Encyclopedia of
Drugs and Alcohol

Encyclopedia of Drugs and Alcohol

VOLUME 2

Jerome H. Jaffe, M.D.

Editor in Chief
University of Maryland, Baltimore

MACMILLAN LIBRARY REFERENCE USA
SIMON & SCHUSTER MACMILLAN
NEW YORK

SIMON & SCHUSTER AND PRENTICE HALL INTERNATIONAL
LONDON   MEXICO CITY   NEW DELHI   SINGAPORE   SYDNEY   TORONTO

Copyright © 1995 by Macmillan Library Reference USA

Macmillan Library Reference
Simon & Schuster Macmillan
866 Third Avenue
New York, NY 10022

Library of Congress Catalog Card Number: 94-21458

Printed in the United States of America

printing number
1 2 3 4 5 6 7 8 9 10

Library of Congress Cataloging-in-Publication Data
Encyclopedia of drugs and alcohol / Jerome H. Jaffe, editor-in-chief.
    p.   cm.
   Includes bibliographical references and index.
   ISBN 0-02-897185-X (set)
    1. Drug abuse—Encyclopedias.   2. Substance abuse—Encyclopedias.
  3. Alcoholism—Encyclopedias.   4. Drinking of alcoholic beverages—
Encyclopedias.   I. Jaffe, Jerome.
   HV5804.E53   1995
   362.29′03—dc20   95-2321
                                              CIP

This paper meets the requirements of ANSI-NISO Z39.48-1992 (Permanence of Paper). ∞ ™

Encyclopedia of
Drugs and Alcohol

ECONOMIC COSTS OF ALCOHOL ABUSE AND ALCOHOL DEPENDENCE

Alcohol abuse and alcohol dependence continue to be major health problems in the United States. The terms *alcohol abuse* and *alcohol dependence* are based on the diagnostic criteria as stated in the American Psychiatric Association's DIAGNOSTIC AND STATISTICAL MANUAL *of Mental Disorders, Third Edition, Revised* (1987). As such, they cost the nation billions of dollars in health-care costs and reduced or lost productivity each year. In 1985, alcohol abuse and dependence cost an estimated 70.3 billion dollars and in 1988 an estimated 85.8 billion dollars (Rice et al., 1990, 1991). Here we estimate the cost of alcohol abuse and dependence for 1990.

EXTENT OF THE PROBLEM

Williams and colleagues (1987) estimated that by 1995, 18.4 million people in the United States ages eighteen and older would be alcohol abusers and therefore at risk for developing alcohol-related health problems. Data for the employed population ages eighteen and over from the 1988 National Health Interview Survey show that 10.3 percent of men and 4.1 percent of women were alcohol dependent, and 3.9 percent of men and 1.0 percent of women were severely dependent (Parker & Harford, 1992).

In 1985, there were an estimated 95,000 alcohol-related deaths in the United States, 4.5 percent of total deaths. Of these, 18.6 percent were caused by conditions, such as cirrhosis of the liver, that are the direct result of alcohol abuse or dependence. The remaining 81.4 percent were alcohol related and accounted for 51 percent of all motor-vehicle crash fatalities, 41 percent of all accidental fall fatalities, 42 percent of all accidental fire fatalities, 26 percent of all suicides, and 75 percent of all deaths due to cancer of the esophagus (Rice et al., 1991). By 1987, the estimated number of alcohol-related deaths rose to 105,095, almost 5 percent of total deaths (Shultz et al., 1990).

COST ESTIMATES FOR 1990

The cost of alcohol abuse and dependence for 1990 is estimated by multiplying the percent changes in socioeconomic indices from 1985 to 1990 by the 1985 cost estimates. (Different socioeconomic indices were used for the different types of costs, see Table 1.) These socioeconomic indices incorporate inflation in the medical-care market as well as the effect of changing demographics and patterns of medical-care utilization.

The total economic costs of alcohol abuse and dependence for 1990 are estimated at $98.6 billion, a 40-percent rise during the five-year period, 1985 to 1990. Table 1 breaks the total costs into core costs—those resulting directly from the illness of alcohol abuse—and other related costs—those related to the secondary nonhealth effects of illness. Within each category are direct costs (for which actual payments are made) and indirect costs (for which resources such as income are lost).

Direct core, or medical-care, costs are estimated at 10.5 billion dollars and account for 10.7 percent of total alcohol-abuse costs in 1990. Morbidity costs—the value of reduced or lost productivity due to illness—amount to 36.6 billion dollars, 37 percent of the total. The costs of alcohol-related deaths amount to 33.6 billion dollars; these mortality costs represent the present value of forgone earnings discounted at 6 percent.

Other related alcohol-abuse costs amount to 15.8 billion dollars, 16 percent of the total economic

costs. The largest component is crime expenditures amounting to 5.8 billion dollars. The value of motor-vehicle crashes due to alcoholism is also high, amounting to 3.9 billion dollars.

CONCLUSION

Alcohol abuse and alcohol dependence are costly to the nation in resources used for care and treatment of persons suffering from these disorders, lives lost prematurely, and reduced productivity. Data show clearly that the measurable economic costs of alcohol abuse continue to be high.

TABLE 1

Estimated Economic Costs of Alcohol Abuse, 1985 and 1990

Type of Cost	1985[1] Amount in Millions	1990[2] Amount in Millions	1990 Distribution
TOTAL	$70,338	$98,623	100.0%
Core Costs	58,181	80,763	81.9
Direct Costs	6,810	10,512	10.7
Specialty organization	2,281	3,469	3.5
Short-stay hospitals	3,017	4,589	4.7
Office-based physicians	141	240	0.2
Other professional services	173	329	0.3
Nursing homes	703	1,095	1.1
Support costs	495	790	0.8
Indirect Costs	51,371	70,251	71.2
Morbidity	27,388	36,627	37.1
Noninstitutionalized population	27,208	36,404	36.9
Institutionalized population	180	223	0.2
Mortality[3]	23,983	33,624	34.1
Other Related Costs	10,546	15,771	16.0
Direct Costs	7,380	10,436	10.6
Crime	4,251	5,807	5.9
Motor vehicle crashes	2,584	3,876	3.9
Fire destruction	457	633	0.6
Social welfare administration	88	120	0.1
Indirect Costs	3,166	5,335	5.4
Victims of crime	465	576	0.6
Incarceration	2,701	4,759	4.8
Special Diseases			
Fetal alcohol syndrome	1,611	2,089	2.1

[1]Rice et al. 1990.

[2]1990 costs are based on socioeconomic indices applied to 1985 estimates.

[3]Discounted at 6 percent.

Use; Homelessness, Alcohol, and Other Drugs; Parents Movement; Partnership for a Drug-Free America; Prevention)

BIBLIOGRAPHY

BOTVIN, G. (1993). A six-year follow up of life skills training. Paper presented at the 7th Annual Drugfree School Conference, Washington, DC.

BOTVIN, G. J., & DUSENBURY, L. (1992). Positive peer groups for homeless adolescents. Research Report for Center for Prevention, New York Cornell Medical School.

BOTVIN, G. J., & TORTU, S. (1988). Preventing adolescent substance abuse through life skills training. In R. Price et al. (Eds.), A casebook for practitioners. Washington, DC: American Psychological Association.

BRY, B., CONBOY, C., & BISGAY, K. (1986). Decreasing adolescent drug use and school failure: Long-term effects of targeted family problem-solving training. Child and Family Behavior Therapy, 8, 43–69.

CENTER FOR HEALTH PROMOTION. (1990). Effective drug education curricula. Palo Alto: Stanford University.

COALITION ON SMOKING OR HEALTH. (1993). Saving lives and raising revenue: Reasons for major increases in state and federal tobacco taxes. Washington, DC: Author.

COMPREHENSIVE HEALTH EDUCATION FOUNDATION. (1990). Here's Looking At You, 2000. Seattle: Author.

CONNELL, D., TURNER, R., & MASON, E. (1985). Summary of findings of the school health education evaluation: Health promotion effectiveness, implementation, and costs. Journal of School Health, 55, 316–321.

DRYFOOS, J. G. (1990). Adolescents at risk. Oxford: Oxford University Press.

FALCO, M. (1992). Making a drug-free America: Programs that work. New York: Times Books.

FEDERAL TASK FORCE ON HOMELESSNESS AND SEVERE MENTAL ILLNESS. (1992). Outcasts on main street. Rockville, MD: National Institute on Mental Illness.

FLAY, B. R., & SOBEL, J. L. (1983). The role of mass media in preventing adolescent substance abuse. In T. Glynn, C. Leukfeld, & J. Ludford (Eds.), Preventing adolescent drug abuse: Intervention strategies. Rockville, MD: National Institute on Drug Abuse.

GARRITY, T. F., & LAWSON, E. J. (1989). Patient–physician communication as a determinant of medication misuse in older minority women. Journal of Drug Issues, 19, 245–259.

GERBNER, G. (1990). Stories that hurt: Tobacco, alcohol, and other drugs in the mass media. In H. Resnik (Ed.), Youth and drugs: Society's mixed messages. Rockville, MD: Office for Substance Abuse Prevention.

HANSEN, J. (1990). Theory and implementation of the social influence model of primary prevention. In K. Rey, C. Faegre, & P. Lowery (Eds.), Prevention research findings: 1988. Rockville, MD: Office for Substance Abuse Prevention.

HAWKINS, D. J., ET AL. (1992). Communities that care. San Francisco: Jossey-Bass.

HAWKINS, J., & LISHNER, D. (1985). Childhood predictors and the prevention of adolescent substance abuse. In C. Jones & R. Battjes (Eds.), Etiology of drug abuse: Implications for prevention. (ADM85-1385). Washington, DC: National Institute on Drug Abuse.

INSTITUTE OF MEDICINE. (1989). Prevention and treatment of alcohol problems: Research opportunities. Washington, DC: National Academy Press.

JOHNSTON, L. D., BACHMAN, J. G., & O'MALLEY, P. M. (1993). Monitoring the future. Rockville, MD: National Institute on Drug Abuse.

LORION, R., BUSSELL, D., & GOLDBERG, R. (1991). Identification of youth at high risk for alcohol or other drug problems. In E. Goplerud (Ed.), Preventing adolescent drug use: From theory to practice. Rockville, MD: Office for Substance Abuse Prevention.

LORION, R., & ROSS, J. G. (1992). Programs for change: Office for substance abuse prevention demonstration models. Journal of Community Psychology. Office for Substance Abuse Prevention, Special Issue.

MARCUS, C., & SWISHER, J. (1992). Working with youth in high risk environments: Experiences in prevention. Rockville, MD: Center for Substance Abuse Prevention.

NORTHEAST REGIONAL CENTER FOR DRUG-FREE SCHOOLS AND COMMUNITIES. (1992). Developing the resilient child. Sayville, NY: Author.

OFFICE FOR APPLIED STUDIES. (1993). Drug abuse warning network. Rockville, MD: Substance Abuse and Mental Health Services Administration.

OFFICE OF THE INSPECTOR GENERAL. (1992). Youth access to tobacco. OEI 02-91-00880. Washington, DC: U.S. Department of Health and Human Services.

PENTZ, M. A., ET AL. (1989). A multicommunity trial for primary prevention of adolescent drug abuse. Journal of the American Medical Association, 261, 3259–3266.

PRICE, R. H., ET AL. (EDS.). (1988). Fourteen ounces of prevention: A casebook for practitioners. Washington, DC: American Psychological Association.

RESNIK, H., & WOJCICKI, M. (1991). Reaching and retaining high risk youth and their parents in prevention programs. In E. Goplerud (Ed.), Preventing adolescent *drug use: From theory to practice.* Rockville, MD: Office for Substance Abuse Prevention.

SWISHER, J., & ASHBY, J. (1993). Review of process and outcome evaluations of team training. *Journal of Alcohol and Drug Education, 39,* 66–77.

SWISHER, J., ET AL. (1993). An evaluation of student assistance programs in Pennsylvania. *Journal of Alcohol and Drug Education, 39,* 1–18.

U.S. DEPARTMENT OF LABOR. (1991). *What works: Workplaces without alcohol and other drugs.* 282-148/54629. Washington, DC: U.S. Government Printing Office.

U.S. GENERAL ACCOUNTING OFFICE. (1992). *Adolescent drug use prevention: Common features of promising community programs.* A report to the Chairman, Subcommittee on Select Education, Committee on Education and Labor, House of Representatives, GAO/PEMD-92-2. Gaithersburg, MD: Author.

JOHN D. SWISHER
ERIC GOPLERUD

EIGHTEENTH AMENDMENT   *See* Alcohol: History of Drinking; Temperance Movement; Women's Christian Temperance Union.

ELDERLY AND DRUG USE   The number of elderly persons in the United States has doubled since 1950, and currently about 13 percent of the U.S. population is sixty-five or older. This percentage is expected to increase to over 20 percent by the year 2030 when most of the baby boom generation will have reached the age of sixty-five or older. These demographic trends, increased understanding of the biomedical correlates of aging, and reviews of prescription and OVER-THE-COUNTER drug-use patterns have led to increased recognition of substance-use problems of elderly people (Robbins & Clayton, 1989).

ILLICIT DRUG USE

Most Americans aged fifty-five and older today entered adolescence and early adulthood during the postdepression and World War II era when ALCOHOL use was prevalent but illicit drug use was relatively uncommon. Since drug initiation typically occurs in adolescence or early adulthood, few of today's elderly population have ever used illicit drugs, and even fewer currently use illegal substances. In the 1992 NATIONAL HOUSEHOLD SURVEY ON DRUG ABUSE, fewer than 1 percent of adults over age 35 reported past-year COCAINE use, and only 3 percent reported past-year use of MARIJUANA, the most commonly used illicit drug (Substance Abuse and Mental Health Services Administration, 1993). These percentages would be even lower for more refined age categories of older adults (e.g., age 55 and older).

Although illicit drug use by elderly persons is not currently a significant social problem, the aging baby boom generation presents the possibility that illicit drug use by members of the elderly population will become more common in years to come as the generation carries its drug-use habits into old age. However, the maturing process—the tendency to decrease or cease illicit drug use as one assumes adult roles and responsibilities (Yamaguchi & Kandel, 1985)—assures that the historically high levels of illicit drug use by the baby boom and current younger generations will abate as they age. The expectation, therefore, is that illicit drug use in old age will be more prevalent in the baby boom generation than among today's elderly people, but much less than this generation practiced in its middle or younger years.

Perhaps more important than illicit drug use itself is the prospect that future older people will face the social and medical aftermath of their earlier drug use. These consequences include increased susceptibility to cognitive and neurological impairment in old age. In addition, the demographic structure of future generations of elderly persons may be altered by the current AIDS epidemic and drug-related VIOLENCE.

MEDICAL DRUG USE

Polypharmacy, the use of multiple prescription and over-the-counter medications, is cited as the most significant drug-safety issue of the 1990s, according to the National Health Promotion and Disease Prevention Objectives for the Year 2000 (Kahl et al., 1992). Polypharmacy is common among older Americans. Compared with previous generations of Americans, today's elderly people have dramatically increased their prescription drug use. This may seem a paradoxical development in light of the improvements older people have experienced in longevity and health, but it is largely a result of pharmacologic developments of the past several decades. Prescription drugs have been developed and marketed for an ever-widening array of diseases and conditions. The number of prescriptions dispensed by community pharmacies rose from approximately 363 million in 1950 to 1.6 billion in 1985. This increase far outstripped population growth over the same period (Lipton & Lee, 1988).

Older adults use more prescription drugs than do young adults. Although they account for only 13 percent of the population, they are given approximately one third of all prescription medications. The majority of older adults living in the community currently take one or more prescription medications; the average is nearly three prescriptions. Medication use is especially heavy among the institutionalized elderly population. A study in 1987 of Massachusetts nursing home facilities reported an average of eight prescription orders per patient (Kahl et al., 1992).

In addition to taking many prescription medications, older men and women are also heavy users of over-the-counter medications (Lipton & Lee, 1988). Purchases of nonprescription drugs peak for people aged 65 and decline only slightly after then, whereas prescription drug use for the same age group continues to increase. Older adults are therefore most likely to be using both prescription and over-the-counter drugs. Oral ANALGESICS and cough-and-cold medications, which, together, account for nearly two-thirds of nonprescription drugs used, are related to accidental injury, cognitive impairment, bleeding problems, and synergistic drug or alcohol-drug interactions. Consumers are frequently unaware of these hazards of over-the-counter medications; they assume that nonregulated drugs are safe and efficacious. The overall levels of use and consequent health risks may increase with the current trend of changes in medications from prescription to nonprescription status.

Unfortunately, much of the medical drug use among older persons is inappropriate and insufficiently monitored. It has been estimated that elderly people experience more than 9 million adverse drug reactions each year, or about one-quarter of all adverse drug reactions. More than 200,000 older adults are hospitalized annually with adverse drug reactions. Although some of these adverse reactions are a result of medication misuse (noncompliance) by the older person, many can be attributed to errors in prescribing and dosage or to avoidable DRUG INTERACTION (Kahl et al., 1992).

Older adults' high levels of medical drug use are natural outgrowths of physical aging as they experience more acute and chronic conditions that cause them to see physicians and to receive prescriptions more frequently. Adverse consequences of medical drug use, however, can be compounded by the biologic processes accompanying aging (Lipton & Lee, 1988). These changes affect drug absorption, metabolism, distribution, and elimination. For instance, the increase in fatty tissue relative to total body weight in the elderly person may cause fat-soluble drugs, such as diazepam (one of the BENZODIAZEPINES), to have lower concentrations at the site of action and more gradual release into the bloodstream and eventual elimination. This creates a potential for cumulative toxicity and unintended drug interactions. Furthermore, since elderly people often use both prescription and over-the-counter medications, and since drinking problems of this population are often unrecognized, it is believed that the medical drug-use patterns might make older adults especially vulnerable to synergistic drug effects or adverse alcohol-drug interactions. Consumption of alcohol in combination with other substances is the most frequent cause of emergency episodes in the DRUG ABUSE WARNING NETWORK (DAWN) system.

PSYCHOACTIVE MEDICATION

In the 1970s, several social scientists criticized the portrayal of women in PSYCHOACTIVE DRUG advertisements carried by medical publications. These advertisements were accused of portraying the stresses of everyday life as disease states treatable by psychotropic drugs. Ads suggested prescribing TRANQUILIZERS to dissatisfied homemakers, lonely women, and even "the woman who can't get along with her new daughter-in-law." Normative transitions such as menopause, the "empty-nest" syndrome, and widowhood were depicted as cause for medication, and psychoactive drugs were even promoted as a way to manage hypochondriacal or difficult patients. More recently, public concern has focused on the psychoactive drug-use patterns of older men and women. As just noted, drug ads often depicted aging-related events.

Men and women aged forty-five and older are more likely than younger men and women to report past-year use of SEDATIVES and tranquilizers, and studies suggest that elderly people are at risk for inappropriate long-term use of sedatives and antianxiety drugs (Robbins & Clayton, 1989; Swartz et al., 1991). Inappropriate use of psychoactive drugs can result in behavioral abnormalities, intellectual deterioration, and MEMORY loss. These problems in elderly people may be mistaken for senility or mental illness, thereby causing doctors to prescribe even further medications that make the condition worse. These declines can result in hospitalization and even unnecessary institutionalization of elderly patients.

About one-half of all nursing home residents take prescription tranquilizers (Lipton & Lee, 1988). The heavy use of medications, especially tranquilizers, in nursing homes has been described as "pharmacological restraint." This excessive reliance on ANTIPSYCHOTIC drugs and other psychoactive medications was addressed in the 1987 Omnibus Reconciliation Act, which introduced new regulations for reviewing and documenting drug use in Medicare- and Medicaid-certified nursing facilities (Kahl et al., 1992).

Among community-residing elderly persons, white people, women, less educated people, and those who are separated or divorced are most likely to use anti-anxiety prescription drugs (Swartz et al., 1991). Compared to nonusers of these drugs, older people who do use them report more psychological distress, more negative life events, and greater use of health care services.

COMPLIANCE

Medication compliance is defined as taking the prescribed dose of a medication at the proper interval for the appropriate period of time (Coons et al., 1994). Reported compliance rates among patients who are sixty years of age and older range from 26 percent to 59 percent. The most typical pattern of medication noncompliance by elderly people is that of taking less than the prescribed amount. This can occur because patients mistakenly take medication symptomatically rather than as prescribed. For example, a patient may only take hypertension medication when he is feeling "tense," anxious, instead of routinely for his high blood pressure. Another common reason for undermedication by elderly patients is the expense of prescription drugs.

No direct relationship has been found between age and compliance with treatment regimens. Factors associated with medication noncompliance by elderly people include number of doctors seen, multiple-disease pathology, multiple prescriptions, complicated drug regimens, low social support, high psychological distress, high monthly medication costs, and lack of insurance coverage for medication. Studies investigating the relationship of age, gender, living arrangements, or socioeconomic status to medication compliance by elderly people have reported inconsistent results.

(SEE ALSO: *Accidents and Injuries; Advertising and the Pharmaceutical Industry; Aging, Drugs, and Alcohol; Alcohol: Complications; Complications; Drug Interactions and Alcohol; Drug Metabolism; Women and Substance Abuse*)

BIBLIOGRAPHY

CLOSSER, M. H. (1991). Benzodiazepine and the elderly: A review of potential problems. *Journal of Substance Abuse Treatment, 8,* 35–41.

COONS, S. J., ET AL. (1994). Predictors of medication noncompliance in a sample of older adults. *Clinical Therapeutics, 16,* 110–117.

KAHL, A., ET AL. (1992). Geriatric education centers address medication issues affecting older adults. *Public Health Reports, 107,* 37–46.

LIPTON, H. L., & LEE, P. R. (1988). *Drugs and the elderly.* Stanford, CA: Stanford University Press.

ROBBINS, C. A. (1991). Social roles and alcohol abuse among older men and women. *Family & Community Health, 13,* 37–48.

ROBBINS, C., & CLAYTON, R. R. (1989). Gender-related differences in psychoactive drug use among older adults. *Journal of Drug Issues, 19,* 207–219.

SCHUCKIT, M. (1977). Geriatric alcoholism and drug abuse. *The Gerontologist, 17,* 168–174.

SUBSTANCE ABUSE AND MENTAL SERVICES ADMINISTRATION. (1993). *National Household Survey on Drug Abuse: Population estimates 1992.* Rockville, MD: U.S. Department of Health and Human Services.

SWARTZ, M., ET AL. (1991). Benzodiazepine anti-anxiety agents: Prevalence and correlates in a southern community. *American Journal of Public Health, 81,* 592–596.

YAMAGUCHI, K., & KANDEL, D. B. (1985). On the resolution of role incompatibility: A life event history analysis of family roles and marijuana use. *American Journal of Sociology, 90,* 1284–1325.

CYNTHIA A. ROBBINS

EMPLOYEE ASSISTANCE PROGRAMS
(EAPs) An Employee Assistance Program (EAP) consists of employer-sponsored services intended to aid employees with personal problems that may adversely affect their job performance. Initially developed to address alcohol-related problems, over the last fifteen years EAPs have emerged as a common response to the problems of ALCOHOL and drug abuse in the workplace. In addition, they provide a variety of services to help employees and their families resolve health, emotional, marital, family, financial, or legal concerns.

While the exact mix of services provided depends on a number of variables, such as size and type of company, EAPs generally offer, at a minimum, confidential client counseling, problem assessment, and treatment referral. A comprehensive EAP offers

1. assessment and referral—EAPs conduct psychosocial assessments to guide decisions to refer clients to treatment and the choice among treatment alternatives
2. treatment follow-up—client follow-up and reintegration into the workplace is an essential EAP function
3. supervisor, management, and union representative training—training provides the information needed on how and when to use the program and how to best assist employees who use it
4. employee education—information on a broad range of problems and how to use the EAP.

The delivery of EAP services may take several forms, depending on such factors as the organization's size and structure. Large companies and organizations, unions, and employee groups often operate their own programs. These services are most often housed within the human resources or medical departments. Smaller organizations, or organizations with dispersed worksites may find it more advantageous to contract with an independent EAP provider located outside the company. A newer trend among small employers is the development of consortium EAP arrangements in which a number of small employers contract with an external provider to provide EAP services.

In the 1980s and 1990s, the number of EAPs has grown dramatically. The Employee Assistance Professionals Association estimates that by the 1990s, 20,000 EAPs were in place in organizations throughout the United States. The Department of Labor's Bureau of Labor Statistics reported that nearly 12 percent of the nonagricultural establishments they sampled offered EAP services. Further, they found that of those sampled, the probability of an establishment offering EAP services increased as a function of establishment size, ranging from 79 percent of employers with over 250 employees, to 9 percent of employers with fewer than 50 employees.

Rapid growth in the number of EAP progams has led to heightened scrutiny concerning their cost effectiveness; in the current economic climate, EAP programs will experience increased pressure to conduct evaluation studies that provide empirical evidence of their efficacy. More research is needed to identify and improve the most essential program components and to aid in tailoring programs to fit specific needs.

Costs incurred in providing EAP services vary widely, but their presence has been clearly tied to overall savings in a number of areas. For example, the McDonnell Douglas Corporation of St. Louis found that employees utilizing their EAP services between 1985 and 1988 for an initial assessment before being referred to treatment had 44 percent fewer lost work days, 81 percent lower termination rates, and lower total four-year medical claims per person than employees seeking treatment for chemical dependence without first consulting the EAP.

For many companies, the approach taken to minimize the impact of drugs in the workplace incorporates a number of additional elements that complement EAPs and constitute a comprehensive strategy. These include a clearly stated formal policy prohibiting drug use, consequences for violating the policy, and alternative strategies to deter drug use.

The Employee Assistance Professionals Association may be consulted for further information: Suite 1001, 4601 North Fairfax Drive, Arlington, VA 22203.

(SEE ALSO: *Contingency Contracts*; *Drug Testing and Analysis*; *Industry and Workplace, Drug Use in*; *Military, Drug and Alcohol Abuse in the U.S.*; *Productivity*)

BIBLIOGRAPHY

HAYGHE, H. V. (1991). Anti-drug programs in the workplace: Are they here to stay? *Monthly Labor Review*, *114*(4), 26–29.

STEVEN W. GUST

ENDORPHINS Endorphins are a group of peptides with potent ANALGESIC properties that occur naturally in the brain. The word *endorphin* is a contraction from *endogenous* and *morphine*—recommended by narcotics researchers in 1975 as the preferred term for a then-hypothetical natural substance capable of action at RECEPTORS for OPIATES (such as HEROIN). The underlying hypothesis was that an endorphin NEUROTRANSMITTER utilized the receptors at which morphine and related drugs exerted their actions. After extensive and intensely competitive research by many groups, three distinct types of such endogenous opioid peptides were found (*peptides* are segments of linked amino acids that can act as neurotransmitters).

Each type of opioid peptide gives rise to one or more opioid peptide prohormones, which can be changed into active neurotransmitters by enzymes in tissues that break the larger inactive peptides into smaller active ones. The pro-opiomelanocortin type, expressed in the corticotrope cell type of the anterior pituitary gland produces the pituitary hormone adrenocorticotropin; the same type expressed in hypothalamic and medullary neurons (nerve cells) of the brain produces β-endorphin, a 31 amino-acid peptide with the greatest intrinsic opioid activity. Each active natural opioid peptide contains the tetrapeptide tyrosine-glycine-glycine-phenylalanine at its amino terminus. The fifth amino acid is either methionine (resulting in the so-called Met^5 enkephalin) or leucine (resulting in leu-enkephalin). Opioid peptides derived from plants—for example, caseimorphin—have also been described. The opioid peptides, of which the proenkephalin- and prodynorphin-derived peptides are most widespread, are found in specific neurons in the brain.

(SEE ALSO: *Enkephalin*; *Opiates/Opioids*)

BIBLIOGRAPHY

COOPER, J. R., BLOOM, F. E., & ROTH, R. H. (1991). *The biochemical basis of neuropharmacology*, 6th ed. New York: Oxford University Press.

FLOYD BLOOM

ENFORCEMENT STRATEGIES AND TACTICS *See* Drug Interdiction.

ENKEPHALIN Enkephalin is either of two pentapeptides (containing five amino acids) with OPIATE and ANALGESIC (painkilling) activity, occurring naturally in the brain, with marked affinity for opiate receptors. While the term ENDORPHIN was initially acceptable as a peptide NEUROTRANSMITTER in the brain, the research team of Hans Kosterlitz and John Hughes gave their own name, enkephalin (a variant of *en-cephal* ["of the brain"]) to the two opioid pentapeptides they had purified from ox brains (c. 1977). They confirmed their discovery by showing that the effects of synthetic peptides were the same in bioassays using opiate RECEPTORS and that both Met^5 enkephalin and Leu^5 enkephalin were authentic endogenous opioid peptides.

(SEE ALSO: *Opiates/Opioids*)

BIBLIOGRAPHY

COOPER, J. R., BLOOM, F. E., & ROTH, R. H. (1991). *The biochemical basis of neuropharmacology*, 6th ed. New York: Oxford University Press.

FLOYD BLOOM

ENZYME-MULTIPLIED IMMUNOASSAY *See* Drug Testing and Analysis.

EPIDEMICS OF DRUG ABUSE Hearing the word *epidemic*, one often thinks first of the flu, measles, the ACQUIRED IMMUNODEFICIENCY SYNDROME (AIDS), or some other contagious disease spreading through a community. In epidemics with person-to-person spread of infection and disease, people become infected and fall victim to the disease, and in the process they come into contact with other people, who in turn get the infection and disease. Often, what is being spread from person to person is not the disease itself, but rather an agent of the disease—for example, one of the viruses that accounts for influenza, the measles virus, or the human immunodeficiency virus (HIV) that causes AIDS.

In EPIDEMIOLOGY (the study of epidemics), it is not the agent, the person-to-person spread of a disease, or the intentional or unintentional nature of acquiring the infection or disease that defines an epidemic. Instead, an epidemic is defined as an unusual occurrence of an infection, disease, or other health hazard in a population. The contrast between "usual" and "unusual" most often is determined by looking at the number of cases that have been occurring within the population over time. If the number of cases occurring in the population this month (or year) is notably greater than the number of cases that occurred in the population during each of the prior months (or years), then it is legitimate to talk of a growing epidemic.

An epidemic may be most obvious when the number of cases goes from zero to a much greater number in a relatively short span of time. For example, before the middle 1970s, the U.S. population apparently had no cases of HIV infection or AIDS. For those years, the usual number of cases per year was zero. Since then, the country has seen a mounting number of HIV infections and AIDS cases each year, and it has become a raging epidemic. Compared to the previous usual number of cases per year, the United States faces an unusual occurrence of disease in the form of thousands of cases per year.

The same concept can be applied on a smaller scale. In the mid-1990s there still are small cities and communities where apparently no one in the population has yet acquired the HIV infection. Health officers who watch over these populations may speak legitimately of an HIV epidemic once the number of cases occurring in the population begins to mount, and there is no need to wait until there are hundreds or thousands of cases before describing the epidemic situation. This is because epidemics are not defined by the absolute number of cases that are occurring. In the early 1990s, there was an epidemic outbreak of hantavirus infection and hantavirus-related deaths in the southwest United States. Because the usual number of hantavirus-related deaths in this region was zero, the situation was declared to be an epidemic well before 100 cases had occurred. Sometimes an epidemic that is limited to a certain place or time will be called an *outbreak*, but this distinction is not a technical one.

There are also epidemics even when no person-to-person spread is involved. For example, in the middle of the twentieth century, there was an epidemic of infant blindness due to retrolental fibroplasia, induced when premature infants were kept in incubators with excessively high concentrations of oxygen. These very high concentrations of oxygen were not a result of machine failure. Instead, the number of cases of retrolental fibroplasia and asso-

ciated blindness kept growing as ever more hospitals raised the oxygen concentration within incubators in a misguided effort to increase survival of the infants by enriching their oxygen supply. Later, clinical and epidemiologic studies showed that this effort to save lives actually led to the increased occurrence of blindness.

Sometimes people object to the usage of the term *epidemic* as applied to drug dependence because it is believed that people bring drug problems down upon themselves by their careless behavior. Epidemiologists, however, typically do not recognize the distinction between "careless" and "careful" behavior when it comes to epidemics. For this reason, they have no trouble speaking about epidemics of syphilis and AIDS, which in some degree are linked to unprotected sexual behavior, something that many would regard as careless behavior.

In summary, the evenhanded application of the concept of epidemic makes it clearly legitimate to speak of an epidemic of smoking-related lung cancer or emphysema, an epidemic of liver cirrhosis due to drinking of alcoholic beverages, an epidemic of leukemia induced by ionizing radiation, an epidemic of mental retardation due to rubella (German measles) infection during gestation, an epidemic of motor vehicle crashes, and an epidemic of deaths by homicide, as well as epidemics of drug use and drug abuse. In order to use the term *epidemic* to describe the health-related experience of a nation, state, or community, it is necessary to demonstrate an *unusual* occurrence of the condition in the population during some specified span of time, relative to the number or rate of cases that occurred in the population during the immediately prior time spans. There is no need to limit usage of the term to infectious diseases with known agents such as rubella or HIV; nor is there a need to limit its usage to diseases spread by person-to-person contact or to be concerned whether the spread of the disease involves careful or careless behavior.

EPIDEMICS IN THE UNITED STATES

An unusual occurrence of drug use or an unusual occurrence of problems connected with drug use can be referred to as epidemics of drug use and drug abuse. In the mid-1990s in the United States, there were multiple indications that the nation had gone through its second major epidemic of COCAINE use and now was in the end-stages of that epidemic.

The first U.S. epidemic of cocaine use started in the late nineteenth century and early twentieth century when cocaine was marketed widely in a variety of forms, including Coca-Cola, Vin Mariana (a wine containing cocaine), and other cocaine products sold without a doctor's prescription. That epidemic subsided, in part because of increased federal and state restrictions on importation and marketing of cocaine, as well as new labeling requirements for patent medicines and other over-the-counter products.

From 1920 through the early 1960s, cocaine use in the United States was not a usual occurrence outside of relatively small circles of HEROIN users, movie and television stars, jazz musicians, and others who came into contact with illicit suppliers of the drug. In the early 1970s, when the federal government began supporting a series of national and state surveys of illicit drug use, cocaine use was found so rarely that it was difficult to get a reliable impression of the characteristics of the cocaine users—there were too few of them in the survey samples.

By studying the series of survey reports from 1972 through the mid-1990s, it is possible to plot the growth of this second U.S. epidemic of cocaine use from what had been typically low levels of use to increasingly greater numbers of cocaine users. The peak years of the epidemic use seem to have been in the late 1970s, which were followed by declining numbers of cocaine users in subsequent years, notwithstanding a small rally in the mid-1980s in connection with the emergence of crack-cocaine smoking.

Although the number of active cocaine users in the U.S. population has dropped back toward the levels observed in the early-to-middle 1970s, it seems that an epidemic of cocaine dependence is still very much in evidence, if the definition of cocaine dependence is meant to encompass very frequent cocaine use as well as the cocaine dependence syndrome described in the more formal terms of clinical research. That is, as the epidemic of cocaine *use* subsided in the late 1980s and early 1990s, there was no parallel falling off in the numbers of daily or other frequent cocaine users, and there was no clear drop in the number of people actively affected by cocaine dependence. Indeed, in the mid-1990s, the number of active cases of cocaine dependence in the population seems to be greater than it ever has been in the nation's history. Thus, it can be said that the epidemic of cocaine dependence is not yet over, for

there continue to be an unusually large number of cocaine-dependence cases in the population. There is not yet enough evidence to say whether fewer newly occurring cases of cocaine dependence are developing in the U.S. population. Once it can be shown that the new occurrence of cases has fallen off, it can then be said with more confidence that the nation has entered a declining phase in this most recent epidemic of cocaine dependence.

With their attention focused upon a declining number of cocaine *users* in the early 1990s, the American public and politicans seemed to turn their attention away from the nation's cocaine problems. At the same time, the level of support for treatment of drug dependence dropped from relatively high levels of expenditures in the mid-1980s, even though the number of people suffering from cocaine dependence had remained about the same as it was during the late 1980s. This set of circumstances underscores the political importance of drawing a distinction between epidemics of drug use versus epidemics of *drug dependence or drug-related problems.* It is likely that many Americans equated declines in the number of cocaine users with declines in the number of cocaine-dependent persons; they were not aware that the epidemic of cocaine dependence continued even as the epidemic of cocaine use was subsiding dramatically.

OTHER PAST DRUG EPIDEMICS

An epidemic during the third century B.C. of "hanshi" use at the end of the Han dynasty in China and the spread of tea drinking prior to 900 B.C., might be the earliest documented epidemics of PSYCHOACTIVE DRUG use in the world, not counting outbreaks of excessive ALCOHOL use (see ASIA, DRUG USE IN).

In the 1600s, in Europe, there were epidemics of CHOCOLATE (cocoa) consumption, TOBACCO consumption, and COFFEE consumption. These epidemics followed shortly after colonization of the Americas by Europeans and were sustained by ever increasing supplies of these products shipped from the cash-poor colonies.

During the nineteenth century, many Europeans became enthusiastic about the inhalation of ether, an intoxicating volatile substance that was investigated for its medical uses by John Snow, one of the fathers of modern epidemiology. Although definitive statistics are not available, it appears that nonmedical inhalation of ether spread through Ireland in an epidemic fashion during the nineteenth century, as did inhalation of NITROUS OXIDE (laughing gas) in the United States. Also during the nineteenth century, China and several other countries experienced epidemics of OPIUM consumption, especially opium smoking. In part, an increased spread of opium smoking in the Americas prompted passage of anti-opium legislation, which ultimately produced international agreements that curbed the supply and distribution of opium and opium products worldwide.

It has been said that the international agreements on these drugs were less effective than the public-health and punitive actions taken within countries to curb opium smoking. For example, harsh jail sentences were imposed for violation of city, state, and federal laws concerning opium, and a tradition of executing "drug criminals" was started in some countries. In Communist China, according to some stories, capital punishment of drug dealers and drug users account for the virtual disappearance of drug problems in that country. The truth of these stories cannot be known.

About the same time that the international agreements on opium and opium products were passed, the United States experienced an increase in tobacco smoking, ultimately with peak population levels of tobacco smoking occurring during World War II and the following years, before declines occurred in conjunction with the surgeon general's 1962 report on smoking and health and other publicity about the health hazards of smoking. When one considers the social climate of the 1990s, a time when tobacco smoking is not at all a socially approved drug-use practice, it may be difficult to imagine that during World War II Lucky Strikes and other cigarettes were passed out to soldiers as part of their daily food rations. This turned out to be an effective way to sustain the epidemic of tobacco smoking, but one cannot be sure whether the tobacco industry's intent was primarily to boost the morale of soldiers or to create and build market strength for tobacco cigarettes. Someone interested in the history of epidemiology might be able to sort this issue out, if industry rcords from that time were opened for inspection.

A more definitive case can be built for the marketing strategies that have been used to increase and build market strength for smokeless tobacco products such as snuff. There was a tremendous increase in the youthful usage of smokeless tobacco between 1970 and 1985. This increase has been traced to deliberate marketing strategies, including formulation of relatively low-cost, "unit dose" supplies of tobacco snuff that had been flavored to increase palatability.

While tobacco consumption was increasing worldwide, Japan's population was affected by an epidemic of METHAMPHETAMINE use during and especially after World War II; later distribution of this drug was seen throughout other countries of the world, including the Scandinavian nations and the United States. At one point in the 1950s, it was estimated that 2 percent of Japan's population had taken methamphetamines nonmedically. It also has been said that especially harsh jail sentences and other criminal penalties accounted for the termination of the amphetamine epidemic in Japan, but as noted in regard to capital punishment and prior Asian drug epidemics, there is no good evidence on this issue. Between 1945 and 1965, other countries saw amphetamine epidemics come and go without the implementation of especially harsh criminal penalties.

The prevalence of nonmedical STIMULANT use in the 1950s did not reach the 2 percent level in the United States as it had in Japan, but it was sufficiently widespread to yield congressional hearings, that focused especially upon AMPHETAMINE use by long-distance truckers (e.g., those who used the drug to promote vigilance and stamina for lengthy trips) and by homemakers (e.g., those who took amphetamines to curb their appetite or because of their mood-altering effects). In part, these epidemics should be understood in relation to the relatively widespread availability of amphetamines in a context of limited regulation of supplies and distribution. These epidemics resulted in legislation and social action to reduce the supply and control the distribution of the amphetamine drugs. In the United States, two especially relevant pieces of federal legislation were the Drug Abuse Control Amendments of 1965 and the CONTROLLED SUBSTANCES ACT of 1970; these laws were directed at controlling the use of the amphetamines as well as the use of other drugs.

The usage of marijuana and the pyschedelic drugs (e.g., LSD) grew during the 1960s and seems to have peaked during the 1970s. In the 1990s, there have been conflicting reports of increasing consumption of these drugs, especially LSD. By some accounts, the nation is facing a new epidemic of LSD usage. It appears, however, that this nationwide increase cannot be detected in population estimates from the NATIONAL HOUSEHOLD SURVEY on Drug Abuse, and it is possible that the apparent nationwide epidemic actually is quite limited in scope.

Several noteworthy developments occurred in relation to HEROIN and the OPIOID drugs during the late 1960s and early 1970s. One important clinical and epidemiological research group based at the University of Chicago developed important innovative strategies for community-level intervention directed at outbreaks of heroin use and heroin dependence. An important element in the group's intervention plan was to employ outreach workers, including staff in recovery from heroin dependence, who would spend enough time on the street corners to identify both new and old users of heroin and to help them get into treatment and stay in treatment. In addition, in Britain, Richard de Alarcon adapted classical methods of epidemiologic research to study the diffusion of injecting drug use (especially injecting heroin use) as an epidemic phenomenon, by plotting the person-to-person spread of the epidemic over time and across the cities of that country.

In 1971, President Richard M. Nixon declared a "war on drugs" following a period of increased heroin use in the United States; he did this partially in association with the return of Vietnam veterans, many of whom had become users of heroin and other opioid drugs during their overseas tours of duty. This epidemic of the late 1960s and early 1970s was documented most readily by examining statistics on clients entering treatment for heroin dependence, including the lag of several years that separated users initial injection of heroin to their first admission for treatment. Despite the war on drugs, a decline in heroin use in the early 1970s was followed by another smaller epidemic of heroin use or dependence during the mid-1970s, followed by apparent decreases in the occurrence of heroin dependence during the late 1970s and early 1980s. The early decrease appears to have coincided with the decrease in importation of heroin to the United States from supplier countries such as Turkey and the mid-1970s increase with the emergence of Mexico and Southeast Asia as suppliers of illicit opiates.

When heroin is the drug of choice and heroin availability declines, users often take other drugs that provide the same functions—either opiate drugs derived from the opium poppy such as morphine or synthetic opioids derived in the chemistry lab and not requiring cultivation products from poppy fields. One example of a synthetic opioid is the so-called China White, which spread through the United States, especially on the West Coast. The number of overdoses linked to China White and related synthetic opioid drugs seemed to increase until the mid-1980s. Since then, there have been declines in the incidence of this type of overdose, possibly because of the increased supplies and street-level purity of poppy-derived heroin.

In addition to the cocaine epidemics already mentioned, there has been a cocaine epidemic in the late twentieth century, which might have been sustained by the introduction of CRACK-cocaine, another unit-dose formulation of a psychoactive drug that reduces cost to a level that can be afforded (at least, initially) by many people. Other articles in this encyclopedia discuss reasons that crack-cocaine smoking might have helped sustain the epidemic of cocaine use, including differences in the pharmacologic, pharmacokinetic, and reinforcement profiles of crack-smoking versus nasal insufflation of cocaine hydrochloride powder. In this context, it is interesting to note that the epidemics of crack smoking and cocaine use ended when they did, during a period of widespread availability of cocaine in a low-cost formulation. In epidemiologic terms, this development carries three very important implications. First, given widespread availability, many Americans had opportunities to smoke crack or take cocaine powder and did not do so. In some important way, these were Americans who were not susceptible to widespread media publicity and other conditions that otherwise might have promoted the use of crack or other forms of cocaine.

Second, for many Americans who tried crack or cocaine powder, the use of these drugs did not compete well with alternative behaviors that were as readily available to them in their home and community environments. They found that there were other, more reinforcing ways with which to occupy their daily lives. This signifies that within the population, for those who have used cocaine, there are differences in the users' susceptibility to becoming cocaine dependent.

Third, within the American population, the balance of these several kinds of susceptibility must have changed over the course of the 1980s. For example, during many other epidemics of contagious disease, as the balance of susceptibility changes, the people who are more susceptible become surrounded by people who are less susceptible. Sometimes, the balance of susceptibility changes without any active and organized public health intervention, as in the case of a typical influenza epidemic in an elementary school population. Sometimes, the balance of susceptibility is changed quite deliberately by organized public-health action, as in the successful worldwide effort to eradicate deadly smallpox by making sure that susceptible persons were immunized against smallpox, and by making sure that infected individuals were surrounded by those who were not susceptible by virtue of either immunization or past infection.

In the case of a drug epidemic, as the more susceptible individuals in the population start to become surrounded by people who will not or do not take the drug, it must be increasingly difficult for them to come into contact with the drug at an individual level, even when the drug supply is great at the societal levels. Furthermore, as the balance of the several kinds of susceptibility changes within the population, there must be an evolution of the social-influence processes that promote the spread of drug use from person to person: Fewer people are being pressured by peers to use the drug; fewer people are talking about the drug in favorable terms; more people are talking about how they had a chance to use it, but it just didn't seem worth it; more people are talking about how they have used the drug but it just didn't do very much for them.

This sort of process must have taken place with regard to the cocaine epidemic for the balance of susceptibility to have changed within the population; otherwise, the epidemic of cocaine use would have persisted. Because we do not have an effective biological vaccine that would reduce susceptibility to cocaine use the way the smallpox vaccine reduced susceptibility to smallpox infection, this change in the balance of susceptibility had to have been caused by something else. *Before* the epidemic of cocaine use had started to decline, the social demographer K. Singh hypothesized that it would decline simply because of demographic changes in the U.S. population caused by a declining birth rate fifteen to twenty-five years earlier. Singh apparently reasoned that, numerically, there would be fewer and fewer people aged fifteen to twenty-five, and this by itself would change the balance of susceptibility in the population because the developmental period from

age fifteen to twenty-five is one that is at especially high risk for starting illicit drug use.

Later, and *after* the epidemic of cocaine use had started to decline, two other main hypotheses emerged. One of these took note of the demographic changes to which Singh had pointed but also drew on three other interrelated epidemiologic observations, namely that (1) cocaine use almost always starts after MARIJUANA use has started; (2) a history of marijuana use probably is the strongest indicator of susceptibility for trying cocaine; and (3) most marijuana users try cocaine once or a few times but do not go on to become dependent upon it (i.e., they are in the second kind of susceptibility group already mentioned). These three epidemiologic observations were also linked with an observation from ethnographic research: When a young person is presented with an opportunity to try marijuana or cocaine, it very often is a slightly older person with a history of marijuana use who presents the opportunity. It might thus have happened that the cocaine epidemic had stopped growing and had started to end once the supply of cocaine had increased to a level where a large proportion of former and current marijuana users had been presented with an opportunity to use cocaine. When these marijuana users either declined to use cocaine or tried and then stopped using cocaine, they then no longer could serve as sources of diffusion to younger persons. That is, the change in balance of susceptibility within the population was related to the number of individuals who previously had tried marijuana and to whether they had completed the normative passage of (1) declining to use cocaine when it was offered to them or (2) trying cocaine a few times without becoming dependent upon it, thereby ceasing to be part of the vanguard of cocaine experimenters who in the glow of their first cocaine experiences would enthusiastically be offering cocaine to others.

According to the other main post-epidemic hypothesis, trends in the perceived danger or risk of harm associated with taking cocaine affected trends in cocaine use. Particularly after basketball star Len Bias died after smoking crack, more young people reported that they perceived there to be substantial risks of harm associated with taking cocaine. Concurrently, there were declines in reported levels of cocaine use. For a number of years, as surveys showed more and more young people reporting that they perceived cocaine use to be dangerous, the levels of cocaine use declined even further, despite increasing or stable levels of cocaine availability. These trends gave rise to the optimistic observation that perhaps it was the increases in perceived dangerousness of cocaine use that accounted for the declines in cocaine use. If such an observation were true, society might be able to stop or curb future epidemics by educating youths to perceive the harmfulness of drug use.

DRUG EPIDEMICS IN THE FUTURE

Singh's prediction based on an analysis of demographic changes in the population and the two main hypotheses that emerged after the epidemic of cocaine use had started to decline have historical importance. Although it was not possible to test these hypotheses about the 1975-to-1994 epidemic of cocaine use in the United States in any rigorous fashion, and it cannot be known for certain that any of them is correct, they may help in the plans for coping with future epidemics of drug use and drug dependence; they also offer pointers what kind of societal response might be needed if a rising line is perceived in the plotted curves of new epidemics. Nonetheless, until a more certain knowledge is acquired about the dynamics of epidemics of drug use, it will be premature for politicians or anyone else to ride to glory on the descending line of these curves. There is enough knowledge to take action, but not enough to say what specific combinations of public-health actions will be effective.

The array of public-health actions to stop or curb future drug epidemics have not yet been exhausted. In the 1970s, Dr. Jerome H. Jaffe and other experts suggested developing prevention strategies that would be based on concepts of reducing susceptibility to drug dependence. This might sound like science fiction, but recent new developments in molecular biology, immunology, pharmacology, and neuroscience have made a viable strategy of this type more plausible.

In seeking to understand the future of drug epidemics in society, it will be necessary to complete more thorough studies of some predicted epidemics that did not materialize. For example, some observers have said that the United States would be swept by a new epidemic connected with smoked methamphetamine ("ice"). It was predicted that the ice epidemic would be more difficult to manage than the crack epidemic. Following the 1990/91 Persian Gulf war and 1992 posting of U.S. troops to Somalia in East Africa, it was said that the United States would suffer a khat-cathinone epidemic as soon as the veterans returned with the experience of seeing khat used by the people of the Middle East and Somalia— and when cathinone (one of khat's active ingredients) was extracted or synthesized by underground chemists for distribution. So far, however, the predictions of the nationwide ice and KHAT-cathinone epidemics have been wrong. There have been isolated epidemics in a few communities but apparently no widespread use, and it is not altogether clear what curbed the spread to other communities.

As of the mid-1990s, many countries have conducted epidemiological surveys to estimate the number of drug users in their populations, and some countries maintain substantial surveillance efforts to assess whether and when drug epidemics are occurring. No country, however, has made a substantial investment in the empirical study of drug epidemics. Most of the hypotheses and theories about drug epidemics remain untested against epidemiologic evidence, including a recently stated and fairly elaborate theory that incorporates what might be the necessary conditions for the expansion, the maintenance, and the decline of drug epidemics. It must thus be said that the present stage of applying epidemiology to the study of drug epidemics is a fairly primitive one.

(SEE ALSO: *Adjunctive Drug Taking*; *Alcohol: History of Drinking*; *Amphetamine Epidemics*; *Education and Prevention*; *Epidemiology of Drug Abuse*; *High School Senior Survey*; *Opioids and Opioid Control*; *Prevention Movement*; *U.S. Government Agencies*; *Vulnerability as Cause of Substance Abuse*)

BIBLIOGRAPHY

ANTHONY, J. C. (1992). Epidemiological research on cocaine use in the U.S.A. In *Cocaine: Scientific and social dimensions*. Proceedings of the CIBA Foundation Symposium 166. Chichester, England: Wiley.

BEJEROT, N. (1970). *Addiction and society*. Springfield, IL: Charles C. Thomas.

DE ALARCON, R. (1969). The spread of heroin abuse in a community. *Bulletin on Narcotics. 21*, 17–22.

ELLINWOOD, E. H. (1974). The epidemiology of stimulant abuse. In E. Josephson & E. E. Carroll (Eds.), *Drug use: Epidemiological and sociological approaches*. Washington, DC: Halsted Press/Wiley.

HELMER, J., & VIETORISZ, T. (1974). *Drug use, the labor market, and class conflict*. Washington, DC: The Drug Abuse Council, Inc.

HUGHES, P. H., & CRAWFORD, G. A. (1974). Epidemiology of heroin addiction in the 1970s: New opportunities and responsibilities. In E. Josephson & E. E. Carroll (Eds.), *Drug use: Epidemiological and sociological approaches*. Washington, DC: Halsted Press/Wiley.

HUGHES, P. H., & JAFFE, J. H. (1971). The heroin copping area: A location for epidemiologic study and intervention activity. *Archives of General Psychiatry, 24*, 394–400.

JOHNSTON, L. D. (1991). Toward a theory of drug epidemics. In R. Donohew, H. Sypher, & W. Bukoski (Eds.), *Persuasive communication and drug abuse prevention*. Hillsdale, NJ: Erlbaum.

MILLER, M. A. (1991). Trends and patterns of methamphetamine smoking in Hawaii. In M. A. Miller, & N. J. Kozel (Eds.), *Methamphetamine abuse: Epidemiologic issues and implications*. National Institute on Drug Abuse Research Monograph. Washington, DC: U.S. Government Printing Office.

MUSTO, D. F. (1987). *The American disease: Origins of narcotic control*, 2nd ed. New York: Oxford University Press.

SINGH, K. (1995). Unpublished communication, presented and discussed in S. B. Sells (1977), Reflections on the epidemiology of heroin and narcotic addiction from the perspective of treatment data. In J. Rittenhouse (Ed.), *The epidemiology of heroin and other narcotics*. Washington, DC: U.S. Government Printing Office.

SUWAKI, H. (1991). Methamphetamine abuse in Japan. In M. A. Miller & N. J. Kozel (Eds.), *Methamphetamine abuse: Epidemiologic issues and implications*. National Institute on Drug Abuse Research Monograph. Washington, DC: U.S. Government Printing Office.

CARLA STORR
JAMES C. ANTHONY

EPIDEMIOLOGY OF DRUG ABUSE

One of the best ways to introduce an article on the epidemiology of drug use and drug dependence is to ask some basic questions that epidemiologic studies can answer but laboratory and clinical studies cannot. Here are some examples:

> In the early 1990s in the United States, about how many ages 12 to 17 had used cocaine at least once?

> In the early 1990s, within which U.S. population subgroups were active cocaine users most likely to be found?

> Within the United States in the early 1990s, among those aged 15 to 24 who had used cocaine, what proportion had become dependent on it?

> In the early 1990s, which age group within the U.S. population was most likely to have experimented with cocaine, and which age group was most likely to have developed cocaine dependence?

> For a young adult living in the United States, what is the risk of developing the problem of alcohol abuse or dependence between one year and the next?

> Is the risk for alcohol dependence greater for some young adults than for others?

> Which subgroups of young adults are at especially high risk for alcohol dependence?

> Are these same subgroups of young adults at especially high risk of becoming dependent on psychoactive drugs such as marijuana or cocaine?

To answer questions of this type, it is necessary to step outside the laboratory and clinical settings where drug users receive treatment. This step can be taken during the course of epidemiologic surveys that seek information about all aspects of the population's drug experience; the surveys take into account not only the relatively modest numbers of drug users who have received counseling and treatment, but also those who never have received any kind of health care or social services. The answers to these questions, based on epidemiologic surveys conducted in the United States between 1980 and the present, are as follows:

> In the early 1990s, among those aged 12 to 17 in the United States, and estimated 169,000 to 339,000 had used cocaine at least once. As a proportion, this amounted to about 1 percent of those 12 to 17 in the United States at that time.

> Within the United States in the early 1990s, young adult men aged 18 to 34 were more likely to be active cocaine users than any other population subgroup categorized by age and sex. For example, slightly more than 1.5 percent of those 18 to 34 were active cocaine users, as compared with 0.4

percent of those 12 to 17, 1.4 percent of women aged 18 to 25, and 0.4 percent of women aged 26 to 34.

Within the United States in the early 1990s, among those aged 15 to 24 who had used cocaine, an estimated 25 percent had become dependent upon it. That is, for every four, who had experimented with cocaine, one had become dependent on it.

Within the United States in the early 1990s, people of the 25 to 34-year age group were most likely to have experimented with cocaine; within this age group, about 30 percent of men had tried cocaine at least once, and about 21 percent of women had tried cocaine at least once. Cocaine dependence also was most prevalent in this age group; it affected about 4 percent of all persons aged 25 to 34. Among cocaine users aged 25-34, an estimated 16 percent had become dependent upon it.

For those 18 to 29 living in the United States, the best available estimate for the risk of developing alcohol abuse or dependence between one year and the next is about 2 to 4 percent.

The risk of succumbing to alcohol abuse or dependence for males aged 18 to 29 is an estimated 6 percent per year, as compared with about 1 percent per year for females aged 18 to 29.

Males between the ages of 18 and 25 are especially high risk of succumbing to alcohol abuse or dependence.

These same subgroups of young adults are at especially high risk of becoming dependent upon psychoactive drugs such as marijuana or cocaine. When all of the abuse or dependence syndromes attributable to nonmedical use of these drugs are considered, the estimated risk for males aged 18 to 29 of developing clinically recognizable drug problem is estimated at 4.4 percent per year; for females aged 18 to 29, it is about 1.6 percent.

There is, of course, good reason to wonder whether epidemiologic surveys of drug use and drug dependence have sufficient validity to be trusted. On the one hand, especially among young people, there may be a tendency to exaggerate drug taking, and to falsify survey responses in the direction of more drug taking than has really occurred. On the other hand, some people may be hesitant to disclose their histories of drug taking or drug problems; they might not agree to participate in the survey, or they might falsify their answers in the direction of less drug taking or fewer problems than have actually occurred.

There fortunately is a body of methodologic research that provides some general assurance about the accuracy of estimates in epidemiologic surveys. Accuracy of the survey results seems to be enhanced considerably when special care is taken to guarantee confidentiality of responses, to protect the privacy of the survey respondents, and to develop trust and rapport before asking survey questions about sensitive behavior, alcohol and drug problems, or illegal activities. In particular, except in poorly conducted surveys of very young respondents, there seems to be very little exaggeration of drug involvement, and older adolescents and adults rarely report drug use unless it actually has happened. Moreover, the accuracy of the estimates does not seem to be distorted too much when the surveys concentrate on household residents and do not extend their samples to include homeless or imprisoned segments of the population. Even though homeless people and prisoners often have significant and special needs for alcohol- and other drug-dependent treatment services that society cannot ignore without peril, the number of homeless and incarcerated persons is small relative to the considerably larger number of persons living in households.

It also is important to note the relatively large size of the survey estimates obtained in these epidemiologic surveys. For example, in 1993, as part of the HIGH SCHOOL SENIOR SURVEY (Monitoring the Future), more than 16,000 high school seniors were asked to fill out confidential questionnaires about their use of such drugs as marijuana and cocaine; more than 40 percent reported having taken these drugs illegally, 87 percent reported consuming alcoholic beverages, and more than 60 percent reported having consumed alcohol to the point of getting drunk. In 1993, more than 26,000 American household residents aged 12 years and older participated in a U.S. government-sponsored NATIONAL HOUSEHOLD SURVEY on Drug Abuse and were asked to answer an interviewer's questions about the use of these drugs; illegal drug taking was reported by an estimated 18 percent of those 12 to 17 years, 51 percent of those 18 to 25, 61 percent of those 26 to 34, and 30 percent of older adults. Furthermore, between 1990 and 1992, almost 9,000 Americans aged 15 to 54 completed confidential interviews as part of a U.S. government–sponsored National Comorbidity Survey. According to this survey, one in three tobacco smokers had tobacco problems, signs, and symptoms consistent with their having become dependent upon tobacco and one in seven drinkers had alcohol problems, signs, and symptoms consistent with their having developed the clinical syndrome of alcohol dependence. Among those who reported use of marijuana, heroin, or other controlled substances, one in seven reported drug problems, signs, and symptoms consistent with their having become dependent upon these drugs. These survey-based estimates are already high enough to provoke social concern. They would be even higher if corrections were to be made to account for respondents who were hesitant to report either their consumption of these drugs or the problems associated with drug use that they had.

DRUG-SPECIFIC ESTIMATES FOR THE U.S. POPULATION

It may be useful if, bearing in mind these potential limitations in the survey methods, one considers each broad drug class one by one, in order to convey the relative frequency of use of tobacco, alcohol, and other drugs in the United States, and to identify population subgroups within which drug use or drug dependence is most common. (Hereinafter, estimates based on the 1993 survey of high school seniors are labeled MF estimates; those from the 1993 National Household Survey on Drug Abuse are labeled NHSDA estimates; and those from the 1990–1992 National Comorbidity Survey are labeled NCS estimates.) In view of recent attention to the CAFFEINE-dependence syndrome and other health hazards of drinking COFFEE or TEA or consuming other caffeinated products, estimates concerning the use of caffeine and caffeine dependence might seem warranted. There is not yet a stable base of epidemiologic data on caffeine use and caffeine dependence, however; these remain topics that ought to be examined in future epidemiologic studies.

Tobacco Smoking in the Early 1990s. Monitoring the Future (MF) estimates show that about 60

percent of high school seniors have smoked TOBACCO cigarettes at least once. An estimated 30 percent of high school seniors smoked tobacco cigarettes at least once during the month prior to the survey, and 19 percent had become daily tobacco smokers.

According to the National Household Survey on Drug Abuse (NHSDA), which included household residents age 12 years and older, an estimated 69 percent to 73 percent smoked tobacco cigarettes at least once, for a total of 143,662,000 to 151,233,000 smokers. An estimated 28 percent to 31 percent had smoked in the year prior to the survey, for a total of 57,817,000 to 64,213,000 recently active smokers; most of these had smoked in the month prior to the survey (47,227,000–53,120,000).

There was important age and sex-related variation in these estimates. For example, among adults past age 34, males were more likely than females to have been recent tobacco smokers (31.6% versus 23.7%). Among those 18 to 25, within the limits of survey error, there essentially were no differences between the sexes in prevalence of smoking, and both estimates were in a range from 32 percent to 46 percent. Among those 12 to 17, there also were no statistically reliable differences between the sexes, and the estimated proportions were between 16 percent and 23 percent; the best estimate for girls aged 12 to 17, however, was 20.3 percent, which was numerically greater than 18 percent, the best estimate for boys of the same age (NHSDA estimates).

Using data from the National Comorbidity Survey of Americans aged 15 to 54, it has been possible to estimate the proportion of tobacco smokers and other drug users who have developed drug-dependence syndromes, as defined in relation to a set of diagnostic criteria for drug dependence that were developed by the American Psychiatric Association in 1987. Before the diagnoses of drug dependence are made, the survey must produce evidence that drug users experienced signs or symptoms of dependence such as going through withdrawal or taking drugs to avoid withdrawal symptoms. Applied to the tobacco smokers identified in the National Comorbidity Survey, these diagnostic methods indicated that almost one-third of tobacco smokers in the survey population had developed tobacco dependence. That is, for every three tobacco smokers, one had developed tobacco dependence and was found to have met the American Psychiatric Association's diagnostic criteria for dependence on this drug. Of the more than

70 percent of respondents who had smoked tobacco at least once, a truly remarkable proportion of about 24 percent was found to have a history of currently active or former tobacco dependence (NCS estimates).

Smokeless Tobacco Use in the Early 1990s. An estimated 32 percent of high school seniors had tried smokeless tobacco at least once, and about 10 percent had used it during the month prior to the survey (MF estimates). Household survey estimates indicate somewhat lower values, except among males aged 18 to 25. For example, among 12- to 17-year-olds, an estimated 8.7 percent had tried smokeless tobacco, and only about 2 percent had used it in the month prior to the survey. By comparison, almost 20 pervent of 18- to 25-year-olds had tried smokeless tobacco; corresponding estimates for 26- to 34-year-olds and those over age 34 were 18.4 percent and 10.1 percent, respectively. Males aged 18 to 25, however, were more likely to be recent or former smokeless tobacco users; Almost 9 percent had used it during the month prior to the survey, while an additional 25 percent had used it at some time before the past month (NHSDA estimates).

Alcohol Use in the Early 1990s. An estimated 87 percent of high school seniors have consumed ALCOHOL at least once. About 75 percent had consumed alcoholic beverages in the year prior to the survey, and almost 51 percent had done so during the month prior to the survey. About 2.5 percent had become daily drinkers (MF estimates).

An estimated 62.5 percent of high school seniors had been drunk at least once—almost 50 percent during the year prior to the survey and almost 30 percent during the month prior to the survey. About 1 percent reported having become daily drinkers (MF estimates).

Among household residents aged 12 and older, an estimated 82 percent to 85 percent have consumed alcoholic beverages; this represents from 170,184,000 to 176,209,000 individuals. During the month prior to the survey, an estimated 50 percent had consumed alcohol. As might be expected, the prevalence values for 18- to 25-year-olds were somewhat higher than they were for the high school seniors, especially in relation to recent drinking: Almost 60 percent of the 18- to 25-year-olds had consumed alcoholic beverages during the month prior to the survey. The values for 12- to 17-year-olds were lower: About 41 percent in this age group had tried alcoholic beverages at least once, and about 18 percent had con-

sumed alcohol during the month prior to the survey (NHSDA estimates).

An estimated 21.5 percent of respondents of all age groups from 12 years upward reported drinking at least once per week or more during the year prior to the survey. Corresponding estimates for respondents aged 12 to 17, 18 to 25, 26 to 34, and 35+ were 4.0 percent, 21.5 percent, 24.6 percent, and 23.6 percent, respectively (NHSDA estimates).

Alcohol dependence was found to have affected 15 percent of those who had consumed alcoholic beverages: Out of every six or seven persons who had tried alcohol, about one had become dependent on alcohol. In relation to the total survey population that included drinkers as well as abstainers, an estimated 14 percent were found to qualify for the diagnosis of drug dependence, according to the American Psychiatric Association's criteria (NCS estimates).

Other Illicit Drug Use in the Early 1990s. When controlled substances such as MARIJUANA, cocaine, and heroin, as well as INHALANT drugs, were considered, it was found that an estimated 46.5 percent of respondents had used these drugs on at least once occasion, almost one-third during the year prior to the survey. About 20 percent had taken one or more of these drugs during the month prior to the survey (MF estimates).

The most recently available National Household Survey on Drug Abuse reported that an estimated 36 percent to 39 percent of the population aged 12 and older had engaged in illicit drug use at lease once; this amounts to about 74 million to 80 million drug takers. The number of recently active drug takers was lower; they represented 5 percent to 6 percent of the population (NHSDA estimates).

According to the National Comorbidity Survey estimates, out of every seven persons who had tried marijuana, cocaine, or other controlled substances and inhalant drugs, one had developed drug dependence (14.7%). In light of the fact that about 51 percent of this survey population of 15- to 54-year-olds reported a history of illicit drug use, the resulting estimate for the prevalence of dependence on controlled substances was 7.5 percent. That is, in the total population of individuals (including both drug users and never users), about one in fourteen had fulfilled the criteria for drug dependence (NCS estimates).

Cannabis Use in the Early 1990s. An estimated 35 percent of high school seniors had tried marijuana or HASHISH (*Cannabis*) on at least one occasion, and about 26 percent had smoked cannabis during the year prior to the survey. An estimated 11 to 19 percent had smoked cannabis during the month prior to the survey, and an estimated 1.9 percent to 2.0 percent reported daily cannabis use (MF estimates).

Within the age ranges of 12 to 17 and among persons aged 35 and older, there are many individuals who have not yet started to use illicit drugs such as cannabis, as well as many others who never will start to use these drugs. As a result, one might expect lower prevalence values in these age groups as compared to the values for other age ranges. In fact, this is precisely what the national survey estimates indicate. Overall, an estimated 32 percent to 35 percent of respondents reported having tried cannabis, but among 12- to 17-year-olds the estimate was only 11.7 percent, and among those aged 35 years and older it was 26.6 percent. Prevalence of cannabis use was most common among 26- to 34-year-olds (59.2%) and among 18- to 25-year-olds (47.4%). This also was true for recent cannabis use during the month prior to the survey: There was a prevalence of 4.3 percent for the population overall, 4.9 percent for 12- to 17-year-olds, 11.1 percent for 18- to 24-year-olds, 6.7 percent for 25- to 34-year-olds, and 1.9 percent for older adults (NHSDA estimates).

Among cannabis users, about 9 percent were found to have developed cannabis dependence. Among *all* 15- to 54-year-olds (including both users and never users), 4.2 percent had become dependent on cannabis (NCS estimates).

Inhalant Use in the Early 1990s. INHALANTS had been used by an estimated 17 percent of high school seniors—about 7 percent within the year prior to the survey and about 2.5 percent during the months prior to the survey. Very few respondents (well under 1 percent) reported daily inhalant use (MF estimates).

The National Household Survey on Drug Abuse indicated that about 5.3 percent of its survey population had tried inhalants at least once; about 1 percent had done so during the year prior to the survey, and from 0.3 percent to 0.6 percent had used these drugs during the month prior to the survey. It was found, when considering age and sex, that the subgroup most likely to have used inhalant drugs during the month prior to the survey was that of males aged 18 to 25; in this group, 1.7 percent reported recently active inhalant use (NHSDA estimates).

An estimated 2.3 percent to 5.1 percent of the inhalant users have been found to qualify for the diagnosis of dependence upon inhalant drugs. Translated into an overall prevalence estimate for both users and nonusers, this amounts to about 0.3 percent prevalence of inhalant dependence in the total survey population (NCS estimates).

Use of Psychedelic Drugs in the Early 1990s. PSYCHEDELIC drugs (primarily LYSERGIC ACID DIETHYLAMIDE, or LSD) had been used by an estimated 11 percent of high school seniors. Almost three-quarters of these users (7.4%) had used them in the year prior to the survey, and about one-quarter (2.7%) had used them during the month prior to the survey. PHENCYCLIDINE (PCP) users were in the minority within this group of drug users; only 2.9 percent of the high school seniors had ever tried PCP (MF estimates).

Among persons aged 12 years and older, from 7.8 percent to 9.7 percent individuals had tried psychedelic drugs such as LSD, but for the most part these drug experiences were not recent: Only 0.2 percent to 0.3 percent reported taking psychedelic drugs during the month prior to the survey. Peak prevalence values for recent use of the psychedelic drugs were observed in the years of early adulthood; only for 18- to 25-year-olds did these values exceed a threshold of 1 percent (1.3%); otherwise, they were at the 0.5 percent level or lower (NHSDA estimates).

About 5 percent of the users of psychedelic drugs were found to qualify for the diagnosis of a dependence syndrome, defined in relation to the American Psychiatric Association criteria. Thus, about 0.5 percent of the survey population of 15- to 54-year-olds had become dependent upon psychedelic drugs.

Cocaine Use in the Early 1990s. Among high school seniors, an estimated 6.1 percent had tried cocaine; within this group of COCAINE users, less than one-half had tried CRACK-cocaine. About 3 percent of high school seniors had used cocaine (including crack) during the year prior to the survey, and just over 1 percent had used it in the month prior to the survey. In the MF sample of about 16,000 high school seniors, daily cocaine smoking was too rare to estimate precisely (MF estimates).

An estimated 10 percent to 13 percent of the National Household Survey's population reported having tried cocaine or crack smoking (or both) at least once. The corresponding value for 12- to 17-year-olds was only 1.1 percent, and there was age-related variation: 12.5 percent of the 18- to 25-year-olds had taken cocaine (including crack); 25.6 percent of the 26- to 34-year-olds had done so, and the prevalence estimate for older adults was 8.5 percent. Translated into absolute numbers, an estimated 21 million to 26 million Americans aged 12 and older had tried cocaine or crack smoking. Recent use was substantially less common: Only 0.5 percent to 0.8 percent of the survey population reported having used these drugs during the month prior to the survey; this represented about 1 million to 1.7 million recently active cocaine users in the survey population.

By the early 1990s, the second American epidemic of cocaine use had peaked and waned. Crack smoking had sustained the epidemic for a time, but in the early 1990s it became clear that crack smoking had not diffused broadly through the U.S. population. The relatively low prevalence values for crack smoking among high school seniors was reflected in the National Household Survey on Drug Abuse, which found that only 1.5 percent to 2.1 percent of its survey population had tried crack smoking; this amounted to 3.1 million to 4.4 million individuals. The age groups with most crack-smoking experience were the 18- to 25-year-olds, with a prevalence value of 3.5 percent, and the 26- to 34-year-olds, with a prevalence value of 4.2 percent. Prevalence of crack smoking during the month prior to the 1993 survey was uniformly under 1 percent for all age and sex groups under study (NHSDA estimates).

For every six individuals who had tried cocaine at least once, one had developed cocaine dependence. That is, among these cocaine users, an estimated 15.2 percent to 18.2 percent had become sufficiently dependent upon cocaine to qualify for the American Psychiatric Association diagnosis. In relation to all persons in the survey population, whether they had tried cocaine or not, an estimated 2.7 percent qualified for the diagnosis of cocaine dependence (NCS estimates).

Use of Non-Cocaine Stimulants in the Early 1990s. The nonmedical use of stimulants other than cocaine (such as AMPHETAMINES) was actually more prevalent than cocaine use among high school seniors. An estimated 15.1 percent of high school seniors had taken these stimulant drugs without any doctor's orders; between 7 percent and 10 percent had done so in the year prior to the survey, and 2.5 percent to 5.0 percent had done so during the month prior to the survey (MF estimates).

For reasons not well understood, the Monitoring the Future sample of high school seniors yields prevalence estimates for non-cocaine stimulant usage that are considerably larger than corresponding estimates from the national household survey. Overall, the household survey population estimate for nonmedical use of these stimulatnt drugs was 6.0 percent, and the age group with the highest prevalence value was that made up of 26- to 34-year-olds, at 10.5 percent. Nonetheless, within the survey population, recent use of the stimulant drugs was found to be 2 percent to 4 percent for the 18- to 25-year-olds, the age group whose level of use most resembled that of the high school seniors (NHSDA estimates).

Slightly more than 11 percent of the persons who had used these stimulant drugs were found to have become dependent upon them. This number of stimulant-dependence cases represents about 1.7 percent of all persons in the survey population aged 15 to 54 (NCS estimates).

Use of Anxiolytic, Sedative, and Hypnotic Drugs in the Early 1990s. About 6 percent of high school seniors had used tranquilizers (anxiolytic) or SEDATIVE-HYPNOTIC (e.g., BARBITURATE) drugs without a doctor's orders. About 3.5 percent had done so during the year prior to the survey, and slightly more than 1 percent had done so during the month prior to the survey (MF estimates).

About 4 to 5 percent of the national household survey population reported nonmedical use of tranquilizers or anxiolytic drugs, while 3 percent to 4 percent reported nonmedical use of sedative-hypnotic drugs without a doctor's orders. For tranquilizers, this amounted to 8.2 to 10.9 millions of nonmedical users. For sedative-hypnotics, the total was 6.1 to 8.3 millions of nonmedical users. The estimated number of recently active users was less substantial; they represented less than 0.5 percent of the survey population for tranquilizers (under 500,000 nonmedical users) and for the sedative-hypnotics (under 1 million nonmedical users).

Grouping the users of the tranquilizer or anxiolytic drugs together with the users of the sedative and hypnotic drugs, the National Comorbidity Survey team found that about 9 percent of these drug users had become dependent upon them. In considering this prevalence value, it is important to note that in this survey nonmedical drug use was defined to include not only use of the drug to get high, but also taking more of the drug than was prescribed or in ways not consistent with accepted medical practice. Overall, the prevalence of dependence upon these drugs was at a level of 1.2 percent in the survey population (NCS estimates).

EPIDEMIOLOGY OF DRUG USE AND DRUG DEPENDENCE OUTSIDE THE UNITED STATES

Each year, the United States allocates more resources to epidemiologic surveys of drug use than does any other country in the world. For this reason, it has been possible to assemble a wealth of epidemiologic survey data on the prevalence of drug use and drug dependence within the United States. Other countries also have conducted surveys of this type, and have produced valuable evidence about their experience with tobacco, alcohol, and other drugs. (See the bibliography for some references that can be consulted to gain more information about the results of these surveys.)

OTHER ASPECTS OF EPIDEMIOLOGY AS APPLIED TO DRUG USE AND DRUG DEPENDENCE

A broad range of research questions must be answered in order to gain a complete understanding of the epidemiology of drug use and drug dependence. The focus in this article has been upon *quantity*: How many people in the population (or what proportion) have been affected by drug use and by drug dependence? Although many epidemiologists now devote their research careers to surveys that are needed to answer this kind of basic question, more stress ought to be placed on the other central questions for epidemiology, especially when the answers to these questions can guide society toward effective strategies for prevention of drug use and drug dependence. These questions are:

Where in the population are the affected cases located (in which subgroups, in which places, during which seasons, years, or epochs)? This is a question of *location*.

What accounts for some people becoming affected while others do not become affected? This is a question about CAUSES.

By what processes or sequence of conditions do people become dependent on drugs? This

is a question about *mechanisms* and linked sequences of causal conditions.

What can we do to prevent and reduce the suffering? This is a question about *prevention* and *amelioration*.

At its best, epidemiology provides critically important answers to each of these questions, and it works to ensure that new findings are translated rapidly into effective strategies for prevention. This is the future agenda for epidemiologic research on drug use and drug dependence.

(SEE ALSO: *Amphetamine Epidemics*; *Comorbidity and Vulnerability*; *Diagnosis of Drug Abuse*; *Diagnostic and Statistical Manual*; *Drug Abuse Warning Network*; *Epidemics of Drug Abuse*; *Social Costs of Alcohol and Drug Abuse*; *Vulnerability As Cause of Substance Abuse: An Overview*)

BIBLIOGRAPHY

ANTHONY, J. C. (in press). International databases on drug dependence. In L. Eisenberg & R. DesJarlais (Eds.), *International behavioral and mental health: A sourcebook*. Cambridge: Oxford University Press.

ANTHONY, J. C., & HELZER, J. E. (in press). Epidemiology of drug dependency. In M. Tsuang, M. Tohen, & G. Zahner (Eds.), *Textbook of psychiatric epidemiology*. New York: John Wiley.

ANTHONY, J. C., WARNER, L. A., & KESSLER, R. C. (1994). Comparative epidemiology of dependence on tobacco, alcohol, controlled substances, and inhalant drugs: Basic findings from the National Comorbidity Survey. *Experimental and Clinical Psychopharmacology, 2*, 1–24.

HELZER, J. E., ET AL. (1990). Alcoholism—North America and Asia. *Archives of General Psychiatry, 47*, 313–319.

JOHNSTON, L. D., O'MALLEY, P. M., & BACHMAN, J. G. (1994). *National survey results on drug use from the Monitoring the Future Study, 1975–1993*. Vol. 1, *Secondary school students*. NIH Publication no. 94-3809. Washington, DC: U.S. Government Printing Office.

MURRELLE, L. (1990). Epidemiologic report on the use and abuse of psychoactive substances in 16 countries of Latin America and the Caribbean. In *Drug abuse*, World Health Organization. Pan American Health Organization Scientific Publication no. 522. Washington, DC: Pan American Health Organization.

PETERSEN, R. C. (ED.). (1978). *The International Challenge of Drug Abuse*. National Institute on Drug Abuse Research Monograph. DHEW Publication no. ADM-

78-654. Washington, DC: U.S. Government Printing Office.

U.S. DEPARTMENT OF HEALTH AND HUMAN SERVICES. (1994). *National Household Survey on Drug Abuse: Population estimates 1993*. DHHS Publication no. (SMA) 94-3017. Rockville, MD: Substance Abuse and Mental Health Services Administration.

JAMES C. ANTHONY

EQUANIL   *See* Meprobamate.

ETHANOL/ETHYL ALCOHOL   *See* Alcohol: Chemistry and Pharmacology.

ETHCHLORVYNOL   This is a complex alcohol that causes depression of the central nervous system (CNS). It is a SEDATIVE-HYPNOTIC drug typically used on a short-term basis to treat insomnia and is prescribed and sold under the name Placidyl. Because of its depressant effects on the brain, it can impair the mental and/or physical abilities necessary to operate machinery, such as an automobile.

Continued use of etchlorvynol can result in TOLERANCE AND PHYSICAL DEPENDENCE leading to abuse. Since the risk of abuse is not very great, it is included in Schedule IV of the CONTROLLED SUBSTANCES ACT. Withdrawal signs, not unlike those seen after ALCOHOL (ethanol) or BARBITURATES, occur upon termination of its use in addicts. Etchlorvynol should never be combined with other CNS depressants, such as ethanol or barbiturates, because their depressant effects are additive.

(SEE ALSO: *Withdrawal*)

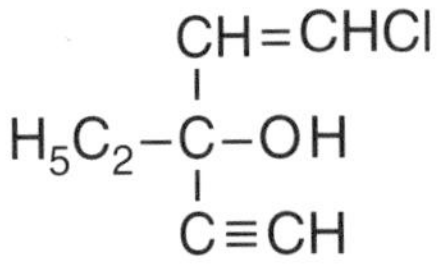

$$H_5C_2-\underset{\underset{C\equiv CH}{|}}{\overset{\overset{CH=CHCl}{|}}{C}}-OH$$

Figure 1
Ethchlorvynol

BIBLIOGRAPHY

RALL, T. W. (1990). Hypnotics and sedatives: Ethanol. In A. G. Gilman et al. (Eds.), *Goodman and Gilman's the pharmacological basis of therapeutics*, 8th ed. New York: Pergamon.

SCOTT E. LUKAS

ETHINAMATE   This is a short-acting SEDATIVE-HYPNOTIC drug typically used to treat insomnia. It is prescribed and sold as Valmid. Structurally, it does not resemble the BARBITURATES, but it shares many effects with this class of drugs; the depressant effects of ethinamate are, however, generally milder than those of most barbiturates. Continued and inappropriate use of ethinamate can lead to TOLERANCE AND PHYSICAL DEPENDENCE, with withdrawal symptoms very similar to those of the barbiturates.

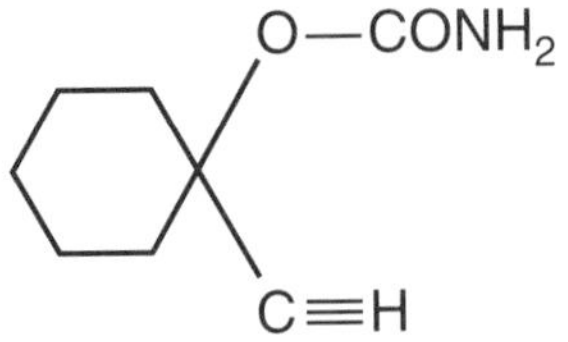

Figure 1
Ethinamate

(SEE ALSO: *Withdrawal*)

BIBLIOGRAPHY

RALL, T. W. (1990). Hypnotics and sedatives: Ethanol. In A. G. Gilman et al. (Eds.), *Goodman and Gilman's the pharmacological basis of therapeutics*, 8th ed. New York: Pergamon.

SCOTT E. LUKAS

ETHNIC ISSUES AND CULTURAL RELEVANCE IN TREATMENT   The HAIGHT ASHBURY FREE CLINICS, Inc. (HAFCI)/Glide Memorial Methodist Church African-American Extended Family Program (AAEFP), described in detail in Reverend Cecil William's book, *No Hiding Place*, represents an important collaboration that has made possible an effective intervention in the inner-city crisis of CRACK-cocaine use.

THE AFRICAN-AMERICAN EXTENDED FAMILY

The key to this intervention has been the adaptation of TWELVE-STEP principals of supported recovery to the AFRICAN-AMERICAN inner city culture. In the HAFCI/Glide program, the basic practicalities of recovery are utilized in a model that is uniquely meaningful in terms of the African-American experience.

The "Big Book" of ALCOHOLICS ANONYMOUS uses the terms "spiritual experience" and "spiritual awakening," manifesting in many different forms, to describe what happens to bring about a personality change sufficient to induce recovery. While some of these may involve an "immediate and overwhelming God consciousness," most are what William James called an "educational variety" of revelation, developing slowly over time. According to a "Big Book" appendix titled "Spiritual Experience," the core of this process is the tapping of an "unexpected inner resource" by members who identify this resource with "their own conception of a Power greater than themselves."

Many members of the black community afflicted with crack-cocaine addiction have been raised in the church. There is a tradition of revelation; many who have been "saved" now believe they are sinners because they have used and sold crack-cocaine to their own people. God has been described in a strict denominational sense. Spiritual awakening in a recovery model within a church program may produce conflict with traditional religious definitions, particularly the third step: "Made a decision to turn our will and our lives over to the care of God *as we understood him.*" Religious leaders, such as Reverend Williams, have played a role in presenting a model of recovery theology that helps mobilize the church as a sleeping giant to better respond to the nation's drug epidemic. In his model, Williams employs self-definition within a spirituality of recovery.

The Black Extended Family Project grew out of an initial collaboration between HAFCI and the Institute on Black Chemical Abuse (IBCA) of Minnesota, headed by Dr. Peter Bell. In 1988, Bell was looking for a West Coast site to replicate IBCA pro-

grams, based on a cultural interpretation of the Minnesota model that had been adapted to black chemical abusers.

In keeping with the IBCA's African-American cultural approach, it was generally agreed that the best site for the new program would be a church. In a Glide conference panel debate on religion and spirituality, Richard Seymour pointed out that under the best of conditions, religion equals spirituality plus culture. This is particularly true in the African-American community, within which the church provides a point of cohesion and a center for both spiritual and community values, and thus a common ground for positive community activity. For a number of reasons, the clear choice was Glide Memorial Methodist Church in San Francisco's Tenderloin, a neighborhood that, though it includes a number of ethnic minorities, is predominantly African-American, low-income, and hard hit by the onslaught of dealing and abuse of crack-cocaine.

Under the leadership of Reverend Williams, Glide had been providing services for indigent and homeless residents, including addicts, for 25 years. Because of his growing concern over the crack-cocaine problem, Reverend Williams and his wife, Jan Mirikitani, executive director of Glide, attended a twelve-step recovery conference conducted by David Smith and Millicent Buxton. Following this conference, they decided to develop a culturally specific recovery program at Glide Church because of the resistance of people of color to participating in the twelve-step process. In 1988, as a preamble to the Black Extended Family Project (BEF), Glide initiated a Facts on Crack education program as an innovative approach to address the drug problem in the African-American community. Reverend Williams embraced the concept, and the BEF was launched that same year with funding from a Pell Foundation grant supplemented by a telethon hosted by KMEL, a San Francisco radio station. Rafiq Bilal, an African-American author and counselor at HAFCI's drug program, was selected as the project's first director. He and other key individuals then received several weeks of training at IBCA in Minnesota.

Problems and Priorities: Programs and Solutions. Specific problems of the African-American target population included the following:

- Low self-esteem
- Late introduction into recovery

- Focus on short-term abstinence rather than long-term recovery
- Dialect of African-Americans
- Institutionalized racism
- Internalized racism
- A unique, often dysfunctional family structure: many classical African cultures have been matrilineal, and look to the "grandmother" for spiritual direction and values. African-Americans developed a matriarchal family structure to survive during slavery. This structure has proved unable to address problems of alcohol and other drug addictions. America is based on a patriarchal family structure, the opposite of the African-American model. It is therefore difficult for African-Americans to relate to systems and to address dysfunctional families when their model is not the norm. The most extreme injury is seen in children being taken from mothers by the system.

The first and foremost priority was bringing to intervention and recovery an approach and nature that African-Americans could identify and live with. Culturally responsive activities needed to be identified and developed.

Many of the BEF's day-to-day activities are structured as workshops. Each component is important in attracting and sustaining the interest of African-American clients into and through recovery:

- Support groups: Daily groups begin with the singing of a selected spiritual. The meeting that follows is structured around the message from the song and a set of commitments developed by African-Americans in recovery. Drug addiction and use are related to slavery. Singing together creates a more relaxed and familiar cultural context for the meeting.
- Women's meetings: For those who have lost children, the comparison between now and the capture of children in Africa during the slave trade is made. Particular emphasis is placed upon the role of women in the more matriarchal African-American family. For many the most positive role model is a grandmother who

passed on the traditions of the family and represents a "higher power."
- Intervention meetings: Addicts and alcoholics are exhorted to stop using and drinking, join the program, and enter into recovery. Active members of the support group participate in this event.
- Fun day: The BEF provides fun days on the first and fifteenth of each month; on those days the program and building become a sanctuary for persons with sums of money in their pockets and intense urges to slip from their recovery. The program provides movies, talent shows, and other entertainment.
- African history classes: Each week a special course is taught about African history and African thought, particularly as it relates to recovery.
- Development of a generation in Glide's BEF: Each support group of men and women is invited to become a generation of addicts born into recovery at Glide. The generation attends support groups, classes, fun days, and other activities together, bonding and graduating together.
- Graduation of generations of addicts in recovery: On a prescribed Sunday, each generation experiences a ritual that involves the entire congregation at Glide. On that day, services at the church are devoted to the members of the graduating generation and their accomplishments and realizations over the past 80 days. Members of the generation speak before the congregation. The whole generation is literally given birth by the congregation, passing through a "placenta," composed of previous generations, that surrounds the graduating group.

Developing Culturally Relevant Approaches to Treatment. In the 1980s, the BEF was showcased in three successive annual national conferences, presented by Reverend Williams, on making use of community resources to combat crack-cocaine and other addictions.

The first two national conferences presented by Glide, entitled "The Death of a Race" and "Rebirth of a Race," focused primarily on drug problems within African-American communities. The third conference, "To Heal a Wounded Soul: Prevention/

Recovery Values and Spirituality," was designed to give emphasis to general cultural diversity and spirituality, and focused on how recognition and utilization of that diversity and spirituality could help all the diverse cultural communities become an extended human family. Although initiated as a means of dealing with addiction, the conference goal extended well beyond the specific problems of addiction to explore cultural differences as a positive value and derive insights of mutual utility from cultural experiences of combating addiction, abuse, misuse, humiliation, degradation, shame, and guilt.

Implicit within these activities is the recognition that treatment is more than the prescribing of medication or the providing of basic and generic counseling based on a homogeneous model of what constitutes addictive disease.

Addiction: A Multicultural Problem in Need of Multicultural Solutions. Just as addiction is a global, rather than a national or regional, phenomenon, so addiction problems in the United States are multicultural. The whole fabric of successful treatment needs to be woven around cultural realities. In this society, twelve-step fellowships, such as ALCOHOLICS ANONYMOUS (AA), NARCOTICS ANONYMOUS, and COCAINE ANONYMOUS, are increasingly seen as the primary means to ensuring long-term abstinence and sobriety through addiction recovery.

It has been suggested that twelve-step fellowships and their success provide credibility to addiction treatment as the bridge between active addiction and active recovery. While this may be increasingly true for the mainstream of white, European-American cultures, it may be less true for other cultures.

Countering the Perception That Twelve-Step Fellowships Have an Exclusively White, Male, Christian, Middle-Class Focus. From its beginnings, elements within the "group conscience" of AA began working to broaden the scope and flexibility of their fellowship. AA may have had its specific beginnings in the Christian Oxford Movement and the personal interaction between its cofounders, Bill W. and Doctor Bob, but its basic tenets reflect a spectrum of cultural antecedents. Throughout history and within various cultures, attempts have been made to deal with addiction and associated human problems. The most generally successful of these have involved in some way the development of individual spiritual maturity within a supportive environment. In that context, the Twelve Steps developed by AA and adapted by other twelve-step fellowships can be seen as a blueprint for developing spiritual maturity, that is similar in intent to the Buddhist Four Noble Truths and Eightfold Path, the Hindu *Vedas*, and the Zen Oxherding Panels.

Individuals with certain religious backgrounds may have particular problems relating to certain tenets of the Twelve Steps. Many Buddhists, for example, venerate the Buddha as a fully enlightened being to be followed and emulated, but do not see him as a "higher power." Not utilizing a concept of God or a higher power in their cultural background, they see their faith as a philosophy and a way of life rather than as a religion. Points of reference need to be established in order for twelve-step recovery to become meaningful for these individuals.

The African-American Extended Family Program is a good example of how the precepts of twelve-step recovery can be adapted to the needs of African Americans. In it, African-American cultural mores and traditions are taken into consideration and made primary to recovery. Ten "Terms of Resistance" are repeated at Glide recovery meetings just as the Twelve Steps are at traditional AA meetings:

1. I will gain control over my life.
2. I will stop lying.
3. I will be honest with myself.
4. I will accept who I am.
5. I will feel my real feelings.
6. I will feel my pain.
7. I will forgive myself and forgive others.
8. I will rebirth a new life.
9. I will live my spirituality.
10. I will support and love my brothers and sisters.

Culture and Spirituality in Twelve-Step Fellowships. While there are many meetings that have a distinct Christian orientation that goes far beyond joining hands and reciting the Lord's Prayer, there are many others that do not. Definitions of God and a "higher power" can and do include an open range of options. Essentially, religiosity and a belief in God as represented in any particular religion are unnecessary for the workings of twelve-step recovery. However, belief in a power outside oneself that is capable of bringing one to sanity in terms of one's addiction is necessary, even if this power is characterized as the meeting group.

From a recovery standpoint, addiction can be seen as a disease of self-centered fear that depends on isolation and deeply held convictions regarding the nature and effects of the addicts' drugs of choice.

That isolation renders the addict incapable of understanding the disease and its personal effects. This is the basis of denial. So long as the addict attempts to fight the addiction through personal willpower alone, he or she is fighting a losing battle, trapped in emotional gridlock in a state of "white knuckle sobriety," where increasing anxiety from the stress will inevitably result in relapse. The reason for this is that the convictions about use are buried within the individual's spiritual belief system, where they can be reached only if the addict is willing to accept that there is something outside his or her own immediate being that can lead him or her to sanity—a power higher than oneself.

Surrender and Powerlessness. The concept of surrender, given its many war-related connotations of occupation, rape, loss of freedom, and so on, is hard enough for anyone to accept, but particularly hard for cultural groups that have, over time, suffered more than their share of occupation, rape, loss of freedom, and so on. African-Americans, for example, may feel that they have been in a state of individual and cultural powerlessness for many generations, and have no desire for further surrender. Native Americans have similar difficulties with that aspect of twelve-step recovery because it runs counter to tribal mores of self-reliance and stoicism. Adolescents, although their cultural cohesion is transitory, are in the process of developing their own individuality and are often loath to appear to be giving up something they have so recently gained. Muslims may have the least problem with the concept of surrender. "Islam" literally means "submission to God's will."

In explication, and to some degree expiation, of the term "surrender" as it is used in recovery, members of the community speak in such terms as "joining a winning team," and urge newcomers to "hang out with the winners." In admitting powerlessness over the disease, addicts are in effect gaining the power, through enlisting the support of their higher power and the fellowship itself, to be responsible for their own recovery. A misunderstanding of this process can lead to an interpretation that people in twelve-step recovery are somehow "copping out" from personal responsibility. The point is that while the addict may not be responsible for having a disease that involves physiological and possibly genetic, psychological, and environmental components, in twelve-step fellowships the addict is most certainly responsible for his or her own recovery.

Outside the United States there is strong professional resistance to both the DISEASE CONCEPT and twelve-step recovery. In France, for example, where the *toxicomanes*, physicians dealing with chemical dependency, are heavily invested in a psychotherapeutic approach, there is professional denial that twelve-step programs exist or, if they do, are effective with French clients. Several *toxicomanes* maintained that even if they, themselves, championed twelve-step recovery and attempted to refer clients to such programs, the French, with their heritage of individual freedom and idiosyncratic behavior and belief, would never abridge their freedom by joining such fellowships as AA.

Health professionals in such wine-producing and -consuming countries as ITALY, Spain, and France also express concern over the issue of addicts needing to abstain from all psychoactive substances. Wine, they maintain, is a food, and should not be included in such a blanket prohibition.

Acceptance of disease-concept-related treatment and recovery outside the United States has differed from culture to culture, from country to country, in some cases from community to community. In Scandinavia, for a studied example, Finland, Iceland, and SWEDEN have experienced phenomenal multiplication of AA groups since the 1970s, whereas Denmark and Norway have experienced a decline in groups over the same period. With the advent of glasnost, narcologists in the former Soviet Union discovered AA Since that time, treatment has been increasingly linked with recovery in Russia and other republics.

Overcoming Points of Resistance and Concern. The distance between cultures may seem like a chasm at times, but it is being bridged by such projects as the AAEFP that provide both recovery and a means to developing cultural parity. Society is changing rapidly, and fortunately, recovery has the flexibility to change along with it. Many groups within AA have learned that if there is no meeting that fits their special need, they can form their own meetings. The challenge is to adapt the process of treatment and recovery to all cultures and races, to counter stereotypes that recovery works only with certain groups.

(SEE ALSO: *Chinese Americans, Alcohol and Drug Use among; Ethnicity and Drugs; Hispanics and Drug Use; Rational Recovery; Sobriety; Treatment, History of; Women for Sobriety*)

BIBLIOGRAPHY

ALCOHOLICS ANONYMOUS. (1976). *Alcoholics Anonymous: Third edition.* New York: Alcoholics Anonymous World Services.

ALCOHOLICS ANONYMOUS. (1957). *Alcoholics Anonymous comes of age: A brief history of A.A.* New York: Alcoholics Anonymous World Services.

ALCOHOLICS ANONYMOUS. (1952). *Twelve steps and twelve traditions.* New York: Alcoholics Anonymous World Services.

BUXTON, M. E.; SMITH, D. E.; & SEYMOUR, R. B. (1987). Spirituality and other points of resistance to the 12-step process. *Journal of Psychoactive Drugs, 19*(3), 275–286.

HAFCI. (1990). Cocaine: Treatment & recovery: African American perspectives on crack. Vol. 2, tape 2. In Darryl Inaba and William E. Cohen (Eds.), *The Haight Ashbury training series.* San Francisco/Ashland, CA: Haight Ashbury Drug Detoxification, Rehabilitation & Aftercare Project and Cinemed.

JAMES, W. (1969). *The varieties of religious experience.* New York, Crowell-Collier. (Originally published in 1902).

SEYMOUR, R. B. (1992). Panel presentation at "To heal a wounded soul." Conference at Glide Memorial Church, San Francisco.

SEYMOUR, R. B., & SMITH, D. E. (1987). *Drugfree: A unique, positive approach to staying off alcohol and other drugs.* New York: Facts on File.

SMITH, D. E., ET AL. (1993). Cultural points of resistance to the 12-step recovery process. *Journal of Psychoactive drugs, 25*(1), 97–108.

STENIUS, K. (1991). Introduction of the Minnesota model in Nordic countries. *Contemporary Drug Problems, 18,* 151–179.

WILLIAMS, C. (1992). *No hiding place: Empowerment and recovery for our troubled communities.* San Francisco: HarperCollins.

ZIMMERMAN, R. (1988). Alcoholism treatment—Soviet style. *American Medical News,* November, 21–22.

DAVID E. SMITH
RICHARD B. SEYMOUR

ETHNICITY AND DRUGS In national statistics for the United States, it is common to see information about different segments of the population. For example, data from the U.S. Census and many national surveys on drug use often are subdivided in relation to four racial groups: (1) white, (2) black, (3) Asian or Pacific Islander, (4) American Indian or Alaska native. In concept, "racial heritage" refers to biologically inherited origins, but most people appreciate that these categories of race are determined more by social ideas and customs than by sharp genetic distinctions among these four groups. Some people even change their racial affiliation as they change their social perceptions.

In some national statistics and survey data, it also is common to see subdivisions in relation to "ethnic heritage," which sometimes refers to a person's country of origin but more generally refers to shared social and cultural characteristics. For example, people with recent or distant family origins in Spain or Portugal, or former colonies of Spain and Portugal (e.g., Mexico, Brazil), are called Iberian, Hispanic, or Latino; in North American statistics, it has been typical to subdivide the racial groups in relation to ethnicity as well: (1) White-not Hispanic, (2) Black-not Hispanic, (3) White-Hispanic, (4) Black-Hispanic, and so on. Here, too, the designation of Hispanic or Latino refers more to a social characteristic than to a specific family-genetic background. For example, American Indians from Mexico may be classified as Hispanic-American on the basis of their Mexican ancestry or as Native American on the basis of their North American Indian ancestry. The utility of these classifications of ethnicity and ethnic heritage depends on the degree to which they reflect sameness of social customs and learned behavior. People who are being compared within different ethnic groups ought to exhibit similarities in social customs and learned behaviors, and sometimes a shared sense of affiliation with that particular group. People across different ethnic groups ought to demonstrate more variation in social customs and learned behaviors than are to be seen among people within these groups.

There are many reasons for national reports to present statistical data on the population classified in relation to racial and ethnic heritage. Anyone reading historical documents for the period during and preceding the nineteenth century will find it difficult to escape a conclusion that these classifications were motivated in part by prejudice and racist thinking. Since the nineteenth century—from the earliest days of the U.S. Census—government officials have been interested in knowing the ethnic origins, as well as the size, of different racial and ethnic groups within the population for various policy and planning purposes.

Despite their somewhat questionable origins and uses, racial and ethnic classifications are important measures of social and historical phenomena in the United States. For example, in the area of public health, when national statistics on liver cirrhosis are examined, it can be seen that Americans who describe themselves as African-American are more likely to develop liver cirrhosis compared with Americans of predominantly European heritage. This type of information can guide public health action directed at preventing and treating liver cirrhosis. It is a help in targeting early detection and intervention efforts intended to reduce the suffering associated with liver cirrhosis. It may help identify specific environmental conditions such as poor nutrition or infectious diseases that might account for the higher risk of liver cirrhosis in the African-American segment of the population.

National statistics on alcohol and other drug use in relation to racial and ethnic heritage also have helped the nation's policymakers to see that some segments of the population have a greater need than others for alcohol and drug treatment and prevention services. Through block grants and other funding mechanisms, the federal, state, and local governments can provide support for services that target the special population groups with more needs for these services.

Although statistics on ALCOHOL and other drug use in relation to race and ethnic heritage can be used for the benefit of the population, it must be said that this topic has been understudied and the evidence often misrepresented. On the one hand, the topic is understudied in the sense that differences can be observed in alcohol and other drug use across racial and ethnic subgroups of the population, but it is not known whether they are due to differences in inherited predispositions or to other differences. On the other hand, the evidence of racial and ethnic differences in alcohol and drug use can be misrepresented and interpreted prejudicially as data showing one group to be inferior to another.

The complicated nature of this topic can be illustrated by considering liver cirrhosis among African Americans in the United States. In part, the occurrence of liver cirrhosis is determined by long-term heavy drinking of alcoholic beverages, but liver cirrhosis is also caused by prior infections or by autoimmune reactions, and vulnerability to alcohol-related liver cirrhosis is also influenced by cofactors such as poor nutrition. In the United States, African

Americans historically have been at great social disadvantage. On average, they are not as wealthy as other Americans, and, in addition, they more often live in poverty, with associated poor nutrition, underutilization of health care services, and compromised health status. Hence, it might be these socioeconomically related conditions that account for the excess occurrence of liver cirrhosis among African Americans rather than any inherited characteristics or personal characteristics related to drinking.

Within the United States, many other racial and ethnic minority groups also live with social disadvantages similar to those endured by African Americans. For this reason, it is easy to misinterpret national statistics on alcohol and drug use among racial and ethnic minority groups if they are taken strictly at face value. Instead, one must look beneath the surface and ask whether social or economic conditions might account for the statistics.

While studying racial and ethnic differences in CRACK smoking and other COCAINE use, some public health scientists have attempted to hold constant the social and neighborhood conditions that also could explain these differences. Once social and neighborhood characteristics had been taken into account, these studies found very little evidence to support the idea that African Americans or Hispanics were more likely to smoke crack or to take cocaine.

Although the importance of social and environmental influences in people's use of alcohol and other drugs has been clearly illustrated, it is important to keep in mind that biological factors may also play a role in determining one's preference for alcohol or particular drugs. For example, Asian Americans, as a group, consume less alcohol than any of the other racial or ethnic groups. Their lower drinking rates have been attributed, in part, to the fact that a majority of Asians possess a particular form of an alcohol-metabolizing enzyme whose action results in unpleasant side effects after drinking alcohol.

It also is interesting to find variation *within* large racial and ethnic groups, because this draws attention to the fact that not all African Americans are alike, nor all Hispanic Americans, Native Americans, Asians, or Pacific Islanders. For example, studying occurrence of alcohol abuse and dependence in different countries of Asia, epidemiologists found that men in South Korea had an extremely high prevalence of these conditions but men in Taiwan an extremely low prevalence. In addition, epidemiologists found more crack smoking among Hispanic Americans whose behavior showed that they had become acculturated to mainstream customs (e.g., by choosing to speak English rather than Spanish) and less crack smoking among other Hispanic Americans (e.g., those who chose to speak Spanish rather than English). This relationship was more pronounced among Hispanic Americans from Mexico than among those from Cuba, however, and this is an additional indication of variation within the large and growing Hispanic segment of the U.S. population.

Studies conducted on alcohol and other drug use by Native Americans provide another example of the variation that can exist within a large racial group. For instance, there is considerable variation in alcohol and other drug experiences from tribe to tribe, from one part of the country to another, and even from one residential location to another (e.g., boarding school students versus other young people). It becomes difficult, therefore, to summarize the alcohol and drug experiences of Native Americans in a few sentences. For many Native-American young people and adults living in urban environments, and sometimes on reservation lands as well, the use of alcoholic beverages and also INHALANT drugs is associated with several social and health problems. Researchers have speculated that the disintegration of Native-American culture has contributed to high rates of STRESS and that this in turn is related to a disproportionately high use of alcohol among this segment of the American population. These statistics alerted the attention of public health workers and government officials, and through their efforts many programs have been initiated to draw Native Americans with alcohol abuse problems into treatment.

Racial and ethnic patterns of alcohol and other drug use and related problems vary by age, gender, and drug. National surveys of high school seniors conducted since the early 1970s, and more recent surveys that included eighth- and tenth-graders, reveal that some minority youth use less alcohol and other drugs than Caucasian youth. Specifically, Caucasians, Native Americans, and Mexican Americans have the highest frequency of reported alcohol use whereas African Americans and Asian Americans have the lowest. Because these surveys include only in-school youth and not children who have dropped out of school, it may be that the true proportions of alcohol and other drug use have been underestimated.

In general, males report using drugs more frequently than females, and this gender difference cuts across racial and ethnic boundaries. For example, African-American males and Caucasian males are more likely than African-American and Caucasian females to use alcohol. It is also true that people in different age groups vary in relation to their reports of using alcohol and other drugs. When researchers carefully divide different racial and ethnic groups by age, some interesting trends in alcohol-use patterns appear. For Caucasian adults, drinking tends to increase until mid to late life, with older people drinking less as a group than younger adults. African Americans, however, tend to be heavier drinkers later in life and to exhibit more alcohol-related health problems (e.g., cirrhosis, esophageal cancer). For some drugs other than alcohol, a similar picture exists. For example, Caucasians and Hispanic Americans report using cocaine earlier in life whereas African Americans report using it later in life. Cigarette SMOKING is more common among young Caucasians (12–17 years old) than it is among Hispanic Americans or African Americans of the same age; however, a higher proportion of the latter groups report smoking later in life.

It is sometimes difficult to interpret findings that point to differences in drug use between minority and nonminority subgroups within the U.S. population. It must be kept in mind that socially shared environmental conditions (e.g., availability of drugs, neighborhood conditions, economic resources) rather than race or ethnic identity may be underlying patterns of drug use. Other factors such as social status and community norms for coping with life stresses may account for reported racial or ethnic differences in drug use.

Continued research is needed to track patterns of alcohol and other drug use in the population and to find out the mechanisms or the reasons that put some groups at higher risk than others for problematic involvement with alcohol and other drugs. Some of the most current information is limited. For instance, minority intravenous drug users are known to have higher rates of exposure to HIV than Caucasian drug users, but no clear explanation for this observation has been determined. Perhaps learning more about barriers to obtaining treatment for intravenous drug use in certain minority populations will contribute to an understanding of this problem.

Researchers, as well as policymakers, need to be culturally sensitive; that is, they must appreciate the social, cultural, and economic conditions that underlie racial and ethnic differences in alcohol and drug use. It is important to realize that racial and ethnic identification can serve as a source of strength to those who design targeted prevention and intervention programs for certain segments of the population.

(SEE ALSO: *Asia, Drug Use in; Causes of Substance Abuse; Chinese Americans, Drug and Alcohol Use among; Epidemiology of Drug Abuse; Ethnic Issues and Cultural Relevance in Treatment; Families and Drug Use; Injecting Drug Users and HIV; Poverty and Drugs; Vulnerability as Cause of Drug Abuse; Women and Substance Abuse*)

MARSHA LILLIE-BLANTON
AMELIA ARRIA

ETHNOPHARMACOLOGY This branch of pharmacology studies the use and lore of drugs that have been discovered and developed by sociocultural (or ethnic) groups. It involves the direct observation and report of interactions between the societies and the drugs they have found in their natural environments and the customs that have evolved around such drugs, whether ceremonial, therapeutic, or other. These drugs, usually found in plants (hence similar study by ethnobotanists as well as ethnologists), are described—as are their effects within the customs, beliefs, and histories of a traditional culture or a specific society.

Examples include descriptions of the use of coca leaves (*Erythroxylon coca*) by indigenous populations of Colombia and Peru, for increased strength and endurance in high altitudes; the ceremonial use of PEYOTE (*Lophophora sp.*) by Native Americans of the Southwest and Mexico; and the use of KAVA (*Piper methysticum*) in ceremonial drinks by the indigenous populations of many South Pacific islands.

(SEE ALSO: *Asia, Drug Use in; Dover's Powder; Plants, Drugs from*)

BIBLIOGRAPHY

EFRON, D. H. (Ed.). (1967). *Ethnopharmacological search for psychoactive drugs*. Public Health Service Publication No. 1645. Washington, DC: U.S. Department of Health, Education, and Welfare.

NICK E. GOEDERS

ETHYL ETHER *See* Inhalants.

EUROPE AS TRANSIT AREA FOR ILLICIT DRUGS *See* International Drug Supply Systems.

EXCLUSIONARY RULE In legal proceedings, the exclusionary rule prohibits the use of any evidence obtained in contravention of the U.S. Constitution. The rule is frequently invoked when government authorities seize evidence in violation of the Fourth Amendment's prohibition against unlawful searches and seizures. Evidence may be illegally obtained when government officials do not have a warrant to search an individual's premises or the warrant is defective. Law-enforcement officers may also lack sufficient probable cause to arrest a person. In addition, the courts may invoke the exclusionary rule when they find a violation of an individual's Fifth Amendment right against self-incrimination or a violation of a defendant's Sixth Amendment right to counsel. Courts often refer to evidence obtained in violation of the Fourth, Fifth, or Sixth Amendment as "tainted" or "the fruit of a poisonous tree."

The exclusionary rule has developed from 1960 to the present based on its initial judicial articulation in the case of *Mapp v. Ohio*. The U.S. Supreme Court extended the rule under the due process clause of the Fourteenth Amendment to require states to exclude evidence obtained from an unconstitutional search or seizure. The Court has often cited an individual's right to privacy and the deterrence of unreasonable police conduct as the primary reasons for excluding evidence obtained from an unreasonable search and seizure.

A number of exceptions to the exclusionary rule have emerged to reduce the effects of the doctrine, such as a police officer's "good-faith" belief that an otherwise defective warrant is valid, evidence obtained in "hot pursuit," or evidence seized in "plain view" of the law-enforcement officer's sight and reach.

IMPORTANCE IN DRUG CASES AND ENFORCEMENT

The exclusionary rule prohibits the introduction of constitutionally tainted evidence. The effect of the doctrine has often been the exclusion of evidence that might be used to convict a suspected drug trafficker or abuser. Courts have excluded evidence of drug PARAPHERNALIA or supplies illegally seized, admissions obtained by coercion or without notifying the party of the right to remain silent, and evidence obtained in violation of a defendant's Sixth Amendment right to counsel, such as a lineup identification. The Supreme Court has determined that it is preferable to allow a drug trafficker to go free than to permit law-enforcement officers to violate a citizen's constitutionally protected rights.

(SEE ALSO: *Drug Laws: Prosecution of; Seizures of Drugs*)

ROBERT T. ANGAROLA
ALAN MINSK

EXECUTIVE OFFICE OF THE PRESIDENT *See* U.S. Government: The Organization of U.S. Drug Policy.

EXISTENTIAL MODELS OF ADDICTION *See* Values and Beliefs: Existential Models of Addiction.

EXPECTANCIES The beliefs a person has about the effects a drug will have are called *expectancies*. The study of expectancies began with the employment of the experimental balanced-placebo design in alcohol research in the early 1970s (see Marlatt & Rohsenow, 1980, for a review). Research on people ranging from light drinkers to inpatient alcoholics revealed that expectancies are predictive of some of the behaviors exhibited when people use a drug. These studies revealed that both the beliefs an individual has—about whether a drink contains ALCOHOL and the specific outcomes that individual expects from consuming alcohol—are in many cases more predictive of subsequent behavior than the pharmacological effects of the drug.

EXAMPLES OF RESEARCH STUDY

An example of research using balanced-placebo design is as follows: In a simulated bar setting, half the participants in a study are told they will receive a drink containing vodka and tonic, and half are told they will receive a drink containing only tonic. After

this expectation is established, half of each group does receive vodka and tonic, while the other half receives only tonic, resulting in four groups: (1) those who expect vodka and tonic and receive vodka and tonic, (2) those who expect vodka and tonic and receive only tonic, (3) those who expect tonic and receive vodka and tonic, and (4) those who expect tonic and receive tonic. Thus, some of the people who expect alcohol receive only tonic, and some who expect only tonic receive a mix containing alcohol.

Behavioral observations following this manipulation reveal that the most powerful predictor of behavior after consuming the assigned drink is not whether the person actually receives alcohol, but whether that person *believes* he or she is drinking alcohol: People who expect alcohol in this experimental situation consume significantly more drink than those who are not expecting alcohol, regardless of whether or not they do receive alcohol in their drink. With the discovery of this phenomenon, even in people who are considered dependent on alcohol, this finding has been interpreted as providing contrasting evidence to the disease model's notion that "loss of control" is caused exclusively by the pharmacological effects of alcohol; the findings introduced the idea that cognitive factors are influential in a person's drug-related behavior. The presence of expectancy effects have also been identified in research on drugs other than alcohol, including TOBACCO and MARIJUANA (Marlatt & Gordon, 1985).

Most of the research on expectancies during the 1970s and 1980s was conducted on college students, with samples ranging from light to heavy social drinkers who were primarily Caucasian. This research has shown that the effect of a person's expectancies depends on whether the behavior involved is socially mediated: Stronger expectancy effects are found for social behaviors (e.g., aggression or sexual arousal) than for nonsocial behaviors (e.g., beliefs concerning motor coordination or memory skills); they are stronger for outcomes that are perceived as positive (e.g., sexual arousal) than as negative (e.g., poor motor coordination).

For socially mediated behaviors, expectancy research has revealed that college students of both sexes show less anxiety in social situations if they believe they have consumed alcohol. In addition, males show heightened sexual arousal when exposed to an erotic environment if they believe they have consumed alcohol (Marlatt & Gordon, 1985). Men and women of college age have also both been found to respond more aggressively when provoked after they believe they have consumed alcohol. Sex differences have been found on the effects of alcohol on anxiety with persons of the opposite sex: Women of college age have shown more anxiety in the company of an unfamiliar man when they believe they have consumed alcohol, while men of college age have shown reduced anxiety when in the company of an unfamiliar female. The results have been interpreted as reflecting gender differences regarding the acceptability of alcohol in social situations with a stranger of the opposite sex.

OTHER STUDIES

Other experimental work has revealed that specific outcomes can vary with the personal beliefs an individual holds regarding alcohol and with the phase of intoxication of an individual (Southwick et al., 1981). Overall, the results based on expectancy research point to the likelihood that people may have established cultural beliefs regarding the effects of alcohol in social situations and that these beliefs play some role in the behavioral effects of alcohol.

Research has also found that expectancies do predict drinking behavior over a one-year period for early adolescents (Christiansen et al., 1989); that expectancies tend to crystallize in people at a young age and that they tend to be resistant to change (Miller, Smith, & Goldman, 1990). Other studies on Caucasian adolescents and young adults have found that those who have mostly positive and only few negative outcome expectancies tend to experience more alcohol-related problems than those whose outcome expectancies are more evenly divided between positive and negative effects (Brown, Christiansen, & Goldman, 1987).

Since the late 1980s, researchers have begun to examine ethnic and racial differences in the expectancy variable. One study of college-age students (Daisy, 1989) revealed that Native-American students had significantly stronger expectancies for the positive social and physical effects of drinking than did Asian-American students. Caucasian students were found to have stronger positive expectancies for social and physical effects than did Asian-American students, but less than did Native-American students. These beliefs concerning the effects of alcohol were also found to be highly associated with the drinking patterns of the study participants: those people whose drinking pattern was considered heavy had stronger beliefs in the above expectancies than individuals who drank less. The study strongly suggests that ethnic differences exist in alcohol-related expectancies, and it confirms that expectancies are related to the amount of alcohol consumed.

The association between expectancies and drinking pattern has been consistent in the research and has therefore become targeted in substance-abuse treatment. Expectancies have been found to influence the way a person copes with high-risk situations after treatment aimed at abstinence (Marlatt & Gordon, 1985; Condiotte & Lichtenstein, 1981). In RELAPSE PREVENTION, positive-outcome expectancies are viewed as the source of urges or cravings for a substance. Treatment according to this perspective therefore includes changing a client's outcome expectancies: If a person believes that drinking will provide immediate relief from stress, then treatment focuses on helping that person consider the long-range implications of drinking—helping the person by adding the negative outcomes of drinking to the anticipated positive results of drinking—and thereby changing the composition of the person's outcome expectancies.

Self-efficacy expectancies, or how effectively one feels he or she can cope with a high-risk situation, are also examined in treatment. If a client lives a stressful lifestyle and believes that only alcohol provides relief from that stress, the therapist helps the client develop and utilize alternative methods for coping with stress. For example, clients can be taught to look forward to meditation or exercise or other positive-reward situations to help cope with stress and to reduce urges and the resulting temptation to drink. Treatment focuses on developing alternative coping strategies for a client's individual high-risk situations, and therefore includes an ongoing assessment of each client's high-risk situations.

Self-efficacy differs from overall motivation to quit or reduce substance use, since perceived control will vary across situations. In research on relapse prevention, self-efficacy has been found to be predictive of the first use of the substance after abstinence-based treatment: Those people who do not believe they can cope with either a specific situation or cope, in general, with the temptation to use a substance are more likely to relapse in the face of a high-risk situation than are people who believe that they are able to maintain their goal of abstinence in the same situation (Condiotte & Lichtenstein, 1981).

(SEE ALSO: *Coping and Drug Use*; *Disease Concept of Alcoholism and Drug Abuse*; *Ethnicity and Drugs*; *Prevention*; *Treatment*; *Women and Substance Abuse*)

BIBLIOGRAPHY

BROWN, S. A., CHRISTIANSEN, B. A., & GOLDMAN, M. S. (1987). The alcohol expectancy questionnaire: An instrument for the assessment of adolescent and adult expectancies. *Journal of Studies on Alcohol, 48*(5), 483–491.

CHRISTIANSEN, B. A., ET AL. (1989). Using alcohol expectancies to predict adolescent drinking behavior after one year. *Journal of Consulting and Clinical Psychology, 57,* 93–99.

CONDIOTTE, M. M., & LICHTENSTEIN, E. (1981). Self-efficacy and relapse in smoking cessation programs. *Journal of Consulting and Clinical Psychology, 49,* 648–658.

DAISY, F. (1989). Ethnic differences in alcohol outcome expectancies and drinking patterns. Ph.D. dissertation, University of Washington.

MARLATT, G. A., & GORDON, J. R. (1985). *Relapse prevention: Maintenance strategies in the treatment of addictive behaviors.* New York: Guilford Press.

MARLATT, G. A., & ROHSENOW, D. J. (1980). Cognitive processes in alcohol use: Expectancy and the balanced placebo design. In N. K. Mello (Ed.), *Advances in substance abuse.* Greenwich, CT: JAI Press.

MILLER, P. M., SMITH, G. T., & GOLDMAN, M. S. (1990). Emergence of alcohol expectancies in childhood: A possible critical period. *Journal of Studies on Alcohol, 51*(4), 343–349.

SOUTHWICK, L., ET AL. (1981). Alcohol-related expectancies: Defined by phase of intoxication and drinking experience. *Journal of Consulting and Clinical Psychology, 49,* 713–721.

ALAN MARLATT
MOLLY CARNEY

F

FAMILIES AND DRUG USE One major debate in the area of families and drug use continues to be whether dysfunctional family life creates drug addiction or whether drug addiction produces dysfunctional families. In other words, are ALCOHOLISM and other drug addictions diseases of individuals or are they products of disorganized families and other social systems? The former is an "individual-focused" view, often held by drug counselors who favor SELF-HELP groups such as AA, Al-Anon, NA, and the like. The latter is a "systemic" view held by professionals who prefer to treat drug addictions by working with families, in order to change family systems into more healthy environments.

Whatever one's position in this debate, almost everyone agrees that the family is the primary socializing agent in society. However, Glick (1988), a senior family demographer, observed that during the past fifty years American families have been undergoing significant transformations. Social acceptance of various forms of families is steadily replacing the older, normative view of a family as comprising only two parents and their children, with the father as a breadwinner and the mother as a homemaker. In the 1960s and 1970s, decades of social protests, Americans witnessed increasing numbers of cohabiting couples, families being maintained by single parents, and many adults living alone. As a result, divorce, single-parenthood, childlessness, and living alone have become more acceptable. Significant transformation has also occurred in gender attitudes, which moved toward greater egali-

tarianism and resulted in increased percentages of young men and women who perceived fatherhood as a fulfilling experience (Lewis, 1986; Thornton, 1989).

These changes continued to occur until the early 1980s when they began to level off, and by 1987 a quarter of all children under eighteen years of age no longer lived with both of their parents. Eighty-two percent of these children lived with stepfathers, whereas only 18 percent lived with stepmothers. The late eighties and early nineties, however, seem to have been a period of stabilization, during which all trends flattened (Glock, 1988; Thornton, 1989).

No systematic analysis has been conducted to assess the association between these social and demographic changes in the family and trends in drug abuse. If one looks at the statistics closely, however, one sees that the trends in families and in drug use look similar. A dramatic increase in the abuse of all kinds of drugs by all age groups was observed during the early 1970s to early 1980s. These trends in drug use also flattened in the early eighties and, as was observed in 1988, are beginning to drop significantly, especially among youth aged twelve to seventeen years.

This line of reasoning is not meant to suggest that the changes in attitudes toward families and the changes in family structures and forms in the last three decades directly caused the current trends in drug use. It may suggest, however, that the instability of families either allows there to be or imposes greater stresses upon individuals and society. Similarly, the stabilization of families provides more secure environments for individuals, who may then more effectively cope without the abuse of substances.

There is nevertheless some evidence and much speculation about a reciprocity between an individual's drug addictions and "family illnesses," since the latter often appear to be passed from one generation to another.

Although recent reductions in the use of illicit drugs present a somewhat optimistic picture of the future of American families, the overall number of drug casualties is still grim and the consequences are debilitating. Every year, 100,000 Americans die as the result of drug abuse. That number should increase with the spread of AIDS. Alcohol, nicotine, and illicit drug abuse are number-one health problems, especially among the young. Life expectancy has steadily risen over the past seventy-five years in all age groups except that for youth aged fifteen to twenty-four, who now have a higher death rate because of injuries and disappearances related to drug use. Long-term substance abuse is associated with DEPRESSION, hostility, malnutrition, lower social and intellectual skills, broken relationships, mental illness, economic losses, and growing CRIME rates.

FAMILY PREDICTORS OF DRUG ABUSE

Family factors that predict drug use may be put into three interrelated categories: structural, historical, and interpersonal. The structural factors pertain to family composition, such as single- or two-parent families, the number of children, sibling spacing, and gender composition. Family historical factors specifically refer to intergenerational patterns, such as the extent and influence of drug usage in the family of origin. Finally, interpersonal factors relate to interpersonal dynamics in the family, such as those reflected in the quality of marital relationships or the quality of parent-and-child or sibling relationships.

Family Structural Factors. Three structural factors—parental composition, family size, and birth order—are the most often included variables referred to in drug and family research. Although these factors seem to contribute to the etiology of drug abuse, one needs to look at the findings more critically to try to evaluate the extent of their influence.

The literature on drug abuse is replete with findings that suggest that, compared with traditional nuclear families, disorganized, especially single-parent, families are more vulnerable environments for children. These families are associated with an earlier onset and greater degree of drug and alcohol abuse. Information regarding the role of family size and birth order, however, is currently insufficient. According to the data, there are very limited indications that an only child is the least at risk, whereas families with seven or more children are at greater risk for drug abuse. However, there seem to be fewer cases of drug abuse involving first-born children compared with the number of cases involving subsequent, especially last-born, children (Barnes, 1990; Glynn, 1984).

Stanton (1985), Hawkins et al. (1987), and Wells and Rankin (1991) have argued that family structural factors do not contribute much to our understanding of drug-abuse behavior. More important risks for children, they suggest, lie in family processes and the quality of family environments.

Divorce, for example, may be a healthy way of ending a hostile marital relationship. The separation of parents may only be the culmination of hostile relationships, painful negotiations, and the draining of family resources prior to the family breakup. Sessa and Steinberg (1991) argue that the most important impact of divorce on children is how much it disturbs the children's developmental tasks—for example, their autonomy. Most children experience relatively brief adjustment problems following a divorce, but continued development of the adjustment process depends on many more factors, such as the age of the children, the gender, the custodial parent, and the quality of life in the home after the divorce.

Different forms of families may possess varied abilities to exercise certain parenting practices, like monitoring and supervision. Dishion, Patterson, and Reid (1988) found interesting linkages between living in a single-parent family, poor parental monitoring, and greater adolescent involvement with drug-abusing peers. In a supportive family relationship, however, parental composition is not a predictor of adolescent drug use.

Variations in family size may impose certain restrictions and may afford opportunities for the utilization of family resources, such as parental support and finances. Birth order seems to expose each child to different opportunities for social learning (e.g., in regard to role models) and different behavioral expectations, depending on one's family traditions. It is therefore important to look at family processes and the quality of family environments as well as at the family structure.

Family History Factors. Some well-established evidence indicates that drug use by any member of the family is related to drug use by other family members. In couple relationships, the initiation of a female partner into illicit drug use and her progression toward drug dependency are related to patterns of drug use in the male partner, whereas illicit drug use by the male partner is more independent of spousal drug use (Weiner, Wallen, & Zankowski, 1990).

Parental and sibling drug use have consistently been found to be associated with ADOLESCENT drug-abusing behavior (Hawkins et al., 1986). The transmission of the problem behavior, however, is perceived differently by different scholars. Although there is an increasing fascination with GENETIC explanations, more research is needed to validate genetic assumptions (e.g., Cadoret, 1990; Searles, 1990, 1991). In their view of the literature, Hawkins et al. (1986) concluded that the evidence from behavioral genetic research was limited to male ALCOHOLISM and the lack of convergent evidence from adoption, twin, and biological response studies. Similar criticism has been presented by Searles (1990, 1991), who also argued that only 20 percent of children of alcoholics become alcoholics and that half of all alcoholics do not have a family history of alcoholism. Research on the family clustering of OPIATE and ALCOHOL abusers indicates that a genetic explanation is inadequate when it is considered that the community or environment affects the choice of the substance of dependence.

A systemic (family) approach presents more compelling explanations. Research focusing on the role of parental attitudes and values has revealed a high congruence between parents' and adolescents' perceptions of the use and abuse of drugs (Barnes, 1990). When parents use drugs such as CIGARETTES and alcohol, it indicates to the children that such use is expected (or at least allowed) in the family.

Heavy drug use in the family, especially by parents, also disrupts functional properties of the family system (e.g., care and support, problem solving, etc.), and this, in turn, provides a conducive environment for drug use and abuse by other members of subsequent generations (Steinglass et al., 1987). Dishion and Loeber (1985) argued that parental drug use diminishes parental ability to exert effective monitoring and supervision, thus allowing children to mingle with peers who abuse drugs frequently. Clinical observation also suggests that parental drug use blocks effective communication, alters modes of interpersonal relations, and is associated with all kinds of child abuse (Barnes, 1990; Leonard & Jacob, 1988).

Interpersonal Factors. There are at least two broad dimensions of interpersonal dynamics in the family—support and control—and one facilitating dimension—communication (Barber, 1992; Rollins & Thomas, 1979). The support dimension refers to the positive affective experience associated with relationships, such as acceptance, encouragement, security, and love. The control dimension pertains to the extent to which children's behavior is restricted by the caregiver(s), and this ranges from establishing rules and discipline to varieties of physical coercion (e.g., hitting and yelling). Familial support is regarded as the most robust variable in the prevention of all kinds of delinquent behaviors in children and adolescents (Baumrind, 1991; Gecas & Seff, 1990). Different aspects of support have recently been identified, such as general support, physical affection, companionship, and sustained contact (Gecas & Seff, 1990), all of which are negatively associated with socially unacceptable behaviors. Coombs and Landsverk (1988), for example, found consistent evidence that maintaining a rewarding parent-child relationship deters substance abuse during childhood and adolescence (see also reviews by Glynn, 1984; Hawkins et al., 1986). Parental praise and encouragement, involvement and attachment or perceived closeness, trust, and help with personal problems are all characteristics of the families of abstainers, whereas parental rejection, conflicts, manipulative relations, and overinvolvement are related to the earlier onset and continued use of drugs (Baumrind, 1991; Hawkins et al., 1986).

The control dimension is more complex than the support dimension, since one needs to differentiate between types of control. Baumrind (1987, 1991), for example, distinguished between authoritative and authoritarian controls. The first is characterized by a combination of warmth, supervision, and opportunity for negotiation; this type of control is associated with positive outcomes. In her study of drug-abusing adolescents, Baumrind found that authoritative control characterized the families of abstainers and soft experimental drug users. Authoritarian control, on the other hand, is based on force, threats, and physical punishment; this is the type of control that characterized the families of dependent drug users. Other studies have revealed that sexual abuse and physical abuse are prevalent in the families of drug abusers.

It has been especially well documented that families with inconsistent or no clearly defined rules also have adolescents who abuse drugs (see Baumrind, 1987; Coombs & Landsverk, 1988; Hawkins et al., 1986; Volk et al., 1989). The constantly changing rules in some families jeopardize parental ability to monitor and supervise children and make it difficult for the children to adapt to family expectations.

In order to function within these two dimensions, families must rely on their communication mechanism. To give support or exert control over others, it is necessary to communicate one's intents. Watzlawick, Beavin, & Jackson (1967) believe that when people communicate, the communication also defines their relationships with other persons. They also believe that to be able to define the relationship, those who communicate should be able to understand each other's perceptions regarding what they talk about and regarding their relationship. In a family where drug use is prevalent, communication is heavily loaded with interpersonal misperception and exchanges of negative affect. Studies also indicate that communication in these families is frequently blocked either by the use of drugs or feelings of not being understood (Hawkins et al., 1986; Jurich et al., 1985; Piercy et al., 1991).

The Family and Other Systems. The peer group and school are two other systems to be considered when the adolescent member of the family who is involved in drug abuse. These systems intervene with their own parenting practices, because they provide much of the environment for learning VALUES, attitudes, and norms as far as expected behaviors are concerned (behaviors that may or may not be expected by the adolescent's family).

It is well known that most new drug users are introduced to drugs by peers and that peers help maintain patterns of use, including greater dependent use. To assess the influence of peers, one should assess the following indicators (Agnew, 1991): (1) time spent with peers, (2) the degree of attachment to peers, and (3) the extent of peer delinquency or drug use.

Although researchers find consistent evidence of the relationship between school DROPOUTS, low performance and underachievement in school, and drug abuse, it is not known when school factors become developmentally salient as possible predictors of drug abuse (Hawkins et al., 1986). Some research indicates that a low grade-point average and dropping out of school are strongly associated with children's involvement with drug-abusing peers. It is clear, on the other hand, that parental involvement in children's schoolwork and activities reduces the changes of a child being seriously involved in drug use.

Hawkins et al. (1987) documented limited evidence with regard to the association of drug use and the social isolation of the family. The 1990 NATIONAL HOUSEHOLD SURVEY indicates that current drug users are concentrated within underprivileged families of lower social economic status and within communities of color.

IMPLICATIONS FOR PREVENTION

In the last ten years, those responsible for drug-PREVENTION efforts have discovered that (1) the most effective programs are multilevel programs; (2) it is most cost-effective to target youth aged twelve and younger; (3) the family is the most influential context within which to set programs, especially with drug users who are younger and female; and (4) LIFE-SKILL programs rather than knowledge-oriented programs are most effective in preventing drug abuse.

In the assessment phase, one can determine the risk status of a family by looking at the intergenerational history of drug usage, reported child abuse, the children's academic performance, the degree of parental involvement in schools, and the characteristics of the community in which the family lives (e.g., population density, extent of economic and social deprivation, rates of criminal activity and drug abuse behavior).

In the program development phase, one may well consider issues embedded in (1) individual and family development (Baumrind, 1991; Steinglass et al., 1987), (2) culture and gender (Weiner et al., 1990), and (3) health and economy, both of which affect the individual and the family (Bush & Iannotti, 1987; Conger et al., 1991). One could also determine how these issues are interconnected in order to come up with the best possible program for specific populations.

In the implementation phase, matching of staff and target group and the ways in which the programs are delivered may affect the outcomes. It may be wise to staff prevention programs delivered in cultures other than the mainstream culture with personnel of similar backgrounds or with those who have an adequate knowledge of that specific culture. Positive and nonthreatening approaches that combine both information and life-skill building are most effective. Parental or significant-other involvement with involvement by the school give programs the most credibility to youth.

FAMILY TREATMENT

As described earlier, dysfunctional family life is one potential contributor to the development of drug addictions in family members. The reciprocal nature of addictions and disorganized families, however, is evident in that not only may dysfunctional families produce addictive behaviors in their members, but these addictions, in turn, may affect the quality of family life, thus negatively impacting the behavior of family members and devitalizing or fracturing family relationships. The most demoralizing aspect of this reciprocity is that drug addictions are often passed from earlier generations to later generations, unless this pattern can be ended by successful treatment or intervention.

Until the mid-1980s, very few drug treatment programs *directly* utilized spouses, parents, or other family members in their treatment of the identified patient. After that time, family therapy became the treatment of choice for most drug abusers, especially in the area of alcoholism treatment. A growing body of research findings has shown that family-centered drug interventions are very effective in getting family members off drugs and keeping them off (Lewis & McAvoy, 1984).

There is evidence, for example, that family groups given systemic family interventions have a higher treatment success rate—that is, decreased drug dependence and less recidivism (Stanton & Todd, 1982). In contrast, if adolescents are treated individually and their family system has not changed, they often return home to resume the same roles and behaviors that had earlier fostered their addictive behaviors.

The inclusion of other family members in an adolescent's drug treatment does add to the complexity of the treatment. Yet this addition often gives a fam-

ily therapist greater leverage for sustained and successful drug treatment (Lewis & McAvoy, 1984), because of the drug abuser's wish to maintain family love and relationships. Strengthening family relationships may therefore help to reduce or eliminate an individual's addictive behaviors.

Some of the better known interventions currently used in the field of alcoholism treatment are treatments based on family systems. For instance, research has revealed that the spouses of alcoholics often play roles that support their spouse's addiction (through co-dependency). Changes in the spouse's behavior and roles, however, can also contribute to the effective treatment of the spouse's alcoholism (Steinglass et al., 1987).

Systemic family treatment has also been widely utilized in the treatment of adolescents' drug abuse, according to the successful research conducted by Stanton and Todd (1982) with adult heroin addicts. In this programmatic research, one of the best controlled studies of family therapy, the researchers found a significant decrease in the heroin usage of young adults when family-focused therapy was employed.

A longitudinal study of 136 adolescents (Lewis et al., 1989) also documents the relative effectiveness of a family therapy program as compared to a family education program and treatment-as-usual (i.e., individual counseling). In this study, the two brief family-based drug interventions together reduced the drug use of nearly one-half (46%) of the adolescents who received them. This success is thought to be due primarily to the fact that both of these outpatient interventions focused on the systemic treatment of entire family groups. In contrast, the family therapy intervention seemed to have been more effective in significantly reducing adolescent drug use for a greater percentage of the adolescents (54.6% compared with 37.5%). Thus family-based interventions (especially family therapy) can be potent and viable drug-treatment programs.

The best drug treatment, however, may be a combined treatment (Lewis, 1989), in which individual treatment focuses on the teaching of social skills and strategies for coping with stress, whereas the emphasis of the family treatment component is on increasing the nurturance and parenting skills of other family members. It is at the intersection of these two approaches that much of the current creativity seems to be taking place. Even though their focus and methods may differ, it is good for these two arenas of inquiry to become better known to each other, since each has a wealth of understanding to contribute to the other.

(SEE ALSO: *Adjunctive Drug Taking; Codependence; Comorbidity and Vulnerability; Conduct Disorder; Ethnic Issues and Cultural Relevance in Treatment; Ethnicity and Drugs; Poverty and Drug Use; Treatment Types; Vulnerability As Cause of Substance Abuse*)

BIBLIOGRAPHY

AGNEW, R. (1991). The interactive effects of peer variables on delinquency. *Criminology, 29,* 47–72.

BARBER, B. K. (1992). Family, personality, and adolescent problem behaviors. *Journal of Marriage and the Family, 54,* 69–79.

BARNES, G. M. (1990). Impact of family on adolescent drinking patterns. In R. L. Collins, K. E. Leonard, & J. S. Searles (Eds.), *Alcohol and the family: Research and clinical perspectives.* New York: Guilford Press.

BARNES, G. M. (1984). Adolescent alcohol abuse and other problem behaviors: Their relationships and common parental influences. *Journal of Youth and Adolescence, 13,* 329–348.

BAUMRIND, D. (1991). The influence of parenting style on adolescent competence and substance abuse. *Journal of Early Adolescence, 11,* 56–95.

BAUMRIND, D. (1987). A developmental perspective on adolescent risk taking in contemporary America. In C. E. Irwin, Jr. (Ed.), *Adolescent social behavior and health.* San Francisco: Jossey-Bass.

BUSH, P. J., & IANNOTTI, R. (1987). The development of children's health orientations and behaviors: Lesson for substance use prevention. In C. L. Jones & R. J. Battjes (Eds.), *Etiology of drug abuse: Implications for prevention.* Rockville, MD: National Institute on Drug Abuse.

CADORET, R. J. (1990). Genetics of alcoholism. In R. L. Collins, K. E. Leonard, & J. S. Searles (Eds.), *Alcohol and the family: Research and clinical perspectives.* New York: Guilford Press.

CONGER, R. D., ET AL. (1991). A process model of family economic pressure and early adolescent alcohol use. *Journal of Early Adolescence, 11,* 430–449.

COOMBS, R. H., & LANDSVERK, J. (1988). Parenting styles and substance use during childhood and adolescence. *Journal of Marriage and the Family, 50,* 473–482.

DISHION, T. J., PATTERSON, G. R., & REID, J. R. (1988). Parent and peer factors associated with sampling in early adolescence: Implications for treatment. In E. R. Rah-
dert & J. Grabowski (Eds.), *Adolescent drug abuse: Analyses of treatment research.* Rockville, MD: National Institute on Drug Abuse.

GECAS, V., & SEFF, M. A. (1990). Families and adolescents: A review of the 1980s. *Journal of Marriage and the Family, 52,* 941–958.

GLICK, P. C. (1988). Fifty years of family demography: A record of social change. *Journal of Marriage and the Family, 50,* 861–873.

GLYNN, T. J. (1984). Adolescent drug use and the family environment: A review. *Journal of Drug Issues, 14,* 271–295.

HAWKINS, J. D., ET AL. (1987). Delinquents and drugs: What the evidence suggests about prevention and treatment programming. In B. Brown & A. Mills (Eds.), *Youth at high risk for substance abuse.* Rockville, MD: National Institute on Drug Abuse, 1987.

HAWKINS, J. D., ET AL. (1986). Childhood predictors of adolescent substance abuse: Toward empirically grounded theory. *Journal of Children in Contemporary Society, 18,* 11–48.

JURICH, A. P., ET AL. (1985). Family factors in the life of drug users and abusers. *Adolescence, 20,* 143–159.

LEONARD, K. E., & JACOB T. (1988). Alcohol, alcoholism, and family violence. In V. B. van Hasselt et al. (Eds.), *Handbook of family violence.* New York: Plenum.

LEWIS, R. A. (1989). The family and addictions. *Family Relations, 38,* 254–257.

LEWIS, R. A. (1986). What men get out of marriage and parenthood. In R. A. Lewis & R. E. Salt (Eds.), *Men in families.* Newbury Park, CA: Sage Publications.

LEWIS, R. A., & MCAVOY, P. (1984). Improving the quality of relationships: Therapeutic intervention with opiate-abusing couples. In S. Duck (Ed.), *Personal relationships: Vol. 5. Repairing personal relationships.* New York: Academic Press.

LEWIS, R. A., ET AL. (in press). The Purdue Brief Family Therapy Model for adolescent substance abusers. In T. Todd & M. Selekman (Eds.), *Family therapy approaches with adolescent substance abusers.* New York: Gardner Press.

PIERCY, F. P., ET AL. (1991). The relationships of family factors to patterns of substance abuse. *Family Dynamics of Addiction Quarterly, 1,* 41–54.

ROLLINS, B. C., & THOMAS, D. L. (1979). Parental support, power, and control techniques in socialization of children. In W. R. Burr et al. (Eds.), *Contemporary theories about the family* (Vol. 1). New York: Free Press.

SEARLES, J. S. (1991). The genetics of alcoholism: Impacts on family and sociological models of addiction. *Family Dynamic of Addiction Quarterly, 1*(1), 8–21.

SEARLES, J. S. (1990). The contribution of genetic factors to the development of alcoholism: A critical review. In R. L. Collins, K. E. Leonard, & J. S. Searles (Eds.), *Alcohol and the family: Research and clinical perspectives.* New York: Guilford Press.

SESSA, F. M., & STEINBERG, L. (1991). Family structure and the development of autonomy during adolescence. *Journal of Early Adolescence, 11,* 38–55.

STANTON, M. D. (1985). The family and drug abuse. In T. E. Bratter & G. G. Forrest (Eds.), *Alcoholism and Substance Abuse: Strategies for Clinical Intervention.* New York: Free Press.

STANTON, M. D., & TODD, T. C., & ASSOCIATES. (1982). *The family therapy of drug addiction.* New York: Guilford Press.

STEINGLASS, P., ET AL. (1987). *The alcoholic family.* New York: Basic Books.

THORNTON, A. (1989). Changing attitudes toward family issues in the United States. *Journal of Marriage and the Family, 51,* 873–893.

VOLK, R. J., ET AL. (1989). Family systems of adolescent substance abusers. *Family Relations, 38,* 266–272.

WATZLAWICK, P., BEAVIN, J. M., & JACKSON, D. D. (1967). *Pragmatics of human communication: A study of interactional patterns, pathologies, and paradoxes.* New York, W. W. Norton.

WEINER, H. D., WALLEN, M. C., & ZANKOWSKI, G. L. (1990). Culture and social class as intervening variables in relapse prevention with chemically dependent women. *Journal of Psychoactive Drugs, 22,* 239–248.

WELLS, L. E., & RANKIN, J. H. (1991). Families and delinquency: A meta-analysis of the impact of broken homes. *Social Problems, 38*(1), 71–93.

R. L. LEWIS
M. S. IRWANTO

FAMILIES IN ACTION   *See* Prevention: National Families in Action.

FAMILY VIOLENCE AND SUBSTANCE ABUSE   Substance abuse has a profound impact on Americans of all ethnic groups. Many people are concerned about substance abuse, especially because it is believed that it has the major consequence of increasing rates of crimes such as robbery and "drive-by" homicides. Yet the physiological, psychological, and social effects of substance abuse extend well beyond acts by individuals against strangers; substance abuse has especially adverse effects on families.

Most individuals' illicit drug use occurs between the ages of eighteen to thirty-five, the childbearing years (National Institute on Drug Abuse, 1993). About 10 million children reside in households that have a substance abuser (Blau et al., 1994), and a minimum of 675,000 children per year are neglected or abused by drug- or alcohol-dependent caretakers (Bays, 1990). At the same time that substance abuse increased, foster care placements increased by 30 percent between 1986 and 1989 (Kelley, 1992).

The extent of spousal abuse by substance abusers is more difficult to document. Although there is much more focus on men as perpetrators and women as victims, women in conjugal relationships do assault their male partners (Halford & Ogarsby, 1993). Recent estimates suggest that annually about 10 percent of married women experience some level of assault (Dutton, 1989) and that between 12 percent to 25 percent experience more serious assault such as being hit or kicked (Andrews & Brown, 1988; Randall, 1990). Physical abuse has been identified as the main reason that between 20 percent and 33 percent af all women seek treatment in emergency rooms (Randall, 1990). Rates for violence against men by their female partners are similar to those reported for violence by men against female partners, but whereas women are believed to commit about 10 percent of murders of nonspouses, they commit 48 percent of murders of husbands and partners (Strauss & Gelles, 1990). Thus, domestic violence by women against men appears much more likely to be lethal when it does occur, whereas domestic violence by men appears more likely to result in severe injuries. Few studies, however, have inquired as to whether either the perpetrator or the victim was a substance abuser or was under the influence of alcohol or drugs at the time of a precipitating incident.

Public awareness of child abuse and neglect has increased dramatically since the mid-1980s, but awareness of spousal abuse has lagged behind. Until recent years, adult victims rarely acknowledged their predicament, attributed signs of physical abuse to other causes, excused perpetrators, and resisted recommendations that they use the legal system to try to deter perpetrators. There are several reasons for reluctance to prosecute. In many instances, wives are dependent on their male partners for economic support, fear loss of their children as a result of custody suits, or conceal abuse to avoid criticism by family, friends, or the community. The still popular notion that women "deserve" abuse prevails and will only diminish as popular beliefs are replaced with information about the complex circumstances facing abused women.

There are few reliable estimates of abuse of elderly people by family members (Pillemer & Suitor, 1988). Many cases may go unreported. One survey reported that 1.5 million elderly persons in the United States were abused in 1989, but others estimate that the range could be somewhere between 4 percent to 10 percent of the elderly population (Boudreau, 1993). Low rates of spousal abuse (3.3%) have been noted for persons over the age of sixty-five, but only 55 percent of this population is married (Strauss & Gelles, 1990). Since women live longer than men, study of the abuse of elderly people by their children or children's spouses focuses mainly on the abuse of mothers. In relationships between adult children and their parents that have become abusive, predisposing factors include health status, dependency status, social isolation, intergenerational transmission of violent behavior, and external stressors. Anecdotal reports indicate that in 30 percent to 45 percent of cases reported to service providers, perpetrators have mental health or substance abuse problems, but the topic requires more systematic study, especially for rates in the general population.

Most studies of family violence involving children have focused on intergenerational relationships. Much less information is available about abuse among siblings or by other children. For example, research emphasis in studies of childhood sexual abuse has examined characteristics of adult male perpetrators who are stepfathers or other relatives, with sexual abuse by brothers identified as the least frequent occurrence.

SUBSTANCE ABUSE AND FAMILY LIFE

It has been estimated that abuse is associated with psychological disorders in about 20 percent of cases (Stark & Flitcraft, 1988). The family plays an important role in factors relating to the development, maintenance, and treatment of substance abuse. The fundamental significance of families as dynamic systems has been recognized and studied (Wolin et al., 1980). Today, treatment plans for substance abusers typically involve family members or significant oth-

ers. The disorganizing impact of alcoholism on families is perhaps the addiction that has been best delineated, but information about the impact of other drug use is increasing (Kosten, Rounsaville, & Kleber, 1985; Bernardi, Jones, and Tennant, 1989).

Disrupted family dynamics can occur irrespective of socioeconomic status and ethnic group membership. Research involving a large cross-sectional sample found that offspring of substance abusers were more likely to experience marital instability and psychiatric symptoms, especially if they had experienced physical and sexual abuse (Greenfield et al., 1993), and it has also been found that alcohol abuse often co-occurs with domestic violence (Fagan, Barnett, & Patton, 1988; Dinwiddie, 1992). Construction of "family trees," or genograms, are now in common use as clinical tools to depict the degree to which abuse of various substances has had effects on several generations in a family, the extent that support is available from family members, and the emotional "valence" of kinship relationships (Lex, 1990). Background factors significant for women include childhood violence experiences, violence from a cohabiting partner, and presence of concurrent antisocial and/or borderline personality disorders (Haver, 1987).

Substance abuse and child abuse may co-occur under similar family conditions and dynamics, or substance abuse can lead to child abuse (Kelley, 1992). Mediating factors, such as social support and education, income, alternative sources of nurturing, and parents' own histories of familial substance abuse and histories of neglect and abuse are important. It is likely, however, that when mothers who use drugs or alcohol are primary caregivers, they will be unable to fulfill some aspects of their children's emotional or physical needs (Tracy & Farkas, 1994).

One typical factor in family lives of substance abusers is the absent father, who usually is affected in some way by substance abuse and whose familial role has had to be reallocated among other relatives (Bekir et al., 1993; Hayes & Emshoff 1993). Often this pattern is transmitted from the grandparental generation to the parental generation. Involuntarily or out of necessity, the missing role is frequently assigned to a child, who has to assume responsibilities inappropriate to his or her age and generation (that is, to act as a spouse or parent). Some children recall having had to raise themselves, since their parents neglected to nurture them or abused or scapegoated them or controlled their activities excessively. Children's responses can include acting out through anger, antisocial behavior, and estrangement, or compliance and assumption of housekeeping, care for siblings, and other domestic tasks. In adulthood, resentment because of the burdens of these childhood role reversals can promote depression in individuals and affect their adjustment to adult roles, and it can, in turn, damage their relationships with their own offspring. In some cases, the onset of substance abuse in children occurs at the age or lifecycle stage when a parent began substance abuse. Substance abusers often appear to expect parental unconditional love from their spouses that includes unquestioned acceptance of their substance abuse and irresponsible behavior (Bekir et al., 1993). Unstated expectations and other communication difficulties occur when the moods and behaviors of substance abusers are closely tied to those of family members (McKay et al., 1993). "Low autonomy" (emotionally dependent) substance abusers, however, appear to respond well to treatment if family members provide more nurturing and support. Conversely, male substance abusers whose attitudes and actions are independent and detached from family concerns seem to exhibit a pernicious individualism that is associated with a poor outcome in treatment.

CONSEQUENCES OF ADDICTION IN CHILDREN

Infants exposed to drugs in utero can present problems for caretakers, such as the consequences of prematurity, low birth weight, retarded intrauterine growth, and developmental delays (Blau et al., 1994; Scherling, 1994). Cocaine-exposed infants can be irritable and easily overstimulated, exhibit increased muscle tone, and resist attempts at soothing (Kelley, 1990). There is also a large literature on alcohol effects in utero, which may affect at least 2.6 million infants annually (for review of this literature, see Finnegan & Kandall, 1992). For drug-dependent mothers, these babies sometimes present overwhelming challenges that are often interpreted as "personal" rejection. Mothers' emotions can include guilt about exposure of their child to drugs as well as anger that their efforts at parenting hyperactive babies with feeding difficulties and abnormal sleep patterns seem unsuccessful and only generate more stress. The attachment between mother and child may be disrupted because mothers experience these infants as being highly demanding and ignore and withdraw

from them or continue to use drugs. All too often, the consequences of disrupted attachment lead to child neglect and abuse.

PRECIPITATING FACTORS

Alcohol, Drugs, and Aggression. It is popularly believed that alcohol use facilitates the commission of violent acts. Although there is an association between alcohol (and drug) use and aggression, it is not appropriate to attribute all family violence to substance abuse, and substance abuse does not inevitably result in violence (Hayes & Emshoff, 1993; Taylor & Chermack, 1993). Individual, familial, and environmental factors are all implicated in family violence. Controlled studies in research laboratories constitute one means of disentangling the important interrelationships of these factors. One series of laboratory experiments that used electric shocks between competitors as a proxy for aggressive behavior (see Taylor & Chermack, 1993) showed that both the quantity of alcohol that has been consumed and the social environment encouraging aggression are two major contributing factors. Results should be interpreted cautiously, since the extent to which controlled laboratory conditions, and the stimulus of a shock, can be generalized to the events in daily domestic life in households with a person who meets the diagnostic criteria for substance dependence or abuse remains to be demonstrated (Leonard & Jacob, 1988).

Experiments were designed to identify factors that could instigate aggression in persons intoxicated with alcohol. In an interactive setting, research subjects were tested while sober and while intoxicated (i.e., about .10 blood alcohol level, or the limit for intoxication while driving in many jurisdictions). Since actual violence could not be condoned ethically, the experiment could only give the illusion that a subject would compete with an "opponent" who could signal intention to send a shock of intense magnitude.

Unless their opponents indicated willingness to administer a strong shock, 80 percent of the sober subjects and 40 percent of the intoxicated subjects were reluctant to retaliate by increasing the magnitude of the shock presumably to be received by the opponent. An additional important factor was pressure from bystanders. In another experiment, two accomplices of the experimenter encouraged both sober and intoxicated subjects to use high-magnitude shocks against their opponents. Under this condition, escalation of shock strength occurred for 10 percent of sober subjects and 50 percent of intoxicated subjects. Once escalation had occurred, however, intervention by a third party was generally ineffective. Instead, the strategies best suited to averting aggression in intoxicated persons were to show the opponent to be nonthreatening, to announce a conventional limit on aggressive behavior (in this instance, magnitude of shocks), or to divert attention from aggression to more socially acceptable behaviors. Although intoxicated subjects expected opponents to be more aggressive than did sober subjects, using a video camera to project an image of the sober opponent's behavior diminished the aggressive responses.

Effects of other drugs on aggression also were evaluated by using this type of laboratory experiment. These studies are important because some tranquilizers are prescribed for anxiety and irritable behavior (Ratey & Gordon, 1993). Low doses of marijuana could result in aggressive behavior, but high doses suppressed it. The use of low doses of benzodiazepines increased aggression, but amphetamines did not augment aggression, and these results were contrary to prevailing expectations. Other studies showed that pretreatment with nicotine, dextroamphetamine, or propranolol (which lowers blood pressure) inhibited aggressive behavior. Furthermore, when individuals were evaluated on an aggression rating scale, the nonaggressive group did not respond to provocation while intoxicated with alcohol, but persons in the moderate- and high-aggression groups responded with aggression.

Thus, pharmacological action of drugs, dosage, characteristics of the consumer, and the social factors surrounding drug taking are all important factors contributing to aggressive behavior. Disturbance of higher-order information processing, or reasoning, appears to be the factor that best explains escalation in aggression while intoxicated. Intoxicated subjects were likely to continue aggressive behavior once it had begun, unless they were strongly prompted to engage in self-reflection. Weak suggestions to limit aggressive behavior apparently are not perceived. Having crossed a behavioral boundary may make it easier to continue to do so.

It also should be noted that alcohol and other drugs have a pharmacological effect on sexual arousal and sexual behavior. Among men, alcohol can cause secondary impotence and heroin use can delay ejaculation. There also is evidence to support the notion that cocaine use can increase sexual interest for men and women, and marijuana use has become associated with uninhibited sexual activity. Some women find that heroin use by their partner prolongs intercourse, and once heroin is used as an adjunct to sexual activity, couples are prone to relapse to drug use (Lex, 1990).

Pharmacological effects of alcohol and drugs can also distort communication. For example, large doses of alcohol consumed in short periods of time can result in blackouts, or disrupted short-term memory. A person in a blackout is unlikely to remember what was said and done during the episode. Excessive cocaine consumption can result in suspicion, hostility, and paranoia. A person in a state of withdrawal from alcohol or drugs can be irritable, and oscillation between withdrawal and intoxication distorts communications, thereby leading to inconsistency, unpredictability, and mistrust (Hayes & Emshoff, 1993).

Social Context of Domestic Violence. Many sociologists have assumed that domestic violence is a relatively rare event, and until the 1980s anthropologists had only a limited perspective on the occurrence of family violence in other societies. In a major analysis of data from ninety societies (Levinson, 1987), it was found that wife beating was nearly ubiquitous and predictably associated with social and cultural factors. The frequency of wife beating was analyzed, and societies were classified according to whether wife beating was absent or rare, occurred in less than half of households occurred in more than half of households, or was present in almost all households. Using these criteria, it was found that wife beating occurred in 84 percent of the societies in the sample. Occurrence of this behavior was best explained by both social acceptance of violence and economic dominance of men. In a restudy by Erchak and Rosenfeld (1994), additional societies were selected for analysis and when wife beating was coded as simply being either present or absent; it was found that that it occurred in 80 percent of the sample. However, social isolation occurred in 47 percent of societies without wife beating, in contrast to occurrence in 94 percent of nonisolated societies. Socially isolated societies were typically smaller, and their members need to be mutually interdependent for the purposes of survival. In comparison, societies where raiding or warfare against outsiders was common—that is, where disputes with outsiders were resolved by physical force—had a wife-beating rate of 85 percent, versus 29 percent for societies without warfare. In societies that strongly emphasized men's role as warriors, rates of wife beating were 94 percent, in contrast to rates of 56 percent in societies lacking these attitudes and behaviors. Neglect or abuse of children co-occurred with wife beating. Other associated values were beliefs about women's inferiority, the lack of value of women's lives, and a widow's ability to choose a new spouse. Additional associated behaviors included tolerance for homosexuality, control of female sexuality, and competition for economic resources. Thus, the current prevailing desire of women for equality between men and women in the United States may be counterproductive and result in more violence, because of increased economic competition between the sexes and increased confusion about appropriate gender-related social behaviors (Erchak & Rosenfeld, 1994).

For impoverished members of minority groups, attributes of the community and neighborhood can adversely affect family life (Wallace, Fullilove, & Wallace, 1992). In a number of urban areas, deterioration of housing, decreases in levels of services such as housing inspections and response by fire-fighting and arson units, and diminished police presence have permitted the dynamics of urban decay to operate. As buildings deteriorate, are further damaged by vandalism, and are destroyed by fire, the impact is much like the spread of a contagious disease. Adjacent buildings may be affected as landlords abandon housing stock and businesses leave or fail. Whole blocks may be damaged, and, finally, entire districts of a city may deteriorate completely.

The quality of life diminishes accordingly. Abandoned buildings are taken over by substance users and sellers or used for other illicit activities such as prostitution. Adolescents can gain ready access to drugs and alcohol, and their behavior may go unchallenged. As people move away, there remain fewer persons available to notice children's behavior, and more unsupervised locations become available where children can engage in disapproved acts. When an area lacks former types of social control, such as sanctions from neighbors, acts such as smoking tobacco cigarettes may escalate to greater deviance, such as using marijuana or crack cocaine. As a consequence, antisocial behaviors may go unchecked, and feelings of anger and hostility can grow. It should be noted, however, that urban settings are not the only locations in which deviance

can increase. Contexts that permit anonymity, including ready accessibility of transportation, can also separate perpetrators from persons who know them or would report deviance to authorities.

Perpetrators of Domestic Violence. Much recent attention has been focused on the psychopathology of both perpetrators and victims. One review (Dinwiddie, 1992) suggested that perpetrators had poor communication skills, higher levels of hostility, and, predictably, less control over their anger. Perpetrators studied for personality problems were more likely to be antisocial, passive-aggressive, or narcissistic. The picture is less clear regarding substance abuse, although men meeting criteria for alcohol abuse or dependence (American Psychiatric Association, 1980) were more likely to hit or throw objects at their wives. Studies of community samples have generally found that perpetrators also meet the criteria for diagnoses of depression and antisocial personality disorder.

In one study, rates of spousal abuse and other problem behaviors were studied in 380 married male relatives of alcoholics (Dinwiddie, 1992). Only 16 percent of the men were self-reported spouse abusers, and 30 percent of these were separated or divorced at the time of the interview, in contrast with 14 percent of the nonabusers. When effects of single diagnoses were examined, alcoholism was the most commonly diagnosed psychological disorder (87%) and was associated with an almost fourfold increase in likelihood of abuse. Diagnoses of antisocial personality disorder (46%) or major depression (33%) were associated with an almost double increased likelihood of spousal abuse. Only four abusers (7%) had no psychological disorder. Most abusers, however, had more than one diagnosis of psychological disorder. Antisocial personality disorder or depression usually co-occurred with alcoholism. Among nonabusers, 65 percent were alcoholic, 23 percent were drug dependent, 20 percent had major depression, and 31 percent had an antisocial personality disorder. Aggressive childhood behaviors were poor predictors of abuse in adulthood, but as adults 95 percent of all abusers reported having physical fights, about half reported marital infidelity, 23 percent had been divorced one or more times, and 17 percent had made attempts at suicide.

Alcohol problems and marital distress appear to be highly interrelated (Halford & Osgarby, 1993). Drinking outside of the home increases marital dissatisfaction, and marital disputes can provoke a relapse in abstinent alcoholics. Divorce rates for alcoholics are thought to be highest among persons with psychological disorders, and divorce or marital problems diminishes the likelihood that alcohol treatment will succeed for individuals. Treatment efforts directed at increasing marital stability, however, can successfully promote abstinence (McCrady et al., 1979). Accordingly, many therapists who treat people for alcoholism suggest conjoint treatment for alcoholism and marital problems. In contrast, few marital therapists address issues of alcohol abuse (Halford & Osgarby, 1993).

A sample of eighty-four women and fifty-six men seeking marriage counseling were identified in a marriage guidance clinic (Halford & Osgarby, 1993). All subjects were still married and cohabiting. The subjects were mainly in their thirties, had about two children, and had been married about nine years. One-third were involved in second or later marriages. The subjects completed questionnaires that probed for information about amounts of alcohol consumption, occurrence of physical violence, and frequency of disputes about alcohol use. About half of the men, but less than 20 percent of the women, met the criteria for a diagnosis of alcoholism. More than 80 percent of the entire sample reported having repeated arguments about alcohol intake, and almost 70 percent reported the occurrence of physical violence. Men and women taking steps leading to divorce were more likely to report disagreements about alcohol use. Women mentioned male violence as a factor in marital dissatisfaction, but men who had been abusive were more likely to seek divorce. In this sample, alcohol abuse was significantly associated with couples taking steps toward divorce, but few other common sources of marital dissatisfaction, such as allocation of household tasks, communication, finances, use of leisure time, and parenting issues, were reported to any significant extent. At the very least, these data suggest that marital therapists should routinely screen their clients for alcohol intake and alcohol-related problems, and that they should assess the extent to which these factors interact with domestic violence. It also is possible that abuse by a husband signals a desire to terminate the relationship rather than to exert greater control over the wife's behavior within the context of marriage.

Disentangling cause-and-effect sequences between alcohol or drug abuse and family violence is an important and necessary step in understanding factors that promote or maintain any interrelation-ships. There are several ways of approaching these questions, and researchers with competing theories have attempted to explain the relevant issues (Fagan et al., 1988; Strauss & Gelles, 1990). One theory termed "deviance disavowal" has argued that drinkers are not responsible for their actions while they are intoxicated (McAndrew & Edgerton, 1969). Drunkenness is used as an excuse, and it is possible that some persons seek an intoxicated state so as to be able to engage in violent behaviors (Gelles, 1974). According to another theory, alcohol acts on the central nervous system to create a "disinhibition" that releases aggression. Although this reflects a popular belief about the effects of alcohol, it is the social environment promoting or discouraging aggression that is an important contributing factor (Strauss & Gelles, 1990; Taylor & Chermack, 1993). Social learning theory has been applied to a wide variety of behaviors, and the proponents of this theory argue that social meaning becomes attached to behaviors, such as alcohol use, with the result that people come to expect certain behaviors in association with alcohol. Researchers who support a more focused approach have suggested that drinking and violence become associated within the family context, and that discussion of drinking behavior acts as a cue or trigger that escalates verbal hostility and culminates in physical aggression (Fagan, Barnett, & Patton, 1988).

Characteristics of Perpetrators and Victims. One study used a Relationship Abuse Questionnaire to assess levels of marital violence among abusive and control subjects, including happily married men, maritally dissatisfied men, and men convicted of a violent offense who had not committed acts of domestic violence (Fagan, Barnett, & Patton, 1988). Men in the marital-violence group were young males from minority groups, with limited education and a high rate of unemployment. All members of these groups had been married for an average of four years, had about two children, and were between one to two years older than their wives. Maritally violent men were more likely to consume whiskey and beer, drink daily, drink at lunch on workdays, and drink at home—after work and in the company of their children or by themselves. In addition, maritally violent men indicated that their female partners also drank, but to a lesser degree than they did. These men in the maritally violent group reported that they drank to "deaden the pain in life," to "cheer up a bad mood," to "relax," to "celebrate special occasions," to "forget worries," "to forget everything," and to allay feeling "tense and nervous." They said their female partners drank to "celebrate special occasions" and to "be sociable." Maritally violent men reported that drinking accompanied abuse about one-third of the time but occurred without drinking occasionally, about one-fourth of the time. Female partners were said to drink on about one-fourth of occasions when abuse occurred. Maritally violent men were most likely to report that in the aftermath of violence they felt "sexy" or "wanted to make love," "tried to stop abuse through reasoning," or "took drugs/had a drink." In sum, these men drank more, drank in many social contexts, perhaps continuously but in low amounts, drank to "escape" unpleasant emotions and events, and had female partners who also drank. Drinking or drug taking could be an outcome, however, rather than the cause of a violent episode. It also should be noted that a violent episode could precipitate sexual activity.

A classic study (Kantor & Strauss, 1989) investigated whether drug or alcohol use by victims increased the likelihood of assault by their partners. Information about violence was obtained from 2,033 married or cohabiting women who responded to the 1985 National Family Violence Survey. Research was stimulated by empirical observations that cultural acceptance of violence was the strongest factor in violence directed at wives. This study was designed to test the hypothesis that victims of violence might in some way precipitate violent episodes. Several studies had indicated that people were more likely to attribute blame for violent episodes to women who had violated the cultural attitude that fosters disapproval of women who are intoxicated and another culturally shaped attitude that excuses intoxicated men from the consequences of their alcohol use, including violence. Specific questions included in the interview asked whether women's alcohol or drug use increased the risk of violence from male partners, whether drinking or drug use by male partners increased the risk of violence, whether intervening variables, such as socioeconomic status, explained the occurrence of violence, and whether minor violence and severe violence had different antecedents.

Events were classified as nonviolent, minor violence (throwing objects, pushing, slappping, or grabbing), and severe violence (kicking, hitting, beating, choking, threatening with knives or guns, or using knives or guns). Subjects also were asked whether they used drugs to the extent of being "high" and alcohol to the extent of being "drunk." Predictably, high rates were obtained for alcohol use. Among nonviolent couples, 16 percent of wives and 31 percent of husbands were reported to use alcohol to the extent of being drunk. In contrast, 36 percent of women and 50 percent of men involved in minor-violence episodes used alcohol, and 46 percent of women and 70 percent of men involved in severe-violence episodes had used alcohol. Correlation of violence with drug use (marijuana) was less than half that of alcohol, but the illegal status of marijuana might have encouraged underreporting. Among nonviolent couples, only 4 percent of wives and 5 percent of husbands were reported to use marijuana. In contrast, 14 percent of women and 18 percent of men involved in minor-violence episodes had used marijuana, and 24 percent of women and 31 percent of men involved in severe-violence episodes had used marijuana. Minor-violence episodes were related to the husband's use of marijuana and to violence in the family of origin of the victim. Drunkenness by the wives and by their husbands, low income, and the wives' acceptance of male violence were significant factors, but wives' marijuana use was unimportant. Severe-violence episodes showed a more restricted pattern. Violence in the women's families of origin and husbands' drunkenness were somewhat stronger factors than husbands' marijuana use. Income level, wives' acceptance of abuse, and wives' drunkenness or being high did not affect the severity of violence. In this study, pregnancy or employment status were not relevant factors.

Some have argued that pregnancy is a factor in the precipitation or escalation of abuse episodes. A recent study examined the extent of physical abuse in a multiethnic sample of pregnant women (Berenson et al., 1991). Of 501 women using services at a prenatal clinic, about 20 percent reported physical abuse, and of this group, 29 percent had been abused while pregnant. However, only 19 percent had ever sought medical help, thus indicating that emergency-room statistics might seriously underreport the prevalence of physical abuse. Abuse occurred typically within the context of a primary relationship, with 92 percent of women reporting abuse by only one person, usually (83% of the time) a male partner. Women who had been abused were more likely to report having a partner who abused alcohol or drugs. Abused pregnant women had significantly more pregnancies and more living children than other pregnant women. Across ethnic groups, white non-Hispanic women were 3.5 times more likely than Hispanic women and 1.6 times more likely than black women to experience physical abuse. Substance abuse increased risk of abuse for white non-Hispanic women to two times that of non-abused women, but for black women, almost four times. Other characteristics were important. Traditional values, as exemplified by speaking Spanish, appeared to be a protective factor for Hispanic women. Divorced or unemployed black women, however, were at higher risk for abuse than either Hispanic or white women. Thus, alcohol or drug use are important factors in the abuse of pregnant women, but black women appear to be at highest risk for abuse when these factors were involved.

There is no single cluster of characteristics that typify men who abuse women. Some studies, however, have indicated that witnessing violence in the family of origin may have taught men to use violence as a coping mechanism. Others have argued that alcoholic abusers also may have had a family history of alcoholism, thereby blurring the relationships between causes and effects in families of origin. In a study of men in a treatment program for family violence (Hamberger & Hastings, 1991), comparisons of marital adjustment, coping with conflict, and personality characteristics were made among alcoholic and nonalcoholic men in treatment and control subjects drawn from the community. The average age of the men was about thirty-five, and they had similar education levels. Nonalcoholic men were more likely to be employed and less likely to have witnessed violence in their families of origin. Alcoholic men who had abused their wives were more likely to have been abused as children, but parental alcohol abuse and parental alcoholism appeared to have no direct role in provoking violence by adult abusers who were alcoholic. As might be predicted, the alcoholic abusers had significantly higher personality-disorder scores for avoidant (passive-aggressive) behaviors, aggression, and negativism, and lower scores for conformity. Both alcoholic and nonalcoholic abusers had a large number of symptoms of pathology, thus scoring high on scales measuring anxiety, hysteria, and depression. Alcoholic abusers had the highest scores on psychotic thinking, psychotic depression, and borderline behaviors. As predicted, abusers had higher scores for personality disorders, and alcoholic abusers had the highest scores in this regard. Alcoholic abusers had witnessed more violence in their families of origin and had themselves been victimized by abusers in their families of origin. Overall, alcohol abuse was significantly related to psychopathology as well as to the degree of harm conferred by abuse. Unemployment as a factor operated in some unknown way to bring abusers to the attention of authorities, but the effect of socioeconomic status was not included in the characteristics examined in this study. Clearly, alcoholic abusers identified through agencies had more severe problems, thus suggesting that treatment programs should carefully assess referral sources of clients. A finding of co-morbidity with depression, anxiety, borderline behaviors, and thought disorders suggests that a program focused on abuse alone would be less successful than a more comprehensive approach that offered services for severe psychological disorders.

In another line of investigation, researchers examined women's histories of victimization and their alcohol use together with characteristics of their partners. The reasoning behind this approach was the consideration that when abusive behavior was modeled, excused, or condoned, children would perpetuate these behaviors as being appropriate to gender roles. Thus boys would devalue women and consider abuse a conventional way to deal with conflict, and girls would expect to be devalued and would tolerate abuse. One study investigated these background factors among forty-nine abused women and eighteen male abusers (Bergman & Brismar, 1992). Abusers were not identified through their female partners, since many of the women were afraid to permit contact with them and many of the abusers refused to participate. Abusers were selected from men who had been sentenced to prison for assault and battery of their female partners. The extent of injuries inflicted by the selected men and experienced by the women were comparable as a result of matching reports from the abused women and those from the convicted abusers. It was intriguing to find that both the men and the women reported having been raised without fathers in their families of origins, that about half of the absent fathers were alcoholic, and that most of the mothers were abstainers. As children, about 80 percent of both men and women had witnessed domestic violence in their families. Moreover, 29 percent of the women and 11 percent of the men had experienced sexual abuse as children. As adults, almost all of the women (94%) had experienced previous abuse, and 49 percent had been abused by former partners. About half of the men and one-fourth of the women had used marijuana, 62 percent of the women and 44 percent of the men had used sedative-hypnotic prescription drugs, and 55 percent of the women and 61 percent of the men acknowledged that both partners had been drunk at the time of the precipitating episode of abuse (only 20% of the women and 11% of the men had been sober). Roughly two-thirds of the men and of the women indicated that the abusive incident probably would not have happened in the absence of alcohol. Transgenerational perpetuation of abuse patterns seemed likely, since 25 percent of episodes were witnessed by the children of the women and the rate of the parents' alcohol and drug abuse was high. Thus, information about histories of alcohol and drug abuse as well as exposure to domestic violence should be evaluated for each partner in a couple involved in domestic violence.

Less information is available about drug use (see Miller, 1990). Abuse is not uniformly associated with drug use, however. Psychopharmacological factors have been implicated in domestic violence in the case of some drugs, such as cocaine (Maher & Curtis, 1992), and for economic reasons, such as when a drug abuser resorts to appropriation of family funds to purchase drugs. Systemic violence, related to the hazards of illicit transactions, may spill over into the domestic area if a drug abuser is concerned or suspicious that a partner may be an informer or may be adulterating drugs. Female drug users may find themselves devalued on the basis of both their gender and their behavior, and because some women are involved in prostitution to obtain drugs for themselves or their partners, their risk of exposure to violent behavior is increased substantially. Intoxicated women also may be more verbally aggressive and thus violate the cultural norm that values the "soft-spoken" woman (Miller, 1990).

Studies of alcohol abuse as it is associated with the abuse of women have not been able to identify a sequence of cause and events. More definitive studies are needed, but one informative study of alcohol and drug abuse by eighty-two male perpetrators and victims sought important linkages. The perpetrators were parolees, and data about psychological disorders, substance abuse, modes of conflict resolution,

and frequency of violent events were obtained from them and their female partners. About three-quarters of the perpetrators, and a surprising 56 percent of their female partners had alcohol problems, and 73 percent of perpetrators and 40 percent of their partners acknowledged using illegal drugs. Similarly, 78 percent of parolees and 72 percent of their female partners reported perpetrating a moderately violent episode, and 33 percent of parolees and 39 percent of their female partners reported perpetrating a severely violent episode at least once during the three months before the interview. About one-third of the episodes were considered severe, and about three-fourths were considered moderate. Neither alcohol nor drug use was involved independently, but concurrent use contributed significantly to violent events, and the separation of drugs into different classes by pharmacological action did not change the effect of alcohol and drug interaction. When combined, however, cocaine and alcohol had a strong effect on violence. In addition, couples with more substance abuse–related problems had a higher incidence of violent episodes, but, overall, alcohol problems most strongly increased the likelihood that violence would occur. Additional studies of women with concurrent alcohol and drug abuse problems are needed to clarify the temporal relationships.

TREATMENT FOR ABUSERS

Shame, guilt, and denial are powerful emotions that impede both the recognition of problems and the admission of the need for help. It is popularly believed that perpetrators enter treatment only under coercion and with considerable reluctance. Given the strong association between substance abuse and marital violence in some individuals, questions arise as to whether treatment of alcohol or drug abuse alone will concomitantly diminish violent acts. Behavioral marital therapy teaches improved communication skills and has been used to improve the marital relationships of patients as their drinking abates (O'Farrell & Murphy, 1994). This treatment modality, however, does not directly address the problem of violence. A comparison was made between eighty-eight couples with a newly abstinent husband and a nonalcoholic control sample of eighty-eight couples undergoing marital therapy. The study covered the year before treatment and the year after it. Acts of domestic violence occurred between four to six times more frequently during the year before treatment. Rates for violent episodes during the year after treatment remained elevated for both men and their wives, and they were higher than the rates among control couples. In instances of relapse, rates were higher than those for couples who had not relapsed. In turn, rates for couples who had not relapsed were comparable to those for controls. Consequently, effective treatment for alcoholism appears to reduce the frequency of domestic violence, although a study that uses a control group of conjugal pairs not receiving behavioral marital therapy is needed for conclusive results. The cause-and-effect relationships between the release of emotions and relapse still need to be disentangled, however, since the former may provoke the latter or have an additive effect.

Another study examined rates of violent acts among seventy-four persons who completed a treatment program for spousal-abuse abatement and thirty-two who relapsed from this program. Men were referred by themselves or the courts, but neither source of referral nor amount of criminal activity had an effect on outcome. Alcohol problems persisted in 32 percent of the men who completed this program successfully, but 56 percent of recidivists had persistent alcohol problems. Recidivists also had higher levels of drug abuse and less empathy as measured on standardized scales. Recidivists also were found to be significantly more narcissistic (self-centered) and gregarious. These findings suggest that alcohol and drug abuse must be addressed when they occur among perpetrators of domestic violence.

COMMENTARY

Numerous studies that use standardized criteria generally support the prediction that substance abuse and domestic violence co-occur in the majority of violent episodes. Roughly one-fourth to one-fifth of episodes, however, occur without substance abuse as a possible co-factor or precipitant. Some additional studies suggest that verbal hostility can escalate domestic conflict to domestic violence (Lindman et al., 1992), but some episodes of verbal hostility may stem from response to life stress and others may be a result of social learning. In other instances, conflict over a child's or a partner's alcohol or drug consumption may prompt the substance

abuser to "protect" the behavior through vehement denial, thereby leading to an escalation of hostility that spirals out of control.

Although any suggestion that women's behaviors might contribute to abuse may seem to take the currently unacceptable position of blaming the victim, there is some evidence that women who express aggression verbally may have had abusive families of origin, and that alcohol abuse may have played a role in fostering a climate of tension and hostility within their households (Gomberg, 1993; Hayes & Emshoff, 1993). This pattern may emerge when women who feel devalued have no behavioral alternative through which to express their frustration. Unfortunately, many potentially interesting and informative laboratory experiments that investigate aggressive behaviors are conducted with undergraduate college students and thus may not disclose important information about effects that stem from income level, social class, educational level, or ethnicity.

Data from alcoholic and drug-abusing women in treatment suggest that younger women may be more verbally aggressive, thus reflecting society-wide changes in gender-role behavior. Other data (Miller, Downs, & Testa, 1993) reveal that women who were victimized as children are more likely to develop alcohol and drug problems in adolescence and adult life. In contrast to women with other psychological disorders, women who require substance-abuse treatment recall more abuse during their childhood. Some contribution to this outcome could be diminished self-esteem and increased alienation from typical childhood socialization processes, as well as limited development of social skills for negotiation and compromise.

It is also possible that the contexts of substance-abuse treatment generate a social expectation that a client must have a family history of substance abuse as well as a background that includes emotional, physical, or sexual abuse. It is clear that additional research is needed and that subject samples need to be drawn from different sources, with different prevalence rates of various types of violence. Longitudinal research that would follow a cohort of children through adolescence, young adulthood, and marital life might hold sorely needed answers. Lacking the answers obtained from definitive research, it is reasonable to continue to screen abuse victims and perpetrators for substance-abuse problems, and to screen substance abusers for perpetration of or victimization through family violence. Because both substance abuse and family violence engender denial that anything is wrong, careful assessment is a prerequisite for effective prevention, intervention, and treatment.

BIBLIOGRAPHY

ANDREWS, B., & BROWN, G. (1988). Marital violence in the community. *British Journal of Psychiatry, 153*, 305–312.

BAYS, J. (1990) Substance abuse and child abuse. *Pediatric Clinics of North America, 37*, 881–904.

BEKIR, P., McLELLAN, T., CHILDRESS, A., & GARITI, P. (1993). Role reversals in families of substance misusers: A transgenerational phenomenon. *The International Journal of the Addictions, 28*, 613–630.

BERENSON, A., STIGLICH, N., WILKINSON, G., & ANDERSON, G. (1991). Drug abuse and other risk factors for physical abuse in pregnancy among white non-Hispanic, black, and Hispanic women. *American Journal of Obstetrics and Gynecology, 164*, 1491–1499.

BERGMAN, B., & BRISMAR, B. (1992). Can family violence be prevented? A psychosocial study of male batterers and battered wives. *Public Health, 105*, 45–52.

BERNARDI, E., JONES, M., & TENNANT, C. (1989). Quality of parenting in alcoholics and narcotic addicts. *British Journal of Psychiatry, 154*, 677–682.

BLAU, G., WHEWELL, M., GULLOTTA, T., & BLOOM, M. (1994). The prevention and treatment of child abuse in households of substance abusers: A research demonstration progress report. *Child Welfare, 7*, 383–394.

BOUDREAU, F. (1993). Elder abuse. In R. Hampton et al. (Eds.), *Family violence prevention and treatment.* Newbury Park, CA: Sage Publications.

DINWIDDIE, S. (1992). Psychiatric disorders among wife batterers. *Comprehensive Psychiatry, 33*, 411–416.

DUTTON, D. (1988). Profiling wife assaulters: Preliminary evidence for a trimodal analysis. *Violence and Victims, 3*, 5–29.

ERCHAK, G., & ROSENFELD, R. (1994). Societal isolation, violent norms, and gender relations: A reexamination and extension of Levinson's model of wife beating. *Cross-Cultural Research, 28*, 111–133.

FAGAN, R., BARNETT, O., & PATTON, J. (1988). Reasons for alcohol use in maritally violent men. *American Journal of Drug and Alcohol Abuse, 14*, 371–392.

FINNEGAN, L., & KANDALL, S. (1992). Maternal and neonatal effects of alcohol and drugs. In J. Lowinson, P. Ruiz, R. Millman, & J. Langrod (Eds.), *Substance abuse: A comprehensive textbook,* 2nd ed. Baltimore: Williams & Wilkins.

GELLES, R. (1974). *The violent home: A study of physical aggression between husbands and wives.* Beverly Hills, CA: Sage Publications.

GOMBERG, E. (1993). Alcohol, women and the expression of aggression. *Journal of Studies on Alcohol, 11*, 89–95.

GREENFIELD, S., SWARTZ, M., LANDERMAN, L., & GEORGE, L. (1993). Long-term psychosocial effects of childhood exposure to parental problem drinking. *American Journal of Psychiatry, 150*, 608–613.

HALFORD, W., & OSGARBY, S. (1993). Alcohol abuse in clients presenting with marital problems. *Journal of Family Psychology, 6*, 345–354.

HAMBERGER, L., & HASTINGS, J. (1991). Personality correlates of men who batter and nonviolent men: Some continuities and discontinuities. *Journal of Family Violence, 6*, 131–147.

HAVER, B. (1987). Female alcoholics: The relationship between family history of alcoholism and outcome 3–10 years after treatment. *Acta Psychiatrica Scandinavica, 76*, 21–27.

HAYES, H., & EMSHOFF, J. (1993). Substance abuse and family violence. In R. Hampton et al. (Eds.), *Family violence prevention and treatment.* Newbury Park, CA: Sage Publications.

HOLMES, S., & ROBINS, L. (1988). The role of parental disciplinary practices in the development of depression and alcoholism. *Psychiatry, 51*, 24–36.

KANTOR, G., & STRAUS, M. (1989). Substance abuse as a precipitant of wife abuse victimization. *American Journal of Drug and Alcohol Abuse, 15*, 173–189.

KELLEY, S. (1992). Parenting stress and child maltreatment in drug-exposed children. *Child Abuse & Neglect, 16*, 317–328.

KOSTEN, T. R., ROUNSAVILLE, B. J., & KLEBER, H. D. (1985). Parental alcoholism in opioid addicts. *Journal of Nervous and Mental Diseases, 173*, 461–469.

LEONARD, K., & JACOB, T. (1988). Alcohol, alcoholism, and family violence. In V. Van Haaelt et al. (Eds.), *Handbook of family violence.* New York: Plenum Press.

LEVINSON, D. (1987). *Family violence in cross-cultural perspective.* Newbury Park, CA: Sage.

LEX, B. W. (1990). Male heroin addicts and their female mates: Impact on disorder and recovery. *Journal of Substance Abuse, 2*, 147–175.

LINDMAN, R., VON DER PAHOLEN, B., OST, B., & ERIKSSON, C. (1992). Serum testosterone, cortisol, glucose, and ethanol in males arrested for spouse abuse. *Aggressive Behavior, 18*, 393–400.

MacAndrew, C., & Edgarton, R. (1969). *Drunken comportment: A social explanation.* Chicago: Aldine Press.

Maher, L., & Curtis, R. (1992). Women on the edge of crime: Crack cocaine and the changing contexts of street level sex work in New York City. *Crime, Law, and Social Change, 18,* 221–258.

McCrady, B. S., Paolino, T. J., Longabaugh, R., & Rossi, J. (1979). Effects of joint hospital admission and couples treatment for hospitalized alcoholics: A pilot study. *Addictive Behavior, 4,* 155–165.

McKay, J., Longabaugh, R., Beattie, M., Maisto, S., & Noel, N. (1993). Changes in family functioning during treatment and drinking outcomes for high and low autonomy alcoholics. *Addictive Behaviors, 18,* 355–363.

Miller, B. (1990). The interrelationships between alcohol and drugs and family violence. In M. La Rosa, E. Lambert, & B. Gropper (Eds.), *Drugs and violence: Causes, correlates and consequences.* Research Monograph no. 103, DHHS Publication no. (ADM) 91-1724. Rockville, MD: National Institute on Drug Abuse.

Miller, B., Downs, W., & Testa, M. (1993). Interrelationships between victimization experiences and women's alcohol use. *Journal of Studies on Alcohol, 11,* 109–117.

National Institute on Drug Abuse. (1993). *National household survey on drug abuse: Population estimates 1992.* Washington, DC: U.S. Government Printing Office.

O'Farrell, T., & Murphy, C. (in press). Marital violence before and after alcoholism treatment. *Journal of Consulting and Clinical Psychology.*

Pillemer, K., & Suitor, J. (1988). Elder abuse. In V. Van Haelt et al. (Eds.), *Handbook of family violence.* New York: Plenum Press.

Randall, T. (1990). Domestic violence intervention calls for more than treating injuries. *Journal of the American Medical Association, 264,* 939–940.

Ratey, J., & Gordon, A. (1993). The psychopharmacology of aggression. *Psychopharmacology Bulletin, 29,* 65–73.

Scherling, D. (1994). Prenatal cocaine exposure and childhood psychopathology: A development analysis. *American Journal of Orthopsychiatry, 64,* 9–19.

Stark, E., & Flitcraft, W. (1988). Male Batterers. In V. Van Haelt et al. (Eds.), *Handbook of family violence.* New York: Plenum Press.

Strauss, M., & Gelles, R. (1990). *Physical violence in American families.* New Brunswick, NJ: Transaction Publishers.

Taylor, S., & Chermack, S. (1993). Alcohol, drugs and human physical aggression. *Journal of Studies on Alcohol, 11,* 78–88.

Tracy, E., & Farkas, K. (1994). Preparing practitioners for child welfare practice with substance-abusing families. *Child Welfare, 73,* 57–68.

Wallace, R., Fullilove, M., & Wallace, D. (1992). Family systems and deurbanization: Implications for substance abuse. In J. Lowinson, P. Ruiz, R. Millman, & J. Langrod (Eds.), *Substance abuse: A comprehensive textbook,* 2nd ed. Baltimore: Williams & Wilkins.

Wolin, S., et al. (1980). Disrupted family rituals: A factor in intergenerational transmission of alcoholism. *Journal of Studies on Alcohol, 41,* 199–214.

Barbara Lex

FERMENTATION Fermentation is a natural metabolic process that produces energy by breaking down carbohydrates (such as sugars) in the absence of oxygen. It occurs in many microorganisms (such as yeasts), and the end product can be either ethyl alcohol (ethanol) or lactic acid; energy is typically given off in the form of heat. The chemical reaction of this process was first described in 1810 by the French chemist Joseph Louis Gay-Lussac. Fermentation is important to the production of many foods and beverages, the most popular of which are bread, butter, cheese, beer, and wine.

Fermented foods first occurred naturally, when stored or forgotten caches were found to be altered but edible. In ancient times, wheat and barley were domesticated, farmed, stored, and used to make breads and porridges—some of which fermented and formed brews. Since that time, the process of fer-

Figure 1
Grapes

mentation has been used worldwide. Industrial means provide huge quantities of fermented foods, as well as alcohol, which is obtained by DISTILLATION from fermented juices of fruits, grains, vegetables, and other plants.

(SEE ALSO: *Beers and Brews*)

SCOTT E. LUKAS

FETAL ALCOHOL SYNDROME (FAS)

This is a constellation of effects that result in the newborn from prenatal alcohol exposure. Diagnosis is made by a specially trained physician and is based on the following criteria: growth deficiency; a pattern of distinct and specific facial abnormalities; and central nervous system (CNS) damage. In addition, there are many other physical abnormalities noted in children with FAS or those more mildly affected, often described as having Fetal Alcohol Effects (FAE). The characteristics listed above and discussed later in this entry must occur in conjunction with confirmed maternal alcohol consumption. Racial, genetic, and familial influences must also be considered when such a diagnosis is made.

HISTORY

The term Fetal Alcohol Syndrome was first used in 1973 to describe the physical problems seen in the offspring of alcoholic women. There have been admonitions against women drinking during PREGNANCY for literally thousands of years—in biblical verses and in the writing of the ancient Greeks. The physical and social implications of women drinking during pregnancy first became highly noticeable during the gin epidemic of the 1750s. At that time, gin became a cheap and easily accessible beverage among low-income women. It was noted that there was a correlation between women who were consuming large amounts of gin and problems among their offspring.

A formal study was conducted in the 1890s by an English physician named Sullivan. He identified the offspring of 120 female "drunkards" in the Liverpool jail and compared them to the offspring of their non-drinking female relatives. From this study Sullivan noted a perinatal mortality rate that was two and one half times higher in the offspring of the female alcoholics.

In 1968, Dr. Paul Lemoine published a study on the children of women alcoholics in a French medical journal. This article did not receive much attention until the landmark articles published in the *Lancet* by Jones, Smith, Ulleland, and Streissguth in 1973. Since 1973, more than 3,000 articles have been published detailing the effects of prenatal alcohol exposure from birth through middle age. There can be no doubt that alcohol is a powerful teratogen (causative agent in fetal malformations) with lifelong aftereffects (sequelae).

DISTRIBUTION

The prevalence of FAS ranges widely from community to community and is determined by the number of women consuming alcohol in any particular community. It is estimated that FAS is now the leading cause of mental retardation in the United States, surpassing Down's syndrome and spina bifada. The prevalence estimates for FAS range from 1 in 600 to 1 in 750 births. However, few prevalence studies have been conducted and many experts have differing views as to the accuracy of those prevalence figures that are available.

PHYSICAL EFFECTS

Scientific research indicates a likelihood that there is no level of alcohol consumption guaranteed free from risk for any period during pregnancy. Individuals react very differently to alcohol and it is difficult, if not impossible, to predict which women will produce a child with FAS. The exception to this is the woman who has already given birth to a child with FAS or FAE. If this woman continues to drink at the same or an increased level, it is highly likely that her subsequent pregnancy will be affected to the same or a greater degree.

Different effects are produced by drinking alcohol during pregnancy and they depend on *when* the alcohol is consumed. During the first trimester, there is a chance of major physical abnormalities and central nervous system (CNS) damage. During the second trimester, alcohol consumption leads to an increased rate of spontaneous abortion and CNS damage, as well as more subtle physical abnormalities. During the third trimester, alcohol consumption can lead to pre- and postnatal growth retardation and CNS damage. These characteristics are detailed below.

As was mentioned above, three major indices are used in diagnosing FAS. First are the common facial abnormalities: These include short palpebral (eye-slit) fissures; a flat midface; a short nose; a long smooth philtrum (upper lip groove); thin upper lip; ptosis (drooping eyelid); strabismus (crossed eyes); epicanthal (eyelid) folds; and ear anomalies. Cardiac (heart) malformations and defects, pectus excavatum (hollow at the lower part of the chest due to backward displacement of xiphoid cartilage); clinodactyly and camptodactyly (permanent curving or deflection of one or more fingers); fusion of the radius and ulna at the elbow, scoliosis (lateral curvature of the spine); kidney malformations, and cleft lip and palate are among the other FAS defects frequently noted.

Growth deficiency in FAS is noted in three parameters—height, weight, and head circumference. Many of the prepubescent patients have growth retardation; their appearance is generally one of being short and skinny. Significant changes in weight are noted as the female patients enter puberty; although the growth deficiency remains in height and head circumference across the lifespan, the girls frequently gain weight and appear plump. The male patients seem to remain fairly short and slender, although a few have gained weight in their late twenties or thirties.

CNS damage is frequently manifested in cognitive and memory deficits, sleep disturbances, developmental delays, hyperactivity/distractibility, a short attention span, an inability to understand cause and effect, lower levels of academic achievement, impulsivity, and difficulty in abstracting. The difficulties noted in infancy and early childhood are often precursors to psychosocial deficits in later life.

PSYCHOSOCIAL AND EDUCATIONAL ISSUES

Ages Birth to Five Years. Diagnosis of alcohol-related birth defects is possible at birth but many physicians are either not trained to identify FAS or do not consider this a possibility. Perinatal behavioral manifestations of FAS include the following: poor habituation, an exaggerated startle response, poor sleep/wake cycle, poor sucking response, and hyperactivity. Failure to thrive, alcohol withdrawal, and cardiac difficulties have become medical concerns frequently noted in this patient population.

Developmental delays in walking, talking, and toilet training are often seen. Concerns such as hyperactivity, irritability, difficulty in following directions, and the inability to adapt to changes are commonly reported. The damage done the brain makes it problematic for children with FAS to learn in a timely and consistent fashion. The more abstract the task, the more apparent this learning gap becomes.

Recommended interventions at this age focus on the family as well as the child. Many children with FAS are removed from the care of the biological mother owing to abuse, neglect, and/or premature maternal death. Newborns and infants with FAS often have trouble feeding; when this is coupled with a mother who may be deeply involved in substance abuse(s) and not attentive to the needs of her infant, it can lead to medical crises. Therefore, it is necessary to provide the following services and interventions:

- Monitoring of health and medical concerns;
- A safe, stable, structured residential placement with services provided to the mother, father, patient, and other family members, such as substance-abuse treatment and parenting training;
- Directions given to the caregivers in a simple, concrete fashion, one at a time; directions given to the child in similar fashion;
- Adaptation of the external environment to fit the child's level of ability to handle stimulation; and
- Setting by caregivers of appropriate goals and expectations for their child.

Ages Six to Eleven Years. Some of the problems noted earlier become less severe as others become more severe—with greater implications for negative social functioning. These are hyperactivity, impulsivity, memory deficits, inappropriate sexual behavior, difficulty predicting and/or understanding the consequences of behavior, difficulties in abstracting abilities, and poor comprehension of social rules and expectations. Children with FAS may show decreasing ability to function in school as they get older. The abstracting deficits become more apparent when the child reaches third and fourth grades and is expected to perform multiplication and division. A summation of suggested interventions at this stage include the following:

- Safe, stable, structured residential placement;
- Establishment of reasonable expectations and goals;
- Use of clear, immediate, and concrete consequences for behavior;
- Listing of chores and expectations in writing;
- Structuring of leisure time and activities;
- Education of parents, caregivers, and the patient regarding age-appropriate sexual and social development;
- Appropriate educational placement that focuses on an activity-based curriculum, development of communication skills, development of appropriate behavior, and basic academic skills embedded with functional skills.

Ages Twelve to Seventeen Years. Children with FAS have the same emotional needs as others this age. Adolescents with FAS may show cognitive deficits, impulsivity, low motivation, lying, stealing, DEPRESSION, suicidal ideation and attempts, and significant limitations in their adaptive behavior skills. Other concerns include faulty logic, pregnancy/fathering a child, and the loss of residential placement. Social deficits noted encompass financial/sexual exploitation and substance abuse. It is frequently difficult for people with FAS to articulate their feelings and needs.

Despite these problems and deficits, adolescents with FAS should not be infantilized. In addition, this is commonly the time where they reach their academic ceiling. The following are some suggested interventions to help them reach their social, emotional, and adaptive potential:

- Changing the focus from academic to vocational and daily-living skills training;
- Structuring of leisure time and activities, such as involvement in organized sports and social activities;
- Education of the patients, parents, and caregivers regarding sexual development and the need for birth control or protection against sexual exploitation and sexually transmitted diseases (STDs);
- Planning for future vocational training and placements, financial needs, and residential placement; and

- Increasing responsibility based on the patient's skills, abilities, and interests.

Ages Eighteen through Adulthood. The problems, deficits, and difficulties seen prior to the age of eighteen are precursors to those seen in young adulthood and into middle age. An additional problem experienced by people with FAS is the increased expectations placed on them by others. Not only can people with FAS often not meet these expectations but their impulsivity and poor judgment have more serious consequences than during their younger years. Issues such as poor comprehension of social rules and expectations, aggressive and unpredictable behavior, and depression coupled with impulsivity, may lead to suicide attempts, antisocial behavior, hospitalization, and/or incarceration.

Other concerns noted in adults with FAS include social isolation and withdrawal; difficulties in finding and sustaining employment; poor financial management; problems accessing and paying for medical care or child care; and a need for help with social/sexual exploitation and unwanted pregnancy. The hyperactivity and distractibility seen in small children with FAS manifests in the adult with not being able to learn job skills or to meet the requirements of many jobs. The following is a brief outline intended to help adults with FAS deal with problematic issues in a productive fashion:

- A guardianship for or systematic help with whatever funds may be received, since arithmetic skills in this population seldom exceed the third grade;
- Subsidized residential placements to help ensure physical safety;
- Medical coupons for care, along with birth-control planning;
- Homebuilders or community housing to help them live as independently as possible;
- Child-care and parenting classes, as needed; and
- Education to others about FAS, including its limitations and skills, to foster acceptance.

SUMMARY

FAS is a preventable birth defect; once it exists it has life-long consequences. Special programs involving planning for future vocational, educational, and residential needs should be implemented as early in childhood as is possible. Education on the harmful effects of alcohol use, focusing on young women and men of childbearing years, is critical to help prevent, or at least reduce, this significant public-health problem.

(SEE ALSO: *Addicted Babies; Alcohol: History of Drinking; Attention Deficit Disorder; Conduct Disorder in Children; Fetus: Effects of Drugs on; Pregnancy and Drug Dependence*)

BIBLIOGRAPHY

ABEL, E. L., & SOKOL, R. J. (1991). A revised conservative estimate of the incidence of fetal alcohol syndrome and its economic impact. *Alcoholism: Clinical and Experimental Research, 25*, 514–524.

CLARREN, S. K. (1981) Recognition of fetal alcohol syndrome. *Journal of the American Medical Association, 245*, 2436–2439.

DORRIS, M. (1989). *The broken cord.* New York: Harper & Row.

JONES, K. L., & SMITH, D. W. (1973). Recognition of the fetal alcohol syndrome in early infancy. *Lancet, 2*, 999–1001.

JONES, K. L., ET AL. (1973). Pattern of malformation in offspring of chronic alcoholic mothers. *Lancet, 1*, 1267–1271.

LADUE, R. A. (1993). Psychosocial needs associated with fetal alcohol syndrome: Practical guidelines for parents and caretakers. Seattle: Fetal Alcohol and Drug Unit, University of Washington.

MALBIN, D. (1993). *Fetal alcohol syndrome: Fetal alcohol effects.* Center City, MN: Hazelden.

MAY, P. A., ET AL. (1983). Epidemiology of fetal alcohol syndrome among American Indians of the Southwest. *Social Biology, 30*, 374–387.

OLSON, H. C., BURGESS, D. M., & STREISSGUTH, A. P. (1992). Fetal alcohol syndrome and fetal alcohol effects: A lifespan view with implications for early intervention. *Zero to Three, 13*, 24–29.

STREISSGUTH, A. P. (1991). What every community should know about drinking during pregnancy and the lifelong consequences for society. *AMERSA: Substance Abuse, 12*, 114–127.

STREISSGUTH, A. P., ET AL. (1991). Fetal alcohol syndrome in adolescents and adults. *Journal of the American Medical Association, 265*, 1961–1967.

STREISSGUTH, A. P., LADUE, R. A., & RANDELS, S. P. (1988). *A manual on adolescents and adults with fetal alcohol syndrome with special reference to American Indians.* Washington, D.C.: US Department of Health and Human Services.

STREISSGUTH, A. P., SAMPSON, P. D., & BARR, H. M. (1989). Neurobehavioral dose-response effects of prenatal alcohol exposure in humans from infancy to adulthood. In D. E. Hutchings (Ed.), *Prenatal abuse of licit and illicit drugs.* New York: Annals of the New York Academy of Sciences.

ROBIN A. LADUE

FETUS: EFFECTS OF DRUGS ON

The pregnant drug-dependent woman subjects her developing infant to a host of problems. When assessing the effects of drugs, especially illicit drugs, on newborn infants (neonates) and young children, two factors must be considered: (1) the duration and concentration of the drug exposure on the developing fetus and (2) any preexisting medical complications in the mother. These factors are interactive and together will influence in varying ways the eventual capabilities of the child. Therefore, the long-term outcome of children exposed to drugs during fetal development should be assessed.

EFFECTS ON THE NEWBORN

Nearly 60 percent of pregnant women or their newborns show evidence of acute infection—for example, hepatitis, ACQUIRED IMMUNODEFICIENCY SYNDROME (AIDS), tuberculosis, and sexually transmitted diseases (STDs). Infected mothers are likely to deliver prematurely.

The placentas of HEROIN-exposed infants, for example, show microscopic evidence of oxygen deprivation. The infants are small for their gestational age, with all their organs affected. In heroin-dependent women, a significant portion of the medical complications seen in their newborns is due to that prematurity and the low birthweight. Therefore, they evidence immature lungs, difficulties in breathing at birth, brain hemorrhage, low sugar and calcium levels, infections, and jaundice.

Women on METHADONE MAINTENANCE (an oral NARCOTIC used for the treatment of heroin addiction) are likely to give birth to normal- or almost normal-sized babies. Because they are in treatment, the complications in their infants are not as severe and generally reflect (1) the amount of prenatal care the mother has received; (2) whether the mother has suffered any complications, including hypertension or infection; and (3) most important, any multiple drug use that may have produced an unstable intrauterine environment for the fetus, perhaps complicated by WITHDRAWAL(s) and/or OVERDOSE.

Multiple drug use may cause a series of withdrawals, when the pregnant woman cannot get to the drug she needs. This series of extreme physical conditions in the pregnant woman can severely affect the oxygen and nutrients that feed the developing fetus, causing various birth defects, depending on when in each trimester the withdrawals occur. If the mother overdoses, a decreased oxygen supply to the fetus can cause aspiration pneumonia—if the mother survives the overdose to give birth.

Laboratory and animal studies have shown that narcotics (OPIOIDS) may have an inhibitory effect on enzymes that influence oxygen metabolism. They also alter the passage of oxygen and nutrients to the fetus by constricting the umbilical vessels and decreasing the amount of oxygen delivered to the developing fetal brain. Such metabolic side effects may cause a derangement in the acid/base balance (acidosis). In contrast, increased maturation of organ systems and certain enzymes have been seen in heroin-exposed infants, including maturation of the lungs, tissue-oxygen unloading, sweat glands, and liver enzymes. The stressful life of the pregnant woman probably contributes to this enhanced maturation in heroin-exposed infants.

The genetic risks to the offspring of addicts on heroin *and* methadone include an increase in the frequency of chromosome abnormalities; infants exposed predominantly to methadone in utero do not. The adverse environmental factors that may contribute to the abnormal findings in heroin-exposed infants may be less prominent in methadone mothers, since drug addiction is compounded by poor maternal nutrition, extreme STRESS, infectious disease, and a lack of early and consistent prenatal care. It is impossible to isolate either methadone or heroin as agents linked to GENETIC damage, however, in the absence of specific clinical abnormalities.

Given the obstetrical and medical complications, the lack of prenatal care, and the prematurity of the infants at delivery, it is not surprising that the death rate for ADDICTED BABIES is higher than for infants born to nonaddicts.

NEONATAL OPIOID WITHDRAWAL SYNDROME

This syndrome is described as a generalized disorder, characterized by signs and symptoms of central nervous system hyperirritability, gastrointestinal dysfunction, respiratory distress, and autonomic nervous system symptoms that include yawning, sneezing, mottling, and fever. At birth, these infants develop tremorous movements, which progress in severity. High-pitched crying, increased muscle tone, irritability, and exaggerated infant reflexes are all characteristic. Sucking of fists or thumbs is common, yet when feedings are administered, the infants have extreme difficulty and regurgitate frequently—because of an uncoordinated and ineffectual sucking reflex. The infants may develop loose stools and are therefore susceptible to dehydration and electrolyte imbalance. At birth, the blood levels of the drug(s) used by the mother begin to fall, so the newborn continues to metabolize and excrete the drug, and withdrawal signs occur when critically low levels have been reached.

Whether born to heroin-addicted or methadone-dependent women, most infants seem physically and behaviorally normal. The onset of their withdrawal may begin shortly after birth to two weeks of age, but most develop symptoms within seventy-two hours of birth. If the mother has been on heroin alone, 80 percent of the infants will develop clinical signs of withdrawal between four and twenty-four hours of age. If the mother has been on methadone alone, the baby's symptoms usually appear by forty-eight to seventy-two hours.

In summary, various studies have shown that the time of onset of withdrawal in the individual infant will depend on: the type and amount of drug used by the mother; the timing of her dose before delivery; the character of her labor; the type and amount of anesthesia and pain medication given during labor; and the maturity, nutrition, and presence or absence of systemic diseases in the infant.

Studies indicate that more full-term infants require treatment for withdrawal than do preterm infants. Withdrawal severity appears to correlate with gestational age; less mature infants show fewer symptoms. Decreased symptoms in preterm infants may be due to either (1) developmental immaturity of the preterm nervous system or (2) reduced total drug exposure because of short gestations.

The most severe withdrawal occurs in infants whose mothers have taken large amounts of drugs for a long time. Usually, the closer to delivery a mother takes heroin, the greater the delay in the onset of withdrawal and the more severe the symptoms in her baby. The duration of symptoms may be anywhere from six days to eight weeks. The maturity of the infant's own metabolic and excretory mechanisms plays an important role. Although the infants are discharged from the hospital after drug therapy is stopped, some symptoms such as irritability, poor feeding, inability to sleep regularly, and sweating may persist for three to four months.

Not all infants born to drug-dependent mothers show withdrawal symptoms, but investigators have reported that between 60 and 90 percent of infants do show symptoms. Since biochemical and physiological processes governing withdrawal are still not fully understood, and since multiple drugs are often used by the mothers in an erratic fashion—with vague or inaccurate maternal histories provided—it is not surprising to find varying descriptions and experiences in reports from different centers. Seizures, a severe outcome in withdrawing infants, are rare in narcotic-exposed infants. One report found that 5.9 percent of 302 newborns exposed to narcotics during pregnancy had seizures that were attributed to withdrawal. Other reports found even rarer occurrences of seizures.

Drug-exposed infants show an uncoordinated and ineffectual sucking reflex as a major manifestation of withdrawal. Regurgitation, projectile vomiting, and loose stools may complicate the illness further. Dehydration, due to poor intake, coupled with excessive losses from the gastrointestinal tract may occur, causing malnutrition, weight loss, subsequent electrolyte imbalance, shock, coma, and death. Neonatal withdrawal carries a risk of neonatal death when these complications are untreated. The infant's respiratory system is also affected during withdrawal: excessive secretions, nasal stuffiness, and rapid respirations are sometime accompanied by difficulty breathing, blue fingertips and lips, and cessation of breathing. Severe respiratory distress occurs most often when the infant regurgitates, aspirates, and develops aspiration pneumonia.

The increased sensitivity to recognition, the accuracy of clinical and laboratory diagnosis, and treatment have essentially eliminated neonatal mortality attributed to withdrawal per se.

ASSESSMENT AND MANAGEMENT OF NEONATAL OPIOID ABSTINENCE

With proper management, the neonate's prognosis for recovery from the acute phase of withdrawal is good. If symptoms of withdrawal appear, simple nonspecific measures should be instituted, such as gentle infrequent handling, swaddling, and demand feeding. Careful attention to fluid-electrolyte balance and calorie support is essential in opioid-exposed infants undergoing withdrawal, since they display uncoordinated sucking, feed poorly, often develop vomiting and diarrhea, and have increased water losses due to rapid respirations and sweating.

Indications for specific treatment, dosage schedules, and duration of treatment courses have varied widely. As a general guide, if, in spite of nonspecific measures, babies have difficulty feeding, diarrhea, marked tremors, irritability even when undisturbed, or cry continuously, they should be given medication to relieve discomfort and prevent dehydration and other complications. The dosages must be carefully regulated so that symptoms are minimized without excessive sedation. Several drugs appear to be effective in treating neonatal narcotic withdrawal, but there has been little controlled comparison of their safety and effectiveness. Drugs such as PAREGORIC or tincture of OPIUM are effective in treating narcotic withdrawal symptoms in the infant, and PHENOBARBITAL is useful, but less so when opioid exposure has occurred in high doses.

NEUROBEHAVIOR IN THE NEWBORN

The Brazelton Neonatal Assessment Scale has been used extensively for evaluating newborn behavior. This instrument assesses reaction to stimuli such as a light or a bell, responsivity to animate and inanimate stimuli (face, voice, bell, rattle), state (sleep to alertness to crying), the requirements of state change (such as irritability and consolability), and neurological and motor development. When using this scale in evaluating drug-exposed infants, it was noted that they were less able than nondrug-exposed infants to be maintained in an alert state and less able to orient to auditory and visual stimuli, most pronounced at forty-eight hours of age. Drug-exposed infants were as capable of self-quieting and responding to soothing intervention as normal neonates, although they were substantially more irritable. These findings have important implications for caregivers' perceptions of infants and thus may have long-term impact on the development of infant–caregiver interaction patterns.

Abnormalities in the interaction of drug-dependent mothers and their infants, on measures of social engagement, have been shown. Abnormal interaction was explained by less positive maternal attachment, as well as difficult infant behavior, which impedes social involvement. Many of these interactive abnormalities reverted to normal by four months of age, but the need for "parenting training" is obvious.

OPIOIDS AND SUDDEN INFANT DEATH SYNDROME (CRIB DEATH)

Sudden infant death syndrome (SIDS) is defined as the sudden and unexpected death of an infant between one week and one year of age, whose death remains unexplained after a complete autopsy examination, full history, and a death-site investigation. Compared to an incidence of approximately 1.5 per 1,000 live births in the general population, narcotic-exposed infants appear to have an increased risk of SIDS. Other high-risk factors for SIDS, such as low socioeconomic status, low birthweight, young maternal age, black racial category, and maternal smoking are all overrepresented in the drug-using groups that are studied. In a most extensive study, New York City SIDS rates were calculated in 1.2 million births from 1979 to 1989. Maternal opiate use, after control for high-risk variables, increased the risk of SIDS by three to four times that of the general population.

LONG-TERM OUTCOME OF CHILDREN WHO HAVE UNDERGONE IN UTERO EXPOSURE TO OPIOIDS

Despite the fact that a drug-exposed newborn may seem free of physical, behavioral, or neurological deficits at the time of birth, the effects of pharmacological agents (used or abused) may not become apparent for many months or years. Although heroin abuse during pregnancy has been recognized for more than 40 years, and methadone treatment has been employed for more than 20 years, follow-up of opioid-exposed infants is still fragmentary. The difficulties encountered in long-term follow-up of this population include an inability to fully document a mother's drug intake, separation of the drug effects from high-risk obstetric variables, problems in maintaining a cohesive group of infants for study, and the need to separate drug effects from those of parenting and the home environment.

The easiest part of caring for the neonate is actually over when drug therapy has been discontinued and the infant is physically well. The most difficult parts then begin—the care involved in discharge planning and assuring optimal growth and development throughout infancy and childhood. Because there is no standard for the disposition of these infants, some may be released to their mothers, some to relatives, and others placed in the custody of a state agency. Still others may be voluntarily released by the mother to private agencies for temporary or permanent placement.

In the United States, pressure recommending separation of infants from their addicted mothers has been growing. This solution may not be practical in cities where social services and courts are already understaffed and overworked. Decent foster care is expensive and hard to find. Pediatricians basically feel that the mother–infant association should not be dissolved except in extreme situations. Aside from intensive drug rehabilitation and medical treatment, these women need extensive educational and job training—to become the productive citizens and loving mothers who may positively socialize their children. Supportive therapies such as outpatient care or residential treatment may help eliminate some of the medical and social problems experienced by drug-dependent women and their children.

Most of the children evaluated for long-term development have been exposed to methadone. Evaluations have occurred at various intervals—at six, twelve, eighteen, and twenty-four months; then at three, four, and five years of age. Testing procedures utilized have been Gesell Developmental Schedule, Bayley Scales of Infant Development, McCarthy Scales of Infant Abilities, and Stanford-Binet and the Wechsler Preschool and Primary Scale of Intelligence. Infants have shown overall developmental scores in the normal range but a decrease in scores at about two years of age—which suggests that environment may confound long-term infant outcome: low socioeconomic groups suffer from this factor particularly, because of poor language stimulation and development.

The developmental scores in these early years, although useful in identifying areas of strength and weakness, may not predict subsequent intellectual achievement. More and more studies have proposed multiple-factor models to assess infant outcome following intrauterine drug exposure. One such postnatal influence involves maternal–infant interaction. Drug-exposed infants are often irritable, have decreased rhythmic movements, and may display increased muscle tone (tensing) when handled. Such behaviors may be interpreted by the mother as "rejecting" behavior, leading to inappropriate maternal caretaking and possible neglect of the infant. Studies of mother–infant interactions show that: (1) infants born to narcotic-addicted women show deficient social responsiveness after birth; (2) this deficient mother–infant interaction persists until the infants' treatment for withdrawal is completed; and (3) maternal drug dosage may affect that interaction.

Based on available data, at five years of age, children born to women maintained on methadone, in contrast to heroin-exposed babies, appear to function within the normal range of their mental development. In addition, no differences in language and perceptual skills were observed between them and children of mothers not involved with drugs and of comparable backgrounds. Difficulty in following large cohorts of drug-exposed infants has led to the study of very limited samples, however.

Positive and reinforcing environmental influences can significantly improve drug-exposed infant development. Women who show a caring concern for their infants are most likely to pursue follow-up pediatric care and cooperate in neurobehavioral follow-up studies. Lacking a large data base, there is an obvious need for comprehensive studies assessing the development of large populations of drug-exposed infants.

COCAINE

The effects of the maternal medical and obstetrical complications seen in opioid-exposed infants are similar to those of COCAINE exposure—although cocaine is a stimulant, not a depressant drug (like the opioids). The infants are frequently small in weight, length, and head circumference as a result of preterm birth and/or retardation of fetal growth. The effects of blood-vessel constriction, a characteristic pharmacologic effect of cocaine, is one of the main reasons for adverse effects—since it results in lack of oxygen and nutrients to the fetus. This predisposes the infant to growth problems, brain hemorrhage, abnormal organ development, and crib death.

The many studies on cocaine effects in the newborn need further clarification because of inadequate sample size, research methodology, and actual drug intake; these include studies that have evaluated brain hemorrhage, structural abnormalities, crib death, and long-term development. Although cocaine-exposed infants have been reported to have some irritability and perform poorly on neurobehavioral tests in the first few days of life, no evidence shows that they have a withdrawal syndrome as described previously in infants exposed to opioids. The symptoms have been related to a cocaine toxicity reaction rather than to a withdrawal syndrome. Infants with opioid *and* cocaine exposure, as compared to opioid exposure alone, have had milder symptoms. This may be a result of interactions between the depressant *and* stimulant properties of these drugs. No treatment has been found necessary to alleviate the symptoms of infants exposed to cocaine, whereas opioid-exposed infants may need treatment in about 40 to 50 percent of cases.

Although a number of reports in the medical literature have described babies who have structural abnormalities related to cocaine exposure, an equal number of studies have found no increased incidence of abnormalities. The abnormalities reported have been those of the urinary tract, intestines, and extremities—all of which are related to the vascular disruption caused by cocaine's ability to constrict blood vessels. The most recent review of the clinical studies describing abnormalities in cocaine-exposed infants shows a very low incidence of occurrence.

Studies evaluating cocaine's effects on the occurrence of SIDS (crib death) have shown diverse results. Although inadequate methodologies and small numbers have accounted for these differences, cocaine-exposed infants have also experienced most of the factors that predispose any child to SIDS. These include low birthweight, POVERTY, neonatal complications, minority ethnicity, low maternal age, and maternal cigarette smoking. When these factors are controlled in the research, cocaine exposure accounts for only a very modest increase in the rate of SIDS.

As with all drugs of abuse, cocaine has properties that permit it to be transmitted through the breast milk. Since a significant portion of drug-using women in the United States may be HIV-positive, until the role of breast feeding in HIV transmission is clarified, breast feeding should be discouraged.

Recent reports indicate that cocaine exposure may even occur in young infants after they leave the hospital. The evidence for the postulated route of cocaine toxicity (passive inhalation of smoked cocaine—"crack") is circumstantial, and the range of occurrences in reported series is 2 to 4 percent. Symptoms involve abnormal neurologic findings, including seizures, drowsiness, and unsteady gait.

Much concern has been voiced regarding the ultimate neurobehavioral outcome of infants following interuterine exposure to cocaine. Based on multiple-risk factors, it appears reasonable to voice these concerns. Commonly, the parents may be of poor socioeconomic status and culturally deprived. The mother may be poorly nourished, may carry medical and sexually transmitted diseases, including AIDS, and may receive little or no prenatal care. After birth, neurologic and neurobehavioral abnormalities may be present in the infant. Stimulation for intellectual growth may be lacking because of prolonged hospital stays, infrequent and inappropriate parental contact, placement in a group-care facility, or discharge to a home in which intellectual nurturing is lacking.

Follow-up studies of large numbers of cocaine-exposed babies are lacking as of the early 1990s. The lay press has reported anecdotal experiences with the first cohort of three- to five-year-old children born of the crack epidemic. Such cocaine-exposed babies have been characterized as showing significant deficits in environmental interactions during play groups and in nursery schools. These babies have also been described as showing less representational play, decreased fantasy play and curious exploration, and lesser quality of play. Others have described these children as "joyless"—unable to fully participate in either structured or unstructured situations, with attention deficits and flat apathetic moods. Developmental evaluations show, however, that the majority of children who were exposed to cocaine in utero and who now have stable environments score in the normal range.

(SEE ALSO: *Complications: Route of Administration; Fetal Alcohol Syndrome; Pregnancy and Drug Dependence; Substance Abuse and AIDS*)

BIBLIOGRAPHY

FINNEGAN, L. P., & KANDALL, S. R. (1992). Maternal and neonatal effects of drug dependence in pregnancy. In J. Lowinson et al., *Comprehensive textbook of substance abuse*, 2nd ed. Baltimore: Williams & Wilkins.

HADEED, A. J., & SIEGEL, S. R. (1989). Maternal cocaine use during pregnancy: Effect on the newborn infant. *Pediatrics, 84*, 205.

KALTENBACH, K., & FINNEGAN, L. P. (1988). The influence of the neonatal abstinence syndrome on mother–infant interaction. In E. J. Anthony & C. Chiland (Eds.), *The child in his family: Perilous development: Child raising and identity formation under stress.* New York: Wiley-Interscience.

ZUCKERMAN, B., ET AL. (1989). Effects of maternal marijuana and cocaine use on fetal growth. *New England Journal of Medicine, 320*, 762.

LORETTA P. FINNEGAN
MICHAEL P. FINNEGAN
GEORGE A. KANUCK

FINANCIAL ANALYSIS IN ENFORCEMENT *See* Drug Laws: Financial Analysis in Enforcement.

FLY AGARIC A poisonous mushroom of Eurasia (*Amanita muscaria*), having typically a bright red cap with white dots. A preparation, consisting primarily of the dried mushroom, is ingested by the people of Siberia as a HALLUCINOGEN. Intoxication by ingestion of several mushrooms moistened with milk or fruit juice leads to a progression of symptoms—beginning with tremors, continuing through

Figure 1
Fly Agaric

a period of visual hallucination that may be interpreted as having religious significance, and finally ending in deep sleep. A similar preparation may be identified with the deified intoxicant *soma* of the ancient Hindus. In some cultures, the urine of intoxicated individuals is ingested by others to induce intoxication, since the active components of the preparation pass unmetabolized through the body.

The active components found in fly agaric are ibotenic acid and several of its metabolites. The predominant metabolite is muscimol, which has agonist properties at a subset of receptors recognizing the NEUROTRANSMITTER GABA. Ibotenic acid itself has agonist properties at certain excitatory amino acid receptors and has been shown to be neurotoxic.

(SEE ALSO: *Plants, Drugs from*)

BIBLIOGRAPHY

HOUGHTON, P. J., & BISSET, N. G. (1985). Drugs of ethnoorigin. In D. C. Howell (Ed.), *Drugs in central nervous system disorders.* New York: Marcel Dekker.

ROBERT ZACZEK

FOOD AND DRUG LAWS   *See* U.S. Government: The Organization of U.S. Drug Policy.

FOREIGN POLICY AND DRUGS   Drug control is a relative newcomer to the list of global issues that are now an integral part of U.S. foreign policy. While arms control and human rights were already important international issues in the 1970s, drug control lagged behind. In 1971–72 some members of Congress tried to use foreign-aid restrictions to stop the entry of Turkish HEROIN, but the government did not want to risk hurting relations with an important defense ally over heroin, which was not considered a "mainstream" drug. The U.S. government found a compromise through diplomatic efforts, which led to the Turkish government severely limiting the cultivation of OPIUM POPPIES (from which heroin is made) and changing the way in which poppies were processed into legitimate medicinal opium. Parallel diplomatic negotiations with MEXICO resulted in cooperation on MARIJUANA eradication efforts with OPERATION INTERCEPT. On the international front, the U.S. government pressed hard

for the ratification of the 1971 UN Convention on Psychotropic Drugs and created the United Nations Fund for Drug Abuse Control (UNFDAC), the predecessor of today's United Nations Drug Control Program (UNDCP). During the rest of the decade, however, drug control gradually declined as a key U.S. foreign policy objective.

Drug control only gained full diplomatic legitimacy in the 1980s when COCAINE use became widespread among entertainers, athletes, and stockbrokers. The government's inability to stop the EPIDEMIC at home prompted Congress to take the issue abroad.

In 1986, in the first of a series of comprehensive international antidrug laws (the Anti-Drug Abuse Act of 1986), Congress placed the burden of halting drug flows on the governments of the drug-producing countries. Using a traditional carrot-and-stick approach, the law required the major drug-producing and TRANSIT COUNTRIES to cooperate fully with the United States in drug matters in order to receive American foreign aid. Half of all assistance was withheld every year until the president certified that the country concerned had met the criteria for receiving aid. Subsequent laws have expanded the requirement, obliging the major drug-producing and transit countries also to comply with the 1988 United Nations Convention Against Illicit Traffic in Narcotics Drugs and Psychotropic Substances. Countries that do not comply not only lose U.S. assistance but incur U.S. opposition to loans from the World Bank and other international financial institutions. For many countries in the developing world, losing access to these loans is an even greater hardship than losing U.S. assistance.

In earning its diplomatic legitimacy, drug control has had to overcome the same obstacles encountered by other global issues, such as human rights or nuclear nonproliferation. The U.S. foreign-policy establishment favors strategic issues affecting vital U.S. national-security or trade interests over law enforcement or scientific endeavor. It has been reluctant to allow multilateral "functional" questions to affect traditional bilateral negotiations. Congress, however, has left no doubt that it intends to keep drug control high on the list of U.S. foreign-policy issues. By denying virtually all forms of aid—excluding humanitarian and drug-control assistance—to countries that refuse to cooperate, Congress has devised an effective form of leverage over drug countries. Since the law also allows the president to waive sanctions when clearly stated national interests are at stake, the Congress had made it difficult for foreign-policy agencies to evade their drug-control responsibilities.

RESPONSIBLE AGENCIES

The U.S. Department of State is responsible for formulating international drug policy. Since 1989, formal coordination authority has rested with the White House Office of National Drug Control Policy (ONDCP) and the National Security Council. Drug control programs, however, involve a broad spectrum of government agencies including the Central Intelligence Agency, the Department of Defense, the U.S. CUSTOMS SERVICE, the Coast Guard, the Department of Treasury, the Justice Department, the DRUG ENFORCEMENT ADMINISTRATION, and the Department of Health and Human Services.

THE REALITIES OF DRUG CONTROL

As presidential administrations have discovered, an effective drug policy is easier to design than to carry out. The drug issue is a typical chicken-and-egg problem. Does supply drive demand or vice versa? The drug-consuming countries traditionally blame the suppliers for drug epidemics, while drug-producing countries allege that without foreign demand, local farmers would not be growing the drug crop at all. Planners must therefore strike the right balance between reducing drug supply and demand. In theory, eliminating drug cultivation in the source countries is the most economical solution, since it keeps drugs from entering the system and acquiring any value as a finished product. Few SOURCE-COUNTRY governments—all of which are in developing nations—will however deprive farmers of a livelihood without substantial compensation from abroad. And the price they seek is usually more than the U.S. government is prepared to pay.

In practice, less than 5 percent of the U.S. drug-control budget is spent on international programs. The bulk of the money goes to domestic law enforcement, drug treatment, and public education. In Fiscal Year 1993, only $523.4 million (slightly more than 4%) of the $12.3 billion U.S. drug budget went to international programs. Approximately $1.5 billion (12.3%) went to interdiction efforts at home and abroad; the bulk, $10.2 billion (84%), was devoted to law enforcement and demand reduction. Even including the interdiction portion, the United States spent only 16 percent of its available drug-control funds on international activities, a small proportion in light of the economic power of the drug trade.

THE NATURE OF THE THREAT

Today's illegal drug trade is one of the most lucrative and, therefore, powerful criminal enterprises in history. Drugs generate profits on a scale without historical precedent—especially given their abundance and low production costs. On the streets of the United States, the average wholesale price of a gram of cocaine or heroin is $100. By diluting (cutting) it with other substances, dealers double its value. Thus, a metric ton (1 million grams) of cocaine is worth between $100 million and $200 million retail. To put these sums in perspective, 5 metric tons of cocaine sold at retail (street) prices would pay for the *entire international portion* of the U.S. drug-control budget for Fiscal Year 1993. Similarly, the 108 metric tons of cocaine seized by the United States in the same fiscal year would have sold for between $10 billion and $20 billion—more than the legal gross domestic product (GDP) of some of the drug-producing countries. Such financial resources, which are well beyond those of most national budgets, give drug traffickers the means to buy sophisticated arms, aircraft, and electronic and technical equipment available to few countries. More importantly, illegal drug revenues allow trafficking organizations to buy themselves protection at almost every level of government in the drug-producing and drug-transit countries, where drug-related corruption remains the single largest obstacle to effective control programs.

As for the drugs themselves, there is a super-abundance. Opium is in especially great supply. In Southeast Asia, Myanmar (formerly Burma) could supply the world's needs several times over with 257.5 metric tons annually. Estimates of heroin consumption in the Unites States range only between 6 and 20 metric tons, less than 10 percent of Myanmar's potential output. In South America, the coca fields may yield as much as 900 metric tons of cocaine annually, enough to satisfy world demand twice over. This surplus is so large that the drug trade easily absorbs losses inflicted by drug-control authorities and still makes enormous profits.

Traffickers have the option of expanding cultivation of drug crops into new areas. Although, for example, coca plants are currently confined to Latin

America, coca once flourished in Indonesia and could do so again if market conditions were right. Opium poppy cultivation is spreading into nontraditional areas, including South America. Gambling on the resurgence of expanding heroin use in the 1990s, South American cocaine-trafficking organizations have been diversifying into opium poppy cultivation. There are now an estimated 20,000 hectares of opium poppy growing in COLOMBIA, with evidence of experimental cultivation in BOLIVIA, Ecuador, and Peru. Without active government antidrug programs, production will continue until the new expanding market is saturated.

CURRENT POLICY

The U.S. government's first priority is to stop the flow of cocaine, which still poses the most immediate threat to potential drug users. Because of rising heroin use promoted by the new, cheaper Latin American producers, the United States must also focus on opium-producing countries. The United States goal is to limit the cultivation of drug crops to the amount necessary for international medical applications. Since all the cocaine that enters the United States comes from coca plantations in three countries— Peru (56%), Bolivia (24%), and Colombia (20%)— the U.S. government has active drug-control programs in the three countries. Political and economic instability, however, limit the effectiveness of these programs. Opium control is more difficult than coca suppression, since most of the world's opium poppy grows in countries where the United States has minimal diplomatic influence (Myanmar, Afghanistan, Laos, Iran, etc.). There also appears to be increasingly important opium poppy cultivation in China, Vietnam, and the Central Asian countries. Left unchecked, this opium expansion will make effective heroin control virtually impossible in drug-consuming countries, as Europe is already aware.

AN INTERNATIONAL APPROACH

Since bilateral programs seldom provide solutions to global problems, the United States has been an active proponent of collective action under the 1988 UN Convention, to which, as of February 1994, ninety-nine countries and territories were parties. This latest agreement covers not only the traditional aspects of drug production and trafficking, but requires signatories to control drug-processing chemicals and outlaw drug-money laundering. The MONEY-LAUNDERING provisions are critical innovations, since they target the enormous international cash flows that sustain the drug trade. As astronomical as drug profits may be, drug money is useless unless it can enter the international banking system. The major industrialized countries are therefore pressing for uniform laws and regulations to exclude drug money in all key financial centers. If honestly implemented, strict money-laundering controls, along with better use of existing programs to suppress drug supply and decrease consumption, offer the hope of reducing the drug trade from an international threat to a manageable concern.

(SEE ALSO: *Crop-Control Policies; Drug Interdiction; Drug Laws: Financial Analysis in Enforcement; Golden Triangle as Drug Source; International Drug Supply Systems; Opioids and Opioid Control: History; Terrorism and Drugs; U.S. Government Agencies*)

BIBLIOGRAPHY

EHRENFELD, R. (1990). *Narco-terrorism*. New York: Basic.

MACDONALD, B., & ZAGARIS, B., (EDS.) (1992). *International handbook on drug control*. Westport, CT: Greenwood.

SIMMONS, L. R. S., & SAID, A. A., (EDS.). (1974). *Drugs, politics, and diplomacy: The international connection*. Beverly Hills: Sage.

TAYLOR, A. H. (1969). *American diplomacy and narcotics traffic, 1909–1939*. Durham, NC: Duke University Press.

U.S. CONGRESS. SENATE COMMITTEE ON THE JUDICIARY. (1975). *Poppy politics*. Hearings before the Subcommittee to Investigate Juvenile Delinquency. Washington, DC: U.S. Government Printing Office.

U.S. DEPARTMENT OF STATE. (1994). *International narcotics control strategy report, April 1, 1994*. Washington, DC: U.S. Government Printing Office.

W. KENNETH THOMPSON

FORFEITURE LAWS *See* Legal Regulation of Drugs and Alcohol; Mandatory Sentencing.

FREEBASING The illicit practice of smoking COCAINE is generally referred to as freebasing. The hydrochloride form of cocaine (powder) is highly soluble in water and, therefore, is efficiently absorbed by the mucous membranes when taken intranasally (snorted) or via blood when injected intravenously (shot up). This form of cocaine is, however, destroyed when it is heated to the temperatures required for smoking it. Therefore, the cocaine alkaloid, called "CRACK" or "freebase," is the form that is smoked. Although not always differentiated, freebase actually refers to cocaine in the base state with all the adulterants removed (Inciardi, 1991). Cocaine hydrochloride is combined with an alkaline substance, such as sodium hydroxide or ammonia, to remove the hydrochloride. The *free* cocaine *base* is then dissolved in ether, and pure cocaine-base crystals are formed. It has been estimated that approximately 560 milligrams of cocaine freebase can be extracted from one gram of street cocaine hydrochloride (Siegel, 1982). Cocaine freebase has a melting point of 208°F (98°C) and is volatile at temperatures above 194°F (90°C), therefore providing an active drug for smoking. Crack, in contrast, although also in the base state and used for smoking (or freebasing), does not have the adulterants of the street cocaine removed. Cocaine base is soluble in alcohol, acetone, oils, and ether—but is almost insoluble in water.

Cocaine freebase is usually smoked in a water pipe containing fine mesh screens, which trap the heated cocaine as it melts. A temperature of 200°F (93°C) is the most efficient. Although the amount of cocaine absorbed by the smoker varies—depending on the kind of pipe used, the temperature of the heat source, and the inhalation pattern of the user—under optimal conditions approximately 30 to 35 percent of the cocaine placed on the mesh screen is absorbed by the smoker.

Although in use since the mid-1970s, freebasing cocaine became popular in the United States in the early 1980s. The popularity of this route of administration was responsible for the rise in U.S. cocaine use during the mid-1980s. When cocaine is smoked, it is rapidly absorbed and reaches the brain within a few seconds. Thus, users get a substantial immediate rush and an almost instant "high," comparable to that after intravenous cocaine. This is in contrast to intranasal use of cocaine, which engenders a high with a much slower onset. Freebasing is thus a convenient way of taking cocaine, with the possibility of repeated and substantial doses. Since the likelihood of abuse is related to the rapidity with which a drug reaches the brain, smoking cocaine makes it more likely that use will lead to abuse than does snorting the drug. Despite losses of more than half of the cocaine when it is smoked, sufficient cocaine rapidly reaches the brain, providing an intense drug effect—which users repeat, often to toxicity. The danger of freebasing, in addition to the inherent danger of cocaine use, lies in what some users perceive to be the greater social acceptability of a route of administration that requires minimal PARAPHERNALIA and can achieve toxic levels of cocaine with relative ease.

(SEE ALSO: *Amphetamine Epidemics; Coca Paste; Complications: Cardiovascular System; Methamphetamine; Pharmacokinetics*)

BIBLIOGRAPHY

INCIARDI, J. A. (1982). Crack-cocaine in Miami. In S. Schober & C. Schade (Eds.), *The epidemiology of cocaine use and abuse*. NIDA Research Monograph no. 110. Rockville, MD: National Institute on Drug Abuse.

PEREZ-REYES, M., ET AL. (1982). Freebase cocaine smoking. *Clinical Pharmacology and Therapeutics, 32,* 459–465.

SIEGEL, R. (1982). Cocaine smoking. *Journal of Psychoactive Drugs, 14,* 271–359.

MARIAN W. FISCHMAN

FREE WILL See *Disease Concept of Alcoholism and Drug Abuse; Values and Beliefs: Existential Models of Addiction*.

FRENCH CONNECTION See *Drug Interdiction; International Drug Supply Systems*.

FREUD AND COCAINE Sigmund Freud (1856–1939), Austrian neurologist and founder of PSYCHOANALYSIS, became interested in COCAINE in the early 1880s. At the time he was in his late twenties and was a medical house officer at the Vienna hospital called the Allgemeine Krankenhaus. He was able both to gain access to the literature about cocaine and, at some expense, to the substance itself (which was not illegal at that time). There had been articles in the American medical literature describing cocaine used in the treatment of various ills and for drug dependencies as almost a panacea. The ability of cocaine to fend off fatigue and enhance mood also came to Freud's attention. He was particularly taken by suggestions that cocaine might be an adjunct to, or even a cure for, ALCOHOL or OPIOID dependencies. His interest was heightened because one of his close teachers and friends, Ernst von Fleischl-Marxow, had become an opiate addict. Using cocaine, Freud treated him with almost disastrous results. At the time, there was no opprobrium attached to the use of cocaine and relatively little concern about any adverse effects.

Freud performed a number of cocaine experiments on himself and reported the results in his experimental paper, "Contribution to Knowledge of the Effects of Cocaine." These were reasonable studies that provided useful data about the physiological and psychological effects of cocaine. Biographies of Freud, such as Ernest Jones's *The Life and Work of Sigmund Freud*, have tended to disparage his experimental paper and other works on cocaine. Although his work was done on himself and was limited in its scope, it has been confirmed in modern replications. Freud was initially skeptical about the possible "addictive" properties of cocaine in normal individuals, but later, in the face of evidence and criticism, he was less vehement on the subject. He became, in later life, very sensitive to criticism of his earlier views on cocaine.

From 1884 to 1887 Freud wrote four papers concerning cocaine, including a definitive review ("Über Coca") in 1884. He obviously felt comfortable in both taking cocaine and writing about it in his letters. He mentions and discusses his use of and dreams about cocaine in the *Interpretation of Dreams* (1889). The true extent and duration of his self-experiments is not known, since access to his correspondence has been severely restricted.

Freud is sometimes credited with the discovery of local anesthesia because of his proposal in his co-

caine review paper that the substance could be used for this purpose. He also claims suggesting the idea to both Koenigstein and Carl Koller prior to their experiments in ophthalmology, which led to the initial papers on local or topical anesthesia. There is a semantic problem in understanding these claims. Almost all investigators of cocaine had noticed the numbing properties of the drug when placed on the tongue. The idea that this property had a practical use in ophthalmological surgery does belong to Carl Koller, a friend and colleague of Freud's, who did the proper experiments and published them promptly. The controversy about the discovery between Koller and Koenigstein with Freud's mediation is well covered in the article by Hortense Koller Becker, "Carl Koller and Cocaine," in *Psychoanalytic Quarterly*.

Extreme viewpoints that attribute Freud's behavior and writings to the influence of the toxic effects of cocaine are unsubstantiated by evidence. Clearly, he used cocaine as a psychotropic agent on himself and this experience led to his faith in its relative safety. Despite this, there is no real support for a viewpoint that he was an addict or that his thought was markedly affected by his drug usage. The combined notoriety of both Freud and cocaine has led to speculative exaggerations that make better newspaper headlines than history.

(SEE ALSO: *Abuse Liability of Drugs; Epidemics of Drug Abuse; Pharmacotherapy*)

BIBLIOGRAPHY

BECKER, H. K. (1963). Carl Koller and cocaine. *Psychoanalytic Quarterly, 32*, 309–343.

BYCK, R. (1974). *Cocaine papers: Sigmund Freud.* (Edited, with an introduction by R. Byck, M.D.). New York: Stonehill; New American Library edition, 1975.

JONES, E. (1953–1957). *The life and work of Sigmund Freud.* 3 vols. New York: Basic Books. (*See* Volume I, Chapter VI, The Cocaine Episode [1884–1887]).

MALCOLM, J. (1984). *In the Freud archives.* London: Jonathan Cape.

ROBERT BYCK

G

GABA *See* Gamma-Aminobutyric Acid.

GAMBLING AS AN ADDICTION Human beings have indulged in games of chance since before recorded history. Archeological sites in both the Old World and the New World yield gambling bones, dice, and counters. The Old and New Testaments mention the casting of lots to determine the distribution of property, presumably as an expression of God's will. In addition, the classical literature of both Eastern and Western cultures includes many accounts of gambling, often with dramatic consequences. Lotteries have been popular in Asia and Europe for centuries. The first European government-sponsored lottery was established by Queen Elizabeth I in sixteenth-century England. The thirteen American colonies and the early American universities—including Harvard, Yale, Princeton and Columbia—were all supported in part by lotteries.

Most societies have recognized the popularity of gambling and its potential for generating social good and personal harm. Therefore, governments have sought ways to regulate gambling. Some governments have prohibited all gambling, while others have established laws limiting the availability of gambling to particular locations, establishing a minimum age, specifying types of games allowed, and regulating the gaming industry to prevent fraud and raise revenues. In the United States, government attitudes toward legalizing gambling have changed radically over time. By the mid-twentieth century, some state governments increasingly looked to state lotteries as a fertile source of revenues. In addition, casino and riverboat gambling, sports betting, card rooms, and bingo games were variously legalized, taxed, and regulated. By 1994, some form of gambling was legal in all states but Hawaii and Utah, and several American Indian nations were operating gambling establishments on tribal land. In 1992, an estimated 330 billion dollars was wagered annually in the United States, generating a profit of more than 30 billion dollars—with the vast majority of the total bet legally. Illegal gambling has its own special set of subcultures—with rules, limits, and penalties for its devotees.

For most people gambling is a pleasurable, if not very profitable, occasional recreation. For a significant minority, however, gambling has the potential to become a compulsive behavior and a ruinous destructive problem. Compulsive gambling has also been known for centuries. The classic Hindu epic, *The Mahabharata*, tells the story of a wise and just king whose single flaw, the inability to control his gambling, leads him to gamble away his wealth and kingdom in a dice game. Still unable to stop, he gambles his brothers, his wife, and himself into slavery. This critical game of chance sets off a train of events that mark the beginning of division and strife in human society.

Famous people among the ranks of compulsive gamblers include sports figures, entertainers, and artists. Fyodor Dostoyevsky, who wrote his novella,

The Gambler, to restore his finances, was a self-described compulsive gambler. Sigmund Freud's 1928 essay about Dostoyevsky was one of the first attempts to understand compulsive gambling as a psychopathological process. This conceptualization and its further development guided the treatment of compulsive gamblers with psychoanalytic therapies.

Until 1980, the term *compulsive gambling* was used to describe the syndrome of apparent loss of control in gambling. At that time, the American Psychiatric Association published the third edition of its DIAGNOSTIC AND STATISTICAL MANUAL (DSM-III). For the first time, the DSM-III established standard criteria to diagnose this disorder, which was renamed *pathological gambling*. The term was coined to avoid confusion with other diagnoses in which the word "compulsive" appeared, such as obsessive-compulsive disorder and obsessive-compulsive personality disorder; these disorders were thought to be unrelated to compulsive gambling. Pathological gambling was grouped under the heading Impulse Control Disorders Not Elsewhere Classified, along with such diagnoses as kleptomania (shoplifting) and pyromania (arson). In 1987, the American Psychiatric Association's *Diagnostic and Statistical Manual* was again revised (to be abbreviated DSM-III-R). In this revision, the term *pathological gambling* and its classification as an impulse-control disorder were retained, but the diagnostic criteria were significantly altered in response to new knowledge about the disorder. Likewise, the fourth edition of the *Diagnostic and Statistical Manual* (DSM-IV) has additional changes reflecting additional research.

THE ADDICTION MODEL OF PATHOLOGICAL GAMBLING

The early psychoanalytic literature often referred to compulsive gamblers as ADDICTS, but it was not until the founding of Gamblers Anonymous (GA) in 1957 that the addictive-disease model became a basis for recovery. GA was initiated through the efforts of a recovering alcoholic who was both an ALCOHOLICS ANONYMOUS (AA) member and a compulsive gambler. GA adapted the TWELVE STEPS of AA, the fellowship's traditions, its spiritual base, and the general format of its meetings to aid in the recovery of gambling addicts. Gam-Anon, a twelve-step group for the friends and families of compulsive gamblers, modeled on Al-Anon Family Groups, was established shortly afterward. Local chapters of Gamblers Anonymous are increasingly available in U.S. communities as well as in treatment units, work settings, and prisons.

The growth of the alcoholism- and drug-addiction-treatment system in the 1960s gave rise to a variety of professional program models that incorporated a cooperative working relationship with twelve-step groups such as Alcoholics Anonymous. In 1971, using one of these models, Dr. Robert Custer developed the first inpatient addiction-oriented treatment unit for compulsive gamblers at the Brecksville, Ohio, Veterans Administration Hospital. Custer's approach proved useful and has been adopted with various modifications by other mental-health and addiction-treatment facilities.

COMMON CHARACTERISTICS WITH OTHER ADDICTIVE DISORDERS

The addiction model conceptualizes pathological gambling as a disease characterized by a dependence on what gamblers refer to as "being in action." The term describes their aroused euphoric state—experienced while gambling. Pathological gamblers who are also users of other drugs compare being in action to the "high" derived from COCAINE or other STIMULANTS. The addiction model is also supported by the many similarities between pathological gambling and substance dependence in risk factors, symptoms, the course of the disease, the nature of relapse triggers, treatment goals, and the process of recovery. A core symptom for both types of disorder is a loss of control over the substance use or gambling behavior. There is also an important comorbidity between the various addictive disorders. For example, a 1986 study of 458 adult inpatients admitted for alcohol and other drug (AOD) dependence to South Oaks Hospital in New York found that 9 percent satisfied diagnostic criteria for pathological gambling and an additional 10 percent had some gambling problems. These rates are many times higher than are found among the general public. In a parallel study of 100 younger AOD inpatients (average age 17), 14 percent met criteria for pathological gambling and an additional 14 percent had some gambling problems. Failure to recognize and address gambling problems during treatment for alcohol or other drug dependence often leads to relapse to substance use in a gambling situation. Less frequently, the result is a switch of addictions from alcohol or another drug to gambling.

EPIDEMIOLOGY OF PATHOLOGICAL GAMBLING

Epidemiological studies conducted during the 1980s in New York, New Jersey, Maryland, and Quebec yielded similar estimates. Approximately 1.5 percent of adults were found to be probable pathological gamblers and an additional 2.5 percent were found to have some gambling-related problems. In contrast, a lower prevalence was found in Iowa. Unlike the other jurisdictions studied, in which legal gambling was well established, Iowa had just initiated a state lottery at the time of the survey. Research has yet to determine whether or not Iowa will experience an increase in gambling problems as a result of this legalization.

In general-population studies in the United States, males outnumber females among probable pathological gamblers by a ratio of about two to one. This is in sharp contrast to male to female ratios observed in treatment programs and GA groups, which are closer to eight or nine males to one female. Some general-population studies in the United States have also found an overrepresentation of nonwhite adults (blacks and Hispanics) among probable pathological gamblers; but these groups, like women, are also underrepresented in treatment and GA populations.

Although less is known about the prevalence of pathological gambling among adolescents than among adults, several surveys of high school students revealed that the vast majority gamble to some extent and that many have problems. For example, a New Jersey study of nearly 900 students found that over 90 percent had gambled at some time in their lives and about 35 percent did so at least weekly. Approximately 5.7 percent of these eleventh- and twelfth-grade students—9.5 percent of boys and 2 percent of girls—were classified as probable pathological gamblers.

Established risk factors for pathological gambling include being male, having a family history of heavy or problem gambling, or of parental alcoholism, and early interest and participation in gambling activities.

CLINICAL CHARACTERISTICS

Gambling usually begins in adolescence, although women may begin gambling later in life. Pathological gambling often develops in three phases, originally described by Custer (1985): (1) the winning phase; (2) the losing phase; and (3) the desperation phase. Female pathological gamblers tend to have a later onset of the illness than males, and may never experience a winning phase.

The Winning Phase. Pathological gamblers often start as winners. Also, in a minority of cases, a significant upsurge in gambling activity begins with a "big win"—a sum equal to half a year's income or more. With or without the big win, individuals developing a dependence on gambling often begin with some success. In this context, they develop an intense interest in gambling and derive an increasing proportion of their self-esteem from feeling smart or lucky. The high derived from being in action becomes a major source of pleasure, a solution to life problems, a remedy for boredom, anger, anxiety, depression, and other uncomfortable feeling states. Bets must be gradually increased in size, in frequency, and sometimes in riskiness to produce the desired psychological effects. This phenomenon parallels the development of tolerance in the substance-dependent patient who must continue to increase the alcohol or drug dosage to reach the preferred feeling state. At this stage of the illness, the gambler devotes a great deal of time and effort to handicapping, studying the sports page, selecting a lottery number, or following the stock market, as well as to the gambling itself. As one gambler put it, "When I'm not occupied with gambling I'm preoccupied with it." Even if the gambler is winning more often than losing, time and emotional investment are withdrawn from friends, family, work, and other interests. The gambler's spouse often senses that something is wrong, but may not identify gambling as the problem. Marital counseling is sometimes sought.

An unreasonable attitude of optimism is also common during the early phase of pathological gambling, sustained by concentrating on wins and making excuses for (or even denying) losses. Because of this denial, the gambler often cannot account for money claimed to have been won. Pathological gamblers who begin with a winning phase are often those who state they gamble for excitement or stimulation.

The Losing Phase. All gamblers know that when on a losing streak it is wise to stop wagering, at least temporarily. For the compulsive gambler, however, losses are experienced as a severe injury to self-esteem. This produces an intense drive to continue gambling in an effort to recoup the money that has been lost, called *chasing losses*. Chasing losses is an important characteristic of this disease and an example of the pathological gambler's impaired control of gambling behavior. Chasing losses accelerates the gambler's losing and initiates a downward spiral. As the gambling debts mount, the pathological gambler will use any and all money available—take out loans, sell property, and gamble with money meant for family necessities. When these sources are exhausted, extended family members or friends may be approached for a "bailout," in the form of a loan or gift to relieve immediate financial pressure. In return, the pathological gambler often promises to give up gambling. However, part of the bailout money is usually gambled in the hope of another big win, and the downward spiral resumes. Although there are both wins and losses during the losing phase, the overall result is mounting emotional and financial distress as well as interference with social, vocational, and family functioning. Serious depression and a variety of stress-related somatic disorders are often experienced. Pathological gamblers report insomnia, gastrointestinal symptoms, dizziness, headache, hypertension, palpitations, chest pains, and breathing problems. Medical help may be sought, but again the connection to gambling behavior is seldom recognized. Family problems become more intense and divorce often results. Alcohol and other drug abuse may accompany gambling and/or function as a substitute when gambling is temporarily interrupted.

Pathological gamblers also describe a WITHDRAWAL syndrome when they are prevented from gambling. Symptoms include craving, restlessness, irritability, insomnia, headache, weakness, gastrointestinal symptoms, shakiness, and muscle aches.

Those pathological gamblers who do not experience a winning phase often describe themselves as gambling for "escape" (from life problems that seem insoluble).

The Desperation Phase. The desperation phase often begins when all legitimate sources of funds are exhausted. The gambling now takes on a desperate quality. The gambler's behavior during this phase may be characterized by activities inconsistent with the individual's previous moral standards, such as lying, embezzling, larceny, and forgery. These activities are justified as temporary expedients until the next big win. Pathological gamblers are often imprisoned both for white-collar crime and for illegal gambling activities such as bookmaking. Violent crime is less common. Studies of prison populations have found gambling problems in 15 to 30 percent of inmates.

An irrational belief in the inevitability of a big win sustains hope to some degree during this phase. Family problems become more intense. Mood swings are common. Severe anxiety, major depression, and suicidal behavior are increasingly noted during the late stages of the disease. Manic or hypomanic states are also seen in some cases. Most pathological gamblers who enter treatment or Gamblers Anonymous do so in the desperation phase. Surveys of Gamblers Anonymous have reported suicide attempts by 17 to 24 percent of members.

PATHOPHYSIOLOGY

Several studies have examined neurochemical changes in pathological gamblers. One study measured levels of NEUROTRANSMITTERS and their metabolites in the body fluids of male pathological gamblers, comparing these to levels in normal male subjects. The researchers found an elevated level of a NOREPINEPHRINE metabolite in the gamblers' urine and cerebrospinal fluid, presumably caused by an increased production of the neurotransmitter norepinephrine within the brain. Furthermore, a psychological measure of extraversion in the gamblers was correlated with levels of norepinephrine and its metabolites in their body fluids. A single study of beta ENDORPHINS in pathological gamblers found lower baseline levels in those who bet on horse races than those who played poker-machines or those who were not gamblers. Although research on the pathophysiology of this disease is still preliminary, commonalities with other addictions through central nervous system mechanisms are being sought.

IDENTIFICATION AND TREATMENT

Since 1987 a valid and reliable paper-and-pencil test, the South Oaks Gambling Screen (SOGS), has been available for screening general or clinical populations for gambling problems. The maximum score on this screening test is 20. A score of 5 or more indicates probable pathological gambling, while a score of 1 to 4 signals some gambling problem. Following screening a formal diagnosis must be established. A thorough assessment of physical, psychiatric, addictive, family, social, financial, and legal problems is also necessary because multiple problems are common. Alcohol and drug dependencies, psychiatric disorders and physical problems are most effectively treated at the same time as the gambling addiction.

Treatment may be provided in both inpatient and outpatient settings. Psychoeducation, individual and group therapies, psychodrama, relaxation training, family counseling and RELAPSE PREVENTION training are commonly used treatment techniques, usually combined with an introduction to Gamblers Anonymous. Family treatment and long-term follow-up are important as well. Abstinence from all forms of gambling is one of the treatment goals, along with improved physical and psychological well-being.

Addiction model treatment may be organized either in a separate facility or as part of a combined substance-dependence and pathological-gambling program. Studies of patients involved in both models of the addiction program have yielded positive outcomes, with gambling abstinence in 56 to 64 percent of the patients who were followed, and improvement in many other aspects of their lives.

OTHER MODELS OF PATHOLOGICAL GAMBLING

Pathological gambling has been explained using models other than addictive disease. It has been considered, for example, a symptom of some other psychiatric disorder; a behavior disorder; learned behavior that can be "unlearned"; a moral problem; or the result of a faulty gambling strategy. The addiction model has, however, proved a useful framework for research, intervention, treatment and self-help. As future research clarifies the neurophysiological mechanisms that underlie alcohol and other drug addiction, both the neurochemical basis of pathological gambling and a "common pathway" of addiction in the brain may also be discovered.

(SEE ALSO: *Addiction: Concepts and Definitions; Addictive Personality*)

BIBLIOGRAPHY

BLUME, S. B. (1991). Gambling problems in alcoholics and drug addicts. In N. Miller (Ed.), *Comprehensive handbook of drug and alcohol addiction*. New York: Marcel Dekker.

CLOTFELTER, C. T., & COOK, P. J. (1989). *Selling hope*. Cambridge, MA: Harvard University Press.

CUSTER, R., & MILT, H. (1985). *When luck runs out: Help for compulsive gamblers and their families*. New York: Facts on File.

GALSKI, T. (1987). *The handbook of pathological gambling*. Springfield, IL: Charles C. Thomas.

GAMBLERS ANONYMOUS. (1984). *Sharing recovery through Gamblers Anonymous*. Los Angeles: Author.

LESIEUR, H. R., & BLUME, S. B. (1988). The South Oaks Gambling Screen (SOGS): A new instrument for the identification of pathological gamblers. *American Journal of Psychiatry, 144*, 1184–1188.

ROY, A., ET AL. (1988). Pathological gambling: A psychobiological study. *Archives of General Psychiatry, 45*, 369–373.

SHEILA B. BLUME

GAMMA-AMINOBUTYRIC ACID (GABA)

This is an amino acid derived by a single-step decarboxylation from GLUTAMATE. GABA is the most abundant (in micromolar concentrations/mg of protein) inhibitory NEUROTRANSMITTER—and it is found throughout the animal kingdom. Its role as a neurotransmitter was first defined for the inhibitory nerve in lobster muscle, where GABA accounted for the total inhibitory potency of nerve extracts. A central inhibitory neurotransmitter role for GABA was securely established only when selective ANTAGONISTS, such as bicuculline, discriminated GABA receptors and pathways from glycine, a related inhibitory amino acid neurotransmitter. GABA actions and receptors for GABA have been linked to central nervous system sedatives such as ALCOHOL and BENZODIAZEPINES.

(SEE ALSO: *Research*)

BIBLIOGRAPHY

COOPER, J. R., BLOOM, F. E., & ROTH, R. H. (1991). *The biochemical basis of neuropharmacology*, 6th ed. New York: Oxford University Press.

FLOYD BLOOM

GANGS AND DRUGS

Youth gangs have been part of the U.S. urban landscape for over 200 years. From the earliest mentions of gangs in the social commentaries of post–Revolutionary War America, gangs have been linked to the use and trafficking of illicit intoxicants. In the late eighteenth century, for example, gangs such as the Fly Boys, the Smith's Vly gang, and the Bowery Boys were well known in the streets of New York City (Sante, 1991). As European immigration increased in the early nineteenth century, gangs such as the Kerryonians (from County Kerry in Ireland) and the Forty Thieves formed in the overcrowded slums of the Lower East Side (of New York City). Gangs proliferated quickly in that time, with such colorful names as the Plug Uglies, the Roach Guards, the Hide-Binders (comprised mainly of butchers), the Old Slippers (a group of shoemakers' apprentices) and the Shirt Tails. Many of these gangs were born in the corner groceries that were the business and social center of the neighborhoods. These groceries also hid the groggeries that were important features of neighborhood life, and guarding them provided a steady income for the gangs. Although not involved in theft, robbery, or the unsavory professions of GAMBLING or tavern-keeping, these gangs warred regularly over territory with weapons—including stones and early versions of the blackjack. They occasionally joined forces to defend their neighborhood, and nearly all were united in their opposition to the police.

Throughout the nineteenth century, gangs emerged in the large cities of the Northeast, in Chicago and in other industrial centers of the Midwest. In the early twentieth century, gangs also formed in the Mexican immigrant communities of California and the Southwest. In what still is widely regarded as the classic work on youth gangs, Thrasher (1927) identified over 1,300 street gangs in the economically disadvantaged neighborhoods of industrial Chicago in the 1920s. He interpreted the rise of Chicago's gangs as symptoms of deteriorating neighborhoods and the shifting populations that accompanied industrialization and the changing populations that lived in the interstitial areas between the central city and the industrial regions that ringed it. Wherever neighborhoods in large cities were in transition, gangs emerged, and their involvement in drinking and minor drug use was a regular feature of gang life.

In the 1990s, gangs are present in large and small cities in nearly every state. They reflect the ethnic and racial diversity of American society (Klein, 1992). Gangs are no longer colorful, turf-oriented groups of adolescents from immigrant or poor neighborhoods. Whereas gangs in the past were likely to claim streetcorners as their turf, gangs today may invoke the concept of turf to stake claims to shopping malls, skating rinks, school corridors, or even cliques of women. Gangs use graffiti and "tagging" to mark turf and communicate news and messages to other gangs and gang members (Huff, 1989). The participation and roles of young women in gangs has also changed. Through the 1960s, women were involved in gangs either as auxiliaries or branches of males gangs, or they were weapons carriers and decoys for male gang members. Today, female gangs have emerged that are independent of male gangs. Fights are common between the new female gangs. There also is some evidence of sexually integrated gangs, where females fight alongside males (Taylor, 1993).

Traditionally, stealing and other petty economic crimes have long been the backbone of gang economic life. For example, Saint Francis of Assisi commented that nothing gave him greater pleasure than stealing in the company of his friends. English common law in the 13th century accorded especially harsh punishments to the roving bands of youths who moved across the countryside stealing from farmers and merchants. The House of Refuge, the first U.S. residential institution for boys, opened in New York City in 1824, largely in response to the unsupervised groups of youths who roamed the city stealing and drinking. For some contemporary gangs, however, entrepreneurial goals, especially involving drug selling, have replaced the cultural goals of ethnic solidarity and neighborhood defense that historically motivated gang participation and activities. A few gangs have functional ties to adult organized crime groups. Other gangs have become involved in drug selling and have developed a corporate structure that has replaced the vertical organization that in the past regulated gang life.

This article examines recent data on the drug and alcohol involvement of street gangs. Recent changes in the social structure of cities has led to a new generation of gangs and gang cities. We look to these changes in cities and neighborhoods to explain the new patterns of substance use and drug distribution among gangs. Changes in the conception of work, the institutionalization of drug selling, and cultural shifts in gangs and ganging, have influenced gang involvement in drugs and alcohol. This article discusses the relationship between political and economic factors that shape the social structure of communities, the neighborhood effects that result from those forces, and the mediating effects of neighborhood processes on the formation of gangs and their use of substances.

DRUG AND ALCOHOL USE AMONG YOUTH GANGS

ALCOHOL and MARIJUANA use have always been, and continue to be, the most widely used substances among both gang and non-gang youths (Fagan, 1989, 1990; Sheley, Smith, & Wright, 1992). Drinking and other drugs (primarily marijuana) consistently are mentioned as a common part of gang life throughout the gang literature. For instance, Short and Strodtbeck's (1965) study of Chicago gangs showed that drinking was the second most common activity of gang members of all races, exceeded only by hanging out on the streetcorner. Although CO-CAINE may be trafficked by some gang members, it is not often used in either its powder or smokeable forms (Fagan, 1990).

Ethnographic studies of gang life (Hagedorn, 1988; Campbell, 1990; Stumphauzer, Veloz & Aiken, 1981; Vigil, 1988; Padilla, 1992; Moore, 1978, 1992a, 1992b; Taylor, 1993) also show the common-place occurrence of drinking and its place in a broad pattern of substance use. Dolan and Finney (1984) and Campbell (1990) illustrated the commonplace role of drug use in gang life among both males and females. Stumphauzer et al. (1981) noted that use patterns varied within and among Los Angeles gangs, but changed for individuals over time. MacLeod (1987) noted high rates of drinking among white gang members but only occasional beer use among the Brothers, a predominantly black (but somewhat integrated) gang. Sanchez-Jankowski (1991) found that all members of all gangs drank regularly, using gang proceeds for collective purchases. Although they used drugs in varying patterns, alcohol was mentioned consistently. But Sanchez-Jankowski also mentioned that the Irish gangs least often used illicit drugs, since access was controlled by nonwhites with whom they did not want to engage in business.

Vigil (1985, 1988) described a variety of meanings and roles of substances among Chicano gang members in East Los Angeles, from social "lubricant" during times of collective relaxation to facilitator for observance of ritual behaviors such as *locura* acts of AGGRESSION or VIOLENCE. In these contexts, drug use

provided a means of social status and acceptance, as well as mutual reinforcement, and was a natural social process of gang life. Vigil (1988) notes that these patterns are confined to substances that enhance gang social processes—alcohol, marijuana, PHEN-CYCLIDINE (PCP), and CRACK-cocaine. There is a sanction against HEROIN use among Chicano gangs. Heroin involvement is seen as a betrayal of the gang and the barrio; one cannot be loyal to his addiction and the addict ("tecato") culture while maintaining loyalty to the gang. Vigil noted that gang members prepared for imminent fights with other gangs by drinking and smoking PCP-laced cigarettes. During social gatherings, the gang members used the same combinations to "kick back" and feel more relaxed among one another. Evidently, gang members had substantial knowledge about the effects of alcohol (and its reactivity to PCP), and they had developed processes to adjust their reactions to the mood and behaviors they wanted.

Feldman et al. (1985) observed three distinct "styles" among Latino gangs in San Francisco that in part were determined by the role and meaning of substances in gang social processes. The "fighting" style included males in gangs who were antagonistic toward other gangs. They aggressively responded to any perceived move into their turf by other gangs or any outsider. Drinking and drug use were evident among these gangs, but this was only situationally related to their violence through territoriality. Violence occurred in many contexts unrelated to drug use or selling and was an important part of the social process of gang affiliation. The "entrepreneurial" style consisted of youths who were concerned with attaining social status by means of money and the things money can buy. They very often were active in small-scale illegal sales of marijuana, pill amphetamines, and PCP. While fighting and violence were part of this style, it was again situationally motivated by concerns over money or drugs. The last style was evident in gangs whose activities were social and recreational, with little or no evidence of fighting or violence but high rates of drinking and marijuana use.

Padilla's (1992) study of a Puerto Rican gang in Chicago described how alcohol and marijuana often accompanied the rituals of induction and expulsion of gang members. These ceremonies often were tearful and emotional, with strong references to ethnic solidarity. Padilla described how emotions intensified as the ceremony progressed, and drinking was a continuous process during the events.

Drinking or drug use also is disallowed in some youth gangs, regardless of the gang's involvement in drug selling. Chin (1990) found that intoxication was rejected entirely by Chineses gangs in New York City. Although they used violence to protect their business territories from encroachment by other gangs, and to coerce their victims to participate in the gang's ventures, "angry" violence was rare; violent transactions were limited to instrumental attacks on other gangs.

Taylor (1990a) and Mieczkowski (1986) described organizations of adolescent drug sellers in Detroit who prohibited drug use among their members but tolerated drinking. Leaders in these groups were wary of threats to efficiency and security if street-level sellers were high and to the potential for co-optation of its business goals if one of its members became involved with consumption of their goods. The gangs were organized around income and saw drug use (but not alcohol) as detracting from the selling skills and productivity of their members. Expulsion from the gang resulted from breaking this rule, but other violent reprisals also were possible. However, gangs in both studies accepted recreational use of substances by members, primarily alcohol, marijuana, and cocaine in social situations not involved with dealing.

In the Mieczkowski study, the sellers particularly found danger in being high on any drug while on the job. Gang superiors enforced the prohibition against heroin use while working by denying runners their consignment and, accordingly, shutting off their source of income. Violence was occasionally used by superiors (crew bosses) to enforce discipline. Gang members looked down on their heroin-using customers, despite having tried it at some point in their lives, which in part explains the general ideology of disapproval of heroin use.

Buford (1980) depicted crowd violence among English football (soccer) "supporters" as an inevitable consequence of the game's setting and the dynamics of crowds of youths. Expectancies of both intoxication and violence preceded the arrival of the "lads" at drinking locations surrounding the stadiums. The expectancies were played out in crowd behavior through rituals that were repeated before each match. Alcohol consumption before and during episodes of unrestrained crowd violence was an integral part of the group dynamic, but Buford does

not attribute to alcohol an excuse function, nor is alcohol a necessary ingredient for the relaxation of social norms. In fact, he pointedly notes that the heaviest drinkers were incapacitated by inebriation and were ineffective rioters, while the crowd leaders were relatively light drinkers. In this context, alcohol was central but hardly necessary to the attainment of the expected behavior—the setting itself provided the context and cues for violence.

GANGS AND DRUG SELLING

In the 1980s, the confluence of social problems involving gangs, violence, and drug trafficking changed both popular and political perceptions of gangs. Beginning with the crack-cocaine crisis, youth gangs have been confounded with new forms of drug distribution organizations that involve young men and women in "underclass" neighborhoods. This terminology has been interchangeably used to describe these groups as gangs. Indeed, the growth in cocaine use in the 1980s did coincide with more visible gang involvement in drug selling and drug-related violence (Huff, 1992). Although drug use and selling have been central features of gang life for decades, gangs were often blamed for many of the new drug problems (Newsweek, 1986; U.S. Department of Justice, 1989; Conley, 1992; Los Angeles County District Attorney, 1992). Once seen as streetcorner groups protecting "turf" and neighborhood, gangs were portrayed in the 1980s by the popular press and criminal-justice officials as nascent organized-crime groups focused on the distribution of drugs, with elaborate intercity networks well-financed by drug income.

Several trends contributed to this changing characterization of gangs. First, gangs became highly visible. Although gangs traditionally have been active in larger cities, gangs emerged in the 1980s in smaller cities—with populations as low as 25,000 (U.S. Department of Justice, 1989). Among the 100 largest U.S. cities, gangs have emerged in 42 cities since 1980 (Klein, 1992). Their recent emergence may belie their stability, however, since gangs are well known to often be temporary groupings with short half-lives (Spergel, 1989). The validity of police reports of gang activity itself remains a difficult measurement problem.

Second, gang violence has become more visible if not more prevalent. Gang-related homicides in Los Angeles grew sharply throughout the 1980s, exceeding 500 annually after 1989 (Los Angeles County District Attorney, 1992). In 1991, record homicide rates were set in both Los Angeles and Chicago, two cities with extensive youth-gang networks (FBI, 1992). Yet the classification of homicides as "gang-related" is quite sensitive to definitions; and this trend, too, may reflect anomalies in definition and measurement of gangs and gang incidents. Maxson and Klein (1989) showed that the Los Angeles (Police Department) rates, based on the gang affiliations of victims or perpetrators, could be halved by applying the motive-based definition used by the Chicago Police Department.

Third, gang involvement in drug trafficking reportedly grew in the 1980s (U.S. Department of Justice, 1989; Spergel, 1990; but see Moore, 1992a). Spergel (1990) found that 75 percent of gang members on probation in San Diego County were convicted of drug offenses at one time or another. Among the 1,200 youths in Chicago, Los Angeles, and San Diego interviewed by Fagan (1990), 32 percent of the gang members reported they were involved in drug selling, compared to fewer than 8 percent of the nongang youths. Based on 45 interviews with California prison inmates, Skolnick et al. (1989) claimed that linkages existed between prison gangs and street gangs around drug distribution. But Klein et al. (1991) showed that adolescent participation in rock-cocaine selling in Los Angeles grew equally for gang and nongang youths.

Drug selling also contributed to changes in the organization and meaning of gangs. Taylor (1990b) and Mieczkowski (1986) illustrated the transformation of Detroit gangs from streetcorner groups protecting territory to highly efficient drug-selling organizations. Padilla (1992) describes how Puerto Rican youths in a Chicago gang refocused the gang to drug dealing as the primary source of income. Hagedorn (1991) showed that twenty-two of thirty-seven African-American gang members in Milwaukee went on to become involved in adult drug organizations.

Fourth, reports of gang migration contributed to the perception of gangs as highly disciplined entrepreneurs intent on establishing intercity drug networks to expand their profits. Gang members are alleged to have set up "franchises" or branch offices in remote cities for selling drugs. Huff (1992) reports arrests of Los Angeles gang members in large and small Ohio cities and Detroit gang members in Cleveland (1989); members of Los Angeles gangs have been arrested on drug charges in cities as far east as Columbus (Ohio) (U.S. Department of Justice, 1989). This diffusion of gang activity from the major gang centers of Los Angeles and Chicago is often cited as a bellwether of the evolution of gangs to a new and more dangerous form.

The temporal proximity of these trends led to their confounding in the popular and sociological literatures, with hasty assumptions that they were causally linked. Few phenomena have been stereotyped as easily as gangs, violence, and drug use, especially in conjunction. These perceptions were amplified in the popular culture through movies and hip-hop music depicting gang life as a stew of violence, drug money, police repression, and the exploitation of women (Taylor, 1992). These perceptions were fueled by gang-related violence in the theaters at the opening of recent films depicting gang life (*Colors, Boyz'n the 'Hood*), as well as reports of violence at rap concerts and in local clubs specializing in "house" and "hip-hop" music (Huff, 1992). These cultural vehicles falsely signaled a transformation of youth gangs from streetcorner groups to more sophisticated crime groups reaping great profits from drug distribution and specializing in lethal violence. In the context of the crack crisis of the mid-1980s and the violence that accompanied it, these portrayals also created a perception that there was an increase in gangs—with more youths in gangs and more violent gangs in urban centers throughout the United States.

The 1980s' changes in drug-use trends, together with earlier changes in labor markets, income dynamics, and demographics in urban centers, suggest that the drug-gang nexus is part of a larger, more complex process of urban change tied to the economic and social transformation of cities. This was a significant era because of the sharp reduction in wholesale cocaine prices, the emergence of crack, and the expansion of street-level drug markets for cocaine and crack distribution (Fagan, 1992a). Reports from law-enforcement agencies suggest that by 1980 gangs formed in smaller cities not traditionally known as gang centers (Klein, 1992), and having little to do with drugs (Fagan & Klein, 1992). This was also an era marked by the emergence or persistent poverty in urban centers and the growth of an "urban underclass" (Wilson, 1987, 1991; Jargowsky & Bane, 1990; Ricketts & Sawhill, 1988; Jencks, 1991). In fact, the emergence of gangs in the 1980s was motivated by broad changes in economic and social conditions during the 1970s, changes that reflected deindustrialization and growing social and economic isolation in large U.S. cities (Jackson, 1991; Hagedorn, 1988; Sullivan, 1989; Fagan & Klein, 1992).

Despite the historically uneven relationship between gangs and drug use or selling (Klein, Maxson, & Cunningham, 1991; Spergel, 1989; Fagan, 1989), recent studies contend that the lucrative and decentralized crack markets in inner cities have created a new generation of youth gangs (Skolnick et al. 1989; Taylor, 1990b). Young drug sellers in these gangs have been portrayed as ruthless entrepreneurs, highly disciplined and coldly efficient in their business activities, and often using violence selectively and instrumentally in the service of profits. This vision of urban gangs suggests a sharp change from the gangs of past decades, and much of the change is attributed to the dynamics of the inexpensive, smokeable cocaine market.

The empirical data suggest otherwise (Fagan, 1989, 1990; Klein et al. 1991; Vigil, 1988; Padilla, 1992; Moore, 1992b; Hagedorn, 1988). Drug selling has always been a part of gang life, with diverse meanings tied to specific contexts and variable participation by gangs and gang members (Fagan, 1990). For example, Fagan (1989) found diverse patterns of drug selling within and across three cities with extensive, integenerational gang traditions, while Klein et al. (1992) reported variability within and across Los Angeles gangs in crack selling.

GANGS, DRUGS, AND NEIGHBORHOOD CHANGE

What are the changes that occurred in cities and communities to explain variation and change in gang participation in drug selling? Two factors have in particular contributed to changes in gangs and the substitution of instrumental and monetary goals for the cultural or territorial affinities that unified gangs in earlier decades. First, cocaine markets changed dramatically in the 1980s, with sharp price reductions. Before cocaine became widely available, drug distribution was centralized, with a small street-level network of heroin users responsible for retail sales (Curtis, 1992; Johnson et al., 1985). The heroin markets from the 1970s were smaller than the mid-1980s crack market, both in total volume of sales and the average purchase amount and quantity. Street-level drug selling in New York City, for ex-

ample, was a family-centered heroin and marijuana business until the 1980s, when new organizations developed to control the distribution of cocaine (Curtis, 1992; Johnson et al., 1990). The psychoactive effects of HEROIN (a depressant) and its methods of administration (by injection) limited its sales volume and number of users.

Cocaine was different in every way—a stimulant rather than a depressant, ingested in a variety of ways (nasally, smoked, or injected), and with a shorter half-life for the "high." The price declined as cocaine became widely available, and the discontinuity in distribution systems across successive drug eras created new opportunities for drug selling, and may even have encouraged participation in it. The sudden change in cocaine marketing, from a restricted and controlled market in the 1970s to a fully deregulated market for crack, spawned intense competition for territory and market share (Fagan, 1992a; Williams, 1989). Law-enforcement officials in New York City characterized the crack industry as "capitalism gone mad" (New York Times, 1989).

In inner-city neighborhoods that since the 1970s had grown more socially isolated, and where legal economic activity was declining quickly, drug selling became a common form of labor market participation. Young men began to talk about drug selling and crime as "going to work" and the money earned as "getting paid" (Padilla, 1992; Sullivan, 1989). In the closed milieu of these neighborhoods, the tales of extraordinary incomes had great salience and were widely accepted, even if the likelihood of such riches was exaggerated (Bourgois, 1989; Reuter, MacCoun, & Murphy, 1990; Fagan, 1992a, 1992b). The focus of socialization and expectations shifted from disorganized groups of adult males to (what was perceived as) highly organized and increasingly wealthy young drug sellers. Many other sellers kept one foot in both licit and illicit work, lending ambiguity to definitions of work and income (Reuter et al., 1990; Fagan, 1992a, 1992b).

Second, profound changes in the social and economic makeup of cities (Tienda, 1989; Wacquant & Wilson, 1989) combined to disrupt social controls that in the past mediated gang behavior (Curry & Spergel, 1988). In this context, gangs became less concerned with cultural or territorial affinities and instead became focused on instrumental and monetary goals (Taylor, 1990; Padilla, 1992; but see Moore, 1992a). The interaction of these two trends provided ample opportunities for gangs to enter into the expanding cocaine economy of the 1980s.

As drug selling expanded into declining local labor markets, it became institutionalized within the local economies of the neighborhoods. Whether in storefronts, from behind the counters in *bodegas* (groceries), on streetcorners, in crack or "freak" houses, or through several types of "fronts," drug selling was a common and visible feature of the neighborhoods (Hamid, 1992). Young men and, increasingly, women had several employment options within drug markets—support roles (lookout, steerer), manufacturing (cut, package), or direct street sales (Johnson et al., 1990). Legendary tales, often with little truth, circulated about how a few dollars' worth of cocaine could be turned into several thousand dollars within a short time. Such quick riches had incalculable appeal for people in chronic or desperate poverty.

THE IMPACT OF DRUGS ON GANGS AND GANG CULTURE

The transition from streetcorner group to ethnic enterprise profoundly shaped the social organization of youth gangs. Money became the driving force and organizing principle for these groups. Greed was elevated to a set of beliefs, expressed consistently among gangs and gang members in the neighborhoods with extensive drug markets (Padilla, 1992). The use of the language of work ("getting paid," "going to work") to describe drug selling signals an ideological shift in the social definition of work and the confounding of illegal and legal means of making money. For the young men using this language, there was no particular meaning assigned to drug selling: they pursued commodities that offered instrumental value as signs of wealth (Sullivan, 1989; Padilla, 1992). Any high-demand contraband consumable commodity would likely have inspired the same behavior (as for example weapons, guns).

Not surprisingly, "materialism" is evident in the motivations expressed by young people participating in drug selling—the attainment of wealth as a manifestation of individual power and achievement. Within the isolated, concentrated poverty areas in inner cities, the absence of mediating social definitions allowed the pursuit of material wealth to become transformed into the very substance of societal bonds and conventional values. Americans always

have looked up to the Horatio Algers, whose "self-made" business success defied social odds. As these models were elevated to societal icons, the attainment of wealth seemed to supersede the importance of law or the collective societal good (Wall Street Journal, 1989).

The interaction of drug selling, violence, and material goals often are combined in an emerging set of sociocultural processes within gangs. Padilla (1992) describes how older gang members reoriented the gangs to become business organizations to fuel increases that were disproportionately taken by the older members. In effect, the older members became local employers themselves. Since the older members were no longer the keepers of the culture and regulators of gang organization, they used traditional appeals to ethnic and neighborhood loyalties to recruit and motivate younger gang members. However, they added money incentives to the mix to strengthen their controls over young gang members.

The emphasis on money, individual gain, and quick wealth was so strong that gang members in Detroit (Taylor, 1990a) and Padilla's Chicago neighborhood themselves regarded low-level drug sellers, even their own "homeboys," as "working stiffs" who were being exploited by other gang members. In the past, such denigration of gang work was heretical: Entry level jobs in the service of the gang typically would be seen as serving the gang's collective interest. Padilla describes how the new pattern of exploitation of lower-level workers (street sellers) in the gang was obscured by appeals to gang ideology (honor, ethnic solidarity, and neighborhood loyalty) combined with the lure of income to control them. Taylor (1990a) also talks about the use of money as social control within Detroit drug-selling gangs—if a worker steps out of line, he simply is cut off from the business, a punishment far more salient than threats to physical safety. Moore (1992b) describes similar age-related exploitation within chicano gangs in Los Angeles but with little involvement in drug selling.

The exaggerated, almost ideological emphasis on money and material wealth interacts in a very complex fashion with ethnicity and local context. It marks a dramatic shift from the gangs of 1970 to 1985. It is difficult to disentangle the order of events. Did drugs bring in more money, and did money take on greater importance (raised the stakes) because of the economic transformation of the cities? Or did the loss of economic structures make drugs more salient? Did the increased stakes/money bring in guns, which in turn increased the lethality of gang violence? Or did the guns come as a manifestation of power for those who are rejected from any other source of economic or personal power?

In Chicago (Padilla, 1992) and Detroit (Taylor, 1990a, 1990b, 1992), gangs superficially are ethnic enterprises, but more substantively serve as economic units with management structures oriented toward the maintenance of profitability and efficiency. For the African-American gangs in Detroit, there was little concern with neighborhood or the traditional meaning of gang life. Although forms of internal control varied, money was manipulated along with appeals to ethnic solidarity to maintain loyalty and discipline within groups that otherwise had evolved from gangs or streetcorner groups to become economic organizations. Among the "Diamonds" in Chicago, appeals to Puerto Rican solidarity were used by older gang members to maintain order and motivation within the gang, while these older members kept the lion's share of the gang's profits from drug sales.

GANG MIGRATION

The appearance of Crips, Bloods, Vice Lords, Black Gangster Disciples, Latin Kings, and other well-known gang names in new gang cities across the country has created concerns that gangs are expanding and migrating. Migration is a term that actually includes several distinct patterns: franchising, opening "branch offices," or acquiring and operating local subsidiaries. Gang migration was virtually unknown until the 1980s, when law enforcement and media reports claimed that gang members were setting up illegal businesses in other cities to expand their drug-selling territories.

There are few instances of gangs operating directly in other cities. Migration seems to be concentrated along interstate highway routes, such as I-75 ('Caine Lane, named for its volume of cocaine traffic) connecting Detroit with Ohio cities, or the I-5 route from Los Angeles through California's Central and San Joaquin valleys (Huff, 1992). Others (Waldorf, 1992) found no evidence of gang migration among San Francisco gangs, either in-migration from Los Angeles gangs or reports of gang members doing gang "business" in other cities.

More often, what appears as migration reflects natural social dynamics of residential relocation, court placements, mimicry, and other forms of gang diffusion. Gang migration also has been confused with the enterprising behavior of individual gang members. There have been sporadic incidents of deliberate migration, isolated among specific gangs in specific cities. But most often, local gangs are composed of local youths who may have adopted the names, graffiti, and other symbols of established gangs from the larger cities.

There are few documented instances of gang migration. Hagedorn (1988) reported that Milwaukee gangs adopted the name of the Vice Lords, a Chicago gang, but had little contact with them. Some Crip or Blood members relocating from Los Angeles may have organized small crews to sell drugs, but law-enforcement officials interpreted this as evidence that Crip chapters had opened in their cities. Chicago gang graffiti appeared in Mississippi as young males were sent away to live with relatives to escape gang violence; but this event was viewed as signs of Chicago gang expansion into the South (Lemann, 1991).

Critics suggest federal initiatives and funds to control gangs have created incentives (and funds) for zealous law-enforcement agencies to identify streetcorner groups or drug gangs as interstate gang conspiracies. Indeed, there have been isolated instances of what Carl Taylor (1990b) calls gang "imperialism," where gangs have established business locations in other cities. Most often, this includes drug selling—and nearly always among entrepreneurial or corporate gangs. Their motives appear to be simply market expansion and increasing profits. Chicago gangs have influenced the gang scene in nearby Evanston. Chinese street gangs operate both regionally throughout the New York metropolitan area and in cities in the Northeast including Philadelphia, Albany, and Hartford (Chin, 1990). The Chinese gangs are not involved in drug trafficking, but their multiple enterprises include extortion and the smuggling of illegal aliens.

SUMMARY

Few phenomena have been stereotyped as easily as gangs, violence, and drug use, especially in conjunction. Drug use has always been a part of gang life, as has peddling of small quantities of whatever street drugs were popular at the time. Many gangs also adopted codes prohibiting drug use, fearing that loyalty to one's drug habit conflicts with loyalty to the gang or efficiency in drug selling. The cocaine and crack crises of the 1980s created opportunities for gang and nongang youths alike to participate in drug selling and increase their incomes. There is little evidence that gang members have become involved in drug selling more so than nongang adolescents. Malcolm Klein and his colleagues, based on police arrest reports following the appearance of crack in Los Angeles in 1985, found no evidence that gang members were arrested more often than nongang members for crack sales, or that drug-related homicides were more likely to involve gang members than nongang members.

Among gangs, involvement in the drug trade varies by locale and ethnicity. Chicano gangs in Los Angeles do not sell cocaine but sell small quantities of other drugs. The crack and cocaine trades are dominated by African-American youths, both gang members and nongang youths. Crack sales began in Chicago more than five years after Los Angeles gangs began selling drugs. As in Los Angeles, both gang and nongang youths are involved. Crack sales in New York flourished beginning in 1986, but there was no discernible street gang structure that participated in drug selling. Instead, loosely-affiliated selling crews provided an organizational structure for drug sales. Chinese gangs have remained outside the cocaine and crack trades. However, some members (but not the gangs themselves) have been involved in transporting or guarding heroin shipments from Asia.

Not all gang members sell drugs, even within gangs where drug selling is common. Drug-selling cliques within gangs are responsible for gang drug sales. These cliques are organized around gang members who have contacts with drug wholesalers or importers. Among the "Diamonds," Padilla (1992) describes how drug selling is a high-status role reserved for gang members who have succeeded at the more basic economic tasks of stealing and robbery. Despite public images of gang members using drug profits for conspicuous consumption of luxury items, drug incomes in fact are quite modest for gang members who sell drugs. Drug incomes are shared within the gang, but the bulk of the profits remain with the clique or gang member who brought the drugs into the gang. The profits from drug selling, combined with the decline in economic "exits" from gang life, provide some incentive for older gang members to remain in the gang.

(SEE ALSO: *Adolescents and Drugs; Crime and Drugs; Ethnicity and Drugs; Poverty and Drug Use*)

BIBLIOGRAPHY

BUFORD, B. (1991). *Among the thugs: The experience, and the seduction of, crowd violence.* New York: Norton.

CAMPBELL, A. (1990). *The girls in the gang,* 2nd ed. New Brunswick, NJ: Rutgers University Press.

CHIN, K. (1990). *Chinese subculture and criminality: Non-traditional crime groups in America.* Westport, CT: Greenwood.

CONLEY, C. (1992). Street gangs: What do we know and what should we do about them? Monograph. Cambridge, MA: Abt Associates, Inc.

CURRY, G. D., & SPERGEL, I. A. (1988). Gang homicide, delinquency and community. *Criminology, 26,* 381–405.

CURTIS, R. A. (1992). Highly structured crack markets in the southside of Williamsburg, Brooklyn. In J. Fagan (Ed.), *The ecology of crime and drug use in inner cities.* New York: Social Science Research Council.

DOLAN, E. F., & FINNEY, S. (1984). *Youth gangs.* New York: Simon & Schuster.

FAGAN, J. (1992a). Drug selling and licit income in distressed neighborhoods: The economic lives of drug users and dealers. In G. Peterson & A. Harrell (Eds.), *Drugs, crime and social isolation.* Washington, DC: Urban Institute Press.

FAGAN, J. (1992b). The dynamics of crime and neighborhood change. In J. Fagan (Ed.), *The ecology of crime and drug use in inner cities.* New York: Social Science Research Council.

FAGAN, J. (1990). Social processes of drug use and delinquency among urban gangs. In C. R. Huff (Ed.), *Gangs in America.* Newbury Park, CA: Sage.

FAGAN, J. (1989). The social organization of drug use and drug dealing among urban gangs. *Criminology, 27,* 633–669.

FAGAN, J., & KLEIN, M. W. (1992). Social structure, social isolation and gang formation. Unpublished. Newark, NJ: School of Criminal Justice, Rutgers University.

FELDMAN, H. W., MANDEL, J., & FIELDS, A. (1985). In the neighborhood: A strategy for delivering early intervention services to young drug users in their natural environments. In A. S. Friedman & G. M. Beschner (Eds.), *Treatment services for adolescent substance abusers*. Rockville, MD: National Institute on Drug Abuse.

HAGEDORN, J. (1991). Gangs, neighborhoods, and public policy. *Social Problems, 38*, 529–542.

HAGEDORN, J., WITH P. MACON. (1988). *People and folks: Gangs, crime and the underclass in a rustbelt city*. Chicago: Lake View Press.

HAMID, A. (1992). Flatbush: A freelance nickels market. In J. Fagan (Ed.), *The ecology of crime and drug use in inner cities*. New York: Social Science Research Council.

HUFF, C. R. (1992). Gangs in America. In A. P. Goldstein & C. R. Huff (Eds.), *The gang intervention handbook*. Champaign, Il: Research Press.

HUFF, C. R. (1989). Youth gangs and public policy. *Crime and Delinquency, 35*, 524–537.

JACKSON, P. I. (1991). Crime, youth gangs, and urban transition: The social dislocations of postindustrial economic development. *Justice Quarterly, 8*, 379–397.

JARGOWSKY, P., & BANE, M. J. (1990). Ghetto poverty: Basic questions. In L. Lynn & M. McGeary (Eds.). *Inner city poverty in the United States*. Washington, DC: National Academy Press.

JENCKS, C. (1991). Is the American underclass growing? In C. Jencks & P. E. Peterson (Eds.), *The urban underclass*. Washington, DC: Brookings Institute Press.

JOHNSON, B. D., ET AL. (1990). Drug abuse and the inner city: Impacts of hard drug use and sales on low income communities. In J. Q. Wilson & M. Tonry (Eds.), *Drugs and crime*. Chicago: University of Chicago Press.

JOHNSON, B. D., ET AL. (1985). *Taking care of business: The economics of crime by heroin abusers*. Lexington, MA: Lexington Books.

KLEIN, M. W. (1992). Twenty-five years of youth gangs and violence. Presented at the Annual Meeting of the American Association for the Advancement of Science. Washington, DC.

KLEIN, M. W., MAXSON, C. L., & CUNNINGHAM, L. C. (1991). Crack, street gangs, and violence. *Criminology, 29*, 623–650.

LEMANN, N. (1991). *The promised land: The great black migration and how it changed America*. New York: Knopf.

LOS ANGELES COUNTY DISTRICT ATTORNEY. (1992). *Gangs, crime and violence in Los Angeles*. Los Angeles: District Attorney's Office.

MACLEOD, J. (1987). *Aint no makin it: Leveled aspirations in a low-income neighborhood*. Boulder, CO: Westview.

MAXSON, C. L., & KLEIN, M. W. (1989). Street gang vio-

lence. In N. A. Weiner & M. E. Wolfgangs (Eds.), *Pathways to criminal violence*. Newbury Park, CA: Sage.

MIECZKOWSKI, T. (1986). Geeking up and throwing down: Heroin street life in Detroit. *Criminology, 24*, 645–666.

MOORE, J. W. (1992a). *Going down to the barrio: Homeboys and homegirls in change*. Philadelphia: Temple University Press.

MOORE, J. W. (1992b). Institutionalized youth gangs: Why White Fence and El Hoyo Maravilla change so slowly. In J. Fagan (Ed.), *The ecology of crime and drug use in inner cities*. New York: Social Science Research Council.

MOORE, J. W. (1978). *Homeboys*. Philadelphia: Temple University Press.

NEW YORK TIMES. (1989). Report from the field on an endless war. March 12, 1989, section 4, p. 1.

NEWSWEEK. (1986). Kids and cocaine. March 17, 1986, 58–68.

PADILLA, F. (1992). *The gang as an American enterprise*. New Brunswick, NJ: Rutgers University Press.

REUTER, P., MACCOUN, R., & MURPHY, P. (1990). *Money from crime*. Report R-3894. Santa Monica, CA: The Rand Corporation.

RICKETTS, E., & SAWHILL, I. (1988). Defining and measuring the underclass. *Journal of Policy Analysis and Management, 7*(2), 316–325.

SANCHEZ-JANKOWSKI, M. (1991). *Islands in the street*. Berkeley: University of California Press.

SANTE, L. (1991). *Low life: Lures and snares of old New York*. New York: Farrar, Straus & Giroux.

SHORT, J. F., JR., & STRODTBECK, F. (1965). *Group process and gang delinquency*. Chicago: University of Chicago Press.

SKOLNICK, J. H., ET AL. (1989). *The social structure of street drug dealing* (monograph). Sacramento: State of California, Office of the Attorney General, Bureau of Criminal Statistics.

SPERGEL, I. A. (1990). Youth gangs: Problem and response. Unpublished. Chicago: University of Chicago.

SPERGEL, I. (1989). Youth gangs: Continuity and change. In N. Morris & M. Tonry (Eds.), *Crime and justice: An annual review of research, volume 12*. Chicago: University of Chicago Press.

STUMPHAUZER, J. S., VELOZ, E. V., & AIKEN, T. W. (1981). Violence by street gangs: East Side story? In R. B. Stuart (Ed.), *Violent behavior: Social learning approaches to prediction, management, and treatment*. New York: Brunner-Mazel.

SULLIVAN, M. (1989). *Getting paid: Youth crime and unemployment in three urban neighborhoods*. New York: Cornell University Press.

TAYLOR, C. S. (1993). *Girls, gangs, women and drugs*. East Lansing MI: Michigan State University Press.

TAYLOR, C. S. (1992). The ecology of crime and drugs in Detroit. In J. Fagan (Ed.), *The ecology of crime and drug use in inner cities*. New York: Social Science Research Council.

TAYLOR, C. S. (1990a). *Dangerous society*. East Lansing, MI: Michigan State University Press.

TAYLOR, C. S. (1990b). Gang imperialism. In C. R. Huff (Ed.), *Gangs in America*. Newbury Park, CA: Sage.

THRASHER, F. M. (1927). *The gang: A study of 1,313 gangs in Chicago*. Chicago: University of Chicago Press.

TIENDA, M. (1989). Puerto Ricans and the underclass debate. *The Annals of the American Academy of Political and Social Science, 501*, 105–119.

U.S. DEPARTMENT OF JUSTICE. (1989). Community-wide responses crucial for dealing with youth gangs. Bulletin, Office of Juvenile Justice and Delinquency Prevention. Washington, DC: Author.

VIGIL, J. D. (1988). *Barrio gangs*. Austin, TX: University of Texas Press.

VIGIL, J. D. (1985). The gang subculture and locura: Variations in acts and actors. Paper presented at the Annual Meeting of the American Society of Criminology, San Diego, CA.

WACQUANT, L. D., & WILSON, W. J. (1989). The costs of racial and class exclusion in the inner city. *Annals of the American Academy of Political and Social Science, 501*, 8–25.

WALDORF, D. (1992). When the Crips invaded San Francisco: Gang migration. Unpublished. San Francisco: Institute for Scientific Analysis.

WALL STREET JOURNAL. (1989). In the war on drugs, toughest foe may be that alienated youth. September 8, p. 1.

WILLIAMS, T. (1992). *Crackhouse*. Reading, MA: Addison-Wesley.

WILSON, W. J. (1991). Studying inner-city social dislocations: The challenge of public agenda research. *American Sociological Review, 56*, 1–14.

WILSON, W. J. (1987). *The truly disadvantaged*. Chicago: University of Chicago Press.

JEFFREY FAGAN

GANJA   Ganja is a Hindi word (derived from Sanskritic) for the HEMP plant, *Cannabis sativa* (marijuana); the term *ganja* entered English in the late seventeenth century. Ganja is a selected and potent preparation of MARIJUANA used for smoking.

The hemp plant was introduced into the British West Indies by indentured laborers from India who arrived in Jamaica in 1845. Considered to be a "holy" plant, ganja is often used in religious ceremonies in both countries. The Indian Hemp Drug Commission traced the origin of ganja use to India.

Although usually smoked, *Cannabis* may also be mixed with foods or drinks; it is considered a remedy for many ailments in herbal medicine. A medical-anthropological study of ganja users in Jamaica was conducted in 1972; the results revealed little evidence of a deleterious effect among users, as compared with nonusers. These conclusions were criticized, however, by investigators who claim that the tests of maturation and mental capacity that were used were not sensitive enough to detect decrements in higher level mental skills or motivation.

(SEE ALSO: *Bhang; Plants, Drugs from*)

BIBLIOGRAPHY

RUBIN, V. & COMITAS, L. (1975). *Ganja in Jamaica: A medical anthropological study of chronic marihuana use*. The Hague: New Babylon Studies in the Social Sciences.

LEO E. HOLLISTER

GATEWAY DRUGS   *See* Adolescents and Drugs.

GATEWAY FOUNDATION   In 1968, Gateway Houses Foundation was incorporated as a not-for-profit corporation and became the first THERAPEUTIC COMMUNITY in Illinois. It was in part the outcome of a four-year effort by a group of concerned citizens (Council for the Understanding and Rehabilitation of Addicts [CURA]) to learn about and fight growing drug addiction. Modeled on DAYTOP VILLAGE, from which its initial key clinical staff was recruited, it was established with an operating grant of $180,000 from the Illinois Drug Abuse Program as a residential setting in which former drug addicts could help other drug abusers find a way to live drug-free, responsible, and useful lives in the community.

Gateway's first two houses were located in Chicago; they provided drug-free self-help programs typical of traditional therapeutic communities. Because of their success and the demand for treatment, a 50-acre site in Lake County was acquired in 1970. By 1972, a center was added in Springfield, and another residence and outpatient center began operations in Chicago; in 1973, the two original Chicago houses closed because of zoning difficulties.

Gateway's programs were shaped by TREATMENT successes, RESEARCH, experience, and the need for public accountability. Education and the contributions of the medical and mental health fields to treatment reinforced Gateway's staff expertise. The professionalism of a maturing staff and the challenge of changing client needs produced effective treatment alternatives that were incorporated into the traditional therapeutic community structure.

The early years of treatment experience demonstrated that not all of those entering Gateway needed long-term residential treatment and that it should be reserved for the neediest clients. Programs were devised or modified to fit the specific needs of the men, women, and youngsters served. The agency adopted the name Gateway Foundation in 1983 to better symbolize the array of services offered. To extended care (residential, long-term treatment), Gateway added outpatient (both intensive and basic), detoxification, and short-term treatment, as well as community-based EDUCATION and PREVENTION PROGRAMS.

As of the mid-1990s, Gateway Foundation offers adult residential centers in Chicago, Lake Villa, Caseyville (southern Illinois), and Springfield (central Illinois) and special Foundation for Youth centers in Lake Villa and Springfield. Youth care services will be available at Chicago's new 150-bed residential center in Chicago's underserved West Side, which opened in the fall of 1993. The therapeutic community remains the core of Gateway's programs and it is one of the few agencies in Illinois offering that service. Participation in TWELVE-STEP support groups (see AA) are the client's mainstay during and after treatment. Outpatient services are offered at two Chicago locations and in Springfield, Belleville, Caseyville, and Waukegan, Illinois. Prevention programs have been broadened to include not only schoolchildren and parents, but Chicago's entire Lake View community and part of the Austin community. In twenty-five years, Gateway has served more than 50,000 residents of Illinois and is supported by both the public and private sectors.

Gateway Foundation's successful treatment center within the Correctional Center of Cook County (the largest U.S. county jail), resulted in treatment programs for men and women in other Illinois correctional programs: Dwight Correctional Center for Women, Sheridan, Graham, Logan, Lincoln, and Jacksonville, Illinois, and Crossroads (a community-based work-release program). The success of the Illinois corrections programs prompted the State of Texas to request that Gateway operate programs in four of its correctional sites in Sugarland, Richmond, Marlin, and Rosharon, Texas. Gateway also operates a therapeutic community in Houston.

Client profiles differ from program to program, but the average Gateway extended care client is indigent. Ninety-five percent have been involved with the criminal-justice system; 70 percent are male, 30 percent female. Seventy-five percent abuse ALCOHOL as well as other drugs; 50 percent are high-school DROPOUTS and have had prior treatment elsewhere. Gateway Foundation's current client body is 60 percent African-American, 35 percent white, and 5 percent Hispanic. Erratic work histories in low-level jobs are common, and only 10 percent have some type of usable job skill. Treatment for all Gateway clients includes work and social-skills development, continuing education, and employment counseling.

Treatment outcomes of the Gateway Foundation are comparable to those found in a twenty-year study, TREATMENT OUTCOME PROSPECTIVE STUDY (TOPS), *Drug Abuse Treatment, A National Study of Effectiveness*, by the Research Triangle Institute, published by the University of North Carolina Press in 1989. The study, in which Gateway participated, found that average abstinence rates one year after treatment were 40 to 50 percent. Gateway is currently participating in the institute's follow-up study, Drug Abuse Treatment Outcomes (DATOS).

(SEE ALSO: *Appendix*, Volume 4)

BIBLIOGRAPHY

HUBBARD, R. L., ET AL. (1989). *Drug abuse treatment, a national study of effectiveness*. Chapel Hill and London: The University of North Carolina Press.

SLOTKIN, E. J. (1972). *Gateway, the first three years*. Funded by the Illinois Law Enforcement Commission.

MICHAEL DARCY

GENDER AND COMPLICATIONS OF SUBSTANCE ABUSE

Does gender have an influence on whether a drug has complications? There is limited research available to answer this question, for many studies include men only. In general, women drink less often and in smaller amounts than men do, and they suffer fewer ALCOHOL-related problems and less dependence (WITHDRAWAL) symptoms. Women use illicit drugs less often than men do, although women have a higher consumption of prescription tranquilizers, sleeping pills, and over-the-counter drugs. Thus, the differences seen between the genders in complications largely reflect the differences in the respective patterns and prevalence of their alcohol and drug use.

The effects of the drugs are relatively similar between men and women. For example, in a heavy drinking and heavy SMOKING sample population, there is little difference in the mortality rates between men and women. Alcohol- and drug-using women are more likely to have partners who are alcohol and drug users. Such women are often victims of violence. Illicit-drug-using women frequently support their drug habits by prostitution, putting themselves at risk for sexually transmitted diseases (STDs) including HUMAN IMMUNODEFICIENCY VIRUS (HIV) and hepatitis B, even if they are not needle users. Accidents and trauma related to substance abuse are more common in men. The skid-row lifestyle is more common in men. Men report DRINKING AND DRIVING more often than women.

ALCOHOL

Women appear to be more susceptible than men are to alcohol-related LIVER damage. For women, cirrhosis may develop with consumption of 20 grams of alcohol (1–2 drinks) per day—as compared to 80 grams (6 drinks) per day for men. Women alcoholics have death rates 50 to 100 percent higher than their male counterparts. Women develop hypertension, obesity, anemia, malnutrition, and gastrointestinal hemorrhage at lower alcohol consumption levels and with a shorter time course of drinking. Women become intoxicated after drinking smaller quantities of alcohol than do men. For an equivalent dose of alcohol corrected for body weight, women absorb alcohol faster and reach a higher peak BLOOD ALCOHOL CONCENTRATION compared to men. These differences can be explained, in part, by the lower total body water of women compared to men. With a higher percentage of fat and lower water content, there is less volume in which to dilute the alcohol, and its concentration is therefore increased. Women also produce less stomach alcohol dehydrogenase—the enzyme responsible for breaking down alcohol. This leads to higher blood alcohol levels, since less is metabolized as it passes through the wall of the stomach and, therefore, as compared to men, more alcohol gets into the bloodstream. There may also be some hormonal or immune effects that account for the increased damage in women.

TOBACCO

Women are at risk for all the same health complications of smoking as are men. The differences seen in the 1990s largely reflect the lower prevalence of women smokers in past generations. For example, as smoking rates have increased in women, lung cancer rates have also increased.

REPRODUCTION

A woman's drinking pattern may be influenced by the mood changes associated with the phases of the menstrual cycle, and her blood alcohol level actually measures higher during the premenstrual period for any given amount of alcohol. This may make it difficult for a woman to predict the effects of her drinking. Oral contraceptives interact with cigarette smoking in contributing to coronary heart disease in women. Cigarette smoking is also correlated with an earlier onset of menopause. In her role as childbearer, a woman's substance use may have harmful effects on the FETUS and newborn. These effects may be related to her lifestyle, such as poor nutrition and poor prenatal care, or to the toxic effects of the drugs themselves resulting in fetal growth retardation, at-birth neonatal abstinence syndrome (withdrawal), and neurobehavioral abnormalities in the child.

Alcohol, tobacco, and illicit drugs like COCAINE and HEROIN are all associated with decreased fertility, increased rate of spontaneous abortion (miscarriage), and decreased birthweight in the newborn. The severely dependent woman may stop menstruating altogether. Menses resume, however, when abstinence or stabilization on methadone maintenance is achieved.

(SEE ALSO: *Fetal Alcohol Syndrome; Pregnancy and Drug Dependence; Women and Substance Abuse*)

BIBLIOGRAPHY

KALANT, O. J. (ED.). (1980). *Alcohol and drug problems in women.* New York: Plenum.

WILSNACK, S. C., & BECKMAN, L. J. (EDS.). (1984). *Alcohol problems in women.* New York: Guilford Press.

JOYCE F. SCHNEIDERMAN

GENETICS *See* Causes of Substance Abuse: Genetic Factors; Vulnerability as Cause of Substance Abuse: Genetics.

GINSENG Ginseng is the most revered and well-known plant of Chinese herbal medicine; it is sold over the counter in Asian apothecaries and groceries worldwide. This plant of the family Araliaceae grows on both sides of the Pacific, with *Panax schinseng* the Asian form and *Panax quinquefolius* the North American form. It is a perennial herb with five-foliate leaves, and its fleshy aromatic root is valued as a tonic and a medicine.

The root has been used by Native Americans, Siberians, Chinese, and other Asians for millennia. Usually it is taken as a tea—once a day as a general preventative tonic, more frequently for therapeutic purposes. Since the North American form is considered the most potent, it is now grown in ASIA along with the local variety. American ginseng is also exported to Asia, then sometimes reimported into the United States as a Chinese or Korean herbal. Both the wild and cultivated forms are used. Roots older

Figure 1
Ginseng

than five years are needed for good effect, and the older and larger the root (seven to fifteen years is prized), the more the ginseng costs. Dried roots are heated and sliced thinly to make tea, but pieces may be kept in the mouth, sucked, and eaten. The many ginseng products now sold (sodas, candies, etc.) have no real tonic or therapeutic value.

Ginseng has a bittersweet aromatic flavor, contains ALKALOIDS, and is said to be good for mental arousal and general well-being. It has not been established in Western medicine and pharmacology, although it contains properties that might be isolated and used pharmacologically.

(SEE ALSO: *Plants, Drugs from*)

MICHAEL J. KUHAR

GLUE/GLUE SNIFFING *See* Inhalants.

GLUTAMATE Glutamate (GLU) is a dicarboxylic aliphatic amino acid. Chemically symbolized as $COOH-CH_2-CH_2[NH_2]-COOH$, it is abundant (micromolar concentrations/mg protein) in NEURONS (nerve cells) as well as in almost all other cells of the body. Its role as the major excitatory NEUROTRANSMITTER in the brain was recognized reluctantly; its universal ability to excite all neurons was considered too nonspecific for a neurotransmitter, so it awaited the development of drugs that antagonized GLU and the specific neuropathways from which it was released.

Its source for this special role in NEUROTRANSMISSION is unknown, but the synaptic vesicles of glutamatergic neurons have a selective ion-exchange mechanism to compartmentalize GLU from other metabolic pathways. Excessive GLU-receptor activation can lead to neuronal death.

(SEE ALSO: *Research*)

BIBLIOGRAPHY

COOPER, J. R., BLOOM, F. E., & ROTH, R. H. (1991). *The biochemical basis of neuropharmacology*, 6th ed. New York: Oxford University Press.

FLOYD BLOOM

GLUTETHIMIDE Glutethimide was introduced into clinical medicine in 1954. It was prescribed to treat insomnia and sold as Doriden. It was first acclaimed as a safer "nonbarbiturate" hypnotic—implying that it was free of the problems of abuse, addiction, and withdrawal that were, by then, recognized drawbacks of the older barbiturate SEDATIVE-HYPNOTICS. Within ten years, however, it was recognized that, in most respects, its actions are like those of the BARBITURATES and it shares the same disadvantages.

Glutethimide is structurally related to the barbiturate drugs and, like the short-acting barbiturates, it depresses or slows the central nervous system. Side effects from its proper use are relatively minor, but a rash is often seen. Like barbiturates, it can produce intoxication and euphoria; TOLERANCE and DEPENDENCE can result with daily use. Glutethimide is metabolized somewhat differently than barbiturates, and OVERDOSE is often far more difficult to treat than barbiturate overdose; fatalities are not uncommon. As a consequence of this and its ABUSE POTENTIAL, glutethimide is included in Schedule III of the CONTROLLED SUBSTANCES ACT. Since the introduction of the BENZODIAZEPINES to treat short-term insomnia, the use of glutethimide has decreased considerably.

(SEE ALSO: *Barbiturates*; *Complications*; *Sedatives*)

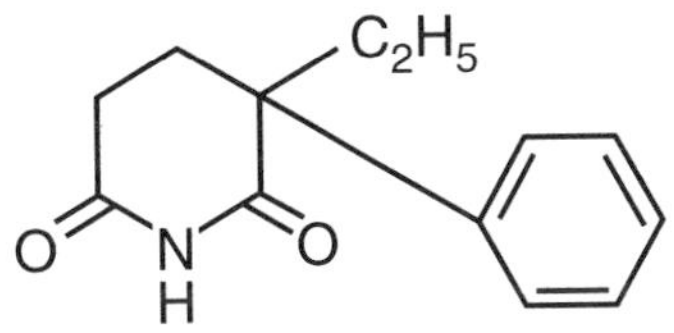

BIBLIOGRAPHY

HARVEY, S. C. (1975). Hypnotics and sedatives: Miscellaneous agents. In L. S. Goodman & A. Gilman (Eds.), *The pharmacological basis of therapeutics*, 5th ed. New York: Macmillan.

SCOTT E. LUKAS

GOLDEN TRIANGLE AS DRUG SOURCE The world's largest illicit OPIUM-growing area is the Golden Triangle—an area in Southeast Asia of some 150,000 square miles (388,500 sq km). The Golden Triangle extends from the Chin hills in the west of Myanmar (formerly Burma), north into China's Yunnan province, east into Laos and Thailand's northern provinces, and south into the Kayah state of Myanmar. It encompasses all the Shan state in Myanmar and supplied some 35 percent of the HEROIN used in the United States since the

Figure 1
Glutethimide

1960s. By 1990, Myanmar, Thailand, and Laos supplied about 56 percent of the heroin consumed in the United States. By 1991, the Southeast Asian opium crop was increasing (heroin is processed from opium).

The United States Government has supplied millions of dollars to Myanmar and Thailand in an effort to reduce OPIUM-POPPY (*Papaver somniferum*) cultivation and interdict heroin destined for the United States. As they have done for years, disenfranchised tribal people cultivate opium as a medicinal and cultural product, as a cash crop to buy food and supplies and improve living conditions, and as a means to procure weapons. Political events in Southeast Asia are complex and are changing constantly. The U.S. government has managed a limited success in helping to reduce opium cultivation in Laos and Thailand; it is anxious over the increased production in Myanmar and the increasing flow of heroin exiting that country via China to Hong Kong, through Rangoon toward Malaysia and Singapore, and through India and Bangladesh.

THE OPIUM SUPPLY

The Golden Triangle had favorable weather throughout 1991, which resulted in record opium crops. By far the largest producer is Myanmar, which until 1988, had attempted to reduce illicit opium production, strike at illicit refineries, and interdict shipments of illicit drugs. Late in 1988, however, the military government shifted its police and military away from drug-control efforts, to suppress domestic political opponents.

Laos, isolated and largely ignored by the West since 1975 when the Communist Pathet Lao seized power, cultivated opium in its nine northern provinces—about 20 percent of Myanmar's production. Partly because of the 1990 collapse of the Soviet Union, Laos's principal trading partner and ally, the Laotian government has entered into a number of cooperative agreements with Western nations.

Opium production leveled off or declined slightly in 1991, due in part to the incipient success of three crop-substitution projects.

Although Thailand is more important as a TRANSIT COUNTRY for Myanmar's opium and heroin, Thailand also cultivated about 10 percent of Laos's production. A traditional producer of opium since the mid-1800s and a net importer of heroin, Thailand's opium is grown in the northern highlands by nomadic hilltribes who are not tied to Thailand culturally, religiously, or politically. Opium cultivation in Thailand remains illegal, so the government has sponsored both eradication and crop-substitution efforts in the north.

In its 1992 *International Narcotics Control Strategy Report* (INCSR), the U.S. government stated that opium cultivation may be emerging as a major problem in China. China has become a major narcotics transit point because of its open border with Myanmar, its location adjacent to the Golden Triangle, and its excellent transportation and communication links with the trade ports of Hong Kong and Macao. Much of the heroin processed from opium by the Kokang Chinese in the Golden Triangle transits through Yunnan, Guangxi, and Guangdong provinces by road to Hong Kong for overseas distribution. In 1991, Chinese law enforcement seized more drugs and investigated more cases than at anytime since the Communist takeover. Accompanying the increased heroin flow in China is an increase in heroin consumption.

CULTIVATION CONDITIONS

A number of factors have contributed to the thriving opium economy of the Golden Triangle—and the complex politics surrounding and sustaining it. First, the topographical and climatic conditions are ideal for opium cultivation. The demographic conditions also provide a division of labor conducive to an economic system rooted in drug cultivation, processing, and trafficking. The area under cultivation is largely mountainous, ranging from about 5,000 feet (1,500 m) to more than 9,850 feet (3,000 m), with four major river systems supporting the transportation networks and any ongoing economic-development efforts. The remote harsh terrain has fostered great efforts

to topple the central governments and to capitalize on the economic opportunities offered by the opium trade.

Second, the ethnography of the region is complex. The region is inhabited by a multitude of ethnic groups, possessing a diversity that defies simple classification. Burman, Shan, Kachin, Thai, and Yunnanese are broad categories that contain widely varied ethnic subgroups. At least twenty-five mutually unintelligible dialects are spoken among the Kachin people. Moreover, there are numerous other groups who do not belong to the larger ethnic division—such as Ahka, Hmong (Miao), Lisu, Lahu, Karenni, and Wa, to name a few. Most of these groups are nomadic—not geographically localized; therefore, little basis exists for territorial political organization. Yet, national boundaries have paid little heed to this fact and have often cut apart ethnic groups, fueling insurgency as the dominant form of politics in the region.

Cultivating opium in the Golden Triangle has been a way of life since the mid-1800s and has represented an important source of income for impoverished, nomadic hill tribes.

(SEE ALSO: *Crop Control Policies; Foreign Policy and Drugs; International Drug Supply Systems; Source Countries for Illicit Drugs; Transit Countries for Illicit Drugs*)

BIBLIOGRAPHY

BUREAU OF INTERNATIONAL NARCOTICS MATTERS, U.S. DEPARTMENT OF STATE. (1990). *International narcotics control strategy report (INCSR)*. Washington, DC: Author.

LINTNER, B. (1990). *Land of jade: A journey through insurgent Burma*. Edinburgh: Kiscadale White Lotus.

LINTNER, B. (1983). Alliances of convenience. *Far Eastern Economic Review* (April).

MCCOY, A. W. ET AL. (1990). *The politics of heroin*. New York: Harper & Row.

WALKER, W. O., III. (1991). *Opium and foreign policy*. Chapel Hill and London: The University of North Carolina Press.

WIANT, J. A. (1985). Narcotics in the Golden Triangle. *The Washington Quarterly* (Fall) 125–140.

JAMES VAN WERT

H

HABIT *See* Addiction: Concepts and Definitions.

HABITUATION *See* Addiction: Concepts and Definitions.

HAGUE OPIUM CONFERENCE OF 1911 *See* Britain, Drug Use in; Opioids and Opioid Control.

HAIGHT-ASHBURY FREE CLINIC The Haight-Ashbury Free Clinic was founded in June 1967 by David E. Smith, M.D., with the help of other physicians from the University of California Medical School at San Francisco and community volunteers to provide medical services for the waves of young people who came to San Francisco during the "Summer of Love." The Haight-Ashbury neighborhood had become a center for a "counterculture" that espoused spirituality, communal living, freedom from past mores, and the use of PSYCHEDELIC drugs. These young people, variously called hippies and flower children, often lived in crowded, unhygienic conditions and were highly vulnerable to the spread of respiratory, skin, and sexually-transmitted diseases. Even when infected or suffering from consequences of their drug use, they felt unable to seek medical care because they felt they could not afford it and found themselves philosophically at odds with established care providers. The Free Clinic offered an alternative to an established medical care system that members of the Counterculture saw as difficult to access, dehumanizing, unresponsive, and often judgmental about their nontraditional lives. The clinic's philosophy included beliefs that health care is a right, not a privilege, and that it should be free, nonjudgmental, demystified, and humane. The free-clinic concept spread nationally, and at one point in the early 1970s, there were more than 600 free clinics in the United States.

The Haight-Ashbury community proved to be a center for drug experimentation, while the free clinic became a primary source of innovative drug-abuse treatment, where many health professionals received their early field training, and treatment approaches were developed for the DETOXIFICATION of OPIOID, SEDATIVE-HYPNOTIC, stimulant, and PSYCHOACTIVE drug abusers. Haight-Ashbury Free Clinic pioneers included John Frykman and George Gay in HEROIN treatment, Donald Wesson and David Smith in sedative-hypnotic and POLYDRUG abuse, Stephanie Ross and Richard Seymour in physician training, Bill Pone and Craig Whitehead in the use of ACUPUNCTURE for detoxification, and John Newmeyer in the development of epidemiological research on drug abuse and HIV (HUMAN IMMUNODEFICIENCY VIRUS) disease.

Today, Haight-Ashbury Free Clinics, Inc., provides a full spectrum of community medical services to an ethnically mixed population of the working poor, the unemployed, and the HOMELESS. Under its current director, Darryl Inaba, the Drug Detoxification, Rehabilitation and Aftercare Program is the clinic's largest division and includes specialized care for inner-city CRACK addicts, pregnant addicts, prison parolees, and the HIV-positive, as well as detoxification from a full range of substances. Other clinic programs provide care for alcoholics and prisoners, and a social-model detoxification for women multiple-drug abusers. HIV and drug research, training for professionals in drug treatment, and correct prescribing practices, including annual specialized conferences, have an ongoing national impact on the drug-abuse field. The clinic continues to provide medical services at the rock concerts and has expanded into residential programs for homeless substance abusers and church-based treatment for African-American crack abusers, such as the model program at Glide headed by the Reverend Cecil Williams.

Parallel to the clinic's growth has been an evolution in funding. In the beginning, the clinic subsisted on personal donations and occasional rock music concert benefits, promoted by rock impresario Bill Graham. All staff was volunteer and funds were used to cover rent, materials, and medications that were not donated. Starting in 1969, federal grants became the primary source of funds and the clinics were staffed by paid full-time employees and part-time volunteers. In the 1990s, the clinic enjoys a wide range of funding sources—the bulk of its income is from City and County of San Francisco treatment contracts, followed by federal, state, and local grants. These are supplemented by a vigorous fundraising within the private sector, but with ever expanding services, much use is still made of volunteers. Of the over $9 million current annual budget of the Haight-Ashbury Free Clinic, all but a small administrative overhead goes to pay for direct client services. The clinic has also evolved into a recovery oriented treatment facility serving as a site for fifty TWELVE-STEP recovery group meetings throughout its twenty-two different sites in the San Francisco Bay area.

(SEE ALSO: *Amphetamine Epidemics; Ethnic and Cultural Relevance in Treatment; Lysergic Acid Diethylamide; Treatment/Treatment Types*)

BIBLIOGRAPHY

PERRY, C. (1984). *The Haight-Ashbury: A history.* New York: Random House.

SEYMOUR, R. B., & SMITH, D. E. (1986). *The Haight-Ashbury Free Clinics: Still free after all these years.* Sausalito: Westwind Associates.

SMITH, D. E., & LUCE, J. *Love needs care.* Boston: Little, Brown.

DAVID E. SMITH
RICHARD B. SEYMOUR
JOHN NEWMEYER

HAIR ANALYSIS AS A TEST FOR DRUG USE

Hair analysis has the potential of providing a historical record of drug exposure that dates back in time for months to years. During hair growth, administered drugs are incorporated into the hair shaft. Although some drug may be leached out by various hair treatments, residual drug may be found for the life of the hair. Since hair grows at a relatively constant rate, segmental analysis of hair strands could localize the time of drug exposure to within a few weeks. This unique property of hair may provide us the new long-term means of detection for drug exposure.

GROWTH AND DEVELOPMENT

Hair is nonliving tissue composed primarily of a sulfur-rich protein material called keratin. Growth occurs at a rate of 0.3 to 0.4 millimeters per day from the follicle, a saclike organ in the skin. Hair does not grow continually but in cycles of active growth followed by transition to a resting phase. For an adult, approximately 85 percent of scalp hair is in the growing stage at any time. Two sets of glands are associated with the follicle: The sebaceous glands excrete sebum, a waxy substance; the apocrine glands excrete an oil that coats the hair (giving it an oily appearance when hair goes unwashed for several days). The color of hair is derived from the genetic programming for varying amounts of pigment, melanin, which is synthesized in hair cells called melanocytes.

DRUGS IN HAIR

How drugs enter hair is not known. Entry may occur from the capillaries, which supply nutrients in the bloodstream to the follicles during the growth stage. Alternatively, drugs may be excreted in the sebum, oil, or sweat that bathe the hair shafts. A more troubling possibility is that drugs might contaminate hair from environmental exposure. Recent evidence indicates that externally applied drug is difficult to distinguish from drug residue within hair—present as a result of self-administration. To overcome the problem of possible environmental contamination, researchers have proposed various washing methods to remove externally applied drug. An alternative method is to perform hair analysis for unique drug metabolites that would be present only as a result of active drug use.

A variety of drugs of abuse have been identified in hair including HEROIN, COCAINE, AMPHETAMINES, PHENCYCLIDINE, MARIJUANA, NICOTINE, BARBITURATES, and their respective metabolites. Analysis is performed by chemical or enzymatic removal of drug from the protein matrix, followed by IMMUNOASSAY or assay by gas chromatography/mass spectrometry. Some technological issues remain unresolved in hair testing; these include (1) selection of appropriate cutoffs that determine whether a sample is positive or negative (2) development of control materials that can be used for assay standardization, and (3) identification of the correct analyte in hair that should be targeted for analysis.

SIGNIFICANCE OF HAIR TESTING

Hair analysis may offer some advantages over other forms of DRUG TESTING. Once drug is embedded in hair, it appears to be stable indefinitely, although concentration diminishes somewhat over time. Cocaine metabolite has, for example, been detected in pre-Columbian mummy hair that is more than 500 years old. Certainly, hair analysis offers a longer window to detection than urinalysis. Urine testing is useful for drug detection within two to thirty days, depending on the drug and individual use patterns. Hair analysis could detect drugs within five years after use (depending upon hair length). Hair samples are easily collected and stored for analysis. If additional analysis is required, another sample may be obtained. These properties make hair testing attractive as a means of assessing possible maternal/fetal drug exposure and as a means of validating self-reports of drug usage. A disadvantage of hair analysis is that it may not reveal drug use as recent as the last week prior to testing, since hair does not grow out quickly enough to show this. Another disadvantage is that with current technology, hair analysis is far more costly than testing for drugs in urine. Urinalysis may cost as little as a few dollars, and the results can be available in minutes. Hair analysis takes much longer and costs much more.

As of the early 1990s, additional controlled clinical studies are needed to validate hair testing—and to define its advantages and limitations.

(SEE ALSO: *Industry and Workplace, Drug Use in; Military, Drug and Alcohol Abuse in the U.S.*)

BIBLIOGRAPHY

HARKEY, M. R., & HENDERSON, G. L. (1989). Hair analysis for drugs of abuse. In R. C. Baselt (Ed.), *Advances in analytical toxicology*. Chicago: Year Book Medical Publishers.

EDWARD J. CONE

HALF-LIFE *See* Dose-Response Relationship.

HALFWAY HOUSES Although the term is of recent origin as used in connection with alcohol or drug treatment, the basic idea of the *halfway house* is almost 200 years old. It designates a residential facility that provides a drug-free environment for individuals recovering from drug or alcohol problems but not yet able to live independently without jeopardizing their progress. By definition, halfway houses are not located in hospitals or PRISONS, but they vary in the extent to which they are integrated with local community life, and in size, sponsorship, sources of financial support, regulatory status (licensed or unlicensed by a state agency), treatment philosophy, and the degree of legal coercion to which residents are subject. Some specialize in alcohol abusers or drug abusers, while some serve both; some focus on specific population groups like offenders, ADOLESCENTS, or WOMEN, while others are inclusive. Some will accept only those with at least a few days of abstinence; others provide DETOXIFICATION services. Some are loosely structured and rely for staff on recovering people; others provide formal treatment and employ a professional staff.

In sum, the term covers a lot of ground and has no stable meaning. Indeed, its meaning in any given state depends on that state's licensing provisions, and whether these make any distinctions among halfway houses, recovery homes, and other similar forms of residential treatment. At a minimum, however, the term implies a group of people with alcohol and/or drug problems living together in a formal, therapeutic arrangement and abiding by the rule of abstinence. In 1987, there were more than 1,300 such programs in North America, many of them members of the National Association of Halfway Houses.

Although there is increasing interest in establishing residential forms that tolerate off-site consumption that does not disrupt facility life, these would not be considered halfway houses in the common use of the term. Further, because the halfway house is a sponsored, therapeutic program, however informally operated, it is everywhere subject to special zoning ordinances that regulate the location of therapeutic agencies. Thus, the halfway house is distinct from what is called "alcohol and drug-free (or sober) housing." The latter is designed to be part of a locality's ordinary housing stock and to be exempt from such regulation.

The Federal Anti-Drug Abuse Act of 1988 (Public Law 100-690) included a provision to encourage the development of ALCOHOL- AND DRUG-FREE HOUSING. Each state that receives federal block grant funds for drug and alcohol programs must establish a 100,000 dollar revolving fund to make start-up loans for such facilities. Although this money can be used to develop halfway houses, as we have defined them, the revolving fund has in practice been used to stimulate less formal approaches to housing recovering people.

(SEE ALSO: *Homelessness and Drugs, History of; Oxford House; Treatment: History of*)

BIBLIOGRAPHY

GERSTEIN, D. R., & HARWOOD, H. J. (EDS.). (1990). *Treating drug problems.* Washington, DC: National Academy Press.

RUBINGTON, E. (1977). The role of the halfway house in the rehabilitation of alcoholics. In B. Kissin & H. Begleiter (Eds.), *Treatment and rehabilitation of the chronic alcoholic.* New York: Plenum.

WITTMAN, F. D. (1993). Affordable housing for people with alcohol and other drug problems. *Contemporary Drug Problems, 20,* 541–609.

JIM BAUMOHL
JEROME H. JAFFE

HALLUCINATION The word *hallucinate* is derived from the Greek *halyein*, meaning "to wander in mind." Hallucinations are perceptions that occur in the absence of a corresponding external sensory stimulus. They are experienced by the person who has them as immediate, involuntary, vivid, and real. They may involve any sensory system, and hence there are several types of hallucinations: auditory, visual, tactile (e.g., sensations on the skin), olfactory (smell), and gustatory (tastes). Visual hallucinations range from simple (e.g., flashes of light) to elaborate visions. Auditory hallucinations can be noises, a voice, or several voices carrying on a conversation. In command hallucinations, the voices often order the person to do things that at times involve acts of violence.

Hallucinations have been a hallmark of mental illness throughout history. They are an important clinical feature of several psychiatric conditions in which psychosis can occur, such as SCHIZOPHRENIA, manic-depressive illness, major DEPRESSION, and dissociative states. WITHDRAWAL from ALCOHOL can cause visual as well as other sensory hallucinations. In alcoholic hallucinosis, a person dependent on alcohol develops mainly auditory hallucinations that can persist after the person has stopped drinking. Hallucinations may be induced by illicit drugs, such as COCAINE, AMPHETAMINES, and LSD. These hallucinations are usually visual, but they can also be auditory or tactile, as in the sensation of insects crawling up the skin (an example of a haptic hallucination). Occasionally, after repeated ingestion of drugs, some people experience "flashbacks"—that is, spontaneous visual hallucinations during a drug-free state, often months or years later.

The cause of hallucinations is not known, but it is likely to be multifactorial through a combination of physiological, biological, and psychological variables. Numerous hypotheses have been proposed. According to a perceptual release theory, hallucinations develop from the combined presence of intense states of psychological arousal and decreased sensory input from the environment (e.g., sensory deprivation) or a reduced ability to attend to the sensory input (e.g., in delirium). This leads to the emergence of earlier images and sensations that are intepreted as originating in the environment. Other researchers suggest that abnormalities in brain cell excitability or in the information processing system of the central nervous system cause hallucinations. Biochemical theories implicate brain NEUROTRANSMITTERS such as DOPAMINE. Drugs that block brain dopamine activity (ANTIPSYCHOTICS) alleviate hallucinations, whereas drugs that stimulate dopamine release induce hallucinations.

Hallucinations can occur in people who are not mentally ill. In acute bereavement, some people report seeing or hearing the deceased. Sensory, SLEEP, food, and water deprivation can produce hallucinations, as can the transition from sleep to wakefulness and vice versa (called hypnopompic and hypnogogic hallucinations, respectively). These hallucinations can occur as side effects of prescribed medications, such as drugs that treat cardiac conditions, or in various medical disorders (e.g., migraines, Parkinson's Disease, infections). They have been described in persons with hearing loss and blindness; in these instances, it has been hypothesized that they may be due to chronic sensory deprivation.

The treatment of hallucinations is part of the treatment of the entire psychotic syndrome. Antipsychotic medications (e.g., haloperidol, chlorpromazine) are effective in reducing and often eliminating hallucinations. When the hallucinations are part of a medical disorder, it is necessary to correct the underlying condition, or remove the causative agent, in addition to prescribing antipsychotic medication.

(SEE ALSO: *Complications: Mental Disorders; Delirium Tremens; Hallucinogenic Plants; Hallucinogens*)

BIBLIOGRAPHY

ASAAD, G., & SHAPIRO, B. (1986). Hallucinations: Theoretical and clinical overview. *American Journal of Psychiatry, 143*(9), 1088–1097.

YAGER, J. (1989). Clinical manifestations of psychiatric disorders. In H. I. Kaplan & B. J. Saddock (Eds.), *Comprehensive textbook of psychiatry*, 5th ed., vol. 1. Baltimore: Williams & Wilkins.

MYROSLAVA ROMACH
KAREN PARKER

HALLUCINOGENIC PLANTS

Literally hundreds of hallucinogenic substances are found in many species of plants. For example, a variety of mushrooms contain indole-type HALLUCINOGENS, the most publicized being the Mexican or "magic" mushroom, *Psilocybe mexicana*, which contains both the hallucinogenic compounds PSILOCYBIN and psilocin, as do some of the other *Psilocybe* and *Conocybe* species. The PEYOTE cactus (*Lophophra williamsii* or *Anhalonium lewinii*), which is found in the southwestern United States and northern Mexico, contains MESCALINE. The seeds of the MORNING GLORY, *Ipomoea*, contain hallucinogenic LYSERGIC ACID derivatives, particularly lysergic acid amide. Many of these plants and plant by-products were and are used during religious ceremonies by Native Americans and other ethnic groups.

Some plant substances may contain prodrugs, that is to say, compounds that are chemically altered in the body to produce PSYCHOACTIVE substances. For example, NUTMEG contains elemicin and myristicin, whose structures have some similarities to the hallucinogen mescaline as well as the psychostimulant AMPHETAMINE. It has been hypothesized that elemicin and myristicin might be metabolized in the body to form amphetamine- and/or mescaline-like compounds, but this has not been proven. The fact that hallucinogenic substances are found in nature does not mean that they are safer or purer than compounds that have been synthesized in the laboratory. Some common edible mushrooms that can be purchased in any supermarket may be sprinkled with LSD or other hallucinogens to be misrepresented as magic mushrooms. In addition, serious problems—even death—may occur when species of hallucinogenic plants are misidentified and people mistakenly ingest highly toxic plants, such as poisonous mushrooms.

(SEE ALSO: *Ayahuasca; Ibogaine; Jimsenweed; Plants, Drugs from*)

BIBLIOGRAPHY

EFRON, D. H., HOLMSTEDT, B., & KLINE, N. S. (EDS.) (1979). *Ethnopharmacologic search for psychoactive drugs.* New York: Raven Press.

SIEGEL, R. K. (1989). *Intoxication.* New York: Dutton.

WEIL, A. (1972). *The natural mind*. Boston: Houghton Mifflin.

DANIEL X. FREEDMAN
R. N. PECHNICK

HALLUCINOGENS The term *hallucinogen* literally means producer of HALLUCINATIONS. A variety of drugs and medicines as well as various disease states can lead to the development of hallucinations. They can occur during a high fever, after acute brain injuries, or as part of a DELIRIUM, accompanied by confusion in judgment, intellect, memory, emotion, and level of consciousness. The patient is said to be "out of it"—not in touch with reality. In fact, many infections affecting the brain, conditions that disrupt the availability of nutrients essential for brain function, or direct brain injury can cause transient or prolonged delirium. Disease states not directly involving the brain also can alter brain function. For example, the overproduction of thyroid or adrenal hormones in endocrine disease can cause psychotic mental symptoms. In addition, poisoning or other toxic reactions can produce hallucinations.

Some drugs used to treat certain illnesses, although not prescribed for their behavioral effects, may be PSYCHOACTIVE and cause auditory and/or visual hallucinations in some but not all patients. High doses of the adrenal hormone, cortisone, which is prescribed to reduce inflammation in arthritis or allergies, can produce elation or depression and mood-related hallucinations. Similarly, the administration of thyroid hormones for the treatment of thyroid gland deficiencies can cause restlessness, nervousness, excitability and irritability, and psychotic mental symptoms. Drugs derived from the belladonna plant, such as atropine and SCOPOLAMINE, have many uses in clinical medicine but in high doses can cause memory lapses and illusions. Delirium also may result from the sudden withdrawal after the chronic administration of certain drugs, especially ethanol (ALCOHOL) and SEDATIVE drugs of the BARBITURATE class. The vivid hallucinations of DELIRIUM TREMENS (DTs) during the WITHDRAWAL from alcohol have been vividly portrayed in the cinema and television.

Many drugs that affect behavior can alter the level of consciousness or the perception of the environment. PHENCYCLIDINE (known as PCP or "angel dust") can produce a state of altered consciousness in which sensations from the body and relationship to the environment are misinterpreted. The subject may experience numbness in the limbs and feel as though they are removed from their bodies. These distorted perceptions of the real world can lead to confusion, delusions, and hallucinations—and violent behavior can occur with the slightest provocation. There is controversy as to whether these varied reactions are psychotomimetic (imitating mental illness with psychoses), but not about the extent to which, depending on the dose, subjects are out of it. High and/or frequent doses of stimulants such as AMPHETAMINE, METHAMPHETAMINE ("speed" or "ice"), or COCAINE can cause paranoid thought or delusions. Moreover, high doses of MARIJUANA or HASHISH can lead to dreamy illusions or hallucinations. Thus, many drugs under certain conditions can cause hallucinations as part of the production of a complex behavioral syndrome, which may include a general alteration of the level of consciousness and the disruption of the ability of the brain to process information and appreciate the real world.

The term *hallucinogens* has come to mean a group of compounds that reliably, temporarily, and universally alter consciousness without delirium, sedation, excessive stimulation, or any intellectual or memory impairment as prominent effects. Indeed these altered mental effects are the main effects of such drugs. There are a number of synonyms for drugs that produce hallucinations that occur with clear consciousness, but the term *psychedelic* has come into wide use. In the 1960s the term was coined by Humphrey Osmond, a British psychiatrist who came to North America to continue studies of the psychiatric effects of MESCALINE and LSD, and was enthusiastic about their use in enhancing insight in psychotherapy. The term *psychedelic* was invented from greek roots to mean "mind manifesting," from *psyche* (mind, soul) + *deloun* (to show). This refers to the convincing clarity with which a subjective experience is compellingly revealed to the subject who has taken a hallucinogen. What is seen, thought, and felt is vivid—contrasting sharply with the normally ordered perceptions of the world in which we move about and perform our practical tasks. Key to the hallucinogenic experience is that drab everyday reality, while clearly perceived in this drug state, has simply lost its importance in favor of vivid subjective sensations and perceptions and interpretations of them that absorb attention. A door is recognized but not simply for its utility; rather the grain of the wood and its fine detail becomes fascinating, and the grain of the wood seems to move and flow. Thus, during the hallucinogenic experience, it is not the utility of what is seen but rather some aspect of shapes and colors and passing thoughts or memories that take on a life of their own, commanding attentive interest. The color of an object is more important than the object. The subjective impact is that thoughts and sights have some uncanny, undeniable, but inexplicit meaning. The sense of great truth is present, but not an urge to test the truth of these images. Rather, one is a spectator of a "TV show in the head." These events are not only clearly "seen" but remembered without confusion. This has been called "consciousness expanding," implying control over a vast span of experiencing. That is wrong, since judgment is *not* enhanced. Rather, the effect is of enhancing the sense of importance of normally unimportant subjective experiences of sensations and perceptions.

Since with hallucinogens everything—even the most familiar scenes—seems novel and is seen in a new way, the experience is in startling contrast with our normal view of the world. Such effects invite many uses. The intrinsic effects of hallucinogenic drugs not only shift perceptions, making the old new, but evoke a loosening of emotions and thoughts. Hence there were efforts to use hallucinogenic drugs therapeutically—to stimulate and enhance new ways of examining problems. But in spite of the alluring promise, no lasting improvement in learning or problem solving has been found after numerous studies. Similarly, the effects produced by hallucinogens seem so significant and strikingly different from everyday life that they can readily be used to enhance mystical thought and belief. Some Native American groups thus use the hallucinogen PEYOTE in religious ceremonies. The intent is to dispose the celebrants to higher thoughts (to be "in the mind of God"); they are told not to attend to the odd perceptions and rather to relax and contemplate higher thoughts. Because with hallucinogens one is not interested in tracking detail, there is greater suggestibility and dependence on structure, on a leader, on a prior belief, or on the flow of music to guide, interpret, or "carry" one through the experiences.

Whether these drugs produce actual hallucinations or, more commonly, illusions (which the subject usually *feels* are very real but *knows* are not) has

sometimes been debated, but not the fact that these perceptions occur. Seeing geometric abstract designs is not unusual. A characteristic effect is the experience of sound triggering color and of the mixing rather than the clear separation of different sensory modalities—called *synesthesia*. For example, sounds may be "seen," or colors "heard." What has just been seen—say, a wall clock—sometimes persists as one focuses on a face. Rather than suppressing a previous perception as we normally do, it may linger. Perceptual boundaries are thus loosened.

The commonly abused hallucinogenic substances can be classified according to their chemical structure. All these hallucinogens are organic compounds, and some are found in nature. Hallucinogenic drugs can be placed in two major groups. The first is known as the indole-type hallucinogens. This family of hallucinogens has in common some structural similarities to the NEUROTRANSMITTER SEROTONIN, suggesting that their mechanism of action could involve the disruption of or some alteration in neurotransmission in NEURONS (nerve cells) that use serotonin as the chemical messenger. The indole-type hallucinogens include lysergic acid derivatives such as LYSERGIC ACID DIETHYLAMIDE (LSD) and other compounds that have structural similarities, such as DIMETHYLTRYPTAMINE (DMT), PSILOCYBIN, and psilocin (see Figure 1).

The second major group of hallucinogens is the substituted phenethylamines (see Figure 2). These are MESCALINE, 2,5-dimethoxy-4-methylamphetamine (DOM or STP), 3,4-methylenedioxyamphetamine, (MDA), and 3,4-methylenedioxymethamphetamine (MDMA or ecstasy). These hallucinogens are structurally related to the phenethylamine-type neurotransmitters, NOREPINEPHRINE, epinephrine, and DOPAMINE. As with the indole-type hallucinogens, the structural similarities of the phenethylamine-type hallucinogens to these natural neurotransmitters may indicate that at least some of their effects involve interactions with systems that use these neurotransmitters. DOM, MDA, and MDMA are synthesized compounds that have structural similarities to the psychostimulant AMPHETAMINE. Thus, they also have some stimulant properties aside from their hallucinogenic activity. They have inaccurately been called psychotomimetic amphetamines, and they are sometimes referred to as stimulant-hallucinogens. It should be pointed out that there are literally hundreds of analogs of the above compounds that

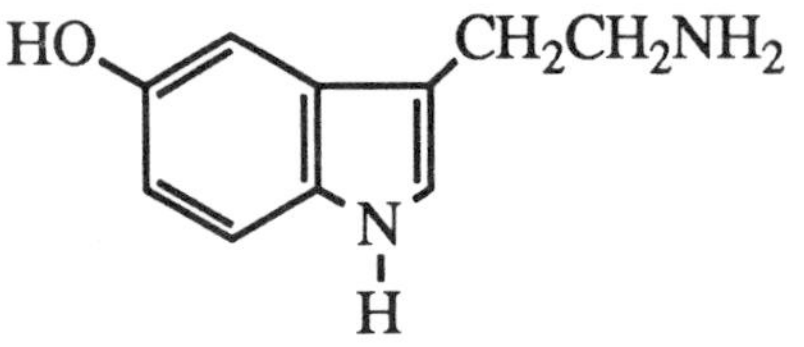

SEROTONIN
(neurotransmitter)

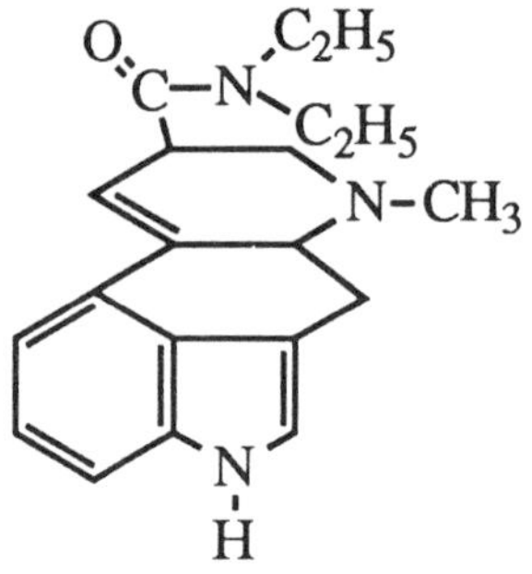

LSD

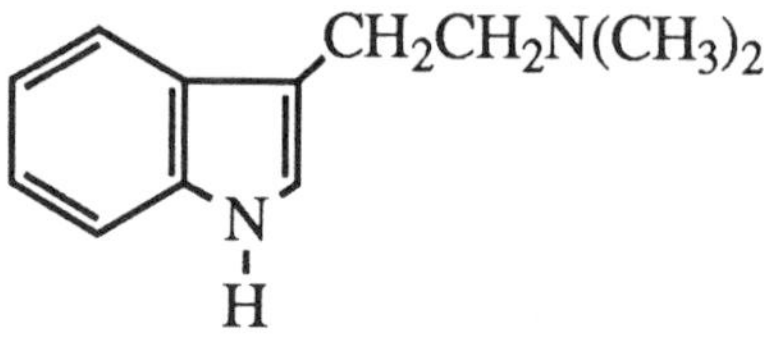

DMT

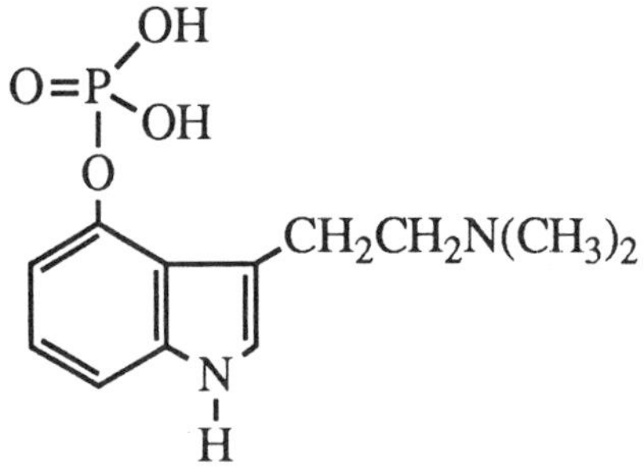

PSILOCYBIN

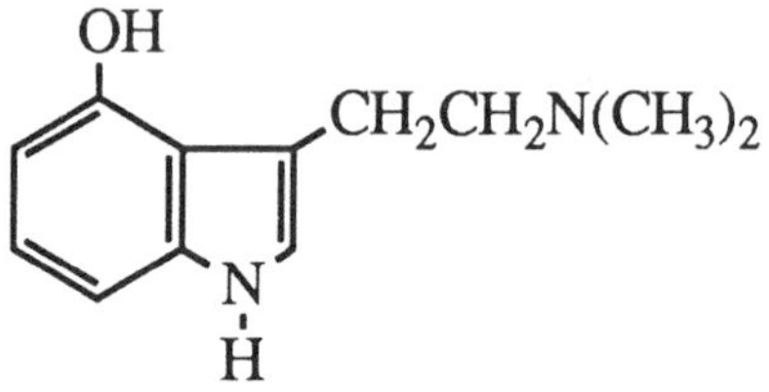

PSILOCIN

Figure 1
Indole-type Hallucinogens

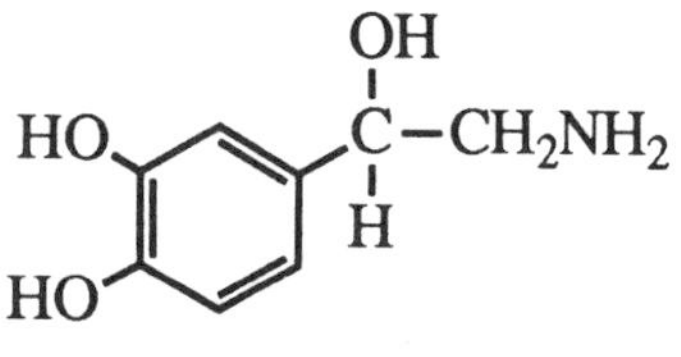

NOREPINEPHRINE
(neurotransmitter)

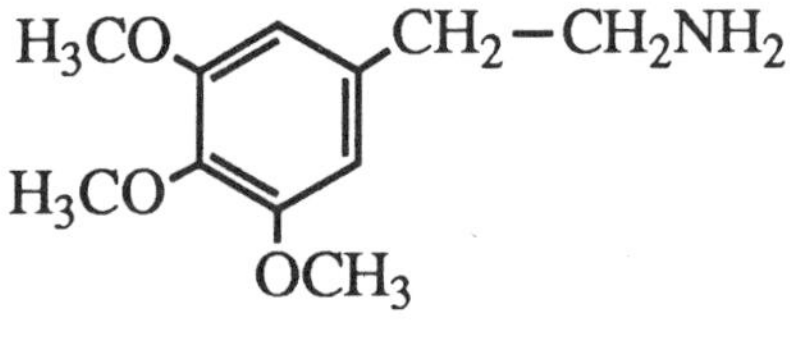

MESCALINE

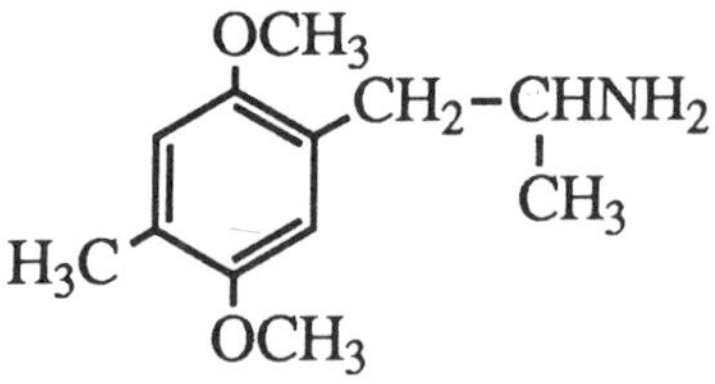

DOM

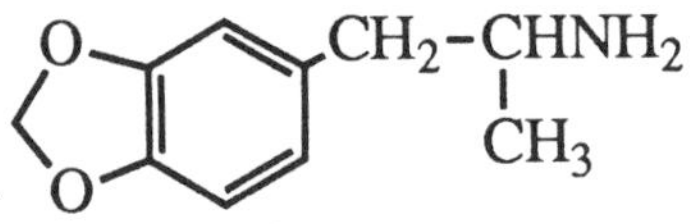

Figure 2
Substituted Phenethylamines

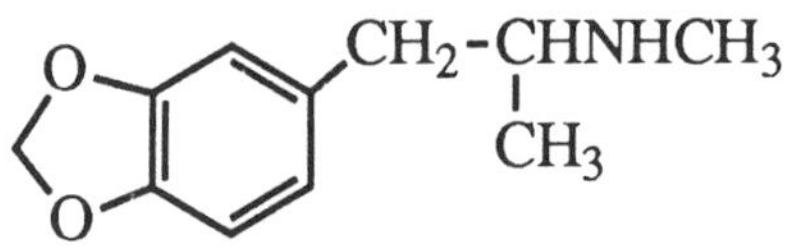

MDA

MDMA

have been synthesized and sometimes are found on the street, the so-called DESIGNER DRUGS.

The overall psychological effects of the hallucinogens are quite similar—but the rate of onset, duration of action, and absolute intensity of the effects can differ. As the various hallucinogens differ widely in potency and in the duration of their effects, some of the apparent qualitative differences between hallucinogens may be due, at least in part, to the amount of drug ingested. Aside from their behavioral effects, the hallucinogens also possess significant autonomic activity, meaning that they can affect the sympathetic and parasympathetic nervous systems. The autonomic effects can include marked pupillary dilation and exaggerated reflexes. There may be increases in blood pressure, heart rate, and body temperature. Some of the hallucinogens may initially cause nausea. These autonomic effects of the hallucinogens are variable and may be due, at least in part, to the anxiety state of the user. Acute adverse reactions include panic attacks and self-destructive behavior.

(SEE ALSO: *Ayahuasca*; *Complications: Mental Disorders*; *Hallucinogenic Plants*; *Ibogaine*; *Plants, Drugs from*)

BIBLIOGRAPHY

FREEDMAN, D. X. (1986). Hallucinogenic drug research—if so, so what?: Symposium summary and commentary. *Pharmacol. Biochem. Behav., 24,* 407–415.

GLENNON, R. A. (1987). Psychoactive phenylisopropylamines. In H. Y. Meltzer (Ed.), *Psychopharmacology: The third generation of progress.* New York: Raven Press.

GRINSPOON, L., & BAKALAR, J. B. (1979). *Psychedelic drugs reconsidered.* New York: Basic Books.

JACOBS, B. L. (1987). How hallucinogenic drugs work. *American Scientist, 75,* 386–392.

JACOBS, B. L. (1984). *Hallucinogens: Neurochemical, behavioral and clinical perspectives.* New York: Raven Press.

JAFFE, J. H. (1990). Drug addiction and drug abuse. In A. G. Gilman et al. (Eds.), *Goodman and Gilman's the pharmacological basis of therapeutics,* 8th ed. New York: Pergamon.

SHULGIN, A., & SHULGIN, A. (1991). *PIHKAL: A chemical love story.* Berkeley, CA: Transform Press.

SIEGEL, R. K. (1977). Hallucinations. *Scientific American, 237,* 132–140.

SIEGEL, R. K., & WEST, L. J. (EDS.). (1975). *Hallucinations: Behavior, experience and theory.* New York: Wiley.

WEIL, A. (1972). *The natural mind.* Boston: Houghton Mifflin.

DANIEL X. FREEDMAN
R. N. PECHNICK

HARM REDUCTION *See* Needle and Syringe Exchanges; Netherlands, Drug Use in the; Policy Alternatives.

HARRISON NARCOTICS ACT OF 1914

The first international drug-control initiative, the 1909 SHANGHAI OPIUM COMMISSION, brought the international community together in efforts to curb the illicit traffic and consumption of OPIUM, a NARCOTIC drug. The Shanghai Commission encouraged participants to enact national legislation that would address the problem of narcotics in their own countries. Representatives of several countries met at the Hague at conferences in 1911 and 1913.

During this period, the U.S. Congress became aware of public opinion favoring PROHIBITION of all "moral evils," especially alcohol and drugs. New York Representative Francis B. Harrison, encouraged by both the Shanghai Commission's directive to enact national legislation to curb narcotics and the reformists in the Progressive movement in the United States who wanted to eradicate drug use completely, introduced two measures—one to prohibit the importation and nonmedical use of opium and one to regulate the production of opium in the United States. Congress enacted the Harrison Act in December 1914 with minimal debate because public opinion considered its passage necessary to combat the "evils" of drugs.

PROVISIONS OF THE HARRISON ACT

Congress regulated drugs by imposing licensing requirements on manufacturers, distributors, sellers, importers, producers, compounders, and dispensers. The Harrison Act required these parties to register with the director of Internal Revenue, within the Treasury Department, and to pay a gradually increasing occupational tax. Congress wanted to monitor the flow of opium and COCA leaves so that government authorities would have records of any transaction involving these drugs. They would be allowed only for limited medical and scientific purposes. Those individuals found in violation of the act faced a maximum penalty of five years in jail, a 2,000 dollar fine, or both.

TREASURY DEPARTMENT REGULATIONS

Congress intended the Harrison Act to generate revenue by imposing taxes on parties involved in the trade, sale, and distribution of drugs. As a result, Congress entrusted enforcement responsibility to the Treasury Department, in particular the Internal Revenue Service and subsequently the Narcotics Unit of the Bureau of Prohibition. The Treasury Department attempted to limit narcotics to medical and scientific use and prevent their illegal diversion by physicians and druggists. The Harrison Act required pharmacists to review prescriptions to determine whether the quantity was unusually large—that is, a suspicious or coerced prescription.

Sales and transfers of narcotics could only be made pursuant to official order forms obtained from the director of Internal Revenue. District offices of the Internal Revenue Service maintained these records for two years. The act permitted a few notable exceptions to form filings. For example, qualified practitioners (physicians, dentists, and veterinarians) could prescribe or dispense narcotics to patients without completing the order forms but were required to maintain records of all the substances distributed. Druggists could also fill lawful prescriptions without completing order forms.

The Treasury Department interpreted the Harrison Act to prohibit drug addicts from obtaining narcotics. Addicts were prohibited from registering and could receive narcotics only through a licensed physician, dentist, or veterinarian. The Treasury Department regulations also prohibited physicians from maintaining a patient-addict on narcotics, a practice frequently used to help addicts avoid severe WITH-DRAWAL pain while they were gradually weaned from narcotic DEPENDENCE. The Treasury Department interpreted possession of narcotics as prima facie evidence of a Harrison Act violation, and the burden of proof shifted to the suspect, who had to document that the narcotics were obtained legally.

The Treasury Department enforced the Harrison Act primarily through warnings. At times, however, the department charged physicians and druggists with conspiracy when authorities arrested an individual who possessed narcotics without a prescription made in good faith, and a connection could be made that the physician or the druggist provided the narcotics.

THE HARRISON ACT AND U.S. DRUG POLICY

Many critics of the Harrison Act argue that the legislation created more problems than it solved. In particular, they charge that the measure failed to eradicate the narcotics problem, primarily because it failed to prohibit the sale and distribution of MARI-JUANA. In addition, detractors argue that the act did not resolve the issue of whether drug addicts should be treated as criminals or as patients requiring medical treatment. They also contend that the courts hampered the Treasury Department's enforcement authority. Specifically, courts prohibited the Treasury Department from seizing narcotics, interpreting the Harrison Act to serve as a revenue, rather than as a penal, measure. After passage of the Harrison Act, illicit use of narcotics increased initially as a result of these omissions or ambiguities.

Despite these criticisms, the Harrison Act is significant because it led to a national focus on the dangers of narcotics and drug abuse. Most important, the Harrison Act served as the impetus for further legislation, such as the 1970 Controlled Substances Act, all of which attempt to combat the illegal sale, distribution, and consumption of narcotics and other abusable substances in the United States, while ensuring their availability for medical purposes.

(SEE ALSO: *Anslinger, Harry J., and U.S. Drug Policy; Britain, Drug Use in; Legal Regulation of Drugs and Alcohol; Opioids and Opioid Control; Psychotropic Convention; Rolleston Report; Treatment: History of*)

BIBLIOGRAPHY

ANSLINGER, H. J., & TOMPKINS, W. F. (1953). *The traffic in narcotics.* New York: Funk & Wagnalls.

McWILLIAMS, J. C. (1990). The history of drug control policies in the United States. In J. A. Inciardi (Ed.), *Hand-*

book of drug control in the United States. New York: Greenwood Press.

MUSTO, D. F. (1973). *The American disease: Origins of narcotic control.* New Haven: Yale University Press.

ROBERT T. ANGAROLA
ALAN MINSK

HASHISH   Hashish is the Arabic word for a particular form of CANNABIS SATIVA; it came into English at the end of the sixteenth century. Hashish is the resin derived principally from the flowers, bracts, and young leaves of the female hemp plant. The resin contains cannabinoids—the one of major interest being TETRAHYDROCANNABINOL (THC). The THC content will vary depending upon the composition of the hashish, but often it is about 4 percent or more. Usually the resinous portion is sticky enough to allow the material to be compressed into a wafer or brick. Some preparations contain only the resin and are known as hashish oil. Similar preparations of the resinous material and flowering tops of the plant have been given a variety of names in different regions—*charas* in India, *esvar* in Turkey, *anascha* in areas of the former USSR, *kif* in Morocco and parts of the Middle East.

One of the ways in which hashish is prepared is to boil *Cannabis* leaves in water to which butter has been added. THC, being extremely fat-soluble, binds with the butter, which can then be used for making various confections, cookies, and sweets; these are eaten to obtain the effects of the drug. Although hashish is often taken by mouth, it can also be smoked, just as MARIJUANA is.

Hashish was introduced to the West in the middle of the nineteenth century by a French psychiatrist, Moreau de Tours, who experimented with the drug as a means of understanding the phenomenon of mental illnesses. He not only experimented on himself but on a coterie of friends of considerable literary talent. These included Theophile Gautier, Alexander Dumas, and Charles Baudelaire. This group named themselves "Le Club des Haschschins" or "The Club of Hashish-Eaters." The lurid descriptions of the drug effects by these talented writers no doubt helped popularize the drug. Most of their accounts dwelt on beautiful HALLUCINATIONS and a sense of omnipotence. Doses must have been large, since the effects described are more characteristic of

TABLE 1
Net Hashish Production, in Tons

Source Country		U.S. Measure	Metric Measure
Lebanon	1990	110	100
	1991	600	545
Pakistan	1990	220	200
	1991	220	200
Afghanistan	1990	330	300
	1991	330	300
Morocco	1990	94	85
	1991	94	85
TOTAL	1990	754	685
	1991	1,244	1,130

SOURCE: *International Narcotics Control Strategy Report 1992.*

HALLUCINOGENIC DRUGS than effects experienced by present-day users (smokers) of marijuana.

Hashish was introduced into England at about the same time, by an Irish physician, O'Shaughnessy, who had spent some time in India, where he had become familiar with it. The material was soon hailed as a wonder drug, being used for all sorts of complaints: PAIN, muscle spasms, convulsions, migraine headaches, and inflamed tonsils. As most of the preparations were weak and the doses used were small, any beneficial effects might be attributable to a placebo effect.

A preparation, Tilden's Extract of Cannabis Indica, became a popular remedy in the United States in the 1850s. An amateur pharmacologist, Fitz Hugh Ludlow, used this preparation for self-experiments in which he was able to explore its hallucinogenic properties. He may have become somewhat dependent on hashish but finally gave it up. His descriptions of the effects of the drug were similar to what had previously been experienced by Asian users: euphoria and uncontrollable laughter; altered perceptions of space, time, vision, and hearing; synesthesias and depersonalization.

Hashish is currently the most potent of all *Cannabis* preparations: A lot of drug effect is packed into a small parcel. Regulation of the dose is difficult because of its variable potency, and labels for street drugs are notoriously unreliable, however. What may be sold as hashish may often be closer to ordinary marijuana in potency.

(SEE ALSO: *Amotivational Syndrome; Creativity and Drugs; Epidemics of Drug Use; Marihuana Commission; Plants, Drugs from*)

BIBLIOGRAPHY

MOREAU, J. J. (1973). Hashish. In H. Peters & G. G. Nahas (Eds.), *Hashish and mental illness.* New York: Raven Press.

NAHAS, G. G. (1973). *Marihuana—deceptive weed.* New York: Raven Press.

LEO E. HOLLISTER

HAZELDEN CLINICS

HAZELDEN CLINICS Hazelden was established in 1949; it was one of the pioneering programs that developed the approach to treatment that is now widely known as the MINNESOTA MODEL. Today, the Hazelden residential rehabilitation program (Center City, Minnesota, with a smaller clinic in West Palm Beach, Florida) provides Minnesota Model treatment for 2,000 adult alcoholic, drug-dependent men and women each year. Hazelden-Pioneer House, in Plymouth, Minnesota, treats 300 ADOLESCENTS and young adults annually. A private, nonprofit organization, Hazelden Foundation serves a national and international clientle through its TREATMENT programs (inpatient, outpatient, intermediate, extended care, and family programming), literature publication, counselor training, continuing education, and consulting services.

Residential treatment consists of an open-ended stay lasting an average of twenty-eight days. Clients are placed in the detoxification unit for a minimum of twenty-four hours, following admission, where they receive a physical examination and are monitored for withdrawal. There are two female units and five male units of twenty-two patients each at the Center City facility. Primary rehabilitation is done by a staff of trained counselors who are also working their own programs of recovery. On a unit, clients are at various stages of rehabilitation with new clients transferring in while others are being discharged. During the first week of primary rehabilitation, the staff concentrates on problem identification, guided by assessments of psychological, spiritual, health, social activities, and chemical-use profiles. Clients are encouraged to relate to the unit peer group and look at consequences of their chemical use. After the client's problem is identified, an individual treatment plan is formulated both for and with the client.

Goals, objectives, and methods are identified in the treatment plan and progress in meeting these expectations is monitored. Clients must participate in treatment unit activities and relate to other clients by helping and being helped in meeting treatment goals. Candidness, honesty, and solidarity are encouraged among the clients. Those nearing program completion model expected behaviors for clients more recently assigned to the unit.

Clients are taught to recognize the DISEASE CONCEPT of chemical dependency, to accept the need for abstinence from all mood-altering chemicals and to become familiar with ALCOHOLICS ANONYMOUS (AA) through reading AA's "Big Book" and focusing on the first five of AA's TWELVE STEPS. Ability to change and the AA program of living with its attention to personal and spiritual discovery are emphasized. AA's Serenity Prayer is shared after all group meetings. AA's fourth and fifth steps are completed before the staff approves the client's discharge. The staff works with clients to develop a mutually agreed on aftercare plan that will promote recovery; then client postdischarge arrangements are made.

The Center City adult population is primarily middle to upper-middle class, although all socioeconomic levels are represented. Approximately 35 percent are college graduates and the same proportion are married. Almost 60 percent of the clients remain abstinent from alcohol and drugs during the year following treatment. Of those not abstaining, 10 percent have an isolated slip but return to abstinence. Posttreatment AA participation and an active religious relationship are predictors of posttreatment abstinence.

(SEE ALSO: *Betty Ford Center; Sobriety; Treatment*)

BIBLIOGRAPHY

LAUNDERGAN, J. C. (1982). *Easy does it: Alcoholism treatment outcomes, Hazelden and the Minnesota Model.* Center City, MN: Hazelden Educational Services.

McELRATH, D. (1987). *Hazelden: A spiritual odyssey.* Center City, MN: Hazelden Educational Services.

J. CLARK LAUNDERGAN

HEART DAMAGE

HEART DAMAGE *See* Alcohol; Cocaine; Tobacco: Medical Complications.

HEMP In the narrow sense, hemp refers to a fiber derived from certain strains of CANNABIS SATIVA, a bushy herb that originated in ASIA. In the broader sense, it also denotes the other use of the plant, as a source of MARIJUANA. Although *Cannabis sativa* is generally considered to be a single species, two genetic strains show considerable differences. One is used for fiber production and has been so used for centuries to make rope, floor coverings, and cloth. Hemp plants have been grown for this purpose as commercial crops in Asia and even in colonial America; during World War II, they were grown in the midwestern United States when the Asian supply was unavailable.

Figure 1
Hemp Plant

The other strain of the hemp plant produces a poor fiber but has a relatively high drug content; it is used for its PSYCHOACTIVE effect. Near the end of the nineteenth century, the Indian Hemp Drug Commission (1895) produced one of the first major assessments of *Cannabis* as a drug, finding it not a major health hazard. Consequently, it remains in legal use in India for both medicinal and social purposes, where it is called BHANG.

(SEE ALSO: *Plants, Drugs from*)

LEO E. HOLLISTER

HEROIN MORPHINE was first identified as the pain-relieving active ingredient in OPIUM in 1806. But morphine was not free of the habit-forming and toxic effects of opium. By the late nineteenth century, the idea of modifying molecules to change their pharmacological actions was well established. It

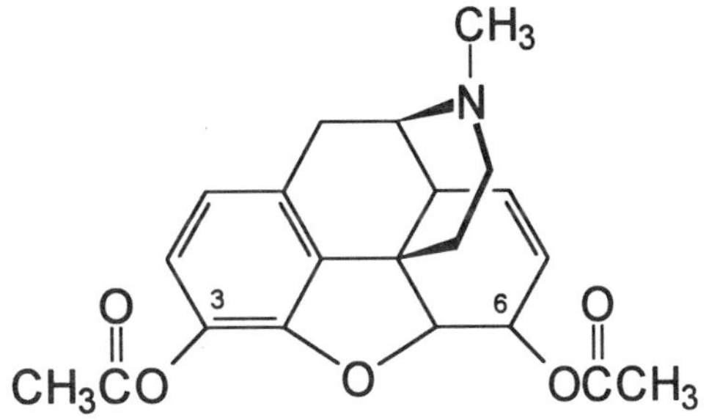

Figure 1
Heroin

seemed quite reasonable to use this approach to develop new chemical entities that might be free of the problems seen with morphine. In Germany, in 1898, H. Dresser introduced such a new drug—3,6-diacetylmorphine—into medical use; it was named there by the Bayer Company, which produced and marketed it, named it heroin (presumably from *heroisch*, meaning "heroical"), because it was more potent than morphine.

Although heroin is structurally very similar to morphine, it was hoped that it would relieve PAIN without the tendency to produce ADDICTION. Turn-of-the-century medical writings and advertisements, both in Europe and the United States, claimed that heroin was effective for treating pain and cough. Many suggested that it was less toxic than morphine and was nonaddictive. A few even suggested that heroin could be a nonaddicting cure for the morphine habit. Clearly, this was not the case, and within a year or two of its introduction, most of the medical community knew so. By the 1920s, heroin had become the most widely abused of the OPIATES.

PHARMACOLOGY

Heroin is a white powder that is readily soluble in water. The introduction of just two esters onto the morphine molecule changes the physical properties of the substance such that there is a significant increase in solubility, permitting solutions with increased drug concentrations. A more subtle advantage of heroin is its greater potency compared to morphine. The volume of drug injected may be particularly important when high doses are used. Thus, 1 gram of heroin will produce the effects of 2 to 3 grams of morphine; by converting morphine to heroin, producers increase both the potency and the value of the drug.

Following injection, heroin is very potent, with the ability to cross the blood—brain barrier and enter the brain. This barrier results from a unique arrangement of cells around blood vessels within the brain, which limits the free movement of compounds. Many factors contribute to the barrier—in general, the less polar a drug, the more rapidly it enters the brain. Heroin, however, has a very short half-life in the blood (amount of time that half the drug remains). It is rapidly degraded by esterases, the enzymes that break ester bonds. The acetyl group at the 3-position of the molecule is far more sensitive to these enzymes than the acetyl group at the 6-position. Indeed, the 3-acetyl group is attacked almost immediately after injection and, within several minutes, virtually all the heroin is converted to a metabolite, 6-acetylmorphine. The remaining acetyl group at the 6-position is also lost, but at a slower rate. Loss of both acetyl groups generates morphine. It is believed that a combination of 6-acetylmorphine and morphine is responsible for the actions of heroin.

MEDICINAL USE

The pharmacology of heroin is virtually identical to that of morphine. This probably reflects its rapid conversion to 6-acetylmorphine and morphine. Detailed studies comparing the actions of heroin and morphine in cancer patients with severe pain have shown very little difference between the two agents, other than simple potency. Heroin may have a slightly more rapid onset of action than morphine and it is certainly two to three times as potent (presumably due to its greater facility in crossing the blood—brain barrier). This difference in potency is lost with oral administration. The pain relief (analgesia) from both agents is comparable when the doses are adjusted appropriately. At equally effective ANALGESIC doses, even the euphoria seen with heroin is virtually identical to that of morphine. From the clinical point of view, there is little difference between one drug and the other. Both are effective analgesics and can be used beneficially in the treatment of severe pain. Heroin is more soluble, which makes it somewhat easier to give large doses by injection, with smaller volumes needed. Many of the similar semisynthetic agents, such as HYDROMORPHONE, however, are many times more potent than heroin and offer even greater advantages.

One widespread use of heroin in the United Kingdom was in the early formulations of Brompton's Cocktail, a mixture of drugs designed to relieve severe pain in terminal cancer patients. The heroin employed in the original formula is now typically replaced with morphine without any loss in effectiveness. For many years, some groups have maintained that heroin is more effective in the relief of cancer pain than morphine is. Careful clinical studies show that this is not true, but the most important issue is using an appropriate dose. Thus, heroin offers no major advantage over morphine from the medical perspective.

STREET HEROIN

Since heroin has no approved medical indications in the United States, it is only available and used illicitly. The marked variability of its purity and the use of a wide variety of other substances and drugs to "cut" street heroin poses a major problem. This inability to know what is included in each drug sale makes the street drug more than doubly dangerous. Typically, heroin is administered intravenously, which provides a rapid "rush," a euphoria, which is thought to be the important component of heroin's addictive properties. It can be injected under the skin (subcutaneously, SC) or deep into the muscle (intramuscularly, IM). Multiple intravenous injections leave marks, called tracks, in a much-used injection site, which often indicate that a person is abusing drugs; but heroin can also be heated and the vapors inhaled through a straw (called "chasing the dragon"). It can also be smoked in a cigarette. While the heat tends to destroy some of the drug, if the preparation is pure enough, a sufficient amount can be inhaled to produce the typical opiate effect.

Heroin use is associated with TOLERANCE AND DEPENDENCE. Chronic use of the drug leads to a decreased sensitivity toward its euphoric and analgesic actions, as well as to dependence. Like morphine, the duration of action of heroin is approximately 4 to 6 hours. Thus, addicts must take the drug several times a day to prevent the appearance of WITHDRAWAL signs. Many believe that the need to continue taking the drug to avoid withdrawal enhances its addictive potential.

Patients taking opiates medicinally can be taken off them gradually, without problems. Lowering the opiate dose by 20 to 25 percent daily for 2 or 3 days will prevent severe withdrawal discomfort and still permit rapid taper off the drug. Abrupt withdrawal of all of the drug is very different—and leads to a well-defined abstinence syndrome that is very similar for both heroin and morphine. It includes eye tearing, yawning, and sweating after about 8 to 12 hours past the last dose. As time goes on, people develop restlessness, dilated pupils, irritability, diarrhea, abdominal cramps, and periodic waves of gooseflesh. The term *cold turkey* is now used to describe abrupt withdrawal with the associated gooseflesh. The heroin withdrawal syndrome peaks between 2 and 3 days after stopping the drug, and symptoms usually disappear within 7 to 10 days, although some low-level symptoms may persist for many weeks. Babies of mothers dependent on opiates are born dependent, and special care must be taken to help them withdraw during their first weeks. Medically, although miserable, heroin withdrawal is seldom life threatening—unlike withdrawal from alcohol, which can sometimes be fatal.

OVERDOSE

Overdosing is a common problem among heroin addicts. The reason is not always clear, but wide variation in the purity of the street drug can make it difficult for the addict to judge a dose. Some impurities used to cut the drug may be toxic themselves. With OVERDOSE, a person becomes stuporous and difficult to arouse. Pupils are typically small and the skin may be cold and clammy. Seizures may occur, particularly in children or babies. Breathing becomes slow, and cyanosis—seen as a darkening of the lips to a bluish color—may develop, indicating inadequate levels of oxygen in the blood. With respiratory depression, blood pressure may then fall. These last two signs are serious, since most people who die from overdose, die from respiratory failure. Complicating the problem is the fact that many addicts may have taken other drugs, used alcohol, and so on. Some of them may have been taken on purpose, and some may have been a part of the street drug.

NALOXONE can readily reverse some opiate problems, since it is a potent opiate ANTAGONIST. This drug binds to opiate RECEPTORS and can reverse morphine and heroin actions. The appropriate dose may be a problem, however, since naloxone can also precipitate a severe abstinence syndrome in a dependent person.

(SEE ALSO: *Addiction: Concepts and Definitions; International Drug Supply Systems; Methadone Maintenance; Opioids: Complications and Withdrawal; Treatment: History of*)

BIBLIOGRAPHY

JAFFE, J. H. (1990). Drug addiction and drug abuse. In A. G. Gilman et al. (Eds.), *Goodman and Gilman's the pharmacological basis of therapeutics*, 8th ed. New York: Pergamon.

JAFFE, J. H., & MARTIN, W. R. (1990). Opioid analgesics and antagonists. In A. G. Gilman et al. (Eds.), *Goodman and Gilman's the pharmacological basis of therapeutics*, 8th ed. New York: Pergamon.

GAVRIL W. PASTERNAK

HEROIN: THE BRITISH SYSTEM What is sometimes referred to as the British "system" of drug control is not really a system; rather, it is a set of principles and programs that represent one form of societal response to HEROIN use and OPIATE DEPENDENCE. The principles encompass the idea that government ought to offer public-health and medical programs that will help contain Britain's heroin problem, in addition to its response in the form of law enforcement. In BRITAIN, the concept of punishing heroin-dependent individuals for dependence as such is as alien as punishing people for becoming infected with syphilis or needing insulin for diabetes.

A key element in this system is allowing medical practitioners to provide maintenance doses of OPIATES or opioid drugs (sometimes including heroin as well as METHADONE and other opioids) when a diagnosis of heroin dependence can be substantiated. The initial programmatic efforts allowed for the prescribing of such drugs by general medical practitioners; but more recently, responsibility for treatment of opioid-dependent persons has shifted to government-run specialized Drug Dependency Units.

BACKGROUND

Drug control in Britain was established between 1910 and 1930, with a solid grounding in public health and medical practice. This British approach to drug problems as public-health problems seemed especially attractive as an alternative to U.S. drug prohibition policies, even when the heroin problem in the United States was relatively small, back before 1960. Thus, beginning in the late 1940s, some Americans started to advocate the use of the British system in the United States—that is, a nonpunitive, public-health approach to the treatment of drug dependence, especially dependence on heroin.

In 1960, the drug problem was essentially a non-issue in the political life of Britain, although the structures for control in the two countries remained very different. In the United States, a prohibitionist policy continued in place whereby criminal penalties were imposed for heroin possession and use—and sometimes for being addicted to heroin. Physicians rarely treated opiate addicts and could not legally provide a known addict with opiates on a maintenance basis. As a result, from early in the twentieth century, virtually all heroin addicts purchased supplies from illegal heroin sellers. With the exception of a brief time during which maintenance programs were available, relatively few addicts sought drug treatment from doctors, and treatment for heroin dependence often was available only at two federal narcotic hospitals and select public and private facilities. In NEW YORK and CALIFORNIA, in particular, large numbers of heroin abusers were arrested and imprisoned for heroin sales, for possession, or for other crimes sometimes committed to gain funds to purchase illegal heroin (e.g., robbery, burglary).

In contrast, by 1960, Britain had had many years of experience with a "medical" or "public-health" policy for controlling heroin and opiates (originating with the ROLLESTON REPORT of 1926). Fewer than 100 heroin addicts and fewer than 500 abusers of all drugs were known in Britain in 1960. Persons identified by a doctor as being addicted to heroin or other dangerous drugs could be (and usually were) treated by a private practitioner. The physician was required to notify the Home Office of the names of the addicts but was at liberty to prescribe heroin or opiates for them in any amounts for long time periods. Their treatment became funded by the National Health Service after World War II, like any other medical service. No other treatment (at a clinic, hospital, or nonmedical facility) was available. Penalties for the illegal sale of heroin or opiates carried sanctions of less than a year and were rarely imposed. Few British prisoners were heroin addicts.

British drug policy has been and continues to be set primarily by Home Office staff in collaboration with leading physicians and addiction specialists. British law-enforcement and criminal-justice practitioners were largely excluded from policymaking—whereas their counterparts in the United States have a primary role in formulating American drug policy. Following the Rolleston precedent, several special committees issued reports establishing the basic directions of British drug policy. The first Brain Committee (1958) reaffirmed the Rolleston recommendation to provide heroin and allow maintenance doses of opiates; it opposed U.S.-sponsored proposals to prohibit heroin manufacture in Britain.

CHANGING MEDICAL POLICIES ON DRUG CONTROL

The situation changed in the early 1960s, however, and, based on recommendations of the second Brain Committee (1964), clinics for controlling and containing the heroin problem were implemented under the Dangerous Drug Act Regulations in 1968. Responsibility for the treatment of addicts generally was shifted from general practitioners (GPs) to Drug Dependency Units (DDUs). When a heroin abuser seeks treatment from a GP, however, the doctor can refuse treatment, refer the patient to a DDU, or provide declining methadone doses over six months (called long-term detoxification in the United States) or provide regular methadone maintenance (although this is rarely done by a GP).

The DDUs or drug clinics provide a range of services funded by the National Health Service. In 1989, 35 DDUs operated in Britain and were directed by consulting psychiatrists who specialized in addiction treatment and prescribing. In smaller towns without clinics, one or two GPs can be licensed by the Home Office to provide treatment for addicts in the area. New applicants are interviewed and their urine tested to verify opiate use. The clinic physician develops a treatment plan with the patient, arranges weekly conferences, and mails the prescription directly to a local pharmacy; it will be filled for the client on a daily basis. The Home Office also convenes meetings with several DDU directors to discuss common policies and practices, and to recommend approval or removal of licenses, when necessary, for physicians to prescribe dangerous drugs.

When the DDUs opened, most clinics made decisions to shift patients receiving prescriptions for injectable heroin onto injectable methadone. The pharmacist dispensed needles, syringes, and ampoules of methadone.

Over the period 1975 to 1983, many clinic directors shifted most patients from injectable to oral methadone maintenance. In the early 1980s, as illegal supplies of heroin became common in British cities, many clinics shifted away from oral methadone maintenance. Instead, the treatment policy at several clinics was to provide gradual withdrawal (detoxification in the United States); rarely were patients provided with long-term maintenance doses. As AIDS was tied to shared needles and syringes by injecting addicts, prevention became an important subgoal of drug treatment; however, new emphasis was then placed on oral methadone maintenance. In the early 1990s, the DDUs had heroin-abusing clients, many of whom received gradual reduction (detoxification) and others who received maintenance on methadone. Relatively few received prescriptions for injectable methadone or heroin, even though DDU doctors could legally and appropriately provide such services.

A continuing controversy within Britain in the 1990s has been whether the clinic system could stem or contain the heroin problem, and whether the clinic's shift away from prescribing heroin and injectable drugs contributed to the growth of black-market heroin. In discussion groups, some experts argued that many black-market heroin users would seek treatment if the clinics returned to prescribing injectable heroin or methadone. Such a policy also might reduce addict crime and prevent transmission of the AIDS virus. This, however, would change the profile of patients: Clinic directors would have to deal with addicts who have no intention of stopping heroin use.

The British have amended the Dangerous Drug Act several times since 1960, thereby making the illegal sale of heroin, cocaine, and marijuana criminal offenses. Although the vast majority of drug arrestees are only "cautioned," even after repeated instances of offense, many illegal sellers and heroin abusers arrested for robbery, burglary, and theft can be and are imprisoned. Thus, an increasingly larger proportion of British prisoners are heroin addicts. Between 1979 and 1984, seizures of illegal drugs went up tenfold, incarcerated drug offenders went up fourfold, and the consumption of heroin increased by 350 percent—but heroin prices decreased by 20 percent.

Rise of Nonmedical Drug Treatment. The increase in black-market heroin, substantial increases

in heroin abusers who avoid the DDUs, apparent increase in penal sanctioning, and a host of complex issues have led to dissatisfaction with the original British System, with its medical model of drug treatment. Influenced by U.S. therapeutic communities and outpatient local programs that promote a drug-free environment, British social service agencies have begun developing similar programs thereby "reaching out" to clients and providing alternative services in a context that is different from the practice settings dominated by the consulting psychiatrists at DDUs.

Other emerging British programs are increasingly built around a philosophy of "harm reduction." This emphasizes informing people of safer ways to take drugs for those who will continue to do so, helping addicts recognize drug-related problems (e.g., infections or diseases), and making sterile injection equipment and/or drug treatment available with minimal restrictions. The program's staff also suggests alternative ways of altering consciousness or seeking pleasure.

AIDS Prevention. Since the years 1984 to 1985, the British have been international leaders in devising innovative programs to reduce the spread of the AIDS virus. Because of the legal provision of opiates by physicians and DDUs, the sale of syringes was never prohibited nor seriously constrained. Addicts using black-market heroin could always purchase sterile needles cheaply as well as receive instructions on safe injection practices although in some areas pharmacists might refuse to sell them to addicts.

Gerry Stimson, a sociologist who had conducted studies of heroin addicts from 1960 through the 1970s, became a leading government consultant in the 1980s in formulating British AIDS prevention policies. Together with other experts, he recommended establishing syringe exchanges to promote safe disposal of used needles (possibly infected with the AIDS virus) and to reach injecting drug users who avoid the clinics. His subsequent research established the facts that untreated addicts could be attracted to these exchanges but that retention rates were low. Possibly as a result of these efforts, the AIDS infection rate in Britain is much lower than that in many cities of the United States.

Heroin Abuse. After 1960, several major increases in heroin use and abuse occurred in Britain. In the early 1960s, a few British physicians began prescribing large amounts of legal heroin to private patients, some of whom resold it to other people. The number of known heroin abusers grew to 2,240 in 1968 and then increased slowly during the 1970s. In the early 1980s, however, a major increase in illegal importation of heroin to Britain was followed by an epidemic of heroin use in that country—thus, 12,500 heroin abusers were reported to the Home Office in 1984. In the mid-1990s, many heroin abusers avoid clinics and doctors and are not reported to the Home Office. Therefore, the actual number of regular heroin abusers in Britain now is estimated to be between 50,000 to 100,000.

CONCLUSION

Since the 1960s, the British system of drug control has evolved and changed in many important ways. Although the heroin problem expanded dramatically in the 1980s, the major policy decisions of the Rolleston Report have continued to govern the British approach. The British government continues to collaborate closely with medical and public-health experts. Treatment practices have been refined by experience and practical considerations, but not because of imposition by government fiat. Prohibition of heroin did not occur and punishment of drug abusers remains a secondary consideration in British policymaking (but is still a dominant consideration in the United States). Since 1960, the British heroin problem has grown and become complex. Drug-policy and treatment response have become diverse and, therefore, there is less of a clear "system."

In comparison with the situation in the United States, British policymakers and the general public favor public-health considerations over other moral concerns. Some British newspapers do promote "dope fiend" images and demand punitive responses—and the American "drug free" and "just say no" philosophies are often articulated. Nonetheless, British drug policy and funding are primarily directed by medical and public-health specialists. This means that heroin addicts and drug abusers are not as heavily stigmatized as they are in the United States.

The British public accepts the idea of providing heroin and methadone as medicine, has few moral qualms about addicts, and little fear of needles. Lacking the harsh and punitive moral consensus against drugs that prevails in the United States, the British government has considerable latitude to experiment with differing policies, to shift treatment practices to accord with practical experience, and to keep modifying its policy responses to the ever-changing drug scene. Whether the British system could work in the United States, which is much larger and more populous than Great Britain, remains an open question.

(SEE ALSO: *British System of Drug Addiction Treatment; Needle and Syringe Exchanges and HIV/AIDS; Policy Alternatives*)

BIBLIOGRAPHY

PARKER, H., BAKX, K., & NEWCOMBE, R. (1988). *Living with heroin.* Stony Straford, England: Open University Press.

PEARSON, G. (1991). Drug problems and policies in Britain. In J. Q. Wilson & M. Tonry (Eds.), *Crime and justice series* (vol. 14). Chicago: University of Chicago Press.

ROUSE, J. J., & JOHNSON, B. D. (1991). Hidden paradigms of morality in debates about drugs: Historical and policy shifts in British and American drug policies. In J. A. Inciardi (Ed.), *The drug legalization debate.* Beverly Hills: Sage.

BRUCE D. JOHNSON

HEROIN EPIDEMICS *See* Epidemics of Drug Abuse.

HEROIN TREATMENT *See* Methadone Maintenance; Treatment.

HEROIN WITHDRAWAL *See* Opioid Complications and Withdrawal.

HEXANE *See* Inhalants.

HIGH *See* Slang and Jargon.

HIGH SCHOOL SENIOR SURVEY The use of illegal drugs by large numbers of young people in the United States became an issue of considerable concern during the late 1960s and early 1970s. At that time, there were few accurate data available to assess the extent of use on a national basis. As stated by Lloyd Johnston, one of the original architects of the study *Monitoring the Future*, "at present there is relatively little information from national samples on drug behavior and attitudes" (1973:21). Because of this lack of information, beginning in 1975 J. G. Bachman and L. D. Johnston of the University of Michigan initiated *Monitoring the Future: An Ongoing Study of the Lifestyles and Values of Youth*.

One of the major purposes of the study was (and is) to develop an accurate picture of the nature and extent of drug use among young people. An accurate assessment of the amount and extent of illicit drug use in this group is a prerequisite for rational policymaking. Reliable and valid data on prevalence are necessary to determine an appropriate allocation of resources and to prevent or correct misconceptions. Reliable and valid data on trends allow for early detection of emerging problems and make it possible to assess the impact of external events, including historical events and deliberate policies.

In addition, the study was designed to monitor factors that might help explain the observed changes in drug use—that is, it was intended to serve both an epidemiological function (to learn how many young people use drugs) and an etiological function (to find out why young people use drugs). The factors measured included attitudes toward drugs, peer norms and behaviors in regard to drugs, beliefs about the dangers of drugs, perceived availability of drugs, religious attitudes, and various life-style factors. The monitoring of these factors has, among other things, provided the country with valuable information, thus helping it to answer a central policy-making question in its war on drugs—that of the relative importance of supply reduction versus demand reduction in bringing about some of the observed declines in drug use.

STUDY DESIGN

The core feature of the design, an annual survey of each new high school senior class, began with the class of 1975. Each year, approximately 17,000 seniors are surveyed in approximately 130 public and private high schools that have been scientifically selected to provide an accurate, representative cross section of high school seniors throughout the coterminous United States. One limitation of the design is that it does not include in the target population the young men and women who drop out of high

school before graduation, and who make up between 15 and 20 percent of each age group nationally, according to U.S. Census statistics. The omission of high school dropouts does introduce biases in the estimation of certain characteristics of the entire age group, but, because the dropouts are only a small proportion of the entire group, the bias due to their omission is small. Furthermore, since the bias resulting from exclusion of the dropouts usually remains constant from year to year, their omission should introduce little or no bias in estimates of change or trends.

Annual follow-up surveys of high school graduates constitute a second feature of the study design. Beginning with the graduating class of 1976, each class was followed up annually after high school on a continuing basis. Since then, a representative sample of 2,400 individuals from each senior class has been chosen for follow-up. The 2,400 selected respondents are randomly assigned to one of two matching groups, each made up of 1,200 graduates; one group is surveyed on even-numbered calendar years and the other group is surveyed on odd-numbered years. The effect of this approach is that each previous high school class is represented every year, but each individual is asked to participate only every other year. All follow-up surveys are accomplished via mailed questionnaires. In 1991, the project was expanded to include nationally representative samples of students from the eighth and tenth grades as well as from the twelfth grade. It was planned that selected panels of students from these grades would be followed longitudinally, primarily in order to provide a basis for studying the extent to which those who dropped out of school differed from those who remained to graduate. This design expansion is still too recent to have yielded definitive findings.

A question that always arises in the study of sensitive behaviors such as drug use is whether honest reporting can be secured. Considerable inferential evidence, however, strongly suggests that the self-report questions, as used in this study, produce largely valid data. This evidence includes the following points: large proportions of respondents report using illegal substances; various drugs exhibit trends in different ways over time; there are very few missing data in response to questions on drug use, even though respondents are instructed not to answer questions they would prefer not to answer; the high correlations with other behaviors such as grades, delinquency, religious attitudes, and truancy indicate a

high degree of construct validity; a high degree of consistency can be noted over time in individuals' reports (that is, the responses are reliable); and other factors that are discussed in detail in other publications (see Johnston, O'Malley, & Bachman, 1992).

MAJOR FINDINGS

Illicit Drugs. The most dramatic finding that has emerged from the Monitoring the Future surveys has been the decrease between 1980 and 1991 in young Americans involved in the use of illicit drugs.

Annual use of any illicit drug (that is, any use of an illicit drug in the past twelve months) peaked among high school seniors in 1979, when more than half (54%) of all high school seniors reported having used such a drug. This peak occurred following a rise in the late 1970s—from 45 percent in 1975, when the first reliable national data were collected. By 1991, the proportion had fallen to 29 percent, representing less than a third of all seniors but still a remarkably high figure.

The statistics for lifetime prevalence are also dramatic. In the peak year of 1982, 66 percent of the

TABLE 1
Trends in Annual Prevalence of Various Types of Drugs for Twelfth Graders, 1975–1994

										Percent who used in
	Class of 1975	Class of 1976	Class of 1977	Class of 1978	Class of 1979	Class of 1980	Class of 1981	Class of 1982	Class of 1983	Class of 1984
Approx. N =	9400	15400	17100	17800	15500	15900	17500	17700	16300	15900
Any Illicit Drug[a,b]	45.0	48.1	51.1	53.8	54.2	53.1	52.1	49.4	47.4	45.8
Any Illicit Drug Other Than Marijuana[b,c]	26.2	25.4	26.0	27.1	28.2	30.4	34.0	30.1	28.4	28.0
Marijuana/Hashish	40.0	44.5	47.6	50.2	50.8	48.8	46.1	44.3	42.3	40.0
Inhalants[d]	—	3.0	3.7	4.1	5.4	4.6	4.1	4.5	4.3	5.1
Inhalants, Adjusted[d,e]	—	—	—	—	8.9	7.9	6.1	6.6	6.2	7.2
Amyl/Butyl Nitrites[f,g]	—	—	—	—	6.5	5.7	3.7	3.6	3.6	4.0
Hallucinogens	11.2	9.4	8.8	9.6	9.9	9.3	9.0	8.1	7.3	6.5
Hallucinogens, Adjusted[h]	—	—	—	—	11.8	10.4	10.1	9.0	8.3	7.3
LSD	7.2	6.4	5.5	6.3	6.6	6.5	6.5	6.1	5.4	4.7
PCP[f,g]	—	—	—	—	7.0	4.4	3.2	2.2	2.6	2.3
Cocaine	5.6	6.0	7.2	9.0	12.0	12.3	12.4	11.5	11.4	11.6
Crack[i]	—	—	—	—	—	—	—	—	—	—
Other Cocaine[j]	—	—	—	—	—	—	—	—	—	—
Heroin	1.0	0.8	0.8	0.8	0.5	0.5	0.5	0.6	0.6	0.5
Other Opiates[k]	5.7	5.7	6.4	6.0	6.2	6.3	5.9	5.3	5.1	5.2
Stimulants[b,k]	16.2	15.8	16.3	17.1	18.3	20.8	26.0	20.3	17.9	17.7
Crystal Meth. (Ice)[l]	—	—	—	—	—	—	—	—	—	—
Sedatives[k,m]	11.7	10.7	10.8	9.9	9.9	10.3	10.5	9.1	7.9	6.6
Barbiturates[k]	10.7	9.6	9.3	8.1	7.5	6.8	6.6	5.5	5.2	4.9
Methaqualone[k,m]	5.1	4.7	5.2	4.9	5.9	7.2	7.6	6.8	5.4	3.8
Tranquilizers[k]	10.6	10.3	10.8	9.9	9.6	8.7	8.0	7.0	6.9	6.1
Alcohol[n,o]	84.8	85.7	87.0	87.7	88.1	87.9	87.0	86.8	87.3	86.0
Been Drunk[l]	—	—	—	—	—	—	—	—	—	—
Cigarettes	—	—	—	—	—	—	—	—	—	—
Smokeless Tobacco[l]	—	—	—	—	—	—	—	—	—	—
Steroids[l]	—	—	—	—	—	—	—	—	—	—

NOTES: Level of significance of difference between the two most recent classes: s = .05, ss = .01, sss = .001. '—' indicates data not available.
SOURCE: The Monitoring the Future Study, the University of Michigan.

graduating class reported having used an illicit drug at some point in their lifetime. By 1991, that percentage was down by one third, to 44 percent; however, this means that even in the class of 1991, four of every nine graduating seniors admitted using illicit drugs. In the early 1980s, two thirds of the graduating classes had had experience with one or more illicit drugs. This makes for an unprecedented contrast with the drug use by graduation classes of previous decades—for example, the early 1960s.

Among the various illicit drugs, marijuana is the most prevalent. The use of marijuana, as indicated by its annual prevalence, peaked among high school seniors in 1979, when a majority (51%) reported that they had used it in the past twelve months, and it has steadily declined since then, reaching a low of 24 percent in 1991. The annual prevalence, thus cut in half, declined from one in two seniors in the class of 1979 to one in four seniors in the class of 1991. A particularly striking trend in marijuana use occurred between 1975 and 1978, when the proportion of seniors who reported using marijuana on a daily or near-daily basis in the past thirty days increased from 6.0 percent to an unprecedented 10.7 percent. This figure subsequently came down by more than 80 percent and stood at 2 percent in 1991. (On the other hand, considerably more seniors—about one in eleven [9%]—reported in 1991 they had smoked

last twelve months

Class of 1985	Class of 1986	Class of 1987	Class of 1988	Class of 1989	Class of 1990	Class of 1991	Class of 1992	Class of 1993	Class of 1994	'93–'94 change
16000	15200	16300	16300	16700	15200	15000	15800	16300	15400	
46.3	44.3	41.7	38.5	35.4	32.5	29.4	27.1	31.0	35.8	+4.8sss
27.4	25.9	24.1	21.1	20.0	17.9	16.2	14.9	17.1	18.0	+0.9
40.6	38.8	36.3	33.1	29.6	27.0	23.9	21.9	26.0	30.7	+4.7sss
5.7	6.1	6.9	6.5	5.9	6.9	6.6	6.2	7.0	7.7	+0.7
7.5	8.9	8.1	7.1	6.9	7.5	6.9	6.4	7.4	8.2	+0.8
4.0	4.7	2.6	1.7	1.7	1.4	0.9	0.5	0.9	1.1	+0.2
6.3	6.0	6.4	5.5	5.6	5.9	5.8	5.9	7.4	7.6	+0.2
7.6	7.6	6.7	5.8	6.2	6.0	6.1	6.2	7.8	7.8	0.0
4.4	4.5	5.2	4.8	4.9	5.4	5.2	5.6	6.8	6.9	+0.1
2.9	2.4	1.3	1.2	2.4	1.2	1.4	1.4	1.4	1.6	+0.2
13.1	12.7	10.3	7.9	6.5	5.3	3.5	3.1	3.3	3.6	+0.3
—	4.1	3.9	3.1	3.1	1.9	1.5	1.5	1.5	1.9	+0.4
—	—	9.8	7.4	5.2	4.6	3.2	2.6	2.9	3.0	+0.1
0.6	0.5	0.5	0.5	0.6	0.5	0.4	0.6	0.5	0.6	+0.1
5.9	5.2	5.3	4.6	4.4	4.5	3.5	3.3	3.6	3.8	+0.2
15.8	13.4	12.2	10.9	10.8	9.1	8.2	7.1	8.4	9.4	+1.0
—	—	—	—	—	1.3	1.4	1.3	1.7	1.8	+0.1
5.8	5.2	4.1	3.7	3.7	3.6	3.6	2.9	3.4	4.2	+0.8s
4.6	4.2	3.6	3.2	3.3	3.4	3.4	2.8	3.4	4.1	+0.7s
2.8	2.1	1.5	1.3	1.3	0.7	0.5	0.6	0.2	0.8	+0.6s
6.1	5.8	5.5	4.8	3.8	3.5	3.6	2.8	3.5	3.7	+0.2
85.6	84.5	85.7	85.3	82.7	80.6	77.7	76.8	76.0	—	
								72.7	73.0	+0.3
—	—	—	—	—	—	52.7	50.3	49.6	51.7	+2.1
—	—	—	—	—	—	—	—	—	—	
—	—	—	—	—	—	—	—	—	—	
—	—	—	—	1.9	1.7	1.4	1.1	1.2	1.3	+0.1

FOOTNOTES: For superscripts "*a*" to "*o*" see *Table 2*, pp. 552–553, Long-Term Trends in Thirty-Day Prevalence of Various Types of Drugs for Twelfth Graders, 1975–1994.

TABLE 2
Long-Term Trends in Thirty-Day Prevalence of Various Types of Drugs for Twelfth Graders, 1975–1994

										Percent who used in
	Class of 1975	Class of 1976	Class of 1977	Class of 1978	Class of 1979	Class of 1980	Class of 1981	Class of 1982	Class of 1983	Class of 1984
Approx. N =	9400	15400	17100	17800	15500	15900	17500	17700	16300	15900
Any Illicit Drug[a,b]	30.7	34.2	37.6	38.9	38.9	37.2	36.9	32.5	30.5	29.2
Any Illicit Drug Other Than Marijuana[b,c]	15.4	13.9	15.2	15.1	16.8	18.4	21.7	17.0	15.4	15.1
Marijuana/Hashish	27.1	32.2	35.4	37.1	36.5	33.7	31.6	28.5	27.0	25.2
Inhalants[d]	—	0.9	1.3	1.5	1.7	1.4	1.5	1.5	1.7	1.9
Inhalants, Adjusted[d,e]	—	—	—	—	3.2	2.7	2.6	2.5	2.5	2.6
Amyl/Butyl Nitrites[f,g]	—	—	—	—	2.4	1.8	1.4	1.1	1.4	1.4
Hallucinogens	4.7	3.4	4.1	3.9	4.0	3.7	3.7	3.4	2.8	2.6
Hallucinogens, Adjusted[h]	—	—	—	—	5.3	4.4	4.5	4.1	3.5	3.2
LSD	2.3	1.9	2.1	2.1	2.4	2.3	2.6	2.4	1.9	1.5
PCP[f,g]	—	—	—	—	2.4	1.4	1.4	1.0	1.3	1.0
Cocaine	1.9	2.0	2.9	3.9	5.7	5.2	5.8	5.0	4.9	5.8
Crack[j]	—	—	—	—	—	—	—	—	—	—
Other Cocaine[j]	—	—	—	—	—	—	—	—	—	—
Heroin	0.4	0.2	0.3	0.3	0.2	0.2	0.2	0.2	0.2	0.3
Other Opiates[k]	2.1	2.0	2.8	2.1	2.4	2.4	2.1	1.8	1.8	1.8
Stimulants[b,k]	8.6	7.7	8.8	9.9	12.1	15.8	10.7	8.9	8.3	17.7
Crystal Meth. (Ice)[l]	—	—	—	—	—	—	—	—	—	—
Sedatives[k,m]	5.4	4.5	5.1	4.2	4.4	4.8	4.6	3.4	3.0	2.3
Barbiturates[k]	4.7	3.9	4.3	3.2	3.2	2.9	2.6	2.0	2.1	1.7
Methaqualone[b,m]	2.1	1.6	2.3	1.9	2.3	3.3	3.1	2.4	1.8	1.1
Tranquilizers[k]	4.1	4.0	4.6	3.4	3.7	3.1	2.7	2.4	2.5	2.1
Alcohol[n]	68.2	68.3	71.2	72.1	71.8	72.0	70.7	69.7	69.4	67.2
Been Drunk[l]	—	—	—	—	—	—	—	—	—	—
Steroids[l]	—	—	—	—	—	—	—	—	—	—

NOTES: Level of significance of difference between the two most recent classes: s = .05, ss = .01, sss = .001. '—' indicates data not available.

SOURCE: The Monitoring the Future Study, the University of Michigan.

[a]Use of "any illicit drugs" includes any use of marijuana, hallucinogens, cocaine, or heroin, or any use of other opiates, stimulants, barbiturates, methaqualone (excluded since 1990), or tranquilizers not under a doctor's orders.

[b]Beginning in 1982 the question about stimulant use (i.e., amphetamines) was revised to get respondents to exclude the inappropriate reporting of nonprescription stimulants. The prevalence rate dropped slightly as a result of this methodological change.

[c]Use of "other illicit drugs" includes any use of hallucinogens, cocaine, or heroin, or any use of other opiates, stimulants, barbiturates, methaqualone (excluded since 1990), or tranquilizers not under a doctor's orders.

[d]Data based on four questionnaire forms in 1976–1988; N is four-fifths of N indicated. Data based on five questionnaire forms in 1989–1994; N is five-sixths of N indicated.

[e]Adjusted for underreporting of amyl and butyl nitrites.

[f]Data based on a single questionnaire form; N is one-fifth of N indicated in 1979–1988 and one-sixth of N indicated in 1989–1994.

[g]Question text changed slightly in 1987.

Past twelve months

Class of 1985	Class of 1986	Class of 1987	Class of 1988	Class of 1989	Class of 1990	Class of 1991	Class of 1992	Class of 1993	Class of 1994	'93–'94 change
16000	15200	16300	16300	16700	15200	15000	15800	16300	15400	
29.7	27.1	24.7	21.3	19.7	17.2	16.4	14.4	18.3	21.9	+3.6sss
14.9	13.2	11.6	10.0	9.1	8.0	7.1	6.3	7.9	8.8	+0.9
25.7	23.4	21.0	18.0	16.7	14.0	13.8	11.9	15.5	19.0	+3.5sss
2.2	2.5	2.8	2.6	2.3	2.7	2.4	2.3	2.5	2.7	+0.2
3.0	3.2	3.5	3.0	2.7	2.9	2.6	2.5	2.8	2.9	+0.1
1.6	1.3	1.3	0.6	0.6	0.6	0.4	0.3	0.6	0.4	−0.2
2.5	2.5	2.6	2.2	2.2	2.2	2.2	2.1	2.7	3.1	+0.4
3.8	3.5	2.8	2.3	2.9	2.3	2.4	2.3	3.3	3.2	−0.1
1.6	1.7	1.8	1.8	1.8	1.9	1.9	2.0	2.4	2.6	+0.2
1.6	1.3	0.6	0.3	1.4	0.4	0.5	0.6	1.0	0.7	−0.3
6.7	6.2	4.3	3.4	2.8	1.9	1.4	1.3	1.3	1.5	+0.2
—	—	1.3	1.6	1.4	0.7	0.7	0.6	0.7	0.8	+0.1
—	—	4.1	3.2	1.9	1.7	1.2	1.0	1.2	1.3	+0.1
0.3	0.2	0.2	0.2	0.3	0.2	0.2	0.3	0.2	0.3	+0.2
2.3	2.0	1.8	1.6	1.6	1.5	1.1	1.2	1.3	1.5	+0.3
6.8	5.5	5.2	4.6	4.2	3.7	3.2	2.8	3.7	4.0	+0.1
—	—	—	—	—	0.6	0.6	0.5	0.6	0.7	+0.1
2.4	2.2	1.7	1.4	1.6	1.4	1.5	1.2	1.3	1.8	+0.5s
2.0	1.8	1.4	1.2	1.4	1.3	1.4	1.1	1.3	1.7	+0.4s
1.0	0.8	0.6	0.5	0.6	0.2	0.2	0.4	0.1	0.4	+0.3
2.1	2.1	2.0	1.5	1.3	1.2	1.4	1.0	1.2	1.4	+0.2
65.9	65.3	66.4	63.9	60.0	57.1	54.0	51.3	51.0	—	—
								48.6	50.1	+1.5
—	—	—	—	—	—	31.6	29.9	28.9	30.8	+1.9
—	—	—	—	0.8	1.0	0.8	0.6	0.7	0.9	+0.2

b Adjusted for underreporting of PCP.

Data based on a single questionnaire form in 1986; N is one-fifth of N indicated. Data based on two questionnaire forms in 1987–1989; N is two-fifths of N indicated in 1987–1988 and two-sixths of N indicated in 1989. Data based on six questionnaire forms in 1990–1994.

Data based on a single questionnaire form in 1987–1989; N is one-fifth of N indicated in 1987–1988 and one-sixth of N indicated in 1989. Data based on four questionnaire forms in 1990–1994; N is four-sixths of N indicated.

k Only drug use which was not under a doctor's orders is included here.

Data based on two questionnaire forms; N is two-sixths of N indicated. Steroid data based on a single questionnaire form in 1989–1990; N is one-sixth of N indicated in 1989–1990.

m Data based on five questionnaire forms in 1975–1988, six questionnaire forms in 1989, and one questionnaire form in 1990–1994. N is one-sixth of N indicated in 1990–1994.

n Data based on five questionnaire forms in 1975–1988, six questionnaire forms in 1989–1992, three of six questionnaire forms in 1993 (N is one-sixth of N indicated in 1993), and six questionnaire forms in 1994.

o In 1993, the question text was changed slightly in some forms to indicate that a "drink" meant "more than a few sips." The data in the upper line for alcohol came from forms using the original wording, while the data in the lower line came from forms using the revised wording.

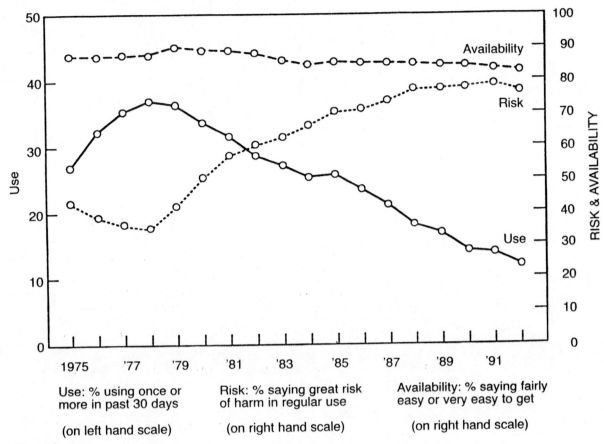

Figure 1
Marijuana: Trends in Perceived Availability, Perceived Risk of Regular Use, and Prevalence of Use in Past Thirty Days for Twelfth Graders

marijuana daily for at least a month at some time in their lives.)

Never as common as marijuana, COCAINE became the drug on which the most attention was focused during the mid-1980s, when the national concern about the drug EPIDEMIC was at its peak. The concern with cocaine was well founded because its use, unlike that of marijuana, had not begun to decline in the very early 1980s. As with marijuana, the daily use of cocaine had increased substantially between 1975 and 1979: Annual prevalence doubled from 5.6 percent to 12.0 percent. Several years followed during which there was little change, with annual prevalence reaching a peak of 13 percent in both 1985 and 1986. A period of decline then ensued during which annual use declined to 3.5 percent in

1991; this was the lowest value recorded since reliable data had begun to be collected in 1975. These data refer to the use of any form of cocaine, including crack cocaine. CRACK cocaine first appeared in the early 1980s and became a significant factor among the illicit drugs in the mid-1980s. It was first assessed on a national basis in 1986, and its annual prevalence among high school seniors at that time was recorded at a disturbingly high 4.1 percent. That first reading turned out to be a peak level, and the use of crack cocaine declined thereafter, reaching 1.5 percent in 1991. Its lifetime prevalence reached a peak of 5.4 percent among the high school class of 1987 but declined to 3.1 percent by 1991.

Although not necessarily illicit drugs, INHALANTS are sometimes used illicitly for the purpose of getting

high. This particular behavior is generally more often seen among younger students rather than among high school seniors. The use of inhalants showed a trend that was slightly upward from its lowest level of 3.0 percent in 1976 (when it was first assessed), to a peak level of 6.9 percent in 1987 and 1990; the figure for 1991 was 6.6 percent. Thus, the use of this class of substance, unlike the use of illicit drugs in general, has not declined much at all. HALLUCINOGENS are the other major class of illicit (or illicitly used) substances that did not evidence declines in the late 1980s and the early 1990s. LSD (LYSERGIC ACID DIETHYLAMIDE) in particular is a very significant exception; its use hardly changed among high school seniors, remaining at the annual prevalence of about 5 percent from 1987 to 1991 after a period of some decline. (The highest recorded value for the use of LSD among this group was 7.2 percent in 1975; the lowest recorded value was 4.4 percent in 1985.) Substances that generally showed declines during the period from the 1970s to the early 1990s include HEROIN, OPIATES other than heroin, AMPHETAMINES, BARBITURATES, and TRANQUILIZERS.

Thus, five classes of illicitly used drugs had a particularly important impact on appreciable proportions of young Americans in their late teens: marijuana, cocaine, amphetamines, LSD, and inhalants. In 1991, they showed annual prevalence rates among high school seniors of 24 percent, 4 percent, 8 percent, 5 percent, and 7 percent, respectively. LSD in particular has become relatively more important with time because its use did not decline while the use of the other substances declined measurably.

Alcohol and Tobacco. The history of the use of the major licit drugs—ALCOHOL and TOBACCO—is rather different than that of the use of most illicit drugs. One significant difference is the extent of the use of alcohol and tobacco. The daily use of cigarettes is far greater than the daily use of any other substance. In 1991, almost one in five (18.5%) high school seniors had smoked one or more cigarettes per day in the past thirty days. About one in thirty (3.6%) seniors had drunk alcohol daily or almost daily, and one in fifty (2.0%) had smoked marijuana that frequently. All other drugs were used on a daily basis by 0.2 percent or less. Although the daily use of alcohol is relatively infrequent among high school seniors, episodic or periodic drinking is more frequent. In 1991, nearly one third (29.8%) of seniors

reported they had had five or more drinks in a row at least once during the past two weeks. (Drinking five or more drinks "in a row" is clearly enough to render the average teenager intoxicated.) This behavior showed some declines in the late 1980s and early 1990s. From 1975 through 1988, the figure for such drinking had been between 35 percent and 41 percent, or consistently more than one in three high school seniors. Between 1988 and 1991, it declined to 30 percent, which represents an encouraging downward trend, although the absolute level remains impressively high.

The daily use of cigarettes peaked in 1977, when 29 percent of high school seniors smoked daily. Most of the decline since then had occurred by 1981, when the figure stood at 20 percent. Throughout the succeeding decade, the figure declined only slightly, to 18.5 percent in 1991. A measure of heavier smoking, the percent of high school seniors who smoked a half pack or more of cigarettes per day, showed a similar trend; it peaked in 1977 at 19 percent, declined to 14 percent by 1981, but was down only to 11 percent in 1991. Thus, although the 1980s showed some declines in cigarette smoking among young Americans, these declines were far more modest than one might have expected. Given the large increases in antismoking legislation, restrictions as to where smoking is allowed, and the general spread of antismoking attitudes, the declines were surprisingly small.

DEMOGRAPHIC DIFFERENCES

Drug use among several groups is monitored in the surveys according to gender. college plans, parental education (an indicator of socioeconomic status), geographical region, population density, and racial or ethnic identification.

Gender. Male adolescents are more likely than female adolescents to use most illicit drugs, and the differences tend to be largest at the higher frequency levels. In 1991, for example, 3.0 percent of male high school seniors reported that they were using marijuana daily, versus 0.9 percent of female seniors. Female ADOLESCENTS are as likely as (or slightly more likely than) male adolescents to use amphetamines, barbiturates, and tranquilizers. As of 1991, there was very little difference between the genders in annual use of alcohol. Large gender differences, however, are seen among high school seniors in the

prevalence of occasions of heavy drinking (38% for male adolescents versus 21% for female adolescents in 1991); thus, as with heavy use of illicit drugs, heavy use of alcohol is more likely among male adolescents than it is among female adolescents. This gender difference is somewhat smaller than the one obtained in 1975, when the figures were 49 percent and 26 percent, respectively. The narrowing of the difference is primarily attributable to the greater decrease in heavy drinking among male adolescents than among female adolescents. In general, there is not much difference between male and female high school seniors where cigarette smoking is concerned. As with most drugs, the greater difference is seen among heavy smokers, but even so the difference is rather small: in 1991, 12 percent of male seniors reported smoking at the rate of a half pack or more per day, versus 10 percent of female seniors.

College-Bound versus Non-College-Bound. Non-college-bound seniors are more likely than college-bound seniors to use any of the licit or illicit drugs. In general, the two groups do not differ greatly when it comes to having tried the various illicit drugs. In 1991, for example, 28 percent of non-college-bound seniors had used marijuana during the past twelve months, compared to 22 percent of college-bound seniors (for a ratio of 1.25). Again, more frequent use of the drug shows more of a difference: non-college-bound seniors are more than twice as likely to use marijuana daily than are college-bound seniors (3.3% versus 1.4%). The use of cocaine, crack cocaine, and amphetamines is about one and a half to two times more frequent among non-college-bound students. A striking difference shows up between college-bound and non-college-bound seniors in terms of cigarette smoking rates. For example, smoking a half pack or more a day is more than two and a half times more prevalent among the non-college-bound than it is among the college-bound seniors (19% versus 7%). Non-college-bound seniors are also considerably more likely than their college-bound counterparts to drink alcohol on a daily basis (5.4% versus 2.9%), and somewhat more likely to report having had five or more drinks in a row in the past two weeks (34% versus 28%).

Parental Education. Among high school seniors there is (perhaps surprisingly) rather little association between parental education and use of illicit drugs. To the extent that there is an association, it is primarily the lower level or lower two levels that stand out as having somewhat higher use rates

than the others. Cigarette use does show a negative monotonic relationship with parental education, going from 16 percent daily use in the highest education group to 21 percent in the lowest.

Geographical Region. As of 1991, the overall use of illicit drugs among high school seniors has been lower in the South than in the other three regions. In the South, 25 percent of seniors had used an illicit drug in the past twelve months, compared with 31 percent in the North Central states, 32 percent in the Northeast, and 33 percent in the West. Both the South and the West tend to exhibit lower rates of licit-drug use among high school seniors than the Northeast and the North Central states. The prevalence of heavy-drinking occasions (that is, five or more drinks in a row on at least one occasion in the past two weeks) among the seniors was found to be 33 percent and 35 percent in the Northeast and North Central states, respectively, compared with 26 percent in both the South and the West. Smoking at the rate of a half pack of cigarettes or more per day was found to be 13 percent in the Northeast and 14 percent in the North Central states, compared with 9 percent in the South and 7 percent in the West.

Population Density. As of 1991, the differences in high school seniors' use of illicit drugs by population density have been surprisingly small. This lack of large differences reflects the fact that illicit-drug use has spread widely throughout the nation. One substance that has shown some significant difference by population density over time is the use of cocaine. The substantial increase in cocaine use in the late 1970s, and the continuing high levels of use until the mid-1980s, was primarily an urban phenomenon. The annual prevalence rates for cocaine were nearly twice as high among high school seniors in the large standard metropolitan statistical areas as they were for seniors in the more sparsely populated areas.

Racial or Ethnic Identification. Because of the design of most national surveys, it is difficult to make definitive statements about the larger minority groups such as African Americans and HISPANICS; it is virtually impossible to make definitive statements about other minority groups. Even Hispanics, who constitute a large segment of the population in many areas, often cannot be accurately represented because there are many important subgroups among the several Hispanic groups (e.g., Mexican, Puerto Rican, Cuban, and Latin American, among others). Nevertheless, certain findings appear to be reliable.

Among high school seniors, African-American students report less use of virtually all substances than do white or Hispanic students (Bachman et al., 1991). The rates of use among Hispanic students appear to fall between those for white and African-American students. Some of the differences observed in twelfth-graders, however, could be affected by differential dropout rates. Hispanic youngsters in particular have distinctly higher dropout rates than do either African-American or white adolescents. As of 1990, for example, about 10 percent of white nineteen- to twenty-year-olds had dropped out of high school, compared with 16 percent of African-American youth of the same age, and 34 percent of Hispanic youth (U.S. Dept. of Education, 1992). Data from national samples of eighth-, tenth-, and twelfth-grade students in 1991 support the following interpretation: The drug-use rates among Hispanic eighth-graders for nearly all drugs are higher than the drug-use rates among white eighth-graders. It thus appears that lower drug use among Hispanic seniors, relative to white seniors, is in part due to differential rates of dropping out of school. African-American students, however, report less drug use than do white students at all grade levels, so that in this instance differential dropping out cannot explain the differences. It is also unlikely that differential reporting biases are likely to be the cause (Wallace & Bachman, 1993).

SUMMARY

Between 1975 and 1991, appreciable declines were found in the use of a number of illicit drugs among high school seniors, but not in all drugs. LSD and inhalants were the notable exceptions. Moreover, relatively slight declines have been seen in the early 1990s in alcohol use and even smaller declines in cigarette use. Although the overall picture for reduced drug use among high school seniors is good, the amount of illicit as well as licit drug use in this age group is still a matter of grave concern, as the following facts demonstrate:

- As of the early 1990s, about 44 percent of young Americans had tried an illicit drug by the time they had neared the end of their last year of high school; this proportion included about 27 percent who had tried some illicit drug other than marijuana.
- By the senior year of high school, one in twelve (8%) had tried cocaine, and about one in every thirty seniors (3.1%) had tried crack cocaine.
- A small but significant number of high school seniors in 1991 smoked marijuana daily (2%), and about one in eleven (9%) reported that they had smoked marijuana daily for at least a month at some time in their life.
- Almost a third (30%) of high school seniors in 1991 had had five or more drinks in a row at least once in the prior two weeks.
- More than a quarter (28%) of seniors had smoked cigarettes in the month prior to the survey, and 19 percent already smoked daily.

In addition to providing basic epidemiologic information on prevalences, trends, and demographic differences, the Monitoring the Future study also contributes information on the reasons for the trends and differences. The study's demonstration that attitudes and beliefs affect drug-use trends (especially in the case of marijuana and cocaine) is particularly important. By virtue of its cohort-sequential design, the study has been able to distinguish among the several possible types of competing changes associated with trends in use—specifically, age, period, and cohort (or birth group) effects (O'Malley, Bachman, & Johnston, 1988). In addition, the study has been able to provide important data with which researchers could evaluate the effects of changes in the laws dealing with marijuana (Johnston, O'Malley, & Bachman, 1981) and alcohol (O'Malley & Wagenaar, 1991). All of these contributions have been vital in the continuing debates about policy regarding the use of licit and illicit drugs.

(SEE ALSO: *Epidemiology; Ethnicity and Drugs; National Household Survey; Vulnerability*)

BIBLIOGRAPHY

BACHMAN, J. G., ET AL. (1991). Racial/ethnic differences in smoking, drinking, and illicit drug use among American high school seniors, 1976–1989. *American Journal of Public Health, 81*, 372–377.

JOHNSTON, L. D. (1973). *Drugs and American youth.* Ann Arbor: Institute for Social Research.

JOHNSON, L. D., O'MALLEY, P. M., & BACHMAN, J. G. (1992). *Smoking, drinking, and illicit drug use among American secondary school students, college students, and young*

adults: 1975–1991: Vol. 1. Secondary school students (DHHS Publication No. NIH 93-3480); *Vol. 2. College students and young adults* (DHHS Publication No. NIH 93-3481). Rockville, MD: National Institute on Drug Abuse.

JOHNSTON, L. D., O'MALLEY, P. M., & BACHMAN, J. G. (1981). Marijuana decriminalization: The impact on youth, 1975–1980 (Monitoring the Future Occasional Paper No. 13). Ann Arbor: Institute for Social Research.

O'MALLEY, P. M., BACHMAN, J. G., & JOHNSTON, L. D. (1988). Period, age, and cohort effects on substance use among young Americans: A decade of change, 1976–1986. *American Journal of Public Health, 78,* 1315–1321.

O'MALLEY, P. M., & WAGENAAR, A. C. (1991). Effects of minimum drinking age laws on alcohol use, related behaviors, and traffic crash involvement among American youth: 1976–1987. *Journal of Studies on Alcohol, 52,* 478–491.

U.S. DEPARTMENT OF EDUCATION, NATIONAL CENTER FOR EDUCATIONAL STATISTICS. (1992). *The condition of education.* Washington, DC: U.S. Government Printing Office.

WALLACE, J. M., JR., & BACHMAN, J. G. (1993). Validity of self-reports in student-based studies on minority populations: Issues and concerns. Pp. 167–200 in M. R. DeLaRosa & J. L. R. Adrados (Eds.), *Drug abuse among minority youth: Advances in research and methodology.* NIDA Research Monograph 130. (DHHS Publication No. NIH 93-3479) Rockville, MD: National Institute on Drug Abuse.

PATRICK M. O'MALLEY

HISPANICS AND DRUG USE, IN THE UNITED STATES

Hispanics in the United States are a large, growing, diverse group. More precisely, 1990 U.S. Census figures put the total at 22 million—of these, 63 percent are Mexican in origin, 11 percent Puerto Rican in origin, and 5 percent Cuban in origin. These three groups are the largest, yet another 14 percent of Hispanics are from the various Central and South American countries; still another 8 percent are classified as "other Hispanic" by the U.S. Bureau of the Census. In this essay the terms *Hispanic* and *Latino* are used interchangeably. *Hispanic* is commonly used in official statistics, and *Latino* is more widely used within the population itself.

The rapid growth of the Latino population within the United States also is noteworthy. It grew by 53 percent between 1980 and 1990. A high birth rate and continuous new immigration fuels this growth.

On average, Hispanics are younger than other minorities and other American population groups. When youthfulness is combined with POVERTY or discriminatory practices, the result sometimes is a disproportionate degree of conflict with law enforcement, especially in connection with drug abuse and drug dealing. The media coverage of these conflicts may lead many into a prejudicial belief about Latinos and drug use.

Although there are many notable exceptions, most Hispanics live in cities in the United States and, lacking other options, they are steadily crowding into the poorest areas of New York, Los Angeles, Chicago, and other large cities. In 1990, 25 percent of Latinos in the United States lived in poverty compared with 31 percent of black families and 13 percent of all other Americans. Poor education, difficulty with the English language, and urban concentration can compound this impoverishment—as it has for the other immigrant minorities in the United States— thereby contributing to the complexity of modern urban problems that they must face daily.

All segments of this highly diverse group are changing rapidly. Documented and undocumented new immigration combined adds about 500,000 arrivals each year, and this flow is increasing. Many of the newcomers crowd into old barrios, and this reduces the quality of life for older residents. Great pressure is therefore exerted on local educational services, health resources, job sources, and job-training services—a pressure that is compounded by problems of acculturation. Many Mexican-American communities predate the Mexican–American War of the 1840s, but other Latino communities have become established in significant numbers only since World War II. Puerto Ricans, for example, settled mostly in the large cities of the Rust Belt in the late 1940s and early 1950s, forming a particularly large concentration in New York City. Like Mexican Americans (Chicanos), they have been sharply affected by recent shifts in the American economy that relegate poorly educated workers to poorly paid service jobs. Central and South Americans are found in diverse locations, with concentrations in New York,

Houston, and Los Angeles, tending to work at the bottom of the labor market. Cubans, who are concentrated primarily in Miami, have been helped both by a vigorous enclave economy (with Cubans owning many of the enterprises and hiring fellow Cubans) and by Miami's emergence as a center for Latin American trade.

HISPANICS AND ILLICIT DRUGS

Latinos often are typecast as drug users (see Helmer, 1975). Such stereotypes persist partly because there is little research information. National statistics about Hispanics mask important variations within the population, not only in ethnicity but in class and culture. Drug problems of the community are treated principally as criminal phenomena, and indeed, in many states a disproportionate number of Latinos are imprisoned for drug-related offenses. The context for drug use is little studied.

What then is really known about drug use by Hispanics? Specifically, 1991 figures from the annual survey of the National Institute on Drug Abuse (NIDA) show that Hispanics are generally less likely to use drugs in their lifetime than either blacks or the white-majority population. However, Hispanics are most likely to have used COCAINE, and next most likely (after blacks) to have used CRACK cocaine. National surveys do not report on HEROIN, an illicit drug that has posed major problems for Latinos, particularly in New York and the Southwest. Heroin use has been studied in several southwestern communities, in particular in the context of peer group and FAMILY in Los Angeles barrios.

The aggregate figures also conceal significant subgroup differences. Puerto Ricans are especially likely to use cocaine, for example, and Cubans are notably less likely to use any drug. (However, clinical data indicate that Cuban drug use is actually higher than survey data show.)

The aggregate figures conceal geographic differences as well. Studies of persons arrested for crimes, for example, show that more than two-thirds of Hispanic arrestees in Chicago, New York, Philadelphia, and San Diego were using drugs but that proportions were far lower in most other cities (U.S. Department of Justice, 1991). Finally, drug-use patterns may change rapidly, even in a high-risk population: for example, 68 percent of San Antonio's Hispanic arrestees were using some drug in 1988, but by 1991

only 47 percent were, according to U.S. Department of Justice figures (1990). Glick (1990) has analyzed the shifting drug-use patterns in Chicago's Puerto Rican community.

Differences in drug use by males and females are sharper for Hispanics than for other ethnic or cultural groups. Mexican American and Puerto Rican boys and girls are socialized very differently to alcohol and drug use—that is, there is more parental and community disapproval for girls and more permissiveness for boys. Yet research on drug use among Hispanic women is scarce. Among the available research, of particular interest is the finding that sedatives and prescription drugs are used differently by women than they are by men (Gonzalez & Page, 1991). There is also research showing that most female heroin addicts usually begin to use heroin with a male friend, spouse, or common-law partner, thus suggesting that the use depends on a relationship. Hispanic women appear to be greatly influenced by traditional ideas about the role of women, even under the pressures of urbanization, acculturation, and poverty (Moore, 1990).

As to adolescents, the most susceptible group, there is little information about how adolescent Hispanic groups differ from other adolescent groups in drug use. National surveys of high school seniors discover only small differences, but the surveys omit dropouts, who are often the adolescents most at risk, and Hispanic adolescents have very high dropout rates. Most studies confirm that the same risk factors that are important for other youth are important for Hispanics: above all, a disruptive family environment; availability of drugs; peer influences; and patterns of unconventional behavior (such as low school achievement, rebelliousness, early sexual activity). These influences (plus the degree of acculturation and individual judgments of the adolescent) seem to be related, in a general way, with beginning drug use and a steady use of drugs (Booth, Castro, & Anglin, 1990). One notable fact is that gender differences are less significant for adolescent Hispanics than they are for adult Hispanics (Gilbert, 1985).

A special factor that affects Latinos is the overriding importance in the culture of the family. This influence has both positive and negative effects. The extended family among Puerto Ricans in New York may limit drug use by protecting and controlling youngsters in both single- and two-parent households (Fitzpatrick, 1990). In Cuban families, by contrast, illicit drug use may occur when the family structure is severely disrupted, often by the trauma of refugee migration, and researchers argue that the very cohesiveness of the Cuban family may be associated with parental overprotectiveness and adolescent rebellion, sometimes accompanied by drug use as a symptom (Rio et al., 1990).

Recent research suggests that Hispanic clients achieve only mixed success in treatment, but that finding needs qualification, because of the limitations of available treatment programs. Because of poverty and residence in blighted areas, a disproportionate number of Latino heroin users, for example, are enrolled in programs that simply administer blocking drugs (e.g., methadone), with virtually no other treatment. Urban drug treatment programs generally face chronic shortages of money and personnel. When drug abusers do get access to broader treatment, failure can often be blamed upon the absence of culturally sensitive therapies (Rio et al., 1990). Fitzpatrick (1990) has suggested that Puerto Ricans in New York City show an "extraordinary" ability to cope with a community saturated with drugs and that efforts should be made to build on this ability.

HISPANICS AND ALCOHOL

Among Hispanic and many other groups, ALCOHOL use has been easier to study than the use of illicit drugs; many of its patterns are similar to and may shed light on drug use. As they do with drugs, Hispanics use less alcohol over their lifetimes than do "Anglos" (i.e., non-Hispanic white U.S. inhabitants in general, not just those of English ancestry), and their usage is only very slightly more than that of blacks. Again as with drugs, there are sharp gender differences in alcohol use, which are especially noteworthy among immigrants. Among Mexican Americans, the gap between male and female drinking narrows but never disappears in succeeding generations, and much recent research focuses on this acculturation effect, so critical in a large new immigrant population (Canino, 1994). Among younger women, the narrowing gap seems to reflect both acculturation and upward social mobility. Even within one city, Mexican-American drinking habits vary greatly by class (Trotter, 1985). But Gilbert found that Mexican Americans in California also speak of family, financial, and job problems as factors in abusive drinking; they tend to recognize alcoholism not

as a medical problem but as a failure of will (Gilbert, 1985). Certainly there is no one set of beliefs, behaviors, and norms associated with Latinos and drinking. Lifestyle diversity within Latino subgroups suggests the need for a corresponding diversity of treatment approaches. The failure of such standard treatments as ALCOHOLICS ANONYMOUS among Hispanics in certain areas should be noted.

Finally, as noted before in regard to drugs, there are important differences in drinking behavior between subgroups of Hispanics. Mainland-dwelling Puerto Ricans' use of both alcohol and drugs is comparatively high wherever studied (Gordon, 1985). (Pentecostal church groups have had notable success in influencing the drinking behavior of some Puerto Ricans, although some clinicians have expressed the view that Puerto Ricans are reluctant to use treatment services.) Cuban drinking patterns are generally moderate: Cultural values of self-control forbid discernible drunkenness for both men and women. With increasing acculturation, there is gradually increasing alcohol usage but reduced reliance on minor TRANQUILIZERS by Cuban women. All the (scanty) information available on the subject stresses the importance of individual ethnic experience.

(SEE ALSO: *Ethnic Issues and Cultural Relevance in Treatment; Ethnicity and Drugs; Families and Drug Use; High School Senior Survey; Inhalants: Extent of Use and Complications*)

BIBLIOGRAPHY

AUSTIN, G. A., & GILBERT, M. J. (1989). Substance abuse among Latino youth. *Prevention Research Update, 3,* 1–28.

BOOTH, M. W., CASTRO, F. G., & ANGLIN, M. D. (1990). What do we know about Hispanic substance abuse? A review of the literature. In R. Glick & J. Moore (Eds.), *Drugs in Hispanic communities.* New Brunswick, NJ: Rutgers University Press.

CANINO, G. (1994). Alcohol use and misuse among Hispanic women. *International Journal of the Addictions, 29,* 1083–1100.

FITZPATRICK, J. P. (1990). Drugs and Puerto Ricans in New York City. In R. Glick & J. Moore (Eds.), *Drugs in Hispanic communities.* New Brunswick, NJ: Rutgers University Press.

GILBERT, M. J. (1985). Mexican Americans in California: Intercultural variation in attitudes and behavior related to alcohol. In L. A. Bennett & G. M. Ames (Eds.), *The American experience with alcohol*. New York: Plenum.

GLICK, R. (1990). Survival, income and status: Drug dealing in the Chicago Puerto Rican community. In R. Glick & J. Moore (Eds.), *Drugs in Hispanic communities*. New Brunswick, NJ: Rutgers University Press.

GONZALEZ, D. H., & PAGE, J. B. (1991). Cuban women, sex role conflicts and the use of prescription drugs. *Journal of Psychoactive Substances, 20*.

GORDON, A. (1985). Alcohol and Hispanics in the northeast: A study of cultural variability and adaptation in alcohol use. In L. A. Bennett & G. M. Ames (Eds.), *The American experience with alcohol*. New York: Plenum.

HELMER, J. (1975). *Drugs and minority oppression*. New York: Seabury Press.

MOORE, J. (1990). Mexican American women addicts. In R. Glick & J. Moore (Eds.), *Drugs in Hispanic communities*. New Brunswick, NJ: Rutgers University Press.

RIO, A., SANTISTEBAN, D., & SZAPOCZNIK, J. (1990). Treatment approahces for Hispanic drug-abusing adolescents. In R. Glick & J. Moore (Eds.), *Drugs in Hispanic communities*. New Brunswick, NJ: Rutgers University Press.

TROTTER, R. (1985). Mexican-American experience with alcohol: South Texas examples. In L. A. Bennett & G. M. Ames (Eds.), *The American experience with alcohol*. New York: Plenum.

U.S. DEPARTMENT OF JUSTICE. (1991). *Drug use forecasting: Drugs and crime (1990 Annual Report)*. Washington, D.C.: National Institute of Justice.

U.S. DEPARTMENT OF JUSTICE. (1990). *Drug use forecasting: 1988 drug use forecasting annual report*. Washington, D.C.: National Institute of Justice.

JOAN MOORE

HISTORY OF TREATMENT *See* Treatment, History of, in the U.S.

HIV *See* Injecting Drug Users and HIV; Needle and Syringe Exchanges and HIV/AIDS; Substance Abuse and AIDS.

HOMELESSNESS, ALCOHOL, AND OTHER DRUGS Because of the amount of media attention given homelessness in the 1980s and into the 1990s, as well as to the extreme visibility of homeless people on city streets, it has been observed that homelessness is the "social problem of the decade." A more true statement might be that "the homeless have always been with us." While homelessness has been described as a complex issue needing complex solutions, the basic concept is simple— homeless people are those who, for whatever reason, are unable to maintain housing in a limited housing market. There is not enough low-rent housing to meet the needs of the growing number of poor in the United States. Homeless individuals are part of the underclass, the bottom tier of poverty, and often they reach that status because of a problem with alcohol and/or other drugs.

Estimating the size of the U.S. homeless population has presented some of the best research methodologists with numerous challenges. Problems include varying definitions of homelessness, the high cost of conducting methodologically sound studies, and the anonymity and transiency of the homeless condition itself. Coming up with a viable estimate of homeless people who experience problems with alcohol or other drugs is even more challenging since this subgroup includes individuals who are among the most transient and most hidden.

To arrive at a figure that is useful for developing policies and programs, the U.S. government funded an evaluation of the best available prevalence studies. Based on this analysis, it is estimated that 40 to 50 percent of homeless people have alcohol-use disorders, while 28 to 37 percent have drug-use disorders. Additionally, 10 to 20 percent of the homeless population suffer from the co-occurrence of a mental health and substance-use disorder.

In addition to prevalence rates, it is important to look at other characteristics of the homeless population. While the chronic public inebriate or traditional skid-row alcoholic—most often single, white, and male—is still very much a part of today's homeless population, new groups have appeared on the scene. These include growing numbers of women with children, the mentally ill, runaway youth, and members of minorities. Research has shown that all of these subgroups of the homeless population have significantly higher rates of alcohol- and other drug-use disorders than the population in general. In addition, the core group of homeless single males has also changed in the 1980s and early 1990s—overall seen as younger (pre-middle age), with a tendency to abuse illegal drugs, most notably CRACK-cocaine, as well as ALCOHOL.

CHALLENGES TO THE TREATMENT SYSTEM

Homeless individuals and families present significant problems to the alcohol- and other drug-treatment system. There are often few or no treatment slots available for people without insurance or other means to pay for treatment. The few programs that are open to this group often cannot accommodate either women with children or people who also have a mental illness. Any who manage to find and successfully complete treatment programs are likely to relapse if they must return to life on the streets or to shelters where alcohol and other drug abuse is rampant and there are few supports available to the individual serious about recovery. Even scarce public housing or publicly subsidized housing, if a homeless person is lucky enough to secure it, is a problem for the newly recovering person, since it is often located in neighborhoods with high levels of alcohol and drug abuse. Most importantly, solving an individual's alcohol or other drug addiction does not necessarily solve the other pressing problems—those of coping with society, homelessness, and poverty.

Similarly, homeless individuals and families with alcohol- and drug-use disorders present significant problems to the social-service-delivery system. Individuals who are actively drinking or using drugs are barred from many shelters in an effort to minimize disruptive behavior or violence. Programs that have been developed to help end homelessness, such as job-training or educational programs, often are not equipped to identify and treat alcohol and other drug problems. People with these problems often fail at such programs; they are then dismissed to make room for others who might have a better chance for success.

Service providers throughout the United States who have implemented programs for the homeless with alcohol and other drug problems and who have experience with the changing face of homelessness, have developed models for treatment and recovery. These models are made up of strategies that have evolved from trial and error as well as from a recognition of the wide range of problems that homeless people face that are barriers to their successful recovery from alcohol and other drug addiction. Research in this area is moving in the direction of conducting rigorous evaluations of these models and their various components to facilitate recommendations about the essential components of treatment.

It is currently acknowledged in the treatment field that alcohol and other drug-treatment interventions designed for people who are homeless should have the following components:

Outreach and Engagement—Going out into the community to identify homeless persons in need of alcohol- and other drug-treatment services (outreach) and removing barriers (psychological, financial, physical, and social) so that an individual can fully participate in treatment (engagement).

Alcohol and Other Drug-Abuse Treatment Services—Detoxification and sobering services, alcohol-and-drug education, individual and/or group counseling. Whether provided on an inpatient or outpatient basis, it is generally agreed that for most homeless individuals this treatment should extend beyond traditional 28-day models, with 6 months to a year being common.

Services and Benefits Acquisition and Coordination—Activities to assist homeless persons to access benefits from the public-welfare or veterans-administration systems, such as social-security payments, unemployment compensation, food stamps, and the like. Often this includes help with money management. In many programs, these services are provided by a case manager, who also helps the individual remain in treatment and avoid relapse.

Prevocational and Vocational Training—Vocational rehabilitation and training services, leading to employment.

Health Care—Including general medical services, dental care, and health education.

Alcohol and Drug-Free Housing—Ranging from single room occupancy (SRO) to group homes, halfway houses, and shelter-type settings, this refers to a group-living setting that supports alcohol- and drug-free life.

The above components are not all-inclusive: In designing treatment programs for the homeless, service providers also consider the special needs of the subgroup with whom they are working. Other important services offered include provisions for child care and child development, educational activities, legal services, and transportation.

SUMMARY

Homeless people present unique challenges to the traditional alcohol- and drug-treatment system. They have multiple problems beyond their addiction and abuse issues and, often, they have cycled through various treatment programs without success. Although homeless people represent a small portion of the entire U.S. population with alcohol and other drug problems, they are highly visible and put tremendous strains on public resources.

(SEE ALSO: *Comorbidity and Vulnerability; Homelessness and Drugs, History of; Poverty and Drug Use; Treatment: History of; Treatment Funding and Service Delivery; Treatment Types*)

BIBLIOGRAPHY

INSTITUTE OF MEDICINE. (1988). *Homelessness, health and human needs.* Washington, DC: National Academy Press.

KOZOL, JONATHAN. (1988). *Rachel and her children.* New York: Crown.

NATIONAL INSTITUTE ON ALCOHOL ABUSE AND ALCOHOLISM. (1993). *Community demonstration grant projects for alcohol and drug abuse treatment of homeless individuals: Final report.* NIH Publication 93-3541. Washington, DC: U.S. Department of Health and Human Services.

NATIONAL INSTITUTE ON ALCOHOL ABUSE AND ALCOHOLISM. (1987). *Alcohol health and research world: Homelessness.* Vol. 11, no. 3. (ADM) 87-151. Washington, DC: U.S. Department of Health and Human Services.

MARGARET MURRAY

HOMELESSNESS AND DRUGS, HISTORY OF

The word *homeless* has a long and complex usage. In its most literal meaning of houseless, it has been employed since the mid-1800s to describe those who have slept outdoors or in various makeshifts, or who resided in temporary accommodations like the police-station lodgings of earlier generations or the emergency shelters of the present day. Another old meaning of the word draws upon the absence of a sense of belonging to a place and with the people who live there. This usage is handed down from the largely rural and small-town society of the nineteenth century, in which the coincidence of family and place provided the basis for community and social order, nurturing traditions of mutual aid and the control of troublesome behavior. To be homeless was to be "unattached," outside this web of support and control; it was to be without critical resources and, equally important, beyond constructive restraint. Many of the young men and women who moved from farm to city, or those who emigrated during the nineteenth and early twentieth centuries, were unattached in this respect. Organizations like the YWCA, and YMCA, and various ethnic mutual-aid societies were invented both to help and superintend them by creating surrogate social ties.

HISTORY

By the 1840s, it was common for Americans to link homelessness with habitual drunkenness. In the popular view, habitual drunkards, usually men, drank up their wages and impoverished their families; they lost their jobs, their houses, and drove off their wives and children by cruel treatment. They became outcasts and drifters and their wives entered poorhouses while their children became inmates of orphanages. By the 1890s, the same logic served to explain the downward, isolated spiral of opiate and cocaine "fiends" (as they were called) and the unhappy circumstances of their families.

Until the early years of the Great Depression (which began in 1929), habitual drunkenness, in particular, often was cited as a principal cause of homelessness. Even so, after the financial collapse of 1893 and an ensuing five-year depression of unprecedented severity, most thoughtful observers did not understand heavy drinking or habitual drug use to cause homelessness in any *direct* manner. Although scholarly studies during the first decades of the 1900s were crude by today's technical standards, their explanations of homelessness were not simpleminded. In fact, they foreshadow today's explanations.

Perhaps most important, pre-Depression students of homelessness noted that the ranks of the dispossessed grew and diminished in close relation to economic conditions. They understood that the profound depressions that haunted the economy long before 1929 caused large numbers of people to lose their grip on security. They noted as well that certain occupations were especially affected by seasonal fluctuations in the demand for labor and by technological change—by the 1920s, agricultural workers, cigar makers, printers, and others had high rates of "structural" unemployment. That is, their jobs had been lost permanently to changes in methods of production and distribution.

These scholars also understood the importance of decisions that employers made about hiring and firing. Workers without families to support and those regarded as the least productive were let go first when the economy soured. Usually, these were single young women assumed able to return to their natal families, married women presumed to be working for "pin money" (people who are today known as secondary wage earners), older men, and in particular, single men known to drink heavily. Minority racial and ethnic status also marked people for layoff. Conversely, in times of high demand for labor, employers relaxed their standards for hiring and job performance. In boom times all but the most seriously disabled, and the most erratic and disruptive heavy drinkers and drug users, could find some kind of work. The ranks of the homeless thus thinned considerably.

Pre-Depression observers also emphasized the impact of working conditions, disability, and the absence of income supports on the creation of homelessness. In an era of dangerous work and widespread chronic disease (especially tuberculosis), large numbers of men, in particular, became substantially disabled, often at a young age. In an era before significant public disability benefits or much in the way of welfare or effective medical treatment, they rapidly became abjectly poor, reduced to begging, soup kitchens, and bedding down in mission shelters or the cheapest, most verminous lodginghouses ("flophouses," as they came to be called).

Some of these men were heavy drinkers, and some were habitual drug users, but it was commonly observed that such problems often developed in the context of POVERTY and rootlessness. The miseries and long stretches of boredom endemic to poverty were understood to promote frequent intoxication—even during the Prohibition years (1920–1933), when illicit ALCOHOL could be had by arrangement, as could illicit drugs. Certain "hobo" occupations that virtually demanded rootlessness, and which brought together large groups of men without families, were regarded as especially corrupting and debilitating. Railroad gangers, cowboys, farmworkers, lumberjacks, and sailors, among others, pursued risky occupations and lived in ways that provided both motive and opportunity for dissipation. During the Depression it was widely feared that tens of thousands of homeless young people in the United States would be maimed hopping freights and would learn bad habits on the road that would transform them into lifelong tramps.

Finally, and related to their understanding of homelessness as an insalubrious and demoralizing experience, early observers paid a great deal of attention to the milieu of homelessness, which is to say, the urban areas where homeless people congregated and the constellation of institutions with which they were involved. Commonly called "hobohemias" before the Depression and "skid rows" thereafter, such areas were characterized by a particular way of life and a peculiar set of economic and social resources. They were honeycombed with cheap restaurants, residential hotels and lodginghouses, private and eventually public welfare agencies, and formal and informal labor exchanges that offered casual ("day") work. Skid row (and the segregated satellites that developed in minority communities) was also a world dominated by single men. Such areas were saturated with saloons (later bars) and sex workers. Some were the sites of a vigorous drug trade.

By the 1940s, winnowed by wartime labor demand, skid row was both repository and refuge, mainly for impoverished single men disabled by age, injury, and/or chronic illness. They survived on private charity, meager public welfare allowances, modest pensions, and undemanding work. Note, however, that they were housed. In the most literal sense, skid-row denizens were not homeless, and from the 1940s through the 1970s they were more often described as "unattached" or "disaffiliated." They were homeless in the broader, social sense discussed above. Further, and contrary to the enduring stereotype, the residents of skid row were not usually heavy drinkers or habitual drug users. While perhaps one-third could be so described, and while public intoxication was common and visible, heavy drinking or drug taking was, as today, the exception not the rule.

With the sustained prosperity of the period between 1941 and 1973, and the simultaneous elaboration of the American welfare state, many observers believed that skid row would wither away. The older men would die off, or—helped by federal Old Age Security, and later by Medicare, state and federal disability benefits, and subsidized housing—would move to better neighborhoods. Or they would remain on a skid row that would be uplifted and transformed by urban renewal projects and effective rehabilitation programs for heavy drinkers and drug users.

In a limited sense, these optimists were correct. The expansion of the welfare state dramatically improved the economic circumstances of the elderly, and they are greatly underrepresented among today's homeless. Aided by federal funds, some cities bulldozed their skid-row areas, thus causing their bricks and mortar, at least, to disappear. But homelessness did not disappear; instead, it underwent an astonishing and tragic transformation. If literal houselessness is used as the definition and measure of the problem, only the Depression produced the prodigious dispossession we see today.

As opposed to the domiciled isolation of skid row, something like today's houseless poverty was beginning to be reported in news magazines and the occasional scholarly publication as early as 1973. But it was not until the early 1980s that a new generation of younger homeless people achieved widespread notice. At first, most observers were struck by the apparently very high rates of mental illness, heavy drinking, and drug use among those new homeless people. Explanations of the problem tended to point toward nationwide changes in policies that governed commitment to and retention in mental hospitals and incarceration for public drunkenness and minor drug offenses. During the 1960s and 1970s many states "deinstitutionalized" both mentally ill people and "alcoholics" and "addicts." That is, state hospital patients were discharged in wholesale fashion, and new commitment laws made initial involuntary commitments difficult; they severely limited the duration of involuntary treatment. Many states also "decriminalized" public drunkenness, referring public inebriates to places where they could sober up rather than housing them in jail for thirty days to six months. Similarly, many minor drug offenders were diverted from jails. During the early 1980s many observers, notably those within the Reagan administration, characterized the resurgence of homelessness as a problem related to mental disorder, excessive drinking, habitual drug use, and the new policies which kept people with such problems from their customary lodgings in state hospitals and county jails. Homelessness was described mainly as a problem in the rehabilitation and control of troubled and troublesome people who were not only houseless but

barred from their traditional institutional shelters and estranged from family and friends who might take them in.

CURRENT VIEWS

While not discounting this view entirely, most scholars now find it too simple and not supported by the evidence. Current scholarship has returned—often unwittingly—to themes first sounded a century ago: the relationship of homelessness to changes in the economy and the nature and supply of housing; to the availability (or "coverage") and sufficiency of income supports and medical care; and to the tolerance and support capacity of kin. Heavy drinking, habitual drug use, and mental illness are considered in this larger context. Such problems are understood to be among many "risk factors" which make it more likely that some people will become homeless repeatedly or remain so for a long time. Moreover, current scholars are concerned increasingly with how such experience wears people down, introduces or rekindles bad habits or poor health, and makes "exits" from homelessness less likely or short-lived.

Briefly and simply, current scholarship suggests the following relationship between homelessness and heavy drinking and habitual drug use.

The problem of poverty has worsened considerably since the mid-1970s. Changes in the economy have added high-skill, well-paid technical jobs and low-skill, poorly paid service positions, but these changes have simultaneously produced job losses among semiskilled but highly paid workers, primarily in manufacturing. This process of "deindustrialization"—the historic passage from a manufacturing to a service economy—has been especially hard on those younger members of the huge baby-boom birth cohort (boomers are those born between 1946 and 1964), especially Hispanics and African Americans, who have entered a glutted labor market without the advantage of prolonged higher education or advanced technical training.

At the same time, the 1980s brought startling inflation in rental housing costs and a steep decline in the inflation-adjusted value of federal and state welfare benefits and unemployment insurance. In consequence, poor people have had an increasingly difficult time forming independent households and poor families are increasingly hard put to support dependent adult members. On top of this and simultaneously, the stock of America's most rudimentary housing, the old hotels and lodginghouses of skid row and similar areas, was decimated by urban renewal.

The baby boom's maturation was crucial in another way. While there is no good evidence that the combined *rate* of persistent and severe mental disorder, heavy drinking, or habitual drug use is significantly higher among boomers, neither is there any evidence that it is substantially lower than in previous birth cohorts. However, if a roughly constant rate (similar percentage) is applied to a much larger population, the resulting prevalence of a problem is of much greater magnitude—the numbers are much larger. Therefore, as huge numbers of boomers reached the age of greatest risk for the development of enduring mental-health, alcohol, and drug problems (roughly 18 to 25 years old), the cohort generated an unprecedented number of such casualties. This situation developed just as conditions of material scarcity were becoming acute and the old policies of institutional containment were being dismantled.

CONCLUSION

In sum, poor people have been badly squeezed since the early 1970s. As a consequence, perhaps 3 percent of all American adults, about 5.5 million people, experienced at least one spell of homelessness between the beginning of 1985 and the end of 1990. Some, however, experience frequent and prolonged episodes of homelessness, and it is among these people that rates of heavy drinking and habitual drug use are very high. It is not simply the case, however, that their drinking and drugging have caused their homelessness. The health problems and troublesome behavior often associated with such habits may have played an important role in job loss, familial estrangement, or displacement from housing—but this is not a new phenomenon, as we have seen.

Now, though, the absorptive mechanisms of earlier generations have gone awry. Deinstitutionalization has been a factor in this breakdown, mainly because its presumed consequence of community care never has been equal to the unprecedented generational need. Nonetheless, more important factors in the creation of widespread houseless poverty

among heavy drinkers and habitual drug users have been the disappearance of casual labor, the erosion of public benefits and the capacities of kinship, and the virtual destruction of the tough but viable refuge of skid-row housing. In 1970, impoverished heavy drinkers and habitual drug users could almost always find some port in the storm, often by moving from one decrepit hotel to another, frequently pooling resources to rent a room by the week. Since the 1980s, they can no longer. Thus they have become a large and highly visible proportion of those who inhabit our public places and persist in our shelters month after month.

(SEE ALSO: *Alcohol: History of Drinking; Alcohol- and Drug-Free Housing; Comorbidity and Vulnerability; Halfway Houses; Homelessness, Alcohol, and Other Drugs; Treatment: History of*)

BIBLIOGRAPHY

ANDERSON, N. (1940). *Men on the move*. Chicago: University of Chicago Press.

ANDERSON, N. (1923). *The hobo*. Chicago: University of Chicago Press.

BAHR, H. (1973). *Skid row: An introduction to disaffiliation*. New York: Oxford University Press.

BLUMBERG, L. U., SHIPLEY, T. F., JR., & BARSKY, S. F. (1978). *Liquor and poverty: Skid row as a human condition*. New Brunswick, NJ: Rutgers Center of Alcohol Studies.

HOPPER, K., & BAUMOHL, J. (1994). Held in abeyance: Rethinking homelessness and advocacy. *American Behavioral Scientist, 37*, 522–552.

KEYSSAR, A. (1986). *Out of work: The first century of unemployment in Massachusetts*. Cambridge, England: Cambridge University Press.

MILLER, H. (1991). *On the fringe: The dispossessed in America*. Lexington, MA: D.C. Heath.

RINGENBACH, P. T. (1973). *Tramps and reformers, 1873–1916: The discovery of unemployment in New York*. Westport, CT: Greenwood Press.

JIM BAUMOHL

HONG KONG AND DRUGS See Opioids and Opioid Control: History of.

HYDROMORPHONE   Hydromorphone is a semisynthetic OPIOID analgesic (painkiller) derived from thebaine, an ALKALOID of the OPIUM poppy (PAPAVER SOMNIFERUM). It is one of the most widely used and effective analgesics for moderate to severe PAIN and is often referred to as Dilaudid, one of the brand names under which it is sold. Its potency is almost eightfold greater than is morphine's. Struc-

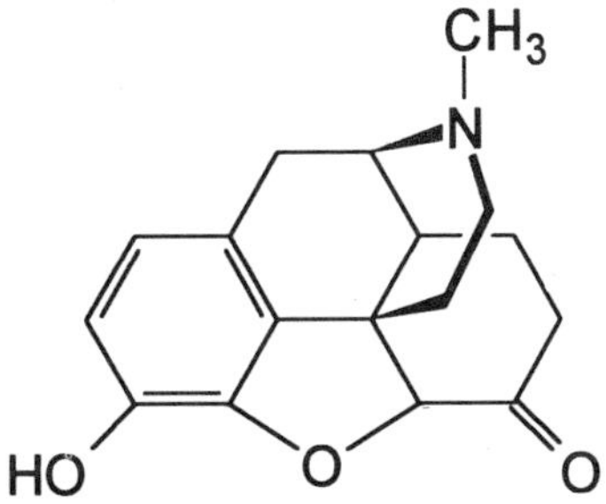

Figure 1
Hydromorphone

turally, it is quite similar to MORPHINE but most like dihydromorphine, differing only in the replacement of the hydroxyl ($-$OH) group at the 6-position with a ketone ($=$O). Thus, it is not surprising that hydromorphone has many of the same side effects—including sedation, constipation, and depression of breathing. Chronic use will produce TOLERANCE AND PHYSICAL DEPENDENCE, much like morphine. This drug is reported to have high abuse potential, perhaps due, in part, to its very high potency.

BIBLIOGRAPHY

JAFFE, J. H., & MARTIN, W. R. (1990). Opioid analgesics and antagonists. In A. G. Gilman et al. (Eds.), *Goodman and Gilman's the pharmacological basis of therapeutics*, 8th ed. New York: Pergamon.

GAVRIL W. PASTERNAK

HYPERACTIVITY   *See* Attention Deficit Disorder; Conduct Disorder in Children.

HYPNOSIS   *See* Treatment/Treatment Types.

HYPNOTICS   *See* Sedative-Hypnotic.

IATROGENIC ADDICTION The potential for ADDICTION or ABUSE influences the licit medical use of many drugs, including OPIOIDS, BENZODIAZE-PINES, BARBITURATES, and others. This influence can be evaluated from two perspectives—(1) the risk that addiction or abuse will result from medical treatment of patients with no such prior history, and (2) the possibility that overconcern about this risk leads to inappropriate undertreatment of certain medical conditions. Although these issues can be discussed with reference to any of these drug classes, the opioids are most illuminating and are emphasized below.

THE RISK OF ADDICTION OR ABUSE

Like any other potential adverse outcome of drug therapy, the prevalence of iatrogenic addiction (drug addiction or abuse during medical treatment) must be determined so that the risk can be assessed by both the practitioner and the patient. An accurate understanding of prevalence, in turn, requires the application of clinically relevant definitions of these phenomena. Unfortunately, there has been little effort to define the addiction syndrome as it occurs in patients, and there is abundant evidence that clinicians commonly use definitions that are inappropriate.

Definition of Addiction in Medical Patients. Accepted definitions of addiction and abuse (Jaffe, 1985; Rinaldi et al., 1988) have been derived from experience with addict populations. These defini-

tions emphasize that addiction is a psychological and behavioral syndrome characterized by psychological dependence on the drug and aberrant drug-related behaviors. There is loss of control over drug use and evidence of compulsive use. Use of the drug continues, and often escalates, despite overt harm to the user or others. The definitions for abuse project a similar sense and stress the persistence of harmful drug use (Rinaldi et al., 1988) or its deviation from accepted societal or cultural norms (Jaffe, 1985).

The validity of these definitions has not been evaluated in medical populations. Although specific behaviors must be used to establish the diagnoses of addiction or abuse, there have been no studies that assess the predictive value of those behaviors that commonly raise concern in clinicians (Table 1). Some behaviors, such as dose escalation, that strongly support a diagnosis of addiction in an individual who does not have an appropriate medical condition or obtains the drug from nonmedical sources may be more difficult to interpret in patients who acquire the drug from a physician to manage an appropriate problem. Some patients with unrelieved cancer pain, for example, have been said to demonstrate pseudo-addiction—behaviors that suggest addiction but disappear as soon as analgesia (pain relief) improves (Weissman & Haddox, 1989).

In the absence of adequate studies of addiction and abuse in medical patients, the evaluation of drug use in the clinical setting is based on observed situations. Although some behaviors may provide compelling evidence (selling prescription drugs), most

TABLE 1
Behaviors that Raise the Suspicion of Addiction or Abuse of Prescription Drugs.

Probably More Predictive	*Probably Less Predictive*
Selling prescription drugs	Aggressive complaining about the need for higher doses
Prescription forgery	
Stealing or "borrowing" drug from another patient	Drug hoarding during periods of reduced symptoms
Injecting oral formulations	Requesting specific drugs
Obtaining prescription drugs from nonmedical sources	Acquisition of similar drugs from other medical sources
Concurrent abuse of related illicit drugs	Unsanctioned dose escalation once or twice
Multiple dose escalations despite warnings	Unapproved use of the drug to treat another symptom
Multiple episodes of prescription "loss"	Reporting psychic effects not intended by the clinician

NOTE: There have been no studies to assess the relative predictive value of these behaviors, but separation into the two categories of "more" or "less" predictive is supported by clinical experience.

will require astute and often repeated assessments. Any suggestion of aberrant drug-related behavior should impel a comprehensive assessment by the clinician of all aspects related to the patient's medical disorder and treatment plan (Portenoy & Payne, 1992).

The Problem of Mislabeling. Clinicians often compound the problem of definition by mislabeling patients as addicts without the evidence to support this diagnosis. Such mislabeling increases the perceived prevalence of iatrogenic addiction and unnecessarily stigmatizes the patient.

The most common type of mislabeling confuses PHYSICAL DEPENDENCE with addiction. Physical dependence is a pharmacologic property characterized by the occurrence of an abstinence syndrome following abrupt dose reduction or administration of an ANTAGONIST. Since physical dependence is not apparent unless an abstinence syndrome occurs, and abstinence can be easily prevented, physical dependence is generally regarded as a minor problem in the clinical setting. Although it has been postulated

that abstinence symptoms can become conditioning stimuli that contribute to the genesis of addiction (Wikler, 1980), it is evident that physical dependence alone does not produce addiction or abuse. Opioid addicts, for example, may or may not be physically dependent, and cancer patients, who are almost certainly physically dependent after receiving high opioid doses for prolonged periods, almost never develop the aberrant drug-related behaviors consistent with addiction or abuse (Kanner & Foley, 1981).

Studies of Addiction or Abuse in Medical Patients. Thus, the risk of iatrogenic addiction or abuse can only be determined if proper definitions are developed and applied to patient populations. Few studies have met these criteria, but those that have are reassuring, indicating a very low risk of these outcomes during medical treatment with drugs of abuse.

Surveys of opioid use are most illustrative. Although older studies of opioid addicts suggested considerable risk of iatrogenic addiction, these data have been replaced by more recent surveys of pain patients. Addiction and abuse are vanishingly rare outcomes of opioid therapy for acute and chronic cancer pain (Kanner & Foley, 1981; Chapman & Hill, 1989). Most experts have concluded that the risk of addiction during opioid treatment for cancer pain is so remote that this outcome should not even be considered in the decision to use these drugs. Similarly, the Boston Collaborative Drug Surveillance Project could document only four cases of addiction among 11,882 patients with no prior history of substance abuse who were administered an opioid during hospitalization (Porter & Jick, 1980); a national survey of burn units could not identify a single case of addiction among 10,000 patients who had no history of substance abuse and received opioids for burn pain. Finally, surveys of selected patients with chronic nonmalignant pain also suggest that aberrant drug-related behavior is distinctly uncommon among those with no such history who are administered opioids on a long-term basis (Portenoy, 1990).

Other drugs have not been evaluated as extensively as the opioids. Recent analyses of BENZODIAZEPINE use, however, conclude similarly that addiction or abuse as defined here is a rare outcome among patients with ANXIETY disorders who are administered these drugs by physicians (Woods et al., 1988; Balter & Uhlenhuth, 1991), although many develop physical dependence.

Together, these data indicate that medical patients with no prior history of substance abuse have a very low risk of iatrogenic addiction or abuse when they are medically administered drugs with a potential for these outcomes. This conclusion is consistent with an understanding of addiction as a disorder related to the use of specific drugs, but not inherent in the pharmacology of any. Addiction is presumably determined by an interaction between the reinforcing qualities of some drugs and a constellation of individual factors, including a genetic propensity, psychosocial aspects, and the specifics of drug availability (Jaffe, 1990, 1992; Chapman & Hill, 1989). The evidence suggests that patients who do not demonstrate a proclivity to addiction or abuse by adulthood are extremely unlikely to develop these outcomes during medical treatment thereafter. Furthermore, it is probable that this small risk could be reduced further by strict adherence to guidelines that set parameters of appropriate patient behavior and follow-up assessments. Such guidelines would also facilitate the identification of those occasional patients who develop any addiction problems.

UNDERTREATMENT

Although the conclusion that iatrogenic addiction and abuse are rare, still this appears to be inconsistent with the attitudes held by many healthcare providers and patients. Fear of addiction is commonplace. Consequently, there is evidence that overconcern about addiction adversely influences prescription practices.

The negative effects on patient care produced by an inaccurate estimate of addiction liability are most clearly documented in pain management—inadequate treatment with opioid drugs results in an unnecessarily high prevalence of unrelieved acute pain, especially cancer pain. Concerns about addiction are among the salient factors that contribute to undertreatment (Portenoy, 1995).

CONCLUSION

The data extant indicate that addiction and abuse are rare outcomes during the therapeutic use of opioids and other drugs in populations with no prior history of substance abuse. The intense concern expressed by clinicians and patients alike and the impact of this concern on prescribing practice appear to be disproportionate to the actual risk. To some extent, this may relate to the difficulties encountered in evaluating addiction and abuse in medical populations, or perhaps more likely to the tendency to mislabel outcomes as addiction that do not fulfill criteria for the diagnosis. Although good clinical practice must recognize the potential for addiction and abuse, optimal therapy depends on an accurate understanding of these phenomena and the limited role they play in clinical practice.

(SEE ALSO: *Abuse Liability of Drugs: Testing in Humans; Addiction: Concepts and Definitions; Controlled Substances Act; Diagnostic and Statistical Manual; Disease Concept of Alcoholism and Drug Abuse; Opioids and Opioid Control; Pain; Prescription Drug Abuse; Vulnerability As Cause of Substance Abuse*)

BIBLIOGRAPHY

BALTER, M. B., & UHLENHUTH, E. H. (1991). Benzodiazepine use/abuse: An epidemiologic appraisal. *New York State Journal of Medicine, 91,* 48S.

CHAPMAN, C. R., & HILL, H. F. (1989). Prolonged morphine self-administration and addiction liability: Evaluation of two theories in a bone marrow transplant unit. *Cancer, 63,* 1636–1644.

JAFFE, J. H. (1992). Current concepts of addiction. In C. P. O'Brien & J. H. Jaffe (Eds.), *Addictive states.* New York: Raven Press.

JAFFE, J. H. (1985). Drug addiction and drug abuse. In A. G. Gilman, et al. (Eds.), *Goodman and Gilman's the pharmacological basis of therapeutics,* 7th ed. New York: Macmillan.

KANNER, R. M., & FOLEY, K. M. (1981). Patterns of narcotic drug use in a cancer pain clinic. *Annals of the New York Academy of Sciences, 362,* 161–172.

PORTENOY, R. K. (1995). Inadequate outcome of cancer pain treatment; Influences on patient and clinician behavior. In R. B. Patt (Ed.), *Problems in cancer pain management: A comprehensive approach.* Philadelphia: Lippincott.

PORTENOY, R. K. (1990). Chronic opioid therapy in nonmalignant pain. *Journal of Pain Symptom Management 5,* S46–S62.

PORTENOY, R. K., & PAYNE, R. (1992). Acute and chronic pain. In J. H. Lowinson et al. (Eds.) *Comprehensive textbook of substance abuse.* Baltimore: Williams & Wilkins.

PORTER, J., & JICK, H. (1980). Addiction rare in patients treated with narcotics. *New England Journal of Medicine, 302*, 123.

RINALDI, R. C., ET AL. (1988). Clarification and standardization of substance abuse terminology. *Journal of the American Medical Association, 259*, 555–557.

WEISSMAN, D. E., & HADDOX, J. D. (1989). Opioid pseudoaddiction: An iatrogenic syndrome. *Pain, 36*, 363–366.

WIKLER, A. (1980). *Opioid dependence: Mechanisms and treatment.* New York: Plenum Press.

RUSSELL K. PORTENOY

IATROGENIC ADDICTION: NONOPIOIDS *See* Prescription Drug Abuse.

IBOGAINE The roots of the shrub *Tabernanthe iboga* first aroused pharmacological interest in 1864 when a French naval surgeon brought some back from Gabon, West Africa. The root was eaten by various Gabonese tribes as part of initiation ceremonies of puberty and was said to produce intoxication, visions, and a reduced need for sleep.

An active alkaloid, ibogaine ($C_{20} H_{26} N_2 O$), was isolated in 1901 from the roots, bark, and leaves of *Tabernanthe iboga*. In the early 1900s, some medical researchers in France recommended ibogaine for use in treating neurasthenia and asthenia (syndromes that would probably be diagnosed in the 1990s as depression or fatigue syndrome). Although the drug was part of a proprietary medication marketed in Europe in the late 1930s and throughout the 1940s, ibogaine attracted little medical or scientific attention until the emergence of interest in indole alkaloids that accompanied the use of reserpine in the 1950s. During the 1960s, when there was considerable research on the use of LYSERGIC ACID DIETHYLAMIDE (LSD) and other psychedelic agents (HALLUCINOGENS) in psychotherapy, ibogaine was also studied, since it appeared to produce mental effects similar in some ways to other hallucinogens. At about the time of these studies, 1967–1968, the World Health Organization and the U.S. Food and Drug Administration (FDA) classified ibogaine as a hallucinogen, along with LSD, MESCALINE, and PSILOCYBIN.

In 1962, Howard Lotsof, who was at the time addicted to heroin, ingested ibogaine in search of a different drug experience. Lotsof came out of a long psychedelic experience, during which he had not taken any heroin, and found that he had no withdrawal symptoms and did not crave drugs. At the time, he noticed that ibogaine had a similar effect on several other heroin addicts. He subsequently remained drug free, completed law school, eventually obtained a patent on the use of ibogaine for the treatment of addiction (brand name ENDABUSE), and became active in seeking funding to further develop the drug and to obtain FDA approval for its medical use in treatment of addiction.

As a Schedule I drug under the CONTROLLED SUBSTANCES ACT, ibogaine is considered to be highly subject to abuse and without any approved medical use. To be approved by the FDA, an agent must be shown to be safe and effective. Throughout the early 1990s the only reports of the efficacy of ibogaine have been anecdotal ones from individuals in Europe who were addicted to heroin, COCAINE, and TOBACCO. Those who take ibogaine are generally highly motivated since the drug is expensive, costing up to several thousand dollars. While many reported a decrease in drug CRAVING after taking ibogaine, relapse to drug use within a few months was also observed.

As a result of pressure from activists, the U.S. government funded animal studies of ibogaine's actions on opioid and cocaine withdrawal, opioid and cocaine self-administration, and neurotoxicity. Studies in animals have not been entirely consistent. High doses of ibogaine reduced some manifestations of opioid withdrawal in monkeys. Studies in opioid-dependent rodents have shown that ibogaine decreases withdrawal, but other studies have not. Some rodent studies have shown a decrease in drug self-administration. Studies of ibogaine toxicity have also produced mixed results. Some studies in monkeys produced no obvious nervous system toxicity, but a study in rats produced damage to neurons in the cerebellum, the part of the brain known best for its role in control and coordination of movement. Other research studies indicate that ibogaine is not similar to opioids such as MORPHINE and heroin nor to hallucinogens such as LSD in terms of actions at drug RECEPTORS.

Despite these inconclusive research findings, in the early 1990s an FDA advisory committee recommended approval of limited trials in humans aimed at establishing safety and efficacy in treating drug dependence. At least one death has been attributed to the use of ibogaine in the treatment of heroin addiction.

(SEE ALSO: *Ayahuasca; Hallucinogenic Plants; Hallucinogens; Pharmacotherapy; Treatment*)

BIBLIOGRAPHY

BLAKESLEE, S. (1993). A bizarre drug tested in the hope of helping drug addicts. *New York Times*, October 27, p. C11.

DEECHER, D. C., ET AL. (1992). Mechanisms of action of ibogaine and harmaline congeners based on radioligand binding studies. *Brain Research, 571,* 242–247.

GOUTAREL, R., GOLLNHOFER, R., & SILLANS, R. (1993). Pharmacodynamics and therapeutic applications of iboga and ibogaine. Translated by William Gladstone. *Psychedelic Monographs and Essays, 6,* 71–111.

JETTER, A. (1994). The psychedelic cure. *New York Times Magazine,* April 10.

KAREL, R. (1993). FDA approves trials of hallucinogen for treating cocaine, heroin addiction. *Psychiatric News,* September 17, p. 7.

RUMSEY, S. (1992). Addiction and obsession. *New York Newsday,* Thursday, November 19, p.1.

SCHULTES, R. E. (1967). The place of ethnobotany in the ethnopharmacologic search for pychotomimetic drugs. In D. H. Efron, B. Holmstedt, & N. S. Kline, (Eds.), *Ethnopharmacologic search for psychoactive drugs.* Public Health Service Publication no. 1645. Washington, DC: U.S. Government Printing Office.

JEROME H. JAFFE

ICD-9/10 *See* International Classification of Diseases.

ICE *See* Methamphetamine; Slang and Jargon.

ILLEGAL/ILLICIT DRUGS *See* Controls: Scheduled Drugs/Drug Schedules, U.S.

IMAGING TECHNIQUES: VISUALIZING THE LIVING BRAIN

Images of the human BRAIN constructed using sophisticated computer systems have proven valuable for studying the effects of abused drugs. Nuclear medicine techniques, such as positron emission tomography (PET) and single photon emission computed tomography (SPECT), allow noninvasive studies of brain function in human volunteers by the administration of small amounts of radioisotopes. These procedures allow visualization and quantification of biochemical processes in the living brain. Functional MRI (magnetic resonance imaging) is a recently developed technique that makes it possible to construct functional brain images without radiation.

PET scanning uses radioisotopes that decay by emitting positrons (positively charged particles), which collide with electrons (negatively charged particles that surround atomic nuclei). In each collision, both the electron and positron are annihilated and energy is released in the form of two photons (quanta of light) that move in opposite directions. The detectors of a PET scanner surround the tissue being studied and register the arrival of photons (Figure 1). The associated computer system can calculate the location of eacch collision and reconstruct an image of the concentration of radioactivity in different parts of the tissue.

The most common applications of PET scanning involve functional measurements of cerebral (brain)

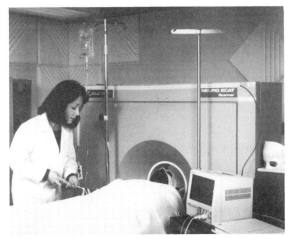

Figure 1
PET scanner with a research subject in position for a study of the human brain.
SOURCE: E. D. London, Addiction Research Center, National Institute on Drug Abuse.

metabolism or cerebral blood flow. PET is also used to map and quantify specific RECEPTORS for drugs and NEUROTRANSMITTERS in the brain. Cerebral glucose consumption (metabolism) and cerebral blood flow both reflect the activity of brain cells. Under normal circumstances, the cerebral metabolism and blood flow are tightly coupled. The most active brain cells require the most glucose, a sugar that is the primary energy source of the adult brain. Brain regions that contain the active cells also require high rates of blood flow for the delivery of nutrients and oxygen. In some conditions, however—including those caused by some drugs—cerebral metabolism and blood flow rates may be dissociated.

Rates of consumption of glucose in the whole brain or in specific brain regions have been measured using fluorodeoxyglucose (FDG) labeled with the positron-emitting isotope fluorine-18 (^{18}F). Cerebral blood flow has been measured using oxygen-15 (^{15}O), either inhaled in $C^{15}O_2$ or injected in ^{15}O-labeled water.

In SPECT, radionuclides that emit single photons are used, including iodine-123 (^{123}I) and technetium-99m (^{99m}Tc), and the photons are measured using a rotating gamma camera. The isotopes used in SPECT have longer half-lives (13 hours for ^{123}I and 6 hours for ^{99m}Tc than those used in PET (110 minutes for ^{18}F and 10 minutes for ^{15}O). Therefore, whereas PET generally requires an on-site cyclotron to produce radioisotopes, SPECT radioactive tracers can be made elsewhere and brought in for use. Although SPECT produces useful images, it does not provide either the quantitative precision or the spatial resolution of PET. Currently available PET scanners can resolve differences in the radioactivity of objects only 4 to 5 millimeters (mm) apart, while the resolution of new SPECT scanners is for 6 to 8 mm.

Before the advent of PET and SPECT, blood flow was measured using xenon-133, given by brief inhalation or intracarotid artery injection. Xenon-133 has a gamma emission with a half-life of 5.27 days, and the radioactivity is monitored outside the skull by an array of detectors that each record a beam of particles from a specific location. Unlike PET, the xenon-133 methods do not provide tomographic information—they do not produce images of "slices" of the brain. Therefore, activity in deep brain structures cannot be measured this way.

Recent advances in magnetic resonance imaging (MRI) technology have permitted functional measurement of cerebral blood volume, which is closely related to cerebral blood flow. Functional MRI assessments are based upon the difference between the paramagnetic properties of oxygenated and unoxygenated hemoglobin. Activation of a brain area causes increased blood flow to the region. Oxygen carried to the activated region is delivered in excess of that which is required by the increased activity. Therefore, it accumulates, as does oxyhemoglobin. Functional MRI produces brain images of very spatial and temporal resolution.

Since researchers are interested in the activity of specific brain structures, data are obtained by functional imaging techniques often adjusted (normalized) to remove the effects of differences between individuals in whole brain activity measurements considered irrelevant to the question under study. Normalized data may be expressed numerically as the quotient of the activity in a region of interest divided by the activity in the whole brain or in the slice containing the region. Such data are not always easy to interpret, since changes in the denominator can obscure the direction and magnitude of change in a region.

ACUTE EFFECTS OF DRUGS

Alcohol. Acute administration of ALCOHOL (ethanol)—a depressant—reduces cerebral glucose utilization, as we learned from measurements taken by the FDG technique. Modest decreases of 15 percent or less are seen in the whole brain in response to a dose of 1 gram/kilogram (g/kg) of ethanol (about 2 oz. of 100 proof whiskey for a 150-lb. person). Slightly more dramatic reductions in metabolism have been noted in the brain's cortex, particularly in the frontal and the occipital regions.

In contrast, acute ethanol administration does not reduce cerebral blood flow. Therefore, ethanol appears to dissociate cerebral blood flow from glucose metabolism. Studies with xenon-133 have indicated that ethanol (0.75 g/kg) increases cerebral blood flow by about 20 percent overall. Furthermore, normalized data obtained by PET scanning, using ^{15}O-labeled water, indicate regional effects of ethanol on cerebral blood flow. The largest changes were noted in the cerebellum (decrease), the prefrontal cortex (increase), and the temporal cortex (increase).

Stimulants. Studies with STIMULANTS have indicated that drugs of this class—including COCAINE and AMPHETAMINE—like the DEPRESSANT alcohol, reduce cerebral glucose utilization. Oral AMPHETAMINE at a dose of 0.5 milligrams/kilogram (mg/kg) decreases cerebral glucose metabolism by an average of about 6 percent of values in the unperturbed state, with no variation in the effect of the drug in different brain regions. A euphorigenic intravenous dose of cocaine (40 mg iv) also reduces cerebral glucose metabolism globally, averaging about a 14 percent decrease overall. The largest reductions occur in the left temporal pole and in the left lateral occipital gyrus.

Benzodiazepines. The effects of diazepam (Valium), a benzodiazepine anxiolytic, on cerebral metabolism and blood flow have also been studied, and results indicate that both of these parameters of brain function are reduced. Glucose metabolism is reduced by taking doses as low as 0.07 milligrams/kilogram orally (about 5 mg, the dose that might be given for anxiety), and the effect does not show regional specificity. Small reductions in cerebral blood flow, as measured with xenon-133, are also seen in response to intravenous diazepam (0.1 mg/kg). The reductions average about 6 percent overall, with the largest reduction seen in the right frontal cortex.

Opioids. The acute effects of HEROIN on cerebral metabolism or blood flow have not been reported, but a euphorigenic intramuscular dose of MORPHINE (30 mg) reduces cerebral metabolism globally, averaging about a 10 percent decrease overall. The largest reduction is found in the left superior frontal gyrus.

Marijuana. The active ingredient in MARIJUANA, delta-9-TETRAHYDROCANNABINOL (THC), produces variable effects on global cerebral glucose consumption but increases normalized metabolism in the cerebellum, as is consistent with the localization of cannabinoid receptors to this region. The metabolic effect is correlated with self-reported intoxication and with the plasma concentration of THC.

Effects of Abused Drugs. Taken together, these results indicate that all drugs of abuse that have consistent effects on cerebral metabolism produce decreases, but the magnitude of the decrease varies. This discrepancy is due, at least in part, to differences in dose and route of administration. The regional distribution of drug effects also varies, but the regional differences in percent change are not large in any of these studies. It seems that drugs of abuse—whether classified as depressants (alcohol), stimulants (cocaine), tranquilizers (benzodiazepines), or ANALGESICS (OPIOIDS)—reduce cerebral glucose metabolism globally.

Effects of abused drugs on global cerebral blood flow are less consistent, with decreases by the tranquilizer diazepam but increases by the depressant alcohol. Differences in regional effects of drugs on cerebral blood flow are minimal or absent, and the effects are generally global. Drugs of abuse may influence cerebral blood flow by direct effects on the cerebral blood vessels. Such direct vascular effects do not reflect changes in blood flow to meet the energy demand of the brain—in contrast, measurements of glucose metabolic rates are less sensitive to vascular responses that are seen as alterations in cerebral blood flow. In this respect, glucose metabolism can be a better measure of brain function than cerebral blood flow.

CHRONIC EFFECTS OF ABUSED DRUGS

Long-term drinking (chronic ethanol abuse) has toxic effects on the brain, and imaging techniques have added to the understanding of these effects. Brain glucose metabolism is decreased in recovering alcoholics (abstinent at least 7 days), even if they do not show brain damage severe enough to be diagnosed as organic brain syndrome. The largest differences from controls were found in frontal lobe structures. Cerebral blood flow, measured using xenon-133, is also decreased in chronic alcoholics, with the largest differences in frontal and temporal lobe structures. To some extent, the changes are reversible with abstinence. Low cerebral blood flow is related to heavy drinking history, with the lowest flow rates in patients with brain damage (organic brain syndrome) due to alcohol.

Chronic use of cocaine has also been associated with persistent effects on functional markers in the brain. Whether measured by PET or SPECT, cerebral blood flow in recovering cocaine addicts (abstinent 4 to 14 days) shows focal abnormalities and lower flow rates than controls, particularly in frontal cortex. The etiology of abnormalities in cerebral blood flow in those with histories of cocaine abuse is not clear. In some cases, focal decrements may be related to the use of alcohol or other drugs of abuse or to the dysphoria related to the withdrawal of cocaine. Heroin addicts showed perfusion abnormalities as measured by SPECT during withdrawal (one week of abstinence), but cerebral blood flow had improved by three weeks of abstinence.

Taken together, studies using imaging techniques suggest that chronic use of alcohol and cocaine may damage certain structures in the frontal lobe of the brain. The frontal lobe is thought to be involved in decision making, planning, and other executive functions necessary for self-control. Thus chronic abuse of these drugs may injure the very brain structures that are required for a person to terminate drug use.

Current imaging techniques offer the promise of delineating the anatomical substrates of the acute and chronic effects of drugs of abuse. Such information may contribute to a further understanding of the causes and the consequences of substance abuse and, ultimately, may lead to more effective prevention and treatment strategies.

(SEE ALSO: *Brain Structures and Drugs; Complications; Reward Pathways and Drugs*)

BIBLIOGRAPHY

ANDREASEN, N. C. (1989). *Brain imaging: Applications in psychiatry.* Washington, DC: American Psychiatric Press.

BELLIVEAU, J. W., ET AL. (1991). Functional mapping of the human visual cortex by magnetic resonance imaging. *Science, 254,* 716–719.

HOLLOWAY, M. (1991). Rx for addiction. *Scientific American,* March, 94–103.

LONDON, E. D. (1984). Metabolism of the brain: A measure of cellular function in aging. In J. E. Johnson, Jr. (Ed.), *Aging and cell function.* New York: Plenum.

LONDON, E. D., ET AL. (1990). Cocaine-induced reduction of glucose utilization in human brain. *Archives of General Psychiatry, 47,* 567–574.

MORGAN, M. J., & LONDON, E. D. (in press). The use of positron emission tomography to study the acute effects of addictive drugs on cerebral metabolism.

ROLAND, P. E. (1993). *Brain activation.* New York: Wiley-Liss.

ROSE, J. S., ET AL. (1995). Cerebral perfusion in early and late opiate withdrawal: A technetium-99m-HMPAO SPECT study. Unpublished manuscript.

JUNE M. STAPLETON
EDYTHE D. LONDON

IMMUNOASSAY Immunology is a laboratory science that studies the body's immunity to disease. The basic mechanism of immunity is the binding of drugs or other chemical compounds to antibodies (large proteins produced by the body's immune system). An assay is a general term for an analytical laboratory procedure designed to detect the presence of and/or the quantity of a drug in a biological fluid such as urine or serum (the fluid component of the blood obtained after removal of the blood cells and fibrin clot). An immunoassay, therefore, is an analytical procedure which has as its basis the principles of immunology—specifically the binding of drugs to antibodies.

Several different types of immunoassay are routinely performed in the laboratory. Although they differ in the types of reagents and instrumentation used, they are all based on the same scientific principle (the binding of drugs to antibodies). The three types of immunoassay that are commonly used for drug testing are the radioimmunoassay (RIA), enzyme multiplied immunoassay (EMIT), and fluorescence polarization immunoassay (FPIA).

It may facilitate the reader's understanding of immunoassay to envision the reactions that occur in the body following a vaccination (e.g., polio). The vaccine contains a weak or a killed solution of (polio) virus. When the vaccine is injected into the body, the immune system recognizes the presence of a foreigner (the polio virus), and it generates antibodies to that virus. These antibodies circulate in the blood, and they constitute the body's protection; if at some later date a live (polio) virus invades the body, the antibodies recognize it by its unique size and shape (similar to the fit of a lock and key); they spontaneously bind to the virus, leading to its inactivation and removal from the body.

This binding of antibodies to drugs forms the basis for immunoassay. In the development of an immunoassay, the first step is to inject an animal (host) with the drug that we ultimately wish to analyze. The host immune system, recognizing the drug as a "foreigner," generates antibodies to this drug, and these antibodies can then be harvested from the serum of the animal. In the test-tube environment of the laboratory (*in vitro*), these antibodies can be recombined with the appropriate drug. Just as it did inside the body (*in vivo*), the antibody will recognize the drug based on the lock-and-key fit and will spontaneously bind to it.

The second step in the development of an immunoassay is to synthesize a "labeled" drug. This involves the chemical addition of a "marker" to the drug. This marker can be small, such as an atom of radioactive iodine, or it can be large, such as an enzyme, which is a fairly large protein. Irrespective of its size, this marker is added in such a way that it does not interfere with the lock-and-key recognition between the antibody and the drug.

Commercially available immunoassay kits contain the antibody (which the company has prepared as described above) and the labeled drug (which has been chemically synthesized) necessary to perform the assay. In the laboratory, a fixed amount of antibody and a fixed amount of labeled drug are placed into a reaction vessel (test tube). If these were the only two ingredients, all the binding sites on the antibody would react with (bind to) the labeled drug. A third ingredient added to the assay is, however, the unlabeled drug (i.e., the urine, saliva, or serum specimen containing the drug that is being measured). Because the label on the labeled drug is placed in a position that does not interfere with binding to the antibody (i.e., it is "hidden"), the antibody cannot distinguish between the labeled and unlabeled drug.

Immunoassays are always designed so that there are fewer antibody-binding sites present in the reaction mixture than there are molecules of (labeled plus unlabeled) drug. Because the labeled and unlabeled drug appear the same to the antibody, they will compete equally for the limited number of available binding sites on the antibody. By measuring the amount of labeled drug bound to the antibody, the analyst can calculate the amount of unlabeled drug in the biological specimen.

All immunoassays work in the same basic fashion. They differ in the types of labels that are added to the labeled drug and in the analytical methods by which the amount of binding of labeled drug to the antibody is measured.

RADIOIMMUNOASSAY

Radioimmunoassay (known as RIA) was the earliest of the immunoassay techniques. It was developed during the 1950s by a pair of research immunologists in New York City, Dr. Solomon A. Berson and Dr. Rosalyn S. Yalow. Their initial RIA was designed to detect very low blood levels of insulin and they published their findings in 1959. Their development of this technique was considered of such importance to science that Dr. Yalow was awarded a Nobel prize in 1977 for their work (since Dr. Berson died in 1972 and Nobels are not awarded posthumously, Berson's contribution was remembered in Yalow's acceptance speech).

In RIA, the marker is an isotope of a radioactive element, hence the name *radio*immunoassay. In most RIAs performed in the laboratory today, the radioactive isotope used as the marker is iodine 125, although tritium (hydrogen 2), carbon 14, and cobalt 57 are used in some assays. RIAs can be used in two different fashions to give information about the drug in a sample: (1) they can be used qualitatively—to determine whether a drug is present or absent (e.g., in urine drug testing); (2) they can be used quantitatively—to determine how much of a drug is present (e.g., to measure serum levels of drugs such as digoxin, a heart medication, or theophylline, an asthma medication).

RIA is an extremely powerful tool. One of its main advantages is the sensitivity that can be achieved. Drug levels in serum and urine that are as low as 10–100 parts per billion are routinely measured. Two of the most sensitive of the radioimmunoassays are the urine LSD assay and the serum digoxin assay, both of which can detect less than 1 part per billion. RIA is also an extremely versatile tool. It is used to measure a wide range of drugs of abuse in blood, serum, saliva, and urine, as well as therapeutic (physician administered) drugs in blood or serum. It is also used as a diagnostic tool to detect and quantify numerous naturally occurring chemicals in human serum and urine. Another characteristic that makes RIA such a powerful tool is the specificity of the assay. The antibodies are highly specific for the drugs analyzed and they rarely make a mistake in recognizing the lock-and-key fit between antibody and drug.

One of the major limitations of the radioimmunoassay is that it generates radioactive waste. To avoid spreading the radioactive compounds and contaminating the environment, the laboratory must conform to very strict regulations, including very elaborate procedures for waste disposal—and undergo frequent inspections. Because of a short half-life for some isotopes, another limitation is that the reagents with a radioactive label have a short shelf life. For instance, the majority are RIAs labeled with iodine

125; they have a shelf life of only approximately 60 days.

Some very sophisticated automated equipment is available for performing RIA or, if need be, the assays can be performed manually. All RIAs require the use of an instrument called a gamma counter, which measures the amount of gamma radiation given off by the radioactive drug bound to the antibody. In the 1990s, gamma counters can be purchased for as little as a few thousand dollars; but the reagents are moderately expensive (costing from less than 50 cents/test to 2–3 dollars/test, depending on the specific assay and the volume of reagents purchased).

ENZYME MULTIPLIED IMMUNOASSAY

The enzyme multiplied immunoassay technique, also known as EMIT™, is a variation of the general immunoassay technique, in which the marker used to prepare the labeled drug is an enzyme, rather than a radioactive isotope. EMIT is a two-stage assay. As in the other immunoassays, the sample, which contains some amount of the drug being measured, is combined with the antibody plus a fixed amount of the enzyme-labeled drug. In the first reaction, the labeled and the unlabeled drug compete for the available binding sites on the antibody (standard immunoassay reaction). A secondary reaction is then performed, which involves only the enzyme portion of the labeled drug. The results of this secondary reaction are used to calculate the amount of enzyme-labeled drug that is bound to the antibody and thus how much (unlabeled) drug there was in the original urine or serum specimen.

As with other forms of immunoassay, the EMIT can be used either qualitatively or quantitatively. In urine specimens, it is used to detect the presence of drugs, such as THC (MARIJUANA), COCAINE, PCP, OPIATES (HEROIN), AMPHETAMINES, and BARBITURATES. In serum specimens, EMIT is used to determine the amount present of drugs used for therapeutic (medical) purposes. Such drugs include acetaminophen (Tylenol), salicylate (aspirin), theophylline (widely used to treat asthma), several drugs used to treat epilepsy, and several drugs used to treat heart abnormalities.

Advantages that the EMIT technology has over the RIA are (1) that no radioactivity is involved, so the waste is more readily disposable; (2) the reagents are relatively stable, which may be particularly attractive to a small laboratory, which runs only a few specimens. The EMIT reagents are also less costly than the RIA reagents. The basic instrumentation requires less capital outlay than does the RIA, however the expense grows as more sophisticated automation is acquired.

Some limitations of the EMIT technique are (1) that it is somewhat less sensitive than the RIA (in particular, the LSD assay requires detection of such minute levels of the drug in urine that it can only be done by RIA); (2) also, EMIT is less specific than RIA and is subject to some interferences that do not affect the RIA—for example, the EMIT assay for amphetamines in urine gives a positive response with several other drugs that are similar in structure to amphetamines.

FLUORESCENCE POLARIZATION IMMUNOASSAY

Fluorescence polarization immunoassay (known as FPIA) is a technique that was developed by Abbott laboratories and marketed under the trade name TD_X. As the name FPIA implies, the marker for the labeled drug is a molecule of a naturally fluorescent compound called fluorescein. The amount of labeled drug that binds to the antibody is measured by a sophisticated instrument called a spectrofluorometer. As with the other immunoassays, this measurement is used to calculate the amount of labeled drug bound to the antibody and thus the amount of drug in the original urine or serum specimen.

The instrumentation necessary to perform the FPIA is only made by Abbott. It is expensive to purchase (upwards of $50,000) but can be leased from the manufacturer. The reagents are more expensive than EMIT reagents, being roughly comparable in cost to RIA reagents. They come in a liquid form and have a more limited shelf life than those for EMIT, but they tend to be more stable than RIA reagents.

The attractiveness of FPIA is in the speed and ease of operation of the instrument. The reagents come in a kit that is bar coded and is placed right into the instrument. All the operator has to do is fill the sample cups with serum or urine, place the reagent pack inside the instrument, and push a button marked "run." The instrument reads the bar code, enters the necessary programs into its memory, performs the assay, and prints out the results. For the routine hospital lab or small drug-testing lab, it is as fast or faster than EMIT or RIA and a lot easier; however, the instrument can only run twenty specimens at a time. For the large drug-testing laboratory, more rapid results can be achieved with the automated instrumentation available for the EMIT or RIA techniques.

FPIA is nearly as sensitive as RIA; digoxin can be run by FPIA, although LSD is still not available. The specificity of FPIA is also comparable to that of RIA.

(SEE ALSO: *Drug Testing and Analysis; Hair Analysis As a Test for Drug Use*)

BIBLIOGRAPHY

BLAKE, C. C. F. (1975). Antibody structure and antigen binding. *Nature* (London), *253*, 158.

CHARD, T. (1990). *Laboratory techniques in biochemistry and molecular biology: An introduction to radioimmunoassay and related techniques*, 4th ed. Amsterdam: Elsevier.

EKINS, R. (1989). A shadow over immunoassay. *Nature* (London), *340*, 599.

HOWANITZ, J. H., & HOWANITZ, P. J. (1979). Radioimmunoassay and related techniques. In J. B. Henry (Ed.), *Clinical diagnosis and management by laboratory methods*, 16th ed. Philadelphia: W. B. Saunders.

NAKAMURA, R. M., TUCKER, E. S., III, & CARLSON, I. H. (1991). Immunoassays in the clinical laboratory. In J. B. Henry (Ed.), *Clinical diagnosis and management by laboratory methods*, 18th ed. Philadelphia: W. B. Saunders.

STEWART, M. J. (1985). Immunoassays. In A. C. Moffat (Ed.), *Clark's isolation and identification of drugs*. London: Pharmaceutical Press.

TIETZ, N. W. (1982). Radioimmunoassay. In *Fundamentals of clinical chemistry*. Philadelphia: W. B. Saunders.

JEFFREY A. GERE

IMPAIRED PHYSICIANS AND MEDICAL WORKERS

Concern about impairment from alcohol and drugs in health-care professionals in the United States and in other countries has waxed and waned during the twentieth century. Until the 1960s, ALCOHOL, the OPIATES, and other PRESCRIPTION DRUGS were the primary concerns. More recently, the concern has been extended to MARIJUANA and COCAINE.

Although there are many estimates of addiction rates among physicians, the prevalence of alcohol and other drug problems within the entire health-care profession is unknown. Brewster (1986) reviewed published estimates of U.S. addiction rates among physicians and found that available reports were not adequate to support firm conclusions about the prevalence rates.

Physicians (since we do have data on them) are as likely as their age and gender peers to have experimented with drugs—both licit and illicit. They are, however, less likely to be current users of illicit substances (Hughes et al., 1992). Self-medication by physicians has changed little since the 1960s, whereas the use of cocaine and marijuana has greatly increased (McAuliffe et al., 1986).

Figure 1 shows the results of three surveys of drug use among U.S. medical students. Substantial numbers of medical students come to medical training having had some experience with illicit drugs.

Disciplinary or diversion actions by health professionals' licensing boards and studies of health professionals who receive treatment for alcohol or drug dependency are additional sources of information about the kinds of problems caused by drugs and alcohol and their relative frequency.

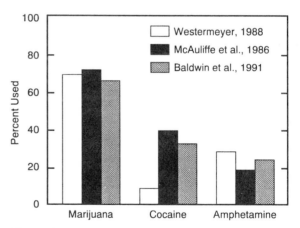

Figure 1

Lifetime Drug Use by Medical Students

Westermeyer (1988) surveyed first-year medical students (*n* = 195) at the University of Minnesota. During 1984–1985, McAuliffe et al. (1986) mailed anonymous questionnaires to a random sample of medical students in New England (*n* = 381). In 1987, Baldwin et al. (1991) mailed questionnaires to senior medical students at 23 schools located throughout the country (*n* = 2,046). (Note: *n* is the number of medical students who returned questionnaires.)

It is widely believed that health-care professionals are especially vulnerable to problems of alcohol and drug abuse because of familiarity with and ready access to drugs; the high STRESS associated with patient-care responsibilities and their own family problems may also contribute. Many physicians self-prescribe medications for relief of PAIN and ANXIETY. A 1989/90 survey of U.S. physicians found that 11.4 percent had used BENZODIAZEPINES and 17.5 percent had used minor opiates during the preceding year without medical supervision (Hughes et al., 1992a).

While the problem of drug addiction among health-care professionals is now widely acknowledged, such awareness has not always been the case. Impaired physicians and other health-care workers have been fearful of seeking help because they have not known how patients and colleagues might respond and because they fear loss of practice privileges and licenses. Physicians often feel uncomfortable about confronting drug or alcohol abuse in a colleague. They want to believe that a colleague in trouble will know when to seek help and will voluntarily seek it. The reluctance of physicians to report colleagues has been called a conspiracy of silence.

In 1972, the American Medical Association (AMA) Board of Trustees accepted the report of its Council on Mental Health and officially ended the conspiracy of silence by making physicians ethically responsible to recognize colleagues' inability to adequately practice medicine—an inadequacy that includes difficulties caused by drug or alcohol abuse. The council recommended a series of steps that should be taken if the impaired physician does not curtail practice: referral of the problem to the medical staff of hospitals in which the physician practices; referral to the state or county medical association; or, if other steps fail, referral to the licensing agency. In 1974, the AMA drafted model legislation allowing states' licensing agencies to require treatment and rehabilitation of impaired physicians as a condition of maintaining licensure. Before that time, the only possible response of the licensing agency was to discipline the physician. Since then, many state medical societies and licensing bodies have established programs for health professionals with alcohol or drug addiction.

In response to increasing malpractice, the U.S. Department of Health and Human Services established the National Practitioner Data Bank, to collect information about malpractice and state-board licensing actions, hospital restrictions, revocation or denial of privileges, or denial of membership by a professional society. The purpose of the data bank is to prevent physicians from moving from state to state and continuing to practice without disclosing previous adverse actions against them. Hospitals must request information from the data bank when a physician applies for clinical privileges. The data bank prevents physicians with untreated alcohol or drug dependencies who have been disciplined in one state from practicing without restrictions in another state.

As a means of detecting drug and alcohol abuse, random urine testing is sometimes proposed for physicians and other health-care professionals. The AMA opposes routine urine testing because it intrudes on personal privacy and because a positive test does not establish impairment. Furthermore, drug- and alcohol-induced impairments are complex psychosocial and neurobehavioral problems that require a comprehensive clinical assessment, and neurobehavioral testing may better reflect the degree of impairment.

Urine testing is useful for other purposes. It is, for example, one of the best ways to document abstinence, which is an indicator of treatment progress.

DIFFERENCES IN PREVALENCE BY SPECIALTY

The choice of a particular drug and route of administration is influenced by accessibility and familiarity. Among anesthesiologists, for example, injectable fentanyl and its analogs are the most frequently abused opioids (see Table 1).

Although opioid addiction—and addiction treatment—among anesthesiologists has received frequent notice, the addiction rate among anesthesiologists may not be higher than among other physicians. Opiate abuse among anesthesiologists may be discovered more frequently because of the hospital environment in which they must practice. Interpersonal stress and the isolation of an office practice are believed to make psychiatrists particularly vulnerable to alcohol and drug abuse. The privacy of a solo office practice also makes detection difficult.

DETECTION OF ADDICTION

In hospitals, drug use by health-care professionals is often uncovered during inventory audits of medications, and the concealment efforts of impaired

TABLE 1
Drugs of Abuse by Anesthesiologists

Drug of Abuse	California Diversion Program n = 42 (percent)	Resident Anesthesiologists n = 132 (percent)
fentanyl/sufentanyl	40	80
meperidine (Demerol) and other opiates	29	17
alcohol	17	11
cocaine	10	3
diazepam (Valium)	2	12
inhalation anesthetics (e.g., nitrous oxide)	0	8
ketamine	0	6
other	2	6

Data from the Diversion Program of the Medical Board of California shows the *primary* drug of abuse of anesthesiologists who were participating in the program during 1989 (Ikeda & Pelton, 1990). The data on resident anesthesiologists show their drug or drugs of abuse (Menk et al., 1990). Some residents abused more than one drug.

health-care professionals are often reflected in their treatment of or attitude toward patients. Some physicians who abuse prescription medications routinely overprescribe opiates or other drugs to patients in an effort to hide their self-prescription; others may prescribe unusually conservatively to avoid drawing attention to themselves.

TREATMENT

Many health professionals are pessimistic about the treatability of substance abuse, and if they develop an alcohol or drug problem, they may discount the value of treatment for themselves. Those who train or work in public-sector hospitals or clinics often observe that the treatment of their patients is rarely successful. Their perception is unduly pessimistic, however, because such clinics often treat recalcitrant, end-stage substance abusers.

The resistance to seeking treatment on their own often necessitates some form of coercion to force health professionals into treatment. One method of breaking down denial and forcing a person to seek treatment is called an *intervention*. The process consists of a group confrontation of the drug-abusing professional by concerned friends, family, and colleagues. A peer professional experienced in conducting interventions often assists in setting up the confrontation. The interventionist rehearses those

who will be involved. When the stage is set, participants each tell the abuser what they have observed concerning the drug abuse and how it has adversely affected them. The confrontation, which may include threats to notify the abuser's employer, hospital, or state licensing board, may motivate an abuser to go for treatment. Such motivation is often fleeting, so it is important for the addict to go immediately into a treatment setting.

TREATMENT MODALITIES

Most treatment for impaired health professionals is drug-free and recovery-oriented, emphasizing follow-up participation in ALCOHOLICS ANONYMOUS (AA), COCAINE ANONYMOUS (CA), or other peer-led groups. Recovering physicians rate participation in AA, for example, as an important factor in their recovery. In most respects, treatment of addiction for health professionals differs little from that used for other middle- and upper-class patients. Health-care professionals who abuse prescription drugs often see themselves as being different from street-drug users. Some programs deal with this form of resistance by insisting on a uniform treatment for all patients. There are, however, special problems that must be addressed. For example, addicted physicians, unlike street addicts, often underreport their degree of PHYSICAL DEPENDENCE on a substance in an effort to project a false sense of being in control. A period of inpatient treatment is often required.

METHADONE MAINTENANCE, which has been employed successfully for some HEROIN (and other OPIOID) addicts, is generally not an option for practicing health professionals, since most licensing boards will not allow them to practice while taking methadone. NALTREXONE has been particularly successful with health-care professionals and is the only medication for treatment of opioid dependency acceptable to most licensing boards. The ingestion of naltrexone reassures licensing boards and hospitals that the recovering health professional is not impaired from abuse of opioids. Its lack of mood-altering effect also fits well with the drug-free treatment philosophy.

WORK REENTRY

Work reentry can be difficult for recovering health professionals. Those who have abused prescription medication face reexposure to their drugs of abuse, which could lead to relapse. Hospital and other professional privileges are not easily regained. Licensing boards often opt to revoke or restrict the impaired health professional's license to practice, and insurance companies often refuse malpractice coverage to recovering addicts.

Some of these obstacles can be overcome with planning and peer support. For example, a nurse may find employment in a blood bank or other setting in which there is no access to drugs. Also, a physician may make arrangements to have a colleague see all the patients that require a NARCOTIC, thus avoiding having to write narcotic-containing prescriptions. Reentry may involve redirecting the health-care practitioner's professional activities to a different location or area of treatment, restricting the recovering health professional's scope of practice, or removing him or her from the previous practice environment altogether. For many health professionals, return to full practice after a period of monitored abstinence and compliance with treatment is possible.

RESPONSE TO TREATMENT

Prognosis for physicians treated for ALCOHOLISM or drug dependency is generally favorable. A study comparing physicians with other middle-class patients similarly treated in an inpatient program showed that physicians did better. The California Physicians' Diversion Program reported a 69 percent success rate for anesthesiologists and an overall success rate of 73 percent. This success is attributed to regular attendance at group meetings, regular testing for sobriety, and immediate corrective action whenever a slip or relapse occurs.

Such high rates of success are not uniformly attained. In a survey of training programs for anesthesiologists, it was found that of the seventy-nine anesthesiology residents who returned to their specialty following treatment, only twenty-seven (34%) did not relapse—and of the fifty-two who relapsed, thirteen (25%) died of drug overdose (Menk et al., 1990).

Some medical specialties are more stringent than others in allowing recovering trainees to return. Minor slips that are often dealt with by additional treatment in some specialties are usually not acceptable in anesthesiology training programs. Therefore, comparison of recovery rates between treatment programs and different subgroups of physicians is difficult to impossible.

(SEE ALSO: *Coerced Treatment; Contingency Contracts; Drug Testing and Analysis; Industry and Workplace*)

BIBLIOGRAPHY

AMA COUNCIL ON MENTAL HEALTH. (1973). The sick physician: Impairment by psychiatric disorders, including alcoholism and drug dependence, *Journal of the American Medical Association, 223*(6), 684–687.

BALDWIN, D. C., JR., ET AL. (1991). Substance use among senior medical students. *Journal of the American Medical Association, 265*(16), 2074–2078.

BREWSTER, J. M. (1986). Prevalence of alcohol and other drug problems among physicians. *Journal of the American Medical Association, 255*(14), 1913–1920.

GALANTER, M., ET AL. (1990). Combined Alcoholics Anonymous and professional care for addicted physicians. *American Journal of Psychiatry, 147*(1), 64–68.

HANKES, L., & BISSELL, L. (1992). Health professionals. In J. H. Lowinson et al. (Eds.), *Substance abuse: A comprehensive textbook*. Baltimore: Williams & Wilkins.

HUGHES, P. H., ET AL. (1992a). Prevalence of substance use among U.S. physicians. *Journal of the American Medical Association, 267*(17), 2333–2339.

HUGHES, P. H., ET AL. (1992b). Resident physician substance use by specialty. *American Journal of Psychiatry, 149*, 1348–1354.

IKEDA, R. M., & PELTON, C. (1990). Diversion programs for impaired physicians. *Western Journal of Medicine, 152*, 617–621.

LING, W., &WESSON, D. R. (1984). Naltrexone treatment for addicted health-care professionals: A collaborative private practice experience. *Journal of Clinical Psychiatry, 45*(9), 46–48.

McAULIFFE, W. E., ET AL. (1986). Psychoactive drug use among practicing physicians and medical students. *New England Journal of Medicine, 315*(13), 805–810.

MENK, E. J., ET AL. (1990). Success of reentry into anesthesiology training programs by residents with a history of substance abuse. *Journal of the American Medical Association, 263*(22), 3060-3062.

MORSE, R. M., ET AL. (1984). Prognosis of physicians treated for alcoholism and drug dependence. *Journal of the American Medical Association, 251*(6), 743–746.

ORENTLICHER, D. (1990). Drug testing of physicians. *Journal of the American Medical Association, 264*(8), 1039–1040.

PELTON, C., & IKEDA, R. M. (1991). The California physicians diversion program's experience with recovering anesthesiologists. *Journal of Psychoactive Drugs, 23*(4), 427–431.

SWEARINGEN, C. (1990). The impaired psychiatrist. *Psychiatric Clinology in North America, 13* (1), 1–11.

TALBOTT, G. D., ET AL. (1987). The Medical Association of Georgia's Impaired Physician Program, review of the first 1000 physicians: Analysis of specialty. *Journal of the American Medical Association, 257*(21), 2927–2930.

VAILLANT, G. E., BRIGHTON, J. R., & McARTHUR, C. (1970). Physicians' use of mood-altering drugs: A 20-year follow-up report. *New England Journal of Medicine, 282*, 365–370.

WESSON, D. R., & SMITH, D. E. (1990). Prescription drug abuse—patient, physician and cultural responsibilities. *Western Journal of Medicine, 152*(2), 613–616.

WESTERMEYER, J. (1988). Substance abuse among medical trainees: Current problems and evolving resources. *American Journal of Drug and Alcohol Abuse, 14*(3), 393–404.

DONALD R. WESSON
WALTER LING

INDIA, DRUG USE IN *See* Asia, Drug Use in; Bhang.

INDUSTRY AND WORKPLACE, DRUG USE IN

The 1991 NATIONAL HOUSEHOLD SURVEY ON DRUG ABUSE (NHSDA) indicated that roughly 10 million employed Americans were "current users" of illegal drugs. Drug use by employees and workers has become an important issue for American business. In the 1990s, employers of all types (large and small businesses, nonprofit organizations, government) are attempting to contain the negative impact of illegal drug use on job performance, PRODUCTIVITY, safety, and health (Walsh & Gust, 1989).

Drug use "in the workplace" is perhaps a misnomer in that today's employer policies focus on drug use by the "worker," whether that use is at the work site or off the job. In the United States, drug abuse in the workplace was recognized as a serious problem in the early 1980s, and in the next decade a slow but progressive response by both labor and management yielded model programs and policies to deal with the issue. Typically, by 1993 most organizations had comprehensive programs that included the following basic components: a written policy; supervisory training; employee education; employee assistance resource; and DRUG TESTING.

With such a comprehensive approach, the workplace has proven to be one of the most effective venues for drug prevention, treatment, and rehabilitation efforts (Gust & Walsh, 1989; Gust et al., 1991). Figure 1 shows the NHSDA data for "current users" (those reporting use of an illegal drug within thirty days prior to the survey), with figures broken down by employment status. Many observers believe that comprehensive workplace-based programs that reach out not only to the workers but also to their spouses and their student children can have an impact on at least 80 percent of all current users.

These endeavors have not been undertaken without controversy, especially with regard to the use of employee "drug testing," but the controversy is perhaps the most interesting part of the story. The development, current status, and future of these policies and programs will be discussed in detail later in this article. The evolution of workplace-based antidrug programs provide insights into the way unique social problems create a need for and the eventual

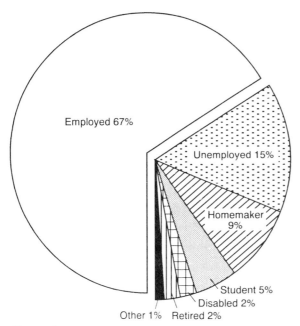

Figure 1
Employment Status of Illegal Drug Users

development of innovative public policy (for additional information, see Walsh & Yohay, 1987).

Workplace "antidrug" policies date back to the 1960s, particularly in the transportation and other safety-sensitive industries. These policies were not then very effective because detection methods were poor and the signs and symptoms of drug use are often subtle and difficult to identify. Not until 1980, when new technology became available that provided reliable, inexpensive detection methods for MARIJUANA and other commonly abused drugs, did workplace detection efforts begin to be effective. Interestingly, the "workplace" that triggered the birth of these antidrug initiatives was the U.S. military.

In 1971, President Richard M. Nixon, as commander in chief, changed the Uniform Code of Military Justice so that testing positive for an illicit drug was no longer a court-martial offense. He ordered the military to start a program of urine testing among U.S. troops in VIETNAM and to offer treatment to those who tested positive for drug use. This urine-testing program was then expanded to service personnel worldwide. In the mid-1970s, the program was discontinued as a result of court challenges. At the time, the only drugs that could easily be tested for were OPIOIDS and some STIMULANTS.

The remarketing and reintroduction of drug-testing technology in 1981 occurred at roughly the same time as the Department of Defense (Burt & Biegel, 1980) and the Congress (House Select Committee on Narcotics, 1981) independently reported the survey results of drug use by MILITARY personnel. The results of these two surveys indicated high rates of drug use by military personnel and brought about considerable congressional scrutiny. The accident on the aircraft carrier U.S.S. *Nimitz* in May 1981, in which drug use was discovered by the postmortems of the crew members, increased political pressure on the military to do something about the drug-abuse problem. The juxtaposition of these events—the availability of drug-testing technology and congressional demands for the Defense Department to address drug taking in the military—was pivotal in justifying the widespread application of drug testing and the formulation of strict policies forbidding the illegal use of drugs on or off the job.

The development of such policies in the military received wide media coverage and generated much discussion in 1981. Shortly thereafter (1982–1983), similar policies began to be adopted in the transportation and utility industries for employees in safety-sensitive positions. The National Transportation Safety Board documentation of drug involvement in railroad and airline accidents more than justified the increasing concerns about workplace drug abuse.

Early military and private-sector policies were punitive in nature; employees found to be using drugs were summarily dismissed. This created a dilemma for many major corporations that recognized they had a drug problem but didn't feel comfortable firing employees, especially when there was no safety or security nexus. The rationale for workplace drug policies and the use of drug testing has evolved considerably since 1981. The philosophy of why to test and what to do with the results changed dramatically during the 1980s and early 1990s. At the outset, the primary purpose of a drug policy was to identify users and to fire them without evaluating the circumstances of the drug use. Subsequently a more positive, helping-hand philosophy evolved.

The basic purpose of today's model corporate drug policy is twofold:

1. to minimize the risk of hiring drug users by denying employment to applicants who use illegal

drugs (as manifested by a positive preemployment drug test); and

2. to provide active programs to get the substance-abusing employee into treatment, to afford the opportunity to get help, and to get the individual back on the job.

This philosophical change to a more politically acceptable, socially responsible policy in dealing with drug abuse evolved about 1986 and allowed major corporations and professional organizations to involve themselves in antidrug workplace initiatives.

The federal government facilitated, encouraged, and in some instances required the development of private- and public-sector workplace antidrug programs. The Federal Railroad Administration began hearings on drug rules for the railroad industry in 1984 and issued regulations requiring written policies and the testing of employees; after a number of legal delays, the regulations went into effect in 1986. In September 1986, President Ronald W. Reagan issued an executive order (EO 12564) that required all federal agencies to develop drug-free workplace programs to ensure that the more than 2 million federal employees were not illegally using drugs on or off the job. In 1988 the Department of Transportation issued regulations for the airline, maritime, trucking, railroad, pipeline, mass transit, and other transportation industries, requiring (1) written policies prohibiting the illegal use of drugs on or off the job and (2) preemployment, reasonable-suspicion, postaccident, and "random" drug testing without cause for employees in specified safety-sensitive occupations. By 1992 the regulations were extended to cover intrastate as well as interstate transportation, a move that increased the number of transportation workers affected by the regulations to 10 million, or nearly 10 percent of the total U.S. workforce. The Nuclear Regulatory Commission also issued regulations requiring written policies and extensive testing of personnel at nuclear sites.

Congress got on the bandwagon and passed the Drug-Free Workplace Act of 1988, which requires all federal grant recipients and federal contractors (whose contracts exceed $25,000) to certify that they will provide a drug-free workplace. The final rules describing the requirements for such grantees and contractors were published in the *Federal Register* on May 25, 1990. In general, the law requires covered employers to:

1. Develop and publish a written policy and ensure that employees read and consent to the policy as a condition of employment;
2. Initiate an awareness program to educate employees about
   - the dangers of drug abuse
   - the company's drug-free workplace policy
   - any available drug counseling, rehabilitation, and employee-assistance programs
   - the penalties that may be imposed on employees for drug-abuse violations;
3. Require that all employees notify the employer or contractor of any conviction for a drug offense in the workplace;
4. Make an ongoing effort to maintain a drug-free workplace.

In 1988, the Bureau of Labor Statistics (BLS) surveyed business establishments throughout the United States about their policies on drug abuse (BLS, 1989). The survey found that half the nation's nonagricultural workforce was employed by organizations with a formal policy on drugs, and that 20 percent of payroll workers were employed in establishments with some type of drug-testing program. More than 90 percent of the establishments surveyed had an EMPLOYEE-ASSISTANCE PROGRAM available to employees. In the years since the BLS survey, the number of corporate and other employers and the employees covered by these policies continued to grow exponentially. The American Management Association (Greenberg, 1993) has surveyed its membership about their workplace drug policies annually since 1987. The 1993 survey indicated that 84 percent of respondents believe that drug testing is an effective way to deal with workplace drug abuse, compared with 50 percent in 1987. The share of surveyed firms that test for drugs rose to 85 percent in 1993. Since 1987, drug testing has increased nearly 300 percent. From the drug-treatment perspective, more than half of all companies (54%) have indicated that employees who test positive are referred for counseling and treatment.

As indicated above, progress in using the work site to intervene in individual substance abuse has not happened without controversy. Generally, employees and workers have no problem with supervisory training or employee education. However, the utilization of drug testing to make employment decisions and the involvement of employee-assistance programs (EAPs) in what many feel is a policing action generates an emotional, gut-level response from both labor and management. The drug testing and EAP components are so critical to any workplace effort that a detailed discussion of the issues is required.

DRUG TESTING

When drug testing is considered, it is important to be familiar with the basic issues with which management and labor have been struggling (a full range of issues are discussed in Walsh & Trumble, 1991). The question of whether to utilize drug-testing technology evokes a complex array of moral, social, ethical, medical, scientific, and legal issues for many Americans. Although most citizens do not condone drug abuse, their concerns about the erosion of civil liberties generate feelings of uncertainty as to whether the end justifies the means. Where will it stop? Where do you draw the line? are questions raised by unions, civil libertarians, and others who worry that employee AIDS testing and pregnancy testing will be the next battlegrounds.

Many Americans view the drug-testing process (i.e., collection of urine) as degrading and dehumanizing. Government employees, unions, and civil libertarians argue strongly that drug testing is an invasion of privacy, that it constitutes an illegal search and seizure (i.e., of body fluids) and therefore violates individual rights guaranteed by the Constitution. In general, the constitutional protections apply only to testing conducted by the government (federal, state, and local). Therefore, testing conducted by private employers is not covered by the constitutional safeguards. However, government-mandated drug testing of private-sector employees—for example, in the federally regulated transportation and nuclear-power industries—must also pass constitutional muster. Although several of these constitutional questions have been brought before the Supreme Court and have generally been upheld, many specific issues may not be resolved by the current cases and will likely continue to be the subject of litigation for some time. This legal uncertainty—whether testing will be upheld and programs go forward or will be found unconstitutional and therefore be restricted—has created confusion for policymakers as well as for employees and unions.

Medical and scientific questions about the accuracy and reliability of drug testing were and are continually raised by those who oppose testing. Concerns have been voiced that many laboratories offering drug-testing services do not have the expertise or capability to perform the assays required. In addition, many employers may be using inappropriate technology and falsely accusing employees of drug use. Congressional support for these concerns has been manifested by the passage of legislation (P.L. 100-71, sec. 503, July 11, 1987) that requires stringent technical and scientific procedures for federal workplace drug-testing programs, as well as standards for the certification of laboratories engaged in drug testing for federal agencies. Similar legislation has been introduced in both the U.S. Senate and House of Representatives that would require such standards and lab certification for the private sector.

In response to concerns about the accuracy and reliability of drug testing, the U.S. Department of Health and Human Services (HHS) issued *Technical and Scientific Guidelines for Federal Drug Testing Programs* (Walsh, 1988). These guidelines are mandatory for federal programs and have rapidly become the gold standard for private-sector programs as well. By 1993, the rigor of the federal standards virtually eliminated concerns regarding accuracy and reliability. The issue of the quality of laboratories has also been addressed by HHS through the establishment of a national laboratory certification program. The College of American Pathologists has also established a forensic urine drug-testing certification program making certified labs available in virtually every state. The use of a certified lab has become the standard by which drug-testing programs are measured. A consensus report from HHS on the scientific issues of drug testing provides detailed information (Finkle et al., 1990).

A discussion of the pros and cons of drug testing provides no clear answers (Walsh & Trumble, 1991). The American Civil Liberties Union (ACLU) has been among the most vocal organizations actively lobbying against drug testing. In addition to constitutional issues, a major concern has been the potential for abuse by managers and supervisors to discriminate against and harass employees. The focus of the ACLU argument is that a positive urinalysis does not prove intoxication or impairment of performance; therefore it cannot be used to draw a nexus between drug use and job performance.

For their part, employers have wrestled with competing objectives and values to develop substance-abuse policies that fulfill multifaceted obligations.

TABLE 1
Some Recent Attempts to Define Alcoholism and/or Drug Dependence

Drug dependence. A state, psychic and sometimes also physical, resulting from the interaction between a living organism and a drug, characterized by behavioural and other responses that always include a compulsion to take the drug on a continuous or periodic basis in order to experience its psychic effects, and sometimes to avoid the discomfort of its absence. Tolerance may or may not be present. A person may be dependent on more than one drug. (*World Health Organization Technical Report Series*, 1969, no. 407, p. 6.) This definition was reaffirmed in the WHO Expert Committee on Drug Dependence Nineteenth Report, *World Health Organization Technical Report Series*, 1973, no. 526, p. 16.

Alcoholism is a chronic, progressive, and potentially fatal disease. It is characterized by tolerance and physical dependency or pathologic organ changes or both, all of

which the direct or indirect consequences of the alcohol ingested. (National Council on Alcoholism/American Medical Society on Alcoholism, 1976.) (See Flavin & Morse, 1991.)

The 1976 definition was revised and broadened in 1991 to include the concept of *denial:*

Alcoholism is a primary, chronic disease with genetic, psychosocial, and environmental factors influencing its development and manifestations. The disease is often progressive and fatal. It is characterized by continuous or periodic impaired control over drinking, preoccupation with the drug alcohol, use of alcohol despite adverse consequences, and distortions in thinking, most notably denial. (National Council on Alcoholism and Drug Dependence/American Medical Society on Alcoholism, 1976) (See Flavin & Morse, 1991.)

On the one hand, many employers feel a moral obligation to do all they can to achieve a drug-free workplace. They have corporate responsibilities to provide a healthy and safe workplace for all employees and to protect shareholders from losses resulting from drug abuse. On the other hand, employers have obligations to their workers—to respect the individual rights and civil liberties of loyal and trustworthy employees (who for the most part are not involved with drugs).

This is an exceedingly difficult balancing act, and, as workplace policies are designed, the balance will shift depending on the individual work site and the nature of the particular job.

EMPLOYEE-ASSISTANCE PROGRAMS

EAPs have become the key component of model workplace policies (Masi, 1984). Although drug testing has provided the major turning point in the evolution of workplace antidrug programs, the EAPs have expanded, grown more sophisticated, and become a vital part of the antidrug initiative. EAP programs were developed in the 1970s to focus on ALCOHOL abuse and to assist employees in dealing with the stresses of employment and personal life. Typically EAP programs provide short-term counseling and serve as a referral source for those employees who need treatment or long-term counseling. So-

called broad-brush EAP programs provide a variety of services, in addition to crisis intervention, including management training and health workshops and seminars (e.g., SMOKING CESSATION, weight reduction).

As managers began to develop antidrug policies, the question was raised: What will we do if we find an employee using drugs? Generally, corporate lawyers and security officers would suggest termination, while corporate medical and EAP staff would recommend treatment. The issue proved difficult to resolve for many corporations when the cost of treatment and the uncertainty of success weighed heavily on the minds of financial officers responsible for making a profit in a bad economy. Fortunately, most corporations have EAP resources to implement the "helping-hand" approach that management sought.

The involvement of the EAP program in the antidrug effort was also not without problems. Initially some EAP providers had difficulty expanding their programs to deal with illegal drug users, a different type of client from ones with whom they had previously worked. The illegal aspect of drug behavior was troublesome in a field where confidentiality is the cornerstone of the therapeutic relationship. Also the advent of drug testing created an ethical dilemma for the EAP provider who was accustomed to being an ombudsman between management and la-

TABLE 2
A Comparison of ICD-10 and DSM-IV Criteria for Dependence

ICD-10 Dependence Syndrome	*DSM-IV Substance Dependence*
A cluster of physiological, behavioural, and cognitive phenomena in which the use of a substance or a class of substances takes a higher priority for an individual than other behaviours that once had greater value. A central characteristic of the syndrome is the desire (often strong, sometimes overpowering) to take psychoactive drugs (medically prescribed or not), alcohol, or tobacco. There may be evidence that return to substance use after a period of abstinence leads to a more rapid reappearance of other features of the syndrome than occurs with nondependent individuals.	A maladaptive pattern of substance use, leading to clinically significant impairment or distress, as manifested by three or more of the following occurring at any time in the same twelve-month period:

Diagnostic guidelines
A definite diagnosis of dependence should usually be made only if three or more of the following have been experienced or exhibited during the previous year:
	(1) tolerance, as defined by either of the following:
	   (a) need for markedly increased amounts of the substance to achieve intoxication or desired effect
	   (b) markedly diminished effect with continued use of the same amount of the substance

(a) a strong desire or sense of compulsion to take the substance;	(2) withdrawal, as manifested by either of the following:
	   (a) the characteristic withdrawal syndrome for the substance . . .
	   (b) the same (or closely related) substance is taken to relieve or avoid withdrawal symptoms

(b) difficulties in controlling substance-taking behaviour in terms of its onset, termination, or levels of use;	(3) the substance is often taken in larger amounts or over a longer period than was intended
(c) a physiological withdrawal state . . . when substance use has ceased or been reduced, as evidenced by: the characteristic withdrawal syndrome for the substance; or use of the same (or a closely related) substance with the intention of relieving or avoiding withdrawal symptoms;	(4) a persistent desire or unsuccessful efforts to cut down or control substance use
(d) evidence of tolerance, such that increased doses of the substance are required to achieve effects originally produced by lower doses (examples are alcohol- and opiate-dependent individuals who may take doses sufficient to incapacitate or kill nontolerant users);	(5) a great deal of time is spent in activities necessary to obtain the substance (e.g., visiting multiple doctors or driving long distances), use the substance (e.g., chain-smoking), or recover from its effects.
(e) progressive neglect of alternative pleasures or interests because of psychoactive substance use, increased amount of time necessary to obtain or take the substance or to recover from its effects;	(6) important social, occupational, or recreational activities given up or reduced because of substance use
(f) persisting with substance use despite clear evidence of overtly harmful consequences, such as harm to the liver through excessive drinking, depressive mood states consequent to periods of heavy substance use, or drug-related impairment of cognitive functioning; determination should be made of the user's actual or expected awareness of the nature and extent of the harm.	(7) continued substance use despite knowledge of having had a persistent or recurrent physical or psychological problem that was likely to have been caused or exacerbated by the substance (e.g., current cocaine use despite recognition of cocaine-induced depression, or continued drinking despite recognition that an ulcer was made worse by alcohol consumption)

Narrowing of the personal repertoire of patterns of psychoactive substance use has also been described as a characteristic feature (e.g. a tendency to drink alcoholic drinks in the same way on weekdays and weekends, regardless of social constraints that determine appropriate drinking behaviour).	Specify if:
	   with physiological dependence: Evidence of tolerance or withdrawal (i.e., either item [1] or [2] is present):
	   without physiological dependence: No evidence of tolerance or withdrawal (i.e., neither item [1] nor [2] is present).

bor. A good percentage of EAP referrals were coming from the drug-testing program in a last-chance situation in which the pressure was on the EAP to "cure" the problem—or management would fire the employee. In the past, many employees using EAP services had sought assistance on their own, and management was never aware of the employee's initiative.

Despite these problems, the EAP field has expanded its efforts to treat substance abuse and has proven to be integral to the entire program. Employers have recognized that EAP programs not only help employees but are cost-effective. New materials, training programs in substance abuse, and certification programs have developed that have made the EAP provider more skilled in dealing with the drug-using employee.

SUMMARY

Although drug abuse in the workplace is still a significant concern of American employers, substantial progress has been made since the early 1980s. Companies with comprehensive programs report significant reductions in accidents, absenteeism, and positive drug tests. There continues to be progressive growth in small and mid-size businesses, as resources for EAP, testing, management training, and legal services are being made available through local business consortia. The business community has developed a consensus that the workplace is an appropriate site for confronting drug abuse and has sent a clear message to the workforce and to the community that drug use will not be tolerated.

For the future, we are likely to see continued growth and expansion of workplace programs. As the country has gained confidence in the accuracy and reliability of drug testing, lower thresholds will be permitted that will make it much more difficult for the casual user to escape detection. We will probably see federal legislation setting additional standards for workplace programs, including standards for testing and for protection of employees.

Educating high school and college students that they must be drug-free to get and hold a job will in the long run contribute significantly to the reduction of drug abuse in the student population. And finally, because the workplace efforts are the most organized drug education, prevention, and treatment initiatives

in the country today, they represent the best prospect for turning around the drug problem in America.

(SEE ALSO: *Accidents and Injuries; Drug Metabolism; Hair Analysis as a Test for Drug Use; Prevention*)

BIBLIOGRAPHY

BUREAU OF LABOR STATISTICS. (1989). *News release* USDL 89-7 (January 11). Washington, DC: U.S. Department of Labor.

BURT, M. R., & BIEGEL, M. M. (1980). *Worldwide survey of nonmedical drug use and alcohol among military personnel.* Bethesda, MD: Burt Associates.

FINKLE, B. S., ET AL. (1990). *Technical, scientific, and procedural issues of employee drug testing: A consensus report.* DHHS Publication no. (ADM) 90-1684. Washington, DC: U.S. Department of Health and Human Services.

GREENBERG, E. (1993). *1993 American Management Association survey of workplace drug testing and drug abuse policies.* New York: American Management Association.

GUST, S. W., & WALSH, J. M. (EDS.). (1989). *Drugs in the workplace: Research and evaluation data.* DHHS Publication no. (ADM) 89-1612. Rockville, MD: U.S. Department of Health and Human Services.

GUST, S. W., ET AL. (EDS.). (1991). *Drugs in the workplace: Research and evaluation data (Vol. II).* DHHS Publication no. (ADM) 91-1730. Rockville, MD: U.S. Department of Health and Human Services.

HOUSE SELECT COMMITTEE ON NARCOTICS. (1981). *Results: Personal drug use survey (study mission to Italy and the Federal Republic of Germany).* U.S. House of Representatives. Washington, DC: U.S. Government Printing Office.

MASI, D. A. (1984). *Designing employee assistance programs.* New York: American Management Association.

WALSH, J. M. (ED.). (1988). *Mandatory guidelines for federal workplace drug testing programs: Final guidelines. Federal Register.* Washington, DC: U.S. Government Printing Office.

WALSH, J. M., & GUST, S. W. (EDS.) (1989). *Workplace drug abuse policy: Considerations and experience in the business community.* DHHS Publication no. (ADM) 89-1610. U.S. Department of Health and Human Services. Washington, DC: U.S. Government Printing Office.

WALSH, J. M., & TRUMBLE, J. G. (1991). The politics of drug testing. In R. H. Coombs & L. J. West (Eds.), *Drug test-*

ing: Issues and options. New York: Oxford University Press.

WALSH, J. M., & YOHAY, S. C. (1987). *Drug and alcohol abuse in the workplace: A guide to the issues.* Washington, DC: National Foundation for the Study of Equal Employment Policy.

MICHAEL WALSH

INHALANTS

INHALANTS   Inhalants are solvents or volatile anesthetics that are subject to abuse by inhalation. Most are central nervous system (CNS) depressants, but some are convulsants. As a class they are characterized by high vapor pressure and significant solubility in fat at room temperature. Vapors and gases have been inhaled since ancient times for religious or other purposes, as at the oracle at Delphi. Experimentation with inhalants did not occur to any significant extent until after the discovery of nitrous oxide and the search for volatile anesthetics commenced in earnest. Arguably the most toxic of abused substances, inhalants can produce a wide range of injuries, depending on the chemical constituents of what is inhaled. Many are very complex mixtures formulated for a specific purpose, or are used because they are the least expensive alternative, or both. Thus their purity and safety are in no way comparable with those achieved by pharmaceutical companies manufacturing medications for human consumption.

Inhalants are typically abused by achieving a high airborne concentration of a substance and deliberately inhaling it. With solvents, this typically involves putting the solvent in a closed container, or saturating a piece of cloth and inhaling through it. Compressed gases are sometimes released into balloons and inhaled; directly releasing these substances into the mouth may freeze the larynx, causing laryngospasm and death by asphyxiation. Once the chemical is inhaled, its uptake and duration of action are determined by its solubility in blood and brain, and by the respiratory rate and cardiac output.

The mechanism of action of this class of agents is less well understood than those of other drugs and medications. As CNS depressants, they have been thought to exert their actions by dissolving in membranes and altering their function in a nonspecific way; the potency of these compounds is frequently related to their solubility in membranes. Many con-sider this relationship to better predict the access of the agent to the site of action, and to be unrelated to the mechanism by which the solvents exert their effects. Solvents impair conduction in isolated nerves, and affect nerves with smaller diameters first. This suggests that parts of the nervous system such as the cortex would be affected before systems consisting of large fibers. There is significant interest in the GABA receptor complex as the site of action of many of these compounds. There is not yet evidence for specific interactions with a receptor, in the sense of a "lock and key" mechanism. However, these agents may "lubricate" or "obstruct" such mechanisms.

Although inhalant abuse has been implicated in a variety of organic diseases, its effects on the nervous system have been of the greatest concern. Such injuries range from paralysis and loss of bowel and bladder control, to permanent impairment of the higher cognitive functions and fine motor control. Those who become involved in inhalant abuse vary across culture and, as in many other types of drug abuse, the vulnerability to becoming dependent on these substances may depend on present economic well-being and perceptions of the possibility of future well-being. Their ability to act as a reward has been demonstrated in laboratory animals, so there is no doubt that they exert powerful actions on the nervous system. Preventive actions are of two types: education about the adverse effects of solvents on bodily function, and the possible formulation of consumer products with less intrinsic toxicity. Some manufacturers have attempted to minimize the abuse of their products by adulterating them with irritants. Intervention strategies for those habitually using inhalants are not different from those employed for other CNS depressant dependence disorders. Frank withdrawal symptoms are rarely seen with organic solvents. They do, however, accumulate under some conditions of use, and can be associated with prolonged delirium and behavioral disturbances.

ALKANES

Alkanes are hydrocarbons of the general formula C_nH_{2n+2}. The potency of this family of straight-chain chemicals increases with the number of carbons. The smaller molecules (methane, ethane, butane, propane) are gases at room temperature; their deliberate inhalation produces cardiac arrhythmias and sudden death. Pentane, hexane, and longer alkanes are liquids that become progressively less volatile. Hexane produces a devastating neurotoxicity. Alkanes are paraffins; cycloparaffins are rings without alternating double bonds; and alkylcycloparaffins have a short substituent on the ring. Alkylcycloparaffins such as methylcyclopentane and methylcyclohexane (hexahydrotoluene) are convulsants.

AMYL NITRITE

Amyl nitrite is a volatile, oily liquid with a sweet, banana-like odor. It is sold by prescription in glass ampules for the treatment of angina pectoris, chest pain caused by the narrowing of vessels in the heart. When the glass ampules are broken, they "pop"; hence they are sometimes called "poppers." Amyl nitrite relaxes the vessels of the heart by relaxing the muscles of the veins as well as all other smooth muscles in the body. When the veins throughout the body dilate, blood pressure falls. Because a minimum blood pressure is required to maintain blood supply to vital organs such as the brain, a reflex protects the brain by increasing heart rate and blood flow. This produces a "rush" as the heart pounds, and there is a throbbing sensation in the head. Users also experience a warm flush as the blood accumulates near the skin because of the dilation of veins. Vision also may "redden" as the retinal vessels dilate. The user may faint if the heart cannot maintain blood flow to the brain. If this occurs, the user falls to the floor, and blood flows to the brain, restoring consciousness. Use in a situation where it is impossible to become horizontal may result in brain damage.

The duration of action of the drug is very brief, and as the effect wears off, the user may experience headache, nausea, vomiting, and a chill. The drop in body temperature occurs because of the loss of heat when the veins dilate and the skin flushes. Use of the drug for prolonged periods, or swallowing the liquid, may produce fatal methemoglobinemia, a "chocolate" blood condition in which the blood is brown and cannot carry oxygen to the brain. The drug produces a thick, crusty brown rash if it is spilled on the skin, and is irritating to the lungs. It is flammable and explosive. Volatile nitrites are converted to nitrosamines in the body, and most nitrosamines are very potent cancer-causing chemicals. There is an association of the use of volatile nitrites with Kaposi's sarcoma, an AIDS-related skin cancer. Volatile nitrites impair the function of the immune system. The physiology of sexual intercourse involves smooth muscle; the nitrites relax those muscles as well and so will affect sexual function.

The prescription requirement for amyl nitrite was eliminated in 1960, and its use became popular; in 1964 prescription requirements were reestablished. "Designer" nitrites, such as butyl and isobutyl nitrites, were then bottled and sold as "room deodorizers" with such names as RUSH, Locker Room, and Aroma of Men, so named because it smelled like a locker room. Since these products were not controlled substances or sold as medicines, they were once legal products.

ANESTHETICS

Anesthetics are used in medicine to permit surgical procedures without pain or consciousness. They are of two types: local and general. A local anesthetic is usually injected near nerves to prevent pain in a limited area, such as a Novocaine injection to anesthetize a tooth. General anesthetics are administered to the whole body and depress the CNS to such an extent that major surgery can be performed without killing the patient from the shock resulting from procedures that otherwise would be unendurable. General anesthetics were developed in the mid-nineteenth century by doctors experimenting, usually on themselves, with the organic solvents available at the time. These experiments were sometimes done by groups of people who inhaled the vapors and described the effects, or passed out. Later, careful experimental work identified volatile chemicals that are used to save lives by permitting surgery that would otherwise be impossible to perform, and that are safe to use and have relatively low toxicity.

Some anesthetics can be given by injection. Short-acting anesthetics are used for brief procedures in medicine and dentistry where inhalation anesthesia is inappropriate or difficult, or for starting anesthesia before longer-acting agents are given to the patient. Drugs used for this purpose include barbiturates such as sodium methohexital and sodium thiopental, and benzodiazepines such as midazolam. Fentanyl and related compounds are used for a longer duration of action. A dissociative anesthetic, ketamine, is used for treating burn patients and small children. These agents affect the brain in a more selective way than other anesthetics, so that there is more muscle tone and better circulation in the head and neck. A related veterinary drug, phencyclidine (PCP), has a longer duration of action; when given to humans, however, it has produced terrifying hallucinations upon recovery. It is subject to abuse.

VOLATILE ANESTHETICS

Volatile anesthetics induce unconsciousness and loss of reflexes for surgical procedures. This CNS depression can be induced by a wide variety of different chemicals; those used in clinical medicine are selected for reasons that include low toxicity, ease of maintaining and adjusting a given depth of anesthesia, and freedom from adverse effects upon recovery. Many compounds were examined in the search for modern anesthetic agents.

The depth of anesthesia depends on how much of the medication is present in the CNS. This in turn depends on how much is in the air, to what extent the anesthetic passes between air and blood, and how much passes from blood to brain (or fat, since the brain is largely fat). An agent that is highly insoluble in blood achieves a plateau, or saturation, concentration very rapidly; an example is nitrous oxide. More soluble agents take a longer time to come to plateau, and take a longer time to be exhaled as well, so recovery from them takes longer. Nitrous oxide and cyclopropane have the same solubility in blood, and take the same amount of time to come to a steady concentration in blood; cyclopropane is more soluble in brain and fat, however, so it takes a much lower concentration to achieve the same effect. (Cycloproane is explosive, and therefore is not used in the operating room.) The way an anesthetic functions in a given individual depends on a number of variables, including the amount of fat in the individual's body, the volume of air inspired per minute, the amount of blood pumped through the lungs per minute, and various preexisting medical conditions.

AROMATIC HYDROCARBON SOLVENTS

Aromatic hydrocarbon solvents have a structure that includes a benzene ring. The simplest form is benzene, a six-membered ring with double bonds and six hydrogen atoms. All other aromatic hydrocarbons have alkyl substituents around the ring; for example, toluene has one methyl group and xylene has two methyl groups.

BENZENE

Benzene is a volatile aromatic hydrocarbon (see above). Its presence in consumer products and in the workplace has been reduced because it causes a form of leukemia. Its chemical formula is C_6H_6; it is a six-membered ring with alternating double bonds and a hydrogen on each carbon. The ring opens when metabolized, causing the formation of reactive and toxic chemicals. Benzine, a name applied to automotive fuel in Europe, is a solvent mixture.

BLACK JACK

This is a trade name for several inhalant products that contain either volatile nitrites or ethyl chloride.

CHLORINATED HYDROCARBONS

These substances comprise a large class of industrial chemicals. Those which are highly volatile are sometimes subject to abuse. Chlorinated hydrocarbons undergo significant metabolism in the body, and these changes in chemical structure usually result in an increase of the solvent's toxicity. Because many of these metabolic products are reactive chemicals, they can produce injuries to the kidneys, the liver, and the blood-forming organs. Chlorinated hydrocarbon inhalation is also associated with lethal disorders of heart rhythm, ventricular arrhythmias.

CHLOROFLUOROCARBON PROPELLANTS

Halogenated hydrocarbons are relatively nonreactive chemicals with very high vapor pressure that have been used to blow products out of containers through a tiny hole. Their widespread use in the early 1960s was followed by an epidemic of aerosol sniffing that led to cardiac arrhythmias and death among young people. The halogens—chlorine, fluorine, and bromine—are used to make various chemicals for purposes ranging from propellants and refrigerants to fire extinguishers. Their use has been severely limited since the recognition that their release into the atmosphere depletes the upper layers of ozone, exposing the earth to excessive amounts of ultraviolet radiation. Freon is a brand name for a family of commercial products.

CHLOROFORM

Chloroform, CHCl₃, was one of the earliest solvents put to use as an anesthetic agent. It has been replaced with agents that are much less toxic. Its use in cough and cold medications is obsolete. Chloroform was widely abused in the nineteenth century.

ETHYL CHLORIDE

This is a local anesthetic, CNS depressant, and refrigerant that has been subject to abuse by inhalation. Ethyl chloride has a very high vapor pressure, and spraying it directly into the mouth may freeze the tissues of the throat and cause fatal laryngospasm (contraction of the muscles of the throat and larynx), and the shutoff of air to the lungs. Ethyl chloride has been sold in canisters and spray cans (e.g., Black Jack). A related chemical, methyl chloride, has similar effects and was used in refrigerators until it was recognized as highly poisonous in closed spaces.

ETHYL ETHER

A volatile anesthetic agent subject to abuse by inhalation, ethyl ether was used as an inhalation anesthetic for many years. It has been supplanted by other agents with fewer recovery side effects, such as headache, nausea, and vomiting. It is explosive. Ethyl ether was drunk during the Whiskey Rebellion of the eighteenth century, when heavy taxes were imposed on whiskey. Consumed by this route, ether "tanned" (hardened dramatically) the soft palate. When swallowed, profound intoxication follows, but recovery is faster than from alcohol. Alcohol is metabolized at a fixed number of grams per hour, except under extreme conditions; ethyl ether is eliminated by exhalation.

FREON

Freon is a brand name applied to a class of aerosol propellants. See Chlorofluorocarbon Propellants, above.

GASOLINE

Gasoline, a fuel that powers internal combustion engines, is a complex petroleum product that is subject to abuse by inhalation. The toxicity produced from gasoline exposure depends on the constituents of the mixture and the route of administration. Oral ingestion of gasoline is usually followed by vomiting; subsequent aspiration of gasoline liquid into the lungs is followed by a frequently fatal chemical pneumonia. Deliberate inhalation of leaded gasoline fumes can lead to brain injury related to absorption of tetraethyl lead, a very toxic chemical.

GLUE

Glues are made by dissolving a sticky or adhesive material in a solvent. When the solvent evaporates, the adhesive material remains attached to the surfaces to which it is applied, sticking them together. Glues are complex mixtures formulated for specific purposes. They are not designed for human consumption. When inhaled, they may produce severe injury or death. Most of the solvents used in glues are flammable, and fires have resulted from their inappropriate use. The solvent mixtures in glues and glue thinners are designed to dissolve the solid glue material and to evaporate evenly at a rate appropriate for the product. Solvents of relatively low industrial purity are used in these products; they are usually complex mixtures whose formulation changes with market price. Their toxicity can be great when concentrated and inhaled. Some manufacturers label their products or add irritants in an attempt to dissuade youths from deliberately inhaling these products.

HEXANE

Hexane is a volatile solvent that contains six carbons in a straight chain and has the chemical formula C₆H₁₄. It can cause severe damage to the peripheral nervous system, producing death of the long myelinated nerves (distal axonopathy). This condition results in an inability to walk, loss of muscle mass in all limbs, and sometimes loss of bowel and bladder control. This injury occurs because hexane is metabolized to a gamma-diketone. Another solvent subject to abuse that undergoes the same change in the body is methylbutylketone.

NITROUS OXIDE

Nitrous oxide is a volatile analgesic and anesthetic agent. It was discovered at the beginning of the nineteenth century by Sir Humphry Davy, who was looking for gases and vapors that might have some therapeutic use. Nitrous oxide quickly produces an inebriation that many found pleasurable, and it rapidly became the subject of much experimentation and merrymaking. Nitrous oxide parties became very fashionable, but could not long be limited to the upper classes. Popular demonstrations were conducted, and at one such demonstration Horace Wells noticed that a participant had injured his leg, yet seemed oblivious to the pain. Although Davy had noted that nitrous oxide deadened the pain of his toothaches, it was Wells who underwent the first tooth extraction using nitrous oxide for pain relief. The first widespread use of nitrous oxide for clinically significant pain relief was its use in childbirth by S. Klikovich. Nitrous oxide inhalation is about as effective as 30 mg of morphine for pain relief.

Nitrous oxide is not very soluble in either blood or brain tissue, and consequently it has a short duration of action and requires very high levels to produce effects, on the order of 15 to 30 percent by volume. Because the use of gases at this high a concentration might result in asphyxiation, special equipment is used to guard against this possibility in medical settings. Because it displaces oxygen, nitrous oxide frequently kills those who inhale it for pleasure in closed rooms or automobiles.

Nitrous oxide was long thought to be a relatively innocuous anesthetic, almost as safe as inert gases. Recent work has demonstrated, however, that its inhalation irreversibly inactivates methionine synthetase, and this enzyme inhibition produces a vitamin deficiency that can injure the peripheral nervous system. This was first observed in dentists and others with access to nitrous oxide and who inhaled it habitually. This nervous system injury is associated with numbness and clumsiness of the hands, and with Lhermitte's sign, a lightning-like shooting sensation that occurs when the patient bends the neck.

Nitrous oxide is used in dentistry because it has both analgesic and anxiety-relieving properties. It is used as a carrier gas and inducing agent in major surgery, facilitating induction of anesthesia maintained by other agents. Because it is not very soluble in blood, oxygen must be provided to patients at the end of the surgery, because the nitrous oxide can displace oxygen as it rushes out of the patient's body (diffusion hypoxia).

PERCHLOROETHYLENE

This chlorinated hydrocarbon solvent, used in the dry-cleaning industry, is also known as PERC (see Chlorinated Hydrocarbons, above).

TOLUENE

Toluene (methyl benzene, toluol) is an aromatic hydrocarbon solvent widely used in industrial processes, fuels, and consumer products. It is among the least irritating of the aromatic hydrocarbon solvents. When inhaled, it can produce CNS depression, like alcohol and other solvents. Its pharmacologic effects resemble those of other CNS depressant drugs, displaying actions like those of medications used for the treatment of epilepsy or for the clinical management of anxiety.

Toluene is removed from the body by exhalation and by metabolism. It is metabolized to methylhippuric acid, and is excreted by the kidneys. Overexposure to toluene can produce distal tubular acidosis of the kidney, an injury attributable to excess acidity that is reversible upon termination of exposure. Toluene has been demonstrated to produce loss of high-frequency hearing in laboratory animals following repeated high exposure, such as occurs during solvent abuse. Toluene also has been implicated in severe injuries to the nervous system in a large number of patients who deliberately inhaled toluene-containing solvents. These injuries are characterized by injury and loss of brain tissue. Patients display flattened emotional responses, impaired cognitive abilities, and a wide, shuffling gait associated with injury to the cerebellum. Animal studies have not yet conclusively demonstrated that toluene alone is responsible for this severe brain injury syndrome; nonetheless, solvent abusers who inhale toluene-containing mixtures run a very high risk of irreversible brain injury.

1,1,1 TRICHLOROETHANE (TCE)

This is a chlorinated hydrocarbon solvent with very high vapor pressure. It is useful in products that need to dry quickly, such as liquid paper products used to cover errors. The deliberate inhalation of these products has been associated with sudden death from ventricular arrhythmias (see Chlorinated Hydrocarbons, above).

TRICHLOROETHYLENE

A chlorinated hydrocarbon solvent used as a degreaser and dry-cleaning agent, it is subject to abuse by inhalation. When alcohol is consumed after exposure to trichloroethylene, profound blushing of the face occurs, the "degreaser's flush." One of the metabolites of trichloroethylene is chloral hydrate, an anesthetic agent used in "Mickey Finns," drinks used criminally to anesthetize robbery victims.

WHIPPETS

Whippets are small canisters of nitrous oxide used at soda fountains to make whipped cream. They have been incorporated into various products, such as balloon inflators, "carburetor pipes," and other drug paraphernalia (see Nitrous Oxide, above).

(SEE ALSO: *Complications; Ethnicity and Drug Use; High School Senior Survey; Inhalants: Extent of Use and Complications*)

RONALD W. WOOD

INHALANTS: EXTENT OF USE AND COMPLICATIONS

About 17 percent of ADOLESCENTS in this country say that they have sniffed INHALANTS—usually volatile solvents such as spray paint, glue, or cigarette lighter fluid—at least once in their lives, according to the National Institute on Drug Abuse (NIDA) in its 1993 MONITORING THE FUTURE study, a national survey of 8th-, 10th-, and 12th-grade students (also called the HIGH SCHOOL SENIOR SURVEY). In fact, results from a number of surveys suggest that among children under 18, the level of use of inhalants is comparable to that of stimulants and is exceeded only by the level of use of MARIJUANA, ALCOHOL, and CIGARETTES (see Figure 1).

Inhalant abuse, however, is a stepchild in the war on drugs. The abuse of inhalants, which include a broad array of cheap and easily obtainable household products (see Table 1), is frequently viewed by the public as a relatively harmless habit and not in the same high-risk category as drugs such as alcohol, COCAINE, and HEROIN. Some people tend to view inhalant "sniffing," "snorting," "bagging" (when fumes

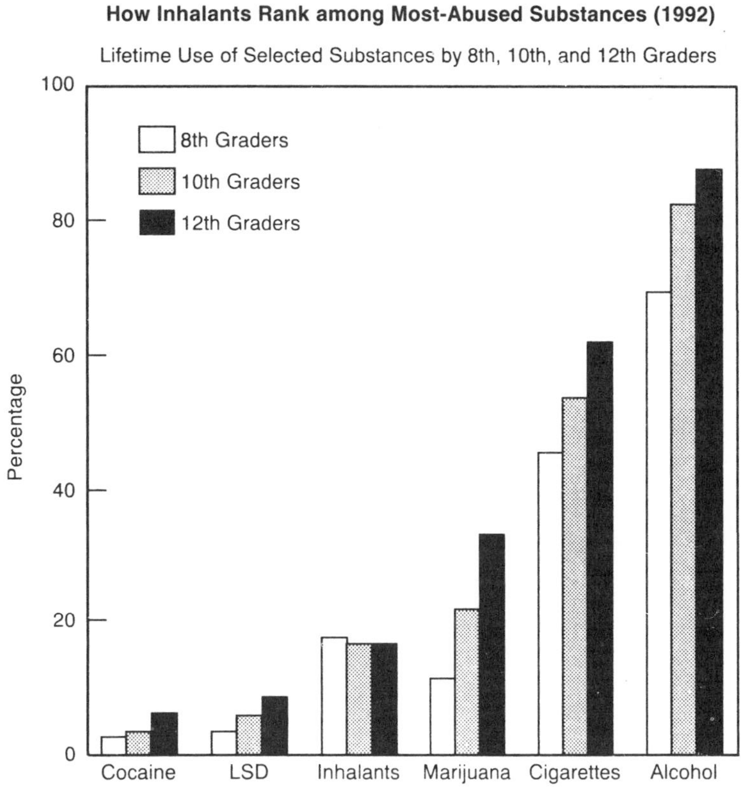

How Inhalants Rank among Most-Abused Substances (1992)

Lifetime Use of Selected Substances by 8th, 10th, and 12th Graders

Figure 1
Experimentation with inhalants is widespread among adolescents. NIDA's 1993 Monitoring the Future survey shows that among 8th, 10th, and 12th graders who have used drugs at least once in their lives, the prevalence of inhalant use is exceeded only by the prevalence of marijuana, cigarette, and alcohol use.

TABLE 1
Inhalants and Their Chemical Contents

VOLATILE SOLVENTS

Products	*Chemicals*
ADHESIVES	
Airplane glue	Toluene, ethyl acetate
Rubber cement	Hexane, toluene, methyl ethyl ketone, methyl butyl ketone
Polyvinlychloride cement	Trichloroethylene
AEROSOLS	
Spray paint	Butane, propane, fluorocarbons, toluene, hydrocarbons
Hair spray	Butane, propane, fluorocarbons
Deodorant, air freshener	Butane, propane, fluorocarbons
Analgesic spray, asthma spray	Fluorocarbons
SOLVENTS AND GASES	
Nail polish remover	Acetone, ethyl acetate
Paint remover	Toluene, methylene chloride, methanol acetone, ethyl acetate
Paint thinner	Petroleum distillates, esters, acetone
Typing correction fluid and thinner	Trichloroethylene, trichloroethane
Fuel gas	Propane
Cigarette lighter fluid	Butane, isopropane
Gasoline	Mixed hydrocarbons
CLEANING AGENTS	
Dry cleaning fluid	Tetrachloroethylene, trichloroethane
Spot remover	Xylene, petroleum distillates, chlorohydrocarbons
Degreaser	Tetrachloroethylene, trichloroethane, trichloroethylene
DESSERT TOPPING SPRAYS	
Whipped cream, whippets	Nitrous oxide ("laughing gas")

NITRITES AND ANESTHETICS

NITRITE ROOM ODORIZERS	
"Poppers" and "rush"	Alkyl nitrite, (iso)amyl nitrite, (iso)butyl nitrite, isopropyl nitrite, butyl nitrite
ANESTHETICS	
Gas	Nitrous oxide ("laughing gas")
Liquid	Halothane, enflurane
Local	Ethyl chloride

SOURCE: Adapted from *Inhalant Abuse: A Volatile Research Agenda*, NIDA Research Monograph 129, 1992.

are inhaled from a plastic bag), or "huffing" (when an inhalant-soaked rag is stuffed in the mouth) as a kind of childish fad to be equated with youthful experiments with cigarettes.

But inhalant abuse is deadly serious. Sniffing volatile solvents, which include most inhalants, can cause severe damage to the brain and nervous system. By starving the body of oxygen or forcing the heart to beat more rapidly and erratically, inhalants have killed sniffers, most of whom are adolescents.

The difficulty people face in recognizing the scope and magnitude of the problem lies in the dearth of documenting information. Survey data on the prevalence of inhalant abuse are difficult to obtain for a number of reasons—and what information does exist may underemphasize the severity of the situation. No one knows how many adolescents and young people die each year from inhalant abuse, in part because medical examiners often attribute deaths from inhalant abuse to heart problems, suffocation, SUICIDE, or ACCIDENTS. What is more, no national system exists for gathering data on the extent of inhalant-related injuries, although medical journals have described the situation as serious. As serious as the situation may be, some researchers warn that doctors and emergency medical personnel are not adequately trained to recognize and report symptoms of inhalant abuse.

SCOPE OF THE PROBLEM

Inhalant abuse came to public attention in the 1950s when the news media reported that young people who were seeking a cheap "high" were sniffing glue. The term *glue sniffing* is still widely used, often to include inhalation of a broad range of common products besides glue.

With so many substances lumped together as inhalants, research data describing frequency and trends of inhalant abuse are uneven and sometimes contradictory. However, evidence indicates that inhalant abuse is far more common among all socioeconomic levels of U.S. youth than is typically recognized by parents and the public. For example, NIDA's Monitoring the Future survey shows that in 1993 one in every six 8th graders, or 19.4 percent, has used an inhalant in his or her lifetime.

Inhalants were used by equally high percentages of 10th and 12th graders, according to the NIDA survey. Lifetime inhalant use among 12th graders, which had increased steadily for most of the 1980s, leveled off somewhat at 17.4 percent in 1993; 10th graders also reported a lifetime inhalant use of 17.5 percent.

Inhalants are most commonly used by adolescents in their early teens, with usage dropping off as students grow older, unlike the case for other drugs. For example, while 5.4 percent of 8th graders reported using inhalants within the past 30 days, known as "current" use, only 2.5 percent of seniors reported current use of inhalants.

One major roadblock to recognizing the size of the inhalant problem is the ready availability of products that are inhaled. Inhalants are cheap, or even free, and can be purchased legally in retail stores in a variety of seemingly harmless products. As a result, adolescents who sniff inhalants to get high do not face the drug procurement obstacles that confront abusers of other drugs. Youthful inhalant abusers can easily buy airplane glue, hair spray, spray paint, cigarette lighter fluid, nail polish remover, or typing correction fluid.

TYPES OF INHALANTS

Inhalants can be divided into three major categories—volatile solvents, nitrites, and anesthetics.

Volatile solvents are either gases, such as butane gas fumes, or liquids, such as gasoline or paint thinner, that vaporize at room temperature. Since the 1950s, the number of common products that contain volatile solvents has increased significantly. Besides gasoline and paint thinner, these products include spray paint, paint and wax removers, hair spray, odorants, air fresheners, cigarette lighter fuels, analgesic sprays, and propellant gases used in aerosols such as whipped cream dispensers.

Volatile solvents produce a quick form of intoxication—excitation followed by drowsiness, disinhibition, staggering, lightheadedness, and agitation. Because many inhalant products contain more than one volatile solvent, it is difficult in humans to clearly identify the specific chemical responsible for subsequent brain or nerve damage or death.

Some volatile solvents are inhaled by abusers because of the effects produced not by the product's primary ingredient but by propellant gases, like those used in aerosol cans of hair spray or spray paint. Other volatile solvents found in aerosol products such as gold and silver spray paint are sniffed not because of the effects from propellant gases but because of the PSYCHOACTIVE effects caused by the specific solvents necessary to suspend these metallic paints in the spray.

Nitrites historically have been used by certain groups, largely gay men, to enhance sexual stamina and pleasure. Often called "poppers" or "rush," some nitrite products are sold as room odorizers. But use of nitrites has fallen off dramatically in recent years. This may be partly because products containing butyl, propyl, and certain other nitrites were banned in 1991, although products using chemical variants of the banned substances are still sold.

For the past thirteen years, NIDA's Monitoring the Future survey has adjusted for the underreporting of nitrite use, recognizing that many survey respondents did not include information about nitrite use when answering survey questions about inhalant abuse. That is because most respondents fail to consider the use of nitrites as a form of inhalant abuse, unless prompted with specific questions mentioning "poppers," "rush," or other nitrite-specific references, researchers say.

Some observers now believe that adjusting inhalant-abuse survey results to combine nitrite use with volatile solvent use can lead to mistaken conclusions when viewing consolidated data over several years. That is because nitrite use is declining while volatile solvent use has been on the rise for a number of years. In combining solvents with nitrites and then adjusting the data, it appears that inhalant use has not changed over the past sixteen years, when, in fact, solvent use has steadily increased for a decade and a half and may now be leveling off (see Figure 2).

Because the current inhalant profile lumping nitrites with volatile solvents leads to misleading data and inferences, many researchers believe that a scientific description of inhalant abuse should distinguish abuse of volatile substances from abuse of nitrites and perhaps anesthetics.

Within the other major category of inhalants, the anesthetics, the principal substance of abuse is nitrous oxide. A colorless, sweet-tasting gas used by doctors and dentists for general anesthesia, nitrous oxide is called "laughing gas" because of its ability to induce a state of merriment, giggling, and laughter. Recent anecdotal reports indicate that nitrous oxide is being sold illicitly to teenagers and young adults at outdoor events such as rock concerts and on the street. Nitrous oxide often is sold in large balloons from which the gas is released and inhaled for its mind-altering effects as well as its ability to arouse gaiety and laughter.

But nitrous oxide is no laughing matter. Inhaling too much of the gas may deplete the body of oxygen and can result in death; prolonged use can result in peripheral nerve damage.

Nitrous oxide should not be confused with helium, which also is often sold in balloons. When inhaled, helium alters the sound of a person's voice, giving ludicrous, cartoonlike intonations to spoken words. Unlike nitrous oxide, helium does not produce euphoric effects. It can be used to treat certain respiratory illnesses and is sometimes mixed with the air supply of deep-sea divers to reduce the danger of decompression sickness, known as bends.

DANGERS OF INHALANT ABUSE

Despite the dangers associated with inhalant abuse, no central system exists in the United States for reporting deaths and injuries from abusing inhalants. A study by Dr. James C. Garriott, the chief toxicologist in San Antonio and Bexar County, Texas, examined all deaths in the county between

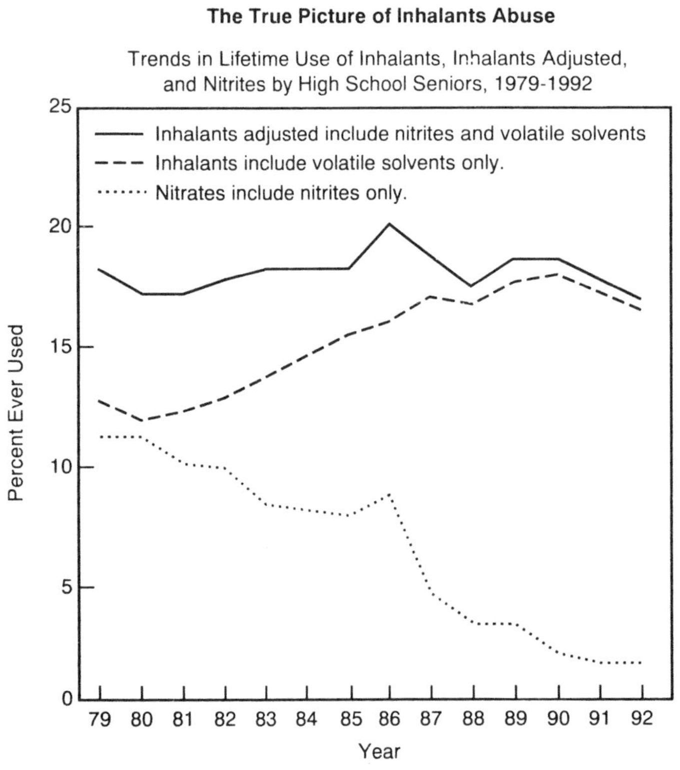

The True Picture of Inhalants Abuse

Trends in Lifetime Use of Inhalants, Inhalants Adjusted, and Nitrites by High School Seniors, 1979-1992

Figure 2
In surveys of drug use, the category "inhalants adjusted" includes the use of nitrites and volatile solvents. Some researchers believe that this category leads to misleading conclusions about trends in the prevalence of inhalant use. For instance, the "adjusted" figures from NIDA's 1993 Monitoring the Future survey indicate that inhalant use by high school seniors has been fairly stable since 1979. But nitrite use has gone down dramatically during that period. The true picture of inhalant use, say some observers, is the category "inhalants," which does not include nitrites and which has steadily increased for several years before leveling off.

1982 and 1988 that were attributed to inhalant abuse. Most of the 39 inhalant-related deaths involved teenagers, with 21 deaths occurring among people less than 20 years old. Deaths of males outnumbered those of females 34 to 5. Many of the abusers met with a violent death possibly related to but not directly caused by the use of volatile solvents. Eleven deaths were caused by suicide (10 by hanging), 9 by homicide, and 10 by accident, including falls, auto accidents, and overdoses.

Most of those people who died in Bexar County had used toluene-containing products, such as spray paints and lacquers, Dr. Garriott reported. The next most frequent cause of death in the Texas study was the use of a combination of chemicals found in typewriter correction fluids and other solvents. Other abused substances that resulted in death included gasoline, nitrous oxide, and refrigerants, such as fluorocarbons (Freon). Freon now has been replaced with butane or propane products in most aerosols.

As reported in the Texas study, the solvent toluene is identified frequently in inhalant-abuse deaths and injuries, because it is a common component of many paints, lacquers, glues, inks, and cleaning fluids. A 1986 study of 20 chronic abusers of toluene-containing spray paints found that after 1 month of abstinence from sniffing the paint, 65 percent of the abusers had damage to the nervous system. Such damage can lead to impaired perception, reasoning, and memory, as well as defective muscular coordination and, eventually, dementia.

In England, where national statistics on inhalant deaths are recorded, the largest number of deaths in 1991 resulted from exposure to butane and propane, which are used as fuels or propellants. Many researchers believe that abuse of butane, which is readily available in cigarette lighters, is on the increase in the United States.

A recent report of this particular inhalant problem in the Cincinnati region indicates that butane gas is the cause of enough deaths to foster national concern about the abuse of fuel gases, whether or not it is a passing form of inhalant abuse. Sniffers seem to go out of their way to get their favorite product; in certain parts of the country, Texas 'shoe-shine'—a shoe-shining spray containing toluene—and silver or gold spray paints are local or current favorites.

Since the banishing of fluorocarbons, the most common sniffing death hazards among U.S. students probably are due to butane and propane. Doctors and emergency room staffs need to be aware that the profile of the teenager who inhales volatile solvents is not limited to the ethnic lower socioeconomic classes. Many sources lead us to believe that abuse of these readily available inhalants has reached epidemic proportions, indicating an urgent need for preventive efforts.

WHO ABUSES INHALANTS?

One possible reason for the increased use of volatile solvents is that more girls are joining boys in sniffing solvents. Studies in New York State and Texas report that males are using solvents at only slightly higher rates than females. Among Native Americans, whose solvent-abuse rates are the highest of any ethnic group, lifetime prevalence rates for males and females were nearly identical, according to 1991 NIDA data.

There is a public perception that inhalant abuse is more common among HISPANIC youth than among other ethnic groups. However, surveys have not found high rates of abuse by Hispanics in all geographic areas. Rates for Hispanics may be related to socioeconomic conditions. Hispanic youths in poor barrio environments may use solvents heavily, but not Hispanic youths in less stressful environments.

In fact, inhalant abuse shows an episodic pattern, with short-term abuse outbreaks developing in a particular school or region as a specific inhalant practice or product becomes popular in a fashion typical of teenage fads. This episodic pattern can be reflected in survey results and can overstate the magnitude of a continually fluctuating level of abuse.

Inhalant abusers typically use other drugs as well. Children as young as fourth graders who use volatile solvents will also start experimenting with other drugs—usually alcohol and marijuana. Adolescent solvent abusers are POLYDRUG users prone to use whatever is available, although they show a preference for solvents. Solvent abuse is held in low regard by older adolescents, who consider it unsophisticated, a childish habit.

It is not just juveniles who are abusing inhalants. Reports in the mid-1990s indicate that college-age and older adults are the primary abusers of butane and nitrous oxide.

(SEE ALSO: *Poverty and Drug Use*)

BIBLIOGRAPHY

HORMES, J. T., FILLEY, C. M., & ROSENBERG, N. L. (1986). Neurologic sequelae of chronic solvent vapor abuse. *Neurology, 36*(5), 698–702.

SHARP, C. W., BEAUVAIS, F., & SPENCE, R. (EDS.). (1992). *Inhalant abuse: A volatile research agenda* (NIDA Research Monograph 129, NIH Pub. No. 93-3475). Washington, DC: Superintendent of Documents, U.S. Government Printing Office.

SHARP, C. W., & ROSENBERG, N. L. (1992). Volatile substances. In J. H. Lowinson et al. (Eds.). *Substance abuse: A comprehensive textbook*, 2d ed. Baltimore: Williams & Wilkins.

SIEGEL, E., & WASON, S. (1990). Sudden death caused by inhalation of butane and propane. *New England Journal of Medicine, 323*(23), 1638.

NEIL SWAN

INJECTING DRUG USERS AND HIV

One of the major risk behaviors for infection by the HUMAN IMMUNODEFICIENCY VIRUS (HIV) is injecting drug use; the others are unprotected male homosexual sex (Centers for Disease Control, 1991a) and unprotected heterosexual sex with an HIV-infected partner. The NATIONAL INSTITUTE ON DRUG ABUSE (NIDA) estimates that there are between 1.1 and 1.3 million injecting drug users (IDU) in the United States (Centers for Disease Control, 1987).

In 1990, 30 percent of the cases of ACQUIRED IMMUNODEFICIENCY SYNDROME (AIDS) were heterosexual injecting drug users; in addition, 30 to 50 percent of new cases identified are related to IDU (Iguchi et al., 1990). Injecting drug use is related to most of the heterosexual transmission of the virus (Centers for Disease Control, 1992). Also, whether directly or indirectly, injecting drug use accounts for 70 percent of AIDS cases among women and children (Centers for Disease Control, 1989). In these cases, either the woman or her sex partner is an IDU (Gayle, Selic, & Chu, 1990). Blacks and HISPANICS are overrepresented in the population of IDUs with AIDS (Battjes et al., 1988). Black and Hispanic women represent 72 percent of women with HIV; HIV infection among such women of childbearing age and their children has almost exclusively been related to intravenous (IV) drug use—in that the women or their sexual partners are IV drug users (Centers for Disease Control, 1990a).

The transmission of HIV among IDUs occurs directly through blood transmission of the virus, as when drug users share used, nonsterilized hypodermic needles and syringes, cotton, cookers, rags, and water that has been contaminated with the infected blood of other users. It is also transmitted when bodily fluids (e.g., semen, saliva, blood) are exchanged during sexual acts. The virus is often transmitted to the fetus by pregnant women who are HIV-positive, but not invariably. However, the risk of transmission to the fetus can be sharply reduced if the HIV-positive woman takes the antiviral drug AZT during pregnancy. Various studies have found that prior to the HIV epidemic, between 70 and 100 percent of IDUs shared injection paraphernalia (Lange et al., 1988; Des Jarlais et al., 1988). These precentages may be decreasing, since the connection with AIDS has been widely publicized since the 1980s.

Historically the most commonly injected drug has been HEROIN; however, the increased availability of COCAINE has resulted in an increased use by IDUs since the late 1980s. More specifically, injecting cocaine has elevated the risk of HIV spread, because the shorter duration of a cocaine "high" leads to more frequent injecting (Gottlieb & Hutman, 1990). It has also been reported that cocaine injectors, when the number of injections was statistically controlled, were at higher risk than other drug-injecting populations for HIV, because cocaine use associated with increased unprotected sexual activity (Chaisson et al., 1989).

BACKGROUND

The prevalence of HIV/AIDS among injectors varies widely from region to region in the United States. The highest rates of IDU and HIV are found along the east coast and west coast, in the southwest, Florida, Puerto Rico, and in the major metropolitan areas. The prevalence of HIV infection is also related to the social context of needle sharing. In areas where injectors go to "shooting galleries"—where anyone using a previously used needle may not know who else used the needle—there are generally high rates of HIV infection. Conversely, in areas where the social network is well known and only a limited number share works with one another, the infection rate is lower (Leukefeld et al., 1991).

The prevalence of HIV infection among IDUs can stabilize at low, moderate, or high rates depending on factors related to the social context of needle sharing. Sharing drug injection works with others, however, regardless of the social network, sets the scene for HIV transmission. Worldwide, some areas of low HIV prevalence have experienced explosive increases (Centers for Disease Control, 1990b).

While IDUs with HIV infection are predominantly males of color (Hispanics and blacks) in their late twenties and early thirties, variations and exceptions are noted and reflect dynamics in individual metropolitan areas. In 1989, the highest prevalence of IDUs in drug treatment centers who tested positive for HIV were in the Middle Atlantic states (New York, New Jersey, and Pennsylvania), where the overall rate for HIV-positive intravenous drug using men and WOMEN in treatment was 44 percent. Prevalence rates vary from this high to lows of 0% in rural midwestern states (Centers for Disease Control, 1990b).

REDUCING RISK-TAKING BEHAVIOR

Drug-abuse treatment and prevention can be effective in controlling the spread of AIDS among IDUs and for reducing the risk of exposure to the HIV virus. The goals of drug treatment and prevention are different. The goal of treatment is to eliminate injecting drug use as a risk factor in the spread of HIV. The goal of prevention is to reduce and eliminate harmful behaviors, like sharing needles, that place the IDU at risk for either becoming infected or infecting others with HIV—without necessarily focusing on changing the drug seeking and needle using behavior. Four areas are considered to be of prime importance: (1) increasing the number of drug abusers in treatment, (2) enhancing the effect of treatment, (3) developing outreach and counseling strategies, and (4) developing prevention strategies for reducing the risk-taking behavior among IDUs.

Drug Treatment. Several organizations and groups have suggested that drug-abuse treatment is important in helping to decrease and prevent the spread of AIDS. These ogranizations include the World Health Organization (WHO), the American Medical Association (AMA), the National Academy of Sciences/Institute of Medicine and the Presidential Commission on the HIV Epidemic.

Drug-abuse treatment can play an important role in preventing HIV transmission. Treatment reduces the number of people engaging in risky behavior. In addition to reducing the number of active drug-using addicts, treatment can reduce the number of people out recruiting new drug addicts (Brown, 1991). Barriers to treatment now exist for IDUs with HIV. IDUs themselves avoid people they suspect have HIV or AIDS, and some treatment programs will not allow the HIV-infected to participate in their programs (Brown, 1991). The most serious barrier to drug-abuse treatment is however, the lack of treatment availability and programs. More specifically, some IDUs, including those known to be HIV infected, are not admitted into drug treatment for as long as six months because of lack of available openings in treatment programs (Gotlieb & Hutman, 1990). In some community-outreach programs designed specifically to target IDUs to prevent HIV transmission, the majority of IDUs contacted have never been in treatment (Schrager et al., 1991). There is evidence that drug-abuse treatment reduces needle sharing by eliminating/reducing the needle using behaviors.

Enhancing Drug-Abuse Treatment. Drug-abuse treatment incorporates several modalities/approaches, which include: (1) drug-free outpatient services, (2) METHADONE MAINTENANCE, and (3) THERAPEUTIC COMMUNITIES (Leukefeld, 1988), as well as a number of programs that do not fit into these categories.

Outreach and Counseling. One way to increase the number of IDUs in treatment is to increase the number of outreach and counseling programs. The National AIDS Drug Abuse Research Demonstration Program is an example of outreach and counseling (National Institute on Drug Abuse, 1988). This demonstration program, initiated in 1987, provided an opportunity to assess the characteristics and risk-taking behaviors of injecting drug abusers not in treatment. Additional purposes included focusing on sexual partners of IDUs at high risk for AIDS, determining and monitoring HIV seroprevalence across cities, and evaluating prevention strategies. The overall goal was to reduce the spread of HIV infection by reducing and eliminating drug-use practices and certain high-risk sexual practices. Counseling and outreach approaches were applied, tested, and evaluated at each community site. Projects were targeted on three levels: (1) high-risk individuals, (2) family and social networks of IDUs, and (3) the larger community. Although intervention components varied across sites, the focus and objectives were similar (Chitwood et al., 1991; Leukefeld, 1988). These projects provided information about protective behaviors, and IDUs were encouraged to enroll in drug-abuse treatment programs. Trained indigenous outreach workers distributed and discussed materials using informal groups or through one-on-one interactions. Sixty-three communities were involved in this demonstration project (McCoy & Khoury, 1990; Leukefeld, 1988).

Strategies for community outreach differ between the IDU, their sex partners, and prostitutes. Reaching the IDU means that outreach workers go to places where IDUs hang out and buy their drugs, as well as going into criminal-justice settings (jails, PRISONS, courts), drug-treatment centers, and the health-care system. Although there is inherent danger in many of these settings, recovering drug users—savvy men and women of the same backgrounds as IDUs—have achieved success in contacting IDUs in these settings (Serrano, 1990; Brown, 1990).

Many male IDUs hang out on the street or can be found in places where other IDUs hang out. The female sex partners of IDUs frequently stay close to home and children, however, and they frequently work (Margolis, 1991). While women may purchase drugs for their partners, they do not generally hang out at those locations. Thus, targeting female partners of IDUs requires different strategies than those used for contacting the IDU. The YES project of San Francisco is an example of a program targeted on female sex partners; it began by supporting high-risk women meet their basic needs by helping them get general assistance, food, clothing, and health care. A second strategy was to rent a hotel room, called "A Room of Her Own" in which education, counseling, and service could be provided to the female partner of the IDU. Another project—serving Bridgeport, Connecticut, San Juan, Puerto Rico, and Juarez, Mexico—contacted the female sex partners of male IDUs; it examined an approach that attracted women to a safe setting established by the program—a clothing boutique where women could pick up new clothes and then stay for an AIDS information video. Another strategy as part of this project was to have outreach staff available in the afternoons and evenings, hours when the women were available (Moini, 1991). In another project in Long Beach, California, a drop-in center was established for youth and women (Yankovich, Archuleta, & Simental, 1991).

Prostitutes, another group at high risk, require strategies appropriate to their setting. Contacting prostitutes can be difficult, since her pimp can severely restrict contact with social-service workers. In one study, contact was made when the pimp was not around and through the Salvation Army mobile canteen that served coffee to prostitutes in the late night/early morning hours (Moini, 1991). Another study reported that prostitutes are aware of AIDS, know how it is transmitted, and are aware that their drug use and unsafe sex behavior are putting them at risk (Shedlin, 1990). However, barriers to behavior change in prostitutes include low self-esteem and low levels of education, also POVERTY, addiction, hopelessness, lack of knowledge, and lack of support services.

Prevention Strategies. Prevention is of central importance in controlling the spread of HIV among IDUs. Abstinence from drug use and needle use is the overall approach for preventing the spread of HIV. Preventing infection is a self-preservation issue (protecting self) while preventing the spread of HIV is an altruistic issue (protecting others) (Moini, 1991). It has been reported that among IDUs there is greater resistance to changing sexual behaviors (using condoms) than drug-use behaviors (sharing needles) (Sorenson, 1990). Thus, it is important to target not only IDUs but also their sex partners and the prostitutes who engage in unsafe sex practices with multiple partners. These people may also be exchanging drugs for sex and may be IDUs themselves (Centers for Disease Control, 1991b). Three prevention strategies have been developed: education, NEEDLE-EXCHANGE PROGRAMS, and community-based interventions.

Education. In addition to the community-outreach programs, three overarching prevention-education strategies have been developed: (1) prevention education for HIV-antibody-negative individuals, (2) AIDS pre- and posttest counseling, and (3) prevention and support for HIV-antibody-positive individuals (Schensul & Weeks, 1991). AIDS prevention education involves delivery of information related to HIV spread, risk behaviors, and preventing the spread of the virus. Educational activities have been targeted on the general public, school-aged populations, and populations at risk, like IDUs. The U.S. Centers for Disease Control (CDC) National Public Information Campaign has produced numerous educational materials for the radio, television, and print media. Education targeted to individuals at risk for HIV infection has included counseling, testing, the teaching of behavioral responses to risky behaviors, and providing support for low or no-risk behaviors (Roper, 1991).

Prevention education for IDUs includes several informational components. Of primary importance to active drug users are issues related to needle sharing as a risk behavior for HIV transmission. Also of critical importance to needle-sharing IDUs in preventing HIV transmission are describing ways to effectively sterilize paraphernalia that will be shared. Of importance to IDUs, the sex partners of IDUs, and prostitutes are safe-sex issues and knowledge of HIV transmission through unsafe sex. Of importance to potential partners—both men and women who have relationships with IDUs and who may be IDUs—is knowledge about the transmission of the virus from mother to fetus (Strawn, 1991). Early to mid-1990s research indicates that the use of AZT (an anti-HIV drug) by pregnant women who are HIV-positive sharply reduces the probability that the baby will be infected with the virus.

Pre- and posttest AIDS counseling is another strategy for HIV prevention. In the early 1980s, at the beginning of the AIDS epidemic, testing was controversial because of the fear of discrimination, concern about the accuracy of tests, the usefulness of the results, and the psychological distress associated with a positive result. With more effective treatment for symptomatic AIDS and early treatment for HIV-infected individuals, however, the resistance is diminishing (Strawn, 1991).

Generally, individuals seek HIV testing for one of two reasons: (1) an agency, like a plasma center or a penal institution, or a medical professional requested it, or (2) the individual seeks to be tested because of identified high-risk behaviors (Roggenburg et al., 1991). HIV testing can represent a crisis in the life of an individual being tested. Receiving the results can be difficult due to the anxiety of the situation, even if the results are negative. Pre- and posttest counseling is necessary to assess the psychological well-being of the individual being tested. Some people believe that one consequence to being informed of a positive test result can be suicide (Strawn, 1991).

Prevention education and support programs for persons who are HIV-antibody-positive are common within the homosexual community, the first to be identified with the virus and the one that responded

politically as well as behaviorally. However, very few support or prevention programs target IDUs.

Needle Exchange. A controversial prevention approach in the United States for preventing HIV infection is the provision of clean needles to IDUs. In needle-exchange programs, a clean needle and sometimes injection equipment (works) are exchanged for used ones. Proponents of these programs argue that needle exchanges help prevent HIV transmission and offer opportunities for education and referral to drug-treatment programs. It has been reported that in areas where needle-exchange programs have been in operation, the incidence of sharing used needles has diminished (Karpen, 1990). Some needle-exchange programs are conducted illegally by AIDS activists (Karpen, 1990). Occasionally, in the United States, needle exchanges are managed legally by health departments. To conduct a needle-exchange program legally, in many regions the PARAPHERNALIA LAWS related to drug-use equipment need to be modified (Wood, 1990).

Opponents of needle-exchange programs point out that the needles and syringes are only two of the many drug-use implements that can be contaminated with blood transmitting HIV. For example, cotton, cookers, and the water used to rinse out syringes can transmit HIV if they have been contaminated with infected blood. In addition, some injecting rituals can transmit HIV even if a clean needle and syringe are used. Sharing an injection can be part of a ritual between addicts. For example, in a "rinse" or a "geezer" one addict injects another person and then injects him- or herself with the remnant in the syringe (Primm, 1990). Few rigorous U.S. studies have examined needle-exchange programs and their effects on HIV transmission.

Interventions. As above, one component of the National AIDS Demonstration Project has been to compare the CDC basic outreach and counseling intervention with an enhanced intervention. The CDC basic intervention includes factual information about AIDS transmission, prevention, and self-assessed risk. Enhanced community-based educational-intervention programs have involved several strategies: counseling individuals, couples, and groups; developing behavioral skills; and using applied ethnography with outreach workers to disseminate information (Chitwood et al., 1991). The rate of sharing between IDUs decreased by up to 59 percent in a five-city study. In the same study, IDU condom use increased up to 16 percent (Iguchi et al., 1990).

CONCLUSION

Preventing the spread of AIDS for IDUs and their sex partners requires a multidisciplinary, multiple-strategy approach. Community-intervention strategies have proven to be partially effective in reducing IDU risk behaviors (Leukefeld, Battjes, & Amsel, 1990). Much remains to be accomplished, however. Targeting HIV-prevention approaches and interventions will receive additional emphasis as the epidemic progresses (Leukefeld & Battjes, 1991). Research needs to continue to examine methods to reduce HIV in IDUs, to reinforce IDU behavior changes, to increase the effectiveness of drug-abuse treatment, and to provide psychosocial and other supports focused on HIV-infected IDUs. Clearly, these activities suggest grappling with controversial issues and related social issues.

(SEE ALSO: *Complications; Heroin: The British System; Prevention: Shaping Mass Media Messages to Vulnerable Groups; Substance Abuse and AIDS; Vulnerability as Cause of Substance Abuse*)

BIBLIOGRAPHY

BALL, J. C., & CORTY, E. (1988). Basic issues pertaining to the effectiveness of methadone maintenance treatment. In C. G. Leukefeld & F. T. Tims (Eds.), *Compulsory treatment: Drug abuse research and clinical practice.*

BATTJES, R. J., ET AL. (1988). AIDS in intravenous drug abuse. *Bulletin on Narcotics.*

BROWN, L. S. (1991). The impact of AIDS on drug abuse treatment. In R. W. Pickens, C. G. Leukefeld, & C. R. Schuster, *Improving drug abuse treatment.* National Institute on Drug Abuse Research Monograph 106. Washington, DC: U.S. Department of Health and Human Services.

BROWN, L. S. (1990). Black intravenous drug users: Prospects for intervening in the transmission of human immunodeficiency virus infection. In C. G. Leukefeld, R. J. Battjes, & Z. Amsel (Eds.), *AIDS and intravenous drug use: Future directions for community based prevention research.* National Institute on Drug Abuse

Research Monograph 93. Washington, DC: U.S. Department of Health and Human Services.

CENTERS FOR DISEASE CONTROL. (1992). The second 100,000 cases of acquired immunodeficiency syndrome—United States, June 1981–December 1991. *Morbidity and Mortality Weekly Report, 41,* 28–29.

CENTERS FOR DISEASE CONTROL. (1991a). Mortality attributable to HIV infection/AIDS—United States, 1981–1990. *Morbidity and Mortality Weekly Report, 40,* 42–44.

CENTERS FOR DISEASE CONTROL. (1991b). Drug use and sexual behaviors among sex partners of injecting drug users—United States, 1988–1990. *Morbidity and Mortality Weekly, 40,* 855–860.

CENTERS FOR DISEASE CONTROL. (1990a). AIDS in women—United States. *Morbidity and Mortality Weekly Report, 39.*

CENTERS FOR DISEASE CONTROL. (1990b). *National HIV seroprevalence surveys: Summary of results,* 2nd ed. Washington, DC: U.S. Department of Health and Human Services.

CENTERS FOR DISEASE CONTROL. (1987). *A review of current knowledge and plans for expansion of HIV surveillance activities: A report to the Domestic Policy Council.* Washington, DC: U.S. Goverment Printing Office.

CHAISSON, R. E., ET AL. (1989). Cocaine use and HIV infection in intravenous drug users in San Francisco. *Journal of the American Medical Association, 261,* 561–565.

CHILDRESS, A. R., ET AL. (1991). Are there minimum conditions necessary for methadone maintenance to reduce intravenous drug use and AIDS risk behaviors? In R. W. Pickens, C. G. Leukefeld, & C. R. Schuster, *Improving Drug Abuse Treatment.* National Institute on Drug Abuse Research Monograph 106. Washington, DC: U.S. Department of Health and Human Services.

CHITWOOD, D. D. (1991). *A community approch to AIDS intervention: Exploring the Miami Outreach Project for Injecting Drug Users and Other High Risk Groups.* New York: Greenwood Press.

DELEON, G., & SCHWARTZ, S. (1984). The therapeutic community: What are the retention rates? *American Journal of Drug and Alcohol Abuse, 10,* 267–284.

DES JARLAIS, D. C., ET AL. (1988). The sharing of drug injection equipment and the AIDS epidemic in New York City: The first decade. In R. J. Battjes & R. W. Pickens (Eds.), *Needle sharing among intravenous drug abusers: National and international perspectives.* National Institute on Drug Abuse Research Monograph 80. Washington, DC: U.S. Department of Health and Human Services.

DOLE, V. P., & NYSWANDER, M. E. (1968). The use of methadone for narcotic blockade. *British Journal of Addiction, 63,* 55–59.

FRAIZIER, L. (1962). Treating young drug users: A casework approach. *Social Work, 7,* 94–101.

GOTTLIEB, M. S., & HUTMAN, S. (1990). The case for methadone. *AIDS Patient Care, 4*(April), 15–18.

IGUCHI, M. Y., ET AL. (1990). Early indices of efficacy in the NIDA AIDS outreach demonstration project: A preliminary report from Chicago, Houston, Miami, Philadelphia, and San Francisco. *Morbidity and Mortality Weekly Report, 39.*

KARPEN, M. (1990). A comprehensive world overview of needle exchange programs. *AIDS Patient Care, 4* (August).

LANGE, W. R., ET AL. (1988). Geographic distribution of human immunodeficiency virus markers in parenteral drug abusers. *American Journal of Public Health, 78,* 443–446.

LEUKEFELD, C. G. (1988). HIV and intravenous drug use. *Health and Social Work,* 247–250.

LEUKEFELD, C. G., & BATTJES, R. J. (1991). The context of HIV prevention. *Journal of Primary Prevention, 12*(1), 3–6.

LEUKEFELD, C. G., BATTJES, R. J., & AMSEL, Z. (1990). Community prevention efforts to reduce the spread of AIDS associated with intravenous drug abuse. *AIDS Education and Prevention, 2*(3), 235–243.

LEUKEFELD, C. G., BATTJES, R. J., & PICKENS, R. W. (1991). AIDS prevention: Criminal justice involvement of intravenous drug abusers entering methadone treatment. *Journal of Drug Issues, 21*(4), 673–683.

MARGOLIS, E. (1991). Evaluating outreach in San Francisco. In *Community-based AIDS prevention: Studies of intravenous drug users and their sexual partners.* Rockville, MD: National Institute on Drug Abuse.

MCCOY, C. B., & KHOURY, E. (1990). Drug use and the risk of AIDS. *American Behavioral Scientist, 33*(4), 419–431.

MCLELLAN, A. T., ET AL. (1982). Is treatment for drug abuse effective? *Journal of the American Medical Association, 247,* 1423–1428.

MOINI, S. (1991). AIDS and female partners of IV drug users: Selected outreach strategies, accomplishments, and preliminary data from one project. In *Community-based AIDS prevention: Studies of intravenous drug users and their sexual partners.* Rockville, MD: National Institute on Drug Abuse.

NATIONAL INSTITUTE ON DRUG ABUSE. (1988). Community research branch AIDS research demonstration project research plan. Bethesda, MD: Author.

PRIMM, B. (1990). Needle exchange programs do not involve the problem of HIV transmission. *AIDS Patient Care, 4*(August), 18–20.

ROGGENBURG, L., ET AL. (1991). The relation between HIV-antibody testing and HIV risk behaviors among intravenous drug users. In *Community-based AIDS prevention: Studies of intravenous drug users and their sexual partners.* Rockville, MD: National Institute on Drug Abuse.

ROPER, W. L. (1991). Current approaches to prevention of HIV infections. *Public Health Reports, 106,* 111–115.

SCHRAGER, L., ET AL. (1991). Demographic characteristics, drug use, and sexual behavior of IV drug users in Bronx, New York. *Public Health Reports, 106,* 78–84.

SCHENSUL, J. J., & WEEK, M. (1991). Ethnographic evaluation of AIDS-prevention programs. In *Community-based AIDS prevention: Studies of intravenous drug users and their sexual partners.* Rockville, MD: National Institute on Drug Abuse.

SERRANO, Y. (1990). The Puerto Rican intravenous drug user. In C. G. Leukefeld, R. J. Battjes & Z. Amsel (Eds.), *AIDS and intravenous drug use: Future directions for community based prevention research.* National Institute on Drug Abuse Research Monograph 93. Washington, DC: U.S. Department of Health and Human Services.

SHELLY, J. (1967). *Daytop Lodge: A two-year report. Rehabilitating the narcotic addict.* Washington, DC: U.S. Government Printing Office.

SIMPSON, D. D., SAVAGE, L. J., & SELLS, S. B. (1978). *Data book on drug abuse treatment effectiveness: Follow-up study of 1969–72. Admissions to the DARP* (Report 78–10). Fort Worth, TX: Texas Christian University, Institute of Behavioral Research.

SIMPSON, D. D., & SELLS, S. B. (1982). *Evaluation of drug abuse treatment outcomes: Summary of the DARP follow-up research.* NIDA Treatment Research Report; DHHS Publication no. ADM 82-1209. Washington, DC: U.S. Government Printing Office.

SORENSON, J. L. (1990). Preventing AIDS: Prospects for change in white male intravenous drug users. In C. G. Leukefeld, R. J. Battjes & Z. Amsel (Eds.), *AIDS and intravenous drug use: Future directions for community based prevention research.* National Institute on Drug Abuse Research Monograph 93. Washington, DC: U.S. Department of Health and Human Services.

STRAWN, J. M. (1991). HIV counseling and testing: Issues in a research context. In *Community-based AIDS prevention: Studies of intravenous drug users and their sexual partners.* Rockville, MD: National Institute on Drug Abuse.

Tims, F. M., & Leukefeld, C. B. (1986). *Relapse and recovery in drug abuse*. National Institute on Drug Abuse Monograph 72. Washington, DC: U.S. Department of Health and Human Services.

Weissman, I. (1971). Drug abuse: Some practical dilemmas. *Social Service Review, 46*, 378–394.

Wood, R. W. (1990). Needle exchange programs stop AIDS! *AIDS Patient Care, 4*(August), 14–17.

Yankovich, D., Archuleta, E., & Simental, S. (1991). A mobile outreach program to intravenous drug users and female sexual partners in Long Beach, California. In *Community-based AIDS prevention: Studies of intravenous drug users and their sexual partners*. Rockville, MD: National Institute on Drug Abuse.

FAYE E. REILLY
CARL G. LEUKEFELD

INSTITUTE ON BLACK CHEMICAL ABUSE (IBCA)

Founded in 1975, the Institute on Black Chemical Abuse (2616 Nicollet Avenue S, Minneapolis, MN 55408; 612-871-7878) is an open-membership organization that provides culturally specific programs and client services for the African-American community. IBCA defines cultural specificity as the creation of an environment that encourages and supports the exploration, recognition, and acceptance of African-American identity and experience, including the unique history associated with being African American in the United States and the role that racial identity plays in drug dependence. Programs are designed to address the devastating effects of the drug-abuse problem on this community. Services are provided in assessment and intervention for outpatient treatment and aftercare, black codependency issues, home-based support, and for pregnant women and young children. Philosophically, IBCA advocates empowering African-American families to achieve a greater state of well-being through a holistic mental health approach.

IBCA's efforts in the community provide training and prevention resources to educate and preserve the people and their families who face the problems of substance abuse. The Technical Assistance Center (TAC) offers training workshops, program consultation, and resource materials on African Americans and substance abuse. TAC also educates and trains clergy members working with these issues in the community. The IBCA prevention programs have involved school and business leaders in social-policy programs aimed at establishing community awareness of substance-abuse issues; the Drug Free Zones program, in particular, has received national recognition. Additional community-outreach programs include volunteer and intern programs and a resource center.

IBCA publishes a newsletter and organizes meetings. The organization is supported by dues from members, profits from activities and services, government and corporate grants, and charitable contributions.

FAITH K. JAFFE

(SEE ALSO: *Ethnic Issues and Cultural Relevance in Treatment; Ethnicity and Drugs; Vulnerability as Cause of Substance Abuse: Race*)

INTERNATIONAL CLASSIFICATION OF DISEASES (ICD)

This is the official classification system of the World Health Organization (WHO). As a general system for the classification of diseases, injuries, causes of death, and related health problems, the ICD is used throughout the world as a common frame of reference for statistical reporting, clinical practice, and education. The ICD is a system of categories to which specific disease entities can be assigned consistently in different parts of the world. Recognizing the growing importance of alcohol and drug misuse, the ninth revision of ICD was published in 1975 (ICD-9), and it introduced the terms *dependence* and *abuse* into the international nomenclature. *Drug dependence* was defined as "a state, psychic and sometimes also physical, resulting from taking a drug, and characterized by behavioural and other responses that always include a compulsion to take the drug on a continuous or periodic basis in order to experience its psychic effects, and sometimes to avoid the discomfort of its absence" (WHO, 1977, 198). *Alcohol dependence* was defined in a similar way. The category Non-Dependent Abuse of Drugs was designed for cases where a person "has come under medical care because of the maladaptive effect of a drug on which he is not dependent and that he has taken on his own initiative to the detriment of his health or social functioning" (WHO, 1978, 43–44).

In 1993, the tenth revision, ICD-10, was introduced—replacing ICD-9 as the official classification system for international use (WHO, 1992a). Chapter

5, which describes mental and behavioral conditions (WHO, 1992b), includes a section for the classification of disorders based on ten kinds of PSYCHOACTIVE substances: ALCOHOL, SEDATIVE-HYPNOTICS, CANNABIS (MARIJUANA), COCAINE, other STIMULANTS, OPIOIDS, HALLUCINOGENS, TOBACCO, VOLATILE SOLVENTS, and multiple drugs. The major disorders associated with these substances are acute intoxication, harmful use, dependence syndrome, withdrawal state, amnesic syndrome, and psychotic disorders (WHO, 1992b). The identification of the substance used may be made on the basis of an interview with the patient, laboratory analysis of blood or urine specimens, or other evidence (such as clinical signs and symptoms or reports from third parties).

Acute intoxication is a transient condition following the ingestion of alcohol or other psychoactive substances. It results in disturbances in consciousness, cognition, perception, mood, or behavior. According to ICD-10, psychoactive substances are capable of producing different types of effect at different dose levels. For example, alcohol may have stimulant effects at low doses, lead to agitation and aggression with increasing dose levels, and produce clear sedation at very high levels. The term *pathological intoxication* in ICD-10 refers to the sudden onset of violent behavior that is not typical of the individual when sober. This occurs very soon after amounts of alcohol are drunk that would not produce intoxication in most people.

A central feature of the ICD-10 approach to substance-use disorders is the concept of a dependence syndrome, which is distinguished from disabilities caused by harmful substance use (Edwards, Arif, & Hodgson, 1981). The *dependence syndrome* is defined as "a cluster of physiological, behavioural, and cognitive phenomena in which the use of a substance or a class of substances takes on a much higher priority for a given individual than other behaviours that once had greater value" (WHO, 1992b, 75). A central characteristic of the dependence syndrome is the strong and persistent desire to take psychoactive drugs, alcohol, or tobacco. Another feature is the rapid reappearance of the syndrome soon after alcohol or drug use is resumed after a period of abstinence. A definite diagnosis of dependence is made only if three or more of the following have been experienced during the previous year: (1) a strong desire or sense of compulsion to take the substance; (2) difficulties in controlling substance-taking behavior in terms of its onset, termination, or levels of use; (3)

a physiological withdrawal state; (4) evidence of tolerance; (5) progressive neglect of alternative pleasures or interests because of substance use; and (6) persisting with substance use despite clear evidence of overtly harmful consequences.

Harmful use, a new term introduced in ICD-10, is a pattern of using one or more psychoactive substances that causes damage to health. The damage may be: (1) physical (physiological), such as fatty liver, injuries associated with alcohol intoxication, or hepatitis from needle-injected drugs; or (2) mental (psychological), such as depression related to heavy drinking or drug use. Adverse social consequences often accompany substance use, but they are not in themselves sufficient to result in a diagnosis of harmful use.

Chapter 5 of ICD-10 is available in several different versions. The *Clinical Descriptions and Diagnostic Guidelines* is intended for general clinical, educational, and service use. *Diagnostic Criteria for Research* is designed for use in scientific investigations and epidemiological studies. A shorter and simpler version of the classification is available for use by primary health-care workers.

(SEE ALSO: *Addiction: Concepts and Definitions; Alcoholism: Origin of the Term; Diagnostic and Statistical Manual; Disease Concept of Alcoholism and Drug Abuse*)

BIBLIOGRAPHY

EDWARDS, G., ARIF, A., & HODGSON, R. (1981). Nomenclature and classification of drug- and alcohol-related problems: A WHO memorandum. *Bulletin of the World Health Organization, 59*, 225–242.

WORLD HEALTH ORGANIZATION. (1992a). *International classification of diseases and related health problems,* 10th rev. Geneva: Author.

WORLD HEALTH ORGANIZATION. (1992b). *The ICD-10 classification of mental and behavioural disorders: Clinical descriptions and diagnostic guidelines.* Geneva: Author.

WORLD HEALTH ORGANIZATION. (1978). *Mental disorders: Glossary and guide to their classification in accordance with the ninth revision of the international classification of diseases.* Geneva: Author.

WORLD HEALTH ORGANIZATION. (1977). *Manual of the international statistical classification of diseases, injuries, and causes of death.* Vol. 1. Geneva: Author.

THOMAS F. BABOR

INTERNATIONAL DRUG CONTROL *See* Anslinger, Harry J., and U.S. Drug Policy; Psychotropic Substances Convention of 1971; Single Convention on Narcotic Drugs.

INTERNATIONAL DRUG SUPPLY SYSTEMS The majority of illicit drugs consumed in the United States are of foreign origin—including all the COCAINE and HEROIN and significant amounts of MARIJUANA. In the early 1990s, the U.S. National Narcotics Intelligence Consumer Committee (NNICC) report estimates that Latin American countries supplied approximately 25 to 30 percent of the heroin, perhaps 60 to 80 percent of the marijuana, and all the cocaine. Southeast Asian and Middle Eastern countries supplied the remaining 70 to 75 percent of the heroin.

Drug use and drug abuse have been a part of many cultures for centuries. Although once considered a problem only for countries with massive demand and consequent loss of labor and life, drugs are now recognized as a policy concern for all countries involved—the producing, TRANSIT, and consuming countries alike. No country is insulated from the destabilizing forces of illicit drugs. For SOURCE (producing) countries, drug trafficking appears to provide short-term economic benefits, but mainly for those involved in the business. Eventually, long-term negative economic consequences ensue, with foreign investment, tourism, and domestic production diminished—and with off-shore money laundering and the concentration of wealth in the hands of a few. The drug trade does not stimulate regional economies through jobs, capital appreciation, and investment.

Since 1971, when modern international drug-control efforts began, a number of major shifts have occurred in the drug-producing capabilities of various countries. For example, in the early 1970s, after the so-called French Connection was broken (Turkish OPIUM was processed into heroin in France), MEXICO replaced Turkey as a major source of U.S. heroin; Pakistan then supplanted Mexico after 1979, when the Islamic political revolution in Iran created a population of refugees. At about the same time, the Soviet Union occupied Afghanistan, and the resistance movements there increased their income-generating opium cultivation practices.

In the 1980s, cocaine production in the Andean countries of Peru, BOLIVIA, and COLOMBIA expanded significantly into nontraditional growing zones (the Bolivian Chapare region and Peruvian Upper Huallaga Valley, or UHV), augmenting the more traditional licit production areas of the Bolivian Yungas and Peruvian Cuzco regions. In the early 1980s, U.S. demand for Mexican marijuana decreased dramatically, because of consumer concern about Mexico's drug-elimination program, where marijuana was sprayed with the herbicide paraquat, some of which is reported to have killed U.S. users. Consequently, Colombia replaced Mexico as the preferred source of high quality marijuana. Colombia and Guatemala also began to cultivate substantial amounts of opium in the early 1990s.

Traffickers have also adjusted their smuggling routes in response to government law-enforcement pressures. For example, in the mid-to-late 1980s, Colombian drug traffickers began to shift their routes away from the Florida peninsula and toward Central America and Mexico. By the early 1990s, the U.S. government estimated that up to 50 percent of the Colombian cocaine consumed in the United States entered via Mexico. Wide variations in source-country response to these shifts in production have also been chronicled, ranging from government complicity and corruption to modest attempts to reduce crop production and trafficking to intensified organized efforts to eliminate or hamper seriously the drug trade.

PRINCIPAL DRUG-PRODUCING COUNTRIES

Coca/Cocaine. As of the early 1990s, all the cocaine, about 30 percent of the heroin, and a significant amount of marijuana entering the United States is produced in the Western Hemisphere—in Mexico, Central, and South America. They are smuggled in through the southern borders of the United States. All of the cocaine consumed in the world is grown and processed in the Andean countries of Peru, Bolivia, Ecuador, and Colombia. Some 60 percent of COCA PLANTS (*Eryroxylon coca*) are cultivated in Peru, about 15 percent in Colombia, and about 25 percent in Bolivia.

Peru. Traditional legal cultivation of coca is licensed for cultivation in Cuzco, Peru, but the majority of

TABLE 1
Cocaine Production Estimates

			U.S. MEASURE				METRIC MEASURE	
Source Country		Net Coca Cultivation (acres)	Estimated Coca Leaf Yield (tons)	Potential Cocaine HCl Capacity (tons)	Source Country	Net Coca Cultivation (hectares)	Estimated Coca Leaf Yield (metric tons)	Potential Cocaine HCl Capacity (metric tons)
Peru	1990	299,611	216,590	627–671	Peru	121,300	196,900	570–610
	1991	298,376	244,970	710–759		120,800	222,700	645–690
Bolivia	1990	124,241	84,480	270–457	Bolivia	50,300	76,800	245–415
	1991	118,313	86,240	275–462		47,900	78,400	250–430
Colombia	1990	99,047	33,310	72	Colombia	40,100	32,100	65
	1991	92,625	33,000	66		37,500	30,000	60

Potential
Cocaine HCl Production
1990 = 969–1,199 tons   1991 = 1,051–1,287

Potential
Cocaine HCl Production
1990 = 880–1,090 metric tons   1991 = 955–1,170 metric tons

NOTE: The figures reflected here are consistent with the *International Narcotics Control Strategy Report (INCSR) 1992*. The INCSR states cultivation in hectares and yields in metric tons. All figures have been converted to acres or short tons, as appropriate, in the chart on the left. A new procedure introduced in the 1991 INCSR is used for calculating coca leaf production. Previous methods did not deduct immature, non-producing fields from net cultivation before calculating production. Multiple harvests of coca do not begin until plants are at least two years old. Here, only mature cultivation was used to calculate production. Estimates included here for 1990 have been revised to reflect the use of the mature cultivation methodology cited. These figures do appear in the NNICC Report 1990 charts in parenthesis.

According to past UHV Reduction Agency (CORAH) and U.S. Agency for International Development (USAID) reports, the dry leaf yield of mature coca in the UHV ranges between 2.0 and 2.7 metric tons per hectare. A mean yield factor of 2.3 metric tons was used for this area. Other areas of Peru have lower yields similar to the Yungas in Bolivia. Last year's reported yield of 1.14 metric tons was for areas of Peru outside the UHV. According to the Bolivia's Coca Eradication Directorate (DIRECO), mature coca leaf yield averages 2.7 metric tons in the Chapare and 1.0 metric tons in the Yungas. The conversion rate for calculating potential cocaine production from dry leaf is 322–345:1 in Peru and 195–330:1 in Bolivia. Production in Colombia is determined by multiplying a yield factor of .8 metric tons per hectare by net cultivation. All Colombian cultivation was assumed to be mature. The conversion rate for Colombia is 500:1.

SOURCE: *International Narcotics Control Strategy Report 1992.*

Peru's illicit crop comes from the Upper Huallaga Valley (UHV), which includes portions of Huanuco, San Martin, and Ucayali departments. Other illicit cultivation occurs in La Convencion and Lares valleys in Cuzco and in the Ayachucho department. Much of the coca leaf is processed into COCA PASTE and cocaine base in crude maceration (steeping) pits positioned near cultivation sites. Clandestine labs then process the paste into base. Normally the base is then shipped to hydrochloride (HCl) laboratories in Colombia, although cocaine HCl production in Peru is rising. Reportedly, traffickers have been moving their laboratories from isolated jungle sites nearer to towns, where corrupt officials can offer protection. The chemicals (kerosene, lime, ether, acetone, hydrochloride acid) needed to process coca leaves into paste, base, and hydrochloride are diverted from legitimate chemical shipments that reach Peru by sea.

Although the Colombian traffickers control most of the cultivation and the processing of coca into paste and base in Peru, some 20 Peruvian trafficking organizations have also been identified. By early 1991, self-limiting by coca growers increased the price for coca derivatives in the UHV; this was largely because of *Sendero Luminoso* (SL—Shining Path), Maoist political insurgents, who demanded a greater share of the cocaine-base profits. The SL extended their area of influence; charged a tax on coca leaf, paste, and base; and attempted to set prices among the Colombian traffickers, growers, and lab operators—therefore, the prices for coca products

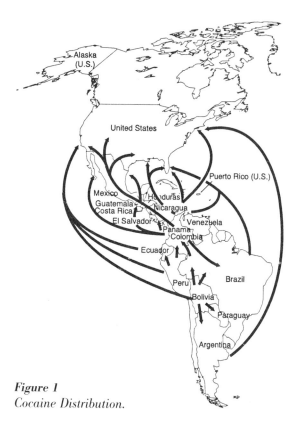

Figure 1
Cocaine Distribution.

varied widely in 1990, showing an average 100 percent increase.

The majority of cocaine base is moved from UHV staging areas by air and by river to Colombia for conversion to cocaine HCl. Drug-control efforts in Peru have been ineffective; violence, political factions, rivalry between the Peruvian police and military, and widespread corruption in a severely depressed economy have contributed to Peru's lack of effectiveness.
Bolivia. By the early 1990s, almost 75 percent of illicit coca was grown in the Chapare, Carrasco, and Arani provinces, in Cochabamba department, Bolivia. Legal cultivation of some 35,000 acres (14,000 ha) occurs only in the Yungas. Small farmers and unemployed migrants cultivate the coca in the Chapare on plots that average one to two acres (less than 1 hectare). When the market price drops below their cost of production, farmers choose not to sell the leaf. Most leaf that is sold to middlemen (*intermediarios*) is processed in the Chapare and then refined into base or cocaine HCl in the Beni, Cochabamba, or Santa Cruz departments. Due to increased enforcement in the early 1990s, some traffickers moved their base of operations to less accessible locations,

and more paste is refined into cocaine base or HCl by about 35 Bolivian trafficking organizations.

Colombian and Bolivian traffickers have integrated some operations vertically, from wholesale paste purchase through cocaine base and HCl refining and export. The U.S. government estimates that as much as 35 percent of Bolivian coca paste may be processed in Bolivia prior to export. Chemicals arrive by truck, train, and aircraft from Brazil, Chile, Argentina, and Paraguay. The base is smuggled to Colombia in private aircraft from the Beni. Increasing its law-enforcement efforts, the Bolivian government eradicated about 10 percent of the cultivation, dismantled a number of laboratories, and disrupted several major trafficking organizations (e.g., Meco Dominquez, Mario Ariaz-Morales, Martin Morales-Daczer).
Colombia. Proximity to a large cash-based U.S. marketplace, powerful criminal organizations, indigenous entrepreneurial spirit, vast tracts of uncontrollable land, and a long tradition of smuggling have made Colombia an ideal source for cocaine. The U.S. government estimated that in 1991 92,000 acres (about 37,500 ha) of the world's 526,500 acres (213,000 ha) of coca were cultivated in Colombia—mainly in the Llanos (plains) region, which encompasses almost 50 percent of eastern Colombia. There is also coca cultivation in Caqueta, Guaviare, Putumayo, and Vaupes departments, with crop expansion into Bolivar department and into south and southwest Colombia. Colombia's drug cartels are the world's leading producers of both cocaine HCL (which is sniffed or snorted) and CRACK (which is smoked).

Colombian cocaine-trafficking organizations are sophisticated and well-organized industries, which derive their strength from control of cocaine laboratories and the smuggling routes to North America. After financing the cultivation of coca plants in Bolivia and Peru, Colombian traffickers often oversee the processing of the leaves into coca paste and sometimes base, which may then be shipped to laboratories in Colombia where the traffickers refine the coca paste—first into coca base and then into cocaine HCl by the ton. Recently, Peru and Bolivia have stopped shipping some of their coca products to Colombia and have begun to refine them into cocaine HCl in laboratories near their own fields, but as of the early 1990s Colombia operates the greatest number of base and HCl labs.

Cocaine is a major threat to weakening Colombia's democratic institutions and directly or indirectly affecting everyone in the country. Colombians increasingly recognize that the violence and corruption that accompany drug trafficking are harming their economy and society. By the early 1990s, under President Cesar Gaviria, the Colombian government security forces began enforcement procedures against cocaine traffickers. The Colombian police have also eradicated virtually all marijuana cultivation in the traditional growing areas along the North Coast and Guajira peninsula. The government of Colombia consequently damaged the leadership structure of the Medellin cartel by jailing its leader, Pablo Escobar. Some feared that jailing Escobar would not curtail his cocaine trafficking, but it did have a symbolic effect on the Medellin cocaine business. (Escobar was later killed after escaping from jail.)

A signatory of the 1961, 1971, and 1988 United Nations International Narcotics Control Conventions, Colombia demonstrates its political will and commitment to investigate and immobilize major cocaine traffickers and to eradicate marijuana and opium. Colombia has also created public-order courts and begun to share evidence, reform its judiciary, and track the substantial money flows into the country—requiring the banking institutions to keep records on cash transactions over $10,000.

In the realm of CROP CONTROL, despite widespread testing of various coca herbicides, the government has not begun a major coca-eradication effort; this is largely because it is not a focus of antidrug efforts—given the location of the fields in terrorist controlled land, it is dangerous for ground forces and almost impossible for air attack. Fearing a new and burgeoning heroin business, in 1992 the Colombian government agreed to spray the common garden herbicide glyphosate (Roundup) to kill the source—the opium poppy fields—after a widespread manual eradication effort in 1991. Since the mid-1980s, marijuana production continues to be minimal because of an effective herbicidal campaign.

The Colombian national police, the military, and the security forces have conducted major operations against the Medellin and Cali cocaine cartels with the assistance of U.S. technical and information support. Colombia's government has, however, paid a heavy price for its action, suffering almost 500 deaths by assassination or during enforcement operations. Colombia has also threatened to use, or has

used, the tool of extradition to incarcerate or immobilize major traffickers. In late 1990, President Cesar Gaviria's offer of amnesty (a plea-bargaining opportunity for major traffickers) resulted in decreased violence throughout the country and the surrender and imprisonment of five traffickers and one terrorist, including Pablo Escobar and the three Ochoa brothers (Jorge Luis, Juan David, and Fabio).

Opium/Heroin. The opium poppy (*PAPAVER SOMNIFERUM*) is the source of heroin. It is grown in three principal geographic regions: Southeast ASIA, Southwest Asia, and Latin America. The Southeast Asian GOLDEN TRIANGLE countries of Myanmar (Burma until 1989), Laos, and Thailand in 1991 cultivated approximately 81 percent of the world's total, 488,000 acres (195,000 ha), yielding 2,500 metric tons of opium, which would yield 250 metric tons of heroin. The Golden Crescent countries of Afghanistan, Iran, and Pakistan cultivated approximately 11 percent, and the Latin American countries of Mexico, Guatemala, and Colombia (plus the Middle Eastern country of Lebanon) produced approximately 8 percent. India is the world's largest cultivator of licit opium, producing about 35,000 acres (14,000 ha) annually for the international medicinal market.

Southeast Asia's Golden Triangle: Myanmar. The largest supply of illicit opium—56 percent of U.S. availability—comes from the Golden Triangle of Southeast Asia. Fields of opium poppy are planted on hillsides that have been prepared by ancient slash-and-burn agricultural methods. Nearly 90 percent of Southeast Asian opium comes from the Union of Myanmar (Burma), where cultivation areas are largely controlled by antigovernment insurgents in the Shan state. Heavy cultivation exists east of the Salween river and in the eastern and southern parts of the Shan state, at an average elevation of 3,300 feet (1,000 m). Fields are small, averaging about an acre (0.5 ha). The climate is ideal for growing poppy. The growers depend on opium for survival, receiving subsistence prices for and selling entire stocks to the political insurgents, who use the proceeds for food, arms, and ammunition. The opium is also consumed locally by large numbers of addicts.

Most processing of opium and heroin in Southeast Asia occurs in Myanmar, with only small amounts in Thailand and Laos. The Shan United Army and the Wa insurgent groups control refineries along the Thai/Myanmar border; the Kokang, Wa

TABLE 2
Illicit Opium Estimates

CULTIVATION				PRODUCTION		
Major Source Country		Net Cultivation (acres)	Net Cultivation (hectares)	Major Source Country	Production (tons)	Production (metric tons)
Burma	1990	370,747	150,100	Burma	1990	2,475	2.250
	1991	395,200	160,000		1991	2,585	2,350
Thailand	1990	8,484	3,435	Thailand	1990	44	40
	1991	7,410	3,000		1991	39	35
Laos	1990	75,533	30,580	Laos	1990	303	275
	1991	73,174	29,625		1991	292	265
Mexico	1990	13,482	5,450	Mexico	1990	68	62
	1991	9,300	3,765		1991	45	41
Guatemala	1990	2,087	845	Guatemala	1990	14	13
	1991	2,828	1,145		1991	19	17
Colombia	1990	Unknown	Unknown	Colombia	1990	Unknown	Unknown
	1991	2,865	1,160		1991	30	27
Afghanistan	1990	30,566	12,375	Afghanistan	1990	457	415*
	1991	42,459	17,190		1991	627	570**
Pakistan	1990	20,266	8,205	Pakistan	1990	182	165
	1991	19,834	8,030		1991	198	180
Lebanon	1990	7,904	3,200	Lebanon	1990	35	32
	1991	8,398	3,400		1991	37	34
Iran	1990	Unknown	Unknown	Iran	1990	330	300
	1991	Unknown	Unknown		1991	330	300
				Total:	1990 = 4.202 tons (3,819 metric tons)	
					1991 = 3,908 tons (3,552 metric tons)	

NOTE: Opium generally converts to heroin hydrochloride at ratio of 10:1. Point estimates cited above reflect a mathematical mean point estimate and are not intended to imply a degree of accuracy or certitude which cannot be obtained due to the nature of illicit drug cultivation and production.

*The U.S. Drug Enforcement Agency believes that multi-cropping and the use of fertilizers in Afghanistan renders the 1990 estimate of 457 tons (415 metric tons) to the lower end of a potentially higher production range in Afghanistan.

**The U.S. Drug Enforcement Agency believes that higher yields may exist in Afghanistan and that production could potentially be above 990 tons (900 metric tons.)

SOURCE: *International Narcotics Control Strategy Report 1992.*

and Kachin ethnic groups also operate large heroin refineries along the China/Myanmar border. Increasing amounts of heroin are smuggled via southern China to Hong Kong, south through Malaysia and Singapore, and west through India and Bangladesh.

With the overthrow of the long-standing government of Burma by a military junta in 1988—and ongoing political strife in the new Union of Myanmar—suspension of aerial opium eradication and diminished enforcement contributed to increases in opium cultivation, heroin refining, and drug trafficking. A signatory to the 1961 SINGLE CONVENTION ON NARCOTIC DRUGS, but not to the 1972 Protocol to the

Convention or the 1971 PSYCHOTROPIC SUBSTANCES CONVENTION, MYANMAR had acceded to the 1988 UN Convention but now disputes the validity of extradition and submission of disputes to the International Court of Justice.

Thailand. Only a small amount of land is used to grow opium in Thailand, but it remains a net importer of opium, consuming far more than it produces. Developed transportation systems make Thailand the primary transit route to the opium/heroin world markets, shipping by air, sea, and overland. Since the mid 1800s, opium has been grown in the northern highlands by nomadic hill tribes, who are

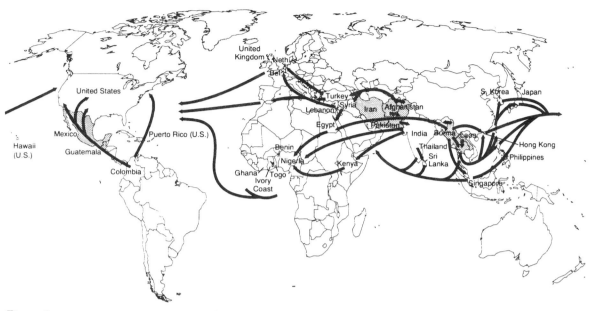

Figure 2
Opium/Heroin Distribution.

not tied to Thailand culturally, religiously or politically. Opium cultivation in Thailand is illegal, so the government has sponsored both eradication and crop-substitution efforts in the north.

Thailand is a party to the 1961 Single Convention on Narcotic Drugs and the 1972 Protocol to the Single Convention. In 1991, Thailand passed conspiracy and asset-forfeiture laws and a new extradition treaty with the United States; both are working on a mutual legal-assistance treaty.

Laos. Recent changes in the world's political order have resulted in cooperation by the Laotian government to reduce opium cultivation. Widespread reports of Lao military corruption and involvement with the traffickers, however, have limited the success. A landlocked country, Laos has been isolated and ignored by the West since 1975 when the Communist Pathet Lao seized power; opium poppies have been grown in its nine northern provinces, yielding in the early 1990s about 20 percent of Burmese production. Three crop-substitution projects have had limited success—one in Houaphanh province, one in Vientiane province, and one in Xiang Khouang province.

The Lao government does not have a mutual legal-assistance treaty or an extradition treaty with the United States, but it does have a formal memorandum of understanding and informal agreements with

U.S. agencies to cooperate more fully in drug-control efforts.

China and the Golden Triangle. In its 1992 International Narcotics Control Strategy Report, the U.S. government stated that opium cultivation may be emerging as a major problem in the People's Republic of China. China has become a major narcotics transit point with its open border to Myanmar, its location adjacent to the Golden Triangle, and its excellent transportation and communication links with the trade ports of Hong Kong and Macao. Much of the heroin processed by the Kokang Chinese in the Golden Triangle travels by road through China's Yunnan, Guangxi, and Guangdong provinces to Hong Kong for overseas distribution. In 1991, Chinese law enforcement seized more drugs and investigated more cases than at any time since the Communist takeover. A spreading domestic opium consumption appears to be accompanying the increased heroin flow.

The Golden Crescent: Pakistan, Afghanistan, and Iran. The Golden Crescent supplied about 21 percent of the heroin consumed in the United States in the early 1990s. In area under cultivation, the Golden Crescent countries produce almost 11 percent of the world's opium.

Pakistan. This is a producer and an important transit country for opiates and HASHISH. The Islamic government of Pakistan maintains a poppy ban in areas under its control and manages to maintain about the same production level from year to year, but cultivation has increased slightly in areas where government control is ineffective or only nominal. Cultivation is both rain fed and irrigated in the northwest and the tribal areas of Kyber, Mohmand, and Bajaur. Once the poppy is harvested, it is processed into opium and heroin in more than a 100 clandestine mobile laboratories in the Northwest Frontier Province (NWFP) bordering Afghanistan, which is controlled by armed tribes who maintain traditional cross-border connections.

Pakistan is party to the 1961 UN Single Convention on Narcotic Drugs, the 1971 UN Convention on Psychotropic Substances, and the 1988 UN Convention Against Illicit Traffic of Narcotic Drugs and Psychotropic Substances. Yet, with widespread corruption and government inaction, Pakistan failed to enforce its counternarcotics laws in the tribal areas, raising questions about its compliance with the 1961 Convention. Pakistan's government does however cooperate with U.S. law-enforcement agencies and has responded positively to extradition requests.

Afghanistan. After Myanmar, Afghanistan is the world's second largest producer of illicit opium. Considered an effective cash crop, opium has been grown for generations in Afghanistan, in the Helmand valley and Nangahar province, and used for medicinal and culinary purposes. The opium is processed into heroin in and smuggled across the borders of Iran through Turkey. Afghanistan's government exerts little control over production or trafficking. Drug revenues continue to finance political resistance operations against the Communist government and provide a livelihood for farmers who depend on the opium crops. Unless the government is willing and able to control opium production in the countryside, both production and domestic consumption will continue to rise. The end of Soviet occupation (1979–1989) has not brought the refugees home, but their return will affect Afghanistan's overall economy and may cause an increase in drug trafficking.

Afghanistan is a party to the 1961 Single Convention but not to the 1972 Protocol amending the Convention. It is a signatory to the 1971 Convention on Psychotropic Substances but not to the 1988 Convention Against Illicit Traffic in Narcotic Drugs and Psychotropic Substances.

Iran. Limited data exist on drug cultivation and trafficking since the Islamic Republic of Iran was established in 1979 under the Ayatallah Khomeini. Iran outlawed opium cultivation in 1980 but growth reportedly occurs in remote areas near the Pakistan and Afghanistan borders. Allegedly, laboratories process heroin from opium in the Kurdish areas of the northwest and the Baluch region in the southeast, with significant Irani and local addict populations consuming the product. The U.S. government estimated that Iran produces about 50 percent of the amount of heroin produced in Afghanistan.

Drug trafficking is increasing along the Afghan-Iran and Afghan-Pakistan borders. Baluch and Pashtun tribesmen from all three Golden Crescent countries smuggle drugs in addition to traditional contraband. Pakistanis and Iranis could increase poppy cultivation to help rebuild their livelihoods that were interrupted by almost twelve years of war.

Mexico. In the 1970s, Mexico began to smuggle significant amounts of heroin into the United States, replacing Turkey as the principal heroin supplier for U.S. addicts. Opium is grown and harvested twice a year—winter and spring—in Mexico's states of Sinaloa, Chihuahua, and Durango. In the 1990s, harvesting has become year round, and cultivation has expanded to include Mexico's west coast from Sinaloa to the Mexican-Guatemalan border. Supplying an estimated 23 percent of the heroin consumed in the United States, Mexican traffickers produce both traditional brown and black-tar heroin, although the predominant type smuggled into the United States is the black-tar type. Conversion from the popular "Mexican brown" in the 1970s to the black-tar variety is a result of traffickers using more cost-effective mobile laboratories. The mobile labs are much harder to detect and can move with the harvesters, as they go from field to field collecting the opium gum and producing the purer black tar preferred by U.S. addicts. Although the mobile labs are found near the fields, Mexican law-enforcement personnel are also finding them near towns and cities, where chemicals and security can be acquired more easily. The administration of President Carlos Salinas (1988–1994) instituted effective law enforcement, including strong measures to combat official corruption, a 40 percent increase in opium eradication, and increased cocaine interdiction.

Mexico is a signatory to the 1988 UN Convention Against Illicit Traffic in Narcotic Drugs and Psychotropic Substances and entered into a Mutual Legal Assistance Treaty (MLAT) with the United States in 1991.

Guatemala and Colombia. These two countries have both begun to cultivate substantial amounts of opium in the late 1980s. By 1991, in Guatemala's western provinces of San Marcos and Huehuetenango, farmers harvested approximately 4,300 acres (1,700 ha) of opium poppy, which had been cultivated in steep mountain valleys on small plots. Mexican traffickers provide the financial, technical, and agricultural support for local growers to harvest three crops per year; the opium, however, is sent to Mexico for processing into heroin. The Guatemalan government has conducted aerial herbicidal eradication with some success, destroying almost a third of the total cultivated, but farmers are relocating their fields to more remote areas. In Colombia, in 1991, over 6,000 acres (2,500 ha) of opium poppy were located in 12 of the 32 states—planted for the most part in the Cauca and Huila departments and financed and controlled by the Cali cartel. Colombia has agreed to begin herbicidal spraying from cropduster aircraft, as it did during its mid-1980s marijuana-eradication program.

Cannabis. A by-product of the HEMP plant CANNABIS SATIVA is marijuana, which is the most commonly used illicit drug in the United States. Both the plant and its PSYCHOACTIVE ingredient TETRAHYDROCANNABINOL (THC) are classified as CONTROLLED SUBSTANCES by the U.S. government, which estimates that Mexico supplies the majority of U.S.-consumed marijuana—perhaps as much as 63 percent. The U.S. supply accounts for another 18 percent, Colombia for 5 percent, Jamaica for 3 percent, and the remaining 11 percent comes from Belize, Laos, the Philippines, Thailand, Lebanon, Pakistan, and Afghanistan. Brazil and Paraguay also cultivate cannabis but the majority is consumed locally or exported to neighboring South American countries.

Mexico. Although *Cannabis* grows throughout the country, major concentrations have been located historically in the western states of Chihuahua, Jalisco, San Luis Potosi, Sinaloa, Sonora, and Zacatecas; it is also found in Mexico's eastern state of Veracruz and, recently, in the southern states of Chiapas, Guerrero, Michoacan, and Oaxaca. Farmers grew two crops per year, traditionally, but in many areas it is grown and harvested year round. *Cannabis* is cultivated by subsistence farmers who often intermingle the crop with corn and beans. Traffickers have introduced irrigation and technological advances to help the farmer (*campesino*) avoid eradication attempts and survive cyclical droughts. The traffickers control the processing and transport of the product into the United States, smuggling the vast majority by road.

Colombia. Once the primary source for marijuana consumed in the United States, in the 1990s Colombia cultivates about 5,000 acres (2,000 ha) in the traditional growing areas of Sierra Nevada de Santa Marta and Serrania de Perija of northeastern Colombia. Since the dramatic success of the Colombian government's 1980s aerial-eradication program, only small amounts of low-quality cannabis have been cultivated in Colombia.

Southeast Asia. This region produces a high-grade marijuana that became popular in the late 1980s; it is cultivated in Thailand and Laos, then shipped to staging points along Thailand's southern coast, to western Cambodia, and to the coast of Vietnam. Moved by ten-wheel trucks, the product is then loaded onto trawlers and taken to motherships in the Gulf of Thailand. Oceangoing vessels, yachts, and sailing boats have all been used to smuggle the product to the United States, with trans-Pacific shipments occurring in the spring and summer. U.S. traffickers usually control the commerce of marijuana into the United States, off-loading their cargo to smaller faster vessels off the U.S. coast.

Two crops a year are generally harvested in Southeast Asia, in December–January and April–May. Cultivators normally press the harvested marijuana into kilogram blocks, using a hydraulic press, and then package the blocks into aluminum foil or plastic bags that are vacuum-packed. They are hermetically sealed with heat and wrapped with nylon-reinforced plastic tape, then stored in tin canisters, burlap sacks, nylon or canvas gym bags, or boxes—all designed to maintain the product's composition, eliminate odor, and prevent mildew.

The THC content of Southeast Asian marijuana can be as high as 9 percent, whereas the average THC content for Mexican or U.S. marijuana is only 2 to 3 percent.

Jamaica. The successful eradication campaigns mounted since 1987 have decreased significantly Jamaica's importance as a supplier of *Cannabis* in the form of ganja. Cultivation has shifted from the accessible wetlands of west-central Jamaica to remote

TABLE 3
Marijuana Production Estimates

U.S. MEASURE

Source Country		Net Cultivation (acres)	Net Production (tons)
Mexico	1990	86,574	21,687
	1991	44,250	8,553
Colombia	1990	3,705	1,650
	1991	4,940	1,650
Jamaica	1990	3,013	908
	1991	2,347	705
Belize	1990	181	66
	1991	133	54
Others	1990	NA	3,850
	1991	NA	4,950
Domestic U.S.	1990	NA	5,500–6,600
	1991	NA	3,977–5,077

METRIC MEASURE

Source Country		Net Cultivation (hectares)	Net Production (metric tons)	Percentage* of Total Supply
Mexico	1990	35,050	19,715	42%
	1991	17,915	7,775	
Colombia	1990	1,500	1,500	8%
	1991	2,000	1,500	
Jamaica	1990	1,220	825	3%
	1991	950	841	
Belize	1990	65	60	>2%
	1991	54	49	
Others	1990	NA	3,500	24%
	1991	NA	4,500	
Domestic U.S.	1990	NA	5,000–6,000	22%
	1991	NA	3,815–4,615	

Summary	1990	1991
Gross Marijuana Available	33,660–34,760 (tons)	19,889–20,989 (tons)
Less U.S. Seizures,** Seizures in Transit, and Losses	3,850–4,950 (tons)	3,850–4,950 (tons)
Net Marijuana Available	28,710–30,910 (tons)	14,939–17,139 (tons)

Summary	1990	1991
Total Marijuana Available	30,600–31,600 (metric tons)	18,080–19,080 (metric tons)
Less U.S. Seizures,** Seizures in Transit, and Losses	3,500–4,500 (metric tons)	3,500–4,500 (metric tons)
Net Marijuana Available	26,100–28,100 (metric tons)	13,580–15,580 (metric tons)

*Percentages were rounded off and reflect midpoints of the quantity ranges in this table. For purposes of calculation and comparison, all the marijuana produced overseas was assumed to be potentially available for import to the United States.

**U.S. seizures included coastal, border and internal (not domestic eradicated sites). Seizures in transit included those on the high seas, in transit countries, from aircraft, etc. The loss factor included marijuana lost because of abandoned shipments, undistributed stockpiles and inefficient handling and transport, etc.

SOURCE: *International Narcotics Control Strategy Report 1992* and the U.S. Drug Enforcement Agency.

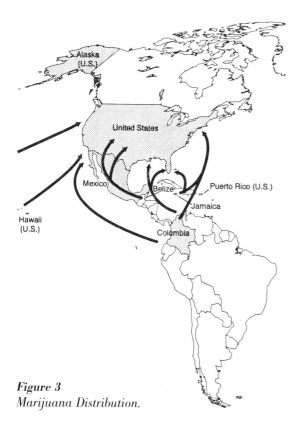

Figure 3
Marijuana Distribution.

sites in the highlands, often in plots smaller than an acre. In the early 1990s, of 4,500 acres (1,800 ha) cultivated, almost 50 percent were reportedly eradicated. The rest was smuggled into the United States in concealed storage areas of pleasure craft, as well as in commercial fishing vessels, cargo ships, and container ships.

POLITICAL AND ECONOMIC SIGNIFICANCE

Although drug cultivators, transportation workers, processors, laboratory workers, middlemen, and smugglers receive their wages, the majority of the money made in the drug business remains in the consuming country or is invested in off-shore banking havens. Drug-producing countries do not normally offer attractive long-term investment opportunities. Countries such as Peru, Bolivia, Myanmar, and Afghanistan have troubled economies, which do not attract traffickers' investment portfolios; rather, traffickers spend money on luxury items, such as foreign real estate and automobiles, race horses, gambling houses, yachts, clothes, and jewels.

In the Golden Triangle and the Golden Crescent, drug production and trafficking offer a primary cash crop for food and the support of political (antigovernment) operations. Resistance groups in Afghanistan and Pakistan and insurgent tribes in the Golden Triangle use the profits from the sale of opium to buy rice and the arms to fight the central governments. Politically speaking, illicit-drug production and trafficking offer a viable means of acquiring wealth, which can be instrumental in buying power and influence. In some countries, the traffickers and insurgent groups may be identical (such as the Wa or the Shan United Army of Burma); in others, insurgency and trafficker goals may be diametrically opposed (such as the Cali cartel and the FARC in Colombia). Most trafficker organizations work to coopt the government and maintain the status quo, buy power and protection, and keep a low profile; insurgent groups, however, seek to be highly visible and wish to change the existing power structure. Despite the opposing objectives of both, traffickers and insurgents often function symbiotically; that is, both need hard currency, security, protection, and armed support to evade detection and apprehension.

HISTORICAL SUPPLY SHIFTS

The 1960s and 1970s. Drug production and trafficking have undergone major shifts since the 1960s. After the so-called French Connection was broken between 1968 and 1972, Mexico began to supply the United States with a low-quality heroin to fill the market demand. As Mexican eradication became more successful in the mid to late 1970s and the Iranian Islamic revolution erupted in 1979, significant amounts of Southwest Asian heroin from Afghanistan and Pakistan were smuggled, often by Iranians, through Western Europe into the United States. Throughout the 1970s, heroin from Mexico, Southeast Asia, and the Middle East was high on the U.S. drug-control policy agenda. No one denied that cocaine and marijuana abuse might be dangerous; indeed, initial attempts were made to initiate bilateral programs with the Andean cocaine source countries in their traditional growing areas, but because policymakers believed that the negative health consequences of heroin consumption were far worse, the U.S. law-enforcement emphasis was placed on cocaine and marijuana.

The 1980s. In the 1980s, targeting heroin gave way to focusing on the reduction of cocaine and marijuana use in the United States, since greater numbers of Americans were using and abusing them, creating large drugged populations. Nongovernmen-

tal institutions became very active in spreading the Just Say No message of the Reagan Administration (1981–1989). Moreover, until the early 1980s, when research had documented the negative health consequences of cocaine, the drug enjoyed a glamour and allure that heroin had never possessed. In some circles, the ability to afford cocaine was a sign that one had succeeded. Most believed that cocaine was not addictive and it became the recreational drug preferred by Hollywood, sports figures, and musicians. The 1980s was the Coke Decade—with cocaine used both by the affluent consumption-oriented yuppie "me generation," as well as the poor disenfranchised underclass who tried to emulate their heros and "make it" in their own world of need and despair. Both the powder and the rock-crystal crack forms found eager markets and ready money in the so-called affluent Reagan years. The stock market crashes of 1987 and 1989 ended the boom in the national and the drug economies.

When Colombia replaced Mexico in the early 1980s as the primary source of U.S. cocaine and marijuana, smuggling vast quantities through the South Florida peninsula, the Reagan Administration turned its focus to cocaine and marijuana control in the Western Hemisphere. In the late 1980s, the Bush Administration (1989–1993) continued the same cocaine policy but decreased federal attention on marijuana supply reduction. Bolivia and Peru quickly expanded their production of illicit coca in nontraditional growing areas of the Chapare and the Upper Huallaga Valley, two areas that remain the major source for the world's illicit coca. Surrounding and potential transit countries also became more involved in the cocaine smuggling enterprise. The Caribbean nations functioned as attractive transit points for both cocaine and marijuana from South America. As the United States placed more enforcement pressure on the Caribbean, the traffickers shifted their routes through Central America and Mexico. In the mid-to-late 1980s, Mexico became a principal transit and smuggling route for an estimated 50 percent of the cocaine entering the United States. In response to smuggler shifts, both Mexico and the United States have increased interdiction efforts along the joint southwest border and the U.S. 1992 Drug Control Strategy focuses its attention in Mexico predominantly on improving cocaine interdiction.

The 1990s. U.S. policymakers are faced with a number of new challenges—namely, increased heroin production and trafficking in Central Asia (China), in Central and South America (Guatemala and Colombia), and in Myanmar and Afghanistan. Another challenge is the growing cocaine business in Bolivia and Peru, where increased processing of coca into cocaine products occurs. Finally, policymakers need to consider the potentially devastating impact of increased cocaine and heroin demand and consumption in the new democracies of Eastern Europe and the new republics of the former Soviet Union.

Beginning in 1991, the U.S. government expressed its concern over an increase in worldwide heroin production, trafficking, and abuse. Record seizures have been made in China's Yunnan province—signaling major changes in trafficking routes out of the Golden Triangle through China, Hong Kong, and Taiwan to the West. Heroin traffickers have begun to use the immense container-shipping industry to smuggle large amounts of heroin from Asia into the United States. In June 1991, the single largest heroin seizure in the world was made in San Francisco, hidden in containerized freight from Taiwan. Colombia became a significant cultivator of opium for the first time, in the 1990s—planting an estimated 6,000 acres (2,500 ha) of opium in 1991. Although opium cultivation has decreased in Mexico and the Golden Crescent, increasing demand in the United States may be met by Colombia and Myanmar.

The cocaine epidemic of the 1980s, as measured by prevalence and incidence indicators, appears to have peaked and is declining for certain cohort populations, but concern continues over the chronic intensive use of the crack form among the predominantly minority underclass; those least able to cope—the uneducated, unemployed, and disenfranchised—are the victims. With processing facilities now closer to source countries least able to implement effective drug-control programs politically and economically, these two problems present daunting challenges for U.S. public policymakers.

Finally, the massive political, economic, and social changes that occurred since 1989—the democratization and political upheaval in the Eastern bloc and the Soviet Union—may result in increased drug demand, driving up drug production, trafficking, and serious negative health consequences. Unfortunately, accompanying the economic difficulties and growing political pains in the fledgling democracies are increases in crime, violence, and drug abuse. The

real challenge for drug policymakers in the 1990s will be to respond effectively to multinational drug organizations that shift their area of focus from a north–south to an east–west trafficking pattern. It will also be important to assess the potential threat of the creation of new markets (e.g., in the face of a saturated U.S. cocaine market), the global sense of laissez faire, and the breakdown of regulatory traditions where societies are in flux.

(SEE ALSO: *Amphetamine Epidemics; Drug Interdiction; Money Laundering; Operation Intercept; Prohibition of Drugs; Terrorism and Drugs; Transit Countries for Illicit Drugs; U.S. Government/U.S. Government Agencies*)

BIBLIOGRAPHY

BUREAU OF INTERNATIONAL NARCOTICS MATTERS, U.S. DEPARTMENT OF STATE. (1991, 1992). *International narcotics control strategy report (INCSR)*. Washington, DC: Author.

HEALY K. (1986). The boom within the crisis: Some recent effects of foreign cocaine markets on Bolivian rural society and economy. In D. Pacini & C. Franquemont (Eds.), *Coca and cocaine: Effects on people and policy in Latin America*. Peterborough, NH: Cultural Survival.

MOORE, M. H. (1988). Supply reduction and drug law enforcement. Remarks made at a symposium at the Castine Research Foundation, Washington, DC.

JAMES VAN WERT

INTERPOL *See* Drug Interdiction; International Drug Supply Systems.

INTOXICATION *See* Addiction: Concepts and Definitions.

INTOXILYZER *See* Drunk Driving.

ITALY, DRUG USE IN In Italy, the impact of illicit drug use was first felt on a broad scale during the mid-1960s. The patterns in Italy were similar to those seen in other European countries. They seemed to be associated with the contestation by young people of existing political and social situations. As in the United States at the same time, this phase was influenced by the cultures of the East—especially those of the Indian subcontinent, Southeast Asia, and the Middle East—in all of which some amount of drug use was not illegal. *Cannabis sativa*, the HEMP plant that produced MARIJUANA, GANJA, HASHISH, and other variants, was particularly unhampered by legislation there and was enjoyed by locals and outsiders, some of whom found ways to smuggle it into the West, where in many instances it was illegal.

In addition, the OPIOIDS (especially HEROIN) began to be used illicitly, and by the 1970s serious consequences ensued. By then, the countercultural movement and its abuse of illicit drugs had lost most of its original idealistic principles. Abusers were simply in search of ever more and ever stronger psychotropic effects. Moreover, criminal organizations took charge of the illicit drug trade, not only to increase their profits but also to control and direct the political and social development of the youth of Italy. For the most part, users became abusers who were physically dependent on their drug, so their behavior could be controlled by the suppliers.

In the 1980s, the drug scene changed, with various control measures and less heroin available. In addition, with less heroin being sold, longer intervals occurred between drug doses for many users. Such modified habits led to decreased tolerance and increased overdosing, with ensuing deaths. For these reasons, the number of heroin addicts in Italy decreased—then, in the mid-1980s, COCAINE emerged as the new illicit drug problem. The CRACK and FREEBASE forms were especially harmful among young ADOLESCENTS. More detailed data are contained in the annual reports of the National Health Council (1985–1991) and the reports of the Department of Social Affairs (1991–1993).

LEGISLATION

At the beginning of the drug-abuse phenomenon of the 1960s, the legislation in force had been passed in 1954. It proved to be insufficient for coping with emerging conditions; it did not take into consideration the political-cultural trends, the scientific knowledge of the day, or the increasingly important role of public health.

New legislation in 1975 was characterized by such innovative elements as the nonpunishment of the addict found to be in possession of a moderate quantity of illicit drugs. The quantity was to be examined and quantified, and it was to be considered in relationship to the physical and psychological needs of the addict. Unfortunately, this individualistic approach was poorly applied, which made the law useless.

The regulations approved in 1990 improved the state's power of both repressive action and intervention, and it defined a daily mean dose to separate administrative offenses from crimes. The objective was to recover and rehabilitate drug addicts. A 1993 referendum, however, repealed the prohibition on

TABLE 1
Treatment Typology Provided by Public Services (1991–1994)

Typology	Addicts (percent)			
	1991	1992	1993	1994
Psychosocial and/or rehabilitating	38.1	37.3	39	41.2
Pharmacologic				
Integrated	24	27	26.5	28.2
(from which "methadone treatment")	(21)	(23.4)	(23)	(14.3)
Methadone treatment (short & long, not integrated)	9.3	9.7	12	27.6
Naltrexone	7.7	7.6	8.2	
Other antagonistic	0.14	0.20	0.13	
Other	19.8	16.3	13.4	

SOURCE: Alcohol and Drug Addiction Central Service, Ministry of Health. Data processed by the Istituto Superiore di Sanità.

TABLE 2
Seizures of Illicit Drugs by Raw Weight in Kilograms

Year	Opioids				Stimulants			Hallucinogens			
	A	B	C	D	E	F	G	H	I	J	K
1980	—	267	197	—	76	—	—	—	4,907	—	—
1981	—	82	141	—	64	—	—	—	11,204	—	
1982	12	0.5	230	1	105	5	4	968	3,908	23	0.1
1983	7	3	314	11	223	71	3	1,018	4,137	23	0.1
1984	2	0.2	457	3	73	0.4	0.4	854	5,185	14	0.3
1985	5	0.9	275	7	107	0.2	0.05	372	1,051	8	0.3
1986	2	7	333	1	129	0.4	0.8	293	15,731	3	0.2
1987	2	0.7	323	9	330	5	1	1,207	11,817	6	0.01
1988	3	1	576	5	612	0.7	0.9	256	6,912	0.0	0.3
1989	1	1	685	0.8	668	0.7	1	239	22,993	0.4	0.1
1990	—	8.8	1,003	2.2	802	0.7	—	182	7,704	1,258	0.23
1991	—	—	1,556	—	1,300	0.7	5,983[a]	499	9,223	—	4,016[b]
1992	—	—	1,353	—	1,367	15.4	22.2	584	22,620	—	12,759[b]

NOTE: A = opium B = morphine C = heroin D = others
E = cocaine F = amphetamine-like G = others[a] (includes MDA, MDMA, in number of units)
H = marijuana I = hashish J = hashish oil K = LSD and others[b] (in number of units)
SOURCE: Drug Enforcement Service, Ministry of the Interior.

personal drug use and canceled references in the regulations to the daily mean dose.

TREATMENT FACILITIES

In accordance with national policy guidelines, a network of facilities was set up and various links were established with rehabilitation, law enforcement, and judicial structures. This process was worked out with public support; the aim has been to sustain every initiative to reduce the availability and demand for drugs.

Of the addicts served by the facilities, almost all are heroin abusers, some not yet dependent. Starting by weaning them from heroin with METHADONE, the facilities provide integrated and custom-designed programs founded mainly on nonpharmacological support.

A wide range of resources are available; 576 public facilities and 276 residential communities and sociorehabilitative structures (public, private, and voluntary—most of them situated in northern Italy). Voluntary services continue to increase in importance both in number and in regional distribution. The effectiveness of the facilities has been proved, since trained personnel and good records provide such statistics on trends (see Table 1).

SEIZURES OF ILLICIT DRUGS

Various trends can be seen by studying the records of seized drugs. Some decreasing trends have been recorded for MORPHINE in 1982, for heroin in 1981 and 1985, and for cocaine in 1984. Irregular trends emerged for *Cannabis* products: a 128-percent increase in 1981, a decreasing trend until 1985, and two huge increases in 1986 and in 1989 (see Table 2). The decision to standardize descriptions of drug seizures by reference to the percent of the primary drug instead of the raw weight of the primary drug seized should improve the accuracy of record keeping (see Table 3).

DRUG ABUSE–RELATED DEATHS

Drug abuse–related deaths also show irregular trends. Most deaths could be attributed to heroin overdoses or to accidents while injecting it. After 1980, two large increases in the death rate occurred, first in 1982 and then in 1984, followed by a steady rise into 1986. From 1986 to 1988, the "empiric" mortality rate nearly doubled; it subsequently remained steady until 1991 and then dropped until 1994 (see Table 4), except among the elderly, for whom the rate increased.

TABLE 3
Seizures of Drugs (Raw and Normalized Weight, average doses*), Italy, 1988

Heroin (kg)			Cocaine (kg)		
raw (37%)	normalized (100%)	doses no.	raw (88%)	normalized (100%)	doses no.
576	213	14,200–8,500	612	539	13,500–9,000

*average dose, heroin = 0.015–0.025 gr
mean dose, cocaine = 0.040–0.060 gr

ALCOHOL ABUSE

ALCOHOL use in Italy strongly differs from drug use for historical, traditional, behavioral, and cultural reasons; supply and distribution are also different, since alcohol is free from legal restrictions. Wine is the most frequently used alcoholic beverage. Although a gradual displacement in wine consumption occurred during the 1980s, with substitution of other liquors and beers, still the total amount of alcohol (percent of ethanol) consumed remained almost constant.

ALCOHOLISM is mainly a problem of chronic abuse by adults over the age of 40. It is mainly a problem in northern Italy. Since the 1980s, however, increasing numbers of young people are abusing liquors and beer. Alcoholism has also become complicated by the combining of alcohol with psychotropics (e.g., tranquilizers), especially by women over 40.

Driving-license regulations have, since 1988, included a test that measures the breath concentration of alcohol. The alcohol level must not be over 8 grams per liter (g/l), approximately that of other countries of the European Economic Community.

(SEE ALSO: *Britain, Drug Use in; Netherlands, Drug Use in; Sweden, Drug Use in*)

TABLE 4
Drug Abuse–Related Deaths (1980–1993) and "Empiric" Mortality Rate (1986–1993)

Year	Deaths	Total Addicts	Rate (percentage)
1980	206	NA	—
1981	237	NA	—
1982	252	NA	—
1983	237	NA	—
1984	392	NA	—
1985	237	NA	—
1986	276	32,000	0.86
1987	516	39,000	1.32
1988	809	47,000	1.72
1989	972	55,000	1.77
1990	1,161	66,700	1.74
1991	1,383	76,200	1.81
1992	1,217	89,000	1.37
1993	888	109,000	0.81

SOURCES: Ministry of the Interior; Ministry of Health.

BIBLIOGRAPHY

AVICO, U., & CAMONI, L. (1994). IX course on "activity aims, valuation of facilities and other structures for addicts' treatment." *Boll. Farmacodipendenze e Alcoolismo, 17*(2):7–120.

AVICO, U., & CAMONI, L. (1994). Multicity study on drug misuse: Rome and Italy: 1993. In Pompidou Group, *Up-to-date report of the epidemiology expert on drug problems.* Strasbourg: Council of Europe.

AVICO, U., ET AL. (1992). *Droga e tossicodependenza* (Drugs and addiction). Brescia: CLAS International. (In Italian).

AVICO, U., ET AL. (1983). Prevalence of opiate use among young men in Italy, 1980 and 1982. *Bull. Narc., 35*(3):63–71.

MACCHIA, T., MANCINELLI, R., BARTOLOMICCI, G., & AVICO, U. (1990). Cocaine misuse in selected areas: Rome. *Ann. Ist. Super. Sanità, 26*(2):189–196.

WHO. PROGRAMME ON SUBSTANCE ABUSE. (1993). Deaths related to drug abuse. Report of a WHO consultation, WHO-PSA/93.14, Geneva.

USTIK AVICO

J

JAILS *See* Prisons and Jails.

JAPAN, DRUG USE IN *See* Amphetamine Epidemics; Epidemics of Drug Use.

JELLINEK MEMORIAL FUND In 1955, the Jellinek Memorial Fund was established to commemorate Dr. E. M. Jellinek's great contribution to the field of ALCOHOL studies. A capital fund was developed and the interest from this fund is used to provide an annual cash award to a scientist who has made an outstanding contribution to the advancement of knowledge in the alcohol/alcoholism field. The first award was presented in 1968.

Each year the board of directors of the Jellinek Memorial Fund designates the specific area of research for which the award will be made and appoints an Expert Selection Committee to review candidates and recommend an appropriate awardee. The awardee may be selected from any country, the sole criterion being the scientific contribution that the person has made within the selected category. The award is traditionally presented at a major international conference and, if necessary, travel and accommodation expenses are provided to permit the winner to attend the conference for the presentation. The following general criteria have been accepted by the board and by previous selection committees as guidelines:

1. The award is to be given to the person deemed to have made, during the preceding years, the greatest scholarly contribution to human knowledge of problems relating to alcohol, in the designated research area.
2. The person selected for the award should be someone who would provide an example and serve as a model for others who might be attracted to work in the field.
3. Only living scientists should be considered for the award.
4. Advanced age or impending retirement would not disqualify someone from candidacy. However, if two or more scientists were considered approximately equal, the one more likely to continue longer in the field would be favored.
5. If the outstanding contribution of a candidate was made more than ten years ago, consideration for the award would require evidence of the candidate's continuing interest and active participation in alcohol research.
6. Other factors being equal, the person would be favored whose primary identification continued to be in the field.
7. If a member of the Expert Selection Committee is deemed eligible for the Jellinek Award, the chair of the selection committee should consult with the president and request the resignation of the committee member.
8. If a previous award winner becomes a candidate and appears equal to or above all other candidates on the basis of unique new achievements,

Year	Recipient	Category
1968	Dr. Jean Pierre von Wartburg University of Bern Switzerland	Genetics and biochemistry
1971	Dr. Kettil Bruun The Finnish Foundation for Alcohol Studies Helsinki, Finland	Sociology
1972	Bill W. Co-Founder of Alcoholics Anonymous	**Honorary**—Self-help treatment
1972	Mr. Robert Popham Addiction Research Foundation Toronto, Canada	Epidemiology and pharmacology
1972	Dr. Harold Kalant Addiction Research Foundation Toronto, Canada	Epidemiology and pharmacology
1974	Dr. Joan Curlee-Salisbury University of Minnesota U.S.A.	Clinical psychology
1974	Dr. Donald W. Goodwin Washington University U.S.A.	Psychiatric genetics
1976	Dr. Charles S. Lieber City University of New York U.S.A.	Clinical medicine
1977	Professor Mark Keller Rutgers University U.S.A.	Documentation
1977	Dr. Werner K. Lelbach University Clinic Bonn, Germany	Clinical epidemiology
1977	Dr. Georges Pequignot National Institute of Health and Medical Research Le Vesinet, France	Clinical epidemiology
1978	Dr. Nancy K. Mello Harvard Medical School U.S.A.	Experimental behavioral studies
1978	Dr. Jack H. Mendelson Harvard Medical School U.S.A.	Experimental behavioral studies
1979	Dr. J. Griffith Edwards Addiction Research Unit Institute of Psychiatry London, England	Socio-behavioral treatment evaluation research
1979	Dr. D.L. Davies Alcohol Education Centre The Maudsley Hospital London, England	Socio-behavioral treatment evaluation research
1980	Dr. Yedy Israel Clinical Institute Addiction Research Foundation Toronto, Canada	Biomedical research—treatment
1981	Dr. Wolfgang Schmidt Addiction Research Foundation Toronto, Canada	Epidemiological research in the alcohol field and, in particular, that relating to physical health consequences of heavy drinking
1981	Mr. R. Brinkley Smithers The Christopher D. Smithers Foundation Mill Neck, New York, U.S.A.	**Honorary**—For his lifelong commitment to the field of alcoholism

Year	Recipient	Category
1982	Dr. Albert J. Tuyns Centre International de Recherche sur le Cancer Lyons, France	Research in clinical medicine in relation to alcohol and alcoholism
1983	Dr. Robin Room Alcohol Research Group Univ. of California at Berkeley U.S.A.	Social-science research having direct implications for the development of preventive strategies
1983	Dr. Klaus Makela Finnish Foundation for Alcohol Studies Helsinki, Finland	Social-science research having direct implications for the development of preventive strategies
1985	Dr. Paul Lemoine Nantes Hospital Nantes, France	Medical and biological research
1985	Dr. Ann P. Streissguth University of Washington U.S.A.	Medical and biological research
1986	Dr. Ting-Kai Li Indiana University Medical Center U.S.A.	Experimental research on alcoholism
1986	Dr. H. David Archibald Addiction Research Foundation Toronto, Canada	**Honorary**—For outstanding achievements in advancing knowledge in the national and international alcohol fields
1987	Dr. Ole-Jorgen Skog National Institute for Alcohol Research Oslo, Norway	Social Sciences—For contributions to the understanding of the collectivity of drinking cultures and for research on the relationship between population drinking and drinking problems
1988	Dr. Boris Tabakoff National Institute on Alcohol Abuse and Alcoholism Bethesda, Maryland U.S.A.	Biological and Medical: clinical—For his major and continuing contributions to our understanding of the neurochemistry of neurotransmitters, receptors, cell membrane response systems and neuropeptide modulators, and their roles in alcohol tolerance and dependence
1990	Dr. G. Alan Marlatt University of Washington U.S.A.	Social and Behavioural Sciences: experimental—For his significant contributions to the systematic study of social and behavioral aspects of alcoholism
1990	Archer Tongue International Council on Alcohol and Addictions Lausanne, Switzerland	**Honorary**—For outstanding contribution to international information exchange
1991	Dr. George E. Vaillant Dartmouth College U.S.A.	Epidemiology and Population Studies—For outstanding contributions to the understanding of the dynamics of drinking careers and the natural history of alcoholism
1992	Dr. Jorge Mardones University of Chile Santiago, Chile	For biological and medical research
1992	Dr. Max Glatt 16 Southbourne Crescent London, England	**Honorary**—For major contribution and pioneer work in treatment of addictions
1993	Dr. Sally Casswell University of Auckland New Zealand	For social and cultural studies
1994	Dr. William R. Miller University of New Mexico U.S.A.	For behavioral (experimental and clinical) studies

he or she should not be ruled ineligible. The chair of the selection committee should consult with the president to ensure that the award is for new achievement and determine if he or she is eligible.

9. The award will normally be made to an individual researcher most highly recommended by the selection committee. In special circumstances, if the selection committee recommends two persons of equal and outstanding merit, a joint award may be made to the two.

(SEE ALSO: *Disease Concept of Alcoholism and Drug Addiction*)

H. DAVID ARCHIBALD

JEWISH ALCOHOLICS, CHEMICALLY DEPENDENT PERSONS AND SIGNIFICANT OTHERS FOUNDATION, INC. (JACS)

This is a nonprofit, tax-exempt, volunteer membership organization located at 426 West 58 Street, New York City, NY 10019, (212) 397-4197. JACS was established as a result of work done by the Task Force on Alcoholism and Substance Abuse of the Federation of Jewish Philanthropies of New York (UJA-Federation). It is supported by dues from its more than 1,000 members, profits from activities and services, charitable contributions, and corporate grants.

JACS provides support programs and conducts retreats enabling recovering Jewish substance abusers and their families to connect with each other, enhance family communication, and reconnect with Jewish traditions and spirituality. The programs are designed to help participants with varying degrees of Jewish background find ways in which Judaism can assist and strengthen their continuing recovery. Informality and personal sharing and disclosure are emphasized. In addition, participants, rabbis, and others with expertise in these areas explore the relationship between Jewish spiritual concepts and TWELVE-STEP PROGRAMS.

In addition to conducting retreats and support programs, JACS provides community outreach programs. These programs disseminate information to educate and sensitize Jewish spiritual leaders, health professionals, and the Jewish community about alcoholism and substance abuse as a disease, and about the effects of ALCOHOLISM and drug dependence on Jewish family life. JACS acts as a catalyst to help Jewish agencies recognize the problems of alcohol and drug abuse in their own programs and communities and to work toward solutions. JACS has a speakers bureau, conducts seminars, and participates in local and national conferences and forums in the Jewish community. The organization is also involved in research, maintains a resource center, holds annual meetings, and publishes a newsletter.

(SEE ALSO: *Ethnicity and Drugs; Jews, Drugs, and Alcohol; Treatment*)

FAITH K. JAFFE

JEWS, DRUGS, AND ALCOHOL

Who hath woe?
Who hath sorrow?
Who hath contentions?
Who hath babbling?
Who hath wounds without cause?
Who hath redness of eyes?
They that tarry long at the wine;
they that go to seek mixed wine.
Look not thou upon the wine when it is red,
when it giveth his color in the cup,
when it moveth itself aright.
At the last it biteth like a serpent,
and stingeth like an adder.
Thine eyes shall behold strange women,
and thine heart shall utter perverse things.
Yea, thou shalt be as he that lieth
down in the midst of the sea,
or as he that lieth upon the top of a mast.
They have stricken me, shalt thou say,
and I was not sick;
they have beaten me, and I felt it not:
when shall I awake?
I will seek it yet again.

—Proverbs 23:29–35

As illustrated by this biblical description of intoxication, alcoholic blackouts, alcohol-related physical and social problems, alcoholic hallucinations, loss of control of drinking, and alcohol dependence was not unknown to the ancient ancestors of today's Jewish population. The Hebrew Bible (called by Christians the Old Testament) includes several illustrations of alcohol-related problems, such as the drunkenness of Noah, which led to family strife, and the incest between Lot and his daughters.

Modern literature about the role of ALCOHOL in the Jewish community displays two very different trends. On the one hand, Jews are regarded as a population with few alcohol problems; and a variety of cultural, spiritual, or physiological explanations are suggested to explain the relatively low rate of ALCOHOLISM among Jews. On the other hand, studies of alcoholism among Jews point out that many cases often go unrecognized, because of the myth of Jewish immunity to alcohol abuse.

Surveys of U.S. drinking practices conducted in the 1960s found that most males who considered themselves Jewish reported drinking to some extent, but few reported alcohol problems. However, the number of Jewish subjects in these studies was small. A more recent study of U.S. male college students and university employees reported that although Jewish and Christian subjects had generally similar drinking patterns, Jews were less likely to drink more than six drinks on any one occasion and less likely to report alcohol problems.

Israel reports a lower per-capita alcohol consumption than countries in Western Europe or the Americas and a lower death rate from cirrhosis of the LIVER. (Cirrhosis mortality is thought to correlate with rates of alcoholism.) The actual rates of alcohol dependence among Jews, either in the United States or in other parts of the world are unknown.

Explanations of Jewish sobriety go back at least as far as the German philosopher Immanuel Kant, who in 1798 theorized that Jews (like women and ministers) avoided drunkenness because their special position in European Christian society was based on the perception that they adhered to a religious law that dictated a higher code of conduct. Intoxication for a Jew would therefore be sinful as well as scandalous. Others have suggested that the traditional use of wine for religious ritual in Jewish life, rather than for hedonistic or social purposes, protects Jews from alcoholism.

In the 1950s, Snyder studied the drinking patterns of seventy-three Jewish men living in New Haven, Connecticut, and also analyzed data from Jewish and non-Jewish male college students. He concluded that SOBRIETY was a positive factor in Jewish identity, as opposed to drunkenness, which was associated with non-Jews. He also concluded that the greater the adherence to Jewish religion and its "ceremonial orthodoxy," the lower the alcohol-problem risk. This finding has led many to theorize that those Jews who do develop alcohol problems are

those who have rejected or left Jewish religious practices, abandoning their Jewish identity.

The finding that genetic factors may predispose to alcoholism has led to speculation that there may also be some hereditary protection for Jews. A 1991 pilot study by Monteiro and colleagues found suggestive evidence that young adult Jewish males were more sensitive to the subjective effects of low levels of blood alcohol than were a control group of Christians. Although this finding awaits replication, the authors theorize that heightened sensitivity in Jews might either deter heavy drinking or help facilitate internal mechanisms for the control of alcohol consumption. Nearly all studies of Jewish sobriety concentrate on male subjects, leaving the applicability of these theories to women unknown.

The most extensive investigation of Jewish alcoholics is the 1980 study by Blume and colleagues, who interviewed 100 Jewish members of Alcoholics Anonymous from the New York City area (58 men and 42 women). The subjects had been abstinent for an average of 4 years. The belief among clinicians that Jewish alcoholics would have a high rate of preexisting psychiatric illness (because they would have to be mentally ill to be so deviant from their cultural group) was found not to be accurate. The Jewish subject group generally resembled their fellow non-Jewish alcoholics in treatment and at ALCOHOLICS ANONYMOUS, with similar family histories of alcoholism, drinking histories, and rates of additional psychiatric diagnoses. They did differ in having an unusually high rate of dependence on prescribed psychoactive medications, a combined result of their attempts to obtain professional help and the frequent failure of their physicians to reach an accurate diagnosis. Although there was evidence of less adherence to orthodox Judaism later in life in these Jewish alcoholics, their subjective feelings of Jewish identity were strong and remained so throughout their alcoholism and recovery.

Many subjects reported that their families, their friends, their physicians and they themselves had dismissed the possibility that they might be suffering from alcoholism, because "Jews can't be alcoholics." They experienced great relief when they finally met another recovering person who was Jewish.

During the late 1970s and 1980s interest in helping Jewish alcoholics grew, both in the United States and in Israel. The Federation of Jewish Philanthropies, based in New York City, organized a task force on alcoholism, which later extended its purview to all addictive diseases, including compulsive GAMBLING. In 1980, the JEWISH ALCOHOLICS, CHEMICALLY DEPENDENT PERSONS AND SIGNIFICANT OTHERS FOUNDATION, INC. (JACS) was organized to serve as a forum for the sharing of recovery by Jewish addicts and their families. Both groups continue to educate the Jewish community and to encourage prevention, treatment, and the opening of synagogues and Jewish community centers to twelve-step groups such as Alcoholics Anonymous, NARCOTICS ANONYMOUS, Gamblers Anonymous, AL-ANON, Nar-Anon and Gam-Anon.

The literature on drug addictions other than alcoholism in the Jewish community has been less divided, because of an absence of long-standing belief concerning Jewish immunity to drug dependence. Nevertheless, denial of drug problems in many Jewish households and communities is an ongoing problem. The rates of drug addiction among Jews, although not known precisely, are very likely similar to those of other ethnic groups of comparable age, sex, and social class. Efforts to promote education, PREVENTION, and TREATMENT of these problems among Jews have gone hand-in-hand with the efforts to fight alcoholism—and have employed the same methods. Self-help fellowships based on the twelve steps of Alcoholics Anonymous can be helpful to alcoholics and to other addicts of the Jewish faith, even though the spiritual base of the twelve steps was originally adapted from the philosophy of a Protestant Christian movement. Several authors have published guides to the twelve steps as related to Judaism.

(SEE ALSO: *Ethnicity and Substance Abuse; Twelve-Step Programs*)

BIBLIOGRAPHY

BLUME, S. B., DROPKIN, D., & SOKOLOW, L. (1980). The Jewish alcoholic: A descriptive study. *Alcohol Health and Research World*, Summer, 21–26.

FLASHER, L. V., & MAISTO, S. A. (1984). A review of theory and research on drinking patterns among Jews. *Journal of Nervous and Mental Diseases, 172*(10), 596–603.

JELLINEK, E. M. (1941). Immanuel Kant on drinking. *Quarterly Journal of Studies on Alcohol, 1*, 777–778.

KANDEL, D. B., & SUDIT, M. (1982). Drinking practices among urban adults in Israel, a cross-cultural comparison. *Journal of Studies on Alcohol, 43*, 1–16.

LANDMAN, L. (1973). *Judaism and drugs*. New York: Federation of Jewish Philanthropies.

LEVY, S. J., & BLUME, S. B. (EDS.). (1986). *Addictions in the Jewish community*. New York: Federation of Jewish Philanthropies.

MONTEIRO, M. G., KLEIN, J. L., & SCHUCKIT, M. A. (1991). High levels of sensitivity to alcohol in young adult Jewish men: A pilot study. *Journal of Studies on Alcohol, 52*(5), 464–469.

MONTEIRO, M. G., & SCHUCKIT, M. A. (1989). Alcohol, drug, and mental health problems among Jewish and Christian men at a university. *American Journal of Drug and Alcohol Abuse, 15*(4), 403–412.

OLITZKY, K. M., & COPANS, S. A. (1991). *Twelve Jewish steps to recovery: A personal guide to turning from alcoholism, drugs, food, gambling, sex*. Woodstock, VT: Jewish Lights.

SELLER, S. C. (1985). Alcohol abuse in the Old Testament. *Alcohol & Alcoholism, 20*(1), 69–76.

SNYDER, C. R. (1978). *Alcohol and the Jews: A cultural study of drinking and sobriety*. Carbondale, IL: Southern Illinois University Press.

SHEILA B. BLUME

JIMSONWEED A poisonous tall coarse annual of the nightshade family (Solanaceae), with rank-smelling foliage and large white or violet trumpet-shaped flowers succeeded by globose prickly fruits. Jimsonweed (*Datura stramonium*) grows in several parts of the world, including North America. A strong intoxicant made from this plant was used

Figure 1
Jimsonweed

by the Woodland tribes of eastern North America. The plant was also used as an ingredient of *wysoccan*, an intoxicant employed in the puberty rites of Native Americans in what is now Virginia.

Smoke from burning jimsonweed was breathed to relieve symptoms of asthma in India, and cigarettes containing jimsonweed have also been used for the same purpose.

Figure 2
Belladonna

As in other members of the Solanaceae family, the psychoactive substances are tropane ALKALOIDS, and the seeds and leaves contain up to 0.4 percent of these compounds. The principal alkaloid found in jimsonweed is the same as in belladonna, *dl*-hyoscyamine (atropine). Atropine acts as a competitive antagonist at muscarinic ACETYLCHOLINE receptors.

While in low doses atropine acts primarily as a peripheral antimuscarinic agent. Large to toxic doses of the drug result in restlessness, irritability, disorientation, hallucinations, and delirium.

(SEE ALSO: *Plants, Drugs from; Scopolamine*)

BIBLIOGRAPHY

HOUGHTON, P. J., & BISSET, N. G. (1985). Drugs of ethnoorigin. In D. C. Howell (Ed.), *Drugs in central nervous system disorders*. New York: Marcel Dekker.

ROBERT ZACZEK

JUVENILE DELINQUENCY AND SUBSTANCE ABUSE   *See* Adolescents and Drugs; Crime and Drugs.

–

K

KAVA A drink prepared from the root of the Australasian pepper shrub *Piper methysticum*. The word *kava*, which is Polynesian for bitter, pungent, is given to the drink because of its strong peppery taste. Several variations of this drink were once used widely as social intoxicants in the islands of the South Pacific, particularly Fiji. The quality of the drink improves with the age of the root, and the roots are generally at least four years old before they are used. After the root is cut and crushed or grated, the active components are extracted by soaking the preparation in water.

Common effects of kava include general muscular relaxation, euphoria, and loss of fatigue. Visual and auditory effects are also common. In large quantities kava can induce muscular incoordination and ultimately stupor.

While no ALKALOIDS or glycosides have been found in kava, several aromatically substituted α-pyrones, including kawain, dihydrokawain, methysticin, and yangonin, have been isolated from the extracted root. Other as-yet-unidentified components of kava may also be important in the effects of the drink.

(SEE ALSO: *Plants, Drugs from*)

BIBLIOGRAPHY

HOUGHTON, P. J., & BISSET, N. G. (1985). Drugs of ethno-origin. In D. C. Howell (Ed.), *Drugs in central nervous system disorders*. New York: Marcel Dekker.

ROBERT ZACZEK

KHAT This is a shrub or small tree that grows wild and has been cultivated in the uplands of Yemen and East Africa. The plant is known under many names; it is called *qat* in Yemen, *tschad* in Ethiopia, and *miraa* in Kenya. The botanical name is *Catha edulis* (family Celastraceae).

The drug is a habituating stimulant containing ALKALOIDS released by chewing the leaves, buds, and sprouts. The leaves are about two to three inches long, with a serrated edge (see Figure 1), are brownish-green, somewhat leathery, and have a glossy upper surface. Since the plant has no unique features, a chromatographic test for its identification has been developed.

Use. Khat leaves are made into a tea or chewed for their stimulating effect. They are thoroughly masticated one by one; the juice is swallowed while their residue is stored in the cheek and later ejected. Young leaves are the most tender and potent; the leaves should be fresh to be effective. A portion is about 100 to 200 grams of leaves, predominantly consumed in a social setting. In Yemen, the sessions are of great importance to social life and the habit is part of the cultural tradition. In East Africa, khat is more recreational, with the leaves being consumed at times with ALCOHOL or other drugs. There is also a tradition of khat use by farmers and craftspeople, who chew it for enhancing work performance and staying alert.

Khat consumption has increased during recent decades; by the mid-1990s, it is about 5 million portions per day. Although the use is limited to the re-

Figure 1
Full-grown Khat Leaf (at about two-thirds natural size)

gion where it grows, small quantities are now imported by air to Europe and North America, where it is sold mainly to immigrants from Yemen and East Africa.

Effects. The pharmacology of khat has been reviewed and its effects are characterized by a moderate degree of central nervous system (CNS) stimulation, resulting in a state of mild euphoria and excitement—often accompanied by talkativeness to excess. High doses may induce restlessness and sometimes manic behavior. Toxic psychosis may result from excess. Khat produces ANOREXIA (loss of appetite) and constipation; it has sympathomimetic effects on the cardiovascular system. Dilation of the pupil and a staring look are indicative of the acute effect of khat. Habitual chewing is usually revealed by a brownish staining of the teeth.

The effects are very similar to those of AMPHETAMINE, and the difference between the two drugs is quantitative rather than qualitative. Accordingly, ha-

bitual khat use may give rise to psychic dependence, which usually is moderate but often persistent. The withdrawal symptoms after prolonged use are slight trembling, lethargy, mild depression, and recurrent bad dreams. Khat use by the habitué is often compulsive, with the necessary supplies obtained at least once a day, even at the expense of vital needs; the socioeconomic consequences of the habit are considerable.

Constituents. Khat contains the alkaloids norephedrine, cathine, and cathinone (see Figure 2). Norephedrine and cathine do not contribute to the psychostimulant action, however, they are probably of importance for the sympathomimetic effects (on the autonomic nervous system). Cathinone is mainly responsible for the stimulant qualities and the dependence-producing effects of khat. Cathinone must be considered a natural amphetamine, since the two substances have the same mechanism of action; however, cathinone has a half-life of only 1.5 hours, much shorter than that of amphetamine. Since cathinone is absorbed gradually during the chewing and is inactivated in the body rather rapidly, the pharmacological effects are usually limited.

(SEE ALSO: *Amphetamine*)

BIBLIOGRAPHY

BRENNEISEN, R., ET AL. (1990). Amphetamine-like effects in humans of the khat alkaloid cathinone. *British Journal of Clinical Pharmacology, 30,* 825–828.

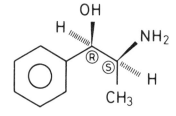

CATHINE   CATHINONE   NOREPHEDRINE

Figure 2
Structure of the Khatamines.
Cathine (S,S(+)phenylpropanolamine or (+)norpseudoephedrine), cathinone (S(−)alphaaminopropiophenone) and norephedrine (R,S(−)phenyl-propanolamine). In an analysis of twenty-two khat samples of different origin, the average concentrations of these alkaloids in 100 grams of fresh khat were found to be 120 milligrams, 36 milligrams, and 8 milligrams, respectively (Geisshüsler & Brenneisen, 1987).

EDDY, N., ET AL. (1965). Drug dependence, its significance and characteristics. *Bulletin of the WHO, 32,* 721–733.

GEISSHÜSLER, S., & BRENNEISEN, R. (1987). The content of psychoactive phenylpropyl- and phenylpentenyl-khatamines in Catha edulis Forsk of different origin. *Journal of Ethnopharmacology, 19,* 269–277.

KALIX, P. (1992). Cathinone, a natural amphetamine. *Pharmacology & Toxicology, 70,* 77–86.

KALIX, P. (1990). Pharmacological properties of the stimulant khat. *Pharmacology and Therapeutics, 48,* 397–416.

KALIX, P. (1987). Khat, scientific knowledge and policy issues. *British Journal of Addictions, 82,* 47–53.

KRIKORIAN, A. (1984). Kat and its use, a historical perspective. *Journal of Ethnopharmacology, 12,* 115–178.

WEIR, S. (1985). *Qat in Yemen: Consumption and social change.* London: British Museum Publications.

PETER KALIX

KORSAKOFF'S SYNDROME   *See* Alcoholism; Complications: Neurological.

L

L-ALPHA-ACETYLMETHADOL (LAAM)

Acetylmethadol (also referred to as *l*-alpha-acetyl-methadol, methadyl acetate, LAAM or L-AAM) is structurally related to METHADONE. LAAM is a potent OPIOID agonist with properties similar to methadone, except for its prolonged half-life. This slow elimination can be useful clinically, since 50–80 milligram doses of LAAM given three times a week are equivalent to daily doses of 50–100 milligrams of methadone in preventing the symptoms of opioid WITHDRAWAL. Thus, addicts on maintenance treatment would need to come to a clinic only 3 times a week for LAAM instead of daily for methadone. Since the early 1970s, methadone has been the only agent approved for use in maintenance-treatment programs for HEROIN addicts, but research has shown that LAAM can be a useful alternative. In 1993, the U.S. Food and Drug Administration (FDA) initiated the legal changes needed to make LAAM available for clinical use.

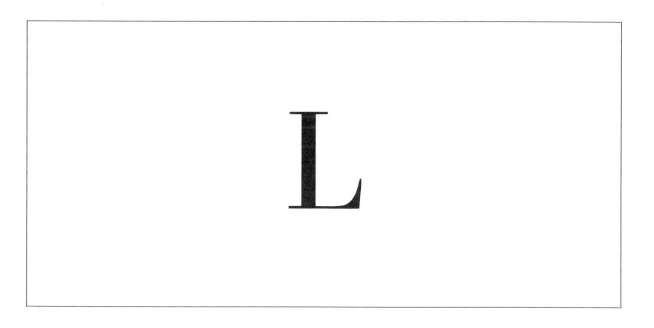

Figure 1
LAAM

(SEE ALSO: *Pharmacotherapy*; *Treatment*)

BIBLIOGRAPHY

GILMAN, A. G., ET AL. (EDS.). (1990). *Goodman and Gilman's the pharmacological basis of therapeutics*, 8th ed. New York: Pergamon.

GREENSTEIN, R. A., FUDALA, P. J., & O'BRIEN, C. P. (1992). Alternative pharmacotherapies for opiate addiction. In J. H. Lowinson, P. Ruiz, & R. B. Millman (Eds.), *Substance abuse: A comprehensive textbook*, 2nd ed. Baltimore: Williams & Wilkins.

GAVRIL W. PASTERNAK

LATIN AMERICA AS DRUG SOURCE

See Bolivia; Colombia as Drug Source; International Drug Supply Systems; Mexico as Drug Source.

LAUDANUM

Laudanum refers to a tincture of OPIUM—an alcoholic extract (about 20%) of opium, which contains approximately 10 milligrams per milliliter of morphine. If used at all during the 1990s, it would be as an antidiarrheal. The solution is more concentrated than PAREGORIC, and smaller volumes are given; however, their actions are almost identical. At standard doses, they rapidly and effectively treat diarrhea without producing euphoria or analgesia. The solution does contain MORPHINE and other opioid alkaloids and, at higher doses, it can be abused—as it was during the late-nineteenth and early twentieth centuries, when it was sold widely as a tonic and cure-all, in shops, by mail order, and by traveling medicine shows. Laudanum use and abuse are often mentioned in novels and plays of and about the period.

(SEE ALSO: *Dover's Powder*)

BIBLIOGRAPHY

GILMAN, A. G., ET AL. (EDS.). (1990). *Goodman and Gilman's the pharmacological basis of therapeutics*, 8th ed. New York: Pergamon.

GAVRIL W. PASTERNAK

LAW ENFORCEMENT   *See* Anslinger, Harry J., and U.S. Drug Policy; Coerced Treatment; Controls; and the Appendix, Volume 4.

LD50   In preclinical studies, the LD50 is the median lethal dose—the dose of a drug that produces death in 50 percent of the experimental animals tested. The LD50 can be estimated from a dose-effect curve, where the concentration of the drug is plotted against the percentage of animals that die. The ratio of the LD50 to the ED50 (the median effective dose) indicates the therapeutic index of a drug for that effect and suggests how selective the drug is in producing its desired effects. In clinical studies, the concentration of the drug required to produce toxic effects can be compared to the concentration required for therapeutic effects in the population to esimate the clinical therapeutic index.

(SEE ALSO: *Research, Animal Model*)

BIBLIOGRAPHY

GILMAN, A. G., ET AL., (EDS.). (1990). *Goodman and Gilman's the pharmacological basis of therapeutics*, 8th ed. New York: Pergamon.

NICK E. GOEDERS

LEGALIZATION OF DRUGS   *See* Policy Alternatives.

LEGAL REGULATION OF DRUGS AND ALCOHOL   Legal regulation can be used in four general ways to influence the incidence, prevalence, patterns, and circumstances of consumption of potentially harmful substances—including ALCOHOL, TOBACCO, and other DRUGS. The most direct mode of legal intervention is to *establish the conditions under which a potentially harmful substance is available.* In doing so, the law can employ either (1) a "prohibitory" scheme, which prohibits the production or distribution of the substance for nonmedical or self-defined uses, or (2) a "regulatory" regime, which permits the substance to be lawfully available for nonmedical or self-defined uses but which may regulate the product, its price, and the conditions under which it is accessible. A completely successful prohibition would prevent any nonmedical consumption of the proscribed substance; however, the more likely consequence of a prohibitory scheme is that an illicit distribution system would arise to respond to whatever demand exists for the substance. In that case, the manner in which the prohibition is enforced can also influence the product, its price, and the conditions under which it is available.

A second mode of legal regulation is to *regulate the flow of information and messages regarding use of the particular substance.* The government may initiate its own informational efforts to influence attitudes, beliefs, and behavior. Government may also attempt to influence private communications, either by proscribing certain messages altogether or by regulating or restricting their content. Such restrictions have generally taken two forms—mandatory warnings and proscriptions of certain types of messages.

A third mode of legal control is *direct regulation of consumption*, either by prescribing and imposing sanctions for undesired behavior or by withholding benefits or privileges to which the individual would otherwise be entitled. Thus, the law may proscribe use of a substance altogether, or it may prohibit such behavior in certain specified circumstances. Examples of total bans include unauthorized possession and consumption of controlled substances and consumption of alcohol by persons under the minimum age. Situational prohibitions include laws against consuming alcohol or smoking tobacco in public areas. Laws that require drug testing of workers and that permit job termination or discipline as a consequence of a positive test illustrate less coercive measures of deterrence.

A fourth use of the law emphasizes its declarative aspects. Whether or not a legal control has a direct impact on the marketplace or on the prevalence of the disapproved behavior, it may *symbolize and express the official government view* of the behavior and may generate derivative effects on behavioral patterns by influencing attitudes and beliefs. To the extent that citizens customarily defer to and respect the law or are influenced by messages of official approval or disapproval, a declaration of illegality may serve an educative, or didactic, role. Specification of a minimum drinking age, regulation of the availability of drug PARAPHERNALIA, and sanctions for possession of illicit drugs may all generate these symbolic effects, even if the direct effects tend to be modest.

AVAILABILITY

The National Commission on Marihuana and Drug Abuse identified four models of availability for psychoactive substances:

The first involves no special controls at all: the substance is treated in the same way as other [unregulated] market commodities. Under the second approach, the substance is subject to special controls but remains lawfully available for self-defined [or nonmedical] purposes. The third model limits availability to specific purposes, generally to medical and research uses only. Under the fourth approach, the substance is not legally available at all except perhaps for narrowly circumscribed use in research.

The first two models can be characterized as regulatory approaches (because the substance is legitimately available for nonmedical or self-defined purposes) and the second two as "prohibitory" approaches (because the substance is not available for self-defined or nonmedical purposes). Tobacco and alcohol are lawfully available for nonmedical uses, but they are subject to variable regulatory controls designed to affect the product, place, and conditions of consumption. (Only the solvents and the INHALANTS—glue, lacquer, thinner, ether, gasoline, nitrous oxide—are essentially uncontrolled.) However, most psychoactive substances (legally denominated controlled substances) are subject to prohibitory controls; with the one minor exception of PEYOTE—which has been available to the members of the Native American Church for sacramental uses—this means their availability is limited by law to medical and research uses.

Alcohol. The availability of alcohol is governed by alcoholic beverage controls (ABC) that vary from state to state. ABC agencies view their primary responsibilities as providing an orderly market for distribution of alcoholic beverages, controlling criminal involvement in the market, and generating tax revenues. Since the 1960s, the trend has been to liberalize restrictions on access to, and availability of, alcohol in order to facilitate private choice, to protect commercial interests, and to raise revenue. Only since the late 1980s have some ABC agencies shown any inclination to use their regulatory authority to influence the prevalence, patterns, and circumstances of consumption. Relevant aspects of ABC regulation include pricing and/or taxation policies, zoning, and rules regarding hours and days of sale.

Direct regulation, under the authority of ABC boards, is not the only method by which the law can influence the conditions under which alcohol is available. For example, one way to discourage retail sellers of alcohol from selling this substance to a person already intoxicated is to hold them legally liable for injuries subsequently caused by the intoxicated consumer, even after leaving the premises. Although the legal theory has changed over the years, the risk of liability for commercial suppliers is relatively well established. Moreover, the courts of several states have extended liability to the hosts of social events who served alcohol to "obviously intoxicated" guests who then cause injuries in their intoxicated condition.

Tobacco. For the most part, the public health dimensions of tobacco regulation have been reflected only in product and package requirements designed to facilitate informed consumer choice. Only since the late 1980s has the federal government moved toward a policy that unequivocally establishes reduced consumption as its goal. Although a national prohibition is unlikely in the foreseeable future, several regulatory initiatives are being undertaken at all levels of government. For example, states will not receive federal money for mental-health and substance-abuse services, unless they implement a plan for enforcing bans against distribution of tobacco products to minors. Many localities have banned vending machines. In addition, several states have raised cigarette excise taxes with the aim of reducing consumption, and the federal excise tax is likely to be increased by a substantial amount, with the dual aim of reducing smoking and raising revenue for health-care reform. Meanwhile, federal subsidies for tobacco producers have come under intensified criticism from public-health advocates.

In addition, smokers or their survivors have sued tobacco companies, without definitive success, seeking damages for smoking-induced disease or death. Obviously, imposition of liability on manufacturers for the adverse health consequences of smoking would have a major impact; even if the conditions of liability were relatively narrow, the threat of liability would introduce a significant new variable into the economics of the industry. In this instance, the indirect regulation of tobacco by the tort system would exert a more potent influence on industry behavior than many direct regulatory alternatives, such as pricing policies, outlet limitations, or tar and nicotine limitations.

Controlled Substances. Manufacture and distribution of OPIATES, COCAINE, CANNABIS (MARIJUANA), stimulants, depressants, and hallucinogenic substances outside medical and scientific channels is unlawful under both federal and state "controlled substance" laws. Production and distribution of these substances within medical and scientific channels are subject to varied levels of restrictions based on their "potential for abuse" and their level of accepted medical use under the CONTROLLED SUBSTANCES ACT of 1970. The wisdom of these prohibitions, especially in relation to cannabis, has been questioned by some on the grounds that suppression of nonmedical use is not a legitimate governmental objective, and if it is, that the costs of the prohibitions exceed the benefits of the reduced consumption they achieve.

INFORMATION REGULATION

A government aiming to discourage what it perceives as unhealthy or unsafe behavior is not likely to be satisfied with the influence of its own messages and may seek to regulate communication by others, within the bounds of the First Amendment, which protects freedom of speech. This can be done in two ways. First, the government may require individuals or organizations to convey the government's desired message. Laws requiring product manufacturers to include information on or with their products have become a standard feature of health and safety regulation. In recent years, mandatory package warn-

ings have been utilized as a means of informing consumers about the dangers of tobacco and, more recently, of alcohol use. Second, government may ban communication of messages that it regards as undesirable. For example, laws banning false or misleading advertising are common, but government may choose to go a step further—to suppress a message because it is thought to encourage unhealthy or socially disapproved drug-, alcohol-, or tobacco-using behaviors. Examples include the federal ban on broadcast advertising of cigarettes and state laws that ban alcohol advertising. Public-health advocates have urged the federal government to prohibit all forms of tobacco advertising. Whether such prohibitions actually affect the level of consumption (as opposed to product choice) remains controversial. Proposals have also been made to move beyond advertising into the content of entertainment programming in ways that would remove messages that portray smoking and drinking in an attractive way. Clearly, such initiatives would raise serious constitutional questions concerning free speech.

Governments have also occasionally attempted to purge the environment of messages that are thought to encourage illicit drug use. For example, one provision of the Model Drug Paraphernalia Act (drafted by the federal drug-enforcement agency as a model for states to enact) specifically bans paraphernalia advertising. In 1973, the Federal Communications Commission (FCC) threatened to revoke the licenses of radio stations whose lyrics were thought to encourage illicit drug use.

DIRECT REGULATION OF
CONSUMER BEHAVIOR

A decision to discourage nonmedical drug use—and to proscribe transactions outside medical channels in order to restrict availability for such use—does not necessarily entail a decision to proscribe and punish unauthorized consumption. Values of individual freedom weigh very differently in the two contexts.

From the perspective of libertarian philosophy, it has been argued that criminalization of private use (and possession for such use) of drugs is categorically illegitimate, and that the criminal prohibition should be limited to behavior that endangers others. This, too, leads to a discussion of the ways in which drugs might affect others. Even if criminalization is not categorically objectionable, the costs of criminalization may exceed the benefits. The National Commission on Marihuana and Drug Abuse relied on such a cost-benefit assessment in 1972, when it recommended decriminalization of possession of marijuana for personal use. A few states have decriminalized possession of marijuana, although they have usually substituted a civil fine. Some of the states that took this action subsequently recriminalized possession. Aside from marijuana, possession of all other controlled substances is a criminal offense in all states as well as under federal law. In addition, possession of alcohol by underage consumers is an offense in most states. Even if possession or use of a substance is not categorically proscribed, prohibitions can be utilized to deter and punish socially harmful behavior or to provide leverage to get individuals into treatment. Public smoking laws and laws prohibiting driving while intoxicated (or while having a certain level of blood alcohol content) provide the prime examples.

DECLARATIVE ASPECTS OF LEGAL REGULATION

Government sends messages by its actions as well as its words. By declaring conduct illegal or by using any of the other instruments of legal intervention described above, the government expresses and formalizes social norms. However, knowledge of the official preferences may actually encourage the disapproved behavior among disaffected, outsider groups. Measuring such symbolic effects is difficult, because of the need to isolate these hypothesized effects from other influences on attitudes and beliefs.

Arguments drawing on the declarative aspects of legal regulation are routinely employed by proponents of restrictive controls over the availability and consumption of alcohol, tobacco, and other drugs. Criminal sanctions against simple possession of controlled substances are frequently regarded as indispensable symbols of social disapproval. Such arguments have been prominent in debates concerning decriminalization of possession of marijuana. Moreover, graded or stratified penalty schemes, which punish possession of "more harmful" drugs more severely than possession of "less harmful" drugs, may be favored because they denote the relative seriousness of these transgressions. Public SMOKING bans and antiparaphernalia laws seem to be particularly designed to reinforce attitudes unfavorable to smoking and recreational drug use.

Statements of legal rules can serve an educative role even if they do not penalize the undesired behavior. Minimum-drinking-age laws (which prohibit distribution of alcohol to persons who are underage) provide a good example, because they denote the norm even if the youthful drinker is not punished. Similarly, bans on alcohol or tobacco advertising might be enacted to erase a possible symbol of social approval even if the proponents did not believe that such bans would directly reduce consumption.

(SEE ALSO: *Advertising; Alcohol; Dramshop Liability Laws; Minimum Drinking Age Laws; Opioids and Opioid Control; Policy Alternatives*)

BIBLIOGRAPHY

BONNIE, R. J. (1986). The efficacy of law as a paternalistic instrument. In G. B. Melton (Ed.), *The law as a behavioral instrument.* Lincoln: University of Nebraska Press.

BONNIE, R. J. (1982). The meaning of decriminalization: A review of the law. *Contemporary Drug Problems, 10*(3), 277–289.

BONNIE, R. J. (1981). Discouraging the use of alcohol, tobacco, and other drugs: the effects of legal controls and restrictions. In N. Mello (Ed.), *Advances in substance abuse*, Vol. 2. Greenwich, CT: JAL Press.

MOORE, M. H., & GERSTEIN, D. R. (1981). *Alcohol and public policy: Beyond the shadow of prohibition.* Washington, DC: National Academy Press.

NADELMAN, E. A. (1988). The great drug debate: I. The case for legalization. *Public Interest, 92*, 3–31.

NATIONAL COMMISSION ON MARIHUANA AND DRUG ABUSE. (1973). *Drug use in America: Problem in perspective.* Washington, DC: U.S. Government Printing Office.

NATIONAL COMMISSION ON MARIHUANA AND DRUG ABUSE. (1972). *Marihuana: A signal of misunderstanding.* Washington, DC: U.S. Government Printing Office.

WALSH, D. C., & GORDON, N. P. (1986). Legal approaches to smoking deterrence. In L. Breslow, J. E. Fielding, & L. B. Lave (Eds.), *Annual review of public health, 7*, 127–49.

RICHARD J. BONNIE

LE PATRIARCHE This is both the name used to denote an organization treating addicts, which was started in Toulouse, France, in 1972, as well as the title used in referring to the founder of this organization, Lucien J. Engelmajor.

Currently, Le Patriarche operates residential treatment centers in Europe, particularly in France (beginning 1972), Spain (beginning 1979), Belgium (beginning 1980), Italy (beginning 1984), Germany (beginning 1986), Portugal (beginning 1988), and Ireland (beginning 1989). These residential treatment centers are primarily located in rural areas on large farming estates. There are, in addition, many small intake and community-interaction units located in the urban hubs throughout most of Mediterranean Europe. Le Patriarche has also opened treatment centers in the Americas—notably in New York City, Miami, San Diego, Toronto, Montreal, and Belize.

Lengths of stay and return to home are not considered the primary program goals—but rather drug-free living with responsibility to others. Interviews with former patients suggest that there is little in organized psychological or behavioral therapies presented but rather the emphasis is placed on work and contributing to Le Patriarche's community. Issues of some controversy have arisen, including the relocation of individuals in Europe from one country to another while withholding their passports and all belongings; the absence of strict rules against the use of force or violence; and situations in some facilities where members have been "forced" back to work or detained against their will.

Further fuel for controversy includes, in Europe, the setting up of special rural environments for some HUMAN IMMUNODEFICIENCY VIRUS (HIV) infected addicts and their families. Negative comments have been made about their coercive containment in a segregated fashion, as well as resentment toward out-of-country HIV-infected members placed in these settings. Positive comments have abounded about their quality of health-care conditions and a quite large computerized database regarding these same members. Estimates of total addict members and HIV-related members in Le Patriarche's diverse centers is about 10,000 worldwide.

(SEE ALSO: *Treatment*; *Treatment Types*)

DAVID A. DEITCH

LIBRIUM   *See* Benzodiazepines.

LIFE SKILLS TRAINING   *See* Prevention; Prevention Programs.

LIFESTYLE AND DRUG-USE COMPLICATIONS   Nonpharmacologic causes of medical complications associated with the abuse of ALCOHOL and other DRUGS may take many forms. While there is no one lifestyle specifically associated with abuse and addiction, there are lifestyle aspects and qualities that may create or contribute to medical COMPLICATIONS. These aspects and qualities may find their way into a variety of general lifestyles through drug use, or they may involve such preexisting socioeconomic complications within specific populations—including ethnic and racial minority populations—as POVERTY, lack of education, and lack of access to mental-health, general-health, and human-service resources. Cultural and even work-related complications may also arise that promote or exacerbate drug use.

Unlike complications caused by mode of administration, or clear-cut medical sequelae, lifestyle related medical complications to substance abuse may not follow a clear cause–effect pattern. More often, lifestyle and abuse patterns are co-effects, working together to create the complications. Life problems may be a leading, or at least supporting, factor in the development of abuse patterns.

Although ethnic and cultural factors may be involved in complications, and certain forms of abuse have been overwhelmingly identified with specific ethnic or cultural groups, these identifications can become stereotypic and misleading, contributing to the behavioral denial of "others." Examples include the identifying of CRACK-cocaine addiction as only an African-American inner-city problem, FETAL ALCOHOL SYNDROME (FAS) as only a problem specific to Native Americans, or ALCOHOLISM as a disease of only the down-and-out. Clearly, others have these same problems and need to face them. Any diagnosis that is based on ethnic bias or cultural considerations can result in incorrect or missed diagnosis for substance abuse. Drug abuse in general is increasingly seen by the treatment community as totally democratic—respecting no ethnic, social, or cultural boundaries. The following, however, represents both sources and medical complications from lifestyle-related factors, including accidents, violence, and poverty.

INJURIES RESULTING FROM ACCIDENTS

One major source of medical complications to alcohol and other drugs is ACCIDENTS—particularly accidents involving automobiles or other machinery. Intoxication, the state professionally referred to as "acute toxicity," often involves impaired judgement, impaired motor skills, and impaired perception and reaction. These impairments make the intoxicated individual highly vulnerable to accidents of all kinds. Driving under the influence of alcohol or other psychoactive drugs produces thousands of injuries and fatalities every year, while many workplace injuries result from operating machinery while intoxicated. The problem becomes particularly acute and many additional lives can be endangered and even lost when the machinery being operated is a commercial airliner, a train, or a bus full of passengers. It is not surprising, therefore, that the U.S. Department of Transportation has taken a leading role in developing protocols and procedures, including drug testing and personnel evaluation, aimed at preventing on-the-job intoxication.

The problem of driving under the influence can be greatly exacerbated by the lifestyle of the user. In some parts of the country, for example, young men still consider that it is their "God-given right" to keep a six-pack or more of beer in their car or in the cab of their truck, and insist that drinking is an integral part of driving. In factory work, construction, and in other lines of work that involve machinery, it is often the custom for a crew to end the day by getting drunk together in a bar. A similar custom for officeworkers is the after-work Happy Hour that goes with free snacks and goes on for hours. This behavior not only contributes to DRIVING UNDER THE INFLUENCE (DUI), but to acute hangovers that can lead to absenteeism or cause errors in judgment or coordination at work the next day.

Not all substance-related accidents involve machinery. Addicts and chronic abusers often develop a history of traumatic injury. For example, a significant number of alcoholics are also victims of traumatic, and often multiple, head injury. With such clients, the addiction and the disability work to-gether to produce an ongoing state of diminished capacity, marked with incidents of reinjury.

Accidents to alcoholics and other drug addicts and abusers need not be major, debilitating, or even traumatic. Patient medical records may show long series of minor accidents at work or in the home. In treating patients who fall within the "accident prone," health professionals do well to suspect and diagnose for chronic abuse or addiction. Many individuals, including health professionals, who finally enter into treatment and recovery look with wonder at their past medical records and cannot understand how their abuse and addiction remained undetected for such long periods. This denial by health professionals highlights another complication due to life-style. Many physicians and other health professionals have had little training or experience in dealing with addictive disease. Victims themselves of "professional denial," these medical professionals may unwittingly ignore clear evidence of a patient's addictive behavior—because they have a mental stereotype of the addict that the patient does not fit or because they may be afraid of their own inexperience in dealing with the matter. The problem of professional denial is being addressed in an increasing number of medical schools and professional training programs through exposure of students to drug- and alcohol-abuse TREATMENT centers and to TWELVE-STEP fellowship meetings where they hear the experiences of recovering addicts from many backgrounds.

INJURIES RESULTING FROM VIOLENCE

Aspects of lifestyle and the effects of certain drugs may combine to promote injuries resulting from VIOLENCE. The same impairment of good judgment that can lead to accidents can also lead to injurious violent behavior. Arguments often erupt among intoxicated individuals, particularly within groups where fighting is an accepted interaction and expected solution to interpersonal problems. Such fights can take place in bars, dance halls, and other social gathering places where alcohol and other drugs may be present; at weddings and other special events; at football games, baseball, soccer and other sporting events where intoxicated groups of rival team fans are apt to get carried away; and any time groups of rival gang members encounter one another. For some groups, and at varying times, fighting becomes a way of life. Recent examples have been the rival English and European soccer fans who rioted at games. Some of these drug-related riots have resulted in multiple fatalities. Such individual and group violence can range from name calling and shoving matches, which may result in only minor injuries, to full-scale and life-threatening melees involving knives, pistols, and even automatic weapons.

Violence in the home is frequently alcohol or other drug related. This includes all varieties of wife, husband, and child abuse. Sexual abuse of spouse or children is often a form of drug-related FAMILY VIOLENCE that can have a wide range of physical and emotional repercussions. It should be noted, particularly in situations involving family members, that not all abuse takes a physical form. Dysfunctional family dynamics often cause long-term psychological damage. A variety of programs and fellowships, such as ADULT CHILDREN OF ALCOHOLICS (ACOA), have been formed to deal with such long-term abuse patterns.

A key factor in alcohol and other SEDATIVE drug-related violence is disinhibition euphoria, or loss of inhibitions coupled with feeling good—disinhibition euphoria is the drug effect most often sought by users. When intoxication is severe, euphoria may turn into dysphoria, or feeling bad or angry, while the intoxicated person is still disinhibited enough to see fighting as an acceptable outlet for his or her negative feelings.

The abuse of stimulant (as opposed to sedative) drugs involves a specific link to violence. Stimulant psychosis, characterized by paranoia with ideas of reference—that is, paranoia that is specific rather than general—and also characterized by rage reactions, is a frequent component of long-term AMPHETAMINE and COCAINE abuse. When the stimulant abuser is also a dealer who is threatened by law enforcement and competing with other paranoid, and frequently armed user/dealers, the situation may be particularly acute. Some of the violence surrounding crack cocaine traffic, including drive-by shootings, executions, and the use of automatic weaponry may be fed by stimulant paranoia.

HIGH-RISK SEXUAL BEHAVIOR

Cultural and lifestyle-related attitudes to sex and sexual activity is another major source of medical complications. The disinhibition euphoria of drugs can contribute to high-risk sexual behavior, leading users to engage in sexual activity with multiple partners and to ignore safe-sex practices in the process. Safe sex is defined here as primarily involving the use of latex condoms and an effective spermicide. Among diseases that result from unsafe sex are those that can also be related to the sharing of needles and other unsafe drug-use practices as well. These include human immunodeficiency virus (HIV), hepatitis, and syphilis.

PREGNANCY

Medical complications to newborns (neonates) result from alcohol and other drug use during PREGNANCY by women, exacerbated by lack of prenatal care, education, or treatment. Often, abusing pregnant WOMEN avoid prenatal care from ignorance, fear of criminal action, or fear of losing the child if their abuse becomes known. FETAL ALCOHOL SYNDROME (FAS) and fetal alcohol effect (FAE) are leading causes of birth defect, and crack-cocaine abuse during pregnancy is an increasing cause of neonate infirmity. While prevention education is beginning to focus on use, particularly drinking and tobacco SMOKING by pregnant women, various lifestyles—both rich and poor—continue to condone drinking and smoking during pregnancy.

POVERTY

Poverty, in and of itself, can give rise to a number of medical complications. These may include malnutrition, resulting from available money going to buy drugs instead of food; acute toxicity or the onset of drug-related disease, resulting from lack of information about drugs and their effects; or even culturally based misinformation about alcohol and other drugs.

LACK OF EDUCATION

Most drug-prevention information is now aimed at schools and the workplace. Many in the depths of POVERTY, often drug-related, neither go to school nor work on a regular basis. They are therefore much more vulnerable to medical complications that result from drug use and risky sexual behavior. An approach model for these individuals was developed in San Francisco and is based on the training of Community Health Outreach Workers (CHOWs), who work among the destitute, HOMELESS, and low-income substance abusers, providing these populations with EDUCATION and information about safe drug use and sexual practices while emphasizing a message of abstinence and recovery.

DIETARY DEFICIENCIES

The dietary deficiencies that may exist in poverty can be greatly exacerbated by alcohol and other drug abuse. Conversely, lifestyle-related dietary deficiencies may contribute to a variety of medical complications. While many drug abusers give a low priority to nutrition, often spending money for their drug of choice rather than eating at all, urban abusers are particularly prone to deep-fried, empty-calorie fast foods when they do eat; and stimulant abusers are prone to high-sugar-content foods when they do eat. Drug- and lifestyle-related dietary deficiencies may lead to medical complications from deficiencies in vitamins, amino acids, and other nutrients. Drug- and alcohol-treatment programs increasingly include dietary information and counseling in their treatment regimen.

OTHER SOCIOECONOMIC COMPLICATIONS

There is a high correlation between HOMELESS-NESS and drug/alcohol-abuse complications. Homelessness and alcoholism have long been linked—although all alcoholics are not homeless and all homeless are not alcoholics or even drug abusers. In a climate of despair it is often difficult to ascertain which came first, the homelessness or the substance abuse. Poverty, homelessness, hopelessness, despair, and abuse all seem to combine in series of interacting correlatives. To these may be added needle abuse, unsafe sex, and generally unhygienic lifestyles that create an explosive potential. All of these factors, exacerbated by substance abuse, have an increasingly negative effect on the immune system, making these individuals particularly vulnerable to all forms of opportunistic disease. Homeless people are exceptionally prone to tuberculosis, for example, with a rate of infection 15 to 20 times greater than that of the general public.

A particularly poignant aspect is the interaction between homelessness, abuse, and related illness found in military veterans. About half of homeless veterans now have alcohol or drug problems. Substance abuse increases the likelihood that those living on the margin will become homeless. Many

veterans turn to alcohol or drugs when post-traumatic stress disorder (PTSD) symptoms or other mental or physical illness go untreated. Veterans are more likely to abuse tobacco and alcohol than the illegal drugs and often seek treatment, only to find that there are waiting lists for treatment programs in nearly every community.

LIFESTYLES OF THE MIDDLE CLASS, RICH, AND FAMOUS

Since the 1960s, illicit drugs joined licit ones in becoming generally available to the U.S. public. Tobacco and alcohol continued to appeal to many—children as well as adults—and various illicit drugs that had always been part of certain lifestyles flooded U.S. markets during the especially rebellious years of the counterculture and VIETNAM war. Since that time, it has been generally considered "cool" to "do" drugs, and many have done them and walked away. Others have not, but they find ways to use them that neither deplete their pocketbooks nor their bodily health.

Not all drug users are down-and-out lost souls, then. Many of the apparently successful and affluent have drug-related lifestyles that may continue for relatively long periods of time with little outward decay. For many, the progress from habituation to addiction was slow and, in fact, may never occur. Not all habituation, not even all abuse, leads to addiction; but given the potential with many drugs for developing physical dependency and the physical toxicity of tobacco and alcohol, a problem may exist whether it is blatant or not.

Among junior high and high school students fads exist—for clothing styles and for drug use—whether the drugs be legal alcohol and tobacco or the illicit varieties. College students often pull all-nighters with CAFFEINE, AMPHETAMINES, and TOBACCO; then move on to MARIJUANA and other illicit drugs (often in combination) for recreational purposes. Many of these use patterns are dropped when they become adults, but some may continue. Similar patterns that develop within the same age groups of the military may also be dropped or may continue beyond that particular lifestyle.

Tobacco use may be declining among U.S. populations, but it and alcohol use continue to be extensive. While the three-martini lunch is a cliché, business executives frequently manage their alcohol-abuse problems in a corporate setting that contributes to denial and institutionalizes alcoholism until the unfortunate individual retires and is then forced to confront his or her addiction. The process is often seen as "late onset alcoholism or addiction" but is more accurately an exacerbation of the social and behavioral symptoms brought on by the loss of a corporate support system.

Many individuals in high-pressure business, entertainment, sports, and politics have turned to COCAINE primarily as a performance-enhancing drug. Compulsive use may continue for long periods of time without apparent loss of control. Cocaine is, however, a major drain on finances, and increasingly potent means of delivery, such as FREEBASING and crack, greatly exacerbate the potential for addiction. Another pitfall is the development of an upper-downer polydrug abuse pattern, coupling cocaine and amphetamine (stimulants) with such downer (depressant) drugs as ALCOHOL, BARBITURATES, and even HEROIN and other opioids.

The habitual use of prescription drugs, such as BENZODIAZEPINES and barbiturates, is a hallmark of anxiety-producing high pressure professions. Newspaper reporters and those in advertising often develop a "post-deadline" drinking and use pattern (Seymour & Smith, 1990).

Statistically, physicians and other health professionals have a high ratio of abuse and addiction. There are many reasons for this, among them, that medicine involves extremely high-stress situations and most doctors, nurses, and so on have relatively easy access to psychoactive medications. In general, use patterns among health professionals involve combinations of alcohol and drugs professionally available to them (Smith & Seymour, 1985).

A number of factors makes it easier for the middle class and affluent to maintain their lifestyles while using drugs, but a steady diet of Bordeaux and Chivas Regal in a cut-glass decanter can produce alcoholism just as readily as cheap Muscatel in a paper bag. Money and connections often place them at one remove from the race for money and drugs that dominates the lives of many street addicts. They also maintain some control over the quality of drugs they use, although such control may be illusory in the illicit drug market.

(SEE ALSO: *Accidents and Injuries; Aggression and Drugs; Creativity and Drugs; Ethnicity and Drugs; Families and Drug Use; Homelessness and Drugs;* *Poverty and Drug Use; Stress; Suicide and Substance Abuse; Vulnerability as Cause of Substance Abuse*)

BIBLIOGRAPHY

NATIONAL COALITION FOR THE HOMELESS. (1991). Still on the front lines: Homeless veterans in America. *Street Sheet*, December.

SEYMOUR, R. B., & SMITH, D. E. (1990). Identifying and responding to drug abuse in the workplace. *Journal of Psychoactive Drugs, 22*(4) 383–406.

SEYMOUR, R. B., & SMITH, D. E. (1987). *Drugfree: A unique, positive approach to staying off alcohol and other drugs.* New York: Facts on File.

SMITH, D. E., & SEYMOUR, R. B. (1985). A clinical approach to the impaired health professional. *The International Journal of the Addictions, 20*(5), 713–722.

DAVID E. SMITH
RICHARD B. SEYMOUR

LIMBIC SYSTEM The limbic system is a group of BRAIN STRUCTURES organized into a functional unit that is important in the expression of emotion and mood states. The term *limbic lobe* and associated terminology can be traced to the French neuroanatomist Paul Broca (1824–1880), who used it first to describe the forebrain structures that encircle the brain stem. The *limbic system* is a broader classification, composed of brain structures that form an integrated circuit surrounding the thalamus—an important relay station between higher brain centers and the hind brain and spinal cord.

The limbic system is thought to be important in emotional behaviors. This was hypothesized on the basis of neuropathological investigations of the brains of individuals displaying bizarre emotional disturbances. These initial clinical observations were followed by animal studies, in which the loss of these structures produced significant changes in emotional responsiveness. As research techniques and methodologies were refined, it became clear that limbic structures had an important and complex role in the expression of behavior. It is now believed that these structures are involved in a number of significant behavioral processes. In particular, the limbic system and related structures are thought to be important in the expression of emotion related to euphoria and

LIMBIC STRUCTURES

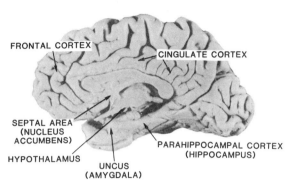

Figure 1

The Limbic System—composed of structures generally located between the brain stem and higher cortical structures. Some of these components are labeled in this sagittal section of the brain. The structures in parentheses lie behind the structures listed above them. The hypothalamus, hippocampus, septal nuclei, nucleus accumbens, amygdala, cingulate cortex, and frontal cortex are components of the limbic system that may have an important role in drug abuse.

feelings of well-being. For these reasons, the limbic system may have an important role in drug abuse.

LIMBIC SYSTEM COMPONENTS

The limbic system that surrounds the thalamus provides an interface between the midbrain and higher cortical structures. The general structure and components of the limbic system are shown in Figure 1. These include the amygdala, the NUCLEUS ACCUMBENS, the olfactory tubercle, the septal nuclei, the hippocampus, the hypothalamus, the cingulate cortex, and the frontal cortex. As can be seen in the figure, these structures are positioned between the brain's major relay station—the thalamus—and higher cortical structures. The separate components of the limbic system are interconnected such that activity initiated in one structure affects other components. One of the hypotheses about the basis of emotion speculated that reverberating neuronal activity in this system was responsible for affective behaviors. Initial animal studies using either direct electrical stimulation or lesions (loss) of various

components of the limbic system substantiated the important role of this system in behavior.

THE ROLE OF THE LIMBIC SYSTEM IN BEHAVIOR

Electrical stimulation or the destruction (lesions) of components of the limbic system alter behavioral processes. Lesions of the hippocampus disrupt memory processes, whereas lesions or stimulation of the amygdala affect emotional behavior and feeding in a manner similar to manipulations of the medial and lateral hypothalamus. Stimulation of the lateral hypothalamus produces aggressive responses, whereas lesions of this area produce a placid behavioral profile. In contrast, lesions of the medial hypothalamus produce a highly excitable and aggressive pattern of behavior, whereas lesions of the amygdala result in placid and nonaggressive behavior. Early studies found that lesions of the lateral hypothalamus can decrease feeding, whereas lesions of the ventromedial region produce excessive levels of feeding resulting in obesity. Recent experimental studies have demonstrated the complex nature of the involvement of hypothalamic cells in feeding and drinking; however, like most complex behaviors, the mechanisms that control hunger and satiety are not simply located in a single brain center.

Some structures of the limbic system are important in REINFORCEMENT processes. The term *reinforcement* applies to processes perceived as rewarding or good, which therefore are repeated, such as electrical self-stimulation. For example, animals will repeatedly emit a response that leads to the delivery of brief electrical stimulation of small electrodes that are implanted in selected brain structures. Humans will also choose to stimulate many of these same brain regions and report positive feelings of well-being and euphoria. The limbic system sites that produce these effects in animals include the lateral hypothalamus, nucleus accumbens, frontal cortex, cingulate cortex, and the brain-stem nuclei believed to be part of the limbic system—these include the substantia nigra and ventral tegmental area, which both contain DOPAMINE neurons that send inputs to many limbic-system components. Measures of brain-glucose metabolism, which directly reflect brain-cell activity, have been used to determine the involvement of specific brain regions in animals electrically self-stimulating three of these brain regions. The stimulation of each of these regions produced significant activation of several limbic-system structures that included the nucleus accumbens, amygdala, hippocampus, and the frontal and cingulate cortices. This area of investigation has led neuroscientists to propose that there are brain circuits dedicated to the behavioral processes related to reinforcement. Drugs of abuse likely produce their positive effects through the activation of these brain circuits.

THE ROLE OF THE LIMBIC SYSTEM IN DRUG ABUSE

A large number of experiments have focused on identifying the brain circuits that mediate the reinforcing effects of abused drugs, because the reinforcing effects are responsible for drug abuse. These experiments have included the use of drug self-administration techniques and sophisticated neurochemical procedures to measure the involvement of specific NEUROTRANSMITTER systems. As of the early 1990s, evidence indicates that limbic structures and brain cells that project to limbic structures play an important role in these processes. It is clear that dopamine-containing neurons that project from the ventral tegmental area to the nucleus accumbens have a critical role in the reinforcing actions of COCAINE and AMPHETAMINE. Removal of these inputs with toxic agents that selectively destroy dopamine-releasing brain cells disrupts intravenous self-administration of these drugs. Additional evidence of the importance of this region in drug abuse comes from glucose-utilization studies. The levels of glucose metabolism are significantly elevated in a number of limbic structures in animals self-administering cocaine intravenously. Other experiments have directly shown dopamine levels in the nucleus accumbens to be increased in animals intravenously self-administering cocaine. Collectively, these data imply an important role for the limbic system in general and specifically for dopamine neurons in the limbic system tied to the brain processes involved in stimulant abuse.

The brain circuits involved in OPIATE reinforcement appear to be very similar to those mediating cocaine self-administration. Limbic structures are clearly implicated in opiate reinforcement, but a central role for dopamine is less obvious. Significant changes in the utilization of some chemicals (neurotransmitters) involved in transmission between brain cells have been shown in the nucleus accum-

bens, amygdala, and the frontal and cingulate cortices of animals intravenously self-administering morphine. However, loss of dopaminergic inputs to the nucleus accumbens does not affect drug intake, whereas a similar loss of serotonergic inputs does. Similarly, nucleus-accumbens dopamine does not appear to be elevated in animals self-administering heroin as it is in animals self-administering cocaine. However, evidence does indicate an important role for limbic structures and chemicals used to communicate between cells of the limbic system in opiate reinforcement.

Limbic structures also appear to be important for ethanol (drinking ALCOHOL) reinforcement. The levels of dopamine appear to be elevated in the nucleus accumbens of rats orally self-administering alcohol. Injections of drugs that antagonize dopamine directly into the nucleus accumbens decrease alcohol self-administration, whereas drugs that enhance dopamine action increase alcohol intake. In addition, animals will self-administer alcohol directly into the ventral tegmental area—an area that contains the cell bodies for the dopamine cells that send inputs to the nucleus accumbens. These data collectively indicate that the nucleus accumbens and dopamine-releasing inputs to the nucleus accumbens are important to alcohol reinforcement.

CONCLUSION

The limbic system plays an important role in behavior. These brain structures appear to be central to the processes that mediate the reinforcing effects of electrical-brain stimulation and of several highly abused drugs. The nucleus accumbens appears to be a structure central to the reinforcing properties of cocaine and amphetamine, but it appears less important to opiate and alcohol reinforcement. A more exact definition of specific neurochemicals and brain-cell pathways in the limbic system that are involved in drug abuse will become clearer as new methodologies are developed.

(SEE ALSO: *Neuron; Neurotransmission; Research; Reward Pathways and Drugs*)

BIBLIOGRAPHY

KOOB, G. F. (1992). Drugs of abuse: Anatomy, pharmacology and function of reward pathways. *Trends in Pharmacological Sciences, 13*, 177–184.

Koob, G. F., & Bloom, F. E. (1991). Cellular and molecular mechanisms of drug dependence. *Science, 242,* 715–723.

ACKNOWLEDGEMENT

Preparation of this entry was supported in part by USPHS Grants DA 00114, DA 01999, DA 03628 and DA 06634.

James E. Smith
Steven I. Dworkin

LIQUOR   *See* Alcohol; Distillation; Distilled Spirits.

LSD   *See* Lysergic Acid Diethylamide and Psychedelics.

LUNG DAMAGE   *See* Crack; Marijuana; Nicotine; Tobacco.

LYSERGIC ACID DIETHYLAMIDE (LSD) AND PSYCHEDELICS

LSD is the abbreviation for lysergic acid diethylamide. It is the most potent member of a group of hallucinogenic substances called the indole-type HALLUCINOGENS. These drugs have structural similarities to another indole, the neurotransmitter SEROTONIN.

HISTORY

LSD was originally synthesized at the Sandoz Pharmaceutical Company, in Switzerland, as part of a long project begun in the 1930s. The aim was to develop useful medicines that were derived from ergot, a fungus (*Claviceps purpurea*) that infects such grasses as rye. Some of these compounds were found to be useful in medicine—such as methysergide, for the treatment of migraine headaches, and ergotamine, which is widely used in obstetrics to induce contractions of the uterus and stop bleeding after the delivery of a baby. These medications do not have hallucinogenic properties.

The chemist in charge of this drug development project was Albert Hofmann. In 1943, he synthesized a compound he called LSD-25, since it was the twenty-fifth compound made in this series of ergot derivatives. He accidentally ingested some of it and within forty minutes had the first LSD "trip." He told his colleagues he was not feeling quite right and got on his bicycle to go home. Later, he carefully described the vividly clear flood of perceptions that are characteristic of the "mind manifesting" or psychedelic drug. This, then, was a complete surprise. Thereafter, the drug and various substitutions of different atoms on the basic molecule were extensively tested for medical uses in the late 1940s and in the 1950s. No specific medical use of LSD or its psychedelic variants has been found.

Because of its potency and the extensive reports of laboratory studies in animals and in the clinic, LSD has become the prototypical hallucinogen, or psychedelic drug. It also became the emblem of a social movement—which, in fact, was a confluence of various movements that had begun in the early 1960s; they peaked in the late 1960s. By 1973, the "acid culture" had subsided into a small but still active subculture of various psychedelic drug devotees seeking meaning and profound insights. The feeling of a "great discovery" about such drugs and the human mind had occurred as early as the nineteenth century; artists and writers, such as Baudelaire and Rimbaud in Paris, had discovered HASHISH and the altered, somewhat dreamy, states of consciousness and euphoria produced by this potent form of MARIJUANA—the active ingredient of which is TETRAHYDROCANNABINOL (THC). For a period, they became absorbed with hashish and wrote about its alluring effects. The drug scene evoked the promise that the human mind must contain remarkable powers. Toward the beginning of the twentieth century, MESCALINE, the active hallucinogenic compound in the PEYOTE cactus, similarly was tried by a few explorers in medicine and in the arts. In New York City, during the early part of World War I, many influential people and intellectuals took either peyote "buttons" (the dried tops of the peyote cactus) or mescaline (the synthesized active ingredient of the buttons) and called it a "dry drunk." Similarly, after World War II, LSD caused a flurry of excitement among some professionals, and its medical value was tested in psychiatric patients. Writers such as Aldous Huxley wrote exciting books about the effects of mescaline and, later, LSD—yet there was still no widely popular movement until 1960.

Then Timothy Leary, a young psychology instructor at Harvard, explored the Mexican or "magic" mushroom, *Psilocybe mexicana*, and its active ingredient, PSILOCYBIN—and later LSD—claiming criminals became loving and peaceful and others more creative. He popularized this on campus and, when he was not reappointed to the faculty, proclaimed himself to be a martyr to his cause. Between 1960 and 1966, the media repeatedly "discovered" LSD—in effect, advertising it. As publicity increased, subcultures experimenting with mushrooms and LSD grew up in East and West Coast cities. Musicians, rock music, the hippie lifestyle, "flower children," and many in the various protest movements against the Establishment and the VIETNAM WAR were loosely joined to Leary's attempt to lead affluent and middle-class youth. Well-publicized festivals celebrated LSD and marijuana, such as the Summer of Love in the Haight-Ashbury section of San Francisco. Leary's challenge was for youth to "turn on, tune in and drop out" with acid. As more and more youth were curious to try experiences their parents had never dreamed of, rebellion led not only to acid experiments but to extensive POLYDRUG ABUSE—the extensive use of marijuana and various street substances. Either LSD or some variant and even heroin were tried. In addition, in the search for new drugs with different and improved characteristics (more or less euphoria, hallucinogenic activity, or stimulant properties), literally hundreds of so-called DESIGNER DRUGS were synthesized (DOM, MDMA, DMT, etc.). Because any drug can have bad effects, the unsupervised use of all of these compounds led to frequent "bad trips" (which fundamentally were panic reactions) that brought people to emergency rooms. This generated widespread concern that all American youth (and, later, those in Europe) would become dreamy and "way-out acid heads." In 1966, the Sandoz Laboratories ceased distribution of the drug because of the often-exaggerated bad reactions and the public concern. As the claims for enduring LSD insights proved transient, research with LSD in humans essentially stopped.

Thus, one of the ways people use the effects of drugs that seem to enhance the clarity of mentation (mental activity) and perception (while not producing confusion, dreamy-euphoria, or oversedation) is to become absorbed in periods of intense exploration with a few others "in the know." Those with such inside information form a kind of cult and then advertise, but they eventually see some bad effects (the wrong people taking the drug in the wrong circumstance with unfortunate consequences) and sooner or later see little real use for the drugs. The minor or major epidemics then die down, only to recur as later generations rediscover the compounds.

EFFECTS

LSD is one of the most potent hallucinogens known; one-billionth of a gram of LSD per gram of brain produces profound mental changes. Although subjective effects occur in some individuals after doses as low as 50 micrograms, typical street doses range from 10 to 300 micrograms—street dosages vary widely. Misrepresentation also frequently occurs; someone will try to purchase synthetic TETRA-HYDROCANNABINOL (THC), the active ingredient of marijuana, and receive LSD. Regular grocery store mushrooms are sprinkled with LSD and marketed as "magic" mushrooms (*Psilocybe*). Thus, the intake of LSD can be accidental as well as intentional, and the lack of quality control in illicit supplies is a hazard. Because of its high potency, LSD can be applied to paper blotters or the backs of postage stamps from which it is dissolved for consumption. Unsubstantiated reports of LSD added to stick-on tattoos for young children have caused alarm, even though absorption through skin would be far too slow to deliver enough drug to the brain to produce and sustain a trip.

The absorption of LSD from the gastrointestinal tract and other mucous membranes occurs rapidly, with drug diffusion to all tissues, including brain. The onset of psychological and behavioral effects occurs approximately 30 minutes after oral administration, peaks in the next 2 to 4 hours, depending on the dose, with gradual return to normal by 10 to 12 hours. The first 4 hours after a 200-microgram dose are called a trip. In the next 4 to 8 hours, when over half the drug has left the brain, the "TV show in the head" has stopped. At this point subjects think the drug is no longer active, but later they recognize that they, in fact, had paranoid thoughts and "ideas of reference" in the last 4 to 8 hours of the trip. This simply means that there is the feeling of being at the center of things, being hyperalert, and having a conviction that everything going on refers to oneself. This is a regular but little publicized aftereffect, which finally dissipates 10 to 12 hours after the dose.

From 12 to 24 hours after the trip, there may be some slight letdown or feeling of fatigue—as if one had been on a long, steep roller coaster ride. After these intense and even frightening moments, the ordinary world might for a time seem drab. There is no craving to take more LSD to relieve this boredom; one trip usually produces satiation for a time, although some may want to repeat the experience. Memory for the events during the trip is quite clear. Those who revisit the experience sooner or later decide they have learned what they can and go on with the practical, daily affairs of living. In one experiment on CREATIVITY, subjects received either LSD or the stimulant amphetamine during a period of pleasant surroundings and music. The only difference between the two groups six months later was a slight tendency for those who had received LSD to buy more recordings! So the promise of lasting insight or creativity was not kept.

Drugs that make one feel different—alcohol being typical—can signal a "holiday from daily reality." The way the effects of such drugs are interpreted is critical. BEER at the Super Bowl means "loudly letting go" and champagne at the White House means a time for graceful speech and feelings. Thus personal and social expectations (called *set*—or how one is set to go) and the surroundings (called *setting*) have much to do with the ultimate effects of drugs. This is distinctively and especially the case with psychedelics. Thus when the chemist Albert Hofmann first ingested the active ingredient of the Mexican mushroom, psilocybin, the perceptions capturing his attention were related to Aztec symbols and art! For some, therefore, the trip may simply be funny and odd—for others it will have special meanings. Set and setting partially determine the character of such trips.

Fundamentally, LSD produces a heightened clarity and awareness of sensory signals—of sights, sounds, touch, lights, and colors. Similarly there is special significance given to thoughts, memories, or verbal interchanges. For example, gestures or inflections of speech or many cues that are normally in the background are felt to be more important than what is being said or usually meant—and in looking at a picture, the central figures may take on a life of their own, the small background details that are normally ignored emerging, capturing attention.

While awareness is strikingly increased, control over what is being attended to is weakened. For all these reasons, unstable surroundings or confused motives at the time of drug ingestion may lead to a less-controlled trip or even a panic-generating trip. Many are aware that the trip is not quite real and fundamentally feel as if they are "spectators" of what they are so intensely experiencing. Many rely on guides, a group, or the rhythm of music to carry them through this period of altered perceptions in which control is diminished. Thus, personal intent and reliable surroundings are major factors affecting the different kinds of experiences that people will have.

While every trip has an individual characteristic, there are regularities in the trips. This has been called a "march of effects" following drug ingestion. Thus, observers note, the first sign of feeling different is like "butterflies in the stomach" or a slight nausea and feeling of "whoops, here we go" as if on a roller coaster. Parts of the body simultaneously feel strange or different. At about the same time (30–40 minutes after drug ingestion), the cheeks are slightly flushed and pupil size begins to increase, maximizing within an hour or two. These changes are due to the effects of LSD on the sympathetic and parasympathetic nervous systems. The pupils react normally but are enlarged. After 4 hours they slowly begin to return to normal size, which finally is achieved at 10 to 12 hours after taking LSD. At the beginning of the trip, all soon note that what is at the periphery of their vision suddenly seems as clear as what is normally at the center of vision. Over the next 90 minutes, there is a feeling that tension is welling up. Laughing or crying will relieve the tension. Often subjects say they are laughing because of what they see or crying because of their feelings. But this is simply based on a need to relieve the fluctuating rise of tension. The trip moves on into the second and third hours when perceptual fluctuations and intensities are mainly noted. People also report perceiving several feelings simultaneously. A common observation is, "I don't know if I'm anxious, thrilled, or terrified." Just as perceptions are in flux so are feelings, and these feelings and emotions may capture center stage in the second and third hours. Throughout the trip, people feel as if they are on the brink of an exhilarating but also dangerous experience. This intensity dies down about 4 hours after the usual dosage. If very large doses of LSD (500–1,000 micrograms) are taken, there is less capacity to be a spectator and far more intense self-absorption and fear. Some call this "dying of the ego" and relate the experience to mystical versions of death and rebirth.

Since the familiar seems novel and is seen in a different way, specialists in perception have been interested in what is called the "breakdown of constancies" that occurs with the drug. Normally we correct for what the retina sees by putting the world into order. We usually suppress the nonessential and focus on what we need to do to get about during the day. Just as with a camera, the retina sees the hand placed 6 to 8 inches in front of the eye as large. But the brain corrects for it and keeps size constant. Under LSD, corrections for constancy do not seem to happen. Many sensations that are normally dampened can thus have free play under the drug and the world will seem far less regular than it does in daily life.

One of the aftereffects in some—clearly not all—people is called "flashbacks." Days, months, or years after tripping, with no particular trigger or with an intense sensation, there may be a sudden few minutes in which subjects feel like they are back under the drug. They also may see flashing lights and other optical illusions. These flashbacks may be very disturbing. Flashbacks can occur after only a single drug experience and unpredictably. There has been no explanation as to why or how flashbacks occur. Scientists cannot predict (by observing a trip) if flashbacks will later occur or who is vulnerable. While these aftereffects are upsetting to some, most people do not experience them or those that do are not bothered. Others simply observe that their dreams may be more intense for a time after the drug experiences. One scientist noted that riding on a train to work, he was distracted from focusing on his newspaper for several months by the telephone poles whizzing by. These were normally at the periphery of his attention as he was reading, but after LSD, he could no longer suppress this irrelevant detail. There were more reports of such phenomena after publicity about them; given the millions of trips with LSD, these aftereffects are certainly infrequent but not rare.

Perhaps the most alarming bad effects of the drug have been the panic states occurring during a trip. Native Americans note that if one is in conflict, the effects of mescaline during religious ceremonies are unpleasant and can evoke terror. They then pray with the panicked person and "talk him down." One cannot predict whether a panic experience will occur. "One good trip does not predict a second one" is the general wisdom concerning this risk. Higher doses lead to less control and more intense effects, but panic states can occur at doses as low as 75 to 100 micrograms. For those who might be at risk for other mental disorders, hallucinogenic experiences may often destabilize them and precipitate some form of mental illness. For others, the experience may lead to a subsequent absorption with the unreal ("dropping out"), rather than coping with the challenges that the tasks of the ordinary world present. Occasional suicides or rare impulsive acting out of odd ideas arising during a trip have led some to loss of control and tragedy.

For most, the experiences have few negative or positive aftereffects. Although it has often been suspected, no permanent change to the cells of the brain (brain damage) has ever been scientifically established. There is no generally accepted evidence that the drug produces chromosomal abnormalities or damage to a developing fetus (although no nonprescription drugs during pregnancy is the only safe rule to follow). The bad effects of a period of diminished control are unpredictable, and in that fact lies the real risk. Thus, it is the intensity of the experience and how well or poorly it can be managed, the unpredictable flashbacks, and how this "TV show in the head" or this "waking dream" gets woven into one's subsequent life that are at issue when hazards are considered.

TOLERANCE

One striking feature of LSD, mescaline, and related psychedelic drugs is tolerance, which is a loss of typical drug effects after repeated doses. In brief, with daily doses the duration and intensity of effects rapidly diminish to the point where no subjective effects are perceived. After 200 micrograms per day of LSD, there is simply no detectable drug effect on the third or fourth day. After three or four days without LSD, the full initial effects can be triggered by the same dose that has been "tolerated." Thus tolerance develops and dissipates rapidly. When subjects are tolerant to LSD, the usual dose of mescaline required for a trip is also no longer effective. This is called cross-tolerance. It is readily seen with similar dosage schedules of psilocybin, LSD, and mescaline. There is no cross-tolerance with the nonhallucinogenic stimulant drug amphetamine. Thus, there must be some common mechanism of action among the psychedelic drugs beyond their structure and similar array of mental effects.

Tolerance is seen both in humans and laboratory animals. The lack of pupil enlargement is a common sign of tolerance. In animals, some drug effects show tolerance and some do not. For example, a heightened sensitivity of rats to mild electric shock persists after daily doses and does not show tolerance. Such persisting drug effects during periods of tolerance have not been studied in humans. How and why a psychedelic drug loses and regains its potency in this fashion is not yet understood, but there is no withdrawal discomfort after stopping a psychedelic drug when it has been taken over several days. This differs from the classic effects described for opioid drugs, where an uncomfortable withdrawal with drug cessation requires more drug for relief. Such physical drug withdrawal phenomena are not found with psychedelics.

LSD AND SEROTONIN

LSD is known to affect many places in the brain where the body's neurotransmitter serotonin naturally has actions and effects, and the biochemical effects of LSD in the brain are mostly linked to those sites related to serotonin. LSD acts as a kind of impostor at receptors that recognize serotonin. LSD is like serotonin but different. Thus with LSD, the receptor signals other parts of the brain that there is too much serotonin, and these parts of the brain respond by tuning down cells that make serotonin. Yet, in fact, the chief effect of LSD is to cause *less* serotonin to be released in the neighborhood of the receptor—rather than too much, there is too little. This is one example of how LSD miscues the systems governing the flow of information between various brain neurons. In fact, overloading the brain with serotonin can reduce the LSD effect, and diminishing brain supplies of serotonin will increase LSD effects. Yet serotonin itself does not cause the scrambled perceptions that LSD does. How this miscue by LSD leads to the vivid effects is still unknown.

LSD, other indole-type psychedelics, and many hallucinogens related to mescaline (but surprisingly not mescaline itself) are known to act especially at a subtype of the serotonin receptor called the $5HT_2$ receptor. In laboratory animals, daily doses of LSD or psilocybin lead to fewer of these receptors, an effect that would be expected to produce tolerance; however, with 3 or 4 days off the drug, the number of $5HT_2$ receptors returns to normal. Both LSD and mescaline act at certain brain neurons, such as the locus coeruleus, and make it more responsive to inputs from the environment—such as a pinch. Researchers speak of such effects as lowering the gates to sensory input. We know the ways by which LSD affects certain brain systems but still far less than we need to know to explain the full panoply of effects.

AMPHETAMINE
(psychostimulant)

LSD

MESCALINE

SEROTONIN
(neurotransmitter)

PSILOCYBIN

Although many of the psychedelic drugs are known to interact with serotonergic $5HT_2$ receptors, and this interaction appears to be of critical importance in producing their hallucinogenic effects, the hallucinogenic drugs can bind to a subtype of serotonin receptors that is located on serotonin nerve-cell bodies and on their terminals (which release serotonin that goes to the adjacent nerves with $5HT_2$ receptors). Interactions with these various receptors can lead to changes in the firing rate of such cells. The designer drugs MDMA and MDA cause the release of both dopamine and serotonin, effects that might contribute to their psychostimulant properties. The differential interactions of the various hallucinogens with multiple sites and systems may underlie the qualitative differences in the experience they produce.

(SEE ALSO: *Cults and Drug Use; Hallucinogenic Plants; High School Senior Survey; Plants, Drugs from; Yippies*)

BIBLIOGRAPHY

FREEDMAN, D. X. (1986). Hallucinogenic drug research—if so, so what?: Symposium summary and commentary. *Pharmacol. Biochem. Behav., 24,* 407–415.

GLENNON, R. A. (1987). Psychoactive phenylisopropylamines. In H. Y. Meltzer (Ed.), *Psychopharmacology: The third generation of progress.* New York: Raven Press.

GRINSPOON, L., & BAKALAR, J. B. (1979). *Psychedelic drugs reconsidered.* New York: Basic Books.

JACOBS, B. L. (1987). How hallucinogenic drugs work. *American Scientist, 75,* 386–392.

JACOBS, B. L. (ED.). (1984). *Hallucinogens: Neurochemical, behavioral and clinical perspectives.* New York: Raven Press.

JAFFE, J. H. (1990). Drug addiction and drug abuse. In A. G. Gilman et al. (Eds.), *Goodman and Gilman's the pharmacological basis of therapeutics,* 8th ed. New York: Pergamon.

SHULGIN, A., & SHULGIN, A. (1991). *PIHKAL: A chemical love story.* Berkeley, CA: Transform Press.

SIEGEL, R. K., & WEST, L. J. (EDS.). (1975). *Hallucinations: Behavior, experience and theory.* New York: Wiley.

WEIL, A. (1972). *The natural mind.* Boston: Houghton Mifflin.

DANIEL X. FREEDMAN
R. N. PECHNICK

M

MADD *See* Mothers Against Drunk Driving.

MAFIA *See* International Drug Supply Systems.

MAGIC MUSHROOM *See* Psilocybin.

MANDATORY SENTENCING Mandatory sentencing laws provide that people convicted of particular crimes receive particular sentences. Examples include laws specifying that people convicted of selling HEROIN or COCAINE within 1,000 yards of a school receive at least a 3-year prison term, or that people convicted of selling more than 4 ounces of heroin or cocaine receive at least a 5-year prison term. Some mandatory sentencing laws require life sentences. A Michigan law, which the U.S. Supreme Court upheld against a claim that mandatory life sentences constitute "cruel and unusual punishment" in violation of the Eighth Amendment to the U.S. Constitution, required life sentences without possibility of parole for people convicted of possessing more than 650 grams of cocaine (*Harmelin v. Michigan*, 49 Cr.L. 2350 [6/27/91]). An Alabama law required life sentences for people who, having previously been twice convicted of felonies, are convicted of a third felony. Laws like Alabama's are sometimes called "habitual offender" or "predicate felony" laws.

ENACTMENT OF MANDATORY SENTENCING LAWS

A historically unprecedented number of mandatory sentencing laws were enacted during the 1970s and 1980s. Most involve drugs, firearms, or both. Between 1978 and 1981, forty-nine states enacted mandatory sentencing laws. Every state and the federal government enacted mandatory sentencing laws during the 1980s. In 1992, approximately 100 separate mandatory minimum penalty provisions were contained in sixty federal statutes.

Despite this proliferation of mandatory sentencing laws, they remain the exception rather than the rule in criminal courts. The laws governing punishment of persons convicted of having committed most crimes set maximum penalties that a judge may impose, but not minimums. For example, a typical robbery law might permit the judge to impose a prison sentence up to twenty years, a fine up to $10,000, or a term of probation up to five years. The judge could then choose, within those limits, what punishment, or combination of punishments, the offender should receive.

Two different kinds of sentencing systems are common in the United States, and mandatory sentencing laws affect each in the same way. In *indeterminate sentencing* states, judges set maximum lengths of prison sentences, and sometime minimums, but parole boards decide when a prisoner will be released. In *determinant sentencing* states, there is no parole board, and the judge decides how long a prisoner will serve. (In both kinds of states, some prisoners also receive time off for good behavior; these "good time" systems are managed by the prison authorities.) Mandatory sentencing laws in either kind of sentencing system limit the judge's discretion to impose a sentence that the judge believes is deserved.

Mandatory sentencing laws have long been controversial. The American Law Institute, an association of lawyers, judges, and law professors that created the *Model Penal Code*, a model law on which the criminal laws of nearly half the states are patterned, opposes enactment of mandatory sentencing laws. So does the American Bar Association. In 1991, a survey of U.S. federal judges showed that 62 percent favored repeal of federal laws calling for mandatory sentences in drug cases.

OBJECTIONS TO MANDATORY SENTENCING LAWS

Opponents of mandatory sentencing laws oppose them for a variety of reasons. Many judges and lawyers believe that mandatory sentencing laws are arbitrary and sometimes require judges to impose sentences that are unduly harsh. They think that justice requires that sentences be individualized to fit the circumstances of the offender and of the crime. They also think that sentences should vary depending on considerations such as whether the offender was a ringleader or a follower; whether the offender played a major role or a minor one; whether he or she was motivated by greed or poverty; whether a seller of drugs was an addict raising money to support a drug habit or a professional drug dealer; and whether the quantity involved was large or small. A law requiring that anyone convicted of selling more than a small amount of heroin receive a five-year prison sentence ignores all such distinctions.

Opponents also complain that mandatory sentencing laws adversely affect court operations. Because prosecuting attorneys decide what charges to file in each case, mandatory sentencing laws shift power from the judge to the prosecutor. Most crimes are not covered by mandatory sentencing laws. Typically, for example, trafficking in drugs is subject to mandatory penalties, but possession of drugs is not. Since nearly every drug trafficker also possesses

drugs, prosecutors can decide which charge to file: a trafficking charge ties the judge's hands; a possession charge gives the judge discretion.

Another objection is that mandatory penalties remove much of the defendant's incentive to plead guilty and thus increase the frequency of trials and lengthen the time required to resolve cases. In most courts, 85 to 95 percent of convictions result from guilty pleas. Many result from plea bargains, in which the prosecutor agrees either to dismiss some charges or to approve a particular sentence if the defendant pleads guilty. If mandatory penalties remove incentives for plea bargains, trials, backlogs, and delays increase.

Yet another objection is that mandatory sentencing laws sometimes result in deceptive practices on the part of judges. To avoid imposing sentences that they believe are too severe, judges sometimes ignore the mandatory sentence law and impose some other sentence or hypocritically acquit defendants of crimes that bear mandatory penalties.

ARGUMENTS FOR MANDATORY SENTENCING LAWS

Supporters of mandatory sentences are not troubled by the harshness of some mandatories or that they shift power from the judge to the prosecutor. One of the goals of these laws is to assure that the mandated sentence will be imposed whether the judge agrees with the sentence or not.

Supporters are troubled by deceptive efforts by judges (and sometimes by prosecutors) to avoid applying them. They argue that judges are wrong to try to circumvent mandatories; if legislatures pass laws, they argue, judges should enforce them even when the judges disagree with them. Finally, supporters say they are sorry if mandatories affect guilty pleas, trial rates, and court delays, but regard those problems as a price worth paying.

Proponents of mandatory sentencing laws make four arguments. First, many citizens are worried about crime, and passage of mandatories is a way for legislators to assure citizens that their concerns are being taken seriously. Second, some people believe that harsh mandatory sentencing laws deter offenders from committing crimes. Third, some legislators think that certain crimes are so serious that people who commit them should be severely punished and that legislators should insist that judges impose severe penalties in such cases. Fourth, some people believe that mandatory sentencing laws are a device for assuring that offenders who commit the same crime will receive the same penalty.

RESEARCH ON MANDATORY SENTENCING LAWS

Evaluations of mandatory sentencing laws offer greater support to their opponents than to their supporters. The Panel on Sentencing Research of the National Research Council, the research wing of the National Academy of Sciences, examined all research on mandatory penalties through 1983. Studies on the deterrent effects of mandatory sentencing laws conclude either that passage of such laws has no deterrent effects or that they have modest deterrent effects that soon disappear. Research on how mandatory sentencing laws affect court operations shows that such laws do shift power from judges to prosecutors, do sometimes result in lower guilty plea rates and higher trial rates, often cause case processing delays, and frequently result in imposition of sentences that the judges and lawyers involved believe are harsher than the defendant deserves. All of these conclusions were reached by the evaluators of the ROCKEFELLER DRUG LAWS in New York State in the mid-1970s.

The conclusions of earlier research were confirmed by the most ambitious and sophisticated study of mandatory penalties ever completed—a report on mandatory penalties in the U.S. federal courts by the U.S. Sentencing Commission. That study concluded that people convicted of crimes subject to mandatory penalties were two and one-half times more likely to be convicted after trials (30% of convictions) than are other federal defendants (12.5%). The study found that "mandatory minimums transfer sentencing power from the court to the prosecutor," that "honesty and truth in sentencing" are compromised by prosecutors' and judges' efforts to work around mandatory sentences, and that "lack of uniform application [of mandatories] creates unwarranted disparity in sentencing."

Thus, on the major empirical issues about which opponents and supporters of mandatory penalties disagree, the great weight of the evidence supports opponents' views. Empirical evidence, however, cannot refute supporters' normative claims that mandatory penalties should be enacted to assure citizens that their concerns about crime are taken seriously or that certain crimes deserve severe punishment and that mandatory sentencing laws should be enacted to increase the likelihood that such punishments will be imposed. Opponents of mandatory penalties do not necessarily disagree that lawmakers should try to respond to citizens' concerns, or that some crimes deserve harsh penalties; they do believe that mandatory penalties are an ineffective way to achieve those goals.

(SEE ALSO: *Civil Commitment; Drug Laws: Prosecution of; Legal Regulation of Drugs and Alcohol; Treatment Alternatives to Street Crime*)

BIBLIOGRAPHY

AMERICAN LAW INSTITUTE. (1962). *Model penal code* (proposed official draft). Philadelphia.

BLUMSTEIN, A., COHEN, J., MARTIN, S., & TONRY, M. (1983). *Research on sentencing—The search for reform*. Washington, DC: National Academy Press.

BLUMSTEIN, A., COHEN J., & NAGIN, D. (1978). *Deterrence and incapacitation—Estimating the effects of sanctions on crime rates*. Washington, DC: National Academy Press.

JOINT COMMITTEE ON NEW YORK DRUG LAW EVALUATION. (1977). *The nation's toughest drug law: Evaluating the New York experience*. Washington, DC: U.S. Government Printing Office.

TONRY, M. (1988). Mandatory penalties. In M. Tonry & N. Morris (Eds.), *Crime and justice—A review of research*, vol. 16 (pp. 243–273). Chicago: University of Chicago Press.

U.S. SENTENCING COMMISSION. (1991). *Mandatory minimum penalties in the federal criminal justice system*. Washington, DC: U.S. Government Printing Office.

MICHAEL TONRY

MARATHON HOUSE   The mid-1960s was a period of significant focus on U.S. social programs. Growing out of President Lyndon B. Johnson's Great-Society strategy was a new way of viewing the community's capacity to take ownership of its social problems, develop collaborative strategies, and heal its own wounds. Toward that end, a new lexicon emerged—*community-based, storefront,* and *streetworker*—to identify but a few terms.

An agency that was very much a product of both its environment and the times was Progress for Providence (Rhode Island). In 1966, streetworkers for that organization began to acknowledge a growing community presence of HEROIN, heroin dealers, and addicts. To their credit, they recognized not only these phenomena but the insufficiency of expertise and resources to address the problem. Consequently, representatives were elected to pursue both knowledge and staffing.

At that time, DAYTOP VILLAGE was conducting training institutes in Swan Lake, New York. Representatives from Progress for Providence attended, seeking both enlightenment and technical assistance to establish a Providence-based initiative. A Daytop alumnus and staff member, James Germano, was deployed to "sow the seeds" for the first New England-based THERAPEUTIC COMMUNITY (TC). Public forums, speaking engagements, and pervasive networking created the foundation of support for Marathon House. This advocacy, charismatic leadership, some seed money from Progress for Providence, and basement space supplied by a church were the ingredients for the establishment of Marathon House in October 1967. In November, then governor, now Senator John H. Chafee of Rhode Island offered a tract of land in Coventry as the temporary site for Marathon House. With dormitories in a converted garage and horse barn, Marathon's first "house" would remain in operation until replaced in February 1991.

Welfare, donations, volunteerism, and the aforementioned seed money were pooled to fund treatment for about twenty recovering heroin addicts. In February 1967, The Rhode Island Foundation (in the early 1990s the seventh largest community foundation in the United States) awarded 10,000 dollars. In early 1968, a formal agreement was reached with the State of Connecticut, which reimbursed $10.50 per Connecticut resident per day treated at the lone Rhode Island facility. Later that year, Marathon was to open its second therapeutic community in Attleboro, Massachusetts, at LaSallette Shrine. It was in late 1968 that the infamous "Daytop split" occurred, which resulted in an exodus of both staff and residents. Many of both found their way to Marathon.

Formal funding from the State of Rhode Island was received in 1969 along with new federal dollars via the NATIONAL INSTITUTE OF MENTAL HEALTH (NIMH). By July 1969, the payroll would grow to 103,000 dollars. The Marathon facility at Gaylordsville, Connecticut, opened in August 1969, under the direction of Richard Mazzochi (who in the early 1990s was vice president for program operations). Soon after the opening, outreach storefronts much like those already operating in Providence opened in a number of Connecticut cities under the Marathon banner.

A facility for ADOLESCENTS in Middletown, Rhode Island, began operating in 1970. While relatively short lived, it laid the groundwork for those modified therapeutic communities Marathon currently operates throughout New England. A contract with the Commonwealth of Massachusetts was signed in 1970, and Marathon's facility in the City of Springfield was opened in 1971. In February 1971, Marathon acquired a historically significant property in Dublin, New Hampshire. Built in 1787 as a farmhouse by Moses Greenwood (1750–1827) upon his return from the Revolutionary War, the property became the Dublin Inn in 1921. In October 1945, the inn was the site of the "Dublin Peace Conference," a gathering of fifty prominent Americans, including John F. Kennedy, who came together to suggest revisions to the United Nations Charter, in an effort to control atomic energy; as a result, the "Dublin Declaration" was published.

In the 1990s, this facility, which is on the National Register of Historic Places, is the center for three distinct Marathon programs. It is, in effect, a campus that houses (1) the original New Hampshire adult therapeutic community, (2) the Lodge at Dublin, a ninety-day residential facility for male adolescents, and (3) the Alcohol Crisis Intervention Unit, a small social-setting detoxification facility.

Since the beginning, Marathon's clientele has been involved with the criminal-justice system. Although there were contracts with the Federal Bureau of Prisons throughout the 1960s and 1970s, there were no formal arrangements. In 1989, however, Marathon began two programs under the aegis of the Rhode Island Department of Corrections. The first is Inside the Walls, an educational and counseling program, which serves all levels of security at Rhode Island's adult correctional institutions, and the second is Outside the Walls, a community-based outpatient counseling service for parolees. The third component of Marathon's Rhode Island criminal-justice effort is a forty-eight bed unit at the minimum security prison.

In Marathon's history, a true milestone was the February 1991 opening of a new seventy-four bed therapeutic community in Exeter, Rhode Island. This replaced the original facility in Coventry and, with expanded bed capacity, allowed all Rhode Islanders in treatment at our other sites to be treated in their home state. In 1993, construction was completed on the Marathon Arts and Fitness Center to abut the Exeter building, which, along with 7.5 acres was deeded to Marathon by the governor of Rhode Island.

Since its founding, Marathon has been actively engaged in research. Moreover, it has participated in national consortium studies examining the effectiveness of therapeutic communities. After the first six years of operation, Research Director Dr. Barry Sugarman discovered that one year after completion, 92 percent of those treated remained drug and alcohol free. It should be noted that the average length of stay per graduate was twenty-six months—extremely lengthy by the standards of the 1990s.

Although by 1982 education on ACQUIRED IMMUNODEFICIENCY SYNDROME (AIDS) was incorporated into the program, not until 1987 did funding become available to study AIDS risk behaviors among needle-using drug abusers. This three-year project, funded by the NATIONAL INSTITUTE ON DRUG ABUSE (NIDA), demonstrated that knowledge, attitude, and risk reduction could be affected with the engagement and education of those not in treatment. Marathon, as contractor, worked with both PROJECT RETURN in New York City and with SECOND GENESIS in the Washington, D.C., area to ensure diversity of sample. Macro International performed the data collection and analysis for the study.

In 1989, again under aegis of NIDA, a five-year research contract was awarded to Marathon to study the correlation between length of time in program and treatment outcome. Subcontracting to Spectrum of Westboro, Massachusetts, and the University of Massachusetts, Project IMPACT examines a six-month versus a twelve-month stay in a therapeutic community as well as a three-month versus a 6-month relapse-prevention model. In 1992, the CENTER FOR SUBSTANCE ABUSE TREATMENT (CSAT) funded an AIDS outreach project, to continue those educational activities begun in 1987 and to encourage, refer, and place those active users not in treatment into the most appropriate and accessible units.

The Marathon organization has grown large. With fourteen sites throughout New Hampshire, Vermont, Massachusetts, and Rhode Island, services range from brief interventions to long-term residential care. The first EMPLOYEE ASSISTANCE PROGRAM (EAP) contract was signed in 1993, launching a new and timely initiative.

Treatment, shelter, and transitional housing are all elements of the Marathon "tapestry." Clients in treatment have ranged in age from thirteen to fifty-seven. Both indigent and affluent co-exist in the facilities, with treatment planning suited to all economic levels. A medical DETOXIFICATION service is in the planning stage to approach a true continuum for those in need.

(SEE ALSO: *Prisons and Jails: Drug Treatment in; Treatment In the Federal Prison System*)

DAVID J. MACTAS

MARIHUANA COMMISSION: RECOMMENDATIONS ON DECRIMINALIZATION

Before 1960, use of MARIJUANA in the United States was generally confined to drug-using subcultures in the inner cities (e.g., jazz musicians) or in rural areas (e.g., migrant farm workers). Sale and use of the drug were prohibited both by federal law and by the laws of every state. Because marijuana was classified in 1937 as a "narcotic drug," along with COCAINE and OPIATES, penalties were severe; simple possession for personal use was a felony in most states. During the 1960s, marijuana smoking suddenly became prevalent on college campuses—for the first time among white middle-class youth of the baby-boom generation. Marijuana use also became associated, as a protest behavior, with dissenters (both adults and youth) against the war in VIETNAM, and by the U.S. MILITARY serving in Vietnam, especially from 1963 to 1973. As use of the drug increased, so did the number of arrests and so did the surrounding controversy. Questions were raised about the actual effects of marijuana on the health and behavior of those who used it and about the wisdom of prevailing social policy.

In response to swirling controversy, many proposals were introduced in Congress for a commission to undertake an authoritative study of the marijuana issue. Eventually, in the Comprehensive Drug Abuse Prevention and Control Act of 1970, Congress established the NATIONAL COMMISSION ON MARIHUANA AND DRUG ABUSE to undertake a two-year study—the first year on marijuana and the second year on the causes of drug abuse in general.

The commission had thirteen members—four from Congress (two each from the House and the Senate) and nine appointed by the president. President Richard M. Nixon appointed Raymond P. Shafer, formerly governor of Pennsylvania, as chairman of the commission, and he appointed Dana L. Farnsworth, M.D., director of Student Health Services at Harvard University, to be vice-chairman. The executive director was Michael R. Sonnenreich, formerly the deputy chief counsel of the Bureau of Narcotics and Dangerous Drugs of the Justice Department.

The commission assimilated the available literature on marijuana use and its effects and also sponsored its own research, including a national survey of use patterns and public attitudes, and a study of enforcement of the marijuana laws in six jurisdictions. In March 1972, the commission issued its first report, *Marihuana: A Signal of Misunderstanding*.

PRINCIPAL FINDINGS

The commission estimated that although 24 million Americans had used marijuana at least once, about 50 percent had simply experimented with the drug out of curiosity and given it up. Among the 50 percent who had continued to use marijuana, most used it only occasionally, once a week or less, for recreational purposes. A small percentage of the more frequent users (about 2% of the total ever-using population—or 4% of the continuing users) used the drug more than once daily. Marijuana use was clearly age-related: about half of the ever-users were 16 to 25 years of age, and 44 percent of those who were currently in college or graduate school had used marijuana at least once.

The commission concluded that there was "little proven danger of physical or psychological harm from the experimental or intermittent use" of marijuana. "The risk of harm," it continued, "lies instead in the heavy, long-term use of the drug, particularly of its more potent preparations." Even this risk was of uncertain dimensions, the commission noted, because the psychological consequences of long-term heavy use were unknown. In light of the fact that 90 percent of marijuana users used the drug only experimentally or intermittently, the commission judged that "its use at the present level does not con-

stitute a major threat to public health." The commission also specifically found that marijuana did not induce physical dependence; did not lead, by virtue of its pharmacology, to use of other drugs; and did not cause criminal behavior.

POLICY RECOMMENDATIONS

The commission's principal policy recommendation was that possession of one ounce or less of marijuana for personal use be "decriminalized." At the same time, the commission rejected outright legalization of the drug and recommended perpetuation of prohibitions against cultivation and distribution for commercial purposes. The commission stipulated that social policy should aim to discourage use of the drug, but it emphasized that the costs of a criminal prohibition against possession far exceeded its benefits in suppressing use.

Although President Nixon disavowed the commission's principal recommendation on marijuana, it won widespread support. In 1973, the National Conference of Commissioners on Uniform State Laws promulgated amendments to the Uniform Controlled Substances Act that codified the commission's recommendation. Some form of decriminalization was endorsed the same year by a variety of national organizations, including the American Bar Association and numerous state and local bar associations, the National Education Association, the Consumers' Union, the National Council of Churches, the American Public Health Association, and the governing board of the American Medical Association.

In 1973, Oregon became the first state to decriminalize possession of small amounts of marijuana. Within the next five years, ten additional states eliminated incarceration as a penalty for simple possession, usually substituting a 100-dollar fine. Five of these states made possession a "civil offense"; in others, it remained a criminal offense although the law typically contained a provision for expunction of criminal records after a specified period of time. Decriminalization of marijuana use was endorsed by President Jimmy Carter in 1977.

Political and legislative support for decriminalization began to wane, however, even during the Carter Administration. The more permissive stance on marijuana use implicit in decriminalization efforts led to mounting public resistance. Some of the strongest opposition came from groups of parents who organized to lobby for more focus on PREVEN- TION efforts. Although these parent groups were generally conservative politically, they gained a re- ceptive ear in the Carter White House. Their argu- ments against decriminalization were bolstered by findings from the National High School Senior Sur- vey showing that, starting in 1975, daily marijuana use had been increasing progressively among high school students. During the Reagan and Bush administrations the parents' movement and their concerns about marijuana use came to have a major influence on national drug policy. In the early 1990s, possession of the drug remained a criminal offense in most states, as well as under federal law.

(SEE ALSO: *Anslinger, Harry J., and U.S. Drug Policy; High School Senior Survey; Legal Regulation of Drugs and Alcohol; Prevention*)

BIBLIOGRAPHY

BONNIE, R. J. (1982). The meaning of "decriminalization": A review of the law. *Contemporary Drug Problems, 10*(3), 277–289.

BONNIE, R. J., & WHITEBREAD, C. H., II. (1974). *The mari- huana conviction.* Charlottesville: University Press of Virginia.

MUSTO, D. F. (1987). *The American disease.* New York: Ox- ford University Press.

NATIONAL COMMISSION ON MARIHUANA AND DRUG ABUSE. (1972). *Marihuana: A signal of misunderstanding.* Washington, DC: U.S. Government Printing Office.

RICHARD J. BONNIE

MARIJUANA In the United States this is the most common term for the HEMP plant *Cannabis sativa* and its PSYCHOACTIVE products. The term derives from the Mexican Spanish *mariguana/ marihuana* (Mary's leaf or plant) or from Maria and Juan (Mary and John). It came into English about 1894 and has become the mainstream term in Amer- ican publication, law, and general usage. The term *cannabis* is sometimes used in medical literature and by the British; it means "hemp" in Latin and is de- rived from the Greek, *kannabis*. In ASIA, where the plant originated, it is grown legally and commercially for both its fiber content and for its drug content— there, it is called BHANG, GANJA, and HASHISH.

Figure 1
Marijuana

BOTANY

Hemp grows easily throughout the tropics, sub- tropics, and temperate regions, varying in size from a few feet to 15 feet (4.6 m) in height. Once estab- lished, it reseeds itself and spreads to neighboring areas; when birds eat the seeds, the defecated seeds may be scattered over considerable distances.

Two genetic strains of hemp are recognized: one produces plants excellent for fiber with very little drug material; the other produces plants with weak fibers but much drug content (TETRAHYDROCANNAB- INOL, THC). To harvest the drug-laden plant, it is simply cut down and chopped into small pieces with all parts included. These clippings resemble lawn cuttings, so one of the slang terms is "grass." The major use of this form in the United States is for illicit marijuana cigarettes.

Since the early 1900s, marijuana has been con- sidered the one drug that might introduce the sus- ceptible to hard drugs, drug abuse, and drug dealing. In the United States until 1937, *Cannabis* had been used in medical practice for a number of condi- tions—but marijuana use for its euphoric effect was relatively uncommon. By 1937, forty-six states had laws against the use of marijuana, and its use had already been made a criminal offense under federal law. Until the 1960s, it was smoked largely by blacks and hispanics in the United States but generally shunned by the white majority. During the 1960s' social and political protests, a change in attitudes al- lowed widespread but illicit marijuana use into all levels of society, along with an increase in the use of several other illicit drugs and a boom in the drug trade that continues into the 1990s.

HISTORY

Various historical allusions to medicinal plants suggest that *Cannabis* was known and used for several millennia. The earliest references to the plant are in ancient Chinese and Indian writings. From India, *Cannabis* use spread to Persia, Assyria, and the rest of the Near East. The Arabs adopted it and spread it through North Africa as they conquered those lands for Islam from the seventh to the fifteenth centuries. Islam forbids the use of ALCOHOL but not explicitly *Cannabis* (since it was adopted after the laws established by the Prophet Muhammad). In Arabic, it is called HASHISH, meaning "grass." When the Arabs crossed Gibraltar into the Iberian peninsula in 711, they ruled there until 1492, but Portugal and Spain did not generally adopt its use. The Spanish conquistadors, however, introduced *Cannabis* into the New World, where it was readily adopted by African slaves, who were already familiar with it because of Arab trade and the spread of Islam into their continent.

CHEMISTRY

Like most plants, *Cannabis* contains many substances, perhaps 200 or more. Those that relate most to the drug effects are a group of chemically similar compounds called cannabinoids. Of these, the most important and plentiful are cannabidiol (CBD), tetrahydrocannabinol (THC), and cannabinol (CBN). The biosynthetic pathway in the plant goes from CBD to THC to CBN. Thus it is possible to identify the maturity of the plant by the relative content of these three cannabinoids. Immature plants show a preponderance of CBD; old plants may contain solely CBN; plants that are at their peak contain all cannabinoids, but mostly THC, which is the agent that produces the psychic effect. Some strains of plants contain variants on the THC structure, which usually have somewhat less drug effect than those with THC. Although some users contend that marijuana has different effects from those of isolated THC, most evidence indicates that virtually all of the psychic effects of marijuana are attributable to the THC content.

The THC content may vary greatly, depending on the genetic strain of the plant, the care in separating the flowers and leaves, and the degree of maturity of the plant. The THC content of plants used for hemp production, such as those that grow wild in the U.S. Midwest, may be negligible to zero; marijuana produced from plants known for high drug production, such as *sensemilla*, may contain 2 to 3 percent THC. Manicured plants, from which the leaves are carefully separated and only the new leaves used for drug effect, may contain 3 to 4 percent THC. Hashish, which represents the ultimate in manicuring, generally contains 4 to 8 percent THC.

THC is sensitive to exposure to air and light. Thus, marijuana that is not protected from such exposure undergoes gradual degradation until the drug content is gone. When protected from air and light, marijuana may retain its activity for many months.

EPIDEMIOLOGY

Marijuana may rank only behind CAFFEINE, alcohol, and NICOTINE as the drug most widely used in the world. It is estimated that between 200 and 300 million people use this material in one way or another. In the United States alone, probably some 20 to 30 million people have used the drug, but the number of regular users is probably far less—still a few million.

In the United States, marijuana is a drug preferred by young people and those who think of themselves as "young"; the rate of use of marijuana is therefore followed among schoolchildren to estimate changing trends in use. Recent surveys of high-school students concerning use of marijuana show quite variable rates of response, both in terms of individuals as well as numbers of individuals from each school. Overall, 3 to 17 percent (median 12%), reported at least a single use of marijuana during the preceding thirty days. Such use compares with that of smoking at least one cigarette (9 to 37%, median 31%) or having at least one drink of alcohol (28 to 64%, median 54%). Thus, it would appear that marijuana is not nearly as widely used as two of our three national drugs. Although these data indicate a trend toward decreased use and greater concern about the drug, this pattern has not held long enough to establish a true trend; it may be simply a minor blip.

A number of factors seem to contribute to increased use of marijuana among young people: Being male, using cigarettes and alcohol, and becoming delinquent are predisposing factors. Coming from a broken home and performing poorly in school are also predictive factors. Among adolescents in Australia and New Zealand, use of stimulants, HALLUCI-NOGENS, NARCOTICS, and SEDATIVES was virtually limited to those young people who used marijuana. Overall, it appears that social factors are more highly predictive of *Cannabis* use than are school factors.

PSYCHOPHARMACOLOGY

Marijuana has a variety of pharmacologic effects that suggest actions like those of AMPHETAMINES, LSD, ALCOHOL, SEDATIVES, atropine, or MORPHINE. Thus, the drug does not fit traditional pharmacologic classifications and must be considered as a separate class.

The expert smoker of marijuana is usually aware of a drug effect after two or three inhalations. As smoking continues, the effects increase, reaching a maximum about twenty minutes after the smoke has been finished. Most effects of the drug usually have vanished after three hours, by which time tests show that plasma concentrations of THC are low. Peak effects after oral administration may be delayed until three to four hours after drug ingestion but may last for six to eight hours.

The early stage is one of being high, characterized by euphoria, uncontrollable laughter, alteration of time sense, depersonalization, and sharpened vision. Later the user becomes relaxed and experiences introspective and dreamlike states, if not actual sleep. Thinking or concentrating becomes difficult, although by force of will the subject can attend.

Two characteristic signs of *Cannabis* intoxication are increased pulse rate and reddening of the conjunctiva (the whites of the eyes). The latter correlates well with the presence of detectable plasma concentrations. Pupil size is not changed. The blood pressure may fall, especially in the upright position (orthostatic hypotension). An antiemetic (decrease in sense of nausea) effect may be present, and muscle weakness, tremors, unsteadiness, and increased deep tendon reflexes (such as the knee jerk) may also be noted.

Virtually any performance test shows impairment if the doses are large enough and the test difficult enough, although no distinctive biochemical changes have been found in humans.

TOLERANCE to *Cannabis* has been demonstrated in virtually every animal species that has been tested. It is apparent in humans only among heavy long-term users. Different degrees of tolerance de-velop for different effects of the drug, with tolerance for the tachycardic effect (increased pulse rate) developing fairly rapidly. A mild WITHDRAWAL SYN-DROME has been noted following very high doses.

HEALTH CONSEQUENCES OF USE

The ambiguity currently surrounding the health hazards of *Cannabis* may be attributed to a number of factors besides those that ordinarily prevail. First, from animal studies, it has been difficult to prove or disprove health hazards in humans. Second, *Cannabis* is still used mainly by young persons in the best of health. Third, *Cannabis* is often used in combination with tobacco and alcohol, among licit drugs, as well as with a variety of other illicit drugs. Finally, the whole issue of *Cannabis* use is so laden with emotion that serious investigations of the health hazards of the drug have been colored by the prejudices of the experimenter, either for or against the drug as a potential hazard to health.

Psychiatric Consequences. *Cannabis* may directly produce an acute panic reaction, a toxic delirium, an acute paranoid state, or acute mania. Whether it can directly evoke depressive or schizophrenic states, or whether it can lead to sociopathy or even to the so-called AMOTIVATIONAL SYNDROME is much less certain.

That *Cannabis* use may aggravate schizophrenia is beyond any question. Such worsening followed acutely after use of *Cannabis* by schizophrenics, despite continued maintenance with antipsychotic drugs, and other adverse reactions were encountered among seventy patients in Sweden—anxiety reactions, flashbacks, dysphoric reactions, and abstinence syndromes.

Whether chronic use of *Cannabis* changes the basic personality of users so that they become less impelled to work and to strive for success has been a vexing question. As with other questions concerning *Cannabis* use, it is difficult to separate consequences from possible causes of drug use.

Automobile Driving. If marijuana is to become an accepted social drug, it would be important to know its effects on driving ability. Fully 50 percent of the fatal auto accidents in the United States are associated with another social drug, alcohol. Neither experimental nor epidemiological approaches to the marijuana question have yet provided definitive answers.

Cardiovascular Problems. For persons with heart disease caused by arteriosclerosis of the coronary arteries or congestive heart failure, the effects of *Cannabis* smoking would be deleterious: tachycardia, orthostatic hypotension, and increased blood concentrations of carboxyhemoglobin.

Clearly, smoking of any kind is bad for patients with angina, but the greater effect of *Cannabis* as compared with tobacco in increasing heart rate makes this drug especially bad for such patients. Thus far, few angina patients have been devotees of *Cannabis*.

Lung Problems. Virtually all users of *Cannabis* in North America take the drug by smoking. As inhaling any foreign material into the lung may have adverse consequences, well proven in the case of tobacco, this mode of administration of *Cannabis* might also be suspect. A formal study has shown that very heavy marijuana smoking for six to eight weeks caused mild but significant airway obstruction.

The issue of damage to lungs from *Cannabis* is somewhat confounded by the fact that many *Cannabis* users also use tobacco. As yet, it is far easier to find pulmonary cripples from the abuse of tobacco than it is to find any evidence of clinically important pulmonary insufficiency from smoking of *Cannabis*.

Endocrine and Metabolic Effects. A review of literature on this subject concluded that no significant effect was found in regard to serum testosterone and that sperm production was decreased but without evidence of infertility. Ovulation was inhibited, and luteinizing hormone was decreased, but cannabinoids had no evidence of estrogenic activity, which had been postulated earlier.

Immunity. A number of in vitro studies, using both human and animal material, suggest that cell-mediated immunity (the capacity of white cells to fight invading bacteria, viruses, or cancer cells) may be impaired after exposure to *Cannabis*. Clinically, one might assume that sustained impairment of cell-mediated immunity might lead to an increased prevalence of opportunistic infections or an increased prevalence of malignancy, as seen in the current epidemic of ACQUIRED IMMUNODEFICIENCY SYNDROME (AIDS). No such clinical evidence has been discovered.

THERAPEUTIC USES

For many centuries, *Cannabis* was used as a treatment, but only during the nineteenth century did a particularly lively interest develop for exploiting its therapeutic potential. *Cannabis* was then reported to be effective in treating tetanus, convulsive disorders, neuralgia, migraine, dysmenorrhea, postpartum psychoses, senile insomnia, depression, and gonorrhea, as well as opium or chloral hydrate addiction. In addition, it was used to stimulate appetite and to allay the pain and anxiety of patients terminally ill with cancer. Few of these claims have even been properly tested in clinical studies.

Antiemetic for Patients in Cancer Chemotherapy. CANCER chemotherapy, especially with the agent cisplatin, produces severe nausea and vomiting, which is extremely difficult to treat with ordinary antiemetic drugs, such as prochlorperazine. This complication is so severe that many patients forgo effective cancer chemotherapy. The antiemetic effects of *Cannabis* had been suggested as early as 1972. In that year a synthetic homolog of THC, nabilone, was developed, which has been tested extensively for antiemetic activity. A crossover study comparing nabilone with prochlorperazine revealed significantly better results (i.e. less nausea and vomiting) following nabilone therapy, although side effects from nabilone were also common.

The potential role of THC as an antiemetic may have been mooted by recent developments. Metoclopramide, a newly developed antiemetic unrelated to the cannabinoids, has been found to be effective when given in high intravenous doses. Lorazepam, dexamethasone, and ondansetron are also useful as antiemetic agents when given by injection. These drugs are often used in various combinations, which meet most requirements. Thus, THC may be superseded even before it has had widespread clinical trial.

Glaucoma. The disease glaucoma causes pressure in the eyeball to increase greatly. If untreated, this can lead to blindness. Discovery of the ability of *Cannabis* to lower intraocular (inside the eyeball) pressure was more or less fortuitous. Intraocular pressure was measured as part of a multifaceted study of the effects of chronic smoking of large amounts of *Cannabis*. Intraocular pressure was found to decrease as much as 45 percent in nine of eleven subjects, thirty minutes after smoking.

This exploitation of cannabinoids for treatment of glaucoma will require much further developmental work to ascertain which cannabinoid will be lastingly effective and well tolerated topically.

Miscellaneous Uses. Cannabinoids have been found to have analgetic (pain relieving) activity and efforts are being made to synthesize new compounds that separate this action from the others. They have also been used as muscle relaxants, for treating bronchial asthma, and as anticonvulsants. Thus far, none of these additional potential therapeutic uses has been fully established.

TREATMENT OF MARIJUANA USE

In general, marijuana users, even those whose use is heavy, do not feel constrained to seek treatment unless such use is complicated by other drugs, such as COCAINE or alcohol. When use is complicated by other drugs, treatment efforts are usually directed toward the complicating drug. Thus, treatment programs directed specifically at marijuana use are rare. A TWELVE-STEP approach, similar to that for alcohol, has been proposed, but neither its feasibility nor efficacy have been tested.

GATEWAY EFFECT

Since about 1950 (but not much prior to that time) in the United States smoking of marijuana has been linked statistically to the use of other illicit drugs, such as heroin and cocaine. Most observers have concluded that the link is sociological rather than biological and that the use of marijuana is a marker for individuals who are more prone to seek new experiences even when these violate social norms and local laws. Further, the process of obtaining illicit marijuana increases the likelihood of contact with dealers and other individuals who have access to drugs such as HEROIN. Consequently, marijuana has been referred to as a "gateway" drug, one whose use often leads to the use of other illicit drugs. Some programs are aimed at preventing even experimentation with marijuana—not only for whatever benefit this may have itself but in the hope that in doing so the movement to other more potentially lethal drugs will be prevented.

LEGAL STATUS

Despite its widespread use, marijuana has not yet been admitted to the company of accepted social drugs. Laws remain that prescribe penalties for its possession, use, and sale. In some jurisdictions, possession and use of small amounts of the drug is a civil crime punishable only by a small fine. Despite this liberalization of the law, these areas have not been overrun with eager marijuana users. Perhaps the reason is that in most other jurisdictions, laws against its use are rarely enforced. Enforcement can be capricious, however, when employed in situations in which more serious crimes cannot be adequately documented.

A new drug application was approved for THC (Marinol) to be used therapeutically for control of the nausea and vomiting associated with cancer chemotherapy. Thus, THC was moved from Schedule 1 of controlled substances (no medical use) to Schedule 2 (medical use despite potential for abuse). The THC homolog, nabilone, used for the same purpose, also has this status.

Thus far, no attempt has been made to establish forensic limits on the amounts of THC in the blood that might be construed as impairing automobile driving. No doubt the issue has not yet appeared to be of enough gravity, since marijuana contributes little to the danger of driving as compared with alcohol.

(SEE ALSO: *Adolescents and Drugs; Cannabis sativa; Complications; Controls; Driving, Alcohol, and Drugs; High School Senior Survey; Marihuana Commission; Yippies*)

BIBLIOGRAPHY

EBIN, DAVID. (1961). *The drug experience.* New York: Orion Press.

HOLLISTER, L. E. (1989). Drugs of abuse. In B. G. Katzung (Ed.), *Basic and clinical pharmacology*, 4th ed. San Mateo, CA: Appleton & Lange.

PETERS, H., & NAHAS, G. (EDS.). (1973). *Hashish and mental illness* by J. J. Moreau et al. (trans. from the French by G. J. Barnett). New York: Raven Press.

LEO E. HOLLISTER

MARIJUANA EPIDEMICS   *See* Epidemics of Drug Use; Yippies.

MAST   *See* Michigan Alcoholism Screening Test.

MDMA   This drug is popularly known as "ecstasy," XTC, and ADAM. It is a synthesized compound and a member of the family of HALLUCINOGENS known as the substituted phenethylamines, which also includes methylenedioxyamphetamine (MDA) and 2,5-dimethoxy-4-methylamphetamine (DOM) (see Figure 1). These hallucinogens are structurally related to the phenethylamine-type NEUROTRANS-MITTERS dopamine, norepinephrine, and epinephrine. Many analogs of these compounds have been synthesized and are sometimes found on the street—the so-called DESIGNER DRUGS.

Controversy exists as to whether MDMA and MDA should be classified with the other hallucinogens. Both MDMA and MDA have structural similarities to the PSYCHOSTIMULANT AMPHETAMINE, and they have amphetamine-like psychostimulant properties. Yet, these designer drugs also have properties in common with LYSERGIC ACID DIETHYLAMIDE (LSD) and MESCALINE; with lower doses, however, they produce fewer perceptual phenomena and less emotional liability, or "keyed-up" feelings and disturbances of thought, than other hallucinogens, and there tends to be a tranquil state with a feeling that tender emotions are meaningful. As doses are increased, the illusions and other LSD-like phenomena are seen. Because of their mixed effects, MDMA and MDA are sometimes referred to as STIMULANT-hallucinogens.

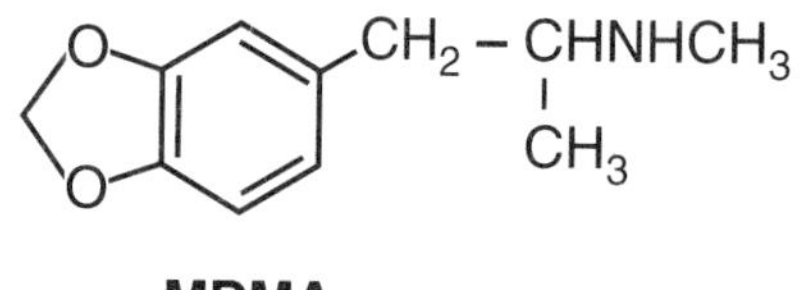

MDMA

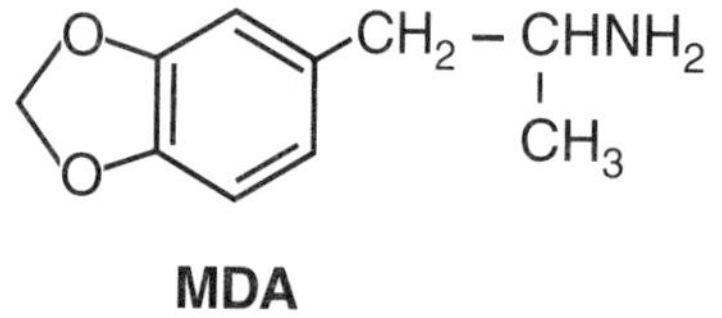

MDA

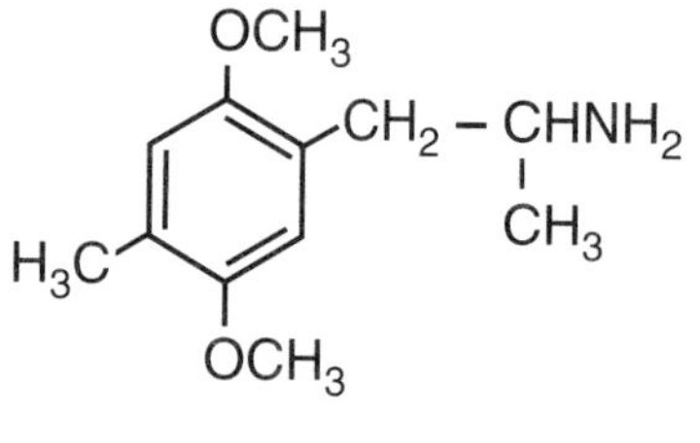

DOM

Figure 1
Phenethylamine Hallucinogens

Unlike LSD, users of MDMA have reported nausea, jaw clenching and teeth grinding, increased muscle tension, and blurred vision, as well as panic attacks. MDMA also causes amphetamine-like stimulation of the autonomic nervous system, producing increases in blood pressure, heart rate, and body temperature. A type of hangover the day after taking MDMA has been described, involving headache, insomnia, fatigue, drowsiness, sore jaw muscles, and loss of balance.

Like the other hallucinogens, the exact mechanisms of action of MDMA are not known. MDMA, like the indole- and phenethylamine-type hallucinogens, binds to receptors for the neurotransmitter serotonin. Thus, many effects might be due to interactions with brain serotonergic systems. MDMA, however, also causes the release of both dopamine and serotonin, so some effects may be related to their stimulant properties.

By the early 1990s, some evidence indicated that MDMA might damage nerve cells. In laboratory experiments, MDMA can produce long-lasting changes in the function of neurons that use serotonin as the neurotransmitter, sometimes causing the death of these cells. Even though LSD also interacts with serotonergic nerve cells, the administration of massive doses of LSD does not damage these cells. In contrast, in experimental animals, a single dose of MDMA approximately three times higher than the typical street dose has been shown to affect brain serotonergic systems for several weeks. In some studies, neurochemical markers did not return to normal until one year after drug administration. Moreover, it is not clear whether there was actual regeneration of neurons or only compensatory changes in the remaining undamaged neurons. In these experiments, the neurotoxic effects of MDMA appear to depend on total exposure. Both the dose taken and the number of times the drug is consumed may be related to brain-cell changes. The exact mechanism of MDMA-induced neurotoxicity is unknown at this time and may be due to MDMA itself, or it could involve the formation of a neurotoxic metabolite.

Although there is controversy whether studies utilizing laboratory animals can be extrapolated to human MDMA users, some evidence suggests that brain function can be altered in humans exposed to MDMA. Although the consequences to behavior and thinking caused by damage to the serotonergic nerve cells in young users are unknown, some effects of MDMA-induced toxicity may become apparent as the users age. Cells die as part of the aging process, and if exposure to MDMA kills or weakens a certain proportion of cells, the effects of normal cell loss due to aging might be exacerbated. Serotonergic systems have been implicated in the control of sleep, food intake, sexual behavior, anxiety, and mood. Thus, serotonergic cell loss could have major consequences.

(SEE ALSO: *Complications: Mental Disorders; Dopamine; Methamphetamine; Serotonin*)

BIBLIOGRAPHY

BARNES, D. M. (1988). New data intensify the agony over ecstasy. *Science, 239*, 864–866.

GLENNON, R. A. (1987). Psychoactive phenylisopropylamines. In H. Y. Meltzer (Ed.), *Psychopharmacology: The third generation of progress*. New York: Raven Press.

RICAURTE, G., ET AL. (1985). Hallucinogenic amphetamine selectively destroys brain serotonin nerve terminals. *Science, 229*, 986–998.

SHULGIN, A., & SHULGIN, A. (1991). *PIHKAL: A chemical love story*. Berkeley, CA: Transform Press.

DANIEL X. FREEDMAN
R. N. PECHNICK

MEDELLIN CARTEL   *See* Colombia As Drug Source.

MEDICATIONS APPROACHES FOR TREATING SUBSTANCE ABUSE   *See* Pharmacotherapy.

MEGAVITAMIN THERAPY   *See* Treatment, History of; Vitamins.

MEMORY AND DRUGS: STATE DEPENDENT LEARNING   The term *state dependent learning* (SDL) refers to the fact that memories acquired while a person is drugged may be forgotten when the drug wears off and not remembered until the person again takes the drug. Conversely, material learned in the undrugged state may be forgotten when a drug is taken; and material learned under one drug may be forgotten when another drug is used. SDL is sometimes called drug dissociation of learning, referring to the fact that material learned while drugged is dissociated from normal consciousness and not able to be retrieved.

Throughout the nineteenth century, there was a high level of public interest in multiple personality, fugue states, and other types of episodic amnesia; SDL was first reported in 1835 by George Combe, an English phrenologist, who viewed it as an analogous phenomenon, perhaps based on similar properties of the brain. SDL became an accepted property of mind during the latter half of the nineteenth century, and was a central theme in the plot of *The Moonstone*, (1868), a well-known mystery novel written by Wilkie Collins. Then, at the beginning of the twentieth century, interest in these dissociative phenomena waned and was replaced by an interest in the amnesias caused by repression, which Freud described. SDL was essentially forgotten.

SDL was rediscovered in the 1960s, this time in experiments using animals, and since then has been a popular topic of research and clinical speculation. Two types of mechanisms are postulated as possibly producing SDL. According to one theory, drugs produce sensory stimuli, subjective sensations—and one's ability to retrieve memories is aided by reinstatement of the stimuli that were present when learning occurred. A second theory suggests that some other property of brain results in memories being most easily retrieved when the conditions of brain excitability that were present during learning are reestablished. Sensory stimuli are not involved in producing SDL, according to this second theory. Thus far it has not been possible to confirm either of these proposed mechanisms experimentally, although the sensory model is more widely accepted.

SDL is produced only by drugs that act on the brain. There are marked differences in the strength of the SDL effects produced by the different centrally acting drugs. For example, BARBITURATES and ALCOHOL produce strong SDL effects, whereas chlorpromazine (Thorazine) produces almost no such effects. SDL is more likely to occur with high doses of drugs, and research on SDL has been severely hampered by the fact that these doses also produce other effects on memorization and retrieval that are difficult to distinguish from SDL effects. Some research suggests that the relative ability of different drugs to produce SDL may differ depending on the type of task that is employed, but this conclusion is not yet well substantiated.

Many consider SDL to be closely related to drug discriminations, believing that the discriminative control exercised by drug conditions is produced by the same drug effects that produce SDL amnesias at higher doses.

After SDL was rediscovered in the 1960s, clinicians feared that the lessons of psychotherapy carried out while a patient was drugged might be forgotten when drug treatment was discontinued. Subsequent studies showed that strong SDL effects typically did not occur except at doses higher than those normally employed during chronic treatment with psychotropic drugs. Some evidence, however, suggests that the stimulant drugs used to treat hyperactive children may produce SDL in those children. There is increasing evidence that some types of learning may take place under general anesthesia, although patients report they remember nothing after the anesthesia wears off. A considerable amount of research is currently focused on the possibility that SDL may block explicit recall of learning under general anesthesia, even though such learning occurs.

Many centrally acting drugs alter moods. A currently active area of research deals with the possibility that emotions act as memory cues and that memories learned in one emotional state may be recalled best when that emotion reoccurs; they may be recalled less easily at other times. Finally, there has been a dramatic increase in the number of reported cases of multiple personality disorder during the past decade. One of the theories used to explain this disorder holds that the process underlying it is similar, at a mechanistic level, to that which produces drug-induced SDL.

(SEE ALSO: *Memory, Effects of Drugs on; Research; Animal Model*)

BIBLIOGRAPHY

COLLINS, W. (1868/1981). *The moonstone.* New York: Penguin Books.

COMBE, G. (1830). *A system of phrenology,* 3rd ed. Edinburgh: John Anderson.

OVERTON, D. A. (1991). Historical context of state dependent learning and discriminative drug effects. *Behavioral Pharmacology, 2,* 253–264.

OVERTON, D. A. (1984). State dependent learning and drug discriminations. In L. L. Iversen, S. D. Iversen, & S. H. Snyder (Eds.), *Handbook of psychopharmacology,* vol. 18. New York: Plenum.

OVERTON, D. A. (1964). State-dependent or "dissociated" learning produced with pentobarbital. *Journal of Comparative and Physiological Psychology, 57,* 3–12.

DONALD A. OVERTON

MEMORY, EFFECTS OF DRUGS ON

Research investigating the effects on memory of ALCOHOL (ethanol) and drugs of abuse is disproportionally small in relation to the widespread use of these substances worldwide. The available evidence clearly indicates that ethanol and abused drugs significantly affect memory processes. Much of current knowledge of the effects of such commonly used substances on memory is based on experiments using laboratory animals. In typical experiments, the animals are trained in a learning task and given a memory retention test after a delay of one day or longer. In experiments on commonly used learning tasks, the animals are trained to acquire responses that provide escape from, or avoidance of, aversive (negative) stimulation. Appetitive motivation (food or water reward) is also used to train animals in mazes and other types of spatial learning.

When investigating acute (single treatment) influences on learning and memory, drugs can be administered before the training, shortly after the training, or before the memory test. When drugs are administered before training, it is difficult to distinguish effects on memory from other influences on sensory, motivational, and motor processes. When administered within a few minutes after training, but not after a delay of several hours, drugs of many classes can enhance or impair memory. Such findings are interpreted as indicating that the drugs can modulate memory-consolidation processes occurring after a training session. The drug effects are typically dose-dependent. For example, drugs that enhance memory when administered in low doses may impair memory when administered in higher doses. Experiments examining the effects of a drug administered prior to memory testing are difficult to interpret, since drugs can affect many processes affecting behavior other than memory. For the same reasons, the alterations in memory performance that are produced by the chronic (long-term) administration of drugs are also difficult to interpret.

ALCOHOL (ETHANOL)

In rats and mice, an acute (a large) dose of alcohol prior to learning usually impairs memory of the training. The effect is heightened by the drug clonazepam, a BENZODIAZEPINE RECEPTOR AGONIST; it is lessened by bicuculline and picrotoxin, drugs that block receptors for the inhibitory NEUROTRANSMITTER GABA (GABA-A receptors). Such findings suggest that ethanol-induced amnesia is mediated by the benzodiazepine/GABA-A receptor complex. These findings are consistent with extensive evidence that benzodiazepines (see section below) induce amnesia in humans as well as in laboratory animals. Memory impairment induced by a large dose of alcohol is also lessened by physostigmine, the acetylcholinesterase inhibitor, suggesting that ethanol influences on memory involve cholinergic mechanisms.

Chronic administration of a high dose of ethanol to rats or mice over time induces memory impairment, accompanied by a decreased function of cholinergic systems in specific brain regions, including the hippocampus and neocortex. The syndrome can be reversed by an implant, into either BRAIN STRUCTURE, of fetal brain tissue that has high numbers of cholinergic cells or by giving oxotremorine, the cholinergic muscarinic agonist, prior to memory testing. Such findings suggest that the memory impairment resulting from chronic ethanol ingestion is associated with a deficit of brain cholinergic function.

Acute or chronic ethanol ingestion produces memory problems in humans. Large amounts of ethanol taken over a short period (hours or days) may cause a severe amnesia—a "blackout" for events occurring during and/or shortly before the period of intoxication. Some alcoholic blackouts may be caused partially by state-dependency—that is, during a later intoxication, individuals may sometimes remember experiences that occurred during a previous blackout. This phenomenon was illustrated in Charles Chaplin's 1931 film *City Lights*, in which the hard-drinking millionaire remembered Charlie only when under the influence of alcohol.

Paradoxically, experiments with human subjects indicate that low doses of ethanol administered immediately after learning enhance retention. Similar results have been obtained in studies using laboratory animals; however, it is not clear that effects seen in animals are due primarily to ethanol effects on brain processes underlying memory. They may reflect, at least in part, the aversive aftereffects of ethanol.

Clinical research shows that chronic ingestion of alcohol can produce three general categories of brain impairment that are associated with memory deficits: the Wernicke-Korsakoff syndrome, alcoholic dementia, and "nonamnesiac" or "non-Korsakoff" disorders. Wernicke-Korsakoff syndrome, the best known, is due to Vitamin B_1 (thiamine) deficiency, resulting from poor food intake during sustained periods of alcohol consumption. It involves an acute phase, with mental confusion and difficulty with eye movements and walking. Most people who recover from this acute phase after treatment with thiamine will have Korsakoff's syndrome, in which impairment of the ability to learn and remember new information (anterograde amnesia) as well as retention of recently acquired information (retrograde amnesia) occur, although apparently normal intellectual function and the ability to acquire and retain skill-based information, such as purely visual/motor tasks, appear to be relatively unaffected. Some improvement in the memory deficits may occur with prolonged abstinence from alcohol.

Alcoholic dementia differs from Korsakoff's syndrome in that it is characterized by severe memory impairment as well as major intellectual deterioration that can be difficult to distinguish from Alzheimer's Disease by clinical examination. Improvements are, however, often seen if patients abstain from alcohol.

It is not known whether the deficits seen in early alcoholic dementia and in Korsakoff's syndrome are accompanied by alterations in GABAergic or cholinergic functioning. The changes seen in late alcoholic dementia, like those of Alzheimer's Disease, involve multiple focal brain lesions, primarily in the temporal lobe but also in other brain regions, and involve deficits in glutaminergic, GABAergic, and cholinergic systems.

The third type of memory problem linked to alcohol ingestion has been variously referred to as "neurologically intact" or "neurologically asymptomatic" and is characterized by subtle impairments in dealing with abstractions, problem solving, and memory. Significant recovery with abstinence is typical.

BENZODIAZEPINES

BENZODIAZEPINES, which are used clinically in the treatment of ANXIETY and the induction of sleep, are among the most widely used (and abused) drugs. It has been known for several decades that benzodiazepines, including diazepam (Valium), triazolam (Halcion), and CHLORDIAZEPOXIDE (Librium) induce anterograde amnesia in humans. Studies using laboratory animals indicate that benzodiazepines impair memory when administered before training, but they generally do not impair memory when administered posttraining. The lack of posttraining effects may be due, at least in part, to the fact that benzodiazepines are absorbed slowly and are slow to reach peak concentrations in the brain following peripheral injections. The anterograde amnesia induced by benzodiazepines is not due either to alterations in sensory or motivational processes affecting learning or to state-dependency.

Benzodiazepines are known to act by modulating GABA-A neurotransmitter receptors on the benzodiazepine/GABA receptor complex. Their effects on memory appear to be mediated primarily by the brain structures designated as the amygdaloid complex and hippocampus. When administered acutely, either systemically or directly into specific brain regions, including the amygdaloid complex and the hippocampus immediately posttraining, retention is enhanced by flumazenil, the benzodiazepine-receptor antagonist, and by the GABA-A-receptor antagonists bicuculline and picrotoxin. Findings indicating that the amnesia induced by peripherally administered benzodiazepines is blocked by GABAergic antagonists administered directly into the amygdaloid complex, as well as by lesions of the amygdaloid complex, provide additional evidence that this brain region is involved in benzodiazepine effects on memory. Although benzodiazepine-like substances are found in the brain, it is not yet known whether they are synthesized in brain cells or derived from food. Evidence that training experiences release these naturally occurring brain substances from synaptic vesicles in neurons suggests that they may play a role in modulating memory-storage processes.

MARIJUANA

In laboratory animals, both acute and chronic administration of marijuana extracts or of their active principles, the TETRAHYDROCANNABINOLS (THC), have been reported to impair the acquisition and retention of a very wide variety of tasks. It is not known whether these effects are due to influences on memory or simply to the sedative influences of the drug. There is some evidence suggesting that acute or chronic use of MARIJUANA impairs human memory. It is not known, however, whether such effects are due specifically to influences on brain processes underlying memory or to other influences on behavior. Cessation of marijuana use typically results in rapid recovery from the drug effects. Little is known about brain influences mediating marijuana effects on learning and memory.

OPIATES AND OPIOID PEPTIDES

The OPIATE drugs MORPHINE and HEROIN, administered posttraining, impair retention in laboratory animals. The memory impairment is not state-dependent: Administration of opiates prior to retention testing does not decrease the impairment. Opiate-receptor ANTAGONISTS, including NALOXONE and NALTREXONE, enhance memory and block the memory impairment produced by opiates. Endogenous opioid peptides (brain peptides that mimic the effect of morphine, heroin, and other opiates) also affect memory. The opioid beta-endorphin is released in the brain when animals are exposed to novel training situations. Memory impairment is induced by posttraining injections of beta-ENDORPHIN as well as by injections into several brain regions, including the amygdaloid complex and medial septum. Opiate antagonists administered into these brain regions enhance memory. Unlike the effects of opiate drugs, the memory impairment induced by beta-endorphin may be due, at least in part, to the induction of state-dependency: Under some conditions beta-endorphin administered (or endogenously released) prior to memory testing may lessen the memory impairment induced by a posttraining injection of the peptide.

Despite the widespread and long-standing use of opiate drugs by humans, there have been no systematic studies on the effect of morphine, heroin, or other opiates on human memory. Chronic opiate users do show memory deficits, but these may result from general deterioration rather than from any specific effect of the opiates. Acute administration of opiates (as in preanesthetic medication, for example) may induce a temporary amnesia. The failure of patients to remember experiences immediately prior to surgery may be due, at least in part, to an amnestic effect of the opiates used for ANALGESIA (PAIN suppression). The effect of opiate antagonists has been explored clinically in the treatment of dementias, but with limited success.

AMPHETAMINE

In laboratory animals, chronic administration of AMPHETAMINE prior to training impairs performance in many types of learning tasks. Such effects are typically obtained in experiments using high doses of amphetamine and complex learning tasks. In contrast, extensive evidence, from studies using a variety of types of training tasks, indicates that acute posttraining injections of amphetamine produce dose-dependent enhancement of memory. Retention is also enhanced by direct administration of amphetamine into several brain regions, including the amygdaloid complex, hippocampus, and caudate nucleus. Amphetamine is known to act by releasing the catecholamines epinephrine, norepinephrine, and dopamine from cells and block their reuptake. Amphetamine effects on memory appear to result primarily from influences on brain dopaminergic systems as well as influences on the release of peripheral catecholamines.

Amphetamine users often report that their "learning capacity" is enhanced by single doses of the substance. Since there are few systematic and well-controlled studies of the effects of amphetamine on memory in humans, however, it is not known whether such reports reflect subjective changes in perception and mood or effects on memory. Chronic amphetamine use is usually accompanied by a deterioration of memory function, an effect that subsides with cessation of use.

COCAINE

Despite the extensive use and abuse of COCAINE, little is known about cocaine effects on memory. Results of studies using rats and mice indicate that acute posttraining administration induces dose-dependent effects comparable to those of amphetamine: Memory is enhanced by low doses and impaired by higher doses. The brain processes mediating cocaine influences on memory have not been extensively investigated. The effects appear to be mediated by influences on adrenergic and dopaminergic systems. Also, as with amphetamine, users of cocaine report that memory is enhanced by acute doses and impaired by chronic use. Systematic, well-controlled studies of the effect of cocaine on human memory are lacking. The effects on memory and intellectual functioning of other drugs—such as PHENCYCLIDINE (PCP), BARBITURATES, NICOTINE, and INHALANTS—are considered in connection with these agents and in separate articles.

(SEE ALSO: *Memory and Drugs; Research: Learning, Conditioning, and Drug Effects; Wikler's Pharmacologic Theory of Drug Addiction*)

BIBLIOGRAPHY

HARDY, J., & ALLSOP, D. (1991). Amyloid deposition as the central event in the etiology of Alzheimer's Disease. *Trends in Pharmacological Sciences, 12,* 383–388.

IZQUIERDO, I., & MEDINA, J. H. (EDS.). (1993). *Naturally occurring benzodiazepines: Structure, distribution and function.* London: Ellis and Horwood.

McGAUGH, J. L. (1989). Dissociating learning and performance: Drug and hormone enhancement of memory storage. *Brain Research Bulletin, 23,* 339–345.

McGAUGH, J. L. (1989). Involvement of hormonal and neuromodulatory systems in the regulation of memory storage. *Annual Review of Neuroscience, 12,* 255–287.

McGAUGH, J. L., INTROINI-COLLISON, I. B., & CASTELLANO, C. (1993). Involvement of opioid peptides in learning and memory. In A. Herz, H. Akil, & E. J. Simon (Eds.), *Handbook of experimental pharmacology: Opioids, part II.* Heidelberg: Springer-Verlag.

U.S. DEPARTMENT OF HEALTH AND HUMAN SERVICES, NATIONAL INSTITUTE ON ALCOHOL ABUSE AND ALCOHOLISM. (1993). *Eighth special report to the U.S. Congress on alcohol and health.* Washington, DC: U.S. Government Printing Office.

WEINGARTNER, H., & PARKER, E. S. (EDS.). (1984). *Memory consolidation.* Hillsdale, NJ: Lawrence Erlbaum.

IVAN IZQUIERDO
JAMES L. McGAUGH

MEPERIDINE   Meperidine is a totally synthetic OPIOID analgesic (painkiller) with a structure quite distinct from MORPHINE, a natural OPIATE. Unlike morphine's rigid fused ring structures, the structure of meperidine is flexible; it is a phenylpiperidine and bends so that the key portions of the

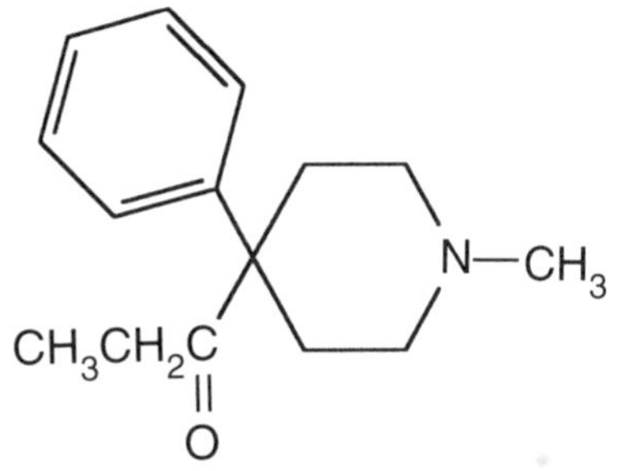

Figure 1
Meperidine

molecule can assume positions similar to those of morphine. A number of other compounds with similar structures are widely used in medicine, including loperamide (used primarily for treating diarrhea) and the extraordinarily potent ANALGESIC agents fentanyl, sufentanil, lofentanil, and alfentanil (for treating PAIN).

Meperidine is a compound with strong analgesic effects similar to morphine's, although greater amounts are needed to produce the same level of analgesia. It is one of the more commonly prescribed opioid analgesics and is better known under one of its brand names, Demerol. Given by injection, 100 milligrams of meperidine equals 10 milligrams of morphine. Meperidine can be administered orally as well as by injection but its potency it not as great following oral administration, so the dose must be increased proportionally. Like morphine, continued use of meperidine is associated with decreased analgesia—TOLERANCE—as well as PHYSICAL DEPENDENCE. As with the other opioids, ADDICTION (defined as a drug-seeking behavior) is not commonly observed with this drug when used for medicinal purposes, but meperidine is highly valued on the street and is widely abused, particularly in its injectable forms.

Medically, meperidine is a significant problem in patients with kidney conditions, where drug-removal from the body is impaired. Metabolized to normeperidine, a closely related compound, it is eliminated by the kidneys. In patients with kidney problems, this metabolite can accumulate to high levels, which can cloud mental processes and even produce convulsions. Since ELDERLY patients often have impaired kidney function, special care must be taken when using meperidine with them.

(SEE ALSO: *Addiction: Concepts and Definitions; Opioid Complications and Withdrawal*)

BIBLIOGRAPHY

JAFFE, J. H., & MARTIN, W. R. (1990). Opioid analgesics and antagonists. In A. G. Gilman et al. (Eds.), *Goodman and Gilman's the pharmacological basis of therapeutics*, 8th ed. New York: Pergamon.

GAVRIL W. PASTERNAK

MEPROBAMATE   This is a SEDATIVE-HYPNOTIC drug that is now typically used to treat muscle spasms. Meprobamate is prescribed and sold as Deprol, Equagesic, Equanil, Meprospan, and Miltown. Because of its abuse potential, it is included in Schedule IV of the CONTROLLED SUBSTANCES ACT. It was first introduced into clinical medicine in 1955 for the treatment of ANXIETY. At the time it was thought to have specific antianxiety effects and to be quite different from other sedative-hypnotics. Also introduced at about the same time were chlorpromazine (Thorazine), which had remarkable ANTIPSYCHOTIC effects, and reserpine, which had tranquilizing as well as blood pressure—lowering effects. These three agents were considered the harbingers of the new era of PSYCHOPHARMACOLOGY and helped popularize the new term *tranquilizer.*

Within a year or two after its introduction, meprobamate had become one of the most widely prescribed drugs in the United States. It was not long however, before its distinction from other sedative-hypnotic agents was reassessed, and within a decade it was recognized that meprobamate shared many of the properties of other central nervous system depressants, such as the BARBITURATES. By the early 1960s, its use for the treatment of anxiety was eclipsed by the BENZODIAZEPINES. Although it is prescribed as a muscle relaxant, the only use currently approved in the United States by the Food and Drug Administration is as a sedative-hypnotic.

Meprobamate has a number of side effects, including tremors, nausea, depression, and various allergic reactions. Continued use of high doses can result in TOLERANCE AND PHYSICAL DEPENDENCE. Convulsions and other signs of withdrawal are reported upon termination of high-dose treatment or inappropriate use.

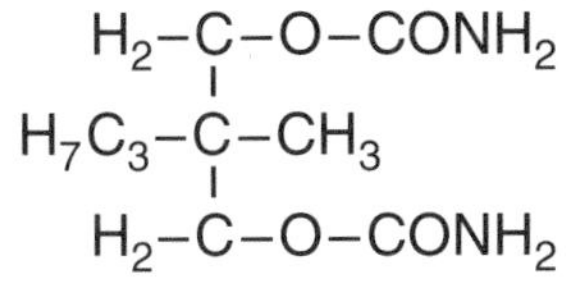

$$H_2-C-O-CONH_2$$
$$H_7C_3-C-CH_3$$
$$H_2-C-O-CONH_2$$

Figure 1
Meprobamate

BIBLIOGRAPHY

RALL, T. W. (1990). Hypnotics and sedatives; Ethanol. In A. G. Gilman et al. (Eds.), *Goodman and Gilman's the pharmacological basis of therapeutics*, 8th ed. New York: Pergamon.

SCOTT E. LUKAS

MESCALINE This is a naturally occurring HALLUCINOGEN, one of the oldest PSYCHEDELIC substances known. It was first obtained from the PEYOTE cactus (*Lophophra williamsii* or *Anhalonium lewinii*), which grows in the southwestern United States and northern Mexico. Peyote buttons, the dried tops of the peyote cactus, were originally used by pre-Columbian Native Americans in those regions as an antispasmodic as well as for highly structured religious rituals; the button was eaten or was steeped to make a drink. It continues to be used in ritual by the Native American Church.

Mescaline is a member of the phenethylamine-type family of hallucinogens, which includes DOM, MDA, and MDMA. The overall behavior effects of mescaline are very similar to those produced by LYSERGIC ACID DIETHYLAMIDE (LSD); however, it is approximately 100 to 1,000 times less potent than LSD, although the effects of mescaline last from 10 to 12 hours.

(SEE ALSO: *Psilocybin; Religion and Drug Use*)

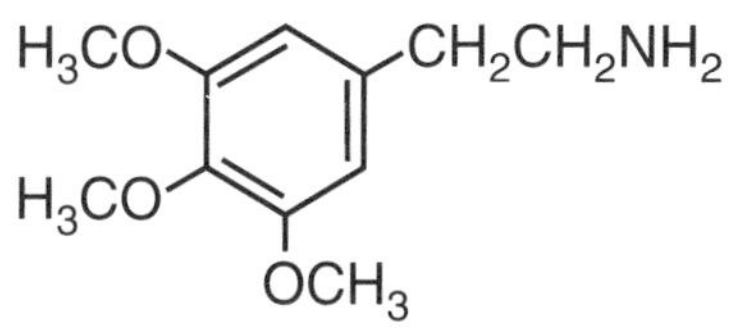

$$H_3CO-\bigcirc-CH_2CH_2NH_2$$
$$H_3CO-\quad OCH_3$$

Figure 1
Mescaline

BIBLIOGRAPHY

EFRON, D. H., HOLMSTEDT, B., & KLINE, N. S. (EDS.). (1979). *Ethnopharmacologic search for psychoactive drugs.* New York: Raven Press.

DANIEL X. FREEDMAN
R. N. PECHNICK

METHADONE Methadone (Dolophine) is a synthesized molecule with pharmacological actions very similar to those of the OPIOID drug, MORPHINE. Methadone serves an important place in the history of opioid ANALGESICS, since it is one of the first synthesized agents (1939). The ability to synthesize opioid analgesics from simple chemicals diminishes our reliance on natural products (such as morphine, CODEINE, and thebaine) to provide the base for many of the currently used opioid analgesics. Structurally, the drug does not look like morphine. Unlike the rigid fused ring structures of morphine, the structure of methadone is extremely flexible. It bends so that the key portions of the molecule can assume positions similar to those of morphine. The structure of methadone is very similar to that of propoxyphene (Darvon), a weaker opiate widely used to treat mild to moderate pain. It has two stereoisomers, but the (−)isomer is far more active than the (+)isomer.

Methadone can be administered orally, intramuscularly, or intravenously. It is well absorbed from the gastrointestinal tract, making it very useful orally. Its oral/parenteral ratio of potency is approximately two. Methadone is threefold more potent than morphine orally, but about equipotent when given by injection. It is metabolized by the liver to a variety of inactive compounds, which then are eliminated by the kidneys.

Pharmacologically, methadone is used in the form of its hydrochloride salt. It has actions quite similar to morphine and works predominantly through *mu* opiate RECEPTORS. As an analgesic, methadone is similar in actions and in potency to morphine. It produces analgesia, as well as many of the side effects associated with morphine use, including respiratory depression and constipation. A major difference between methadone and morphine is methadone's long duration of action. Typically, the drug is given to patients every six to eight hours. This long duration of action can be very advantageous, particularly in patients who require the drug

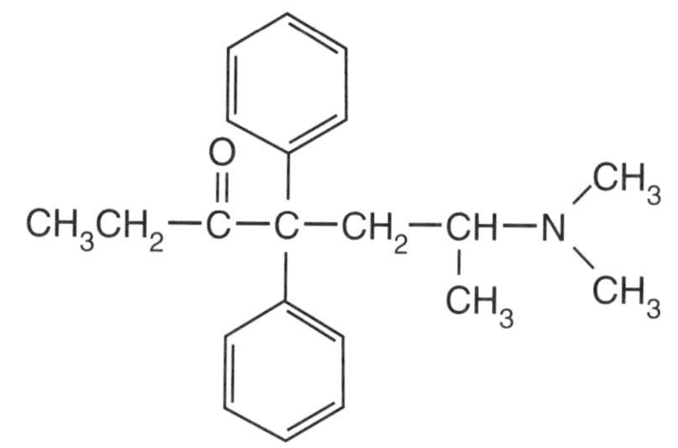

Figure 1
Methadone

for long periods of time, such as cancer patients. However, there are some disadvantages. With a HALF-LIFE ranging from twenty to thirty hours, it may take many days of continued dosing to reach constant (or steady-state) levels of the drug in the body. Thus, the full effect of a change in drug dose may not be seen for three or four days. This may make it difficult to adjust the dose for an individual patient. Increasing the dose too rapidly may even lead to delayed increases in its concentration in the body, far beyond those anticipated and, in some situations, may actually lead to an overdose. Continued administration of methadone will produce TOLERANCE AND PHYSICAL DEPENDENCE. The actions of methadone, like those of morphine, are readily reversed by ANTAGONISTS such as NALOXONE or NALTREXONE; however, these antagonists will also produce an immediate WITHDRAWAL syndrome in physically dependent people.

Despite its clear utility in the control of PAIN, the major use of methadone in the United States is in the treatment of HEROIN addicts. Although methadone must be administered approximately every six to eight hours to maintain analgesia, its slow rate of elimination prevents the appearance of withdrawal symptoms for over twenty-four hours. This slow appearance of withdrawal signs has made this agent very useful in maintenance programs, since it permits once-a-day dosing. With chronic administration of high doses of methadone, addicts become very tolerant, markedly limiting the euphoria an addict might obtain from illicit use of other opiates such as heroin. Thus, methadone minimizes occasional opiate use, is readily tolerated by the addicts, and can be administered once a day, which makes it easily

dispensed. Methadone has been used clinically in maintenance programs and is one of the most effective treatment modalities available for opiate addicts.

(SEE ALSO: *Addiction: Concepts and Definitions; Methadone Maintenance Programs; Pain, Drugs Used in Treatment of; Treatment Types: Pharmacotherapy*)

BIBLIOGRAPHY

JAFFE, J. H. (1990). Drug addiction and drug abuse. In A. G. Gilman et al. (Eds.), *Goodman and Gilman's the pharmacological basis of therapeutics*, 8th ed. New York: Pergamon.

JAFFE, J. H., & MARTIN, W. R. (1990). Opioid analgesics and antagonists. In A. G. Gilman et al., (Eds.), *Goodman and Gilman's the pharmacological basis of therapeutics*, 8th ed. New York: Pergamon.

GAVRIL W. PASTERNAK

METHADONE MAINTENANCE PROGRAMS The history of methadone treatment offers a striking example of the benefits and limits of research findings on public attitudes and policies. To understand methadone maintenance treatment, it is necessary to appreciate the profound stigma attached both to the patients and to the treatment providers. This establishes the context for understanding how a modality with the most extensive research base of anything in the addiction-treatment field can nonetheless engender passionate dispute.

As a TREATMENT modality, methadone maintenance was developed in the mid 1960s by Drs. Vincent Dole and Marie Nyswander, in response to prevailing concerns about HEROIN dependence and related health problems, mortality (especially among young people ages 15 to 35), and the high relapse rate of those who quit use (Lowinson et al., 1992). Methadone itself had been synthesized in Germany during World War II, to be used as an ANALGESIC (painkiller); it was then studied at the U.S. PUBLIC HEALTH SERVICE HOSPITAL in Lexington, Kentucky, after the war. It was approved by the U.S. Food & Drug Administration in August 1947 for medical use in the treatment of PAIN.

Methadone's initial use in the treatment of addiction was to withdraw addicts from heroin. It was subsequently determined to be well suited to long-term maintenance treatment. As a treatment tool, methadone provides a safe and effective way to eliminate heroin CRAVING, WITHDRAWAL, and drug-seeking behavior; it frees most patients to lead productive lives. In conjunction with educational, medical, and counseling services, methadone has been thoroughly documented to enable patients to discontinue or reduce illict drug use and associated criminal activity, improve physical and mental well being, become responsible family members, further their education, obtain and maintain stable employment, and resume or establish a productive lifestyle. Despite three decades of research confirming its value, methadone maintenance treatment remains a source of contention among treatment providers, the public in general, and officials and policymakers in particular. Unlike controversies based on a difference of opinion between informed parties, debate about methadone usually involves several common misunderstandings about the drug and its uses.

COMMON MISUNDERSTANDINGS

Much of the uneasiness about methadone stems from the idea that it is "just substituting one addicting drug for another." Indeed, this is technically correct; methadone treatment is drug-replacement therapy, in which a long-acting preparation that is administered orally is substituted for a short-acting OPIOID that is used intravenously. The long-acting (24 to 36 hours) effect of preventing withdrawal allows most patients to receive a daily dose and function in a stable manner—without the four-hour cycles of euphoria and withdrawal that characterize heroin use. The objection that methadone is "addicting" reflects the recognition that this medication produces DEPENDENCE. Addiction-treatment professionals increasingly distinguish between PHYSICAL DEPENDENCE and ADDICTION; *addiction* is characterized by behavior that is compulsive, out of control, and persists despite adverse consequences. Patients with chronic pain will develop *physical dependence* although their overall functioning improves. If physical dependence is a factor to be considered, then addiction specialists increasingly assess the extent to which the person's functioning and quality of life is improved or impaired when determining whether physical dependence is an acceptable consequence of medication use.

There is also uneasiness about the erroneous belief that "methadone keeps you high." This notion reflects general misunderstanding about the effects of a properly adjusted methadone dose. Once stabilized, most patients experience few subjective effects; heroin addicts will readily state that they seek methadone to avoid "getting sick" (prevent withdrawal effects), not to "get high." While the patient's dose is being stabilized, he or she may experience some subjective effects, but the wide therapeutic window allows for dose adjustment between the extremes of craving and somnolence. Dose adjustment may take some weeks and it may be disrupted by a variety of medical and lifestyle factors, but once achieved, the patient should function normally. Ample scientific evidence shows that the long-term administration of methadone results in no physical or psychological impairment of any kind that can be perceived by the patient, observed by a physician, or detected by a scientist (Kreek, 1973, 1992; Novick, 1993; Lowinson et al., 1992). More specifically, there is no impairment of balance, coordination, mental abilities, eye–hand coordination, depth perception, or psychomotor functioning. Recently, advocacy efforts have been successful on behalf of patients identified through workplace drug testing and threatened with negative consequences. It is anticipated that the Americans with Disabilities Act will further protect patients against such forms of discrimination.

A third point of resistance, objection to long-term or even lifelong maintenance, is better addressed following the presentation of some basics about opioid addiction and the nature of treatment.

HOW DOES METHADONE TREATMENT WORK?

Most addiction specialists agree that addictive disorders are complex phenomena involving the interaction of biological, psychosocial, and cultural variables, all of which need to be considered to make treatment effective. Vincent Dole and Marie Nyswander, who pioneered the use of methadone, held the view that there was something unique about opioid addiction that made it difficult for patients to remain drug free (Lowinson et al., 1992). Although originally intended as a long term treatment for a metabolic defect, many initially hoped that methadone could be used to help heroin addicts make the transition to a drug-free lifestyle and then be discontinued. Research since then indicates that less than 20 percent of those who require methadone maintenance will be able to discontinue methadone and remain drug free. Dole (1988) postulated that a receptor-system dysfunction resulting from chronic opioid use leads to permanent alterations that we do not yet know how to reverse. New brain-IMAGING technology holds the promise of better understanding and, eventually, improved intervention. In the interim it appears that methadone is corrective if not curative for the person severely addicted to opioid drugs. Two important questions for future research are (1) whether a preexisting condition enhances the VULNERABILITY of some patients more than others, and (2) whether long-term opioid addicts can ever recover normal functioning without maintenance therapy.

Studies indicate that methadone is a benign drug that exhibits stability of RECEPTOR occupation, thus permitting interacting systems to function normally; an example of this is the normalization of hormone cycles in heroin-dependent women and the return of regular menstrual cycles (Kreek, 1992; Lowinson et al., 1992). This distinguishes methadone from heroin, a short-acting NARCOTIC, which produces such rapid physiological changes that a stable state of adaptation is impossible. While TOLERANCE develops to most drug effects, even long-term use (20 years or more) of methadone does not produce tolerance to the effects of either reduced craving or to prevention of narcotic withdrawal.

The desired response to methadone depends largely on maintaining relatively stable blood levels at all times. Appropriate doses usually keep the patient in the therapeutic range and produce the stable state so important for rehabilitation. What is referred to as a "rush," or " high," is the result of rapidly rising blood levels; thus once therapeutic levels are achieved and maintained, the patient experiences little of such subjective effect (Kreek, 1972, 1993; Lowinson et al., 1992).

Unfortunately, negative attitudes toward methadone have historically played a significant role in dosing practices, so dose ceilings have been imposed by state or local authorities without regard to medical criteria. Such policies place value on giving as little of the drug as possible—instead of the therapeutic level needed to accomplish the goal—influenced in part by the unsubstantiated belief that lower doses would make it easier to discontinue methadone. It was common in some programs to have dose ceilings of only 40 milligrams (mg) per day. It is now well established that for most patients this dose is inadequate to maintain the necessary plasma concentrations to be effective; the effective

dose range for most patients is between 60 and 120 mg per day, with higher doses strikingly well correlated with reductions in illicit drug use (GAO 1990; Capelhorn and Bell, 1991).

For many years and in many programs, patients on low doses who complained that "my dose isn't holding me" were often dismissed with the assertion that they were "merely engaging in drug-seeking behavior," and when the distressed patient then supplemented the methadone dose with heroin, it was concluded either that the patient was poorly motivated or that the treatment was ineffective. Studies by D'Aunno and Vaughn (1992) show that more than 50 percent of patients in the United States receive doses that are inadequate to prevent continued illicit narcotic use—indicating poor physician training, the attitude of program leaders, or inappropriate invasion by regulatory agencies and legislative policies; in many cases all of these conditions exist.

Initial hopes to use methadone as a drug to help patients progress to a medication-free lifestyle have proven unrealistic for most patients. Studies indicate that although short-term abstinence is common, eventual relapse is the norm for more than 80 percent (McLellan et al., 1983; Ball & Ross, 1991). Clinicians who have worked with this population over the long term believe that although lifestyle changes are essential to successfully discontinuing methadone, such changes in conjunction with high motivation will still be insufficient for most; neurobiological factors remain the deciding factor. Because the current treatment system, overburdened by regulations and inappropriate expectations, is dehumanizing for many, programs usually make efforts to assist the patient wishing to taper off. Many programs attempt to set a tone in which the patient is encouraged to succeed but also to resume methadone treatment promptly rather than return to a heroin-using lifestyle.

METHADONE AND OTHER DRUG USE

Methadone patients may drink COFFEE or ALCOHOL, use COCAINE, smoke TOBACCO and/or MARIJUANA, and use other drugs prevalent in their communities. Methadone is opioid specific; it does not in itself increase or prevent other kinds of drug use. Methadone programs, however, offer the enormous advantage of making the patient accessible to other kinds of intervention. Rules governing take-home medication are intended to reduce diversion of methadone onto the illicit market. At minimum, they mandate that the patient come to the clinic at least once weekly and, in most cases, more frequently. Thus the patient can be exposed to educational presentations, materials, and counseling interventions—as indicated by an individualized treatment plan, a required part of the treatment effort.

Cocaine use has received particular attention, since it has been identified as increasing the dropout rate, slowing progress, and undermining the gains of previously stable patients. Research and training efforts have turned to this problem. Alcohol use remains a problem, particularly since many patients define their difficulty only in terms of *illicit* drug use and are resistant to the notion of giving up drinking. With the blending of the "cultures" of alcohol and drug-treatment providers, counselors are increasingly sophisticated about addressing problem drinking. It is, however, uncommon for methadone programs to define goals for everyone in terms of abstinence from all intoxicants as other parts of the treatment system for drug and alcohol abusers typically do. Nonetheless, some methadone programs have the advantage of being able to dispense Antabuse (DISULFIRAM) with methadone when appropriate. Disulfiram in the patient's system causes a bad reaction to alcohol and is therefore a deterrent to drinking.

TREATING OPIOID-ADDICTED PREGNANT WOMEN

Methadone maintenance has been viewed as an effective treatment for opioid addiction in the pregnant woman since the early 1970s. In addition to the benefits of psychosocial interventions provided by the program, methadone maintenance treatment prevents erratic maternal opioid drug levels, thus protecting the FETUS from repeated episodes of withdrawal. Most programs either provide prenatal care on-site or monitor to see that prenatal care is obtained elsewhere, thus reducing the incidence of obstetrical and fetal complications, in utero growth retardation, and neonatal morbidity and mortality (Finnegan, 1991). Exposure to HIV infection through ongoing needle use is also reduced. Programs typically provide interventions around nutrition, parenting skills, exercise, and other related topics.

Methadone maintained women produce offspring more similar to drug-free controls, in contrast to the poorer health status of offspring born to women using street drugs. It is clear that the most damaging consequences of opioid use during PREGNANCY are caused by repeated episodes of intoxication and withdrawal (Jarvis & Schnoll, 1994). Although women can be stabilized on methadone, body changes specific to pregnancy frequently cause women to develop increasing signs and symptoms of withdrawal as pregnancy progresses. They may need dose increases to maintain their therapeutic plasma levels and to remain comfortable (Kreek, 1992). Splitting the dose so that it can be ingested twice daily often produces better results, both reducing fetal stress and increasing the comfort of the pregnant woman— but local regulatory obstacles often deny a woman access to half her daily dose out of the clinic.

Inconsistent evidence exists for the commonly held belief that the severity of the neonatal abstinence syndrome is proportional to the methadone dose. Many programs do urge the woman to reduce her dose, however, so the "baby won't be born addicted." In fact, the management of neonatal (newborn) abstinence syndrome is relatively straightforward; discomfort usually can be eliminated within hours and withdrawal can be accomplished within 14 to 28 days. No lasting impairment from these experiences has been reported.

ADDRESSING PSYCHOSOCIAL ISSUES

Many existing methadone programs are far short of the resources needed to do an effective job, but extensive research over a long period of time has clarified many of the treatment tasks. The stigma against heroin addicts in general and against methadone patients in particular has created a treatment climate in which both patients and treatment providers may be demoralized about the value of the treatment endeavor. Providers may not be able to obtain access to resources for patients on methadone, who often are isolated from society's mainstream. For example, methadone patients are often excluded from housing or from inpatient residential treatment. Even with such drawbacks, documentation is growing that minimal intervention using methadone does reduce illicit opioid use and, hence, needle sharing; enchanced treatment with regular counseling and other services accomplishes much more (McLellan et al., 1993).

Historically, drug counseling has been provided in methadone clinics by counselors who often have no formal credentials but who are trained on site. This counseling focuses on the patient's management of life problems—specific problems in the area of drug use, physical health, interpersonal relationships, family interactions, and vocational and educational goals. The counselor also performs the role of the case manager and is a liaison between physicians and medical institutions, courts and social services. Properly trained counselors may help the patient to develop COPING strategies for current problems, perform initial screening for medication and other program services, and attend to issues concerning program rules, privileges, and policies. Regulations governing methadone treatment are complex, detailed, and restrictive—maintaining a therapeutic alliance with patients while meeting such obligations is a daunting task for clinical staff.

Studies of the drug-dependent population indicate that more than 50 percent have a comorbid psychiatric disorder (Regier et al., 1990); and among the opioid-dependent population, depression is particularly common. Treatment outcome is therefore improved by adding supplemental psychotherapy by professionally trained staff (Woody et al., 1983) for patients who meet criteria for high psychiatric severity. It is important that such staff be well integrated into the treatment team. Medication for such psychiatric disorders may also be given concurrently with methadone, and the use of ANTIDEPRESSANTS is increasingly common. The interaction effects of the drugs are manageable with consistent monitoring and good staff teamwork. Psychotic conditions are somewhat less frequent, but clinics are likely to see some highly disturbed individuals as part of their population and should be able to recognize and manage such patients. Methadone patients benefit from the structure of frequent clinic attendance combined with the lowest possible psychological intrusiveness.

It is most desirable for vocational interventions to be integrated into a methadone program, although economic conditions in many urban areas necessitate the development of alternatives to effect employment opportunities. Parenting classes, which include information, skill training, and the opportunity to explore issues, are often well received by parents who feel the absence of good role models and skills.

Although negative attitudes toward methadone have been a barrier to the utilization of TWELVE-STEP PROGRAMS, methadone patients have increasingly become involved in them (e.g., NARCOTICS ANONYMOUS [NA]). Some conceal their participation in methadone treatment, others find meetings that are hospitable anyway. Some methadone programs are developing specialized meetings on site, which in turn encourages patients to utilize meetings in the community. Although medical treatment was not incompatible with twelve-step program participation in the minds of the founders of ALCOHOLICS ANONYMOUS (AA) (Zweben, 1991), meeting participants may nonetheless continue to be condemning toward methadone patients who seek to become involved. Narcotics Anonymous and other drug fellowships were designed to deal with this situation and with the various forms of POLYDRUG ABUSE that are beyond the scope of ALCOHOLISM.

HIV/AIDS

A reexamination of the value of methadone treatment has been greatly stimulated by the documentation of its role in reducing the spread of the HUMAN IMMUNODEFICIENCY VIRUS (HIV). Seroprevalence is much lower among those who have been on long-term maintenance, particularly those who entered treatment prior to the 1980s onset of the rapid spread of HIV in the local population (Hartel et al., 1988; Batki, 1988; Kreek, 1992; Lowinson et al., 1992). Methadone clinics provide accessibility to large numbers of intravenous (IV) drug users, making clinics an excellent site for prevention education, screening, testing, and counseling. Because methadone patients have a continuing forum to discuss their life issues, counselors may be able to facilitate behavior change around the disease issues relating to safe sex practices and other high-risk behaviors. Further gains accrue as the patient progresses in treatment, since an abstinent person is in a better position to exercise good judgment than is an intoxicated one. Efforts are being made to integrate HIV/AIDS-related activities as fully as possible into the programs.

THE FUTURE

Methadone maintenance has demonstrated its effectiveness in reducing illicit opioid use and facilitating the transition to a productive lifestyle. Data-based decision making should certainly result in its inclusion into the mainstream as our health-care delivery system goes through any legislated transition. Enhanced services at the beginning of treatment promote good outcomes, but long-term methadone maintenance is needed for most people to retain their gains. Specialists hope that regulatory restrictions will be reduced and that a "medical maintenance" model—for patients who are doing well to be relieved of many of the regulatory obligations—will simplify the process and reduce the costs (Novick, 1993). Other maintenance pharmacotherapies, particularly LAAM and BUPRENORPHINE, have been developed and should broaden the options for effective care. Federally sponsored training efforts have improved the quality of care; they will continue to be essential in disseminating up-to-date information and providing opportunities for skill development. Slowly, patients are beginning to come forward to serve as visible examples of success and role models. Barriers to participation in residential treatment communities (TCs) are beginning to be removed. It is hoped that these gains allow methadone treatment the acceptance it deserves.

(SEE ALSO: *Addicted Babies; Comorbidity; Productivity: Effects of Drug Use on; Professional Credentialing; Substance Abuse and AIDS*)

BIBLIOGRAPHY

BALL, J., & ROSS, A. (1991). *The effectiveness of methadone maintenance: Patients, programs, services, and outcomes.* New York: Springer-Verlag.

BATKI, S. (1988). Treatment of intravenous drug users with AIDS: The role of methadone maintenance. *Journal of Psychoactive Drugs, 29,* 214–226.

CAPLEHORN, J. R. M., & BELL, J. (1991). Methadone dosage and retention of patients in methadone treatment. *The Medical Journal of Australia, 154,* 195–199.

D'AUNNO, T., & VAUGHAN, T. E. (1992). Variations in methadone treatment practices: Results from a national study. *Journal of the American Medical Association, 267*(2), 253–258.

FINNEGAN, L. (1991). Treatment issues for opioid-dependent women during the perinatal period. *Journal of Psychoactive Drugs, 23*(2), 191–201.

HARTEL, D., ET AL. (1988). Methadone maintenance treatment and reduced risk of AIDS and AIDS-specific mortality in intravenous drug users. Paper presented at the Fourth International Conference on AIDS, June 1988, Stockholm, Sweden.

JARVIS, M., & SCHNOLL, S. (1994). Methadone treatment during pregnancy. *Journal of Psychoactive Drugs, 26*(2).

KREEK, M. J. (1992). Rationale for maintenance pharmacotherapy of opiate dependence. In C. P. O'Brien & J. H. Jaffe (Eds.), *Addictive States.* New York: Raven ress.

KREEK, M. J. (1973). Medical safety and side effects of methadone in tolerant individuals. *Journal of the American Medical Association, 223,* 665–668.

LOWINSON, J. H., ET AL. (1992). Methadone maintenance. In J. H. Lowinson et al. (Eds.), *Substance abuse: A comprehensive textbook,* Baltimore: Williams & Wilkins.

MCLELLAN, A. T. (1983). Patient characteristics associated with outcome. In J. R. Cooper et al. (Eds.), *Research on the treatment of narcotic addiction: State of the art.* Rockville, MD: U.S. Department of Health and Human Services.

MCLELLAN, A. T., ET AL. (1993). The effects of psychosocial services in substance abuse treatment. *Journal of the American Medical Association, 269*(15), 1953–1996.

NOVICK, D. M., ET AL. (1993). The medical status of methadone maintained patients in treatment for 11–18 years. *Drug and Alcohol Dependence, 33,* 235–245.

REGIER, D. A., ET AL. (1990). Comorbidity of mental disorders with alcohol and other drug abuse. *Journal of the American Medical Association, 264*(19), 2511–2518.

WOODY, G. E., ET AL. (1986). Psychotherapy for substance abuse. *Psychiatric Clinics of North America, 9*(3), 547–562.

WOODY, G. E., ET AL. (1983). Psychotherapy for opiate addicts—does it help? *Archives of General Psychiatry, 40,* 639–645.

ZWEBEN, J. E. (1991). Counseling issues in methadone maintenance treatment. *Journal of Psychoactive Drugs, 23*(2), 177–190.

JOAN ELLEN ZWEBEN
THOMAS PAYTE

METHAMPHETAMINE Methamphetamine (also called METHEDRINE) is a potent PSYCHOMOTOR STIMULANT with a chemical structure similar to AMPHETAMINE. Methamphetamine's stimulant effects on the central nervous system are more pronounced than those of amphetamine, while its peripheral effects (e.g., cardiovascular and gastrointestinal) are less marked. Like amphetamine, it causes increased activity, increased talkativeness, more energy and less fatigue, decreased food intake, and a general sense of well-being. Injecting the drug intravenously results in the production of a "rush," described by some as the best part of the drug effect. Methamphetamine is more soluble than DEXTROAMPHET-AMINE, and, when available, of this group is generally the illicit user's drug of choice for intravenous injection—although dextroamphetamine dissolves sufficiently to permit intravenous use.

Japan was the first nation to experience a major epidemic of methamphetamine use. Immediately following World War II, large quantities of methamphetamine, which had been produced to keep combat troops alert, were released for sale to the Japanese public. Within a short time there was widespread use and abuse of the drug, much of it intravenously. At the peak of the epidemic, more than a million users were involved. Despite the experience of the Japanese, the belief persisted in the United States that amphetamines did not lead to serious compulsive use, and these drugs were not subject to any special regulatory controls like the ones governing the availability of the opioid drugs until 1964.

The first methamphetamine ("speed") epidemic in the United States began in the 1960s in the San Francisco area. A number of physicians there were prescribing the drug to HEROIN abusers for self-injection—to treat their heroin dependence by substituting methamphetamine. The drug achieved widespread popularity, with increasing numbers of people claiming heroin abuse and requesting prescriptions for methamphetamine. When the sale of intravenous methamphetamine to retail pharmacies was curtailed in the mid-1960s, illicitly synthesized methamphetamine began to appear. By the late 1960s a substantial number of users throughout the United States were injecting high doses of this illicit methamphetamine in cyclical use patterns—resulting in toxic syndromes that included the development of a paranoid psychosis (i.e., amphetamine psychosis).

Although illicit methamphetamine never completely disappeared from street use, its availability was considerably reduced by the 1970s. This trend began to reverse during the 1980s, with pockets of methamphetamine abuse occurring in the United States. Hawaii was the first area of the United States to ex-

perience the most recent methamphetamine outbreak, mostly in the form of smokable methamphetamine. Initial reports of smoking methamphetamine occurred in late 1986, with increases occurring about a year later, and a more sustained increase occurring in 1988 and 1989. Called "ice" or "crystal," this is the same substance as "speed," which was abused several decades earlier.

Methamphetamine, sold as "ice," is a large, usually clear crystal of high purity (greater than 90%) that is generally smoked using a glass pipe with two openings, much like a CRACK-cocaine pipe. Because it is a large crystal, it is difficult to adulterate with inert substances, a property that makes it extremely desirable to purchasers of illicit products. The smoke is odorless and, unlike crack, the residue of the drug stays in the pipe and can be resmoked. The effect is long-lasting, reported by users to be as long as twelve hours, although it is likely that this prolonged effect is due to the use of several doses.

Like COCAINE, methamphetamine abuse occurs in binges, with users taking the drug repeatedly for several hours to several days. During this time the user generally neither eats nor sleeps. Ending a methamphetamine binge is accompanied by fatigue, depression, and other "crash"-related effects. One of the most profound of the toxic effects of repeated methamphetamine use is the development of a paranoid psychosis, often indistinguishable from schizophrenia. With time off the drug, this psychosis generally resolves, although it can reappear if the user returns to methamphetamine abuse. Some Japanese psychiatrists have reported that methamphetamine psychosis may persist for many months.

(SEE ALSO: *Amphetamine Epidemics; Designer Drugs; Epidemics of Drug Abuse; Toxicity*)

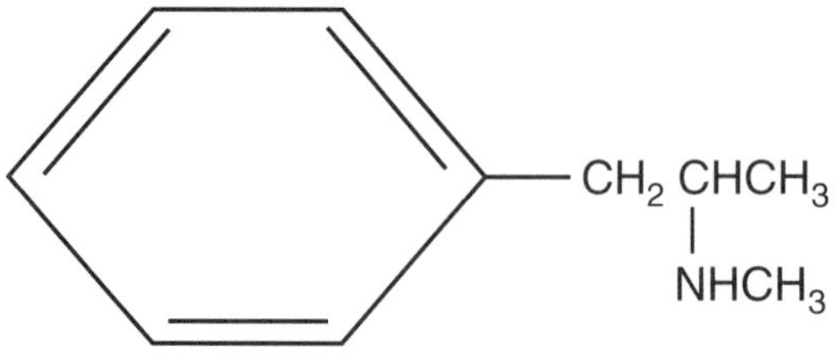

Figure 1
Methamphetamine

BIBLIOGRAPHY

KALANT, O. J. (1973). *The amphetamines: Toxicity and addiction,* 2nd ed. Springfield, IL: Charles C. Thomas.

MILLER, M. A., & KOZEL, N. J. (1991). Methamphetamine abuse: Epidemiologic issues and implications. NIDA Research Monograph Series, no. 115. Washington, DC: U.S. Government Printing Office.

SMITH, D. E., & WESSON, D. R. (EDS.). (1972). The politics of uppers and downers. *Journal of Psychedelic Drugs,* 5, 101–182.

MARIAN W. FISCHMAN

METHANOL

Methanol is the simplest alcohol, containing only one carbon atom, four hydrogen atoms, and one oxygen atom (CH_3OH). It is also called methyl alcohol, WOOD ALCOHOL, carbinol, wood naphtha, wood spirit, pyroxylic spirit, and pyroligneous alcohol or spirit. It is a flammable, potentially toxic, mobile liquid, used as an industrial solvent, in antifreeze, and in chemical manufacture. Ingestion may result in severe acidosis, visual impairment, and other effects on the central nervous system. Methanol does not produce significant inebriation unless a very large amount is consumed.

Methanol itself is not toxic, but it is metabolized by enzymes in the body to create formaldehyde and formic acid—both of which are very toxic substances. The formic acid can cause blindness. Ethanol (ethyl alcohol—drinking alcohol) can be used as an antidote for methanol poisoning, because it competes with the methanol for the enzyme. As a result, there is a delay of formaldehyde and formic acid production, and these toxic substances do not rise to such high levels. Although methanol is frequently added to ethanol-based cleaning solutions, its addition denatures the solution and makes it unsafe to drink. Only desperate alcoholics will drink methanol, but it is sometimes drunk by accident by people experimenting with various alcohol substitutes.

(SEE ALSO: *Alcohol*)

BIBLIOGRAPHY

CSÁKY, T. Z., & BARNES, B. A. (1984). *Cutting's handbook of pharmacology,* 7th ed. Norwalk, CT: Appleton-Century-Crofts.

PERRY, J. H. (1990). *Methanol: Bridge to a renewable energy future.* Lanham, MD: University Press of America.

S. E. LUKAS

METHAQUALONE

This is a nonbarbiturate, short-acting SEDATIVE-HYPNOTIC drug that has been used to treat insomnia. It was originally introduced in 1951 as a treatment for malaria. In the 1960s and 1970s, it became a popular drug of abuse among college students. Frequently called Quaaludes or "Ludes," the drug, like the short-acting BARBITURATES, produced euphoric effects; some users claimed it had APHRODISIAC effects.

It is usually taken in pill form, and depending on the dose, the effects last a few hours. The body eliminates about half of the ingested dose in about ten to forty hours, so that even forty-eight hours after ingestion, some drug may still be present. Prolonged use of methaqualone in high doses can lead to TOLERANCE AND PHYSICAL DEPENDENCE, and abrupt cessation of daily ingestion can result in WITHDRAWAL symptoms that are quite similar to those seen in barbiturate withdrawal. Fatal convulsions have resulted from sudden withdrawal. Fatal overdoses can occur when the drug is used alone, but especially when it is mixed with ethanol (ALCOHOL) and/or barbiturates. Because it was so commonly abused in the United States, the drug was shifted to Schedule I of the CONTROLLED SUBSTANCES ACT in the 1980s. Thus, it can no longer be prescribed and its nonmedical use is subject to severe criminal penalties. Although it is rarely used illicitly in the United States, it is still available in other countries and is a drug of abuse in some.

(SEE ALSO: *Addiction: Concepts and Definitions*)

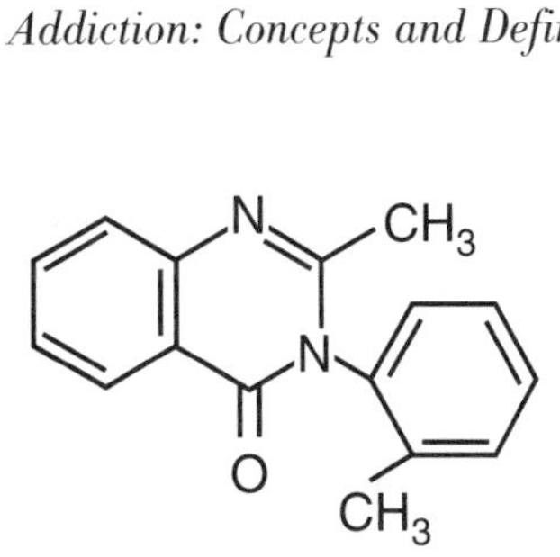

Figure 1
Methaqualone

BIBLIOGRAPHY

HARVEY, S. C. (1980). Hypnotics and sedatives. In A. G. Gilman et al. (Eds.), *Goodman and Gilman's the pharmacological basis of therapeutics*, 6th ed. New York: Macmillan.

SCOTT E. LUKAS

METHEDRINE   Methedrine was the proprietary name given METHAMPHETAMINE hydrochloride by the pharmaceutical company Burroughs Wellcome. It was sold in ampules and until 1963–1964 was readily available by prescription. Methedrine (or "meth") became one of the street names of methamphetamine during the 1960s and early 1970s when high-dose methamphetamine ("speed") was a major drug of abuse. It was a particular problem in northern California where, after the manufacturer withdrew commercially made Methedrine from the market in 1963, large quantities of black-market, illicitly synthesized methamphetamine became available for sale.

(SEE ALSO: *Amphetamine Epidemics; Designer Drugs; Epidemics of Drug Abuse*)

MARIAN W. FISCHMAN

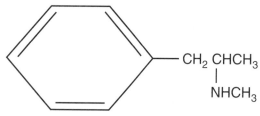

Figure 1
Methedrine

METHYLPHENIDATE   This is a central nervous system STIMULANT, structurally related and with similar effects to AMPHETAMINE. It is used by prescription as Ritalin. It was initially marketed as a mood enhancer in the mid-1950s and described as having less abuse potential than amphetamine; however, within a few years a number of dramatic reports of its abuse and toxicity were published. Methylphenidate is commercially available (by prescription) in pill form, reaching peak effect in one to two hours. Like the amphetamines and other stimulant drugs, methylphenidate is a controlled substance, placed in Schedule II of the CONTROLLED SUBSTANCES ACT to indicate that although it has medical utility it also has substantial ABUSE LIABILITY.

In most people, methylphenidate increases general activity, decreases food intake, produces positive subjective effects (an elevated mood), and can interfere with sleep. With continued use, tolerance can develop to these effects and users will often escalate their doses to achieve the desired effects of their initial doses of methylphenidate. Continued high-dose methylphenidate use can result in toxic consequences similar to those seen after amphetamine use—with ANXIETY, sleeplessness, and eventually a toxic paranoid psychosis. High-dose users often begin with oral methylphenidate use but switch to injecting the drug in order to maximize the effect and achieve the initial "rush" that is typical of intravenous drug abuse. Commercially manufactured methylphenidate pills (the only form available) contain talcum, an insoluble substance, which can cause toxic effects (such as abcesses) when the pills are dissolved in water and injected intravenously or under the skin.

Laboratory animals tested with methylphenidate show increases in locomotor activity after single doses, increased sensitivity to this effect after repeated doses, and the development of stereotyped repetitive behavior patterns after chronic dosing. In addition, these animals remain more responsive to methylphenidate even after the drug treatment has been discontinued. It has been suggested that the continuous repetition of behavior that characterizes the response to chronic methylphenidate treatment is a good model for the human stimulant psychosis and, as in animals, humans who use high doses become increasingly sensitive to stimulants such as methylphenidate, with psychosis increasingly likely at lower doses after its initial appearance. There are, however, no data to support this hypothesis.

In addition to its action as an appetite suppressant, methlyphenidate has been found to have other therapeutic utility. Like *d*-amphetamine, it has been used successfully in the treatment of ATTENTION-DEFICIT HYPERACTIVITY DISORDER (ADHD), a syndrome that first becomes evident during childhood and is characterized by excessive activity and difficulty in maintaining attention. Because of its relatively short half-life, two or three doses of methylphenidate are required each day, although recently a slow-release form of the medication has become available, promising more stable blood levels with only a single daily dosing. Methylphenidate has been shown to alleviate or moderate many of the symptoms of this disorder, although it is not effective in all cases and its long-term efficacy is not well understood.

Side effects of treatment can include insomnia, loss of appetite, and weight loss, all effects of stimulant drugs in general. In addition, concern about the longer lasting effects on learning and cognition in youngsters maintained on this drug for many years has made practitioners cautious and often unwilling to prescribe it. Recent research and practice, however, has supported methylphenidate as the stimulant of choice for treating this disorder. As with the amphetamines, methylphenidate is also effective in the treatment of narcolepsy, in which sudden attacks of sleep can occur unexpectedly.

BIBLIOGRAPHY

GRILLY, D. M. (1989). *Drugs and human behavior.* Needham, Mass: Allyn & Bacon.

KALANT, O. J. (1973). *The amphetamines: Toxicity and addiction,* 2nd ed. Springfield, IL: Charles C. Thomas.

MARIAN W. FISCHMAN

MEXICO AS DRUG SOURCE Drug control in Mexico is unique—the reason both for Mexico's paradoxical success as well as for its ongoing difficulty in managing the issue. Believing that destruction at their agricultural source is the most effective way to reduce supplies and halt trafficking, Mexico began to spray the OPIUM poppy (PAPAVER SOMNIFERUM) and MARIJUANA plant (CANNABIS SATIVA) in late 1975 with the herbicides paraquat and 2,4-D. Plants, not people, became the target in the 1970s and 1980s. Until the early 1990s, the drug-eradication program was the centerpiece of Mexico's program. With the 1990s increase in Colombian cocaine transiting Mexico, the administration of Mexican President Carlos Salinas (1988–1994) increased its efforts to work with the United States in halting COCAINE smuggled through Mexico, sharing intelligence, extraditing non-Mexican nationals, and reducing drug-related corruption. Mexico's principal agency for drug control is the attorney general's office, but the Mexican military has also been involved

with manual crop eradication and operational support for herbicidal spraying. Recently, the Mexican military has also become involved in tactical reconnaissance, interdiction, and destruction of secret landing strips. Historically, the government of Mexico has increased its effectiveness in the drug-control field as a positive response to U.S. diplomatic and enforcement pressures; these include OPERATION INTERCEPT and the diplomatic pressures that followed the 1985 death in Mexico of U.S. Drug Enforcement agent Enrique Camarena.

UNIQUENESS OF MEXICO

At least four factors set Mexico apart from its drug-producing neighbors to the south, creating an environment for drug control. First, Mexico is the only country in the Western Hemisphere that produces significant amounts of opium poppy and HEROIN with little use by its people. Although large numbers of its people abuse marijuana and INHALANTS, Mexico may be the only opium-producing country in the world with almost no domestic market. Yet, Mexico shares a 1,900-mile (3,057-km) border with the U.S.—a country that has one of the world's largest and most lucrative markets for heroin.

Second, powerful drug rings have bought power and influence in several Latin American countries, yet, unlike their Peruvian and Colombian counterparts, Mexican drug traffickers have no symbiotic relationship with ideological, terrorist-oriented, political factions—whose goal is to change the prevailing political order. Mexican drug organizations pose less of an overt threat to government stability than do Colombian and Peruvian drug cartels.

Third, growing opium in Mexico is relatively recent; it has always been illegal and involves only a small number of citizens. Illicit opium and marijuana are grown not on privately owned plots but on open unowned (hence, government) land, largely as extra-cash crops, not as subsistence crops. If these illicit crops were all destroyed, growers would not starve. Unlike coca—which is a legal crop (that can provide the raw material for an illegal commodity) and has been cultivated for centuries in Bolivia and Peru—Mexico's opium crop has never become the center of a social, cultural, or agricultural economy.

Fourth, Mexico is a relatively wealthy country with vast natural and human resources; its strong central government has control over its territory. Unlike much of the Andean region, Mexico has shown its ability to build an effective infrastructure to implement an ongoing drug-control campaign. Beginning in the mid-1970s, Mexico started the world's first successful herbicidal opium-eradication program, which continues today.

CAMARENA MURDER

Drug control has been an important issue between the United States and Mexico for the past twenty-five years, yet the abduction and murder of U.S. DEA agent Enrique Camarena in Mexico in February 1985 elevated the drug issue on the bilateral diplomatic agenda of the two countries. The murder focused public attention on the perhaps decreasing effectiveness of Mexican drug-control efforts and represented a turning point in U.S.–Mexican relations. After Camarena's murder, drugs became a confrontational issue at uncharacteristically high levels of the two governments. Both the U.S. secretary of state and Mexico's foreign minister discussed the murder and subsequent government response as a paramount diplomatic issue; drug control was no longer treated only as a law-enforcement issue between the two countries. In response to continuing U.S. pressure, the Mexican government took a series of actions that resulted in the apprehension and incarceration of the several drug traffickers responsible.

HISTORICAL ROOTS

Mexico's international drug-control efforts have their roots in the SHANGHAI Convention of 1909 and the Hague Opium Convention of 1911–1912. In 1923, Mexico's President Alvaro Obregon prohibited the production of opium and condemned what was then widespread and increasing drug-induced violence. In 1934, President Cardenas del Rio created the first centralized narcotics administrative unit in the government.

After the United States entered World War II in 1941, Mexico was asked to provide opium for the war effort, since it was processed into MORPHINE, a medication used extensively for war-related wounds. In both Mexico and the United States, HEMP was grown to fill U.S. military need for rope and cordage; hemp is processed from *Cannabis sativa*, which is also used

as marijuana. By mid-1943, opium constituted the most profitable cash crop in Mexico's northwestern state of Sinaloa. Despite Mexico's efforts to control the production of these crops after the war, drugs were grown, processed, and smuggled into the United States from Mexico.

In the late 1960s and early 1970s, Mexico soon became the major supplier for the illicit U.S. heroin market when Turkey prohibited opium cultivation and the French Connection had been ended. Consequently, in the fall of 1969, the U.S. BUREAU OF NARCOTICS AND DANGEROUS DRUGS (the predecessor organization of DEA) and the U.S. Customs Service initiated Operation Intercept—a three-week operation that subjected every person crossing the border in the San Ysidro, California, area to intensive baggage and body searches. The economic losses and dismay on both sides of the border prompted termination of the operation—but not before focusing attention on the volume of drugs entering the United States from Mexico. The Mexican government then began to locate and manually destroy the poppy fields—the source, at that time, of all the heroin produced in the Western Hemisphere. Originally, the search for poppy fields was made in small planes that flew over mountain zones where crops were suspected to be growing on remote plots of government land.

Prior to 1975, once the poppy had been spotted and the approximate location registered in official correspondence, military squads were sent to destroy the plants by cutting them down. In 1975, the Mexicans began to use the most modern technology—a system called Multi-Opium Poppy Sensing (MOPS), which used multispectral sensing cameras on board low-flying aircraft to read and print images from the electromagnetic spectrum. In nature, every substance emits its own unique electromagnetic waves that can be read on the color spectrum using special cameras. The fields were then destroyed by aerial application of the contact herbicides 2,4-D and paraquat. By the 1990s, a fleet of nearly 120 aircraft were being used.

MEXICAN GOVERNMENT ORGANIZATIONAL STRUCTURE

Organizationally, the Mexican attorney general's office plans and implements the drug-eradication campaign. Nearly 700 civilian pilots, mechanics, communications experts, and technical personnel make the campaign as effective as possible— working specific zones and sectors, with a coordinating office in each zone. Forward operating bases connect all the zones to Mexico City via a sophisticated communications system. Mexico's military is also used to stop the illicit cocaine that transits Mexico from South America to the United States, exchanging intelligence and training, and destroying clandestine trafficker landing strips.

ERADICATION RESULTS

The drug eradication program has had dramatic results. When heroin availability in the United States is measured through changes in retail price and potency (or purity) and through certain changes in negative health consequences, it is clear that between late 1975 (when the aerial spraying began) and 1980, Mexican heroin became less abundant in the United States. The price increased; retail purity decreased (i.e., the heroin was "cut" more); and overdoses, drug-treatment admissions, and emergency room episodes decreased—indicating that the supply of Mexican heroin had also decreased.

Between 1982 and 1989, Mexico's rapidly deteriorating economy, bureaucratic inertia, technical inefficiency, poor management, low morale, complacency, and corruption led to the decreased effectiveness of the eradication program. Countermeasures by growers who planted smaller fields, at higher altitudes, under cover of foliage, during more than the two traditional growing seasons, further decreased program success. In the mid-1970s, the eradication campaign was managed in large part by specialized organizations of both governments (Mexico's attorney general and U.S. law-enforcement units). In the mid-1980s, the Camarena murder took the campaign out of the strict purview of the specialist law-enforcement agencies and into the diplomatic arena. In the 1990s, the drug-control efforts increased to include interdiction of South American cocaine traveling through Mexico but destined for the United States. In 1991, the Mexican government increased its eradication of opium by 40 percent over 1990; and its eradication of marijuana by 60 percent. The U.S. government estimates that in 1991 Mexico provided approximately 25 percent of the heroin and perhaps as much as 60 percent of the marijuana consumed in the United States.

(SEE ALSO: *Bolivia; Colombia; Drug Interdiction; International Drug Supply Systems*)

BIBLIOGRAPHY

BUREAU OF INTERNATIONAL NARCOTICS MATTERS, U.S. DEPARTMENT OF STATE. (1991, 1992). *International narcotics control strategy report (INCSR)*. Washington, DC: Author.

CRAIG, R. (1987). United States antidrug policy with Mexico: Consequences for American society and U.S.–Mexican relations. Paper presented to the Bilateral Commission on the Future of United States–Mexican Relations, Queretaro, Mexico.

CRAIG, R. (1980). Operation Condor: Mexico's anti-drug campaign enters a new era. *Journal of Interamerican Studies and World Affairs, 22*, 346–347.

CRAIG, R. (1980). Operation Intercept: The international politics of pressure. *Review of Politics, 42*(4), 556–580.

VAN WERT, J. M. (1986). Mexican narcotics control: A decade of institutionalization and a matter for diplomacy. In Gabriel Szekely (Ed.), *Mexico-Estados Unidos 1985*. Mexico City: Colegio de Mexico.

VAN WERT, J. M. (1982). *Government of Mexico herbicidal opium poppy eradication program: A summative evaluation*. Unpublished doctoral dissertation, University of Southern California, Los Angeles.

WALKER, W.-O., III. (1981, 1989). *Drug control in the Americas*. Albuquerque: University of New Mexico Press.

JAMES VAN WERT

MICHIGAN ALCOHOLISM SCREENING TEST (MAST)

This is a brief self-report questionnaire designed to detect ALCOHOLISM (Selzer, 1971). It is widely used in clinical and research settings. The twenty-four scored items assess symptoms and consequences of ALCOHOL abuse, such as guilt about drinking; blackouts; DELIRIUM TREMENS; loss of control; family, social, employment, and legal problems following drinking bouts; and help-seeking behaviors, such as attending ALCOHOLICS ANONYMOUS meetings or entering a hospital because of drinking. Several shorter versions of the MAST have also been developed including the thirteen-item Short-MAST (Selzer, Vinokur, & van Rooijen, 1975) and the ten-item Brief-MAST (Pokorny, Miller, & Kaplan, 1972).

To complete the MAST, individuals answer yes or no to each item. The items are weighted on a scale of 1 to 5, with items concerning prior alcohol-related treatment experiences and help-seeking behaviors receiving higher weights. The total MAST score (range: 0–53) is derived by adding the weighted scores from all items that are endorsed. Studies indicate that the long version of the MAST possesses good internal-consistency reliability, as indicated by Cronbach's alpha coefficients of .83 to .93 (Gibbs, 1983). Therefore, the scale does appear to measure a unitary construct.

Selzer (1971) originally recommended adopting a cutting score of 5 or higher for a diagnosis of alcoholism with the MAST. However, since this cutting score was shown to produce a relatively high percentage of false positives (Gibbs, 1983), Selzer, Vinokur, and van Rooijen (1975) suggested the following cut points: 0 to 4, not alcoholic; 5 to 6, maybe alcoholic; 7 or more, alcoholic. Skinner (1982) recommended that scores of 7 to 24 be regarded as clear evidence of alcohol problems, and that scores of over 25 be considered evidence of substantial alcohol problems. In a recent study, Ross, Gavin, and Skinner (1990) compared scores on the MAST to diagnoses of alcoholism obtained from the National Institute of Mental Health Diagnostic Interview Schedule (NIMH-DIS) (Robins, Helzer, Croughan, & Ratcliff, 1981). In this study, the MAST cutting score that yielded the highest overall accuracy was 13 or greater.

The validity of the MAST has been examined in a number of studies in which MAST scores, or scores from the shorter versions of the instrument, were compared to other measures of drinking status, including diagnostic interviews, physicians' diagnoses, and other self-report instruments. In reviewing twelve of these studies, Gibbs (1983) concluded that MAST diagnoses agreed with diagnoses of alcoholism reached through other assessment procedures in about 75 percent of cases. Where inconsistencies between results were found, it was found that the MAST tended to overdiagnose alcoholism. This probably reflects the fact that a cutting score of 5 or higher on the MAST was used in these studies. By adopting a cutting score of 13, Ross et al. (1990) were able to achieve a greater degree of agreement when comparing MAST scores to DIS-derived diagnoses.

As with any instrument that relies on the veracity of self-report information, the reliability and validity of the MAST is dependent on the willingness of the interviewee to answer the items truthfully. All the items possess high face validity, which means it is relatively easy to answer them so as to appear nonalcoholic. The MAST may therefore not be a useful screening tool with individuals who are motivated to conceal their alcohol problems.

(SEE ALSO: *Addiction Severity Index; Diagnosis of Drug Abuse: Diagnostic Criteria; Diagnostic and Statistical Manual; Disease Concept of Alcoholism and Drug Abuse; Minnesota Multiphasic Personality Inventory*)

BIBLIOGRAPHY

GIBBS, L. E. (1983). Validity and reliability of the Michigan Alcoholism Screening Test: A review. *Drug and Alcohol Dependency, 12*, 279–285.

POKORNY, A. D., MILLER, B. A., & KAPLAN, H. B. (1972). The brief MAST: A shortened version of the Michigan Alcoholism Screening Test. *American Journal of Psychiatry, 129*, 118–121.

ROBINS, L. N., ET AL. (1981). National Institute of Mental Health Diagnostic Interview Schedule: History, characteristics, and validity. *Archives of General Psychiatry, 38*, 381–389.

ROSS, H. E., GAVIN, D. R., & SKINNER, H. A. (1990). Diagnostic validity of the MAST and the Alcohol Dependence Scale in the assessment of DSM-III alcohol disorders. *Journal of Studies on Alcohol, 51*, 506–513.

SELZER, M. L. (1971). The Michigan Alcoholism Screening Test: The quest for a new diagnostic instrument. *American Journal of Psychiatry, 127*, 1653–1658.

SELZER, M. L., VINOKUR, A., & VAN ROOIJEN, L. (1975). A self-administered short version of the Michigan Alcoholism Screening Test (SMAST). *Journal of Studies on Alcohol, 43*, 117–126.

SKINNER, H. A. (1982). *Guidelines for using the Michigan Alcoholism Screening Test.* Toronto: Addiction Research Foundation.

A. THOMAS MCLELLAN

MIDDLE EAST AS DRUG SOURCE   *See* International Drug Supply System.

MILITARY, DRUG AND ALCOHOL ABUSE IN THE U.S.

Both drug and alcohol abuse are strongly opposed by the U.S. armed forces because of their negative effects on the health and well-being of military personnel and because of their detrimental effects on military readiness. In the military, drug abuse or illicit drug use is defined as any use of illegal drugs (e.g., MARIJUANA, COCAINE) or any nonmedical use of legal drugs (e.g., STIMULANTS, TRANQUILIZERS) without a doctor's prescription for purposes other than intended. Alcohol abuse is defined as any ALCOHOL use that results in negative consequences such as adverse effects on work performance, health, or social behavior.

DEVELOPMENT OF MILITARY POLICY

U.S. military policy toward drug and alcohol abuse originated largely in response to reports of widespread use of illicit drugs among U.S. troops in VIETNAM. The Department of Defense convened a task force in 1967 to investigate drug and alcohol abuse in the military and in 1970 formulated a drug and alcohol abuse policy based on task force recommendations. In 1971, the Department of Defense was directed by President Richard M. Nixon to initiate a policy of URINE TESTING for drugs for all troops leaving Vietnam and to provide treatment for those who tested positive for OPIATE drugs. The policy was updated in 1972 and again in 1980 and currently is one of ZERO TOLERANCE. It includes an emphasis on preventing and detecting abuse and either discharging abusers from the military (the approach generally followed for drug abuse) or providing treatment and rehabilitation (the approach generally followed for alcohol abuse).

The reaction to the crash of a jet fighter on the aircraft carrier U.S.S. *Nimitz* in 1981 called public attention to the military's drug-abuse problem, particularly marijuana use. Autopsies of the fourteen Navy personnel killed in the crash showed evidence of marijuana use among six of the thirteen sailors and nonprescription antihistamine use by the pilot. The military reinstituted urine testing for drugs in 1981 as a result of this incident and of other concerns about drug use in the services. (Worldwide military testing had also been done in 1971 to 1974 but was then discontinued.) Urine tests, which are conducted either randomly or when a person is suspected of using drugs, are a major tool for the detection and deterrence of illicit drug use.

WORLDWIDE SURVEY SERIES

To help monitor the extent of drug and alcohol abuse, the Department of Defense initiated a series of worldwide surveys among active-duty military personnel in the Army, Navy, Marine Corps, and Air Force. Personnel in the reserves or National Guard were not included. The first survey was conducted by Burt Associates in 1980 (Burt et al., 1980); and the other four by Robert Bray and his colleagues at Research Triangle Institute in 1982, 1985, 1988, and 1992 (Bray et al., 1983; 1986; 1988; 1992). The goals of the surveys have been to provide data to assess the prevalence, correlates, and consequences of substance abuse and health in the military.

The surveys have all been conducted using similar methods. Civilian researchers first randomly selected a sample of over sixty military installations to represent the armed forces throughout the world. At these designated installations, they randomly selected men and women of all ranks to represent all active-duty personnel. Civilian research teams administered printed questionnaires anonymously to selected personnel in classroom settings on military bases. The few personnel (about 10%) who were unable to attend the group sessions (e.g., were on leave, sick, or temporarily away from the base) were mailed questionnaires and asked to complete and return them. Participants answered questions about their use of illegal drugs (e.g., marijuana, cocaine, HEROIN), about the misuse of prescription drugs (e.g., stimulants, tranquilizers), about alcohol use, and about problems resulting from drug or alcohol use. These data-collection procedures yielded from over 15,000 to nearly 22,000 completed questionnaires for the various surveys. From 77 percent to 84 percent of those eligible to take part actually did so.

TRENDS IN DRUG AND ALCOHOL USE

Figure 1 presents the trends, over the five worldwide surveys, in the percentage of the total active military force who engaged in any illicit drug use or heavy alcohol use during the 30 days prior to the survey. Illicit drug use was measured in terms of consumption, one or more times during the past 12 months and the past 30 days, of marijuana/HASHISH, cocaine, INHALANTS, HALLUCINOGENS, heroin, and the nonmedical use of prescription-type drugs including stimulants, SEDATIVES, tranquilizers, or ANALGESICS. Heavy alcohol use was defined as five or

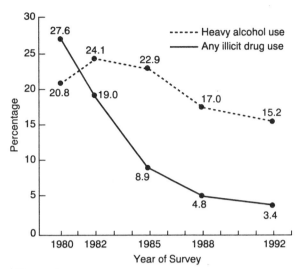

Figure 1

Trends in Any Illicit Drug Use and Heavy Alcohol Use, Past 30 Days, Total Department of Defense, 1980–1992

SOURCE: Bray et al. (1992). *Worldwide survey of substance abuse and health behaviors among military personnel.* Report prepared for the Assistant Secretary of Defense (Health Affairs) and Department of Defense Coordinator for Drug Enforcement Policy and Support, U.S. Department of Defense (Contract no. MDA 903-91-C-0220). Research Triangle Park, NC: Research Triangle Institute.

more drinks per typical drinking occasion at least once a week. As shown in the figure, use of any illicit drug declined sharply, from 27.6 percent in 1980 to 3.4 percent in 1992; heavy drinking declined significantly, from 20.8 percent in 1980 to 15.2 percent in 1992, although the decrease was less dramatic than for drug use. Heavy drinking by itself does not constitute alcohol abuse, but it indicates drinking levels that are likely to result in negative consequences.

EFFECTS OF DEMOGRAPHIC CHANGES

Although the downward trends in illicit drug use and heavy drinking noted in the figure are impressive, the question arises whether they are the result of military programs and policies or of other factors. One possible explanation for the decrease could be shifts in the demographic composition of the armed forces between 1980 and 1992. Members of the military in 1992, for example, were more likely to be older, to be officers, to be married, and to have more education than their counterparts in 1980. These characteristics are associated with less substance use. For example, nearly 63 percent of personnel in 1992 were married, compared with 53 percent in 1980; and 61 percent were older than age 25 in 1992, compared with 43 percent in 1980.

Analyses that adjusted for demographic differences across survey years, from 1980 to 1992, showed that illicit drug use had the same significant decline as found before the adjustment, whereas heavy alcohol use did not. This finding indicates that the declines in illicit drug use shown in the figure *were not* explained by shifts in the demographic composition of the military population, whereas declines in heavy drinking *were* largely explained by demographic changes. Stated another way, if the demographic composition of the military in 1992 were like the composition in 1980, rates of illicit drug use in 1992 would still have been notably lower, but rates of heavy drinking between these two surveys would have been about the same.

MILITARY AND CIVILIAN COMPARISONS

Another possible explanation for the trends in drug and alcohol use observed in figure 1 is that the military may simply mirror similar trends occurring among civilians. Drug use among civilians has declined substantially in recent years (SAMHSA, 1993), whereas declines in alcohol use among civilians have been more moderate (Clark & Hilton, 1991). To address this issue, two studies (Bray et al., 1991; 1992) compared illicit drug use and heavy alcohol use among military personnel and civilians. In the first study, military data were drawn from the 1985 worldwide survey and civilian data from the 1985 NATIONAL HOUSEHOLD SURVEY ON DRUG ABUSE (NHSDA), a nationwide survey of drug abuse. In the follow-up study, military data were drawn from the 1992 worldwide survey and civilian data from the 1991 NHSDA. In both studies, the military and civilian data sets were equated for age and geographic location of respondents, and civilian substance-use rates were standardized (adjusted) to reflect the demographic distribution of the military. Standardized comparisons showed the same pattern of results for the two studies—military personnel were significantly less likely than civilians to have used any illicit drugs during the past 30 days but were significantly more likely to have been heavy drinkers. These findings are illustrated in Figure 2 for data from the 1992 worldwide survey.

The findings indicate that substance-use trends in the military do not simply mirror similar changes among civilians. The lower rates of drug use among military personnel than among civilians suggest

either that military policies and practices deter drug use in the military or that military personnel hold attitudes and values that discourage substance use. In contrast, the higher rates of heavy drinking among military personnel suggest that certain aspects of military life may foster use and/or that military policies and programs aimed at reducing heavy

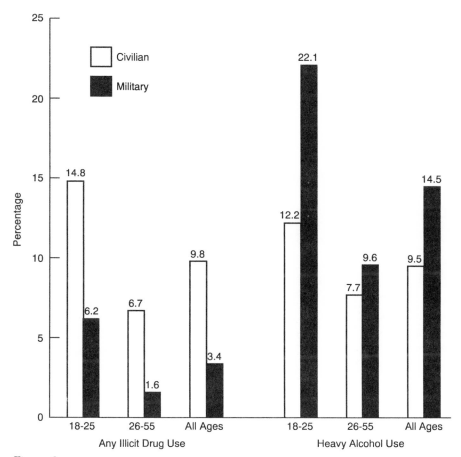

Figure 2

Standardized Comparisons of Military and Civilian Prevalence of Illicit Drug Use and Heavy Alcohol Use, Past 30 Days, 1992 (ages 18–55)

NOTE: Military data for the U.S.-based personnel include those in Alaska and Hawaii. Civilian data have been standardized to the military data by sex, age, education, race/ethnicity, and marital status.

CIVILIAN DATA SOURCE: Substance Abuse and Mental Health Services Administration (SAMHSA). (1993). *National Household Survey on Drug Abuse: Highlights 1991* (DHHS Publication no. SMA 93-1979). Washington, D.C.: U.S. Government Printing Office.

MILITARY DATA SOURCE: Bray et al. (1992). *Worldwide survey of substance abuse and health behaviors among military personnel.* Report prepared for the Assistant Secretary of Defense (Health Affairs) and Department of Defense Coordinator for Drug Enforcement Policy and Support, U.S. Department of Defense (Contract no. MDA 903-91-C-0220). Research Triangle Park, NC: Research Triangle Institute.

alcohol use have not been as effective as similar efforts among civilians.

SUMMARY

Overall, these findings indicate that the military has made steady and notable progress in combating illicit drug use, particularly during the 1980s and early 1990s. In 1992, illicit drug use in the military was at minimal levels, and rates were substantially lower than among civilians. In contrast, the military has made little progress in reducing heavy drinking. In 1992, heavy drinking affected over one-seventh of active-duty personnel and was significantly higher than among civilians. Declines in heavy drinking in the military between 1980 and 1992 were largely explained by changes in the military's demographic composition. In summary, the military appears to have developed an effective formula to reduce illicit drug use and now needs to develop a comparable plan to reduce heavy drinking.

(SEE ALSO: *Epidemiology of Drug Abuse; High School Senior Survey; Industry and Workplace: Drug Use in*)

BIBLIOGRAPHY

BRAY, R. M., ET AL. (1992). *1992 worldwide survey of substance abuse and health behaviors among military personnel.* Report prepared for the Assistant Secretary of Defense (Health Affairs) and Department of Defense Coordinator for Drug Enforcement Policy and Support, U.S. Department of Defense (Contract no. MDA 903-91-C-0220). Research Triangle Park, NC: Research Triangle Institute.

BRAY, R. M., ET AL. (1991). Standardized comparisons of the use of alcohol, drugs, and cigarettes among military personnel and civilians. *American Journal of Public Health, 81,* 865–869.

BRAY, R. M., ET AL. (1988). *1988 worldwide survey of substance abuse and health behaviors among military personnel.* Report prepared for the Assistant Secretary of Defense (Health Affairs), U.S. Department of Defense (Contract no. MDA 903-87-C-0854). Research Triangle Park, NC: Research Triangle Institute.

BRAY, R. M., ET AL. (1986). *1985 worldwide survey of alcohol and nonmedical drug use among military personnel.* Report prepared for the Assistant Secretary of Defense (Health Affairs), U.S. Department of Defense (Contract no. MDA 903-85-C-0136). Research Triangle Park, NC: Research Triangle Institute.

BRAY, R. M., ET AL. (1983). *1982 worldwide survey of alcohol and nonmedical drug use among military personnel.* Report prepared for the Assistant Secretary of Defense (Health Affairs), U.S. Department of Defense (Contract no. MDA 903-83-C-0120). Research Triangle Park, NC: Research Triangle Institute.

BURT, M. A., ET AL. (1980). *Worldwide survey of nonmedical drug use and alcohol use among military personnel.* Bethesda, MD: Burt Associates.

CLARK, W. B., & HILTON, M. E. (EDS.) (1991). *Alcohol in America: Drinking practices and problems.* Albany: State University of New York Press.

SUBSTANCE ABUSE AND MENTAL HEALTH SERVICES ADMINISTRATION (SAMHSA). (1993). *National household survey on drug abuse: Highlights 1991* (DHHS Publication no. SMA 93-1979). Washington, DC: U.S. Government Printing Office.

ROBERT M. BRAY

MINIMUM DRINKING AGE LAWS Before the twentieth century, there were few legal restrictions on the consumption of alcoholic beverages by youth. Early in the twentieth century, laws prohibiting alcohol sales to minors began to be implemented, as part of a broader trend of increasing legal controls on adolescent behavior. The temperance movement succeeded in establishing national Prohibition in 1919 but when it was repealed in 1933, all states implemented legal minimum ages for alcohol purchase or consumption, with most states setting the age at 21.

From the 1930s through the 1960s, the issue received little public attention. In 1970, the 26th Amendment to the U.S. Constitution lowered the voting age in federal elections from 21 to 18. By 1974, all fifty states had lowered their voting ages for state elections to 18. As part of this trend of lowering the "age of majority," twenty-nine states lowered their minimum drinking ages between 1970 and 1975, most setting the age at 18 or 19. In the mid-1970s, studies began to emerge that showed significant increases in the rate of young drivers' involvement in traffic accidents following the reductions in the legal drinking age. The trend toward lower drinking ages was reversed, with Maine being the first state to raise its legal drinking age from 18 to 20 in October 1977. Several other states soon followed, and research studies completed by the early 1980s found significant declines in youth traffic-crash involvement when states raised their legal drinking age. With the support of organized efforts by citizen-action groups such as REMOVE INTOXICATED DRIVERS and MOTHERS AGAINST DRUNK DRIVING, federal legislation was passed in 1984 that called for the withholding of a portion of federal highway-construction funds from any state that did not have a legal drinking age of 21 by October 1986. As a result, all the remaining states with a legal drinking age of below 21 raised their age to 21 by 1988. Thus, all states now have a uniform legal drinking age of 21, although details in regard to the purchase, possession, consumption, sales, and furnishing of alcohol to underage youth vary from state to state.

The legal drinking age became a major issue because of the serious consequences of young people's consumption of alcohol. Most teenagers drink; in addition, almost a third regularly become intoxicated. Damage resulting from the drinking of youth is extensive. Car crashes are the leading cause of death for teenagers (Baker et al., 1992), and one third to one half of the crashes involve alcohol (National Highway Traffic Safety Administration, 1990). Other leading causes of disability and death among youth, such as suicide, homicide, assault, drowning, and recreational injury, involve alcohol in one quarter to three quarters of the cases (Wagenaar, 1992). Injuries are only part of the problem. Early use of alcohol appears to affect multiple dimensions of physical, social, and cognitive development (Semlitz & Gold, 1986). Alcohol use increases the odds of having unprotected sex (i.e., failure to use a condom), which increases the chance of pregnancy and catching sexually transmitted diseases, including the human immunodeficiency virus (HIV), which causes AIDS (Plant, 1990; Strunin & Hingson, 1992). Many "date rape" situations involve individuals who have been drinking (Wagenaar et al., 1993a). Early use of alcohol increases the odds one will move on to using other drugs, such as MARIJUANA, COCAINE, or HEROIN (Kandel, 1989). Finally, the earlier one starts a pattern of regular drinking, the higher the chance of later serious problems with alcohol, including dependence (i.e., getting "hooked" so that it is very hard to quit). Despite the many problems associated with young people's drinking, the most obvious one, and the one that received the most attention in debates on the legal drinking age, is traffic-crash involvement.

EFFECTS OF THE DRINKING AGE ON CAR CRASHES

Seventeen studies of the effects of lowering the legal age for drinking on traffic crashes appeared between 1974 and 1982 (see Wagenaar, 1983, for a review). Although results varied across studies and across states, most studies found significant increases in traffic crashes among youth after the drinking age had been lowered (usually from 21 to 18). Typically, lowering the drinking age resulted in 5 percent to 20 percent increases in fatal and injury-producing crashes likely to involve alcohol, such as single-vehicle crashes occurring at night.

Thirty-nine studies of the effects on traffic crashes of raising the legal age for drinking have appeared between 1979 and 1992 (see Wagenaar, 1993, for a review). Twenty-eight of these studies found significant reductions in the involvement of youth in traffic crashes following increases in the legal drinking age. Typically, raising the drinking age resulted in 5 percent to 20 percent declines in fatal and injury-producing crashes likely to involve alcohol. With the aid of the better-designed studies with longer follow-up periods, it could be estimated that the long-term effects of raising the drinking age to 21 would be a 13 percent decline in single-vehicle nighttime crashes among those whose legal access to alcohol was removed (i.e., 18 to 20-year-olds).

The legal drinking age is probably the most extensively researched policy that is designed to reduce traffic crashes and other alcohol problems. Scientists and professionals in the field agree that lowering the legal age for drinking increased car crashes among youth, and that, subsequently, raising the legal age reversed the effect: It lowered car crashes among youth (United States General Accounting Office, 1987). The National Highway Traffic Safety Administration estimates that, even when counting only those states that raised the legal age after 1982, the U.S. age-21 policy now saves over one thousand lives per year in reduced car crashes alone (Arnold, 1985).

EFFECTS OF THE DRINKING AGE ON OTHER PROBLEMS

Four studies have appeared of the effects on problems other than car crashes of raising the legal age to 21 (see Wagenaar, 1993, for a review). One study found that vandalism was down 16 percent in four states that raised the drinking age, and another found that significant reductions in suicides, pedestrian injuries, and other unintentional injuries were associated with higher legal drinking ages. A study of two Australian states that lowered the legal drinking age found 22 percent to 40 percent increases in trauma-hospital admissions for causes other than car crashes, although another study did not confirm these findings. A Massachusetts study found no reductions in nontraffic trauma, suicide, and homicide deaths after the drinking age had been raised, perhaps because many of Massachusetts' residents lived close to bordering states that had lower drinking ages at the time of the study.

EFFECTS OF THE DRINKING AGE ON ALCOHOL USE

Seven studies examined the effect of the legal drinking age on aggregate alcoholic-beverage sales. Effects were mixed—some studies found that alcohol sales were related to the legal age, but others did not find such a relationship. These studies were difficult to interpret because alcohol sales to young drinkers could not be distinguished from sales to older drinkers.

Surveys of the effects on alcohol use among youth of lowering or raising the drinking age have produced conflicting results. Some have found that there was little effect of the legal drinking age on young people's drinking, whereas others found that lower rates of youth drinking resulted when the legal drinking age was higher (see Wagenaar, 1993, for a review of the fourteen survey studies to date). A major limitation of many of these studies was their use of nonrandom samples of youth from particular high schools, colleges, and local communities rather than samples that were broadly representative of the youth in a state. Surveys of college students, which are usually limited to students in introductory social sciences courses, frequently report finding little effect of the legal drinking age on drinking patterns. In contrast, surveys of random samples of high school seniors and 18- to 20-year-olds across many states, including those entering college and those in the work force, report finding significant reductions in drinking that are associated with higher legal drinking ages (e.g., Maisto & Rachal, 1980; O'Malley & Wagenaar, 1991). It appears, on the basis of the best-designed studies, that raising the legal drinking age results in reductions in young people's drinking.

The age-21 policy, however, by no means eliminates this drinking by youth.

ENFORCEMENT OF THE MINIMUM DRINKING AGE

Although drinking among youth is now significantly down from its peak in 1980, when questioned, 54 percent of high school seniors still reported drinking in the last month, and 30 percent reported having had five or more drinks at a time at least once in the previous two weeks (Figures 1 and 2; data from Johnston, O'Malley, & Bachman, 1991). Among the many reasons that youth continue to drink, one important reason is that alcohol remains easily available to them, despite the minimum drinking age law. A recent study by Wagenaar and associates (1993b) indicated that only two of every one thousand episodes of underage drinking resulted in an arrest of the youth involved. More important, only five of every hundred thousand episodes of drinking by underage youth resulted in any action being taken against a store, restaurant, or bar for selling or serving alcohol to a minor. Because the chance of getting caught was so low, half or more of all alcohol outlets tested sold alcohol to youth without requesting any age identification (Preusser & Williams, 1991; Forster et al., 1993).

CONCLUSIONS

Evidence that showed that raising the drinking age to 21 reduced deaths and injuries in car crashes was a major factor in the debate about the drinking age. Other arguments were also heard, such as: Is it unconstitutional to discriminate solely on the basis of age? Federal courts have ruled that the drinking age is not discriminatory, because (1) drinking is not a fundamental right, (2) age is not an inherently suspect criterion for discrimination, (3) and the higher drinking age has a "rational basis" and is "reasonably related" to a legitimate goal of the state to reduce death and injury from traffic crashes (Guy, 1978). In a democracy, laws should have the support of the governed. Repeated polls have shown that the majority of the public clearly supports a legal drinking age of 21. Even among youth under the age of 21, some polls have shown majority support for the minimum drinking age of 21.

Is it logical to set the legal age of drinking at 21, when other rights and privileges of adulthood (e.g.,

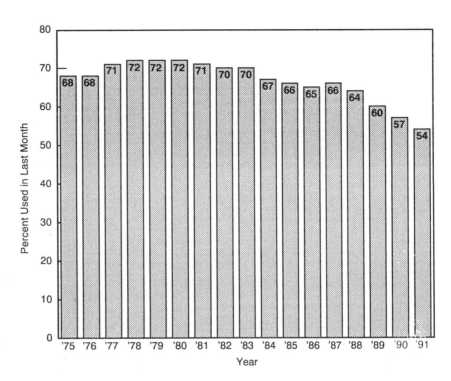

Figure 1
Alcohol Use, U.S. High School Seniors.

voting, signing legally binding contracts) begin at age 18? The answer is yes, because we have many different legal ages, varying from 12 to 21, for voting, driving, sale and use of tobacco, legal consent for sexual intercourse, marriage, access to contraception without parental consent, compulsory school atten-

dance, and so forth. Minimum ages are not set uniformly; they depend on the specific behavior involved, and they are arrived at by balancing the dangers and benefits of establishing the particular age.

Some have argued that a minimum drinking age of 21 will make things worse when young people fi-

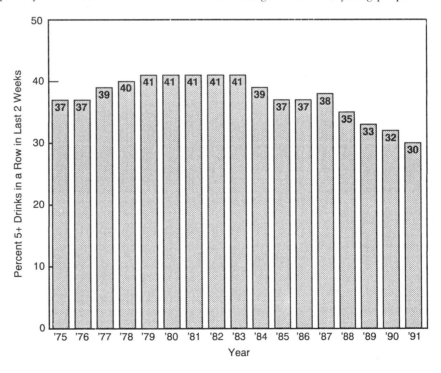

Figure 2
Binge Drinking, U.S. High School Seniors.

nally get legal access to alcohol. This is the "rubber band" theory whereby it is claimed that prohibiting teenagers from drinking will cause a pent-up demand for the forbidden fruit. At 21, they will break loose and drink at significantly higher rates than they would have if they had been introduced to alcohol earlier. This theory is clearly not supported by research. For example, O'Malley and Wagenaar (1991) found just the opposite results in their nationwide study—that is, persons aged 21 to 24 drank at lower rates if they had to wait until 21 to have legal access to alcohol. A frequently heard related argument is that a minimum drinking age of 21 may reduce car crashes among teenagers, but this will only be a temporary effect if it simply delays those problems until the teenagers reach age 21. This argument is also false. The minimum age of 21 significantly reduces car crashes among 18- to 20-year-olds, and those injuries and deaths are permanently saved. There is, furthermore, no rebound effect at age 21; in fact, the higher legal age appears to produce benefits, in terms of reduced drinking, that continue into a person's early twenties.

The debate surrounding the legal age for drinking appears settled in the United States. However, other countries (particularly in Europe where drinking ages are typically set at 18) are now examining the research and experience of the United States with increasing interest. Professionals in the areas of public health and traffic safety, as well as other professionals and citizens, are beginning to see the benefits of the age-21 drinking law in the United States, and they are initiating in their own countries the debate on the most appropriate age for legal access to alcohol.

(SEE ALSO: *Accidents and Injuries from Alcohol; Driving, Alcohol, and Drugs; Driving Under the Influence; Social Costs of Alcohol and Drug Abuse*)

BIBLIOGRAPHY

ARNOLD, R. D. (1985). *Effect of raising the legal drinking age on driver involvement in fatal crashes: The experience of thirteen states*. National Highway Traffic Safety Administration Technical Report (DOT HS 806-902). November.

BAKER, S. P., ET AL. (1992). *The injury fact book* (2nd ed.). New York: Oxford University Press.

FORSTER, J. L., ET AL. (1993). *Alcohol merchants' practices as a factor in underage drinking*. Paper presented at the Society of Behavioral Medicine's Fourteenth Annual Scientific Sessions, San Francisco, CA, March 10–13.

GUY, HON. R., JR. (1978). *Opinion*. Ref. Civil No. 8-73015 and Civil No. 8-73159. Detroit, MI: U.S. District Court, Eastern District of Michigan, Southern Division. December 22.

JOHNSTON, L. D., O'MALLEY, P. M., & BACHMAN, J. G. (1991). *Drug use among American high school seniors, college students and young adults, 1975–1990. Vol. 1. High school seniors*. (U.S. Department of Health and Human Services Pub. no. ADM 91-1813). Rockville, MD: National Institute on Drug Abuse.

KANDEL, D. B. (1989). Issues of sequencing of adolescent drug use and other problem behaviors. *Drugs and Society*, 3, 55–76.

MAISTO, S. A., & RACHAL, V. (1980). Indications of the relationship among adolescent drinking practices, related behaviors, and drinking-age laws. In H. Wechsler (Ed.), *Minimum drinking-age laws*. Lexington, MA: D. C. Heath, Lexington Books.

NATIONAL HIGHWAY TRAFFIC SAFETY ADMINISTRATION. (1990). *Alcohol and highway safety 1989: A review of the state of knowledge*. Washington, DC: U.S. Department of Transportation.

O'MALLEY, P., & WAGENAAR, A. C. (1991). Effects of minimum drinking age laws on alcohol use, related behaviors, and traffic crash involvement among American youth 1976–1987. *Journal of Studies on Alcohol*, 52, 478–491.

PLANT, M. A. (1990). Alcohol, sex and AIDS. *Alcohol & Alcoholism*, 25(2/3), 293–301.

PREUSSER, D. F., & WILLIAMS, A. F. (1991). *Sales of alcohol to underage purchasers in three New York counties and Washington, D.C.* Arlington, VA: Insurance Institute for Highway Safety.

SEMLITZ, L., & GOLD, M. S. (1986). Adolescent drug abuse: Diagnosis, treatment, and prevention. *Psychiatric Clinics of North America*, 9, 455–473.

STRUNIN, L., & HINGSON, R. (1992). Alcohol, drugs, and adolescent sexual behavior. *International Journal of the Addictions*, 27(2), 129–46.

U.S. GENERAL ACCOUNTING OFFICE. (1987, March). *Drinking-age laws: An evaluation synthesis of their impact on highway safety*. Washington, DC: U.S. Government Printing Office.

WAGENAAR, A. C. (1993). Minimum drinking age and alcohol availability to youth: Issues and research needs. *Alcohol and Health Monograph*, no. 25 (Economics and the Prevention of Alcohol-related Problems). Rockville, MD: National Institute on Alcohol Abuse and Alcoholism.

WAGENAAR, A. C. (1992). Alcohol and injuries: Epidemiology and control. Presented at the "Healthy People/ Healthy Environments" Secretary's National Conference on Alcohol-Related Injuries. Washington, DC, March 23–25.

WAGENAAR, A. C. (1983). *Alcohol, young drivers, and traffic accidents: Effects of minimum-age laws.* Lexington, MA: D.C. Heath, Lexington Books.

WAGENAAR, A. C., ET AL. (1993a). Where and how adolescents obtain alcoholic beverages. *Public Health Reports, 108*(4), 459–464.

WAGENAAR, A. C., ET AL. (1993b). *Enforcement of the legal minimum drinking age.* Washington, DC: National Highway Traffic Safety Administration, U.S. Department of Transportation.

ALEXANDER C. WAGENAAR

MINNESOTA MULTIPHASIC PERSONALITY INVENTORY (MMPI)

This is a self-report test containing 550 statements that can be answered true or false (Levitt & Durkworth, 1984). It was first published in 1943 for use in routine diagnostic assessment. As one of the most widely used psychological tests, the MMPI is sometimes given to alcoholics and drug users to evaluate the psychological effects of substance use as well as the personality characteristics of substance abusers.

The MMPI is scored in subunits or scales. Eight scales comprise the main parts of the clinical profile, which is a standard way of describing the patient's personality features in relation to population norms. The clinical scales measure hypochondriasis, depression, hysteria, psychopathic deviancy, paranoia, psychasthemia, schizophrenia, and hypomania.

The MMPI has three main applications to the diagnosis and study of substance-use disorders. First, it has been used to evaluate the effects of alcohol and drug abuse. Several studies (Pettinati et al., 1982; Babor et al., 1988) have found that MMPI clinical scales measuring depression, paranoia, and other psychiatric symptomatology tend to be higher than normal when alcoholics are drinking—but return to the normal range during periods of abstinence. Second, the MMPI has been used to identify subtypes of alcoholics and drug users that might benefit from specialized treatments. For example, several studies have found three types of alcoholics based on their MMPI profiles: neurotic, psychotic, and psychopathic (Conley, 1981; Nerviano & Gross,

1983). Third, the MMPI has been used in the development of screening tests. The MacAndrew scale (MacAndrew, 1965), for example, is used to measure impulsivity, pressure for action, and acting-out potential that may lead to alcoholism and drug abuse. Persons who score high on the MacAndrew scale are therefore considered to be at risk for substance-use disorders.

(SEE ALSO: *Addiction Severity Index; Diagnostic and Statistical Manual; Disease Concept of Alcoholism and Drug Abuse; Michigan Alcohol Screening Test*)

BIBLIOGRAPHY

BABOR, T. F., ET AL. (1988). Unitary versus multidimensional models of alcoholism treatment outcome: An empirical study. *Journal of Studies on Alcohol, 49*, 167–177.

CONLEY, J. J. (1981). An MMPI typology of male alcoholics: Admission, discharge and outcome comparison. *Journal of Personality Assessment, 45*, 33–39.

LEVITT, E. E., & DUCKWORTH, J. C. (1984). *Minnesota multiphasic personality inventory.* In D. J. Keyser & R. C. Sweetland (Eds.), *Test critiques.* Kansas City, MO: Test Corporation of America.

MACANDREW, C. (1965). The differentiation of male alcoholic outpatients from non-alcoholic psychiatric patients by means of the MMPI. *Quarterly Journal of Studies on Alcohol, 26*, 238–246.

NERVIANO, V. J., & GROSS, H. W. (1983). Personality types of alcoholics on objective inventories. *Journal of Studies on Alcohol, 44*, 837–851.

PETTINATI, H. M., SUGARMAN, A. A., & MAURER, H. S. (1982). Four year MMPI changes in abstinent and drinking alcoholics. *Alcoholism, Clinical and Experimental Research, 6*, 487–494.

THOMAS F. BABOR

MINORITIES AND DRUG USE

See Ethnicity and Drugs.

MMPI

See Minnesota Multiphasic Personality Inventory.

MONEY LAUNDERING

A myriad of financially motivated crimes, from DRUG TRAFFICKING and prostitution to securities and insurance fraud, make up the burgeoning underground economy. Estimates of the amounts of cash generated in the so-called illegal sector are generous, but due to enormous measurement problems, are of dubious accuracy. Projections of drug trafficking profits alone have ranged from 60 billion to over 300 billion dollars annually, just for the U.S. portion of the trade. Criminals involved in activities that generate these vast profits take a variety of steps to prevent discovery of their cash accumulations, which stand as red flags alerting wary investigators to the crimes that generated the enormous sums. Drug traffickers and other criminals also seek to avoid punishment for income tax evasion and crimes they commit in the course of continuing to conceal their ownership of illicit income by not reporting the drug-derived profits as income.

Money laundering is a conversion process that takes large amounts of "dirty" cash and transforms it into other mediums of exchange that serve the purpose of disguising ownership, not directly traceable—if traceable at all—to the underlying crimes. Departing from the literal meaning of the metaphor, the process of laundering only gives the appearance that the money (or other asset) is legitimate, or "clean"; it cannot perform the sanitizing function of converting the money into something that is pristine and untainted by the money's illicit origin.

Depending on the objectives of individual criminals for such factors as convenience and security, the laundering process can be effected with as few as one and as many as a dozen discrete steps. In its most familiar form, hundreds of thousands of dollars in drug proceeds are taken to a financial institution and exchanged for a cashier's check, which the trafficker can carry around (or out of the country) with much less suspicion than suitcases full of cash. A slightly more involved scenario entails taking the same cash to the same bank, where it is deposited into an account and then sent by wire transfer to a bank in a foreign country; probably a jurisdiction renowned for the relative secrecy it affords customers like the hypothetical drug dealer.

In even more elaborate schemes, the same funds are wire transferred around a circuit of accounts in different countries, bearing the names of legitimate businesses. After the transfer reaches its final destination abroad, the owner in the United States arranges a sham loan transaction, thereby bringing the funds back into this country as the proceeds of a pur-

ported loan. There are literally countless varieties of laundering schemes, bounded only by limits on the imaginations of criminals and a more widespread impatience with sending one's funds too far away.

In 1970, a law was enacted to help law-enforcement agencies track large cash movements in and out of financial institutions and both currency and monetary instrument movements across the U.S. border. Known as the Bank Secrecy Act, the Congressional intent of this legislation was to virtually force institutions accepting large cash deposits to complete a report, to be filed later with the U.S. Treasury Department, recording the particulars of the transaction and the person making it. Ideally, the government would then have a "paper trail" that begins when such large cash transactions leave the recordless underground economy and surface in banks and other financial institutions. The Bank Secrecy Act does not *per se* punish laundering but merely tries to compel financial institutions to report large cash transactions that may involve currency derived from criminal activity.

The early years of the Bank Secrecy Act were noted for notoriously low compliance by financial institutions. In addition to stiffening the penalty structure, in 1986 Congress passed a law that finally proscribed the act of knowingly converting currency or other crime-derived property to another form of asset. Known as the Money Laundering Control Act, this legislation has enabled federal enforcement agencies to target the financial infrastructure of drug trafficking and organized crime groups, as well as individuals whose laundering transactions attempted to conceal corruption, fraud, and even acts of terrorism.

(SEE ALSO: *Colombia as Drug Source; Drug Laws: Financial Analysis in Enforcement; International Drug Supply Systems; Terrorism and Drugs*)

CLIFFORD L. KARCHMER

MONITORING THE FUTURE *See* High School Senior Survey.

MONOAMINE A monoamine is an amine that has one organic substituent attached to the nitrogen atom (as RNH_2). SEROTONIN is such an amine, one that is functionally important in NEURO-TRANSMISSION. Chemically, monoamines include the catecholamines (derived from tyrosine) and the indoleamines serotonin and melatonin (derived from the amino acid tryptophan). Acetylcholine also has only a single (but trimethylated) amine, while histamine (a diamine formed from histidine) stretches the condition only slightly. Neurotransmitters in this class share several properties—nanomolar concentrations/milligram protein; neurons (nerve cells) that contain thin, generally unmyelinated axons to many brain regions; and their receptors (except for the cholinergic nicotinic receptor and one of the ten or so subtypes of serotonin receptors) employ second-messenger coupled transduction. Monoamine neurotransmitters are often involved in the action of mind-altering drugs and have been well studied.

(SEE ALSO: *Dopamine, Neurotransmitters*)

BIBLIOGRAPHY

SNYDER, S. H. (1980). *Biological aspects of mental disorder.* New York: Oxford University Press.

FLOYD BLOOM

MOOD AND DRUGS *See* Research: Measuring Effects of Drugs on Mood.

MOONSHINE Moonshine (white lightning) is the colloquial term for illegally produced hard liquor—whiskey, rum, brandy, gin, and vodka. The term probably originated around 1785, when it was recorded in a British book on vulgar language—used to describe the white (clear) brandy that was smuggled to the coasts of Kent and Sussex in England. In the New World, moonshine was made in homemade stills, usually from corn, especially in rural areas in the southern United States—before, during, and after Prohibition—and continues to be made today. The ethanol (drinking ALCOHOL) content is usually high, often approaching 80 percent (160 proof). First-run moonshine contains a number of impurities, some of which are toxic, so it is necessary to double and triple distill the liquor to purify it for drinking.

(SEE ALSO: *Alcohol: History of Drinking; Legal Regulation of Drugs and Alcohol; Still*)

BIBLIOGRAPHY

DABNEY, J. E. (1974). *Mountain spirits.* New York: Scribner's.

S. E. LUKAS

MORNING GLORY SEEDS The seeds of the morning glory, genus *Ipomoea* of the family Convolvulaceae, contain many lysergic acid derivatives, particularly lysergic acid amide. The hallucinogenic properties of some of these derivatives are not known. The seeds can be ingested whole; they can be ground and used to prepare a tea; or the active compound can be extracted using solvents. The seeds have also been used as a source of precursors for the synthesis of ·LYSERGIC ACID DIETHYLAMIDE (LSD). Since the seeds contain lysergic acid derivatives, people ingesting morning glory seeds may feel "different"; however, the experience is not identical to an LSD–type "trip," even though the seeds are marketed on the street as an LSD equivalent.

Although morning glory seeds are easy to purchase legally, many varieties (those sold by reputable garden-supply distributors) have been treated with insecticides, fungicides, and other toxic chemicals—as well as with compounds that will induce vomiting if the seeds are eaten.

(SEE ALSO: *Hallucinogenic Plants; Mescaline*)

Figure 1
Morning Glory

BIBLIOGRAPHY

Efron, D. H., Holmstedt, B., & Kline, N. S. (Eds.) (1979). *Ethnopharmacologic search for psychoactive drugs.* New York: Raven Press.

DANIEL X. FREEDMAN
R. N. PECHNICK

MORPHINE Morphine is a major component of OPIUM, a product of the poppy plant (PAPAVER SOMNIFERUM or *P. album*). Named after Morpheus, the Greek god of sleep, morphine is a potent ANALGESIC (painkiller) that is widely used for moderate to severe PAIN. Morphine is one of approximately twenty ALKALOIDS in opium; it was first purified in 1806 and, by the mid-1800s, pure morphine was becoming widely used in medicine. At approximately the same time, the hypodermic needle and syringe was developed, which permitted the injection of the drug under the skin (subcutaneous, S.C.), into muscles (intramuscular, I.M.), or directly into the veins (intravenous, I.V.). Together, these routes of administration are termed parenteral. Injections provide rapid relief of pain and can be used in patients who are unable to take medications by mouth. These advantages led to the wide use of morphine injections during the American Civil War (1861–1865). At that time, the intense euphoria and addictive potential of these agents following injections was not fully appreciated, leading to the addiction of a large number of soldiers. Indeed, morphine was not illegal and was sold over the counter; ADDICTION soon became known as the Soldier's Disease.

Since that time, a major objective of pharmaceutical companies has been to develop, for medication purposes, a nonaddictive analgesic with the potency of morphine. The concepts of PHYSICAL DEPENDENCE and addiction were not clearly differentiated until the late twentieth century, and it is likely that most of those early addicts were attempting to prevent the onset of WITHDRAWAL symptoms. Today very few patients become addicted to opiates, despite the fact that with continued administration all will become physically dependent—this may reflect our better understanding of the drugs plus our ability to take a patient off medications without precipitating withdrawal symptoms.

Morphine produces a wide variety of actions, some desired and others not. The definition of a desired action and a side effect depends on the reason

Figure 1
Morphine

for using the drug. For example, opiates such as morphine can be used to treat diarrhea—but their constipating actions are usually considered an undesirable side effect when they are used to treat pain.

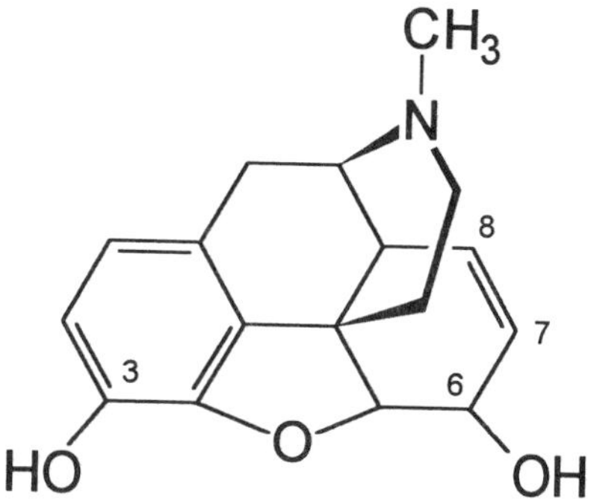

Clearly, the control of pain remains the most important use for morphine. Morphine and other OPIATES relieve pain without interfering with traditional sensations. Patients treated with morphine often report that the pain is still there but that it no longer hurts. Morphine works through *mu* opiate RECEPTORS located both within the brain and the spinal cord. Morphine has a number of other actions as well. Its ability to constrict the pupil is one of the most widely recognized signs of opiate use. In addition, morphine produces sedation and, at higher doses, morphine will depress respiration. Very high doses of morphine will stop breathing entirely—a common occurrence in overdoses.

Morphine also has a major influence upon the gastrointestinal tract, which is the basis for its antidiarrheal effect. Here, morphine decreases the motility of the stomach and intestine, through local actions on the organs themselves, as well as through control systems located within the brain and spinal cord. Other systems can be affected as well. Morphine produces a vasodilation, in which the peripheral blood vessels are relaxed. This can lead to significant drops in blood pressure when a person shifts from a lying to a standing position as the blood is pooled in the legs. This ability to pool blood in relaxed blood vessels can be used clinically to treat conditions such as acute pulmonary edema, an accumulation of fluid within the lungs, which occurs in acute myocardial infarctions (heart attacks). Increasing the capacity of the vascular system by relaxing the blood vessels permits the reabsorption of the lung fluid. Finally, morphine and similar drugs, such as CODEINE, are also effective agents in the control of coughing.

All these effects of morphine can be easily reversed by ANTAGONISTS. NALOXONE is the most widely used antagonist. Given alone, it has virtually no actions; however, low doses of naloxone are able to block or reverse all the actions of morphine described above.

Morphine is given either by mouth or by injection. Oral administration is associated with a significant metabolism of the drug by the liver, explaining its lower potency as compared to that attained by injections. From three to six times more morphine must be taken by mouth to produce the same effects as an injected dose. Thus, higher doses are needed when giving the drug orally. Morphine injections can be given either intramuscularly, subcutaneously, or intravenously. Continuous infusions are also becoming more common, but their use is restricted to physicians expert in the treatment of pain. Morphine has a relatively short half-life in the body, around two hours, and it is usually given to patients every four to six hours. It is extensively metabolized. In the late 1980s, it was discovered that one of these metabolites, morphine-6β-glucuronide, is very potent, far more potent than morphine itself. The importance of this compound with a single morphine dose is probably not great; however, with chronic dosing, the levels of morphine-6β-glucuronide in the blood actually exceed those of morphine—so this metabolite may be responsible for most of morphine's actions. Since this metabolite is removed from the body by the kidneys, special care must be taken when giving morphine to patients with kidney problems.

One common problem associated with morphine is nausea. This is difficult to understand, since nausea does not occur in all patients and often is seen with one drug but not others. This lack of consistency raises questions about whether it is a specific receptor-mediated action or whether it may be a nonspecific side effect.

With chronic use, morphine has a progressively smaller effect, a phenomenon termed TOLERANCE. To maintain a constant action it is necessary to increase the dose. Along with tolerance, morphine also produces physical dependence. Physical dependence (physiological dependence; neuroadaptation) develops as the body attempts to compensate for many of morphine's actions. As long as a person continues to receive the drug, no symptoms are noted. Abrupt cessation of the drug or the administration of an antagonist, such as naloxone, produces a constellation of symptoms and signs termed the withdrawal syndrome. Early symptoms include a restlessness, tearing from the eyes and a runny nose, yawning, and sweating. As the syndrome progresses, one sees dilated pupils, sneezing, elevations in heart rate and blood pressure, and gooseflesh—which is responsible for the term "cold turkey." Cramping and abdominal pains are also common.

Physical dependence (or neuroadaptation) is a physiological response to repeated dosing with morphine and is seen in virtually all patients. Physical dependence, however, is not the same as addiction (drug dependence). Drug dependence (addiction) is defined as drug-seeking behavior, whereas physical dependence is simply a physiological response to the medication. While addiction is common among drug abusers, it is rare when morphine is used for appropriate medical conditions. The reasons for this difference are not clear, and they remain a major issue in understanding and treating morphine addiction.

(SEE ALSO: *Addiction: Concepts and Definitions; Diagnostic and Statistical Manual; Opiates/Opioids; Opioid Complications and Withdrawal*)

BIBLIOGRAPHY

JAFFE, J. H., & MARTIN, W. R. (1990). Opioid analgesics and antagonists. In A. G. Gilman et al. (Eds.), *Goodman and Gilman's the pharmacological basis of therapeutics*, 8th ed. New York: Pergamon.

GAVRIL W. PASTERNAK

MOTHERS AGAINST DRUNK DRIVING (MADD)

This organization works to reduce DRUNK DRIVING and to help the victims of drunk-driving ACCIDENTS. Many of MADD's members are volunteers who have personally suffered from the results of drunk driving. This national organization was founded by Candy Lightner, whose thirteen-year-old daughter, Cari, was killed by a drunk driver on May 3, 1980. Ms. Lightner was outraged to learn that only two days previously the driver had been released from jail, where he had been held for another hit-and-run drunk-driving crash. Although he had been arrested for drunk driving several times before, he was still driving with a valid California license. Candy Lightner decided to begin a campaign to keep drunk drivers off the road, so that other mothers would not have to suffer the anguish that she was experiencing. On September 5, 1980 (Cari's birthday), MADD was incorporated.

Since then, MADD has evolved into an organization with millions of members and hundreds of local chapters across the United States. Chapters have also been started in Canada, Great Britain, New Zealand, and Australia. Membership is not restricted to mothers of victims or to the victims themselves. Everyone who is concerned about the drunk-driving issue is welcome to join. Funding for the organization comes from membership dues and contributions; MADD also applies for and receives grants from federal and state governments and private organizations. Paid staff are employed to provide leadership on the state and national levels. MADD is involved in three major kinds of activity (1) advocacy for stricter drunk-driving laws and better enforcement, (2) promotion of public awareness and educational programs, and (3) assistance to victims.

THE LEGISLATIVE AGENDA

According to MADD, drunk driving is a violent crime. One of its rallying slogans is, "Murder by Car Is Still Murder!" Over the years, MADD members have worked to generate public support for passage of stricter drunk-driving legislation, punitive sanctions, and more consistent enforcement measures aimed at deterring drunk driving. In the 1980s, intense lobbying efforts were undertaken for the passage of laws making 21 the minimum legal age for drinking (now in force in all 50 states). The group believes that this measure has saved thousands of young lives that would have been lost in drunk-driving crashes.

MADD has also lobbied for changes in judicial procedures that would make the system more responsive to victims of drunk driving. For example, in many states victims had been barred from the courtroom during the trial of their own drunk-driving cases, because their testimony (or even their presence) might prejudice the jury. Owing to the efforts of MADD and other groups, victims' rights bills have now been passed in all states. These ensure that victims will be notified about court hearings and, in most states, allowed to testify about the impact of the crime on their lives. Other lobbying efforts have sought to close legal loopholes that drunk drivers

were using to avoid punishment. For example, drivers might have refused to take a breath or blood test for intoxication and have been allowed to plead guilty to a lesser charge. In other cases, drivers were allowed to claim that despite their high blood-alcohol content (BAC), their driving was not really impaired.

MADD has been instrumental in the passage of over 1,000 tougher drunk-driving laws that close these loopholes and institute other deterrence measures, such as mandatory jail sentences for drunk drivers. MADD also supports efforts to require offenders to undergo treatment for alcoholism and/or drug dependency, if this is deemed necessary.

PUBLIC AWARENESS AND EDUCATION

MADD is involved in various efforts to raise public awareness and concern about drunk driving. The "National Candlelight Vigil of Remembrance and Hope" is held in many locations each December, drawing victims together to give public testimony to the suffering that results from drunk driving. During the "Red Ribbon Tie One On for Safety" campaign, which takes place between Thanksgiving and New Year's Day, MADD encourages citizens to attach a red ribbon to their car as a reminder to themselves and others to drive sober. MADD's well-known public awareness campaign of the past used the slogan, "Think . . . Don't Drink and Drive" in public-service announcements on radio and television and in print materials. A more recent campaign, "Keep It a Safe Summer" (KISS) emphasized the need for sobriety during recreational activities that involve driving, boating, or other risky activities. MADD also provides curriculum materials for schools and each year sponsors a poster and essay contest for children on the subject of drunk driving.

ASSISTANCE TO VICTIMS

Programs that provide aid to victims of drunk-driving crashes constitute the heart of MADD's mission. Support groups help victims share their pain with others who understand their feelings. MADD members send "We Care" cards to victims of recent crashes. Specially trained victim advocates offer a one-on-one personal relationship with victims, trying to respond to both their emotional and practical needs. Victims are briefed on their legal rights and on the judicial procedures relevant to their cases. They can call a toll-free number (1-800-GET MADD) for information and for help in case of crisis. MADD also offers death-notification training for police and specialized training for other community professionals, such as clergy and medical workers, who are called upon to assist victims.

20 × 2000

Since the founding of MADD in 1980, the percentage of alcohol-related traffic fatalities has steadily decreased from almost 60 percent to around 50 percent. MADD's current goal "20 × 2000" seeks to reduce that proportion by an additional 20 percent by the year 2000. Intensified efforts will focus on more effective law enforcement, increased sanctions, and prevention programs that include education for youth and more responsible marketing and service practices in liquor establishments.

(SEE ALSO: *Blood Alcohol Concentration, Measures of; Blood Alcohol Content; Breathalyzer; Dramshop Laws; Driving, Alcohol, and Drugs; Driving Under the Influence; Legal Regulation of Drugs and Alcohol; Minimum Drinking Age Laws; Psychomotor Effects of Alcohol and Drugs; Remove Intoxicated Drivers; Students Against Driving Drunk*)

BIBLIOGRAPHY

BLOCH, S. A., & UNGERLEIDER, S. (1988). Whither the drunk driving movement? The social and programmatic orientations of Mothers Against Drunk Driving. In F. B. Dickman (Ed.), *Alcohol and traffic safety.* New York: Pergamon.

LIGHTNER, C. (1987). Youth and the road toll. In P. C. Noordzij & R. Roszbach (Eds.), *Alcohol, drugs and traffic safety.* Amsterdam: Elsevier.

MANN, P. (1985). *Arrive alive.* New York: McGraw-Hill.

SADOFF, M. (1990). *America gets MADD.* Irving, TX: MADD.

DIANNE SHUNTICH

MPTP To circumvent the laws regarding controlled drugs, a chemist attempted to synthesize a derivative of MEPERIDINE. By synthesizing a new derivative not specifically covered by the CONTROLLED SUBSTANCES ACT and existing Drug Enforcement Agency laws and by synthesizing the drug and selling it within the same state, the chemist had hoped to profit but to avoid violation of the laws. This DESIGNER DRUG approach was being widely used to avoid prosecution for selling drugs of abuse — however, in this case a side product was also formed in this reaction, MPTP (1-methyl-4-phenyl-1,2,3,6-tetrahydropyridine). People who bought this mixture on the street quickly developed a neurological syndrome virtually indistinguishable from Parkinson's disease. Initially, the cause of this problem remained unknown. With intense investigation, the blame was placed on the side product in the reaction, MPTP. MPTP had long been used as an intermediate in chemical synthesis and was commercially available. The ability of MPTP to provoke a Parkinson-like syndrome helped explain a report from years ago of a chemist working with this compound suddenly developing a Parkinson-like disease.

The Parkinson-like syndrome is very similar to the symptoms originally described in Parkinson's disease. The most notable aspects of the syndrome are the marked cog-wheel rigidity of the muscles, along with a generalized decrease in movement usually associated with problems initiating the movement. Patients often have difficulty with fine motor skills, such as writing, and with walking, which usually becomes a series of small, shuffling steps termed a "festinating gait"; their greatest problem is starting and stopping. Diminished blinking coupled with a limited facial expression can be very prominent and is termed "masked facies." In Parkinson's disease, patients also have a pill-rolling tremor and a tendency to fall, because of problems with blood pressure and the reflexes important to maintaining posture.

Pathologically, Parkinson's disease is noted for a degeneration of pigmented nuclei within the brain, including the substantia nigra. The loss of the dopaminergic NEURONS in the substantia nigra that project to the part of the brain called the striatum is responsible for the motor problems, while the degeneration of other areas of the brain, including the locus coeruleus, are presumably responsible for the autonomic problems. The cause of Parkinson's disease is still not known; treatment is symptomatic. Early studies demonstrated the ability of anticholinergic medications to help with many of the motor symptoms, especially the tremor. However, the drug of choice in the 1990s is L-dopa, a precursor of DO-

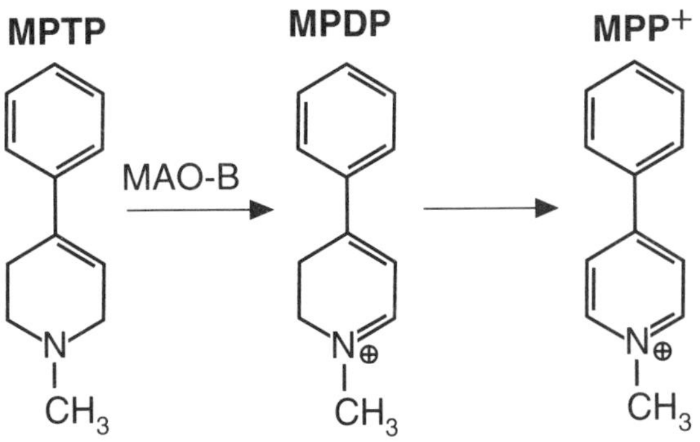

MPTP　　　**MPDP**　　　**MPP⁺**

Figure 1
MPTP Conversion to MPDP and MPP⁺

PAMINE. Unlike dopamine, which does not traverse the blood–brain barrier, L-dopa is readily transported into the brain where it is taken up into neurons and converted to dopamine—thereby helping to reduce symptoms caused by loss of dopamine-containing neurons. Replacement of the dopamine can markedly limit the severity of the motor symptoms; however, the duration of this benefit is often limited to only about five years, presumably due to the continued progression of the disease.

MPTP does not bind to OPIOID RECEPTORS and it has no opioid activity, although it is a side product in the synthesis of a meperidine analog. When ingested, it is taken up into neurons containing a catecholamine transporter, greatly limiting the neurons affected. Once in the cell, the drug is converted by the enzyme monoamine oxidase (type B) in a series of steps to another compound, MPP+, which is believed to be responsible for its toxic actions. The need for the transporter to take up the toxin into the cells partially explains its selective toxicity within the brain. There, this drug destroys the same groups of pigmented catecholinergic neurons affected in Parkinson's disease, including the substantia nigra and the locus coeruleus. The greater sensitivity of pigmented neurons to the toxin is still not completely understood. One hypothesis has been put forward: The color in the neurons is due to the pigment melanin, which actively binds the toxin. Therefore, it has been suggested that this binding results in the accumulation of very high levels of the drug, which persist in the neurons for long periods of time, enhancing its toxicity.

Clinically, MPTP produces a syndrome virtually identical to that seen in Parkinson's disease, but Parkinson's is a progressive degenerative disease, which,

over the period of many years, gradually leads to a variety of difficulties with thought and memory. It is not thought that MPTP produces a similar global, diffuse loss of function. The marked similarity, though, has led to the speculation that Parkinson's may be due to the exposure to a toxin similar to MPTP. Since the toxicity of MPTP depends on its conversion by type B monoamine oxidase (MAO-B), it has been suggested that inhibition of this enzyme may prove beneficial. Seligine is a selective MAO-B inhibitor, and early clinical trials suggest that the progression of Parkinson patients taking this medication may be slower than in the control groups.

BIBLIOGRAPHY

CEDARBAUM, J. M., SCHLEIFER, L. S. (1990). Drugs for Parkinson's disease, spasticity, and acute muscle spasms. In A. G. Gilman et al. (Eds.), *Goodman and Gilman's the pharmacological basis of therapeutics*, 8th ed. New York: Pergamon.

GAVRIL W. PASTERNAK

MULES　　*See* Slang and Jargon.

MULTIDOCTORING　　Multidoctoring, or double-doctoring, is the practice of obtaining medications from more than one physician without informing the other physician(s) involved of any medication already prescribed. Almost always, the medications involved are PSYCHOACTIVE medications, which may then be abused or misused. Individuals who engage in this behavior may be obtaining the medication for their own use or for the purpose of diverting it to sell on the street. People who seek drugs for the purpose of selling them on the street are often very convincing in their appeals and can get the physician to prescribe the particular drug they are after without even mentioning it by name. In Canada and the United States, legislation prohibits people from acquiring a narcotic prescription without informing the physician of other narcotics that have already been for them prescribed that month. Failure to do so results in criminal charges. Physicians can record a patient's response to the question about other prescribed narcotics, and psychoactive drugs in general, as a means of discouraging multidoctoring.

Physicians themselves may be involved at various levels in multidoctoring and the diversion of drugs to the street. These are the physicians termed "script doctors," who willfully prescribe controlled substances to people seeking them, or who prescribe them as a result of being misled or simply uniformed about the prevalence of multidoctoring and the substances involved. Educating the public regarding the risks of prescription-medication abuse and increasing the skills of physicians in recognizing patients engaged in multidoctoring will help to decrease the diversion and misuse of prescription drugs.

(SEE ALSO: *Controls: Scheduled Drugs; Iatrogenic Addiction*)

BIBLIOGRAPHY

AMERICAN MEDICAL ASSOCIATION COUNCIL ON SCIENTIFIC AFFAIRS. (1992). Drug abuse related to prescribing practices. *Journal of the American Medical Association, 247*(6), 864–866.

GOLDMAN, B. (1987). Confronting the prescription drug addict: Doctors must learn to say no. *Canadian Medical Association Journal, 136*, 871–876.

MYROSLAVA ROMACH
KAREN PARKER

NA *See* Narcotics Anonymous (NA).

NALOXONE Naloxone is an OPIOID ANTAGO-NIST (i.e., a blocker of morphine-like agents) commonly used to reverse the actions of drugs such as morphine. In the early 1990s, it was the treatment of choice for reversing the life-threatening effects of opioid overdose. Structurally, naloxone is very closely related to OXYMORPHONE, both compounds being derivatives of the opium alkaloid thebaine. Indeed, the structural differences between oxymorphone and naloxone are minimal; they are restricted to a simple substitution on the nitrogen atom. Oxymorphone has a methyl group whereas naloxone has an allyl substitution. This small substitution changes the pharmacology of the compound dramatically. Whereas oxymorphone is a potent ANALGESIC with actions very similar to MORPHINE, naloxone has no

analgesic actions by itself and instead has the ability to antagonize, or reverse, virtually all the effects of morphine-like drugs. This ability to reverse opiate actions has proven valuable clinically. However, giving naloxone to opiate addicts will immediately precipitate WITHDRAWAL symptoms.

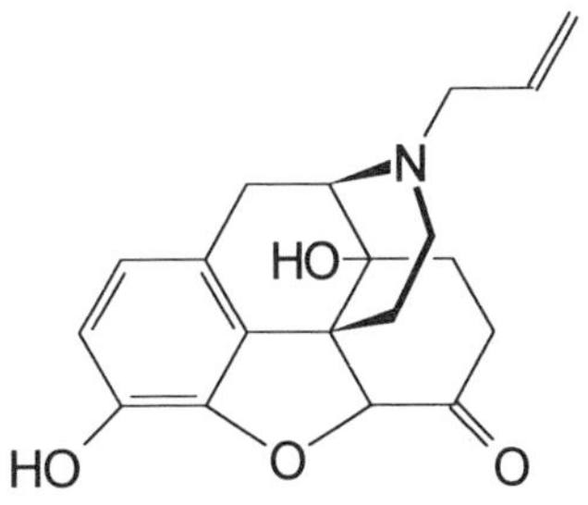

Naloxone is rapidly metabolized in the liver to inactive compounds, resulting in a relatively brief duration of action. When naloxone is used clinically to reverse the actions of morphine and other OPIATES, care must be taken to ensure that the drug being reversed does not last longer than the naloxone. Should that happen, a patient may be revived by naloxone only to relapse back into a coma or even die from the side effects of the initial opioid AGONIST. Despite its effectiveness following injection, naloxone is not very active when given orally; this, together with its short duration of action, prevents its widespread use as a treatment for opioid addiction.

(SEE ALSO: *Naltrexone*; *Naltrexone in Treatment of Drug Dependence*; *Opioids: Complications and Withdrawal*)

BIBLIOGRAPHY

JAFFE, J. H., & MARTIN, W. R. (1990). Opioid analgesics and antagonists. In A. G. Gilman et al. (Eds.), *Goodman and Gilman's the pharmacological basis of therapeutics*, 8th ed. New York: Pergamon.

GAVRIL W. PASTERNAK

Figure 1
Naloxone

NALTREXONE Naltrexone is an OPIOID ANTAGONIST (i.e., a blocker of substances with morphine-like actions), with a structure very similar to another antagonist, NALOXONE. It also closely resembles the potent ANALGESIC (painkiller) OXYMORPHONE. The differences between naloxone and naltrexone are restricted to a simple substitution on the nitrogen atom, with naltrexone having a methylcyclopropyl group, yet this small substitution changes the pharmacology of the compound dramatically. Naltrexone has no analgesic actions by itself and has the ability to antagonize, or reverse, virtually all the effects of morphine-like drugs. Like naloxone, naltrexone will precipitate WITHDRAWAL in physically dependent people.

Naltrexone is rapidly metabolized in the liver, but one of its metabolites is 6-naltrexol, which has some activity and a longer duration of action. In the 1990s, naltrexone has been used to treat opiate addiction. Its greater potency than naloxone, along with its greater and longer activity after oral administration, has made this the antagonist of choice in the treatment of opioid addiction.

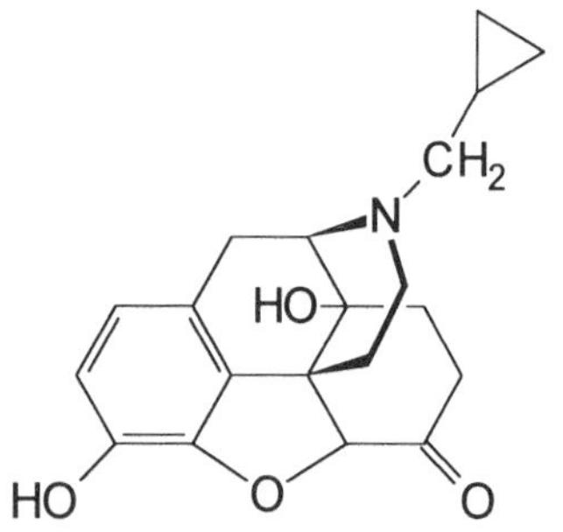

Figure 1
Naltrexone

In the early 1990s, several research groups reported that naltrexone, when given to alcoholic men following detoxification, reduced the likelihood of relapse to ALCOHOL. This finding seemed to support the hypothesis that some of the reinforcing (euphoric) effects of alcohol are due to interactions with naturally occurring opioid systems in the brain.

(SEE ALSO: *Naltrexone in Treatment of Drug Dependence; Treatment: Alcohol; Treatment Types: Pharmacotherapy*)

BIBLIOGRAPHY

JAFFE, J. H., & MARTIN, W. R. (1990). Opioid analgesics and antagonists. In A. G. Gilman et al. (Eds.), *Goodman and Gilman's the pharmacological basis of therapeutics*, 8th ed. New York: Pergamon.

GAVRIL W. PASTERNAK

NALTREXONE IN TREATMENT OF DRUG DEPENDENCE Naltrexone (brand names Trexan[R], Revia[R] [U.S.], Nalorex[R] [France, U.K.]) is a synthetic antagonist of opiate (morphine-like) drugs, which blocks their actions without itself having any opiate effects. Naltrexone differs from most other pure opiate antagonists in having a relatively long duration of action (at least 24 hours) and being effective when taken by mouth. These characteristics have led to its clinical use as a long-term or maintenance treatment for OPIATE and OPIOID dependence after detoxification. Naltrexone is also being studied experimentally as a possible treatment for cigarette smoking and eating disorders, and was approved in 1995 for treatment of alcoholism.

The use of opiate ANTAGONISTS as treatment for opiate dependence was first proposed by William Martin and Abraham Wikler and their colleagues at the U.S. Addiction Research Center in the early 1960s. They hypothesized that chronic administration of an opiate antagonist, by blocking the pleasurable or rewarding effects of opiate drugs, would lead to the extinction of drug-seeking and drug-taking behavior—since the addict would no longer receive any pleasurable effects from taking an opiate. With abstinence from opiates, PHYSICAL DEPENDENCE and any chronic withdrawal syndrome would dissipate, removing important factors that cause craving for opiates. They suggested that antagonist treatment would have several advantages over treatment with an opioid such as METHADONE. Since antagonists do not produce any pleasurable effects, the addict would have little incentive to misuse the medication or divert it to illegal channels. Chronic use of an antagonist would not produce physical dependence, and an overdose of antagonist would not cause life-threatening opiate effects such as suppression of breathing. Use of the antagonist in nondetoxified opioid addicts, however, would cause an acute but not life-threatening withdrawal.

HISTORY

The earliest studies of opioid antagonists were not satisfactory, because of drawbacks in the then available antagonists. For example, NALOXONE was short-acting and not very effective when taken by mouth. Nalorphine and cyclazocine had some kappa-opioid effects (i.e., were not pure antagonists), which produced unpleasant side effects.

Further work was stimulated by the SPECIAL ACTION OFFICE FOR DRUG ABUSE PREVENTION created by President Richard M. Nixon in June 1971 as part of a "war on drugs." The 1972 funding legislation for this office called for research on "long-lasting, nonaddictive, blocking and antagonist drugs . . . for the treatment of heroin addiction." Eventually twenty-two studies with naltrexone (which had been synthesized by Blumberg and Dayton in 1965) were conducted at various treatment programs in the United States. These studies demonstrated the safety and effectiveness of naltrexone after detoxification as a long-term treatment for opiate dependence, leading to its marketing in North America and Europe. Its effectiveness, however, was defined in terms of blocking the effects of HEROIN, not in the success of changing the behavior of heroin users.

TREATMENT

Naltrexone is usually used in conjunction with counseling and other rehabilitation services, as part of a structured and monitored treatment program. The best treatment results tend to occur in highly motivated, psychologically healthy addicts who are employed and well-functioning socially, especially when they face severe economic or legal consequences for failing treatment. For example, addicted health professionals whose treatment is required by their professional licensing boards and monitored as a condition of continued licensure will regularly take naltrexone for several years and remain abstinent from opiates. Some programs have reported five-year success rates as high as 95 percent. Most street addicts (e.g., those with unstable living situations who support their drug use by criminal activity) refuse to take naltrexone or, if started in treatment, quickly drop out. This is believed due to the lack of reward effect. Many such addicts prefer maintenance treatment with the synthetic opiate methadone and others find even methadone nonrewarding, so they relapse.

Fifty milligrams of naltrexone block the effects of 25 milligrams of heroin for 24 hours, so the typical weekly naltrexone dose is 350 milligrams. The actual medication schedule is adjusted to the individual patient and may range from 50 milligrams every day to 150 milligrams every third day. Patients are put on the least frequent medication schedule possible to enhance patient cooperation and reduce the number of clinic visits. To further reduce medication scheduling, researchers are working on a depot form of naltrexone that can be injected once a month and which slowly releases the medication into the body.

Care must be taken to avoid administering naltrexone to individuals still physically dependent on opiates. In opiate-dependent individuals, an antagonist will precipitate an acute opiate withdrawal syndrome. While not life-threatening, this syndrome can be extremely uncomfortable, with symptoms such as abdominal cramps; diarrhea; muscle, joint, and bone pain; runny nose (rhinorrhea); and goose bumps (piloerection). To avoid this situation, naltrexone is not administered to patients until they have been free of opiate drugs for at least seven to ten days to allow dependence to wear off. To confirm the absence of dependence, patients may be challenged with the short-acting antagonist naloxone before starting on naltrexone. To shorten the required opiate-free period, some programs are experimenting with combined administration of naltrexone and CLONIDINE, a medication that reduces symptoms of opiate withdrawal.

Naltrexone was shown to reduce the rate of relapse of full-blown compulsive drinking by detoxified alcoholics, although it did not substantially increase the number who were totally abstinent. In one research study, naltrexone seemed to reduce craving for alcohol. In contrast with opioid addicts, alcoholics were more willing to take naltrexone.

(SEE ALSO: *Treatment/Treatment Types; Wikler's Pharmacologic Theory of Drug Addiction*)

BIBLIOGRAPHY

GINZBURG, H. M., & GLASS, W. J. (1984). The role of the National Institute on Drug Abuse in the development of naltrexone. *Journal of Clinical Psychiatry, 45*(9), 39–41.

GONZALEZ, J. P., & BROGDEN, R. N. (1988). Naltrexone: A review of its pharmacodynamic and pharmacokinetic properties and therapeutic efficacy in the management of opioid dependence. *Drugs, 35,* 192–213.

JULIUS, D., & RENAULT, P. (1976). Narcotic antagonists: Naltrexone progress report. NIDA research monograph no. 9. Washington, DC: U.S. Government Printing Office.

O'MALLEY, S. S., ET AL. (1992). Naltrexone and coping skills therapy for alcohol dependence. *Archives of General Psychiatry, 49,* 881–887.

VOLPICELLI, J. R., ET AL. (1992). Naltrexone in the treatment of alcohol dependence. *Archives of General Psychiatry, 49,* 876–880.

DAVID A. GORELICK

NARA *See* Narcotic Addict Rehabilitation Act (NARA).

NARCANON/SCIENTOLOGY *See* Cults and Drugs.

NARCOTERRORISM *See* Terrorism and Drugs.

NARCOTIC The term derives from the Greek *narkōtikos,* meaning benumbing. It was originally used (since the fourteenth century) to refer to drugs that produced a stupor associated with pain relief (analgesia), primarily OPIUM and its derivatives, the morphine-like strong ANALGESICS, or the opium-like compounds (OPIOIDS)—these, in moderate doses, dull the senses, relieve pain, and induce profound sleep but in large doses cause stupor, coma, or convulsions.

During the nineteenth century, the term was widely used to include a number of agents that produced sleep. Toward the end of the nineteenth century, the term came to imply drugs that could lead to addiction, and so by the turn of the twentieth century, "narcotic" came to describe drugs as diverse as opioids and COCAINE. During the twentieth century, the term became widely used in a legal context to refer to psychoactive drugs and drugs of abuse—those subject to restriction—as "addictive narcotics," whether in fact the agents were physiologically addictive and narcotic or not. This imprecise usage has left the term nebulous, although it is still used extensively in the media and by the general population. The term is no longer used in scientific discourse to categorize drugs.

(SEE ALSO: *Drug Types; Opiates/Opioids; World Health Organization Expert Committee on Drug Dependence*)

BIBLIOGRAPHY

JAFFE, J. H., & MARTIN, W. R. (1990). Opioid analgesics and antagonists. In A. G. Gilman et al. (Eds.), *Goodman and Gilman's the pharmacological basis of therapeutics,* 8th ed. New York: Pergamon.

GAVRIL W. PASTERNAK

NARCOTIC ADDICT REHABILITATION ACT (NARA) Public Law 89-793, the Narcotic Addict Rehabilitation Act (NARA), was passed by Congress in 1966. This legislation was designed to allow the use of the federal courts and criminal-justice system to compel drug addicts to participate in treatment. Several developments provided the context for this legislation. In the early 1960s, the problem of NARCOTIC drug use and ADDICTION were perceived to be increasing. There was also a perception that treatment was not particularly effective and that the RELAPSE rate was high. In response, California, in 1961, and New York, in 1962, passed legislation permitting the CIVIL COMMITMENT of narcotic addicts; that is, they could be compelled to accept treatment even if they had committed no crime but could be shown to be using illicit narcotic drugs. In both of these states the legislatures also provided substantial funds to establish residential facilities where addicts could be treated initially as well as aftercare programs to provide supervision following their release from the residential facilities. Several other states, including Illinois, passed similar civil commitment legislation, but only New York and California launched massive programs to implement compulsory treatment and civil commitment.

In January 1963, the Presidential Advisory COMMISSION ON NARCOTIC AND DRUG ABUSE appointed by President John F. Kennedy made a number of recommendations, including the enactment of a federal civil commitment statute that could provide an alternative to prison for confirmed narcotic or marijuana abusers convicted of federal crimes. The advisory commission also recommended increased assistance to states and municipalities to develop and strengthen their own treatment programs.

As passed by Congress, NARA had four titles, or main parts: *Title I* provided that eligible addicts charged with a federal offense could choose civil commitment or treatment instead of prosecution. After being examined by clinicians at a treatment center, an addict, if found suitable, could be committed to the custody of the surgeon general for thirty-six months of institutional treatment and aftercare. *Title II* provided for civil commitment after conviction. *Title III* stated that even if no federal crime had been committed, an addict or a related individual could petition the U.S. attorney in the district of residence and, if local facilities were unavailable, the U.S. District Court could commit the person to custody of the surgeon general for treatment. *Title IV* provided for funding to states and localities to establish or expand treatment for addicts.

Treatment under NARA began to be provided in 1967. The two U.S. PUBLIC HEALTH SERVICE HOSPITALS—in Lexington, Kentucky, and Fort Worth, Texas—which had been treating both addicted federal prisoners and voluntary patients, were redesignated "Clinical Research Centers" and became the sites for the institutional phase of treatment for addicts committed to the Surgeon General under NARA. Aftercare was provided by local programs supported by contracts with the NARA program administered by the Division of Narcotics within the National Institute of Mental Health (NIMH).

From 1967 through 1973, the two clinical centers admitted more than 10,000 NARA patients, 5 percent under Title I, 2 percent under Title II, and 93 percent under Title III. Women made up 15 percent of admissions. Race and ethnicity were noted for admissions between 1970 and 1973, during which time the designations and distribution were as follows: Anglo 43 percent, black 47 percent, Puerto Rican 1 percent, Mexican American 9 percent.

Many of the patients referred were found "not suitable for treatment" (38% at Fort Worth and 51% at Lexington), a designation that generally meant they were too disruptive or antagonistic. Some of this unsuitability was deliberate. Many of those under Title III, while not being charged with a federal crime, were under court pressure because of state or local crimes; as part of plea bargaining with local courts, they agreed to accept commitment under NARA Title III. They quickly learned that the centers would not require them to stay in residence, nor would NARA officials compel them to stay in aftercare. Once released from the centers as "not suitable," they would find ways to convey to the local courts how motivated for treatment they still were and how puzzled they were not to be offered treatment.

The general approach to treatment during the residential phase was based on THERAPEUTIC COMMUNITY principles, which delegate many responsibilities to former addicts and to patients participating in the program. The average duration of the residential phase of treatment was intended to be about 6 months, but of those admitted for examination, only about 35 percent were discharged to aftercare as having completed the residential phase. A number of studies have been conducted on the effectiveness of the NARA program, including aftercare. One study found that only 38 percent of the 35 percent that completed the residential phase remained in aftercare for the full six months after discharge from residential treatment. Reasons for attrition included death, disappearance, recommitment, conviction, and incarceration. One study by Gold and Chatham in 1971 found that 46 percent of addicts in aftercare had used an illegal drug during the month preceding the interview; about 50 percent were working. Another study found that 87 percent had used narcotics during the first six months after the residential phase; 65 percent had become readdicted.

While this rate of readdiction did not seem as bleak as that seen after the discharge of the early cohorts from Lexington, it was not seen as particularly successful—given the high cost of the six-month residential phase and the high attrition rates. Because of the attrition, the readdiction rate, while not inevitable, was occurring among only the better candidates. Another study by Mandell and Amsel (1973) compared the outcome of those treated compared to those found "not suitable" for treatment. The difference in outcome between the two groups was not significant.

While the legal authority for federal civil commitment remained in effect through the early 1990s, the actual application of NARA fell into disuse in the mid-1970s as more federal prisons developed programs for Title II offenders and as more communities developed their own treatment programs. The use of treatment under civil commitment also declined, because the involvement of courts and expensive legal procedures made it far more expensive than voluntary treatment. In 1971, the Fort Worth facility was closed and turned over to the Bureau of Prisons. The Lexington facility experienced the same fate in 1974.

(SEE ALSO: *California Civil Commitment Program; Civil Commitment; Coerced Treatment; New York State Civil Commitment Program*)

BIBLIOGRAPHY

GOLD, R., & CHATHAM, L. R. (1973). Characteristics of NARA patients in aftercare during June 1971. DHEW Publ. No. (HSM) 73-9054. Washington, DC: U.S. Government Printing Office.

MADDUX, J. F. (1978). History of the hospital treatment programs, 1935–1974. In W. R. Martin & H. Isbell (Eds.), *Drug addiction and the U.S. Public Health Service.* DHEW Publ. No. (ADM) 77-434. Washington, DC: U.S. Government Printing Office.

MANDELL, W., & AMSEL, Z. (1973). Status of addicts treated under the NARA program. Baltimore, MD: School of Hygiene and Public Health, The Johns Hopkins University.

JEROME H. JAFFE
JAMES F. MADDUX

NARCOTICS ANONYMOUS (NA) Even though the origins and strategies of Narcotics Anonymous (NA) are closely intertwined with those of AL-COHOLICS ANONYMOUS (AA), NA has devised its own unique adaptations to them. There is no question that NA's roots were in the AA program, but it soon came to realize its uniqueness and had to give AA's program its own "spin." Briefly sketched—an energetic, relatively new AA member, while doing twelve-step work in 1944, recruited an alcoholic who was also an abuser of MORPHINE (he used this drug to avoid hangovers). The AA program helped the recruit with alcohol, but not with morphine. He soon found himself an involuntary patient in the U.S. PUBLIC HEALTH SERVICE HOSPITAL in Lexington, Kentucky.

In the meantime, his AA sponsor, who was much puzzled by AA's help with alcohol but not other drugs, was transferred to Frankfort, Kentucky, near the Lexington Hospital. He (dubbed "Houston" in a *Saturday Evening Post* article) reportedly repeated to himself, "I was convinced that the TWELVE STEPS would work as well for drugs as for alcohol" (Ellison, 1954:23). As a result, Houston called on Dr. V. H. Vogel, the director of the Lexington Hospital, and told him of his convictions and his partial success with his AA "pigeon." Further, he offered to start a group directed at drugs in the hospital, and Dr. Vogel agreed. The first meeting was on February 16, 1947. Weekly meetings have gone on ever since.

In 1948, an addict known as Dan returned to the hospital from New York City for the seventh time; after a period of severe withdrawal, he began attending the meetings begun by Houston the year before. Dan, Houston, and Houston's former AA "pigeon" spent many hours together apart from the regular meetings. From these discussions Dan experienced a miraculous change, focusing enthusiastically on the twelfth step of AA. In high spirits, he returned to New York hoping to form the first group outside Lexington Hospital—and to call it Narcotics Anonymous. Dan looked up others whom he had known at Lexington and suggested weekly meetings. Only three responded: a barber, a housepainter, and a waiter. No organization was then willing to provide them with a room for a meeting until the Salvation Army provided one. Slips plagued the first few months, but three of the original four remained committed. Slowly, the group grew in size despite disputes over policy—for example, should withdrawal from drugs be done "cold turkey" at home or within institutional care? The group finally decided to encourage the latter.

As NA emerged, it faced a dilemma. On the one hand, it wished to use the basic AA strategies and program that were directed solely against alcohol. On the other hand, it attracted, as did AA itself, many who abused a rather wide variety of drugs besides alcohol. At first, NA attracted mainly HEROIN users; later, abusers of BARBITURATES, AMPHETAMINES, and MARIJUANA began to appear at meetings. As the NA groups spread from New York City to other cities, AA groups began to thrash out a policy on the matter that further encouraged the formation of NA groups. The policy came to be known as "cooperation, but not affiliation" between AA and NA. The result was that AA freely offered their steps and traditions to NA for adaptation but steadfastly clung to their singleness of purpose—namely, to encourage alcoholics only to join. Thus, NA had to deal with a variety of drugs, not a sole prominent one, such as alcohol.

In their meetings, NA members tended to focus on the differences between the various drugs they had abused, thereby creating considerable chaos. Slowly, however, they decided on a radical change in the wording of step one. Rather than "We admitted we were powerless over drugs," they decided on "We admitted we were powerless over our ADDICTION." In other words, what all members had in common was a belief that they suffered from a disease of addiction. They pass on their experiences and hopes to the addict who still suffers; they do not become embroiled in the differing features of the various drugs to which members were addicted. In this respect, they are quite different from Cocaine Anonymous, a group that focuses on only one drug, cocaine.

(SEE ALSO: *Addiction: Concepts and Definitions; Disease Concept of Alcoholism and Drug Abuse; Rational Recovery; Treatment Types: Self-Help and Anonymous Groups*)

BIBLIOGRAPHY

ELLISON, J. (1954). These drug addicts cure one another. *Saturday Evening Post, 277.*

HARRISON M. TRICE

NATIONAL COMMISSION ON MARIHUANA AND DRUG ABUSE

In response to a substantial increase in drug-use patterns in American society during the 1960s and a swirling controversy about changing the marijuana laws to legalize the substances, in 1970, the U.S. Congress established the National Commission on Marihuana and Drug Abuse. The commission was directed to conduct a two-year study, the first on MARIJUANA and the second on "the causes of drug abuse and their relative significance." The commission was composed of thirteen members, four appointed by the Congress (two each from the Senate and the House) and nine appointed by the president. The chair of the commission was Raymond P. Shafer, former governor of Pennsylvania, and the vice chair was Dana L. Farnsworth, M.D., the director of Student Health Services at Harvard University.

In March 1972, the commission issued its first report, *Marihuana: A Signal of Misunderstanding,* which recommended decriminalization of possession of marijuana for personal use. The commission's final report, *Drug Use in America: Problem in Perspective,* was issued in March 1973. The 500-page report was supplemented by 1,000 pages of appendices. In its report, the commission summarized its findings concerning the patterns of drug use in the United States, psychosocial and institutional influences on drug-using behavior, and the social impact of drug dependence and drug-induced behavior. The commission also proposed a framework for policymaking and made specific recommendations in the areas of legal regulation, prevention, treatment and rehabilitation, and research.

The most enduring impact of the commission's final report probably lies in its efforts to revise the vocabulary of the drug field. The commission insisted that ALCOHOL be recognized as the major "drug" problem in the United States; it recommended that the term "drug abuse" be eschewed in favor of more descriptive terminology concerning drug-using behavior. For example, the commission developed a typology of drug-using behavior (experimental, recreational, situational, intensified, and compulsive use) and emphasized the need for different social responses for different patterns of use. In another important contribution, the commission fostered the development of information systems for monitoring changes in drug-using behavior in U.S. society, including national surveys of drug-using behavior among high-school students and in the general population.

The commission strongly endorsed the national treatment strategy, codified in the Drug Abuse Office and Treatment Act of 1972, which aimed to create a national network of treatment services and to establish appropriate incentives for people to seek these services voluntarily. In addition, the commission sought to reorient the role of the criminal law in implementing a policy of discouraging drug use. In the short term, the commission concluded, the criminal sanction should be retained, but should be utilized primarily as leverage for entry into prevention and treatment programs. In regard to government organization, the commission recommended that the law-enforcement and public-health dimensions of national drug-abuse prevention policy be combined into a single agency.

(SEE ALSO: *Commissions on Drugs; Marihuana Commission; U.S. Government*)

BIBLIOGRAPHY

NATIONAL COMMISSION ON MARIHUANA AND DRUG ABUSE. (1973). *Drug abuse in America: Problem in perspective.* Washington, DC: U.S. Government Printing Office.

NATIONAL COMMISSION ON MARIHUANA AND DRUG ABUSE. (1972). *Marihuana: A signal of misunderstanding.* Washington, DC: U.S. Government Printing Office.

RICHARD BONNIE

NATIONAL COUNCIL ON ALCOHOLISM AND DRUG DEPENDENCE (NCADD)

This is the ninth largest voluntary health organization in the United States and the country's major public advocate for the prevention and treatment of alcohol and other drug problems. Working through hundreds of local affiliate councils, state councils, and its New York City and Washington offices, NCADD sponsors prevention and education programs, information and referral services, scientific and clinical consensus development, public policy advocacy, and other related activities.

NCADD was established in 1944 as the National Committee for Education on Alcoholism. As the organization grew, its name and scope enlarged. It became the National Committee on Alcoholism in 1950, was renamed the National Council on Alcoholism in 1957, and assumed its present name in 1990.

The NCADD was the idea of a single individual, Marty Mann; she was its director until her retirement in 1968 and its guiding spirit until her death in 1980. Mrs. Mann was the first woman to recover from alcoholism through the fellowship of ALCOHOLICS ANONYMOUS (AA). During the early years of her recovery, she became increasingly aware that the United States was uninformed about the disease of ALCOHOLISM. The resulting stigma and prejudice kept alcoholics and their families from receiving the medical, social, and spiritual help they needed to recover. The structure and traditions of AA prevented it from becoming a public-health agency similar to those concerned with promoting prevention, treatment, and research for polio, tuberculosis, cancer, and heart disease. With the support of the Yale Center of Alcohol Studies, the council was incorporated and an office was established in the New York Academy of Medicine building in New York City. In 1950, it became independent of Yale. Ruth Fox, a psychiatrist who had helped found the council, became its first medical director in 1958. In 1969, she was succeeded by Frank A. Seixas, an internist.

During its early years, the council's activity consisted mainly of developing literature and presenting lectures to professional and lay groups on the concept of alcoholism as a disease and of organizing local affiliates to pursue this educational process in their own communities. By 1947, a survey of American adults showed that 36 percent believed alcoholism to be a disease, a remarkable increase from 6 percent who held this view in 1943. As interest in alcohol and drug problems expanded, the council developed and then published in 1972 the first set of medical criteria for the diagnosis of alcoholism. In 1976, it sponsored Operation Understanding, in which fifty-two men and women known for their contributions in the areas of government, medicine, industry, science, journalism, and the arts publicly revealed their histories of recovery from alcoholism.

These and other activities have made NCADD an important force in the nation's development of service systems and health policy related to alcohol and other drug problems. NCADD helped establish the first industrial alcoholism programs, the first research society devoted to alcoholism, the first public education campaigns to promote the concept of alcoholism and other drug dependence as diseases, the movement to recognize the special needs of WOMEN with substance-related problems, and the nation's effort to understand and prevent FETAL ALCOHOL SYNDROME (FAS) and other effects in the fetus.

NCADD is also a leader in the U.S. campaign against alcohol-related highway ACCIDENTS and in promoting appropriate treatment services for substance-dependent pregnant and postpartum women and their children. Through its local affiliates, NCADD provides direct services, including education and prevention, in school and community settings, as well as information, intervention, and referral counseling, local alcohol- and drug-awareness campaigns, and other related activities.

(SEE ALSO: *American Society of Addiction Medicine; Association for Medical Education and Research in Substance Abuse; Disease Concept of Alcoholism and Drug Abuse; Parents Movement; Society of Americans for Recovery; Women for Sobriety*)

BIBLIOGRAPHY

LENDER, M. E., & MARTIN, J. K. (1987). *Drinking in America: A history*. New York: Free Press.

MORSE, R. M., & FLAVIN, D. K. (1992). Joint Committee of the NCADD and the American Society of Addiction Medicine: The definition of alcoholism. *Journal of the American Medical Association, 268*, 1012–1014.

MURPHY, W. *NCA's first forty years.* (1984). New York: National Council on Alcoholism and Drug Dependence.

NATIONAL COUNCIL ON ALCOHOLISM, CRITERIA COMMITTEE. (1972). Criteria for the diagnosis of alcoholism. *Annals of Internal Medicine, 77*, 249–258.

C. D. SMITHERS FOUNDATION. (1979). *Pioneers we have known in the field of alcoholism.* Mill Neck, NY: Author.

SHEILA B. BLUME

NATIONAL HOUSEHOLD SURVEY ON DRUG ABUSE (NHSDA)

The National Household Survey on Drug Abuse is the primary source of statistical information on the use of illegal drugs by the population of the United States. Conducted periodically by the federal government since 1971, the survey collects data by administering questionnaires to a scientifically selected sample of persons age twelve and older living in the nation. The primary purpose of the survey is to estimate the prevalence of illegal drug use (i.e., the number of people using illegal drugs) in the United States, and to monitor changes in prevalence over time.

Legal drugs, such as ALCOHOL and TOBACCO, are also covered by the survey. Prevalence rates (the percentage of the population using any type of drug) for various population subgroups and for various types of drugs are generated from the survey data; these rates are compared by analysts to provide insight into which population groups are most prone to illicit drug use—which drugs are most commonly used. These basic statistics are used by the federal government in planning federal policies and funding priorities related to substance abuse. Statistical reports, containing the survey estimates and descriptions of the surveys, have been routinely published. The raw survey data are also available on data tapes, which are widely used by substance-abuse researchers studying the EPIDEMIOLOGY of substance abuse, and the results of these studies are published in professional journals.

HISTORY OF THE NHSDA

The NHSDA traces its origin to a survey conducted by the NATIONAL COMMISSION ON MARIHUANA AND DRUG ABUSE (1970–1972). The commission required baseline data on the public's beliefs, attitudes, and use of marijuana, to satisfy its charge of developing recommendations for legislation and administrative actions in helping to deal with the illicit drug problem. Through a private contractor, they conducted two surveys, in 1971 and 1972. The NATIONAL INSTITUTE ON DRUG ABUSE (NIDA) continued the survey in subsequent years (1974, 1976, 1977, 1979, 1982, 1985, 1988, 1990, and 1991) to satisfy the continuing need for current data. Starting in 1990, the survey was conducted annually. In 1992, sponsorship of the survey was transferred to the newly created SUBSTANCE ABUSE AND MENTAL HEALTH SERVICES ADMINISTRATION (SAMHSA). All the surveys were conducted by private contractors selected by the government.

Expansion of the survey took place in 1985 with the implementation of a new sample design that had larger samples of AFRICAN AMERICANS and HISPANICS (resulting in a sample size of 8,038). Further expansions took place in 1990, with the intensive sampling of the Washington, D.C., metropolitan area as a part of the survey and in 1991, with the addition of five more oversampled metropolitan areas and an increase in the national sample component (for a total sample of 32,594 in 1991). The metropolitan oversampling was continued through 1993, but beginning in 1994 the survey was scaled back to a national sample of about 18,000 interviews. All surveys conducted from 1971 through 1991 were done at a particular time of year, usually spring or fall. In 1992, a continuous data collection design was implemented—with quarterly samples and January to December data collection. A major revision to the survey questionnaire was also implemented in 1994, to improve the validity and reliability of the survey estimates.

DESCRIPTION OF THE SURVEY METHODOLOGY

Since its inception, the NHSDA has undergone various design changes affecting primarily the sample design, as described above. Data collection methods have been consistent in the survey since 1971.

Target Population. Prior to 1991, the NHSDA covered all persons age twelve and older living in households in the forty-eight contiguous states. Beginning in 1991, this was modified so that the survey covers the civilian noninstitutionalized population aged twelve years old and older within the fifty states. In addition to including all household residents (except persons on active MILITARY duty), it includes the residents of noninstitutional group quarters (e.g., shelters, rooming houses, dormitories) as well as residents of civilian housing on military bases. Persons excluded from the target population are those with no fixed address, residents of institutional quarters (e.g., jails and hospitals), and active-duty military personnel.

Sample Selection. A complex multistage sample design is used to select people to be respondents in the survey. The first stage of sampling is the selection of nonoverlapping geographic primary sampling units (PSUs), consisting of counties or metropolitan areas. For the second stage of sampling, area segments (constructed from U.S. Census block groups or enumeration districts) are selected within each PSU. Field staff count and list all dwelling units within sample segments and mark their location on a map. A *dwelling unit* is either a housing unit, such as a house or apartment, or a group-quarters unit, such as a dormitory room or a shelter bed. From these listings, a sample of dwelling units is then selected by sampling staff, and interviewers are assigned to contact these dwelling units.

Prior to arrival at the sample dwelling unit (SDU) an introductory letter is mailed to the SDU, briefly explaining the survey and requesting participation. When the interviewer visits the SDU, a brief screening interview is conducted that involves listing all SDU members on a screening form. Based on a predetermined random selection procedure, the interviewer identifies which SDU member(s) will be asked to participate in the survey.

Questionnaire Administration. Interviewers control the questionnaire administration, but to enhance respondent confidentiality, drug-use questions are answered by respondents on self-administered answer sheets that are not reviewed by interviewers. As the respondent records the answer choices and completes each answer sheet, they are placed in an envelope. At the end of the interview process, all materials are sealed in this envelope by the respondent and mailed to the data-processing site.

Data Processing. All questionnaires are received by mail at a data-processing site, where they are checked and then all data are entered onto a computer data base. Consistency checks and other editing is done, after which statistical tables showing estimates of prevalence rates for various drugs are produced. All questionnaires are destroyed after the data file is complete, and no identifying information (such as addresses) is included in the data file.

STRENGTHS AND LIMITATIONS OF THE NHSDA

Strengths. The major strengths of the NHSDA are its size, continuity, and national representativeness. The survey has a sample large enough to allow comparisons of drug-use prevalence among many different population subgroups each year and over time. The length of the questionnaire and amount of data collected provides a rich data base for examining the characteristics of drug abusers, the relationships of drug use with many demographic and other variables, and the changing patterns of drug use over time. The methodology used, while expensive, has been extensively evaluated and found to be effective (relative to other methodologies) in eliciting valid data from respondents. Through intensive call-back procedures, participation rates in the NHSDA have been excellent. The 1992 participation rate for the screening questionnaire was 95 percent and the participation rate for the main questionnaire was 83 percent.

Limitations. The survey does not cover certain populations likely to have heavy illicit drug use, such as the homeless and prison populations. While these missing populations, because they are small, make little difference in estimating MARIJUANA or ALCOHOL prevalence, rarer behaviors such as HEROIN or CRACK use may be severely underestimated by the NHSDA. Data validity from the survey is also in question because of the self-report methods employed and the voluntary nature of the survey.

MAJOR FINDINGS OF THE SURVEY

The NHSDA has tracked the changing nature of drug abuse since 1971. At the time of the first survey, about 10 percent of the population age twelve and older had ever used illicit drugs. This was estimated to be more than double the rate of lifetime use as of

TABLE 1
Prevalence of Any Illicit Drug Use (1979–1993) by Percent of U.S. Population

	1979	1982	1985	1988	1990	1991	1992	1993
Use in Past Month								
all ages 12 +	13.7%	12.2%	12.1%	7.3%	6.4%	6.3%	5.5%	5.6%
12–17	17.6	12.7	14.9	9.2	8.1	6.8	6.1	6.6
18–25	37.1	30.4	25.7	17.8	14.9	15.4	13.0	13.5
26 +	6.5	7.5	8.5	4.9	4.6	4.6	4.1	4.1
26–34	18.5	19.2	21.1	13.0	9.8	9.0	10.1	8.5
35 +	2.5	3.4	3.9	2.1	2.8	3.1	2.2	2.8
Use in Past Year								
all ages 12 +	19.5	18.7	19.6	14.1	13.3	12.7	11.1	11.8
12–17	26.0	22.0	23.7	16.8	15.9	14.8	11.7	13.6
18–25	49.4	43.4	42.6	32.0	28.7	29.1	26.4	26.6
26 +	10.0	11.8	13.3	10.2	10.0	9.4	8.3	8.9
26–34	26.9	29.5	32.0	22.6	21.9	18.4	18.3	17.4
35 +	4.4	5.6	6.6	5.8	6.0	6.4	5.1	6.3
Use in Lifetime								
all ages 12 +	33.3	32.3	36.9	36.6	37.0	37.0	36.2	37.2
12–17	34.3	27.6	29.5	24.7	22.7	20.1	16.5	17.9
18–25	69.9	65.3	64.3	58.9	55.8	54.7	51.7	50.9
26 +	23.0	24.7	31.5	33.7	35.3	36.0	36.0	37.3
26–34	51.5	57.7	62.2	64.2	62.6	61.8	60.8	61.1
35 +	13.6	13.2	20.4	23.0	25.9	27.3	28.0	29.9

SOURCE: SAMHSA, Office of Applied Studies, National Household Survey on Drug Abuse
NOTE: Any illicit drug use includes use of marijuana, cocaine, hallucinogens, inhalants (except in 1982), heroin, or nonmedical use of sedatives, tranquilizers, stimulants, or analgesics. The exclusion of inhalants in 1982 is believed to have resulted in underestimates of any illicit use for that year, especially for 12–17 year olds.

TABLE 2
Trends in Past Month Drug Use for Ages 12 and Older in U.S. Population

Drug	1977	1979	1982	1985	1988	1990	1991	1992	1993
			Number of Users (in millions)						
Any Illicit Use	—	24.3	22.4	23.0	14.5	12.9	12.8	11.4	11.7
Marijuana	16.4	22.5	20.0	18.2	11.6	10.2	9.7	9.0	9.0
Cocaine	1.6	4.3	4.2	5.8	2.9	1.6	1.9	1.3	1.3
Drug	1977	1979	1982	1985	1988	1990	1991	1992	1993
			Percent of Population						
Any Illicit Use	—	13.7%	12.2%	12.1%	7.3%	6.4%	6.3%	5.5%	5.6%
Marijuana	9.5	12.7	11.0	9.4	5.9	5.1	4.8	4.4	4.3
Cocaine	1.0	2.4	2.3	2.9	1.5	0.8	0.9	0.6	0.6

— = not available
SOURCE: SAMHSA, Office of Applied Studies, National Household Survey on Drug Abuse
NOTE: Any illicit drug use includes use of marijuana, cocaine, hallucinogens, inhalants (except in 1982), heroin, or nonmedical use of sedatives, tranquilizers, stimulants, or analgesics. The exclusion of inhalants in 1982 is believed to have resulted in underestimates of any illicit use for that year, especially for 12–17 year olds.

the early 1960s. By 1992, 36 percent or about 74 million Americans reported using illicit drugs. The reports showed that more than one drug had been used by some of the total 74 million, with a breakdown of this figure as follows: Some 68 million reported using marijuana or HASHISH; 23 million cocaine; 24 million PRESCRIPTION DRUGS nonmedically; 16 million HALLUCINOGENS (such as LSD and PCP); and 10 million INHALANTS.

The 1992 NHSDA estimated that only 11.4 million of the 74 million lifetime illicit drug users were *current users*—meaning that they had used within the past month. Current use of illicit drugs reached a peak in 1979 at 13.7 percent of the population; it decreased to 5.5 percent of the population in 1992.

All the NHSDAs conducted since 1971 have shown that marijuana is the most commonly used illicit drug, with current use at 4.4 percent in 1992. Marijuana use peaked in 1979, when 12.7 percent had used in the past month. Current cocaine use reached a peak in 1985 at 2.7 percent, but the survey showed dramatic declines in cocaine use after 1985, to 0.6 percent in 1992.

The NHSDA has shown varying rates of use in different segments of the population. Use of illicit drugs since 1971 has been most common among young people (under age 35), although recent surveys demonstrate that because of the aging of the now-over-35 cohorts of (continuing) drug users, the proportion of current users age 35 and older shows an increase from 10 percent in 1979 to 23 percent in 1992.

The surveys have also shown that while illicit drug use occurs in all segments of society, prevalence rates have been greatest among males; in metropolitan areas; among the unemployed; and among high-school dropouts. In the early to mid-1990s, prevalence rates show cocaine use to be higher among blacks and Hispanics than among whites. The other "risk factors" that NHSDA data show as correlating with illicit drug use in the under-35 groups include SMOKING cigarettes and drinking alcohol at a young age; having divorced parents; and having parents or older siblings who also use illicit drugs.

(SEE ALSO: *Drug Abuse Warning Network*; *Drug Use Forecasting Program*; *High School Senior Survey*)

BIBLIOGRAPHY

SUBSTANCE ABUSE AND MENTAL HEALTH SERVICES ADMINISTRATION. (1994). *National household survey on drug abuse: Main findings 1992*. Washington, DC: U.S. Government Printing Office.

SUBSTANCE ABUSE AND MENTAL HEALTH SERVICES ADMINISTRATION. (1993). *National household survey on drug abuse: Population estimates 1992*. DHHS Pub. no. (SMA) 93-2053. Washington, DC: U.S. Government Printing Office.

TURNER, C. F., LESSLER, J. T., & GFROERER, J. C. (1992). *Survey measurement of drug use: Methodological studies*. National Institute on Drug Abuse, DHHS Pub. no. (ADM) 92-1929. Washington, DC: U.S. Government Printing Office.

JOSEPH C. GFROERER

NATIONAL INSTITUTE ON ALCOHOL ABUSE AND ALCOHOLISM (NIAAA) *See* U.S. Government Agencies.

NATIONAL INSTITUTE ON DRUG ABUSE (NIDA) *See* U.S. Government Agencies.

NATIONAL PARENTS RESOURCE INSTITUTE (PRIDE) *See* Prevention.

NATIVE AMERICANS, DRUG AND ALCOHOL USE AMONG *See* Alcohol: History of Drinking; Ethnicity and Drugs.

NATURAL HISTORY OF NARCOTIC USE *See* Opioid Dependence: Course of the Disorder over Time.

NEEDLE AND SYRINGE EXCHANGES AND HIV/AIDS The first syringe exchange (SE) program was begun in 1984 in Amsterdam, the NETHERLANDS, out of concern for the spread of hepatitis B among INJECTING DRUG USERS (IDUs). While the hepatitis B virus, hepatitis C virus, and human T

cell lymphotropic virus can all cause fatal illness and are all spread through multiperson use ("sharing") of drug-injection equipment, the threat of human immunodeficiency virus (HIV) has clearly become the dominant force in implementing needle- and syringe-exchange programs throughout the world.

HIV is the causative agent for ACQUIRED IMMUNODEFICIENCY SYNDROME (AIDS). At present (1995), HIV infection is eventually fatal; there is no permanently effective treatment for HIV infection; and a vaccine against infection (if possible) is still well in the future. HIV has now been reported among IDUs in sixty countries, from all continents except Antarctica, and from both industrialized and developing nations.

A disturbing facet of HIV infection among injecting drug users is the potential for the rapid spread of the virus through a local population of IDUs. In Edinburgh, Scotland, HIV spread, after the introduction of the virus, into the local population to infect over 40 percent of the local IDUs within two years (Robertson, 1990). In Bangkok, Thailand, the percentage of HIV-infected IDUs (HIV seroprevalence) increased from 2 percent to over 40 percent in less than one year (Vanichseni et al., 1992). In the state of Manipur, India, over 50 percent of the local population of IDUs were infected with HIV within one year after the introduction of the virus into the group. The rapid spread of HIV among IDUs results from a lack of awareness of HIV/AIDS as a local threat and from mechanisms, such as shooting galleries (places where addicts "shoot up" together) and dealers' works, that allow large numbers of the population to be exposed to the virus through infected needles and syringes (Des Jarlais et al., 1992).

Once HIV becomes well established within a population of IDUs, their homosexual and heterosexual partners and transmission to developing fetuses (perinatal) become additional significant problems. In most developed countries, IDUs are the predominant source for both heterosexual and perinatal transmission of HIV.

The need to reduce HIV transmission among and from injecting drug users has led to a variety of prevention programs. The programs have had differing degrees of effectiveness, although there is evidence that "education-only" programs (i.e., those that do not provide the physical means for behavior change) are the least effective. In almost all industrialized and in some developing countries, increasing legal access to sterile (or uncontaminated) injection equipment has become the most common HIV/AIDS prevention strategy for IDUs. This strategy has included both increased over-the-counter sales of sterile injection equipment and syringe-exchange programs, in which IDUs can turn in used injection equipment for sterile equipment at no cost.

Increasing legal access to sterile injection equipment has been politically controversial in several industrialized countries, notably the United States and Sweden, and in many developing countries. Concerns have been raised as to whether increased legal access would lead to increased injection of illicit drugs and whether increased legal access would appear to "condone" illicit drug use or "send the wrong message" about illicit drug use (Martinez, 1992). The empirical data on these questions will be reviewed below, but first it is important to address operational issues involved in needle-exchange programs—to specify how needle exchanges actually work before addressing evaluations of their outcomes.

ORGANIZATIONAL CHARACTERISTICS OF PROGRAMS

At first glance (and regrettably, in much of the public debate about needle-exchange programs thus far), the operation of a program seems quite simple—one would merely select a location and provide staff who could trade new injection equipment for used. In practice, since the exchanges are service-delivery programs, the organization of the services is critical to their effectiveness. Some programs are heavily utilized—for example, the Amsterdam programs exchange approximately 6 million needles and syringes per year in a city with an estimated 3,000 injection drug users. In contrast, the first legal program in New York City traded fewer than 1,000 needles and syringes per year in a city with an estimated 200,000 injecting drug users.

To date, there have been only two comparative studies of the organizational characteristics of the programs (Stimson et al., 1988; Lurie & Reingold, 1993). According to the Stimson study, the most important aspect of an exchange program is "user-friendliness"—which includes such practical considerations as convenient location and convenient hours of operation but also addresses some of the philosophical issues involved.

Perhaps the most vital element of user-friendliness is the nonjudgmental attitude of the staff toward the participants in the exchange. Participants in a user-friendly program are treated with dignity and respect. They are not stigmatized as morally and psychologically impaired simply because they inject psychoactive drugs. The participants are presumed to care about their health and to be capable of taking actions to preserve their health and the health of others.

User-friendliness also requires that exchanges offer multiple services. Other concerns need to be addressed beyond the provision of sterile injection equipment; the sexual transmission of HIV also needs to be prevented, which includes the distribution of condoms without cost. Moreover, the trusting relationships that gradually develop between staff and participants lead to the discovery of other health and social-service needs, especially the need for drug-abuse treatment. The exchange service should be able to respond positively to such needs, either through referral or through on-site provision of assistance. Failure to do so would undermine the trusting relationships between staff and participants.

There is as yet no consensus as to which additional services should be offered on site and which ones through referral—or even a set of available guidelines for how an individual exchange program should decide which additional services to offer on site and which to offer through referral. However, a broad range of additional services are presently being offered on site, with some programs offering conventional drug-abuse treatment, self-help recovery groups, women's support groups, tuberculosis screening and treatment, and Bible study groups.

The need to provide on-site (or link to other) services means that exchange programs should be considered a part of a system of services for preventing HIV infection among injecting drug users, rather than as self-sufficient HIV prevention programs.

THE EFFECTIVENESS OF THE EXCHANGES

Studying the effectiveness of an HIV prevention program that facilitates sustained risk reduction is extremely difficult. Research ethics require that comparison subjects be provided with some intervention to reduce their chances of HIV infection, and it is not easy to determine an appropriate comparison condition for a program. Should the comparison subjects be told/permitted to purchase sterile injection equipment from pharmacies? Should they be told to purchase sterile injection equipment through an illicit market? or find some method of disinfecting their own injection equipment?

The logical unit of analysis in an exchange evaluation would be the needs of the local population of injecting drug users rather than the needs of individual drug users. If HIV-infected drug users participate in exchanges—returning their needles and syringes to the exchange rather than passing them on to other injectors—those who do not participate in the exchange would then still be protected against HIV infection. Using communities as the unit of analysis in a clinical trial, however, would be extremely expensive, and it is doubtful that many communities would accept random assignment to experimental or control conditions.

No needle-exchange study to date has approached a randomized clinical trial. Most studies have measured HIV risk behavior prior to and after participation in an exchange, or have compared risk behavior among exchange participants with that of some other group of injecting drug users. Conclusions about the effectiveness of needle-exchange programs must thus be drawn from the consistency of findings across many methodologically limited studies, rather than rely on a single or small group of methodologically rigorous tests of needle exchange.

Drug Injection. A common concern expressed by opponents to exchange is that the programs would increase the frequency of illicit drug injection. However, research studies have consistently found that such exchange is not associated with any detectable increase in drug use on either a community or an individual level (Des Jarlais & Friedman, 1992). The most recent review emphasized that "there is no evidence that needle exchange programs increase the amount of drug use by needle exchange clients or change overall community levels of noninjection or injection drug use" (Lurie & Reingold, 1993). Of the eight relevant studies analyzed in this review, three found reductions in injection associated with needle exchange, four found mixed or no effect, and one found an increase in injection compared with the controls. Data from the New York City exchange evaluation (which were not available at the time of Lurie & Reingold's 1993 review) indicate a modest *decrease* in the frequency of injection among participants using needle exchange (Paone et al., 1995).

Moreover, although opponents have often expressed an additional concern—that exchange programs would attract new injectors—the overwhelming number of IDUs participating in exchanges have long histories of drug injection. The mean length of time usually ranges from five to ten years or more. Typically only 1 to 2 percent of exchange participants initiated drug injecting within the previous year. If providing sterile injection equipment had induced large numbers of people to begin injecting drugs, then the numerous studies to date should have observed substantial numbers of new injectors participating in programs.

HIV Injection Risk Behavior. Consistent findings across studies indicate declines in self-reported frequencies of injection with potentially HIV-contaminated needles (Paone et al., 1993). The magnitude of the reduction is difficult to estimate, because studies have used different metrics for risk behavior; some studies have used differences in pre- and post-exchange measurements, while other studies have compared participants with various other groups of drug injectors. Nonetheless, the trend observed from participants in a program has been a reduction in risk behavior, through injection of contaminated equipment, ranging from 50 percent to 80 percent. No studies, however, have shown anything approaching complete elimination of risk behavior among needle-exchange participants.

Exchange programs probably attract drug injectors who are relatively concerned about their health, and it is possible that, even in the absence of exchange programs, these injectors would seek alternative ways of reducing HIV injection risk, such as purchasing sterile injection equipment from pharmacies or on the illicit market. Thus the present data do not permit a conclusion that exchange programs are necessary to reduce risk behavior leading to HIV infection. However, the possibility of alternative methods for reducing injection risk behavior does not imply that an exchange program is not effective in reducing such behavior.

Nevertheless, the fact that very few new injectors participate in exchange programs may be considered a limitation on their current effectiveness. Since IDUs are typically exposed to hepatitis B and C within the first few years of injecting drugs (Hagan et al., 1993), new injectors may already be infected

with these blood-borne viruses before they start to obtain sterile injection equipment from an exchange program. Moreover, in cities with high HIV-seroprevalence. even new injectors may be at high risk for HIV infection. In New York City, the estimated seroconversion rate among new injectors is 6.6 per 100 person-years at risk (Des Jarlais et al., 1994). The new injectors may become infected with HIV before they even begin to participate in an exchange program.

Sexual Risk Behavior. While all exchange programs address sexual transmission of HIV to some extent, fewer studies have examined the effect that the program has had on sexual-risk reduction among participants. Moreover, the findings from these few studies are ambiguous. Very few HIV prevention programs for injecting drug users have had consistent success in changing the sexual behavior of IDUs, particularly those with "regular" sexual partners (Friedman et al., 1994). The one exception might be programs that provide HIV counseling and testing, since drug injectors who know they are infected with HIV are more likely to change their behavior to reduce the chances of transmitting HIV to others (Vanichseni et al., 1993).

Effects on HIV and Hepatitis B Transmission. Research data on exchange programs has produced a body of consistent findings with regard to reduced risk behavior through drug injection. Studies within the programs of HIV seroprevalence and HIV seroincidence tend to validate the self-reported risk reduction. Seroprevalence rates have usually stabilized after a program has been implemented, and the rates of new infections among participants have ranged from zero to less than 1 per 100 person-years at risk to a moderate 4 per 100 person-years at risk in Amsterdam. While there is as yet no definite evidence that participation in a needle exchange reduces the chances of HIV infection, the available HIV seroprevalence and seroincidence data are largely consistent with this hypothesis.

The same behaviors that transmit HIV infection (multiperson use of injection equipment and unprotected sexual behavior) also transmit hepatitis B. The epidemiology of these viruses is similar in most countries, and injecting drug users are at high risk for infection with both viruses.

Studies on the effects of exchange-program participation and new hepatitis B infection among drug users in several cities have shown actual declines

(Hagan et al., 1991), further validating self-reported risk reduction and indicating that exchange programs do have a large-scale effect on AIDS risk behavior among injecting drug users.

Discarded Syringes. Exchange programs create an economic value for used needles and syringes—they can be traded for new injection equipment. Thus exchanges have the potential for reducing the amount of used and damaged equipment that is just discarded in the community. Indeed, the one study that systematically examined the amount of discarded injection equipment before and after implementation of an exchange program found a significant reduction in needles and syringes left on sidewalks and in the streets (Oliver et al., 1992)— where anyone might touch it and become a potential victim.

THE "MESSAGE" OF EXCHANGE PROGRAMS

Objections that exchange programs will lead to increased illicit drug use or that they will not lead to reductions in HIV risk behavior can be addressed through empirical studies. Such studies show consistent findings of *no* increase in illicit drug injection and consistent reductions in HIV risk behavior (although it has not yet been possible to translate the reductions in risk behavior into empirically grounded reductions in HIV transmission rates).

A common objection to the programs, however, is that they "condone" or "send the wrong message" about illicit drug use. The symbolism of a government providing the equipment needed for the injection of illicit drugs seems to contradict society's fundamental disapproval of illicit drug injection; and exchange participants do not misinterpret a need to prevent HIV infection as indicating a reversal of prevailing societal attitudes toward the injection of psychoactive drugs.

The important political message in the programs is not that the injection of drugs like HEROIN and COCAINE is a social good but that previous policies on illicit drug use cannot cope with a public-health catastrophe such as HIV infection among injecting drug users, their sexual partners (and theirs), and their children. The "war on drugs" or "ZERO TOLERANCE" approach focused on reducing the use of illicit drugs. It was clearly impractical, however. Our ability to treat drug users so that they will never take drugs again is also clearly limited, and letting drug injec-

tors, their sexual partners, and their children die of HIV infection is clearly inhumane—and they have potential for spreading HIV into the rest of society.

Needle-exchange programs suggest the possibility of greatly reducing the individual and social harm associated with drug use through means other than simply reducing drug use or the drug supply. Making the distinction between reducing drug-related harm and reducing drug use per se is the fundamental premise of a new approach to drug policy that has been termed "harm reduction" or "harm minimization." Harm-reduction practices existed before HIV/AIDS and exchange programs and extend well beyond HIV/AIDS issues, but they have come to be recognized as a prototype of the harm-reduction approach in general.

The harm-reduction perspective itself is in a period of rapid development, so it is not possible to state its fundamental principles definitively, but there are at least four common assumptions in descriptions of the approach:

1. Pragmatism is valued over idealism. The nonmedical use of both licit and illicit psychoactive drugs is likely to continue indefinitely, so policies should be formulated on a realistic basis rather than on the basis of a utopian drug-free society.
2. Reducing drug use, particularly very heavy (dependent, addictive) drug use, is the most desirable but not the only means of reducing the individual and social harms associated with psychoactive drug use. Exchange programs to prevent HIV infection are a clear example of reducing harm without necessarily reducing drug use. (Designated-driver programs are another example of harm reduction—reducing the harm associated with alcohol use without necessarily reducing alcohol use.)
3. In general, drug-related harm is likely to be reduced through integrating drug users into society rather than stigmatizing them and treating them as social outcasts.
4. While drug addiction clearly restricts an individual's ability to control his or her own behavior, drug users are still capable of making rational choices and should be offered choices among different ways of reducing the harm that drug misuse causes them and society.

The harm-reduction perspective is thus quite different from the war on drugs–zero tolerance perspective. Harm reduction is also distinct from the

LEGALIZATION of all psychoactive drugs. The individual and social harms of drugs are not likely to be minimized by the mass marketing of drugs. NICOTINE/TOBACCO is a prime example of how large-scale harm has been created through uncontrolled merchandising of an addictive drug.

Rather than base policy on a utopian ideal of a drug-free society or the equally implausible ideal of a society that freely uses psychoactive drugs without problems, the harm-reduction perspective calls for basing policy on a flexible pragmatism. Specific harms associated with specific types of drug use can be identified, and concrete steps can be taken to reduce those specific harms. Exchange programs to reduce HIV infection among injecting drug users and their social contacts are a prototypical example of a concrete action for reducing drug-related harm. The message sent by exchange programs thus should not be read as "drug injecting is good" but rather that drug policies should be based on their pragmatic effects instead of on their symbolism.

(SEE ALSO: *Alcohol and AIDS; Complications: Route of Administration; Injecting Drug Users and HIV; Substance Abuse and AIDS*)

BIBLIOGRAPHY

DES JARLAIS, D. C., & FRIEDMAN, S. R. (1992). AIDS and legal access to sterile injection equipment. *Annals of the American Academy of Political and Social Science, 521,* 42–65.

DES JARLAIS, D. C., ET AL. (1994). Continuity and change within an HIV epidemic: Injecting drug users in New York City, 1984 through 1992. *JAMA, 271,* 121–127.

DES JARLAIS, D. C., ET AL. (1992). International epidemiology of HIV and AIDS among injecting drug users. *AIDS, 6,* 1053–1068.

DONOGHOE, M. C., ET AL. (1989). Changes in HIV risk behaviour in clients of syringe-exchange schemes in England and Scotland. *AIDS, 3*(5), 267–272.

FRIEDMAN, S. R., ET AL. (1994). Drug injectors and heterosexual AIDS. In L. Sherr (Ed.), *AIDS and the heterosexual population.* Chur, Switzerland: Harwood Academic Publishers.

HAGAN, H., ET AL. (1993). An interview study of participants in the Tacoma, Washington, Syringe Exchange. *Addition, 88,* 1691–1697.

HAGAN, H., ET AL. (1991). The incidence of HIV infection and syringe exchange programs. *JAMA, 266,* 1646–1647.

HEATHER, N., ET AL. (EDS.). (1993). *Psychoactive drugs and harm reduction: From faith to science.* London: Whurr.

LURIE, P., & REINGOLD, A. L. (EDS.). (1993). *The public-health impact of needle-exchange programs in the United States and abroad: Summary, conclusions, and recommendations.* San Francisco: Institute for Health Policy Studies, University of California.

MARTINEZ, R. (1992). Needle exchange programs: Are they effective? *ONDCP Bulletin, 7,* 1–7.

OLIVER, K., ET AL. (1992). Comparison of behavioral impacts of syringe exchange and community impacts of an exchange. In *Final program and abstracts of the VIII International Conference on AIDS,* Amsterdam (abstract PoC 4284).

PAONE, D., ET AL. (1993). AIDS risk reduction behaviors among participants of syringe-exchange programs in New York City. Presented at the Ninth International Conference on AIDS, Berlin, June (abstract PO-C24-3188).

PAONE, D., ET AL. (1995). *New York City syringe exchange: An overview.* Washington, DC: National Academy of Sciences.

ROBERTSON, R. (1990). The Edinburgh epidemic: A case study. In J. Strang & G. V. Stimson (Eds.), *AIDS and drug misuse: The challenge for policy and practice in the 1990s.* London and New York: Routledge.

STIMSON, G. V., ET AL. (1988). *Injecting equipment exchange schemes: Final report.* London: Monitoring Research Group, Goldsmith's College.

VAN HAASTRECHT, H., ET AL. (1991). The course of the HIV epidemic among intravenous drug users in Amsterdam, the Netherlands. *American Journal of Public Health, 81,* 59–62.

VANICHSENI, S., ET AL. (1992). HIV testing and sexual behavior among intravenous drug users in Bangkok, Thailand. *Journal of Acquired Immune Deficiency Syndrome, 5,* 1119–1123.

DON C. DES JARLAIS
DENISE PAONE

NERVOUS SYSTEM DAMAGE   See Brain Structures and Drugs; Complications: Neurological.

NETHERLANDS, DRUG USE IN THE

The context of drug use in the Netherlands has received much international attention because of the nation's innovative policy initiatives. A popular misconception has been that the Netherlands is soft on drugs and permits a freewheeling atmosphere for illicit drug use. This misconception is based on a lack of understanding of a national drug policy. After the explosion of drug use in the 1960s and 1970s, the government chose to experiment with a shift to different goals rather than a more-of-the-same strategy. The primary goal was to reduce the harm that drugs caused to both the individual and to society. A corollary to this approach was that the efforts to control drugs should not cause more harm than the drugs themselves. During the 1970s and 1980s, the pursuit of this policy by the Netherlands often resulted in bitter controversy with neighboring countries, which complained that the Dutch drug policy was undermining their own drug-control efforts. The 1990s saw a more harmonious situation: Many neighboring countries were adopting Dutch harm-reduction programs while the Netherlands itself was developing a more differentiated policy that increased the penalties for drug trafficking while providing more services for drug users. Elimination of "drug tourism" (that is, the buying and use of drugs by foreign nationals in the Netherlands) has become a top priority, and a number of ways of decreasing this are being tried.

The drug policy of the Netherlands has been characterized by two main principles—the separation of markets and the normalization of drug problems. The separation-of-markets principle is based on the idea that drugs can be classified pharmacologically according to their socially acceptable risks and that drug markets should be controlled on the basis of this classification. For example, in many societies ALCOHOL is a drug regarded to have acceptable risk, and the market for alcohol is legal for adults, with varying degrees of government regulation. The Dutch have decided that cannabis (MARIJUANA, HASHISH) is also a drug of acceptable risk and therefore should be separated from the markets for HEROIN and COCAINE, which have an unacceptable risk. Because of international regulations, however, the cannabis market cannot be equated with the alcohol market. Thus cannabis trafficking still remains illegal in the Netherlands, although it has a low law-enforcement priority in many jurisdictions. The so-called AHOJ-G policy for marketing cannabis requires limited advertising; no hard drugs—cocaine or heroin—are allowed to be sold or on the premises; no social nuisance; no youths under 16 years of age; only small amounts—less than 30 grams—can be sold. This policy regulates the system of cannabis-selling coffeehouses that have sprung up in most Dutch cities. Additional local regulations require that the coffeehouses provide recreational facilities, such as pool tables, so that something more than cannabis is offered to the customers.

The second main principle of the Netherlands drug policy is the normalization of drug problems. This principle recognizes that much of the harm attributed to hard-drug use such as heroin is based on negative definitions that are held by society and internalized by the drug users. The principle of normalization leads to multiple efforts to reintegrate the heroin user into the community and to fight against his or her stigmatization. This is done by an extensive system of METHADONE MAINTENANCE PROGRAMS (a widely used pharmacotherapy for heroin users), counseling, and social-service support. In addition, drug users are encouraged to organize self-help groups and to mobilize for positive changes in their own subcultures, all in the interest of increasing both their participation in and their responsibility for the development of the drug-use context.

The minister of justice of the Netherlands, addressing the United Nations International Conference on Drug Abuse and Trafficking in Vienna in 1987, urged that "international cooperation is indispensable. However, an attempt to reach an internationalization of drug policies in the sense of a single, nondifferentiated approach is bound to be counterproductive for many countries." This official statement at the conference expresses the Netherlands' commitment to find "cooperation in diversity." This commitment has provided the political context for drug use in the Netherlands and has been a significant contributor to the dynamic equilibrium of the international drug-control system. Much like the earlier influence of the British through international organizations, the Dutch have maintained an important role despite a widespread opinion that the Netherlands is somehow in violation of existing international law.

On the contrary, the Netherlands, as a signatory of the 1988 Vienna agreement, supports the pact's mandate (Article 14) to adopt appropriate measures aimed at reducing demand. Although the international emphasis on demand reduction is relatively new, demand reduction has been the deliberate aim of Dutch drug policy for the last two decades. This definition provides the foundation of the principles of the separation of drug markets and the normalization of drug problems.

The specific context of drug use in the Netherlands has not resulted in any remarkable irregularities in drug use. The number of OPIATE users increased sharply in the 1970s but leveled out in the 1980s. According to most official sources, there are between 15,000 to 25,000 heroin-dependent people in the Netherlands. Thus, the prevalence of heroin DEPENDENCE in the Netherlands is between 100 to 133 per 100,000 population. This rate is slightly higher than in Germany (99 to 115 per 100,000) and slightly lower than in Great Britain (106 to 140 per 100,000). An encouraging sign has been the gradual aging of the heroin-using population, indicating that fewer young people are being initiated into the use of this drug.

The 1980s saw an increase in the prevalence of cocaine use in the heroin-using population. In some studies as many as 90 percent of daily heroin users also took cocaine. This trend has made the POLY-DRUG user the most common type seen by social-service agencies. In a 1983 national survey with a sample of Dutch youth and young adults aged 15 to 24 years, 3 percent had used cocaine at least once. The estimate was twice as high (6%) in large Dutch cities. Data drawn from a sample of the general population of Amsterdam showed signs that cocaine use must have decreased between 1987 and 1990.

Overall, cannabis use in the Netherlands seems to have decreased since 1970 but has shown a slight increase since 1980. The prevalence of cannabis use is about the same as in Norway and Sweden and far lower than in the United States, although these countries have a far more restrictive cannabis policy. However, in surveys of Dutch students, major increases in cannabis use were reported since 1984/85. In the mid-1990s, as many as 22 percent of the student population may be experimenting with cannabis. These increases may be interpreted as either the result of the current drug policy or as part of global cultural changes among youth, which promote cannabis use in place of alcohol, cocaine, and heroin. In either case, the changes in cannabis use document the limits of tolerance in the policy of the Netherlands. More attention is being placed on abuses in the coffeehouse system and preventive projects that focus on the harmful effects of excessive cannabis use in the young.

Research continues to play an important role in reformulating the system of Dutch drug use. A number of university, private, and governmental institutions conduct research in almost every area of drug use. In general, this research seems to show that the drug policy of the Netherlands has been functioning positively. For example, it seems that the goal of reducing the secondary effects of drug abuse (for example AIDS, VIOLENCE) is being reached. Studies of cocaine use in nondeviant social groups in Amsterdam and in the general population of Rotterdam provide evidence that patterns of use do not always lead to negative consequences, although it is difficult to say who can use without experiencing harm. A longitudinal study of heroin addicts indicates that the normalization policy has been effective in diverting the career of heroin addicts from criminal to conventional but has been less effective in getting heroin users clean. Ongoing evaluations of clinical programs for addicts—programs that concentrate on outcomes, retention, and the place of psychopathology in the provision of a wide array of treatment services—are yielding results that should help improve the social-services system. Model harm-reduction programs, such as needle exchange as it relates to AIDS prevention, have been shown to be efficacious. Policies to help migrant drug users and drug-using prostitutes have also been evaluated with generally positive results.

Future policies toward drug use in the Netherlands will probably involve the refinement of existing strategies, and the development of new ones, aimed at reducing the risks of drug use to optimal low levels. The social nuisance of drug dealing and drug use will be given an increasing priority. Because the atmosphere toward drug users has been relatively open, friendly, and tolerant, abuses in the system have become readily visible and the subject of public concern. To enhance the quality of life in cities and to accommodate the need for security in the coming years, the problem of drugs as a social nuisance will have to be recognized and dealt with. There are increasing signs that the connection between excessive drug use and crime will be an important issue, and more repressive measures toward drug-using criminals will be accepted. At the same time, PREVENTION projects for youth and more differentiated services to addicts are likely to appear. There is a strong tendency to maintain continuity and to improve the organization of drug policymaking. This approach will require a greater spirit of cooperation between many sectors of society, including the drug users themselves. It will also require more effective distribution of information that aims at developing consensus in drug policy.

If the experience of drug policy in the Netherlands has anything to teach, it is that drug use is intimately related to the social context of drug policy. Society's responses to drug use as well as drug-use patterns seem to combine to determine the outcome and severity of drug problems.

(SEE ALSO: *Needle and Syringe Exchanges and HIV/AIDS; Sweden, Drug Use in*)

BIBLIOGRAPHY

BIELEMAN, B., ET AL. (1993). *Lines across Europe: Nature and extent of cocaine use in Barcelona, Rotterdam and Turin.* Amsterdam/Lisse: Swets & Zeitlinger.

DE KORT, M. (1994). The Dutch cannabis debate, 1968–1976. *Journal of Drug Issues,* 24(3), 417–427.

JANSEN, A. C. M. (1991). *Cannabis in Amsterdam: A geography of hashish and marihuana.* Muiderberg: Dick Coutinho.

LEUW, E., & HAEN, M. I. (EDS.). (1994). *Between prohibition and legalization: The Dutch experiment in drug policy.* New York: Kugler.

VAN DE WIJNGAART, G. F. (1991). *Competing perspectives on drug use (the Dutch experience).* Berwyn, PA: Swets & Zeitlinger.

VAN VLIET, H. J. (1990). Separation of drug markets and the normalization of drug problems in the Netherlands: An example for other nations? *Journal of Drug Issues,* 20 (3), 463–471.

CHARLES KAPLAN

NEUROLEPTIC Any of a group of drugs that are also called ANTIPSYCHOTICS. They are used as medications in the treatment of acute psychoses of unknown origin, including mania and SCHIZOPHRENIA. The prototype neuroleptic drugs are chlorpromazine (Thorazine) and haloperidol (Haldol). The site of action for these drugs is the central nervous system where they produce antipsychotic effects.

These drugs are also used for antianxiety, although other agents are more effective and do not have the long-term therapy side effects that neuroleptics do. Drug therapy alone is not entirely effective in treating psychoses, and it is used in combination with acute and long-term support and medical care. Some neuroleptics are also used in the treatment of nausea, vomiting, alcoholic hallucinogens, neuropsychiatric diseases marked by movement disorders (e.g.,

Huntington's disease and Gilles de la Tourette's syndrome), pruritus, and intractable hiccough.

BIBLIOGRAPHY

BALDESSARINI, R. J. (1990). Drugs and the treatment of psychiatric disorders. In A. G. Gilman et al. (Eds.), *Goodman and Gilman's the pharmacological basis of therapeutics,* 8th ed. New York: Peragamon.

GEORGE R. UHL
VALINA DAWSON

NEURON The gross anatomy of the central nervous system—the brain and spinal cord—was studied in some detail during the seventeenth and eighteenth centuries, but not until the nineteenth century did scientists begin to appreciate that the central nervous system (CNS) was composed of many millions of separate cells, the neurons (also called nerve cells). This discovery had to await technical improvements in the microscope and the development of specialized stains that permitted scientists to observe the microscopic anatomy of the nervous system.

HISTORY

In the 1870s, the Italian anatomist Camillo Golgi developed such a special staining technique, and he and other scientists were then able to observe under the microscope the fine structures of the cells of the nervous system. Yet Golgi may not have fully appreciated that what seemed to be an extended network of nerve tissue in reality were millions of distinct neurons with fine fibrils touching each other. It was the Spanish scientist Santiago Ramón y Cajal who was credited with expounding the neuron theory. In 1906, Golgi and Ramón y Cajal shared the Nobel prize in physiology/medicine for their discoveries on the nature of the nervous system.

Even after the concept of separate neurons was generally accepted, there was controversy for many years about how the separate neurons communicated with each other. At the end of the nineteenth century, many scientists believed they did so by means of electric impulses. Others believed there was a chemical messenger that allowed neurons to influence each other. Around 1920 ACETYLCHOLINE was discovered, the first of many nerve messengers

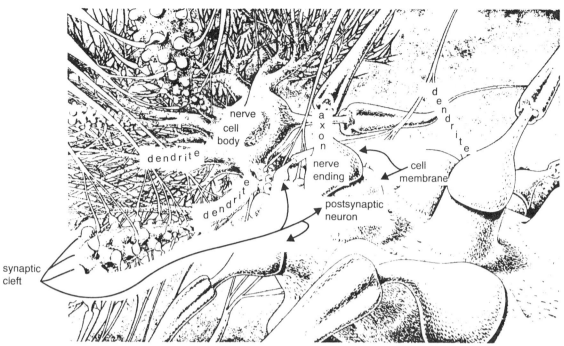

Figure 1

Neuronal Complexity. The complexity of the neuronal network in the brain is demonstrated by this bundle of neurons, which form a vast and ramified structure with their cell bodies, outgrowths, and intercellular contact points.
SOURCE: Modified from Figure 1, in M. J. Kuhar's "Introduction to Neurotransmitters and Neuroreceptors," in *Quantitative Imaging*, edited by J. J. Frost and H. N. Wagner. Raven Press, New York, 1990.

that would be discovered during the subsequent decades.

FUNCTION

The neuron is the basic functional cellular unit of nervous system operations; it is the principal investigational target of research into the actions of addictive drugs and ALCOHOL. An essential feature of the cellular composition of the brain is the high density of extremely varied, heterogeneously shaped neuron groups (see Figure 1). To understand the specialized aspects of neurons and their function, therefore, requires a discussion of the general structural and functional features characteristic to all neurons and the degree to which unique variations form consistent subsets of neurons.

Neurons share many cellular properties that distinguish them significantly from other cell types in other tissues; those changes within the cell's regulatory processes of greatest interest to researchers of addictive drugs, however, depend on features that form distinctions within the class of cells called neurons. Furthermore, the assembly of individual neurons into functional systems, through highly precise circuitry employing highly specified forms of chemical interneuronal transmission, allows for the sensitivity of a brain to addictive drugs.

In some organs of the body—such as the liver, kidney, or muscle—each cell of the tissue is generally similar in shape and function. Within that tissue, all perform in highly redundant fashion to convert their incoming raw material into, respectively, nutrients, urine, or contractions, which establishes the function of the specific tissue. In the nervous system, the variously (heterogeneously) shaped neurons (see Figure 2), supported by an even larger class of similarly (homogeneously) shaped nonneuronal cells, termed *neuroglia*, convert information from external or from internal sources into information ultimately integrated into programs for the initiation and regulation of behavior.

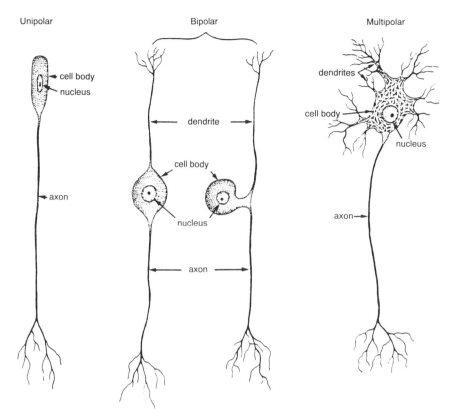

Unipolar — cell body, nucleus, axon

Bipolar — dendrite, cell body, nucleus, axon

Multipolar — dendrites, cell body, nucleus, axon

Figure 2
Three Types of Neurons

This integrative conversion of sensory information into behavioral programs results from the rich interconnections between neurons, and it depends on the extremely differentiated features of neurons—their size and shape; their extended cell-surface cytoplasmic processes (dendrites and axons); and their resultant interconnections that establish the sources of their incoming (afferent) information and the targets of their outgoing (efferent) communication (see also Figure 4).

COMMON FEATURES

As cells, neurons share some features in common with cells in all other organ systems (see Figure 3). They have a *plasma membrane* acting as an external cell wall to form a distinct boundary between the environment inside (intracellular) and outside (extracellular) the cells. The intracellular material enclosed by the plasma membrane is termed the *cytoplasm.* Like all other cells (except red blood cells), neurons have numerous specialized intracellular organelles, which permit them to maintain their vitality while performing their specialized functions.

Thus, neurons have *mitochondria* (singular, mitochondrion), by which they convert sugar and oxygen into intracellular energy molecules, which then fuel other metabolic reactions. Neurons have abundant *microtubules*, thin intracellular tubular struts, by which they form and maintain their often highly irregular cell structure. Neurons are also rich in a network of intracellular membranous channels, the *endoplasmic reticulum*, through which they distribute the energy molecules, membrane components, and other synthesized products required for functioning. Like other cells that must secrete some of their synthesized products for functioning, as neurons do with their neurotransmitters, some parts of the endoplasmic reticulum, the *smooth endoplasmic reticulum*, are specialized for the packaging of secretion products into storage particles, which in neurons are termed *synaptic vesicles.* At the center of the pool of cell material, the *cytoplasm*, neurons possess a *nucleus*, which, as in other nucleated cells, contains the full array of the genetic information char-

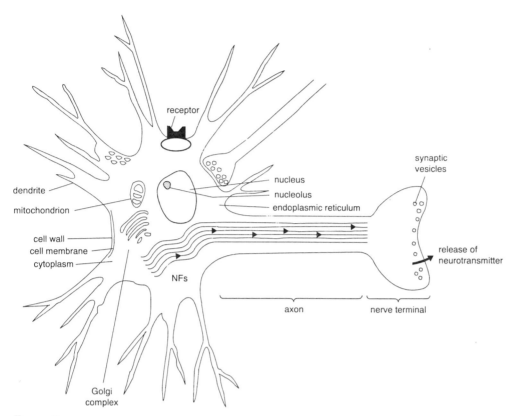

Figure 3
Features of the Neuron

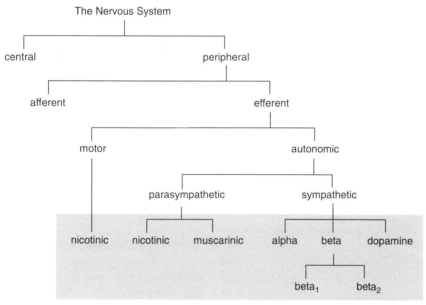

Figure 4
Relationship of Receptor Types. Efferent nerves in the peripheral nervous system. (Receptor subdivisions for alpha and dopaminergic receptors are not included.)

acteristic of the individual organism. From this nucleus, selected subsets of genetic information are expressed to provide for the general shared and the specific unshared features of the cell. The nucleus of the neuron cell is enclosed within a membranous envelope that, as in many other types of cells, exhibits multiple nuclear pores through which information can be conveyed to and from the nucleus.

UNIQUE FEATURES

The plasma membrane of neurons differs from that of non-neuronal cells in that it contains special proteins, termed *voltage-sensitive ion channels.* Such channels are conceptually small tubular proteins embedded in the membrane of the neuron, which, when activated under specific conditions, allow positively charged ions of sodium, potassium, and calcium to enter the neuron. The existence of such electrically sensitive channels permits the neuron to become electrically excitable. The expression and selective distribution (compartmentalization) of such electrically excitable channels along its efferent processes, the axons, permit neurons to conduct signals efficiently for long distances; this also accounts for the bioelectrical activity of the brain assessed by *electroencephalography (EEG).* Similarly, the distribution of such electrically excitable ion channels along the receptive surfaces of the neurons (its dendrites and cell body [soma]) allows them to conduct and integrate signals from all over the extended shape of the neuron.

The smooth endoplasmic reticulum of the neuron is somewhat more elaborate and extensive than other cells that secrete their products; this specialized and extensive smooth endoplasmic system is termed the *Golgi complex* (or *Golgi apparatus).* Discovered accidentally, it was a useful marker for staining the nervous system to distinguish neurons from other cells of the brain when under inspection by microscope.

The nucleus of neurons is often highly elaborated, with multiple creases or infoldings, exhibiting complex configurations, within which are typically dense accumulations of cytoplasmic organelles, and almost always a very distinctive intranuclear clustering of genetic material, the *nucleolus.* Differentiated neurons—neurons whose developmental stage is past the step at which cell-type dedication has occurred—are unable to undergo cell division, in dis-

tinct contrast to comparably metabolically active cells in such complex tissues as liver, kidney, muscle, or skin. As a result, mature neurons can repair themselves, up to a point, but are unable to regenerate themselves or respond to their growth factors in a manner that would in other tissues lead to cell division and replacement.

The most distinctive cellular feature of neurons is the degree to which they express unique patterns of size and shape. In mammals, all neurons have highly irregular shapes; such shape variations are categorized in terms of the number of cell surface extensions, or neuronal processes, that the neuronal subset expresses, as in Figure 2.

Some neurons have only one cellular process extending from the surface of a round or nearly round cell body; this form of neuron, a *unipolar* neuron, is typical of invertebrate nervous systems. Typical unipolar neurons are the cells of the dorsal root ganglia, in which a single efferent axon conducts information toward or away from the cell body through a branched axon.

Most neurons of the central nervous system of mammals are multipolar. That is, in addition to the efferent axon, which may also have many subsets of secondary axons, called *collateral branches,* that stem from the main efferent process axon, elaborations may also be expressed from the cell body surface. The latter elaborations are termed *dendrites,* because their shape resembles the limbs of trees. Dendrites protrude from the cell body, and they as well as the cell body constitute the receptive surfaces of the target neuron onto which the afferent connections make their synaptic connections.

DISTINGUISHING NEURONS

Since neurons come in so many shapes and sizes, early investigators of the brain sought to make distinctions among them, based in part on their locations, their sizes and shapes, and the connections they could be shown to receive or emit. Every scientist who worked in the formative era of brain research sought to describe a unique subset of neurons that were forever after named for their initial describer or the unique property defined. Thus, we have *Betz neurons,* large layer V-VI neurons of the motor cortex, and *Purkinje neurons,* the major output neurons of the cerebellar cortex, as well as neurons named for their shapes and appearance—*pyramidal neurons* of the cerebral and hippocampal cortices, *mitral* and *tufted neurons* of the olfactory bulb, and *granule cell neurons* of the cerebellar, hippocampal, and olfactory cortices. The last mentioned have relatively compact cell bodies, densely packed together, giving the brain a granular appearance by optical microscopy.

Dendrites and axons exhibit highly distinctive morphological patterns. The surfaces of dendrites and axons can be distinctive in the shapes of their branches. This permits fine discrimination among neurons (stellar, or star-shaped, neurons; chandelier neurons; or mossy or climbing axon fibers). Some neurons exhibit dendrites whose surfaces are smooth (aspiny); others are highly elaborated (spiny), which may serve to enlarge the receptive surfaces and enhance the degree to which such neurons may integrate afferent information.

Similarly, the morphology and stability of the axons may also be highly variable. Some neurons direct their axons to highly constrained targets in a more or less direct route; others may be highly branched, with multiple collateral branches to integrate communications from one cell cluster to many divergent targets. To provide the essential support of anabolic and secretory materials within these highly elaborated cellular structures, neurons have evolved an efficient form of intracellular transport, an energy-dependent, microtubule-guided, centripetal and centrifugal process by which organelles are dispensed to and returned from the distal processes (as well as probable macromolecular signals sensed by pinocytotic-like [fluid uptake] incorporation of such signals by distal dendrites and axons). Such signals may serve as local growth-regulatory factors, allowing even the nondividing neurons to alter their shape and connections in response to activity and signals received from their afferent sources.

NEURONAL IDENTITY

An individual neuron may be referred to on the basis of its size (magnocellular, parvicellular). A layer or "nuclear" cluster of neurons may be referred to by shape (pyramidal, mitral), the morphology of its axon terminals (i.e., *basket cells*, whose axon terminals make basket-shaped terminations on their targets), and its position in a sensory or motor circuit. In the latter classification scheme, those neurons closest to the incoming sensory event or to the out-

going motor-control event are termed *primary sensory* or *motor* neurons, respectively, whereas neurons at more distal positions of circuitry from the primary incoming or outgoing event are termed *secondary*, *tertiary*, and so on, depending on their position in that hierarchy.

In addition to these morphological qualities, neurons may also be separately distinguished on the basis of the functional systems to which they are connected (visual, auditory, somatosensory, proprioceptive, attentional, reinforcing, etc.) and on the basis of the neurotransmitters they employ to communicate with the neurons to which they are connected (cholinergic, adrenergic, GABA-ergic, etc.). Each of those features provides for a multidimensional definition of virtually every neuron in the brain.

(SEE ALSO: *Brain Structures and Drugs; Neurotransmission; Neurotransmitters; Receptor: Drug; Reward Pathways and Drugs*)

BIBLIOGRAPHY

CORSI, P. (ED.). (1991). *The enchanted loom: Chapters in the history of neuroscience*. New York: Oxford University Press.

FLOYD BLOOM

NEUROTRANSMISSION NEURONS (nerve cells) communicate chemically by releasing and responding to a wide range of chemical substances, referred to in the aggregate as NEUROTRANSMITTERS. The process of *neurotransmission* refers to this form of chemical communication between cells of the central and peripheral nervous system at the anatomically specialized point of transmission, the SYNAPSE (synaptic junctions). Thus, it is convenient to conceive of "the" neurotransmitter for a specific instance of synaptic connections between neurons in one brain location (the source neurons) and their synaptic partner cells (the target neurons) in another neuronal location. For example, the phrase "dopaminergic neurons of the nigro-accumbens circuit" refers to the DOPAMINE-transmitting synaptic connections between the brain neurons of the substantia nigra and their targets in the NUCLEUS ACCUMBENS. Current concepts of neurotransmission, however, require a broader view; they would consider as neurotransmitters all the chemical substances that a given neuron employs to signal the other neurons to which it is anatomically connected (its synaptic targets) and through which that neuron may also be able to influence other neuronal and nonneuronal cells in the adjacent spatial environment of its circuitry (nonsynaptic targets).

In some cases—more frequent in invertebrate nervous systems, in more primitive vertebrates, and in the embryonic nervous system than in the adult mammalian nervous system—neurons may also communicate "electrically," by direct ionic coupling between connected cells, through specialized forms of intercellular junctions referred to as "gap junctions," or *electrotonic junctions*. Such electrotonic transmission sites are of relatively little direct concern to the actions of addictive drugs and ALCOHOL. In contrast, it is the more pervasive process of chemical neurotransmission that underlies the main molecular and cellular mechanisms by which addictive drugs act—and through which the nervous system exposed to such drugs undergoes the adaptations that may lead to DEPENDENCE, HABITUATION, WITHDRAWAL, and the more enduring changes that persist after withdrawal from the once-dependent state.

The critical characteristic of a substance designated as a neurotransmitter is the manner in which it is made and secreted. To qualify as a neurotransmitter, the release of the substance must be coupled to neuronal activity according to two rather stringent functional rules (see Figure 1).

1. The transmitter substance must be synthesized by the transmitting neuron. In most cases, the substance is made well in advance and stored in small organelles (synaptic vesicles) within the terminal axons of the source neuron, ready for eventual release when called upon.
2. The transmitter substance must be released by that neuron through a special form of activity-dependent, calcium ion (Ca^{2+})–selective, excitation–secretion coupling. Substances released through other nonactivity–coupled and non-Ca^{2+}–coupled mechanisms may be regarded as excretion (as with metabolic byproducts to be degraded), rather than secretion.

The synaptic junction is the site at which the axons of the source neuron physically make most intimate contact with the target neuron to form an anatomically specialized junction; concentrated there are the proteins that mediate the processes of transmitter release (from the presynaptic neuron) and response (by the postsynaptic neuron). Indirect evidence for some neurotransmitter systems has suggested to some scientists a general concept of *nonsynaptic* interneuronal communication, sometimes also referred to as *paracrine* or *volume-transmission* communication, in which the neurotransmitter released by a designated set of presynaptic terminals may diffuse to receptive neurons that are not in anatomic contact. The sets of chemical substances that neurons can secrete when they are active can also influence the non-neuronal cells, such as the cells of the vascular system (the glia) and the inflammatory-immune cells (the microglia).

The activity of neurons can also be modified by substances released from the non-neuronal cells of the central or peripheral nervous system, substances often termed *neuromodulators*. This same term, however, is frequently applied to the effects of neuron-produced transmitter substances whose mechanisms of action and whose time course of effect differ from those of the classic junctional neurotransmitter acetylcholine.

The current research on neurotransmitters and neuromodulators pertinent to drugs and alcohol is devoted to (1) understanding how exposure to addictive drugs may regulate the genes that control the synthesis, storage, release, and metabolism of known neurotransmitters; (2) identifying new substances that may be recognized as neurotransmitters, whose effects may be related to the effects of or reactions to addictive drugs and alcohol; (3) understanding the molecular events by which neurons and other cells react to neurotransmitters in both short-term and long-term time frames (a process often termed *signal transduction*, which cells of the nervous system share with most other cells of the body) and how these processes may themselves be perturbed by the influence of addictive drugs and alcohol; and (4) understanding the operations of neuronal communication in an integrative context of the circuits that release and respond to specific transmitters, and the way in which these neuronal circuits participate in defined types of behavior, either normal or abnormal.

NEUROTRANSMITTER ORGANIZATION

There are three major chemical classes of neurotransmitters.

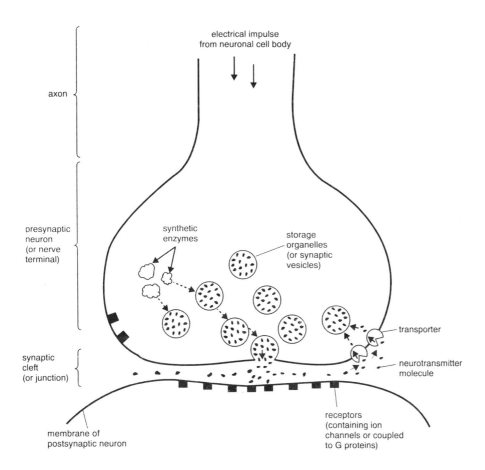

Figure 1

Synapse. Nerve ending from one neuron forms a junction, the synapse, with another neuron (the postsynaptic neuron). The synaptic junction is actually a small space, sometimes called the synaptic cleft. Neurotransmitter molecules are synthesized by enzymes in the nerve terminal, stored in vesicles, and released into the synaptic cleft when an electrical impulse invades the nerve terminal. The electrical impulse originates in the neuronal cell body and travels down the axon. The released neurotransmitter combines with receptors on postsynaptic neurons, which are then activated. To terminate neurotransmission, transporters remove the neurotransmitter from the synaptic cleft by pumping the neurotransmitter back into the nerve terminal that released it.

1. *Amino acid transmitters:* glutamate (GLU) and aspartate are recognized as the major excitatory transmitting signals; GAMMA AMINOBUTYRATE (GABA) and glycine are the major inhibitory transmitters. These transmitter substances occur in concentrations of one millionth part per milligram (μM/mg) protein. Since they are considered the most frequently employed transmitter substances, they

have been linked to many aspects of the actions of addictive drugs.

2. *Aminergic transmitters:* ACETYLCHOLINE, epinephrine (also called adrenaline), NOREPINEPHRINE (also called noradrenaline), DOPAMINE, SEROTONIN, and histamine. The aminergic neurons constitute a minor population of neuronal transmission sites, as reflected in the fact that

their concentrations in the brain are roughly 1/1000th that of the amino acid transmitters or one billionth part per milligram (nM/mg protein). Because of their divergent anatomy (a few clusters of aminergic neurons may project onto literally millions of target neurons in many locations of the brain) and the ability of their synaptic signals to produce long-lasting effects, the aminergic neurons represent a very powerful subset of transmission conditions that is important to the effects of addictive drugs. Of particular relevance are the dopaminergic neurons—for their pertinence to the sites of reward for stimulants, opiates, and certain aspects of ethanol (alcohol) action—and the noradrenergic and serotonergic neurons—for their association with the phenomena of drug adaptation and tolerance.

3. *Neuropeptides:* of which there are dozens. Peptides are molecules containing a specific series of 2-50 amino acids, chemically arranged in a specialized "head-to-toe" chemical linkage known as a peptide bond. The order and number of the linked amino acids determine the linear structure of the peptide. In the nervous system, peptides in general occur in still lower concentrations than do the two prior classes of transmitter, namely at 10–100 trillionth part per milligram (pM/mg) protein. A revolutionary finding has emerged here in concepts of brain system interactions: It would now seem that neuropeptides are almost certainly never the sole signal to be secreted by a central neuron that contains such a signaling molecule, but rather accompany either an amino acid or an amine transmitter (at intrasynaptic terminal concentrations a thousand to a millionfold higher), such sites may even contain a second or third peptide as well.

Neuropeptides are of interest to the molecular and cellular mechanisms of addictive drug and alcohol action, because they may provide the postsynaptic receptors through which the drugs act (as in the case of the opiates and possibly the case for the natural BENZODIAZEPINES) or modify the effects of the presynaptic transmitters (as in the case of the peptide cholecystokinin that accompanies some forms of dopaminergic transmission, through which stimulants act and may modify responses to that amine if cosecreted).

Because of the ability to read the linear sequences of the amino acids, it has become clear that many of the neuropeptides share select small sequences and thus conceptually constitute "families" of peptides. For example, the opioid peptides all share one or more repeats of the amino-acid sequence tyrosine-glycine–glycine-phenylalanine; thus, each of the opioid-peptide genes leads to the expression of a different pre-prohormone by different sets of neurons of the central and peripheral nervous system. The existence of the shared amino-acid sequences implies that at some point in evolution, there may have been only one opioid-peptide signal, which was then duplicated and modified for use by the increasing number of neurons that came with the evolution of the mammalian brain. Such family relationships also exist for other peptide families (oxytocin/vasopressin; the tachykinin peptides; the secretin/glucagon-related peptides; the pancreatic polypeptide-related peptides), whose amino-acid sequences have shown great conservation over large domains of the evolutionary tree, attesting to the high signal quality of these molecules and the transductive mechanisms of their receptors. Other peptides, such as somatostatin and gonadotropin-releasing hormone, have no known family relationships as yet—but the discovery process here is probably not complete.

OTHER TRANSMITTER CANDIDATES

Other kinds of molecules may also be made within neurons to play auxiliary roles in intercellular transmission in the nervous system—from purines like Adenosine Triphosphate, lipids like arachidonic acid and prostaglandins, and steroids similar to those made and released by the adrenal cortex and the gonads. These substances may, in some cases, act as intracellular second messengers to underlie the effects of the aminergic and peptidergic transmitters (see below); they therefore have implicit relevance to the effects of the addictive drugs whether or not they may also serve as primary transmission signals.

Investigators have revealed that under some conditions active neurons may synthesize gaseous signals, such as nitric oxide and carbon monoxide, which can carry rapidly evanescent signals over short distances. The effects of these transmission-related substances will undoubtedly become of increasing importance to the explanations of the mechanisms of action or adaptation to the addictive drugs.

SIGNAL TRANSDUCTION ORGANIZATION

Aside from the chemistry of the neurotransmitter substances, further insight into their role in the actions of addictive drugs arises from the viewpoint of their synaptic physiology and their underlying mechanisms of signal transduction. When neurons respond to neurotransmitters, the ultimate changes in the excitability and metabolic activity of the responding neuron generally require changes into or out of the cell in the flow of ions (natural chemical elements of the extracellular fluid)—some with positive charge (sodium, potassium, and calcium) and others with negative charge (chloride).

As a general rule, it would appear that every neurotransmitter has more than one form of postsynaptic receptor through which its effects are mediated. Before the ability to characterize these receptors through molecular genetics, such receptor subtypes were identified on the basis of the comparative pharmacological potency of synthetic AGONISTS or ANTAGONISTS of the natural transmitter. With the advent of molecular cloning, however, an even finer subtyping would appear to be required, since many of the conclusions on receptor pharmacological patterns were based on analyses of tissue fractions that undoubtedly contained many molecular forms. A major effort in the future will be to link more explicitly the molecular and pharmacological characterization of neurotransmitter receptor subtypes and to determine which of them are most critical to the effects of, and adaptations to, addictive drugs.

Three major formats have been revealed for the transductive process.

1. *Directly regulated ion channels.* Here the ion channel to be opened is formed by the units of the receptor molecule itself, as recently established by direct cloning of several such receptor-ionophores. Such receptors are now known to be the motif of the nicotinic-cholinergic receptors of the neuromuscular junction and the central nervous system, as well as for the three types of glutamate receptor, the several isoforms of the GABA$_A$ receptor, the glycine receptor, and at least one form of a serotonin receptor.

   Common features of these receptors are (a) they are composed of several (3–5) subunits, called monomers, that apparently may be combined in differing ratios (so-called multimeric re-combinations) by various neurons to constitute the "holoreceptor"; (b) each monomer consists of four presumed transmembrane domains; and (c) discrete sections of the receptor monomer, either within the membrane or the cytoplasm, account for their voltage and chemical sensitivity, and for the ease and duration of openings in the ion channel.

2. *Indirectly regulated ion channel-receptors.* This form is based on the similarities between the visual pigment rhodopsin—the molecule used by photoreceptor neurons (rods, cones) to transduce light into signals to other neurons of the retina—and the beta-adrenergic receptor—one of the types of receptors regulated by the amine norepinephrine. This general form of transducing molecule was later found to be the form also used by the cholinergic muscarinic receptor, as well as by most serotonin and all known dopamine receptors, plus all the known peptide receptors.

   The common features of this class are (a) the receptor is a single molecule, with seven transmembrane domains; (b) activation of these receptors by their signaling molecules leads to further interactions of the receptor with other large proteins, some of them enzymes, within or near the plane of the membrane; and (c) the eventual indirect regulation of the ion channel, either the opening or closing of the channel, is then mediated through small molecular intracellular second messengers, such as the calcium ion (Ca^{2+}) or the products of the associated enzymes, yielding intracellular second-messenger molecules, such as cyclic adenosine monophosphate (cAMP), or a lipid such as an inositol phosphate, diacylglycerol, or an arachadonic acid catabolite. The essential common second step of such transduction cascades is that the activated receptor interacts with a guanosine triphosphate (GTP)-binding protein (termed a *G-protein*) composed of three monomer subunits. The G-protein complex dissociates to activate the enzyme making the second messenger and, at the same time, hydrolyses the GTP and reassociates to end the cycle of signal generation. The second messenger consequences of this form of transduction, however, may be more enduring—activating one or more enzymes (protein kinases or phosphatases) that can add or remove phosphate groups on structural proteins or other enzymes, to activate or inactivate them. Such events can significantly shift the metabolic state of the responding cell and eventually regulate the expression of its specific genes. One such gene target is the immediate early genes of the nervous system, the proto-oncogenes, discovered some years ago because of the mutated forms used by oncogenic viruses, which induce cancer in non-neuronal cells.

3. *The receptor-enzyme.* This third major molecular motif of signal transduction has been elucidated recently; although it is already clear that this form does exist in the mammalian brain, it has been studied more in non-neuronal systems. This motif's characteristics are that the receptor for some peptides is itself the enzyme guanylate cyclase, which is directly activated by receptor-ligand binding, leads to an intracellular generation of cyclic guanosine monophosphate, and then to a cascade of events similar to that described for AMP.

SYNAPTIC INTERACTIONS

Most neurons receive synaptic input simultaneously from hundreds of other neurons, each of which employs its own mix of transmitters. The transductive processes underlying these individual events can influence the intensity and duration of the subsequent responses, thereby integrating incoming signals and providing the basis by which activity in assemblies of interconnected neurons results in behavioral output by the brain.

To gain insight into the basis by which the events of neurotransmission can lead to multineuronal programs of interaction, such as those required to initiate responding for an addictive drug, requires knowledge both of the anatomical substrate over which such programs of neuronal activity take place and of the effects of the neurotransmitters at each of the cellular elements of such an interactive ensemble of neurons.

(SEE ALSO: *Addiction: Concepts and Definitions; Brain Structures and Drugs; Limbic System; Tolerance and Physical Dependence*)

BIBLIOGRAPHY

BARONDES, S. H. (1993). *Molecules and mental illness.* New York: Scientific American Library.

BLOOM, F. E. (1990). Neurohumoral transmission in the central nervous system. In A. G. Gilman et al. (Eds.), *Goodman and Gilman's the pharmacological basis of therapeutics,* 8th ed. New York: Pergamon.

COOPER, J. R., BLOOM, F. E., & ROTH, R. H. (1991). *The biochemical basis of neuropharmacology,* 6th ed. New York: Oxford University Press.

KORNEMAN, S. G., & BARCHAS, J. D. (EDS.). (1993). *Biological basis of substance abuse.* New York: Oxford University Press.

WATSON, R. R. (ED.). (1992). *Drugs of abuse and neurobiology.* Boca Raton, FL: CRC Press.

FLOYD BLOOM

NEUROTRANSMITTERS A neurotransmitter is any chemical substance (the first recognized was ACETYLCHOLINE) that NEURONS (nerve cells) secrete to communicate with their target cells (glands, muscles, and other neurons). Neurotransmitters diffuse from their sites of release—from the presynaptic nerve terminal—across the synaptic cleft, to bind to receptors on the external surface of the postsynaptic cell. Activation of these receptors allows for the transmission of commands (excitation, inhibition, and other more complex forms of regulation) from the presynaptic neuron to the postsynaptic cell.

A neurotransmitter is released from a nerve ending, interacts with specific RECEPTORS, and is then either transported back into the presynaptic neuron or destroyed by metabolic enzymes in the synaptic cleft.

Chemically, neurotransmitters are amino acids, amines, or peptides. Peptide transmitters commonly coexist and may be cosecreted with amino acid or amine transmitters.

(SEE ALSO: *Dopamine; Endorphins; Neurotransmission; Norepinephrine; Serotonin*)

BIBLIOGRAPHY

BARONDES, S. H. (1993). *Molecules and mental illness.* New York: Scientific American Library.

BLOOM, F. E. (1990). Neurohumoral transmission in the central nervous system. In A. G. Gilman et al. (Eds.), *Goodman and Gilman's the pharmacological basis of therapeutics,* 8th ed. New York: Pergamon.

COOPER, J. R., BLOOM, F. E., & ROTH, R. H. (1991). *The biochemical basis of neuropharmacology,* 6th ed. New York: Oxford University Press.

KORNEMAN, S. G., & BARCHAS, J. D. (EDS.). (1993). *Biological basis of substance abuse.* New York: Oxford University Press.

WATSON, R. R. (ED.). (1992). *Drugs of abuse and neurobiology.* Boca Raton, FL: CRC Press.

FLOYD BLOOM

NEW YORK STATE CIVIL COMMITMENT PROGRAM

Historically, the New York State Civil Commitment Program has been viewed as the largest and most expensive program of its kind. With the CALIFORNIA CIVIL Addict Program (CAP) serving as a prototype, it was established in the early 1960s in response to the explosion in New York's heroin-addict population. The addicts were accounting for much of the street crime, political candidates were expected to have a policy in regard to the heroin-addiction problem, and many people believed that compulsion was necessary for treatment to be effective (Winick, 1988). In 1962, an initial reaction to the problem came in the form of the Metcalf-Volker Narcotic Addict Commitment Act, which sent arrested addicts to state mental-hygiene facilities for treatment. The failure of this program (90 percent of the patients did not complete the process) prompted New York Governor Nelson Rockefeller to substantially modify and expand the program in 1966 by creating a Narcotic Addiction Control Commission (NACC). NACC was to administer a New York State Civil Commitment Program involving a major statewide network of residential treatment centers.

Six different types of centers handled the following phases of treatment: examination and detention; detoxification, orientation, and screening; residential treatment and rehabilitation; temporary return; indefinite return; and halfway houses. Those who were eligible for treatment at a center included addicted individuals who had been arrested or convicted for a felony or misdemeanor, who had been involuntarily committed by their family or a friend, or who had volunteered to be treated. The treatment process consisted of a period of commitment within the institution, followed by community aftercare. Clients were under the control of the New York State agency for an average of 25 months, of which 10.1 months was spent residing in the institution (Winick, 1988).

THE PROGRAM'S DEMISE

The New York State Civil Commitment Program reached its peak in 1970 when 24 state facilities with 4,100 beds and a staff of over 5,000 provided services to 6,600 addicts. Follow-up studies of the program at this time were few, but they tended to indicate some positive outcomes (Winick, 1988). Inciardi (1988) maintains that the NACC withheld studies with unfavorable results and delivered carefully selected positive statements about the program to the public. After 1970, the program began to lose public support and became a regular political target because of charges of cost overruns, allegations of staff brutality, and questionable administrative procedures (Winick, 1988). There was also a general change in philosophy that drew politicians away from supporting state-run institutions and toward recommending community-based treatment. Governor Rockefeller announced in 1971 that he had lost confidence in the New York program and initiated a two-thirds cutback in budget and clients. By 1975, there were 2,100 occupied beds in 15 facilities; 8 facilities remained by the following year, 3 in 1977, and the last 2 centers shut down in 1979 (Winick, 1988). From 1966 to 1979, the program had cost approximately 1 billion dollars. By the time the program was closed, each resident was costing an average of $29,000 per year, as compared with $8,500 for a resident in a THERAPEUTIC COMMUNITY and $14,500 for a prison inmate (Winick, 1988). In 1980, the state legislature repealed the civil commitment law.

WHY THE PROGRAM FAILED

Inadequacy of planning was a key contributor to the problems of the New York State Civil Commitment Program (Winick, 1988). The first eight facilities opened in less than a year, in response to Governor Rockefeller's campaign pledge of 1966, to get large numbers of addicts off the streets quickly. Under this rapid development, there was little time to select and train the staff. The directors of the treatment facilities had inadequate administrative or clinical experience, since they were mostly political and civil service appointees (Inciardi, 1988). Because there was a need to quickly obtain large, secure buildings, NACC staff purchased underutilized moderate- to maximum-security prisons from the New York Department of Corrections and used them as treatment facilities. Many of the former prison guards were maintained as Narcotic Rehabilitation Officers who performed both a counseling and custodial function. These officers had not received adequate training in drug treatment or counseling, and they often disciplined program participants overly harshly (Inciardi, 1988). The result was an environment that did not offer therapeutic benefits and was not conducive to behavioral change. The screening of candidates for the program, moreover, was not consistent, and the criteria for completion of the program were ambiguous. The reentry and aftercare programs were equally ill equipped to handle the task at hand. The aftercare "officers" had no authority to arrest a client for violation of aftercare conditions, and their caseloads were too large to allow close supervision. As a consequence, a great number of parolees fled or stopped reporting—approximately 30 percent to 40 percent of clients fled at any given time (Winick, 1988).

An analysis of the areas of weakness and failure of the New York State Civil Commitment Program has significant policy-making implications for the treatment of drug abusers. As the policymakers for such states as Texas begin new, large-scale residential treatment programs in the criminal justice system, it will be helpful for them to study and understand the New York experience.

(SEE ALSO: *California Civil Commitment Program; Civil Commitment; Coerced Treatment; Narcotic Addict Rehabilitation Act; Prisons and Jails; Rockefeller Drug Laws*)

BIBLIOGRAPHY

INCIARDI, J. A. (1988). Compulsory treatment in New York: A brief narrative history of misjudgment, mismanagement, and misrepresentation. *Journal of Drug Issues, 18*(4), 547–560.

LEUKEFELD, C. G., & TIMS, F. M. (1988). Compulsory treatment: A review of findings. *Compulsory treatment of drug abuse: Research and clinical practice* (NIDA Research Monograph 86). Rockville, MD: U.S. Department of Health and Human Services.

WINICK, C. (1988). Some policy implications of the New York State Civil Commitment Program. *Journal of Drug Issues, 18*(4), 561–574.

HARRY K. WEXLER

NIAAA *See* U.S. Government Agencies: National Institute on Alcohol Abuse and Alcoholism (NIAAA).

NICOTINE This is a PSYCHOACTIVE chemical substance found in TOBACCO products, including cigarettes, cigars, pipe tobacco, and smokeless tobacco such as chewing (spit) tobacco and oral and nasal SNUFF. The nicotine molecule is composed of a pyridine ring (a 6-membered nitrogen-containing ring) with a pyrrolidine ring (a 5-membered nitrogen-containing ring).

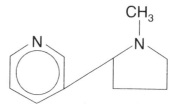

Figure 1
Nicotine

Nicotine can occur in two forms. The active form, called L-nicotine, is found in tobacco plants of the genus Nicotiana. These are chiefly South American plants of the nightshade family (Solanaceae)—annuals cultivated since pre-Columbian times for their leaves, especially *Nicotiana tabacum*. The inactive form, D-nicotine, is not present in tobacco leaves but is formed to a small extent in the combustion of tobacco during smoking. These two forms are stereoisomers, meaning that even though they are both nicotine, they have different three-dimensional structures. In pure form, nicotine is a colorless liquid, but it turns brown on exposure to air.

Nicotine is water-soluble and transfers from tobacco to cigarette smoke readily, because it vaporizes easily. Once it is in the body, conditions are ideal for rapid distribution to blood and tissues because nicotine is a weak base, and when un-ionized under alkaline conditions, such as those found in the blood stream, it crosses cell membranes easily.

The primary natural source of nicotine is the tobacco plant, but nicotine is also found in some amount in related plants. Small amounts are in foods of the nightshade family, such as tomatoes and eggplants. Consumption of nicotine has not been limited to the use of plants in which it naturally occurs. In 1828, the German scientists Posselt and Reiman isolated nicotine from tobacco leaves, and since then it has been added to other products. For example, it is widely used as an insecticide in such products as Black Leaf 40, which contains 40 percent nicotine sulfate.

EFFECTS OF NICOTINE

The first pharmacological studies of nicotine were initiated in 1843 by Orfila. Nicotine is an alkaloid that affects major organs, such as the heart and brain. It also affects the body at the cellular level.

Effects in the Body. The actions of nicotine in a human body are complex. They depend on the amount of nicotine given, the route of administration (e.g., by mouth or intravenously), the time over which the dose is given, and the individual's history of exposure to nicotine. In high doses, nicotine produces nausea, vomiting, convulsions, muscle paralysis, cessation of breathing, coma, and circulatory collapse. Such high doses are seen after accidental absorption of a nicotine-containing insecticide or an overdose of nicotine.

In lower doses, such as those used by people who consume tobacco products, the effects are very different. They include a speed up in heart rate and blood pressure; increased force of contraction of the heart; constriction of blood vessels in the skin, producing cool, pale skin; constriction of blood vessels in the heart; relaxation of skeletal muscles; increased body metabolic rate; and the release of hormones such as epinephrine (adrenaline), NOREPINEPHRINE, and cortisol into the bloodstream. Nicotine's effects on the brain are very complex because nicotine works in part by enhancing the release of chemicals that transmit information from one neuron to another (NEUROTRANSMITTERS) by brain cells. For example, nicotine enhances the release of DOPAMINE, which may produce pleasure; norepinephrine, which may suppress appetite; acetylcholine, which produces arousal; SEROTONIN, which may reduce anxiety; and beta ENDORPHIN, which may reduce pain. The development of addiction to nicotine in tobacco users is attributed in part to many of the effects of nicotine that people find desirable.

Effects of Nicotine in Cells. Nicotine binds (attaches) to RECEPTORS on cell membranes that normally bind a neurotransmitter called ACETYLCHOLINE. Acetylcholine, like other neurotransmitters, is a chemical released by nerve endings in the body that binds to certain receptors on cells and activates them. The activated cells communicate messages to other nerves or produce specific actions on body organs. Nicotine activates only certain of the receptors that bind acetylcholine. These receptors are now called nicotinic cholinergic receptors. Using the selective action of nicotine on cholinergic receptors, scientists are able to observe their activity separately from muscarinic cholinergic receptors, receptors activated by a chemical called muscarine. Nicotinic cholinergic receptors are located at the ganglia in the autonomic nervous system, where there are specialized areas for communications between nerves, in the adrenal gland, at the neuromuscular junctions, where nerves attach to and activate muscles, and in many parts of the brain.

The greatest number of nicotine cholinergic receptors in the BRAIN are found in the hypothalamus, hippocampus, thalamus, midbrain, brain stem, and many parts of the cerebral cortex. Nicotine acts on sensory receptors, including those that mediate pain sensations. The effects of nicotine on these specific receptors have been an important tool in studying the effects of neurotransmitters on cell receptors and on the nervous system as a whole. In addition, these studies provide information about the widespread effects of nicotine introduced into the body during tobacco use.

DEVELOPMENT OF PHYSICAL DEPENDENCE ON NICOTINE

Nicotine is the chemical substance responsible for PHYSICAL DEPENDENCE on tobacco products. During the development of physical dependence on a drug such as nicotine, brain chemistry and function change. They return to normal in the presence of nicotine and come to depend on the drug for normal function.

The change that results in normal function in the presence of nicotine is called neuroadaptation or TOLERANCE. When tolerance develops after a period of use of nicotine, or of any drug, the same dose produces less of an effect than previously. Tolerance develops to many of the effects of nicotine. It is well-known that people smoking their first cigarette often experience nausea and vomiting. However, after repeated exposure to cigarette smoke, these effects disappear. Their disappearance is the development of tolerance to the toxic effects of nicotine in the cigarette smoke. Tolerance also develops to the more desirable effects of nicotine such as pleasure and alertness.

The development of tolerance is associated with changes in the brain, such as an increased number of nicotinic cholinergic receptors found in the brains of smokers studied at autopsy. The changes in the brain correspond to a state in which the tolerant brain comes to depend on nicotine for normal functioning. This state is called physical dependence.

Physical dependence also means that abstinence or WITHDRAWAL symptoms occur when a person who has taken a drug on a regular basis stops taking it. Physical dependence on nicotine has been clearly demonstrated. Thus a person who stops using tobacco after his or her body has adapted to the presence of nicotine will experience withdrawal symptoms in the form of irritability, restlessness, drowsiness, difficulty concentrating, impaired job performance, anxiety, hunger, weight gain, sleep disturbances, slow down in heart rate, and a strong urge for nicotine. In general, withdrawal symptoms are opposite to the effects produced by nicotine when a person who is not tolerant uses it. Thus a person will start using tobacco primarily to experience the desired effects of nicotine, but once the ADDICTION develops, use of tobacco may be chiefly to prevent the emergence of unpleasant withdrawal symptoms. Use of a drug to prevent withdrawal is common in people who are addicted to a drug.

ABSORPTION OF NICOTINE FROM TOBACCO

Nicotine, which is absorbed into the body when tobacco products are used, can be absorbed by different routes and at different rates. Some products deliver nicotine in smoke that is inhaled. In tobacco smoke, nicotine is present in droplets that also contain water and tar. These droplets are carried by gases that include carbon monoxide, hydrogen cyanide, and nitrogen oxides. Such suspended droplets carried by gas are called an aerosol. When the aerosol is inhaled, the droplets are deposited in the small airways of the lungs, from which nicotine is absorbed into the blood stream. After absorption through the lungs, blood containing nicotine moves into the heart and then into the arterial circulation, including the brain. Nicotine reaches the brain within 10 to 15 seconds after a puff on a cigarette. This rapid delivery of nicotine to the brain produces more intensive effects than following slower delivery and provides the close temporal link between SMOKING and the development of addiction.

Nicotine is absorbed into the body in other ways. It can be absorbed in the mouth even if not inhaled in pipe or cigar smoke. In addition, not all tobacco products deliver nicotine through smoke. Chewing tobacco consists of shredded tobacco or plugs of tobacco that are enhanced with licorice and other flavorings. These products are periodically chewed, and the saliva generated is spat out, hence the term *spit tobacco*. Oral snuff is finely cut tobacco. A portion of oral snuff, called a pinch, is placed between the lip and the gum. Nicotine is absorbed from these forms of tobacco more slowly than from inhaled smoke, but the total amount absorbed is similar. Nasal snuff is finely powdered tobacco that is sniffed into the nose, where nicotine is rapidly absorbed.

DOSES OF NICOTINE TAKEN IN TOBACCO

The dose of nicotine absorbed from a cigarette is on average about 1 milligram (mg). The average user smokes about 25 cigarettes a day, an average nicotine intake of 20 to 30 mg daily. The average amount of nicotine absorbed from chewing tobacco or snuff per day is similar to that obtained from cigarettes. A person who smokes 25 cigarettes a day will absorb about 200 grams of nicotine in 20 years of smoking.

NICOTINE-CONTAINING MEDICATIONS

Nicotine is available as a medication, used to assist people in quitting smoking (see articles on NICOTINE DELIVERY SYSTEMS and TREATMENT of smoking and TOBACCO abuse). These medications are meant to provide nicotine to smokers as a substitute for nicotine formerly consumed from tobacco use. Nicotine medications reduce withdrawal symptoms and increase the likelihood that the individual will quit tobacco use. Two forms of nicotine medication are currently available. Nicotine chewing gum (nicotine polacrilex, also known as Nicorette) consists of nicotine in a gum that slowly releases nicotine during chewing. Each gum is typically chewed for about 30 minutes. People chew up to 16 pieces per day when trying to quit smoking.

Nicotine patches are applied to the skin. They release nicotine slowly through the skin over 16 or 24 hours, depending on the patch used.

Both forms of nicotine-replacement medication deliver doses of nicotine equivalent to that taken in by the average tobacco user. Nicotine chewing gum delivers about 1 to 2 mg per piece. Nicotine patches deliver from 5 to 21 mg, depending on the patch and its strength.

ELIMINATION OF NICOTINE FROM THE BODY

Nicotine in the body is eliminated primarily by breakdown by the liver. The rate of breakdown is such that the level of nicotine in the blood falls about one-half after two hours. This rate is also known as a half-life of two hours. The primary breakdown product of nicotine is cotinine. Cotinine levels in the body are about 10 times higher than those of nicotine. The half-life of cotinine is 16 hours, and cotinine persists in the body for 4 days after a person stops smoking. Cotinine levels can be measured as an indicator of how much nicotine a person is taking in.

NICOTINE ADDICTION

Addiction to nicotine is well documented. The development and characteristics of nicotine addiction are described in detail in a report from the U.S. Surgeon General published in 1988. In this report, *The Health Consequences of Smoking: Nicotine Addiction*, the surgeon general presents criteria for nicotine addiction including the following:

1. Highly controlled or compulsive use. Smokers have great difficulty abstaining. Seventy percent of the 45 million smokers in the United States today report that they would like to quit and can't.
2. Psychoactive effects. Nicotine, as described earlier in this article, has pronounced effects on the brain.
3. Drug-reinforced behavior. Tobacco use is motivated by a desire for the effects of nicotine. People do not smoke cigarettes that do not contain nicotine. Very few people choose to smoke cigarettes that deliver very low doses of nicotine. (See also the article on tobacco.)

Other factors lead to the conclusion that nicotine is addictive:

1. It is used despite harmful effects. Most people know that smoking is harmful to their health and continue to smoke. Many people who have nicotine-related diseases are still unable to quit.
2. RELAPSE following abstinence. Most smokers can quit for a few days or even weeks (abstinence), but most of these smokers return to smoking within a month. Typically, it takes four or five attempts before a smoker is successful at quitting permanently.
3. Recurrent drug cravings. Most smokers have an intense craving or urge to smoke when they have not smoked for some period of time.
4. Tolerance
5. Physical dependence
6. Pleasurable effects

The last three factors were described previously.

Smokers carefully regulate nicotine intake to maintain desired levels of nicotine in the body. Such careful regulation is further evidence that nicotine is addictive. Smokers keep the amount of nicotine obtained from cigarettes constant in two ways.

1. When people are given cigarettes that are labeled as low-yield (see tobacco history for detailed discussions of yields), they smoke more intensively to obtain the same dose of nicotine they were used to obtaining from the higher-yield cigarettes.
2. When they are forced to cut down on the number of cigarettes they smoke each day, they will take in more nicotine per cigarette. Thus when smoking is restricted, smokers tend to maintain the nicotine in their bodies at close to levels maintained during unrestricted nicotine intake.

BEHAVIORAL ASPECTS OF TOBACCO ADDICTION

People continue to smoke both because they enjoy the direct drug effects of nicotine and because use of nicotine becomes associated with other pleasures through learning—for instance, when the pleasurable effects of nicotine occur repeatedly in the presence of specific cues or events in the environment. Eventually, those cues and events become a signal to smoke. For example, people often smoke after meals, while drinking a cup of coffee or an alcoholic beverage, during a break from work, while talking on the phone, or while with friends who smoke. After smoking in these situations hundreds of times, the user may find that these situations themselves produce a powerful urge for a cigarette.

There are other learned pleasures that keep people smoking independent of the pharmacological effects of nicotine. Handling of smoking materials, and the taste, smell, or feel of tobacco smoke in the throat, all can become associated with the effects of nicotine and then become pleasurable in themselves. A person who tries to quit must learn to give up not only the pharmacological actions of nicotine but also the aspects of smoking that have become pleasurable through learning. Urges aroused after learning an association between aspects of the environment and the pleasures of smoking prompt relapses in many people who have already overcome withdrawal from nicotine and quit tobacco use.

Smokers report many other reasons for their habit. For example, many smokers, particularly women, smoke to maintain lower body weight. Others seem to use tobacco to control mood disturbances, such as DEPRESSION or ANXIETY.

COMPARISON OF ADDICTION TO NICOTINE AND OTHER DRUGS

Nicotine addiction is similar to and as powerful as addiction to other drugs, such as HEROIN, ALCOHOL, and COCAINE. All these drugs have psychoactivity and produce pleasure. They increase the likelihood that people will spend time looking for them and engaging in rituals while taking them and that users will continue to take them in the face of risk to their well-being and health. The psychoactivity of nicotine is subtle and does not interfere with normal functioning in daily life. Thus nicotine's psychoactivity differs from that of heroin and cocaine, which produces more intense euphoria and may be disruptive to everyday functioning. Despite this difference, nicotine is addictive. A subtle psychoactive effect, especially when experienced with each puff of smoke, taken hundreds of times a day, exerts a powerful effect on behavior over time. The magnitude of effect becomes apparent when each puff of cigarette is considered as a dose of nicotine. A smoker who takes 8 puffs per cigarette and smokes 20 cigarettes per day is receiving up to 160 doses of nicotine per day. The dosing is equivalent to 58,400 doses a year, or 1,168,000 doses after 20 years of smoking.

When difficulty in quitting and relapse after attempting to quit are compared, it becomes apparent that nicotine is even more addictive than other drugs of abuse. Ninety percent of all people who smoke cigarettes are addicted and have difficulty quitting. In contrast, only about 10 percent of people who drink alcohol at all have difficulty controlling use and would be classified as addicted. The percentage of occasional versus addicted users of heroin and cocaine is not known, but when multidrug users are asked about which drug they would have most difficulty giving up, the choice is most commonly nicotine (that is, cigarettes). Relapse rates among adults after cessation of alcohol, heroin, and tobacco use are similar.

NICOTINE ADDICTION IN YOUTH

Ninety percent of all tobacco users begin smoking before the age of 20. The earlier in life one starts smoking, the more likely he or she is to become a regular smoker and the more cigarettes he or she will smoke as an adult. The development of addiction in youth involves a series of steps including

- a trying stage
- experimentation
- regular smoking
- nicotine addiction

The typical interval between trying and addiction is 2 to 3 years.

Initially, young people smoke for social and psychological reasons. The motivations include the influence of parents and friends who are smokers, and the positive images of smoking perpetuated in television and movies and in advertisements in magazines, at music and sports events, and on billboards. Personal factors also play a role. Some include poor school performance, low self-esteem, poor self-image, sensation seeking, rebelliousness, failure to take seriously the adverse effects of tobacco use, and depression or anxiety. While early stages of smoking usually consist of occasional sessions with friends, tolerance develops and withdrawal symptoms are experienced between cigarettes as smoking becomes more frequent. Many youths report withdrawal symptoms and difficulty quitting. They consider themselves addicted to tobacco.

TREATMENT OF NICOTINE ADDICTION

Treatment of nicotine addiction is discussed in the articles entitled *Treatment: Tobacco*. The approach may be summarized as follows. Initial therapy usually does not include drugs. Smokers are encouraged to pick a day and just stop (go cold turkey). Some smokers participate in formal behavioral therapies such as are available in smoking-cessation clinics. Those who are unable to stop on their own or with behavior therapies are more likely to be highly addicted to nicotine and are candidates for pharmacological (drug) therapy. The main drug therapies for smoking are nicotine-containing medications such as chewing gum or transdermal (skin) patches.

(SEE ALSO: *Addiction: Concepts and Definitions; Adolescents and Drugs; Reward Pathways and Drugs; Tobacco: Smokeless; Tolerance and Physical Dependence; Withdrawal: Nicotine*)

BIBLIOGRAPHY

BENOWITZ, N. L. (1988). Pharmacologic aspects of cigarette smoking and nicotine addiction. *New England Journal of Medicine, 319,* 1318–1330.

U.S. SURGEON GENERAL. (1988). *The health consequences of smoking: Nicotine addiction.* Washington, DC: U.S. Government Printing Office.

NEAL L. BENOWITZ
ALICE B. FREDERICKS

NICOTINE DELIVERY SYSTEMS FOR SMOKING CESSATION

Several nicotine-delivery systems have been devised to assist nicotine-dependent cigarette smokers to quit smoking. The aim is to provide temporary relief for SMOKING (NICOTINE) WITHDRAWAL symptoms—such as irritability, anxiety, hunger, restlessness, drowsiness, and CRAVING for cigarettes. Meanwhile, the smoker learns to resist smoking in a variety of situations that have been associated with smoking thousands of times in the past. Eventually, the hope is that the smoker will give up the alternative source of nicotine, which is not as addictive as cigarettes. Quitting smoking, which is such a difficult task for many, is thereby simplified by breaking the process into two steps: (1) giving up the habit of smoking while retaining some of the effects of nicotine, and (2) giving up the nicotine, perhaps weeks or months later. While using an alternative nicotine-delivery system, smokers avoid the intake of hazardous smoke components, such as carbon monoxide and cancer-causing "tar."

NICOTINE CHEWING GUM

Nicotine chewing gum was the first alternative nicotine-delivery system to be approved by the U.S. Food and Drug Administration (FDA) as a SMOKING-CESSATION aid. Nicotine is contained in a gum resin and is slowly released upon chewing. Currently, nicotine gum is available in two strengths; each piece of gum contains either 2 or 4 milligrams of nicotine; however, in many cases only about 1 or 2 milligrams are released on chewing. This is comparable to the amount of nicotine delivered by a single cigarette. Unlike cigarette smoking, which delivers the nicotine rapidly into the bloodstream through the lungs, nicotine from the chewing gum is slowly absorbed through the mucous membrane of the inner cheeks. Most of the nicotine that is swallowed does not reach the general circulation, because after being absorbed from the small intestine, it is destroyed as it passes through the liver. Use of nicotine gum has been shown to double success rates in smoking cessation, but only if it is employed with a behavioral-treatment program. Even so, success rates are only about 20 to 30 percent at one year after treatment. Problems with the gum include bad taste, jaw soreness, stomach upset from swallowed nicotine, and inconsistent levels of nicotine in the bloodstream.

NICOTINE SKIN PATCHES

Partly to overcome the unpleasant side effects of nicotine chewing gum, nicotine skin patches (transdermal patches) were developed. They release a controlled amount of nicotine directly through the skin. Nicotine is easily absorbed through the skin, and it is possible to provide a steady delivery of approximately 21 milligrams per day, equivalent to the amount of nicotine delivered from a pack of cigarettes. However, just as with chewing gum, the nicotine is delivered much more slowly than from cigarettes, and the peak blood levels are thus lower than those obtained from cigarettes. The patch is applied once a day, and typically smokers are advised to use a full-strength patch for four weeks, followed by a two-thirds-strength patch for two weeks and a one-third-strength patch for two more weeks. This provides a gradual reduction of the nicotine dose to minimize WITHDRAWAL symptoms. Sufficient nicotine is delivered at the beginning of treatment to relieve many smoking withdrawal symptoms, and use of the patch has been shown to double or triple success rates in quitting smoking. Unlike the gum, the patch does not produce a bad taste or stomach upset; however, a small proportion of patients (less than 10%) do experience skin irritation from wearing the patches. Also, nicotine patches seem to improve success rates even in the absence of a behavior-therapy program, although behavioral treatment in combination with the nicotine patch further enhances success rates.

NICOTINE SPRAYS

The nicotine-delivery systems just described provide a much slower absorption of nicotine than one gets from cigarettes. Some researchers have speculated that a more rapid absorption of nicotine would more closely simulate the effects of cigarettes desired by smokers and increase success rates in smoking cessation. A nicotine nasal spray and a nicotine aerosol for inhalation that would provide a rapid delivery of nicotine are currently under development. Problems to overcome include the local irritation caused by nicotine itself. Additionally, such devices have the potential of being addictive because of the greater resemblance to smoking.

COMBINATION APPROACHES

Numerous combination approaches have yet to be tested, which may increase success rates beyond those of any one technique alone. This has already been seen in the enhancement of success rates with nicotine chewing gum or skin patches by a behavior-therapy program. Other combinations may include the use of a nicotine spray early in treatment, to make an easier transition from cigarettes, followed by switching to a slower delivery system such as a skin patch. Alternatively, it may be helpful for some smokers to use two or more delivery systems at the same time. A patch might provide a steady baseline level of nicotine, which could be supplemented as the need arises by the use of gum or sprays. These and other possibilities need to be tested in future research, because smoking has proven to be a more formidable adversary, as well as a more tenacious addiction, than many would have suspected initially.

(SEE ALSO: *Addiction: Concepts and Definitions; Tobacco: Dependence*)

BIBLIOGRAPHY

KROGH, D. (1991). *Smoking: The artificial passion*. New York: W. H. Freeman.

POMERLEAU, O. F., & POMERLEAU, C. S. (1988). *Nicotine replacement: A critical evaluation*. New York: Alan R. Liss.

JED E. ROSE

NICOTINE GUM The number of cigarette smokers has been steadily declining in North America since 1964, when smoking was first publicized as the most preventable cause of disease and death. Many treatment programs have been developed to assist people in overcoming nicotine DEPENDENCE. Of the pharmacologic treatments, the most extensively studied has been NICOTINE replacement in the form of an ion-exchange resin chewing gum, which was developed in the 1970s. When chewed slowly, the nicotine is released from the resin and absorbed buccally (through the inner cheek membrane), thus providing plasma levels slightly lower than those attained by smoking but sufficient to control nicotine withdrawal symptoms—one of the major factors for smokers to maintaining smoking behavior.

Because patients who use the gum frequently complain of throat and mouth irritation, nausea, vomiting, hiccups, and stomach upset, compliance is a problem. Nicotine gum is most effective as an adjunct to behavioral therapy in the context of a SMOKING-CESSATION clinic. Transdermal nicotine patches were introduced in the 1990s, and research on nicotine aerosols and nasal solutions continues. Although the majority of smokers quit without undergoing any formal therapy, and there is still much controversy surrounding the issue of substituting another dependence for smoking, nicotine replacement—prescribed in gradually decreasing doses—remains a favorable alternative to continued cigarette smoking.

(SEE ALSO: *Nicotine Delivery Systems for Smoking Cessation; Tobacco: Dependence; Treatment*)

BIBLIOGRAPHY

GARVEY, A. J., HEINOLD, J. W., & ROSNER, B. (1989). Self-help approaches to smoking cessation: A report from the normative aging study. *Addictive Behavior, 14,* 23–33.

HUGHES, J. R., & HATSUKAMI, D. (1986). Signs and symptoms of tobacco withdrawal. *Archives of General Psychiatry, 43,* 11–17.

LAM, W., ET AL. (1987). Meta analysis of randomised controlled trials of nicotine chewing-gum. *Lancet, 2,* 27–30.

U.S. DEPARTMENT OF HEALTH, EDUCATION AND WELFARE. (1979). *Smoking and health: A report of the Surgeon General* (PHS), 70-5006. Washington, DC: U.S. Government Printing Office.

MYROSLAVA ROMACH
KAREN PARKER

NICOTINE PATCH *See* Nicotine Delivery Systems for Smoking Cessation.

NIDA *See* U.S. Government Agencies: National Institute on Drug Abuse (NIDA).

NITRITES *See* Inhalants; Inhalants: Extent of Use and Complications.

NITROUS OXIDE *See* Inhalants.

NOREPINEPHRINE This is the biochemical product of DOPAMINE and the enzyme dopamine-beta-hydroxylase. It is the NEUROTRANSMITTER for the sympathetic nervous system, as well as for several sets of long axon, multiple-branched neurons (nerve cells) of the central nervous system.

The availability of pharmacological agonists and antagonists and the development of histochemical methods in the 1960s and 1970s for direct light mi-croscopic visualization led to many successful studies of this neurotransmitter and the neurons that contain it. Noradrenergic receptors, termed alpha and beta, can act independently or synergistically to mediate the activity of norepinephrine and related drugs. Brain noradrenergic neurons in the nucleus locus ceruleus are activated during withdrawal from addictive drugs.

(SEE ALSO: *Neurotransmission*)

BIBLIOGRAPHY

COOPER, J. R., BLOOM, F. E., & ROTH, R. H. (1991). *The biochemical basis of neuropharmacology,* 6th ed. New York: Oxford University Press.

FLOYD BLOOM

NUCLEUS ACCUMBENS The nucleus accumbens is a structure in the brain that has been demonstrated to have an important role in STIMULANT self-administration. This structure is part of the limbic system and is located near the midline in the frontal region, beneath the frontal lobe. Destruction of brain cells in this structure and inputs from several other regions disrupt stimulant self-administration by rodents. At this time, there is no direct scientific data to show that the nucleus accumbens has the same importance to human stimulant abuse—which likely is the result of the limited experimental methodologies available in human studies.

(SEE ALSO: *Brain Structures and Drugs; Limbic System; Reward Pathways and Drugs*)

JAMES E. SMITH

NUTMEG Nutmeg, the common spice obtained from the aromatic seed of the tree *Myristica fragrans* (native to the Moluccas, the spice islands of the East Indies), has been used for centuries for food and medicinal purposes. It has some HALLUCINOGENIC activity when consumed in large amounts. Since nutmeg is found in most kitchens, including food-preparation areas found in prisons, it has been used by prisoners. Therefore, it has been removed from ready access in prisons to the tighter control of drugs of abuse; Malcolm X wrote about such use.

Figure 1
Nutmeg

Nutmeg contains elemicin and myristicin, whose structures have some similarities to the hallucinogen MESCALINE as well as to the PSYCHOSTIMULANT AMPHETAMINE. It has been hypothesized that elemicin and myristicin might be metabolized in the body to form an amphetamine- and/or mescaline-like com-pound, but this has not been proven. The effects of nutmeg have been reported to have some similarities to those produced by MARIJUANA; however, the large amounts of nutmeg that must be ingested to get be-havioral effects can cause dry mouth and thirst, in-creases in heart rate, vomiting and abdominal pain, severe headaches, agitation, and panic attacks.

(SEE ALSO: *Lysergic Acid Diethylamide and Psyche-delics; Plants, Drugs from*)

BIBLIOGRAPHY

MAX, B. (1992). This and that: The essential pharmacology of herbs and spices. *Trends in Pharmacological Science, 13,* 15–20.

DANIEL X. FREEDMAN
R. N. PECHNICK

NUTRITION, ALCOHOL, AND DRUGS
See Complications: Nutritional.

OBESITY This term derives from the Latin (*obesus*, meaning "to eat up"), and it came into use in English in the early 1600s to mean a condition characterized by excessive bodily fat. Excess body weight is associated with the increased storage of energy in the form of adipose tissue. Standard criteria for obesity are (1) greater than 20 percent above ideal body weight (IDW) for a given height, as determined from actuarial tables, or (2) body mass index (BMI), defined as weight in kilograms divided by height in meters squared (kg ÷ m^2 = BMI), greater than 27 for men and greater than 25 for women.

Obesity represents the upper end of a body-weight continuum, rather than a qualitatively different state. Obesity can derive from a variety of causes, but a significant genetic contribution has been demonstrated.

Being overweight to a statistically significant above-average degree or having proportionately more body fat than average is believed to be due primarily to genetic factors that influence appetite, metabolism, and activity levels. Most notably, obesity is more prevalent in persons whose parents, brothers, or sisters are obese. Studies in identical twins have clearly demonstrated that genetics plays a major role. For example, nonidentical twins raised together were less similar in weight than identical twins raised apart.

The *prevalence* of obesity (in this case defined as having body fat in excess of 20% for males or 28% in females) varies remarkably across ethnic groups and cultures and across age groups. In the United States, obesity is consistently less common among African-American men than among white men across the entire age range; is consistently more common among African-American women than among white women; and tends to be more common among women of Eastern European and Italian ancestry than among those of British ancestry. Socioeconomic factors affect the prevalence of obesity, but men and women are affected differently: It is more common among all women in lower socioeconomic groups, but men in lower socioeconomic groups are leaner than average.

Some researchers and clinicians see similarities among certain patterns of overeating and other excessive behaviors such as drinking too much ALCOHOL, compulsive GAMBLING, engaging in "too much" sexual activity, and even exercising compulsively. While there may be such similarities, the semantics attached to problems of overeating and OBESITY are formidable.

Not all persons whose weight is above average are obese (they may have excess muscle mass); not all who are obese eat excessively; not all who eat excessively become obese; and some individuals who have clinically recognized disorders centered on eating and body weight, such as BULIMIA, may or may not be obese.

(SEE ALSO: *Bulimia Nervosa; Overeating and Other Excessive Behaviors*)

BIBLIOGRAPHY

BALL, G. G., & GRINKER, J. A. (1981). Overeating and obesity. In S. J. Mulé (Ed.), *Behavior in excess.* New York: Free Press.

TIMOTHY H. MORAN

OD *See* Overdose, Drug; Poison Control Centers, Appendix I, Volume 4.

ODALE *See* U.S. Government Agencies: Office of Drug Abuse Law Enforcement (ODALE).

ODAP *See* U.S. Government Agencies: Office of Drug Abuse Policy (ODAP).

ONDCP *See* U.S. Government Agencies: Office of National Drug Control Policy (ONDCP).

OPERATION INTERCEPT Described by government sources as the largest peacetime search-and-seizure operation in U.S. history, Operation Intercept was launched along the United States–Mexico border in September 1969. This unilateral program was instituted, ostensibly, to halt the flow of MARIJUANA, HEROIN, and other dangerous drugs from MEXICO into the United States. However, Intercept's true goal was not to interdict narcotics but to publicize the war on crime promoted by President Richard M. Nixon, who had taken office the previous January, and to force Mexican compliance with Washington's antidrug campaign. Fashioned by well-meaning but shortsighted law-enforcement officers, who all but totally neglected the State Department and knowledgeable border-state residents, Operation Intercept constituted a classic example of international pressure politics and became a serious incident between Mexico and the United States.

On September 16, 1968, presidential candidate Nixon had pledged to an Anaheim, California, audience that, if elected, he would move against the source of drugs and accelerate the development of tools and weapons to deter NARCOTICS in transit. As president, he came face-to-face with the reality of a staggering national drug abuse problem and accelerating drug-related street crime. With the director of his own BUREAU OF NARCOTICS AND DANGEROUS DRUGS contending that the United States had "failed miserably" in controlling narcotics abuse, Nixon chose to couple a highly publicized law-and-order campaign at home with an international offensive against foreign sources of heroin and marijuana. Attorney General John Mitchell was chosen to implement the program, and in April 1969 he assembled a multiagency task force to attack the importation into, and illegal sale and use of illicit drugs in, the United States.

Establishing a linear relationship between marijuana, deteriorating health, heroin usage, and increased crime, the task force turned its attention to the border problem. Mexico was correctly deemed the primary source of high-potency marijuana entering the United States. Officials noted further that (1) a significant percentage of the heroin was of Mexican origin, (2) substantial quantities of European heroin were being smuggled across the southern frontier, (3) Mexico served as an in-transit point for South American COCAINE, and (4) considerable amounts of AMPHETAMINES and BARBITURATES entered the United States surreptitiously from Mexico. In the midst of so much smuggling, Mexico's resources and efforts remained inadequate. Something had to be done to elicit a concerted, sustained antidrug program from Mexico City. That something was Operation Intercept.

On Sunday afternoon, September 21, 1969, at exactly 2:30 P.M. Pacific standard time, "the biggest, broadest-based enforcement task ever mounted" was launched. Noting that the Mexican government had been kept "fully informed" of the operation, a U.S. TREASURY DEPARTMENT news release termed Intercept a "coordinated effort" encompassing the law-enforcement resources of several branches of the federal government. Involving intensified land, sea, and air surveillance along the entire 1,945-mile U.S.–Mexico border, the effort would continue "for an indefinite period," as everything and everyone, no matter their nationality or status, was thoroughly and painstakingly searched.

More than 4.5 million individuals and their belongings were ultimately inspected. Vehicles, their component parts, personal baggage, purses, books, lunch boxes, jackets, toys, and in some cases even blouses and hairdos were searched. The daily routine of life in Mexican border cities was radically altered, as traffic backed up for miles, car radiators boiled over, and tempers, both private and diplomatic, flared. No person or object—including diplomatic and consular officials, their children, possessions, and even their diplomatic cargo—was spared during Intercept's 20-day existence. In the process, the maneuver encompassed some 2,000 personnel, intensified inspections, heightened air and sea surveillance, and the expenditure of some 30 million dollars.

Analyzed solely on the basis of drugs confiscated, Intercept surely was not worth the cost and effort it entailed. Seizures, however, were of minor importance. The primary objective was to "bring the Mexicans around, get them really moving against cultivation and trafficking." In this regard, the operation must be judged a qualified success. Diplomatic outcries notwithstanding, Intercept did play an undeniably important role in energizing Mexico's moribund antidrug program during the 1970s.

Viewed retrospectively, Operation Intercept's basic weakness was embodied in its title, for its purpose was not to interdict drugs at the border but to pressure Mexico through economic denial. Seeking a politically expedient solution to the highly complex problem of domestic drug abuse, the Nixon administration chose Mexico. Unfortunately, the White House failed to recall the salient fact that Mexico is a foreign country, and a friendly one at that.

Neglect of the State Department proved a serious blunder. Overlooked or overpowered by law-enforcement officials during Intercept's crucial formative stage, U.S. diplomats ultimately terminated the ill-advised project before it became an even greater diplomatic disaster. More important, if its supporters had managed to prolong the unilateral maneuver for an extended period, U.S. authorities probably would have never secured the level of cooperation they sorely needed to impair the cultivation of drugs in Mexico and the trafficking of drugs across the border.

Equally damaging was the failure of Intercept officials to gauge the impact of such a blockage on the U.S. border's economy. Highly dependent on Mexican shoppers, American border merchants reacted angrily and effectively through professional and civic groups. Pressure on the administration from border-state members of Congress was intense, and its impact increased as the project was prolonged. Along with diplomatic protests, this proved crucial to Intercept's demise.

Additionally, the operation was poorly timed; it came on the eve of *tapadismo*, the process through which Mexico chooses its next president, but before the Nixon administration's announcement of a Latin American policy. Furthermore, Mexico played host during the Intercept period to a regional meeting of the United Nations Commission on Narcotic Drugs and the thirty-eighth annual assembly of INTERPOL, thereby compounding its embarrassment over the blockade's indignities.

Yet despite its numerous shortcomings, Operation Intercept was not entirely void of accomplishments. Because of the tremendous publicity it engendered, the program made Mexican officials keenly aware of a reality heretofore ignored or slighted—that nation's own burgeoning drug problem. Politicians and journalists became introspective and reluctantly admitted that the availability of domestically produced drugs posed a danger to the health of *nuestra juventud* (our youth) as well as providing an everyday pastime for "gringo jippies" (American hippies).

Intercept also helped spur a previously lagging Mexican campaign against the cultivation, manufacture, and shipment of illicit drugs of all kinds. Since the fall of 1969, the government of Mexico has budgeted ever increasing funds for *la campaña permanente* (the permanent campaign) and is presently conducting (mid-1990s), with U.S. assistance, the world's most comprehensive eradication program against opium poppies and marijuana plants. As a corollary to this effort, cooperation between Mexican and American narcotics officials improved dramatically during the 1970s, only to tail off during the 1980s. Thus, while Intercept proved a short-term diplomatic blunder, it indirectly and somewhat ironically became a long-term catalyst to an accelerated Mexican antidrug campaign and a springboard to more effective international cooperation.

(SEE ALSO: *Border Management; Crime and Drugs; Crop Control; Drug Interdiction; International Drug Supply Systems; Transit Countries for Illicit Drugs; U.S. Customs Service*)

BIBLIOGRAPHY

BARONA, L. J. (1976). *México ante el reto de las drogas.* Mexico City: Impresiones Modernas.

CRAIG, R. B. (October 1980). "Operation Intercept: The international politics of pressure." *Review of Politics, 42,* 556–580.

CRAIG, R. B. (May 1978). *La campaña permanente:* Mexico's antidrug campaign. *Journal of Interamerican Studies and World Affairs, 20,* 107–131.

EPSTEIN, E. J. (1977). *Agency of fear.* New York: Putnam.

GOOBERMAN, L. A. (1974). *Operation Intercept: The multiple consequences of public policy.* New York: Pergamon.

SCHROEDER, R. C. (1975). *The politics of drugs: Marijuana to mainlining.* Washington, DC: Congressional Quarterly.

RICHARD B. CRAIG

OPERATION PAR   Operation PAR, Inc. (Parental Awareness & Responsibility) was founded in 1970 by Florida State Attorney James T. Russell, former Pinellas County Sheriff Don Genung, County Commissioner Charles Rainey, and Shirley Coletti, a concerned parent. In the years since its founding, PAR has developed one of the largest and most comprehensive nonprofit systems of substance-abuse EDUCATION, PREVENTION, TREATMENT, and RESEARCH in the United States.

Operation PAR provides a comprehensive continuum of care to individuals at risk for, or overcoming, CHEMICAL DEPENDENCY. It is dedicated to reducing the incidence of substance abuse in its local service area and to providing consultation and training to other communities. Its services are well known locally and nationally. In 1986, PAR was named the outstanding program in the United States by the Alcohol and Drug Problems Association.

In 1990, PAR served more than 4,800 people in treatment and 30,500 people in case management, criminal-justice diversion, prevention, and education and training programs. About 75 percent of clients are adults. Most are in their early to mid-twenties and have criminal-justice system involvement. Twenty percent are minorities, although the local population is less than 10 percent minority. Almost all ADOLESCENTS are referred through the local juvenile-justice system. Several studies have been conducted on the effectiveness of PAR's adult and adolescent treatment programs, especially its residential services. Results consistently show that more than 80 percent of clients who have completed treatment continue to be drug free one year after completing services.

At present, PAR operates more than twenty-five substance-abuse programs in nineteen locations in three west central Florida counties, including a 149-bed long-term adult residential-treatment program with residential capacity for women and their children; a 30-bed residential program for adolescents; a 10-bed drug-detoxification center; outpatient treatment programs, including METHADONE MAINTENANCE; specialized programs for pregnant and postpartum women, including day treatment, intensive case management, and on-site therapeutic child care; criminal offender assessment and diversion programs; intensive monitoring programs for adult and adolescent offenders; prison education and treatment programs; school-based prevention programs; intervention programs for high-risk elementary and middle school children; drug and alcohol education and treatment programs for teens and their parents; parent programs; diagnostic and evaluation services; extensive community education and training programs; and a licensed urinalysis laboratory. PAR is also involved in several federally funded research projects.

In 1992, Operation PAR had a permanent work force of approximately 310 full-time employees and an annual budget in excess of $12.5 million. It administered grants of more than $10 million from a wide variety of federal, state, and local sources.

PAR THERAPEUTIC COMMUNITY AND PAR VILLAGE

Operation PAR's THERAPEUTIC COMMUNITY (TC) has been in continuous operation since 1974. This 149-bed facility is the largest publicly supported residential drug-treatment program in Florida, serving clients from approximately 33 percent of all Florida counties. The center provides long-term residential treatment services, averaging eighteen to twenty-four months in duration, to adults age 18 and over who have histories of drug problems. The program targets individuals who are severely dysfunctional and who exhibit a wide range of antisocial behaviors as a result of substance abuse. The facility is an important alternative to incarceration for criminal courts throughout central Florida. Approximately 70 percent of clients have histories of significant involvement with the criminal-justice system, including felony convictions and prior incarceration.

Overall services provided by PAR TC include individual and group counseling, counseling groups for special populations, AA and NA support groups, on-site educational services, an extensive vocational training and job placement program, work experience, recreational therapy, parenting therapy and classes, and a wide range of personal-development activities.

In April 1990, services were expanded to include residential living for the children of maternal substance abusers. Known as PAR Village, addicted mothers can bring their children to live with them at the TC. PAR Village can care for fourteen mothers and twenty-two children (ranging in age from birth through age 10). While in treatment, women receive individual and group treatment, parenting-skills classes, vocational and educational training, medical services, and other therapeutic interventions and services. Pregnant women also are eligible to participate in PAR Village. Children are given therapeutic day care at the on-site, state-licensed, professionally staffed, child-care center. Infant, preschool, and afterschool care are provided.

(SEE ALSO: *Amity, Inc.; Coerced Treatment; Pregnancy and Drug Dependence; Prisons and Jails: Drug Treatment in; Treatment Alternatives to Street Crime*)

SHIRLEY COLETTI

OPIATES/OPIOIDS Technically, the *opiates* are drugs that are found in OPIUM or are derived from a substance found in opium, which is the juice of the opium poppy (PAPAVER SOMNIFERUM). In contrast, the *opioids* include the opiates, totally synthetic agents, and naturally occurring peptides that bind to one or more opioid receptors found in a number of animal species. In general usage, both terms are often used interchangeably—but opioids is the larger grouping.

The effects of opium have been known for several thousand years. For most of this time it was not clear which of the ingredients in opium gave it its analgesic (painkilling) and other therapeutic properties. MORPHINE and CODEINE, two of the most abundant constituents of opium, were the first pure opiates isolated—morphine in 1806 and codeine in 1832. Chemical modifications were soon attempted in an effort to eliminate their problematic side effects. One of the first attempts (in the 1890s) produced 3,6-diacetylmorphine, which is known as heroin. This agent did not eliminate the problems of tolerance, dependence, or abuse. Since then, extensive studies of the important components of morphine's structure

have led to the development of a number of different classes of organic compounds. In 1939/40, the first synthetics were discovered. The recent discovery of the opioid peptides have provided even more diversity in drug design.

AGONISTS, ANTAGONISTS, AND PARTIAL AGONISTS

Some drugs have very complex actions and many drugs act at specific RECEPTOR locations on the surface of a cell. All of the drugs that belong to the class of drugs called opioids act at opioid receptors on the surface of cells. Usually these cells are neurons, but there are also opioid receptors on white blood cells. Once a drug binds to a receptor, it can either turn it on (AGONIST) or do nothing (ANTAGONIST). If a compound does nothing once it binds to the receptor, nevertheless it blocks the site and prevents an active compound from binding to the receptor. The situation is much like a key in a lock; some keys fit into the lock but will not turn, and as long as they remain in the lock they prevent the insertion of keys that would turn the lock. Finally, there are drugs that we term *partial agonists*; these compounds bind to the receptor and turn it on but not nearly as well as pure agonists.

Again, using the key analogy, these partial agonists will turn in the lock, but only with some jiggling, lowering the efficiency in opening the door. Pharmacologically, partial agonists have limited effects at the receptor, termed a *ceiling* effect. This means that increasing the dose further will not give a greater response. To further complicate our understanding of these drug actions, it is important to recognize that the opioid receptors (and many other types of receptors as well) are actually families of similar but subtly different receptor types. Some opioids are agonists at one receptor type and partial agonists or even antagonists at another receptor type. These drugs are termed mixed agonist/antagonists and they can have complex pharmacological profiles.

RECEPTORS

Morphine and drugs with similar actions work through specific recognition sites, termed *receptors*, located on the outside of cells (see Table 1). A number of general classes of opioid receptors have now been identified and it is likely that even more will be

TABLE 1
Tentative Receptor Classification

Receptor	Agonists	Analgesia	Other Action
Mu	Morphine		
mu$_1$		Supraspinal*	Prolactin release
			Acetylcholine turnover
mu$_2$		Spinal	Respiratory depression
			Inhibition of gastrointestinal transit
			Guinea pig ileum bioassay
Kappa			
kappa$_1$	Dynorphin A	Spinal	Diuresis
			Sedation (?)
			Rabbit vas deferens bioassay
kappa$_2$	Bremazocine		Pharmacology unknown
kappa$_3$	Nalorphine	Supraspinal	
Delta	Enkephalins	Spinal	Mouse vas deferens bioassay
			Dopamine turnover

*The supraspinal system is far more sensitive than the spinal one.

discovered. The major types of opioid receptors have been designated mu, kappa, and delta. From the clinical perspective, the mu opioid receptors are the most important. This class, comprised of two subtypes, mu$_1$ and mu$_2$, have high affinity for morphine and most of the clinically used agents. Both mu subtypes mediate analgesia but through different mechanisms and locations within the brain and spinal cord. Mu receptors have been implicated in euphoria and mu agonists have often been abused; equally important, activation of mu receptors depresses respiration and inhibits gastrointestinal transit. In addition to analgesia, euphoria, respiratory depression, and decreased activity in the gut, mu agonist opioids produce some actions that are clinically useful, such as cough suppression. However, most of their actions are considered unwanted side effects; for example, they affect endocrine function, constrict pupils, induce sweating, and cause nausea and vomiting. All mu agonist opioids also induce tolerance and physical dependence.

Kappa opioid receptors were defined using keto-cyclazocine, an experimental benzomorphan derivative, and subsequently with dynorphin A, an endogenous opioid, which is believed to be the natural ligand for at least one of the kappa receptor subtypes. Morphine has relatively poor affinity for kappa receptors, but other drugs, such as pentazocine and nalbuphine (analgesics in clinical use), interact with kappa receptors quite effectively. The importance of kappa mechanisms in their actions have only recently been appreciated. The pharmacology of kappa receptors in humans has not been extensively studied; however, animal studies indicate that the kappa receptors also can relieve pain through receptor mechanisms distinct for each of the subtypes. Many of the clinically used drugs active at kappa receptor are mixed agonists/antagonists. Although they are agonists at kappa receptors, they are antagonists or partial agonists at mu receptors. In contrast to mu agonists, which can produce mood elevations and euphoria, drugs that activate kappa agonists appear to produce weird feelings and dysphoria.

The discovery of the enkephalins—endogenous peptides with opioid properties—soon led to the identification of delta receptors. The clinical pharmacology of delta receptors is not well known, primarily because so few agents have been tested in humans. Again, animal work indicates an important role of delta receptors in analgesia, which is supported by a few studies with humans; however, there are no pure delta agonists clinically available yet.

Although all the various receptor subtypes examined can relieve pain, each receptor represents a different mechanism of action. Their sites of action within the brain differ and, most importantly, agents highly selective for a specific subtype do not show cross-tolerance. While tolerance develops with continued activation of any of the various receptors, tolerance to one does not lead to tolerance to another.

For example, tolerance to morphine does not diminish the response to a kappa or delta drug. Similarly, mu agonists produce a characteristic variety of physical dependence, and there is cross-dependence among mu agonists (that is, people dependent on heroin will not experience withdrawal if given methadone.) However, there is no cross-dependence between mu agonists and kappa agonists.

All the various subtypes produce a number of actions other than analgesia. Most of the nonanalgesic actions of opiates can be explained by considering the receptors to which they interact. An excellent example are mu_2 receptors, which mediate respiratory depression and the constipation seen with morphine. Drugs that are agonists at these receptors also produce these side effects while compounds lacking affinity for these receptors do not. The role of multiple receptors is important clinically, primarily since few drugs are specific for one receptor. Even morphine, which is highly selective for mu receptors, interacts with two mu subtypes, and at higher doses with delta receptors as well.

CLASSES OF OPIOIDS

Opioids can be divided into a series of classes based upon their chemical structures, illustrated by prototypic compounds from each group (see Figure 1). These include morphine and its close analogs, the morphinans, the benzomorphans, the phenylpiperidines, and methadone. The pharmacology of agents within each category can be quite varied and often can be predicted from their affinity for various opioid-receptor subtypes. Most of the clinically relevant drugs will interact with more than one receptor. Thus, their actions can be ascribed to the summation of a number of receptor actions.

The importance of various regions of the morphine molecule has been well studied and a number of related compounds are widely used (see Figure 2). Early studies examined small changes in morphine's structure. One of the critical groups is the hydroxyl group at the 3-position on the molecule. Blockade of this position by adding chemical groups markedly reduces the ability of the drug to bind to opioid receptors. Although this may seem at odds with the analgesic activity of codeine, which lacks a free hydroxyl group at the 3-position, evidence indicates that codeine itself is not active and is metabolized to

morphine, which is responsible for its actions. A similar situation exists for OXYMORPHONE and OXYCODONE.

Some changes in the morphine molecule do not adversely effect activity. For example, the double bond between carbons 7 and 8 can be reduced to a single bond without loss of activity, and the hydroxyl group at the 6-position can be converted to a ketone or eliminated completely without sacrificing potency.

The morphine molecule has a single nitrogen atom. The substituent on the nitrogen in these series of opiates can have major effects on activity. Morphine and most of the mu agonists contain a methyl (CH_3-) group on the nitrogen, but a number of other compounds with different substituents have been developed. Replacing the methyl group with an allyl (-$CH_2CH = CH_2$2) or methylcyclopropyl (-CH_2 $CHCH_2CH_2$)group does not have much effect upon the ability of the compound to bind to opioid receptors, but it markedly changes what happens when they do bind. For example, oxymorphone, with its methyl group on the nitrogen, is a clinically useful analgesic many times more potent than morphine. Replacing the methyl group with an allyl group produces NALOXONE. Naloxone is an antagonist, a drug that blocks or reverses the actions of other opiates. Clinically, naloxone is used as an antidote to opiate overdose. It is interesting that such simple changes can influence the pharmacology of these agents so profoundly.

Further investigations revealed that Ring C of morphine can be eliminated, to give the benzomorphans—many of which are potent analgesics. The major drug in this group is pentazocine (Talwin). Even simpler structures produce potent analgesics, such as methadone. The phenylpiperidines comprise another large group of opioids. The first of these to be used clinically was meperidine, which was first described in 1939 and which still is extensively used. Modifications of the phenylpiperidine structure led to a subgroup of drugs, with fentanyl as a prototype. Fentanyl is approximately 80-fold more potent than morphine. Its very short duration of action has required continual infusions. An advantage is that once the infusion is discontinued, the effects of the drug clear rapidly. This ability to quickly turn on or off the drug's actions, along with its great potency, has made this agent a valuable tool in anesthesia. Recently, this high potency has been exploited to de-

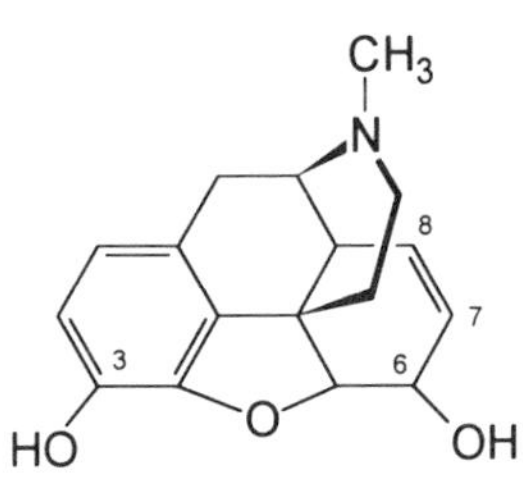

Morphine

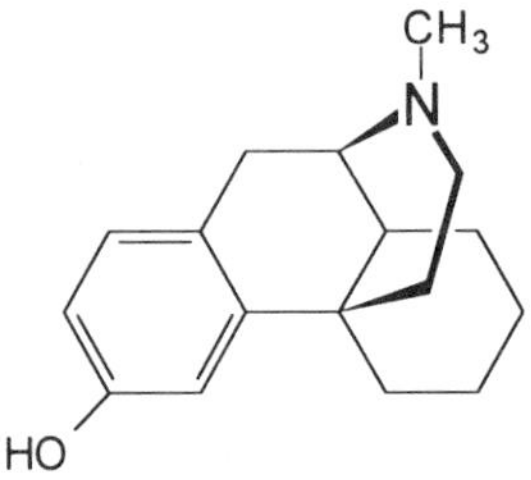

Levorphanol
(morphinan)

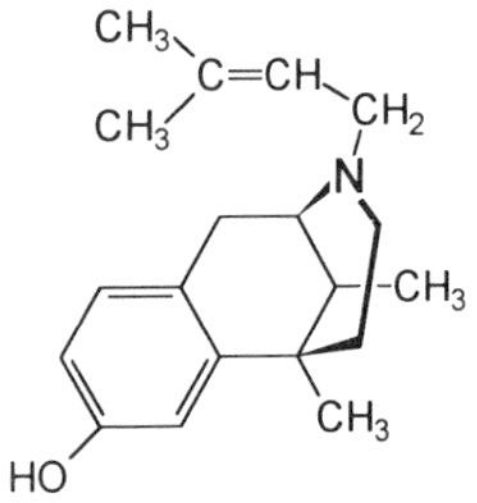

Pentazocine
(benzomorphan)

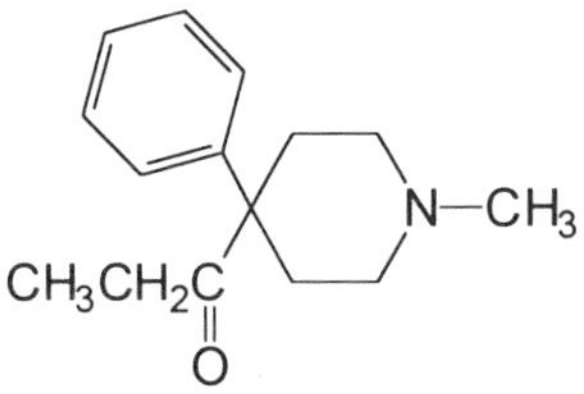

Meperidine
(phenylpiperidine)

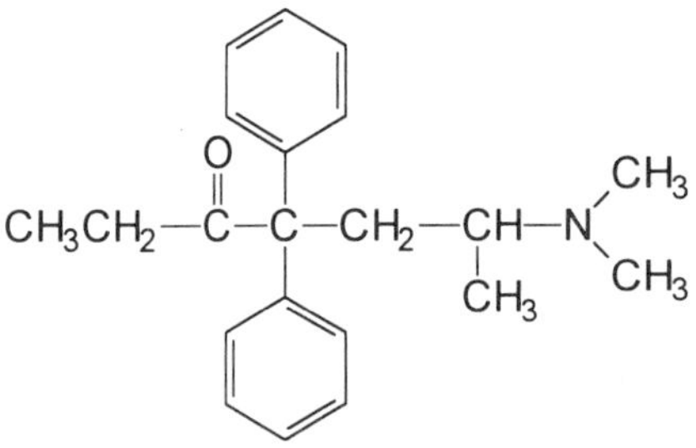

Methadone

Figure 1
The Classes of Opiod Compounds,
Based on Structure

velop skin patches which give a constant release of fentanyl into the body as the drug is absorbed through the skin. Other agents within this series, such as sufentanil and alfentanil, are even more potent than fentanyl. Two other members of this series, loperamide and diphenoxylate, have activity but very poor solubility. This property has led to their use as antidiarrheal agents since they cannot be made soluble and injected and are therefore less likely to be abused.

Together, these structure activity studies reveal that the basic requirements needed for opioid activity are quite simple. However, the wide variety of structures becomes even more intriguing since morphine and the other opioids act within the brain by mimicking naturally occurring peptides—the endogenous opioids. The enkephalins were the first such naturally occurring substances to be isolated and sequenced (Table 2). Initially, these results were somewhat confusing since the two enkephalins—both pentapeptides—contain the identical first four amino acids and differ only at the fifth. The complexity of these peptides become more clear with the subsequent isolation and characterization of β-endorphin, a 31 amino acid peptide derived from a larger protein, which also gives rise to active

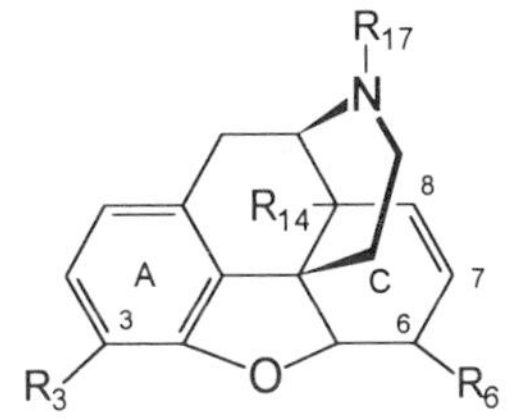

POSITION ON MOLECULE

Drug	R_3	R_6	R_{14}	R_{17}	C_6–C_7	Action
Morphine	OH	OH	H	CH_3	=	Agonist
Codeine	CH_3O	OH	H	CH_3	=	Agonist
Heroin	$\overset{\parallel}{O}$ CH_3CO	$\overset{\parallel}{O}$ CH_3CO	H	CH_3	=	Agonist
Oxymorphone	OH	=O	OH	CH_3	–	Agonist
Oxycodone	CH_3O	=O	OH	CH_3	–	Agonist
Hydromorphone	OH	=O	H	CH_3	–	Agonist
Naloxone	OH	=O	OH	$CH_2CH=CH_2$	–	Antagonist
Nalbuphine	OH	OH	OH	CH_2—◇	–	Ag/Antag
Nalorphine	OH	OH	H	$CH_2CH=CH_2$	–	Ag/Antag

Figure 2
The Morphine Molecule and Some Widely Used Related Compounds, Based on Region of the Molecule

compounds, including ACTH and α-MSH. The first five amino acids in β-endorphin are identical to [met⁵]enkephalin, but [met]enkephalin and β-endorphin derive from different gene products. There are also a series of compounds containing the sequence of [Leu⁵]enkephalin, including dynorphin A, dynorphin B and α-neoendorphin. All these compounds (the ENKEPHALINS, ENDORPHINS, and dymorphine) have distinct genes and are expressed independently from one another. Thus, they comprise a family of similar, but discrete NEUROTRANSMITTERS.

The opioid peptides are only now becoming important clinically. A major difficulty in the use of peptides is the fact that they are broken down when taken by mouth, and thus, most have very limited oral activity. However, new derivatives specifically designed to be more stable have been developed, which will provide new leads. The enkephalins are potent at delta receptors, and it might be expected that their derivatives also would be delta-selective. Many are. Some of the more recent derivatives label delta receptors over 10,000-fold more selectively than others. Yet other peptides are very much like morphine in terms of their pharmacology and receptor binding. Finally, peptides with opioid actions are now being identified in a variety of other tissues; for example, toad skin has dermorphin, a potent and stable opioid peptide.

(SEE ALSO: *Addiction: Concepts and Definitions; Opioid Complications and Withdrawal; Pain*)

TABLE 2
Selected Opioid Peptides

[Leu⁵]enkephalin	**Tyr-Gly-Gly-Phe-Leu**
[Met⁵]enkephalin	**Tyr-Gly-Gly-Phe-Met**
Dynorphin A	**Tyr-Gly-Gly-Phe-Leu**-Arg-Arg-Ile-Arg-Pro-Lys-Leu-Lys-Trp-Asp-Asn-Gln
Dynorphin B	**Tyr-Gly-Gly-Phe-Leu**-Arg-Arg-Gln-Phe-Lys-Val-Val-Thr
α-Neoendorphin	**Tyr-Gly-Gly-Phe-Leu**-Arg-Lys-Tyr-Pro-Lys
β-Neoendorphin	**Tyr-Gly-Gly-Phe-Leu**-Arg-Lys-Tyr-Pro
β_h-Endorphin	**Tyr-Gly-Gly-Phe-Met**-Thr-Ser-Glu-Lys-Ser-Gln-Thr-Pro-Leu-Val-Thr-Leu-Phe-Lys-Asn-Ala-Ile-Ile-Lys-Asn-Ala-Tyr-Lys-Lys-Gly-Glu
Dermorphin	Tyr-D-Ala-Phe-Gly-Tyr-Pro-Ser-NH₂

BIBLIOGRAPHY

JAFFE, J. H. (1990). Drug addiction and drug abuse. In A. G. Gilman et al. (Eds.), *Goodman and Gilman's the pharmacological basis of therapeutics*, 8th ed. New York: Pergamon.

JAFFE, J. H., & MARTIN, W. R. (1990). Opioid analgesics and antagonists. In A. G. Gilman et al. (Eds.), *Goodman and Gilman's the pharmacological basis of therapeutics*, 8th ed. New York: Pergamon.

GAVRIL W. PASTERNAK

OPIOID COMPLICATIONS AND WITH-DRAWAL Opioids are frequently used in medicine, but they are also among the most common drugs of abuse. When taken under medical supervision, opioid drugs have a low level of serious toxicity. The most common side effects are nausea, drowsiness, and constipation—but when self-administered, not under medical supervision, their use is associated with a high incidence of untoward actions and side effects, as well as with a high death rate when used alone or in combination with other drugs (including ALCOHOL).

Table 1 presents estimates of untoward actions of opioids, derived from data collected by the DRUG ABUSE WARNING NETWORK (DAWN), which appeared in the *Annual Emergency Room and Medical Examiner Data, 1992*. As can be seen, opioids account for approximately 16 percent of emergency room and 64 percent of medical-examiner (death) mentions. (Suspicious and accidental deaths are sent to the county medical examiner.) More than 76 percent of the medical-examiner opioid mentions involve death by opioid drugs in combination with either alcohol or COCAINE, while more than 20 percent are in combination with other opioids. It is further estimated that about 67 percent of all such deaths were unintentional overdoses (ODs). Adverse results also occur in patients given opioids for therapeutic reasons, including, although uncommonly, serious respiratory depression.

RESPIRATORY DEPRESSION

It is generally believed that the most common life-threatening complication of opioid use, whether therapeutic or illicit, is respiratory depression (loss of the ability to breathe automatically). Probably the most important action of morphine-like drugs in producing respiratory depression is the lessening of the sensitivity and responsivity of the brain's medullary respiratory center to carbon dioxide (CO_2—the metabolic waste that circulates in the blood, derived from carbonic acid during animal respiration). Therefore, CO_2 becomes an inefficient respiratory stimulant, and automatic breathing ceases.

Administering a specific opioid ANTAGONIST such as NALOXONE to patients with severely depressed res-

TABLE 1
Drug Abuse Warning Network (DAWN): Emergency Room and Medical Mentions 1992

Drug	Emergency Room Mentions	Medical-Examiner Mentions
heroin-morphine	48,003	2,912
d-propoxyphene	6,551	398
oxycodone	3,750	41
hydrocodone	6,105	92
methadone	2,812	431
codeine	1,896	880
meperidine	1,163	42
hydromorphone	615	25
Pentazocine	547	0
total opioid mentions (% of all mentions)	71,442 (16%)	4,821 (64%)

SOURCE: *Annual emergency room data, 1992: Data from the Drug Abuse Warning Network* (1992). National Institute on Drug Abuse Statistical Series, 1:12-A. DHHS Pub. No. (SMA) 94-2080 also add annual medical examiner.

piration frequently produces a dramatic increase in the rate of respiration and the volume of air taken in per breath. This occurs when a partial or completely resensitized respiratory center is confronted with high brain levels of CO_2. When the brain CO_2 levels are dissipated as a consequence of the evoked excessive rate and volume of breathing (hyperpnea), the minute volume (the volume of air breathed per minute) decreases. Yet when brain levels of the antagonist decrease, the respiratory depressant action of the opioid may assert itself again. The drug naloxone is a relatively short-acting antagonist. Patients who, for example, have received an overdose of long-acting opioids (e.g., METHADONE) have experienced a fatal respiratory depression following successful treatment with naloxone.

TOLERANCE AND PHYSICAL DEPENDENCE

Another group of complications associated with chronic use of opioids is the development of tolerance and dependence. Although these phenomena can be easily conceptualized superficially, they become operationally complex when they are analyzed.

Tolerance. The most common concept of TOLERANCE to opioid drugs is that following chronic administration of a drug, its effects are diminished. Several mechanisms have been demonstrated to be involved in the development of tolerance to drugs, and these include (1) the induction of drug-metabolizing enzymes; (2) the development of coping strategies; (3) the exhaustion or depletion of NEUROTRANSMITTERS; and (4) an alteration in the number of active and inactive RECEPTORS. These mechanisms have, by and large, failed to provide adequate explanations for tolerance to opioid drugs. This may stem in part from the complexity of the results of chronic administration of opioids, the involvement of multiple mechanisms, and the influence of the dose, route, and frequency of drug administration.

Opioids, for example, alter the functioning of some body homeostats, and apparent tolerance is related to the establishment of new equilibrium conditions. This is clearly evident in respiratory depression, where opioids depress both the sensitivity and the reactivity of the brain-stem respiratory CO_2 homeostat, causing CO_2 to be a less effective respiratory stimulant. Yet when CO_2 accumulates because of depressed respiration, the increasingly higher concentrations will cause stimulation of respiration to the degree that the altered homeostat dictates. The ability of opioids to constrict pupils is dose-related, and patients receiving opioids frequently have miosis—near-maximally constricted pupils; hence it is difficult to determine if tolerance develops to opioids' miotic effect. This has given rise to the commonly accepted view that tolerance does not develop to the miotic effects of opioids.

In former opioid addicts, morphine-like drugs produce dose-related feelings of enhanced self-image, of being more efficient and effective, and of well-being. These related subjective states form the essence of opioid-induced euphoria, which is produced in patients who are plagued by feelings of inadequacy. This can be quantitatively measured using the Morphine-Benzedrine Group scale of the Addiction Research Center Inventory.

Tests in many normal subjects (nonabusers), who are not suffering from pain indicate that opioids do not produce euphoria—but in sufficiently large doses instead produce feelings of apathy and ineffectiveness, which can be dispiriting (dysphoric). When opioids are administered chronically to addicts, the subjective effects they produce change from feelings of well-being to feelings of being withdrawn, tired, and weak. With regard to these effects of chronic opioid administration, they are not simply diminished but rather changed.

The development of tolerance can be a problem when opioids are used in the treatment of pain. Although some degree of tolerance to ANALGESIC effects is expected when opioid drugs are used repeatedly, in practice there is a great deal of variability among patients. Some patients with CANCER pain appear to derive satisfactory relief from the same dose of MORPHINE or similar drugs over a period of many months. For these patients, a need to increase the dose can be a signal that the disease is progressing. Other patients with terminal disease can develop remarkable tolerance. There are reports of patients who have been given the equivalent of 1000 mg of morphine per hour intravenously. This is an impressively large dose, since the usual starting therapeutic doses of morphine are 10 to 15 mg by injection every 4 to 6 hours, and doses of more than 60 mg by injection can cause potentially fatal respiratory depression in nontolerant individuals. It is not usually of much benefit to change to another opioid that acts at the same receptor. For example, morphine acts at the mu-opioid receptor. When tolerance develops to morphine, other opioids acting at mu receptors will be less effective, a phenomenon referred to as "cross-tolerance."

Physical Dependence/Withdrawal. Closely related to the phenomenon of tolerance is the phenomenon of physical dependence. Subjects given repeated doses of opioid agonists exhibit a syndrome when the drug is withheld or when the subject is administered an opioid antagonist. The resulting group of signs and symptoms is called the WITHDRAWAL or precipitated abstinence syndrome; subjects who exhibit an abstinence syndrome are termed *physically dependent* on the opioid. The degree of physical dependence and the intensity of the abstinence syndrome is related to the dose of the opioid agonist chronically ingested. In general, the intensity of all signs and symptoms covary together.

The abstinence syndrome includes restlessness, weakness, chills, body and joint pains, gastrointestinal cramps, anorexia (loss of appetite), nausea, feelings of inefficiency, and social withdrawal. Signs of abstinence include activation of the autonomic nervous system, lacrimation (tearing eyes), rhinorrhea (running nose), piloerection (gooseflesh), tachypnea (rapid breathing), mydriasis (dilated pupils), hypertension (high blood pressure), tachycardia (rapid heart beat), muscle spasms, twitching, restlessness, vomiting, and diarrhea. The waves of gooseflesh that occur during severe opioid withdrawal reminded some observers of the look of a plucked "cold turkey," a term that has come to be used not only for any abrupt discontinuation of a drug, but also for sudden cessation of any habit or pattern of behavior. The twitching and kicking movements of the lower extremities that can occur during opioid withdrawal have given the English language another widely used term, "kicking the habit," to denote the process of giving up any pattern of behavior or drug use.

The time of onset of opioid abstinence depends on the length of activity for the dependence-producing opioid. The abstinence syndrome of subjects dependent on morphine or HEROIN is well developed within 24 hours after the last dose of the opioid, peaks after 48 hours of abstinence, and gradually subsides thereafter. Signs of abstinence in patients dependent on METHADONE begin to emerge 24 to 48 hours after the last dose and may not peak for 2 weeks.

After this early abstinence syndrome subsides, a protracted abstinence syndrome emerges. The protracted abstinence syndrome becomes manifest 5 to 10 weeks after acute or early withdrawal in humans. It differs from the early abstinence syndrome in some ways but not in others. In subjects who were dependent on morphine or methadone, protracted abstinence is characterized by the following signs: a modest hypotension (low blood pressure), bradycardia (low heart rate), hypothermia (lower than normal body temperature), miosis (small, constricted pupils), and tachypnea. Other signs of protracted abstinence may include an inability to concentrate and a decrease in fine-motor control. Symptoms associated with protracted abstinence in patients who were dependent on methadone include feelings of tiredness and weakness, withdrawal from society, inefficiency, decreased popularity and competitiveness, and loss of self-control. Patients withdrawn from methadone have also exhibited a significant elevation of the Sc (schizophrenia) scale of the MINNESOTA MULTIPHASIC PERSONALITY INVENTORY (MMPI). This elevation of the Sc scale may be related to feelings of social withdrawal that patients in protracted abstinence experience. Protracted abstinence persists for at least 25 weeks after withdrawal. Protracted abstinence following addiction to morphine has also been demonstrated in rats and in dogs.

The patterns of abstinence and time course of symptoms described above are those seen when opioid drugs that have been used for weeks or months are discontinued. However, opioid withdrawal can also be observed when a drug-dependent person is given an opioid antagonist (a drug such as naloxone that competes with opioid agonists for the opioid receptor). In a matter of minutes, this will produce a precipitated abstinence syndrome that can be severe, with vomiting, cramps, and diarrhea. This precipitated abstinence is usually brief, however, because as soon as the antagonist is metabolized (usually less than an hour for naloxone), the opioids still in the body can again attach to the opioid receptors and suppress the abstinence syndrome.

The biological mechanisms that are responsible for the development of opioid physical dependence are set into motion with the very first doses of an opioid drug. If volunteer subjects are given standard doses of morphine (15 to 30 mg) and then, after an interval varying from 6 to 24 hours, they are given naloxone, they report nausea and other feelings of dysphoria and exhibit yawning, dilated pupils, tearing, sweating, and runny nose. Changes in endocrine levels are also seen that are in the same direction, although not as extreme, as those seen when chronically administered opioids are abruptly discontinued.

TREATMENT OF OPIOID WITHDRAWAL (DETOXIFICATION)

The opioid withdrawal syndrome varies in severity depending on the amount of opioid used and the duration of use. For the average user of illicit opioids, such as heroin, withdrawal is rarely severe because the amount of drug used typically is not high. The withdrawal syndrome from such a level of use can be uncomfortable, but it is not life threatening in otherwise healthy individuals. However, death can occur if severe withdrawal is left untreated in individuals who are weakened by other medical conditions.

The process of treating someone who is physically dependent so that acute withdrawal symptoms are controlled and the state of physical dependence is ended is usually referred to as detoxification. For opioid drugs, this process can be managed on an ambulatory (outpatient) basis or in a hospital or other residential (inpatient) setting. The most common approach to easing the severity of opioid withdrawal is to slowly lower the dose of opioid over a period of days or weeks. This can involve giving progressively smaller doses of the opioid that the patient has been using. However, in the United States, if the drug has been heroin, a substitution technique is used instead. Since virtually all of the opioids that are abused act as AGONISTS at the mu-opioid RECEPTOR, any mu agonist could be a suitable substitute, but the only ones approved for this purpose in the U.S. are methadone and LAAM (L-ALPHA ACETYLMETHADOL). These medical agents are effective when taken by mouth. Methadone can completely suppress the opioid abstinence syndrome. This capacity of one opioid to prevent the manifestations of physical dependence from another is called cross-dependence.

Outpatient detoxification using methadone typically involves using doses of 20 to 40 mg per day for a few days and then gradually reducing the dose over several weeks. Because so many patients return to illicit drug use as the dose of methadone approaches zero, government regulations controlling methadone permit a long period (up to 180 days) of slow dose reduction.

When detoxification takes place in a hospital or other residential setting, where the patient is presumably not as likely to be exposed to environmental cues that elicit CRAVING for opioids, dose reductions of methadone can be more rapid (for example, over 8 to 10 days), although the intensity of discomfort will be higher.

Other opioid agonists and partial agonists that have been used satisfactorily to facilitate detoxification include BUPRENORPHINE, a partial mu agonist, and LAAM. The opioid withdrawal syndrome can also be modified and reduced in severity by using agents that do not act at the mu receptors, but instead act on some of the physiological systems that exhibit hyperactivity as part of the syndrome. The use of CLONIDINE is an example.

Because opioid withdrawal is time-limited and rarely life threatening, many nonmedical treatments have also been used, including ACUPUNCTURE and herbal medicines.

NAUSEA AND VOMITING

Nausea and vomiting are common side effects associated with the use of opioid analgesics. These effects are experienced following administration of opioids orally, by injection, or by injection into the spinal canal (epidurally)—they are worsened by movement and the resulting stimulation of the vestibular (inner ear organ responsible for balance). The site and mechanism responsible for these actions of opioids is presumed to be a special area in the brain stem or medulla, the chemoreceptive trigger zone of the area postrema.

CONSTIPATION

Constipation, an often undesirable effect of opioids, is sometimes a useful effect for which opioids can be prescribed. It is undesirable when opioids are used for the relief of pain and in opioid-dependence maintenance therapy.

The oldest of the therapeutic actions of opiates is their antidiarrheal and constipating effects. It is now known that the extrinsic innervation (nerves leading from the central nervous system to the gut) and the intrinsic innervation (nerves within the wall) of the gastrointestinal (GI) tract are complex and vary from species to species. A variety of naturally occurring neurones with diverse neurotransmitters have been identified, including neurones and their process that contain opioid peptides: the enkephalins, B-endorphin, dynorphins, and other ligands derived from pro-opiomelanocortin. Further mu and delta opioid receptors have been identified in the GI tract. The vagus nerve also has fibers that contain enkephalins, and the central nervous system has opioid mechanisms that modulate GI movement (motility).

Several influences must play a role in the constipating effects of opiate agonists—these include increased segmental activity, decreased propulsive activity, and decreased secretory activity. Naloxone, even when administered in high doses for a long period of time in antagonist therapy of opioid abusers, does not produce an overt stimulation of the GI tract resulting in diarrhea. When opioid antagonists are administered to opioid-dependent subjects, however, GI cramps and diarrhea develop as classic opioid withdrawal signs.

PRURITUS

The ability of morphine-like drugs to produce the sensation of itching (pruritus) is well known, and it is a discomforting complication when opioids are administered for therapeutic reasons. Further, many morphine-like drugs (e.g., codeine) release histamine from white blood cells that store it (mast cells and basophils). When morphine is administered intravenously, wheals (hives—raised red lumps) may appear at the site of the injection and along the course of the vein. The wheals may be associated with the sensation of itching. For this reason, morphine may have a peripheral site of action in producing itching. Occasionally, large doses of morphine may produce generalized itching. Rarely does morphine produce pulmonary edema (fluid in the air sacs of the lung), bronchoconstriction (narrowing of the air tubes in the lungs), or wheezing. With the advent of the use of intrathecal and epidural morphine (injection of morphine into spinal fluid or around the lining of the spinal canal) in pain management, the incidence of morphine-induced pruritus has become greater. Under this circumstance, the distribution of itching may be segmental (limited to the part of the spinal cord involved). Itching remains an elusive phenomenon and is harder to define and investigate than pain. It is thought that it may be mediated by a subgroup of nociceptive (pain-carrying) C fibers. Ballantyne and coworkers (1988) argue that opiates produce itching by interacting with opiate receptors in peripheral nerves and with peripheral and central neurones through a stimulatory

action. Further, morphine's histamine-releasing property has been implicated in its ability to produce itching, as histamine does in allergic reactions.

CONVULSIONS

Although most opiates produce convulsions when administered in very large doses, convulsions are most frequently observed when excessively large doses of MEPERIDINE or d-propoxyphene are administered. Emergent meperidine seizures are characterized by tremors and twitching, which may evolve into tonic-clonic (epileptic) convulsions. Tonic-clonic convulsions have been reported to occur without prodromata (preceding twitching or tremors), however. Focal and tonic-clonic seizures have been observed in patients overdosed with d-propoxyphene. The mechanisms whereby opioid drugs produce convulsive phenomena are not well understood and may involve several mechanisms, including (1) direct and indirect dysinhibition of glycine and GABA-mediated inhibition and (2) excitatory actions that are probably mediated by yet-to-be-classified receptors. The convulsant effects of d-propoxyphene can be readily antagonized by naloxone; however, meperidine's convulsant effects may be more resistant. Meperidine probably has a convulsant effect in its own right when administered in very large doses acutely, yet convulsant phenomena seen following the administration of multiple doses of meperidine are produced by the accumulation of a metabolite, normeperidine.

DYSPHORIA, DELUSIONS, AND HALLUCINATIONS

It is rare for morphine-like analgesics to produce psychotic reactions. In patients with severe pain and discomfort and in opiate addicts, single doses of morphine-like drugs most commonly produce feelings of well-being. In normal subjects with no pain or with only modest levels of discomfort, morphine produces feelings of apathy and enervation, which are somewhat dysphoric. The drug d-propoxyphene has been reported to produce bizarre reactions—delusions and hallucinations—particularly when taken chronically in large doses and when used to suppress opioid abstinence. Some agonists-antagonists (e.g., pentazocine, nalorphine, and cyclazocine) produce feelings of apathetic sedation, perceptual distortions, anxiety, delusion, and hallucinations. These effects are thought to be due to interactions with kappa and sigma opioid receptors.

STREET DRUGS

The complications described in the preceding sections are most commonly associated with pure, unadulterated opioids. When street drugs are used, which are typically diluted by the seller with quinine, lactose, or other powdered materials—and injected by the user in an unhygienic manner, in doses that vary significantly—the range of complications widens. These are described fully in the entry on neurological complications, but among the complications of heroin use reported in the medical literature are strokes, inflammation of cerebral (brain) blood vessels, disorders of peripheral nerves, impairment of segments of the spinal cord, and widespread injury to muscle tissue (rhabdomyolysis)—which by releasing muscle protein can denote damage to the kidneys.

In addition to these toxic reactions are those complications and infectious diseases that arise from the bacteria and viruses frequently associated with injection of street drugs. Furthermore, some synthetic opioids produced in clandestine laboratories are contaminated with toxic chemicals, which produce damage to a number of bodily organs.

(SEE ALSO: *Addiction: Concepts and Definitions*)

BIBLIOGRAPHY

Annual emergency room data, 1992: Data from the Drug Abuse Warning Network. (1992). National Institute on Drug Abuse Statistical Series, 1:12-A. DHHS Pub. no. (SMA)94-2080.

Annual medical examiner data, 1992: Data from the Drug Abuse Warning Network. (1992). National Institute on Drug Abuse Statistical Series, 1: 12-B. DHHS Pub. no. (SMA)94-2081.

BALLANTYNE, J. C., LOACH, A. B., & CARR, D. B. (1988). Itching after epidural and spinal opiates. *Pain, 33*, 149–160.

BIOULAC, B., LUND, J. P., & PUIL, E. (1975). Morphine excitation in the cerebral cortex. *Canadian Journal of Physiological Pharmacology, 53*, 683–687.

GILBERT, P. E., & MARTIN, W. R. (1976). The effects of morphine- and nalorphine-like drugs in the nondependent, morphine-dependent and cyclazocine-dependent dog. *Journal of Pharmacol. Exp. Ther., 20*, 66–82.

HAERTZEN, C. A. (1974). Subjective effects of narcotic antagonists. In M. C. Braude et al. (Eds.), *Narcotic antagonists*. Advances in Biochemical Psychopharmacology, vol. 8. New York: Raven Press.

HURLE, M. A., MEDIAVILLA, A., & FLOREZ, J. (1982). Morphine, pentobarbital and naloxone in the ventral medullary chemosensitive areas: Differential respiratory and cardiovascular effects. *Journal of Pharmacol. Exp. Ther., 220*, 642–647.

JAFFE, J. H. (1990). Drug addiction and drug abuse. In A. G. Gilman et al. (Eds.), *Goodman and Gilman's the pharmacological basis of therapeutics*, 8th ed. New York: Pergamon.

JAFFE, J. H. & MARTIN, W. R. (1990). Opioid analgesics and antagonists. In A. G. Gilman et al. (Eds.), *Goodman and Gilman's the pharmacological basis of therapeutics*, 8th ed. New York: Pergamon.

KROMER, W. (1988). Endogenous and exogenous opioids in the control of gastrointestinal motility and secretion. *Pharmacological Review, 40*, 121–162.

LASAGNA, L., VON FELSINGER, J. M., & BEECHER, H. K. (1955). Drug induced mood changes in man. 1. Observations on healthy subjects, chronically ill patients and "post-addicts." *Journal of the American Medical Association, 157*, 1006–1020.

MARTIN, W. R., & JASINSKI, D. R. (1969). Physiological parameters of morphine dependence in man—tolerance, early abstinence, protracted abstinence. *Journal of Psychiatric Research, 7*, 9–17.

MARTIN, W. R., ET AL. (1976). The effects of morphine and nalorphine-like drugs in the non-dependent and morphine-dependent chronic spinal dog. *Journal of Pharmacol. Exp. Ther. 197*, 517–532.

MARTIN, W. R., ET AL. (1973). Methadone—a reevaluation. *Archives of General Psychiatry, 28*, 286–295.

MUELLER, R. A., ET AL. (1982). The neuropharmacology of respiratory control. *Pharmacological Review, 34*, 255–285.

WANG, S. C., & BORISON, H. L. (1952). A new concept of organization of the central emetic mechanism: Recent studies of the site of action of apomorphine, copper sulfate and cardiac glycosides. *Gastroenterology, 22*, 1–12.

WANG, S. C., & GLAVIANO, V. V. (1954). Locus of emetic action of morphine and Hydergine in dogs. *Journal of Pharmacol. Exp. Ther., 111*, 329–334.

WILLIAM R. MARTIN

OPIOID DEPENDENCE: COURSE OF THE DISORDER OVER TIME

Opioid dependence is the modern diagnostic term for narcotic addiction, but the older term is still often used. This entry, however, uses the modern term. The term *opioid* refers to natural and synthetic substances that have morphine-like effects. The term *opiate* is generally used in a more restricted sense to refer to MORPHINE, HEROIN, CODEINE, and similar drugs derived from OPIUM. OPIOID dependence is defined as a cluster of symptoms related to continued use of an opioid drug. One of the prominent features of the disorder is the inability to stop using the drug. Persons with repeated periods of opioid dependence are often called narcotic addicts. Because they are not always dependent (that is, addicted), the term *opioid users* seems more suitable and therefore is used here. During the late nineteenth and early twentieth centuries the principal opioid drugs used were LAUDANUM (a solution of opium in alcohol, taken orally) and morphine (usually injected by needle). During the latter half of the twentieth century, heroin has been the principal drug of opioid users. It is usually taken by intravenous injection, but sometimes by insufflation, that is, by sniffing it into the nasal cavities.

The course of opioid dependence is affected by multiple interacting conditions in the person and in the environment. The combined conditions create thresholds for the onset, continuation, and relapse after remission of opioid dependence. Different methods of investigation (for example, pharmacological, psychological, sociological, psychiatric) have led to different theoretical conceptions of the causal conditions and processes in opioid dependence. These conceptions, however, tend to be compatible and supplementary rather than contradictory. In the following description of the course of opioid dependence, the principal conditions thought to affect its onset and course will be identified.

In the United States legal and medical conditions affecting opioid use and dependence have changed since the nineteenth century. In the nineteenth century many persons regularly used laudanum or morphine that they obtained legally from physicians, retail drug stores, or other sources. Physicians often prescribed or recommended these drugs for treatment of chronic physical PAIN or psychological distress. Although daily use of an opioid drug with consequent dependence on it probably impaired the social performance of many persons, reports exist of persons—including some with distinguished careers—who acceptably filled social roles during years of opioid drug dependence. Though some antisocial persons used opioid drugs, such use itself did not lead to criminal behavior.

In the twentieth century opioid dependence became closely associated with criminal behavior. Enactment and enforcement of federal and state laws to control the production and distribution of opioid drugs (mostly called narcotic drugs in the laws) became prominent features of the twentieth-century environment of opioid use. Physicians could no longer prescribe opioid drugs to maintain dependence, and opioid users now had to obtain their drugs from illicit sources. Furthermore, because the illicit opioid drugs were expensive, users often engaged in illegal moneymaking activities—especially theft, burglary, fraud, prostitution, and illicit drug traffic—to pay for their drugs. In addition, twentieth-century opioid users have often had histories of delinquent behavior that preceded their opioid use.

WHO IS SUSCEPTIBLE?

At the turn of the century, when opioid drugs were legally and easily available to all adults, only a few persons became dependent on them. Although the exact scale of opioid dependence at that time is not known, it probably did not exceed 2 percent of the adult population. An interview survey conducted in the 1970s of a national sample of young men in the United States revealed that 5.9 percent had used heroin at some time in their lives, but only 1.7 percent ever considered themselves dependent on this drug (O'Donnell et al., 1976). Other studies indicate that normal people free from physical pain tend to react to the effects of opioid drugs with indifference or dislike. With rare exceptions, patients who receive opioid drugs to relieve pain after surgery make no effort to continue drug use after they become free from pain. It is now well-known that opioid dependence develops in only a small proportion of those exposed to the effects of the drugs.

The characteristics of persons susceptible to opioid dependence have not been clearly defined, but clinical and other studies point to three personality problems that probably increase susceptibility. First, chronic emotional distress, such as DEPRESSION, tension, ANXIETY, anger, or mixtures of these, is relieved by opioid drugs, and this relief probably prompts repeated use of the drug. Second, impaired capacity to regulate emotional distress increases the urgency of the need for relief. Third, an antisocial attitude makes it easy for the person to perform the illegal actions needed for regular illicit opioid use. The notion that opioid drugs are used to relieve emotional distress is called the self-medication hypothesis. The origins of the personality problems that increase susceptibility probably lie partly in genetic inheritance and partly in adverse psychosocial experience. Modern opioid users often come from dysfunctional parental families.

Environmental conditions in the deteriorated areas of the large cities in the United States place young persons living there at special risk for opioid dependence. Most of the retail illicit drug traffic and much of the opioid use takes place in these areas. Young persons are consequently exposed to available heroin and heroin-using role models and associated criminal behavior. Since these areas are heavily populated by minority groups—primarily African Americans, Puerto Ricans, and Mexican Americans—these groups are at special risk. The experience of POVERTY and adverse discrimination may contribute to emotional distress in members of these groups and thereby increase their susceptibility to opioid dependence. Apart from environmental conditions, ethnic status as such does not seem to affect susceptibility. Men seem to develop opioid dependence more often than women do; the ratio of men to women in treatment programs is about three to one.

ONSET OF OPIOID DEPENDENCE

Opioid use is usually preceded by use of tobacco, ALCOHOL, and MARIJUANA. Before their first opioid use, most users dropped out of school and began to associate with opioid users. Heroin is nearly always the drug of choice. With few exceptions, it is first used within a few years of the user's twentieth birthday. Users report that they were not coerced or urged to use heroin by either their associates or drug dealers. In a typical sequence a person becomes aware of drug use by his friends or relatives, becomes curious about its effects, and asks for the first injection. As already noted, most persons exposed to the effects of heroin do not become regular users.

Susceptible persons rarely become compulsive daily users immediately after first use. A variable period of occasional use—once a month or more often but not daily—usually ensues. Curiosity fades as a motivation; the effects of the drug are what prompt repeated use. The drug users call these effects the "high." The high is not described as exhilaration or excitement but rather as relaxation and mood elevation. Descriptions of the high offered by many drug users suggest that it amounts to relief of the chronic emotional distress mentioned before as a factor in susceptibility. Susceptible persons increase the frequency of use until it reaches once or several times daily. From first use to daily use typically takes about one year, but it may take much longer. In a study of opioid users in San Antonio, one man reported that he first used heroin at the age of sixteen. He did not like it and did not use it again for fifteen years. At that time he felt depressed following the death of a friend and decided to try heroin again. This time the heroin made him feel better, and he quickly became a daily user (Maddux & Desmond, 1981).

With daily or nearly daily use, the user develops physiological DEPENDENCE on the drug. This means that when the drug use is reduced or stopped, the user develops distressing symptoms called the WITHDRAWAL illness. The threat or the onset of withdrawal symptoms provides additional strong motivation to continue daily use of the drug.

In the progression from initial use to daily use, heroin users learn how to inject heroin intravenously, how to acquire the drug and injection equipment, and with some exceptions, how to conduct illegal moneymaking activities to pay for the heroin. Those who began a delinquent career before their initial use of heroin were already oriented toward criminal activity. In some cases, heroin users or dealers provide a regular supply of heroin to their spouses or live-in companions; the latter thus do not have to engage in regular illegal activity to pay for their drug. Another exception to the pattern of illegal moneymaking activity is linked to opioid dependence among physicians and other health professionals. Health professionals rarely purchase heroin from street retailers. They have access to meperidine or other opioids available in pharmacies and hospital supplies, and they use these drugs instead of heroin.

Probably the most serious and disabling feature of opioid dependence is the inability to put a stop to it, also called loss of control of the drug use. After drug use has become daily and physiological dependence has developed, many opioid users try to stop using it and find themselves unable to do so. This inability to stop is a subjective mental state reported by drug users. It probably starts as a mild impairment of control and progresses to complete or nearly complete loss of control.

EARLY TERMINATION OF OPIOID DEPENDENCE

Continued daily use with loss of control depends partly on the availability of the drug and other environmental conditions. American soldiers serving in VIETNAM during the Vietnam War were exposed to an environment in which heroin was easily available and heroin use was common. An interview survey of a sample of returning veterans revealed that 35 percent had tried heroin while in Vietnam and 19 percent (about half of those who tried it) considered themselves addicted to it. During a three-year period after return to the United States, however, only 12 percent of those addicted in Vietnam became readdicted in the United States. These represented about 2 percent of the entire sample interviewed (Robins et al., 1980). Other studies of early termination of opioid dependence in the United States have identified various life events as probable causative factors in the termination. Among these are change of residence, marriage, a drug-related arrest, and death of a friend from overdose. Many persons who terminate their opioid dependence do so without treatment.

CHRONICITY, REMISSION, AND RELAPSE

With continued daily use and physiological dependence, the user's bond to the drug becomes stronger. Drug use, drug seeking, and illegal activity become the dominant activities of the user's life. Psychosocial development is retarded. Those who become dependent during adolescence often fail to complete high school and never develop regular work habits or job skills. With continued dependence, opioid users become impaired marital partners or parents.

Daily use does not continue indefinitely. In some cases, as noted, an important life change leads to cessation of use. In other cases, pressure from family or friends or other sources prompts entry into a treatment program. In still others, arrest, conviction, and incarceration interrupt the daily use. Sometimes conviction leads to probation with treatment as a requirement of the probation. After treatment or incarceration, the majority of chronic users resume

opioid use within six months. The common long-term pattern consists of initial use followed by irregular sequences and varied durations of occasional use, daily use, treatment, abstinence, and incarceration. Remissions enduring for three years or longer followed by relapse are not unusual. Variations in the course of opioid dependence are illustrated in the following case summaries.

An employed man first used heroin at the age of twenty-six and after two months of occasional use began daily use. He continued working but engaged in the illicit heroin traffic to pay for his heroin. Two years after first use, he was arrested and convicted for sale of heroin. In lieu of prison, he was sent to a federal hospital for treatment. Released on parole at age twenty-nine, he remained abstinent for ten years, when he was last interviewed at age thirty-nine. He abstained from heroin, he said, because he did not want to return to "that miserable life."

Shortly after release from an institution for delinquents, a boy had his first injection of heroin at the age of sixteen. He became a daily user in about three months. During the next thirteen years he had two brief periods (each of about five months' duration) of abstinence from heroin. He used heroin occasionally or daily during the remaining time, except for four years in prison. He was murdered by gunshot at the age of twenty-nine.

After dropping out of school, a fourteen-year-old boy learned to make money by selling marijuana and heroin. He tried heroin at age sixteen, liked it, and promptly became a daily user. He used heroin daily for the next twenty years, except for relatively brief periods when he was in prisons and hospitals. Then, at age thirty-six, he was sent to prison for two years. During this period in prison, he felt some change in himself while participating in a THERAPEUTIC COMMUNITY program. After release, he abstained from heroin for the next eight years. He obtained employment as a counselor in a drug-abuse treatment program. He was aged forty-six when last interviewed.

Modern TREATMENT of opioid dependence includes drug withdrawal done as an inpatient or outpatient procedure, residential treatment, therapeutic community, drug-free outpatient treatment, the use of opioid ANTAGONISTS, and METHADONE MAINTENANCE. Prompt abstinence from opioid drugs is the goal of the first five of these types of treatments. Methadone maintenance, in contrast, consists of continued substitution of methadone, itself an opioid drug, for the illicit opioid. In addition to these forms of treatment, self-help groups such as NARCOTICS ANONYMOUS are available as well as special religious programs for drug users.

Opioid users who enter treatment aimed at prompt abstinence reveal mixed motivations for the treatment. They would like to become free of the burden of their drug dependence, but they do not want to give up the effects of the drug. Most leave treatment before completing it. Relapse after treatment is common, but the severity of the dependence is usually reduced, short periods of abstinence are often achieved, and for a small proportion of users, enduring cures of opioid dependence are attained. Methadone maintenance aims for social rehabilitation, with opioid abstinence as a possible distant goal. It has become a major mode of treatment for chronic opioid users and benefits many of them by helping them reduce or stop illicit opioid use and stop their criminal activity. This treatment, however, only infrequently leads to enduring abstinence.

USE OF MULTIPLE SUBSTANCES

In the early twentieth century, many alcoholics were converted from ALCOHOLISM to opioid dependence. If the opioid dependence was terminated, alcohol dependence often replaced it. In the later twentieth century, the patterns of use of other psychoactive substances during the course of opioid dependence have become more complex. Heroin users often substitute alcohol when they become abstinent from opioids, but, in addition, many use alcohol regularly while using heroin daily. They also use TOBACCO, marijuana, and cocaine. In a recent interview study of opioid users in California, 75 percent reported current use of tobacco, 20 percent reported being drunk on alcohol in the previous seven days, 38 percent reported use of marijuana in the previous thirty days, and 18 percent reported use of cocaine in the previous thirty days (Hser, Anglin & Powers, 1993).

WHY DOES OPIOID DEPENDENCE BECOME INTRACTABLE TO TREATMENT?

This important question can be answered only partially and tentatively. The conditions that contribute to the onset of opiod dependence also support the tendency to continued use. These, as previously noted, include chronic emotional distress, drug-using models, an available opioid drug, and withdrawal symptoms. Two other effects of the drug dependence probably contribute to relapse after treatment or incarceration. First, mild withdrawal symptoms such as muscular aching, insomnia, and irritability often persist for six months or longer after the last dose. These symptoms (called protracted withdrawal) are promptly relieved by an opioid drug, and they probably contribute to relapse after treatment. Second, the opioid user becomes conditioned to environmental conditions associated with withdrawal symptoms, so that after a period of abstinence, exposure to a conditioned stimulus will evoke withdrawal symptoms. This conditioned withdrawal probably contributes to relapse.

Three other changes in the mental state of the user probably also contribute to the intractable quality of the disorder, but these have not been as well defined and studied. First, over time the drug-using habit tends to become automatic, requiring no conscious decision to use or abstain. Second, the drug-seeking and the associated criminal behavior seem to become a part of an established lifestyle, and the user becomes enmeshed in a social network that includes illicit drug users and criminals. Third, with repeated relapses after treatment or incarceration, the opioid user comes to a self-perception as an addict with a diminishing capacity for change. This complex of learned attitudes and behaviors amounts to a personality change, which is probably accompanied by change in the brain. Such change may not become permanent, but it tends to endure.

LONG-TERM OUTCOMES

In follow-up studies extending from five to more than twenty years after admission to treatment, the percentages of users reported abstinent from opioid drugs have varied from 9 percent to 21 percent (Maddux & Desmond, 1992). Some of this variation was due to different ways of counting abstinence. In some studies the users were counted as abstinent only if they remained so during the entire period from treatment to follow-up, whereas in others the users were counted as abstinent if they were found so at the time of follow-up. Despite these differences, the studies collectively indicate that only a minority of opioid users are found to be abstinent on long-term follow-up.

Although only small to medium percentages were found to be abstinent, it should not be assumed that the remainder of people were using opioid drugs. Some were dead, some were in prison, and some were in treatment. The death rate of opioid users is about three times the expected rate. Overdose, homicide, suicide, accidents, and liver disease account for many of the deaths. In the 1980s the acquired immunodeficiency syndrome appeared as an additional hazard for drug injectors. A follow-up of opioid users in San Antonio revealed the following different statuses twenty years after first use: 16 percent were abstinent, 29 percent were using heroin, 30 percent were in prison or other institutions, 8 percent were maintained on methadone, and the remaining 17 percent were dead or their status was unknown (Maddux & Desmond, 1981).

WHAT CAN BE DONE?

Since policies and programs to reduce drug abuse are described elsewhere in this encyclopedia, only a brief comment will be offered here. Two broad approaches—supply reduction and demand reduction—have been put in place in the United States. Supply reduction consists of the enactment and enforcement of drug control laws. Although the supply-reduction effort has undoubtedly reduced the supply of illicit opioid drugs, it has failed by far to eliminate them from the environment of susceptible persons.

Demand reduction consists of treatment and prevention. Treatment of opioid dependence produces short-term abstinence and reduces the pool of daily users in the community, but it achieves few enduring cures. Publicly supported treatment programs in the United States are insufficiently financed to provide prompt treatment to all who seek it. A few pilot projects have been developed for reaching out to young persons at risk for opioid dependence and providing special services for them, but more research is needed on this type of preventive effort. Finally, opioid use in the United States seems embedded in a complex matrix of family dysfunction, poverty, undereducation, unemployment, and crime. Anything that reduces these problems would likely reduce illicit opioid use. Easy solutions seem unlikely.

(SEE ALSO: *Addiction: Concepts and Definitions; Britain, Drug Use In; Causes of Substance Abuse; Coerced Treatment; Conduct Disorder and Drug Use; Crime and Drugs; Opioid Complications and Withdrawal; Opioids and Opioid Control: History; Vulnerability; Wikler's Pharmalogic Theory of Drug Dependence*)

BIBLIOGRAPHY

BESS, B., JANUS, S., & RIFKIN, A. (1972). Factors in successful narcotics renunciation. *American Journal of Psychiatry, 128*, 861–865.

BRECHER, E. M., & THE EDITORS OF *Consumer Reports* (1972). *Licit and illicit drugs.* Mount Vernon, NY: Consumers Union.

CHEIN, I., ET AL. (1964). *The road to H: Narcotics, delinquency, and social policy.* New York: Basic Books.

HSER Y. I., ANGLIN, M. D., & POWERS, K. (1993). A 24-year follow-up of California narcotics addicts. *Archives of General Psychiatry, 50*, 577–584.

KOLB, L. (1962). *Drug addiction: A medical problem.* Springfield, IL: Charles C. Thomas.

LETTIERI, D. J., SAYERS, M., & PEARSON, H. W. (EDS.). (1980). *Theories on drug abuse* (National Institute on Drug Abuse Research Monograph 30, DHHS Publication no. [ADM] 80-967). Rockville, MD: U.S. Department of Health and Human Services.

MADDUX, J. F., & DESMOND, D. P. (1992). Methadone maintenance and recovery from opioid dependence. *American Journal of Drug and Alcohol Abuse, 18*, 63–74.

MADDUX, J. F., & DESMOND, D. P. (1992). Residence relocation inhibits opioid dependence. *Archives of General Psychiatry, 39*, 1313–1317.

MADDUX, J. F., & DESMOND, D. P. (1981). *Careers of opioid users.* New York: Praeger.

O'DONNELL, J. A. (1969). *Narcotic addicts in Kentucky* (Public Health Service Publication No. 1881). Washington, DC: U.S. Department of Health, Education and Welfare.

O'DONNELL, J. A., ET AL. (1976). *Young Men and Drugs: A Nationwide Survey* (National Institute on Drug Abuse Research Monograph 5, DHEW Publication No. [ADM] 76-311), Rockville MD: U.S. Department of Health, Education and Welfare.

POWELL, D. H. (1973). A pilot study of occasional heroin users. *Archives of General Psychiatry, 28*, 586–594.

ROBINS, L. N., ET AL. (1980). Vietnam veterans three years after Vietnam. In L. Brill & C. Winick (Eds.), *The yearbook of substance use and abuse* (Vol. 2), pp. 213–230. New York: Human Sciences Press.

Schasre, R. (1966). Cessation patterns among neophyte heroin users. *International Journal of the Addictions, 1,* 23–32.

Vaillant, G. E. (1973). A 20-year follow-up of New York narcotic addicts. *Archives of General Psychiatry, 29,* 237–241.

James F. Maddux
David P. Desmond

OPIOIDS AND OPIOID CONTROL: HISTORY

Throughout recorded history, and in most parts of the world, OPIATES have occupied a central place in medicinals. They have been used popularly against a wide range of ills and to produce calm or well-being. Opiates are renowned for their powerful ability to relieve PAIN. They also have been used for their PSYCHOACTIVE properties and, within the last 100 years, have come to symbolize the problems with attempts to control drug use through legislation and enforcement. (Technically, opiates are a subset of the OPIOIDS, which also include synthetic agents and naturally occurring peptides that bind to opioid RECEPTORS found in certain animal species.)

The OPIUM poppy (*Papaver somniferum*) grows easily in semiarid parts of the Middle East and southern Asia, including dry or steep locales where other crops are difficult to cultivate. For thousands of years, farmers in these regions have grown the poppy as an important staple crop. For traditional poppy farmers, opium is a cash crop that supplements an agricultural livelihood. The entire plant is used: Poppy seeds are baked into breads, or oil for cooking or fuel is extracted from them and the body of the plant is fed to cattle. The labor-intensive aspect of collecting the sap for sale means that whole families are pressed into service at harvest time. The desire for opium in other parts of the world has long made it an important commodity in international trade networks.

References to opium appear in inscriptions and texts of ancient Sumer, Egypt, and Greece. The Greek physician Galen, in the second century A.D., noted that opium cakes were widely sold in Rome. This observation highlights an important difference between drug use before the twentieth century and contemporary drug use. Currently, drug use is divided into medical and nonmedical (or recreational) uses. Nonmedical use for opiates is banned in most countries, and persistent demand fuels a large and vigorous illicit trade. Medical uses are defined exclusively by physicians, and consumption of these drugs is allowed only in the context of treatment by a physician.

The sharp separation between medical and nonmedical uses of drugs is comparatively new in human history, although attempts to control drug use legislatively are not. In the past, physicians constituted only a small group of specially trained professionals who found their clientele primarily among the rich and powerful. A wide range of healers provided different kinds of health care; for example, in Europe from the Middle Ages to about the mid-nineteenth century, apothecaries prepared and sold drugs to anyone seeking treatment. Apothecaries consulted with the patient, helping diagnose an ailment and suggesting a remedy, but they charged a fee only for the sale of the drug.

Opium became an important European drug in the sixteenth century. During the Middle Ages, the severing of ties between Europe and the Middle East meant that large amounts of opium were not shipped to Europe. In the Middle East, however, the ancient Roman and Greek texts remained important sources of knowledge, and medical, as well as scientific and mathematical, theories were developed and debated among scholars like the Arab physician Avicenna. In these Moslem countries, where alcohol was absolutely forbidden, both opium and cannabis were widely used.

During the European Renaissance, renewed ties with the Middle East brought the ancient texts and their Arab interpretations to the attention of European scholars. Galen, who had systematized humoral theory in his writings, was recognized as an important authority in sixteenth- and seventeenth-century Europe. Galen's views were challenged by the sixteenth-century Swiss physician Paracelsus, who favored chemical remedies (such as mercury) to herbal ones. Paracelsus valued opium highly. He devised a mixture of alcohol, opium, and other ingredients that he called "laudanum" (from the Latin for "praise") to suggest its superiority.

Thomas Sydenham, the influential English physician, wrote in 1680: "Among the remedies which it has pleased Almighty God to give to man to relieve his sufferings, none is so universal and so efficacious as opium." This valuation of opium (and later of its derivatives) has been repeated by physicians in the centuries since as ongoing testimony to the drug's central role in medical treatment.

Medical use of opium grew more widespread in eighteenth-century England; for example, the relief of pain at the time of death was seen as an important adjunct to preparing the patient for death in a blessed state of peace. England was an important commercial power in this period, and new kinds of goods from distant parts of the world became increasingly plentiful. Opium was a valuable commodity, and, as such, it was handled commercially like any other. Individuals seeking to treat themselves for aches or ailments, or wanting to relieve drudgery or sleeplessness or persistent coughs, could buy pellets of opium from various merchants, innkeepers, or apothecaries. This pattern persisted through most of the nineteenth century, although by the late eighteenth century a particular effect of chronic opium consumption was described: If a habitual user stopped taking the drug, a clearly recognizable syndrome of symptoms ensued. These included runny nose, tearing, sweating, aches, muscle tremor, vomiting, and diarrhea. These problems were seen as an expected difficulty connected with taking medicines; they were not portrayed as a unique and devastating kind of problem that threatened the social fabric.

In the United States, also, opiates were freely sold. In the first half of the nineteenth century, neither medications nor medical practice were regulated. During the presidency of Andrew Jackson, antimonopolistic sentiment had led many states to repeal licensing requirements for physicians, on the grounds that such licenses created artificial elites. Many people saw no physician at all; they treated themselves or their family members with homemade or purchased remedies. Taking charge of one's own medical care also reflected the kind of broadened democratic spirit that characterized the Jacksonian age. In home treatment, opiates were valued for their wide-ranging effects, including quick and dramatic improvement in how one felt. Physicians also administered opium generously as part of the heroic brand of therapy favored in the nineteenth century. Based on humoral theory, "heroic therapy" sought to provide clear evidence of its effects on body fluids by promoting fluid discharges. Emetics and cathartics were the hallmarks of such practice, but the ability of opiates to produce sweat in addition to their other valuable effects made them a component of heroic therapy.

For individuals who appeared chronically weak, perhaps as the result of lingering fever, opium improved spirits and energy and was considered by many medical practitioners to have a STIMULANT effect (although it is now classed as a DEPRESSANT). Individuals who took the drug to relieve vague feelings of unease, or in the absence of serious medical conditions, were said to take the drug for its stimulant properties.

Rapid industrialization caused profound social shifts in England in the first half of the nineteenth century. People whose families had worked on the land for generations became part of the first large-scale factory work force. Working conditions were brutal; men, women, and children worked fourteen-hour days, six days a week. Working women often had to bring young children to the factory with them. For working people, opium was an easily available source of relief for many complaints of both adults and children.

Early in the nineteenth century, Thomas De Quincey and Samuel Taylor Coleridge wrote about opium-induced reveries. Although their works were widely read, their opium use was treated more as a curiosity than a cause for alarm. The earliest concerns about excessive or indiscriminate opiate use centered on adulteration or on deaths due to accidental OVERDOSE. These were voiced by a new group of professionals, public health workers. Extensive surveys of health conditions in England in the 1840s both pointed up problems and created opportunities for government and professional workers to expand their professional arenas. At the same time, the old three-rank system of health-care givers, in which physicians treated the well-to-do while surgeons and apothecaries met the health needs of those of more modest means, was giving way. Surgeons and physicians joined a unified healing profession while pharmacists prepared and sold drugs without providing diagnostic or therapeutic advice. As physicians worked to increase their professional authority, they sought to gain control over the use of drugs, defining them as medicines that only the medically trained could use or prescribe with safety. Toward the middle of the nineteenth century, a few physicians expressed concern about opium use for its "stimulant" effects. These voices foreshadowed an alarm about nonmedical use of opiates that would transform how this behavior was viewed. In the meantime, the 1868 Pharmacy Act called for precise labeling of any preparation containing opium.

The incidence of addiction also worried some observers, and this phenomenon became increasingly visible in part as a result of new pharmacological discoveries and changing medical technology. In 1806, Frederich Sertürner of Hannover, Germany, announced that he had isolated the chief active component of opium. He named this new drug MORPHINE, after Morpheus, the Greek god of dreams. Morphine was the first drug compound to be isolated from the plant that contained it, and as such it marked the first step in the development of scientific pharmacology. Drug effects could not be precisely described and measured until individual compounds were isolated. The isolation of CODEINE followed in 1832. In time, the systematic modification of molecular structure of such compounds would be an important source of new medications and the basis of the modern pharmaceutical industry.

In the 1840s, the invention of the hypodermic syringe provided a new means of administering drugs. Morphine was among the first drugs to be administered by syringe, and the immediate introduction of the dose into the bloodstream provided stronger and faster drug effects than by swallowing and digesting the drug.

During the American Civil War (1861–1865), the combination of the more potent morphine, the hypodermic syringe, and wartime conditions contributed to widespread hypodermic morphine use. Large numbers of wounded soldiers and relatively few physicians meant that many soldiers were given syringes and supplies of morphine to treat their own pain. Many soldiers inevitably became addicted. Following the war, some of these soldiers phased out opiate use as their wounds healed while others continued their pattern of morphine use for years. In the postwar period of industrial and commercial expansion, a wide variety of preparations containing opium were sold through vigorous advertising in an unregulated market. Physicians prescribed opiates, including morphine and codeine, for a wide variety of conditions. Many preparations were advertised specifically for women's health problems or for children bothered by colic or teething pain.

After 1850, Chinese laborers were brought to the American West to work on railroad building and other forms of gang work. As they moved away from these forms of labor, some Chinese took up placer mining in the Sierra Nevada or settled in Pacific coast cities like San Francisco. There, as white laborers sought to exclude them from the labor market, many opened and operated small businesses. The Chinese brought with them the practice of smoking opium to induce a two-to-three hour state of dreamy relaxation. Prejudice against Chinese people was based largely on fears that they would displace white laborers by accepting wages that white people considered to be below subsistence level; this prejudice focused on Chinese customs such as opium smoking. The U.S. Congress passed several laws in the 1880s to reduce the importation into the United States of opium intended for smoking.

In 1898, the Bayer company of Germany began marketing the newly trademarked drug Heroin, produced by modifying the morphine molecule. At first, HEROIN was valued for its apparent ability to cure morphine addiction; a dose of heroin quickly relieved all symptoms associated with morphine withdrawal. Within a few years, heroin's addictiveness was recognized and physicians stopped prescribing it, despite its effectiveness in relieving pain and coughing.

For many who became addicted through self-medication, addiction was a source of shame that they could not free themselves of. They sought treatment in privately run clinics that promised anonymity and offered little more than a place to rest while they went through withdrawal; or they purchased purported cures that in fact merely contained more opiates. Others continued to take opium or morphine and managed their jobs or other responsibilities as long as their drug supply remained uninterrupted. The initial response to rising rates of addiction was to blame unscrupulous medicine merchants and physicians who administered opiates too readily.

In the United States, the concerns about adulteration, overdose, and addiction associated with an unregulated drug market became acute around the turn of the twentieth century. In the context of Progressive Era reform, the 1906 Pure Food and Drug Act required that any medication containing opiates state their presence and the amounts on the label.

In both the United States and England, what is now called recreational use of drugs emerged in about the 1890s. People began taking opiates for pleasure, or to provide a novel experience, in a social setting with no medical overtones. Rising alarm about drug use as a particularly dangerous kind of social problem dates from this period, which also saw the rising political power of the Temperance Movement and its efforts to enact a total prohibition on the use of alcohol. Unfamiliar drug-use practices provided an additional focus for social anxieties in a time of rapid economic change. A Protestant middle-class ethos helped burgeoning new groups of professionals and business people adjust to new kinds of economic opportunity in an industrial age. Behaviors that challenged that ethos with pleasure seeking, new modes of entertainment, and unfamiliar drug-use practices proved disturbing.

In the 1890s in England and the United States, small numbers of artists and bohemians, seeking to challenge what they saw as restrictive Victorian artistic and social standards, visited Chinese opium dens where they learned to smoke opium. For some Chinese in London or Liverpool, opium smoking provided a means of relaxation from a life of hard work in an alien land. As the existence of opium dens became more widely known, however, images of ghostlike, numb pipe smokers began to appear in popular literature. The middle- and upper-class pleasure seekers who smoked opium prompted a compassionate response, but British working-class use of opium was said to betoken laziness, poor child-rearing habits, or loose morals.

In the United States, the 1880s and 1890s brought waves of new immigrants from southern and eastern Europe—and they brought new customs to the American cities they settled. By the early 1900s, sniffing heroin, for example, had become a practice of some young adults in urban neighborhoods crowded with large immigrant families or for some single adults making their way alone in a new industrial setting.

Rising concern about opiate use in this period was only partly a reaction to incidence of opiate addiction, which, with alcoholism, was classed as a psychiatric condition called inebriety. In the late nineteenth century, many troubling conditions were redefined as diseases, especially as forms of psychopathology, and opiate addiction was among them, although many physicians even decades later saw addiction as resulting from a moral failing.

Worldwide missionary activity also resulted in concerns about opiate addiction. Christian missionaries in China and the Philippines, for example, believed that opiate addiction among the local populations helped explain what they perceived as economic backwardness. Like some temperance advocates in the United States, reformers concerned about addiction portrayed it as a form of slavery that followed a collapse of moral will. In such a frame-

work, opiate addiction appeared as a scourge to be eradicated. Between 1911 and 1914 reformers met at The Hague to urge worldwide control of opiate supplies so as to prevent any nonmedical use of the drugs. Some countries joined in signing and ratifying a treaty that marked the first attempt to develop a coordinated international system for controlling worldwide opiate supplies. The U.S. representatives to these meetings were embarrassed by the lack of any U.S. legislation for controlling access to opiates. A lobbying effort to bring U.S. legislation into line with the goals of The Hague resolutions led to passage of the Harrison Anti-Narcotic Act in 1914, the first U.S. law enacted to control who could buy a drug. The act banned sale of opiates and cocaine except for use by physicians or through a doctor's prescription. The American Medical Association (AMA), sensitive to charges that physicians' overprescribing of opiates was the chief cause of addiction, supported the legislation.

Following implementation of the HARRISON ANTI-NARCOTIC ACT in 1915, health authorities in several American cities were worried that the sudden lack of opiate supplies for addicted individuals would create great personal stress and a possible public crisis. They opened clinics that were intended to dispense opiates to addicts so that they would not go suddenly into withdrawal when legal supplies were cut off. In many cases, the mission of a clinic was unclear: Were patients expected to reduce their doses gradually and wean themselves off opiates, or would some be permitted to continue to maintain their addiction by means of opiates supplied through the clinics? The U.S. Treasury Department, charged with enforcing the Harrison Act, moved vigorously to enforce it by charging certain physicians with the excessive prescription of opiates and by arguing in court cases that the act specifically disallowed addiction maintenance. In 1919, the Supreme Court ruled that the Harrison Act meant that physicians could not prescribe opiates to addicts except as part of a short-term program of detoxification. Again, the AMA agreed. Armed with this legal support, the Treasury Department continued its enforcement activities against the maintenance clinics, and by the mid-1920s all had been closed.

The Harrison Act was envisioned by its proponents as part of a planned worldwide system of treaties in which each country that imported opiates would allow only the amounts needed for medical treatment to cross its borders. Opium-producing countries and the European countries where, in this period, most of the world's opium was refined into drugs like morphine or codeine would also cooperate to limit supplies of the drug. This approach to drug control has characterized the drug policies of most countries ever since.

Meanwhile, morphine and heroin use became part of a new urban social scene that included new kinds of entertainment. Concerns about opiate addiction shifted from compassion for innocent victims of improper medication to alarm about new centers of vice in urban neighborhoods. Inner cities became populated with groups whose social and political behaviors worried some business leaders, middle-class reformers, and workers who felt their jobs were threatened.

The passage of the Harrison Act was followed by the creation of federal enforcement bodies to prohibit unauthorized entry of opiates into the country, and to arrest and convict unauthorized sellers and possessors of opiates. In the 1920s, psychiatric theory held that chronic addicts suffered from personality deficits that caused them to feel inordinate pleasure from opiates and thus become mired in addiction. Opiate addiction was now viewed as both a medical and a criminal problem. The creation of the Federal BUREAU OF NARCOTICS in 1930, and the appointment of HARRY J. ANSLINGER as its head, moved drug enforcement out of the Prohibition Unit that oversaw enforcement of the Volstead Act. Following repeal of alcohol prohibition in 1933, the Federal Bureau of Narcotics continued to carry out the enforcement of the prohibition of opiates and cocaine. Harry J. Anslinger was a skillful administrator with a background in diplomatic service. He oversaw American participation in the activities of the League of Nations' Opium Advisory Committee, which furthered the work on international control of opium supplies that had been initiated through the Hague Opium Treaty. On the domestic front, Anslinger managed an efficient team of nationwide enforcement officials. Believing that harsh and early punishment would be effective deterrents, he supported increasingly severe punishments for drug offenders, including mandatory minimum sentences for first offenders. For decades, the "drug problem" remained in the background of public consciousness as a kind of exotic problem associated with a city world of jazz, marijuana, and beatniks, but the threat carried enough symbolic weight to cause penalties for drug trafficking and possession to be stiffened in

1951 and again in 1956. Anslinger remained the U.S. government's chief drug-enforcewment official until his retirement in 1962, when, in both medical and legal circles, a new generation of observers were urging less punitive responses to drug offenses and greater emphasis on medical approaches to treating addicts.

The British approach to controlling opiate use in the twentieth century proceeded along a policy basis that was different from that of the American approach, despite some similarities in legislation. The Dangerous Drugs Act of 1920, like the American Harrison Act, restricted the use of opiates to legitimate medical needs. However, the British government did not seek to define the limits of those medical needs. The government-appointed ROLLESTON Committee, which met in 1924, recommended regarding addiction as an illness to be treated by physicians. Reacting in part to perceived difficulties in enforcing America's prohibitions of both alcohol and opiates in that period, the Rolleston committee members sought to avoid stimulating an illicit market by banning opiates. Rather, they favored allowing individual physicians to prescribe opiates to selected addicts—that is, they recommended a policy of addiction maintenance. British policy was also conditioned by the demographics of opiate use in Britain, which differed from patterns in the United States. In Britain, opiate use continued to be associated with affluent bohemianism and iatrogenic addiction, and the powerful stigma against addicts that characterized the American scene did not develop to the same degree in Britain. In such an atmosphere, nonpunitive policies appeared appropriate.

In the 1960s, startling new patterns of drug use brought the issue to mainstream consciousness in the United States and throughout Western Europe. Since the nineteenth century, the leaders of American reform efforts aiming to curb drug use had typically couched their rhetoric as concern about use patterns among specific population groups—foreigners (as in opium use by Chinese people) or the working class. Now, illicit drugs were typically being used by young, white, and middle-class persons.

Events of the 1960s prompted a generation of young people raised during the prosperous 1950s to question the ideals of the relatively calm and affluent world that they knew. These events included the ongoing civil-rights movement, the assassinations of President John F. Kennedy, Martin Luther King, Jr., and presidential candidate Robert F. Kennedy, and the escalating war in VIETNAM. As they questioned and challenged the establishment, young people disregarded old prohibitionist messages about illicit drugs; at the same time that they sought to forge new values, they also hoped they could eliminate the superficial and hypocritical aspects in American life. MARIJUANA and PSYCHEDELIC drugs most closely symbolized the new spirit, but young people buying drugs on the illicit market and sharing lore about highs also encountered amphetamines and opiates.

For the young men who went to Vietnam to fight the war, the ready availability of heroin provided one possible avenue of escape from the horrors some of them experienced and witnessed daily (although boredom was often reported as a common motive for use). Southeast Asia remained an important source for the world heroin market, the more so as the trade from Turkey through southern France became hampered by enforcement activity. It was relatively easy for many returning veterans to stop using heroin once they returned to the United States. The men came back, however, after fighting a losing war to a United States deeply divided over the conflict. Receiving little welcome, many veterans had difficulty in readjusting to civilian life; for some of these, continued drug use remained part of a web of problems made up of chronic medical conditions or difficulties in finding work, although opiate use specifically was remarkably uncommon.

In 1972, President Richard M. Nixon was reelected on a platform that included bringing an end to the war and responding to growing American fears about crime. He united these concerns by increasing enforcement resources directed against drug use. In 1971, Nixon had proposed the most significant federal drug-policy initiatives since the passage of the Harrison Anti-Narcotic Act of 1914. He announced the creation of the Special Action Office for Drug Abuse Prevention (SAODAP). This office, administratively located in the White House and headed by Jerome H. Jaffe, M.D., led an expanded federal funding for drug treatment and special programs to identify and treat addicted soldiers returning from Vietnam. Jaffe had been director of an innovative program in Illinois that offered a range of treatment services, including methadone maintenance. The previous U.S. policy toward opiate addiction, which placed emphasis on law enforcement, was for a time replaced by one that emphasized concern for treatment and prevention in addition to control of the drug supply. Beginning in 1963 in New York, Vin-

cent Dole and Marie Nyswander had demonstrated that longtime heroin users, stabilized on daily doses of oral methadone and supported with a range of rehabilitative services, showed reduced criminal activity and improved functioning in social and employment areas. Nixon came to believe that methadone maintenance would provide a cost-effective means of reducing the money-seeking crimes committed by street addicts. Previously viewed as an experimental treatment, methadone maintenance, though subjected to special regulations, was made an accepted element in the treatment of opiate addiction. In the same legislation that created the Special Action Office, Congress included language that authorized the formation of the National Institute on Drug Abuse to coordinate federal funding of treatment services and research on drug abuse.

Meanwhile, in the 1970s, under federal leadership, treatment programs were expanded and new ones created in cities across the United States. Increasingly, those running the programs encountered patients who did not fit the model of the criminally involved longtime heroin addict. Younger patients, more women, and those using a variety of drugs reflected changing U.S. drug-use patterns. As these patterns were recognized, opiates ceased to dominate images of drug abuse in both the popular mind and in policy circles. Rather, opiates became just one group among many that were traded on the illicit market and used for philosophical, lifestyle, political, recreational, and even habitual reasons.

The CONTROLLED SUBSTANCES ACT of 1970, also passed at Nixon's initiative, reformulated how drugs were assigned legal status. The act created five schedules for categorizing psychoactive drugs, ranging from those considered to have no medical use and high risk of abuse to those having important medical use and only a mild risk of abuse potential.

In Britain, as in the United States, drug users in the 1960s and 1970s experimented with a growing range of drugs besides opiates. New patterns of chronic drug use, new, flamboyant behaviors symbolized by the lives of celebrities and rock stars, and a sharp escalation in the absolute numbers of heroin addicts prompted some divisions in Britain's medical community about the wisdom of continuing Britain's nonpunitive maintenance policy toward opiate addiction. Some physicians became unwilling to treat addicts, whereas others remained committed to a purely medical approach to addiction with maintenance as an important component of the policy. In

1968 new laws were passed that limited the role of the general physician in the prescribing of heroin and that established a system of clinics supervised by specialists.

The early 1980s advent of ACQUIRED IMMUNODE-FICIENCY SYNDROME (AIDS) has added a new dimension of concern about drug use by injection, the preferred mode of administration of many heroin users. Because sharing used syringes can transfer the human immunodeficiency virus (HIV) from one person to another, drug use by injection has been named a high-risk behavior for its transmission.

Opiates remain important medication for the treatment of pain, cough, and diarrhea. Recent discoveries that opiates achieve their effects by mimicking compounds occurring naturally in the body (e.g., ENDORPHINS and ENKEPHALINS) have spurred exciting neuroscience reseearch about how the brain works. After millennia of use, then, opiates continue to be one of the most interesting classes of drugs.

(SEE ALSO: *Asia, Drug Use in; Britain, Drug Use in; Chinese Americans, Alcohol and Drug Use among; Dover's Powder; Shanghai Opium Conference; Terry and Pellens Study*)

BIBLIOGRAPHY

BERRIDGE, V., & EDWARDS, G. (1981). *Opium and the people: Opiate use in nineteenth-century England.* New York: St. Martin's Press.

COURTWRIGHT, D. T. (1982). *Dark paradise: Opiate control in America before 1940.* Cambridge: Harvard University Press.

MORGAN, H. W. (1981). *Drugs in America: A social history, 1800–1980.* Syracuse: Syracuse University Press.

MUSTO, D. F. (1987). *The American disease: Origins of narcotic control* (exp. ed.). New York & Oxford: Oxford University Press.

SCHUR, E. M. (1963). *Narcotic addiction in Britain and America: The impact of public policy.* Bloomington: Indiana University Press.

CAROLINE JEAN ACKER

OPIUM The milky juice derived from the unripe seed capsules of the poppy plant *Papaver somniferum* is called opium. This material, which dries to a brownish gum contains a large number of alkaloid compounds. These ALKALOIDS can be cate-

Figure 1
Opium Poppy and Pod

gorized into two major groups—the benzylisoquinolines and the phenanthrenes. The phenanthrene group includes the OPIOIDS, the most important of which is MORPHINE, which constitutes approximately 10 percent of opium. CODEINE is present in far smaller quantities, at 0.5 percent, and thebaine is only 0.2 percent. Both morphine and codeine can be extracted from opium and each crystallized to yield pure compounds. Virtually all morphine is derived from opium, since to synthesize it is complex and expensive. Although morphine and codeine have been used extensively in the clinical treatment of PAIN, thebaine is equally important—it is the starting material for the synthesis of many semi-synthetic opioid analgesics (painkillers). Of these, the most widely used include oxymorphone, oxycodone, and naloxone. Thebaine, itself, has no opioidlike effects.

Opium has a long history of use and abuse. It was initially used for the treatment of diarrhea and then for the relief of pain. Today, opium still has a number of medicinal uses, primarily as tincture of opium, a concentrated alcoholic extract of opium. Although this is occasionally used for extreme diarrhea, most physicians prescribe paregoric, a camphorated opium-tincture preparation containing approximately 0.4 milligrams per milliliter of morphine in 45 percent alcohol. The concentration of morphine in paregoric is far smaller than in opium tincture, so doses are adjusted accordingly. Doses that effectively treat diarrhea typically do not cause euphoria or analgesia—however, excessive doses can be abused and can lead to dependence.

History. The plant grows wild in the Middle East—especially in the Turkish plateau region—and has been known and used since antiquity. Opium was introduced into India by Arab traders of the thirteenth century. By the seventeenth century, along with the spread of TOBACCO use, the Chinese had devised a method of smoking opium—using small sticky balls of opium gum in opium pipes. It is said that by 1900, about 25 percent of the Chinese smoked opium, although it was banned by the emperor. This high level of use was the result of the British East India Company's practice, beginning in the mid-eighteenth century, of shipping opium to China from their conquered lands around Bengal (1750)—one of the major opium-producing areas of the subcontinent. Export of opium to China helped balance the company's trade deficit, caused by tea purchases. After 1780, opium was produced as a monopoly by the company.

China was at that time basically closed to all kinds of outside trade, except for certain port cities, where special concessions were granted by the emperor. Indian opium was auctioned to British traders in Calcutta, who carried it to Southeast Asia and China, often by way of shippers and smugglers off the South China coast and the islands there, including Hong Kong. British concern for the security of their opium trade led to the colonization of Malaysia and Singapore and, eventually, to the Opium War of 1840–1843, with China, where the emperor's troops were outmaneuvered. The series of treaties that ended that conflict "opened" China to trade with the West and to European political and economic domination.

Suppression of the trade began with the concerns of Protestant missionaries and physicians in China—which outweighed the concerns of the emperor for keeping his people producing for him, not enslaved by opium dreams. International bodies were formed in the late nineteenth and early twentieth centuries to restrict the opium trade, but the British refused to move toward any kind of regulation until 1905. The international conferences and conventions of 1909, 1915, and 1930 led to the restriction and prohibition of traffic in opium and opium derivatives—morphine, codeine, and heroin.

In the United States, opium abuse is not anywhere near the problem of HEROIN abuse, as of the early 1990s. Opium smoking and opium eating are the two major forms of abuse. Some immigrants to the United States have brought these customs with them, but on a small scale. When smoked, opium is prepared by heating it over a flame until a small ball of roasted opium gum is formed. The ball is then pushed into a pipe, where it is held over either a flame or a coal and smoked. Opium eating is widely practiced in India and in other countries where the opium poppy is cultivated—Turkey, Afghanistan, Southeast Asia, and so on. It is used as a household remedy for pain and other ailments, much as it has been for hundreds of years. Approximately 50 percent of opium eaters in India, for example, use it for medicinal purposes, taken as a pill or as a solution.

While the legitimate opium trade had slowed by the 1930s, illegal production continued in several places. In Southeast Asia, colonial governments drew revenues from opium monopolies until 1942, when the invading Japanese suppressed it during World War II. With the victory of the Communists in China in 1949/50, steps were taken to eradicate the growing of opium and its use. By 1960, opium production was confined to a few isolated areas of Burma, Laos, and Thailand. During the VIETNAM War, various tribal peoples were encouraged to grow opium by a number of politically motivated groups, resulting in the establishment of the GOLDEN TRIANGLE as one of the major centers of illegal opium production.

(SEE ALSO: *Asia, Drug Use in*; *Shanghai Opium Conference*)

BIBLIOGRAPHY

EMBREE, A. T. (Ed.). (1988). *Encyclopedia of Asian history*. New York: Scribners.

FAY, PETER W. (1976). *Opium war, 1840–1842*. New York: Norton.

JAFFE, J. H., & MARTIN, W. R. (1990). Opioid analgesics and antagonists. In A. G. Gilman et al. (Eds.), *Goodman and Gilman's the pharmacological basis of therapeutics*, 8th ed. New York: Pergamon.

GAVRIL W. PASTERNAK

ORGANIZATIONS FOR EDUCATION, PREVENTION, AND TREATMENT OF SUBSTANCE ABUSE *See* Appendix, Volume 4.

OSAP *See* U.S. Government Agencies: Office of Substance Abuse Prevention (OSAP)

OVERDOSE, DRUG (OD) Administration of a drug in a quantity that exceeds that which the body can metabolize or excrete before toxicity develops constitutes an overdose. Whether it is accidental or deliberate, drug overdose is a significant problem that is encountered by providers of emergency medical care. Accidental overdose is common among users of illegal substances of abuse, since little reliability can be placed on the potency, presence of adulterants, and even identity of the street substance. For example, HEROIN potency has been demonstrated to range from 3 to 90 percent. Overdoses and deaths from heroin are therefore common. The prevalence of comorbid disorders in substance-abusing populations, particularly DEPRESSION, has been found to be high. Thus, deliberate drug overdoses taken in the attempt to commit SUICIDE are frequently encountered in this population. Also, people with a psychiatric illness but no drug-abuse problem most often attempt suicide with a drug overdose. Substances frequently implicated in drug overdose involve nonnarcotic ANALGESICS (painkillers), BENZODIAZEPINES (tranquilizers), OPIATES, or ANTIDEPRESSANTS—often in combination with alcohol.

The treatment of a drug overdose begins by providing basic supportive care (i.e., ensuring that there is adequate ventilation and monitoring the heart), calling 911, an emergency medical service (EMS), or the Poison Control Center (see Appendix I in Volume 4). If little time has elapsed since ingestion, efforts may be made to prevent further absorption of the drug by such means as gastric lavage or by administration of activated charcoal. Other treatments include increasing the rate of excretion through forced diuresis or giving specific antidotes (e.g., NALOXONE for opiate overdose) when the substance is known or can be identified from the presenting clinical syndrome. Obtaining a careful drug history from the patient or accompanying individuals is of paramount importance in effectively treating and minimizing risks from a drug overdose, which often results in death.

(SEE ALSO: *Drug Abuse Warning Network; Drug Interactions and Alcohol*)

BIBLIOGRAPHY

KOSTEN, T. R. (1989). Pharmacotherapeutic interventions for cocaine abuse: Matching patients to treatment. *Journal of Nervous and Mental Disease, 177,* 379–389.

KOSTEN, T. R., & ROUNSAVILLE, B. J. (1988). Suicidality among opioid addicts: 2.5-year follow-up. *American Journal of Drug and Alcohol Abuse, 14,* 257–369.

ROSS, H. E., GLASER, F. B., & GERMANSON, T. (1988). The prevalence of psychiatric disorders in patients with alcohol and other drug problems. *Archives of General Psychiatry, 45,* 1023–1031.

MYROSLAVA ROMACH
KAREN PARKER

OVEREATING AND OTHER EXCESSIVE BEHAVIORS Overeating, a behavior not always limited to persons with BULIMIA, is grouped together with substance abuse and dependence in a superfamily of disorders designated as behavioral (non-substance-related) addictions. The term *impulse control disorders* has been used by some clinicians to describe these behaviors. In this context the notion of ADDICTION centers on the repetitiveness of the behavior and would include such behaviors as compulsive spending, compulsive gambling, pathological overeating (bulimia), hypersexuality, kleptomania (repetitive, compulsive stealing when there is no need), as well as miscellaneous obsessive-compulsive behaviors such as tics and hair-pulling (trichotillomania). Some researchers have pointed out similarities among these disorders and believe that there may be similar brain mechanisms involved in some of them. For example, it has been shown that DOPAMINE levels in certain areas of the BRAIN (such as NUCLEUS ACCUMBENS) are elevated by the ingestion of reinforcing drugs including COCAINE, AMPHETAMINES, OPIOIDS, and, to some degree, NICOTINE. However, increased dopamine levels in these same brain circuits have been shown to occur when animals anticipate food or sexual activity. Also, learning, conditioning, and reinforcement play important roles in these repetitive behavior disorders as well as in the more traditional chemical or substance-abuse and -dependence disorders. It has also been pointed out that treatments for nonchemical "addictive" disorders often follow principles used in substance-abuse disorders; for example, identifying trigger and high-risk situations, teaching alternative coping behaviors, and emphasizing relapse prevention. Self-help groups using AA principles have also been organized, such as Overeaters Anonymous or Gamblers Anonymous. Some pharmacologic agents appear to alter both drug ingestion and obsessive-compulsive behaviors that are not drug related. For example, SEROTONIN UPTAKE INHIBITORS, now used as ANTIDEPRESSANT medications, seem to help alcoholics decrease alcohol consumption and compulsive hair-pullers reduce that behavior.

Such broad definitions of addictive behaviors have disadvantages when they focus too much attention on the commonalities among the diverse behaviors while minimizing the differences and particularities. At a time when rapid progress is occurring in the understanding of the biological processes associated with substance dependence, focusing only on commonalities may obscure the value of therapeutic interventions aimed at specific disorders. For example, nicotine transdermal patches seem to have considerable value in treating tobacco dependence but are probably of no value for cocaine dependence or compulsive gambling.

The way society (or science) chooses to categorize behaviors—desirable or undesirable, repetitive or episodic—is determined in large measure by the objectives of developing the categorization. There are probably some circumstances where it is helpful to think about a broad category of problematic excessive behaviors encompassing everything from substance abuse to television watching. There is also the risk that in doing so we convey the notion that excessive drug use is no more serious or refractory to intervention than watching television or jogging. Certainly at the present time the social costs and medical consequences of the substance-use disorders are so great that we should be cautious about embracing any conceptual scheme that tends to trivialize or make these problems seem less serious than they are.

(SEE ALSO: *Addiction: Concepts and Definitions; Adjunctive Drug Taking; Causes of Substance Abuse: Learning; Obesity; Research, Animal Model: An Overview of Drug Abuse*)

BIBLIOGRAPHY

AMERICAN PSYCHIATRIC ASSOCIATION. (1994). *Diagnostic and statistical manual of mental disorders-4th edition.* Washington, DC: Author.

BALL, G. G., & GRINKER, J. A. (1981). Overeating and obesity. In S. J. Mulé (Ed.), *Behavior in excess.* New York: Free Press.

JAFFE, J. H. (1990). Trivializing dependence. *British Journal of Addiction*, 85, 1425–1427.

LEVISON, P. K., GERSTEIN, D. R., & MALOFF, D. R. (EDS.). (1983). *Commonalities in substance abuse and habitual behavior.* Lexington, MA: Lexington Books.

MARKS, I. (1990). Behavioural (non-chemical) addictions. *British Journal of Addiction*, 85, 1389–1394.

JEROME H. JAFFE

OVER-THE-COUNTER (OTC) MEDICATION

This class of medication can be purchased without a prescription. Which medications require prescriptions and which do not varies widely from country to country. Common examples of OTC medications in the United States include ANALGESICS (aspirin, Tylenol[R]), cough and cold products (Sinutab[R], Drixoral[R]), allergy medications (Benadryl[R], Tavist[R]), gastrointestinal products (Maalox[R]), and antidiarrheals (Imodium[R]).

Prescription medications are labeled with patient-specific instructions determined by a physician whereas OTC products provide general information for use by consumers. OTC products *are drugs*, and as such they may cause side effects or adverse effects, or they may interact adversely with foods, ALCOHOL, or other medications. Some of the more than 500,000 OTC products that are available have the potential to be misused or abused. Antihistamines, hypnotics, decongestants, analgesics, laxatives, and DIET PILLS are often consumed in higher than recommended quantities; they have caused physical and/or psychological dependence. An epidemic of the early 1990s among adolescents has been "baby speed," the combining of OTC CAFFEINE pills with the decongestant pills pseudoephedrine. Handfuls cost only a few dollars and are responsible for overstimulating the heart and central nervous system, causing strokes and death.

An estimated 28 percent of adults in the United States use all kinds of OTC products, often responsibly but also in combination with prescription medications or alcohol. The high cost of visits to a physician and stays in a hospital has generated heightened interest in self-medication, which has increased opportunities for pharmacists to counsel patients. This situation is also contributing to the increased availability of medications as products are transferred from prescription to OTC status. The legislation that controls OTC products is quite recent.

It was in 1951 that the United States first separated drugs into the two categories—prescription and OTC. A drug that is available only on prescription cannot be made available as an OTC product until its relative safety and efficacy have been reviewed by the U.S. Food and Drug Administration.

(SEE ALSO: *Drug Interactions and Alcohol; Legal Regulation of Drugs and Alcohol*)

BIBLIOGRAPHY

GRAEDON, J., & GRAEDON, T. (1985). *Joe Graedon's the new people's pharmacy*. New York: Bantam.

PALUMBO, F. (1991). The impact of the prescription to OTC switch on practicing pharmacists. *American Pharmacy, 4*, 41–44.

MYROSLAVA ROMACH
KAREN PARKER

OVER-THE-COUNTER (OTC) MEDICATION USE AND ABUSE   *See* Poison Control Centers, Appendix, Volume 4.

OXFORD HOUSE

The autonomous self-owned halfway-house movement of the 1990s, Oxford House, Inc., owes its momentum to J. Paul Molloy, who in 1975 established the first Oxford House in Silver Spring, Maryland. The stimulus for this first house was a decision by the state of Maryland to save money by closing a publicly supported halfway house. The men living in it decided to rent and operate the facility themselves. Operated democratically, somewhat like a college fraternity, residents of the house determined how much each would have to pay to cover expenses, developed a manual of operations, and agreed to vote out anyone who returned to alcohol or drug use. The name Oxford House was selected to acknowledge the role that the Oxford movement had played in the original development of ALCOHOLICS ANONYMOUS (AA) in the 1930s. When the first Oxford House found itself with a surplus of funds, the residents decided to use the money to rent another house and expand the concept. Each subsequent house followed suit. There are now separate houses for men and women. By 1988, there were fourteen Oxford Houses in Maryland and the District of Columbia. About 40 percent of the residents were black and 56 percent were white.

While not affiliated in any way with AA or NARCOTICS ANONYMOUS (NA), the principles of these groups are integral aspects of the operation of each Oxford House. Individuals can remain in residence as long as needed to become stably sober—there are no time limits. The average length of stay is thirteen months, with 24 percent of residents staying less than four months and 25 percent staying four to nine months.

Although a recovery house can be self-run and self-supported without being an Oxford House, if it wishes to affiliate, it must file an application for a charter with Oxford House, Inc. (9314 Colesville Road, Silver Spring, Maryland 20907). Oxford House, Inc., a nonprofit corporation, does not itself own property. Incorporated in 1987, its Board of Directors is made up of the presidents of the twelve Oxford Houses in existence at the time of incorporation. Oxford House, Inc., helps groups wanting to start a new house by providing a detailed manual on how the process works, and it helps coordinate the individual Oxford Houses.

(SEE ALSO: *Alcohol- and Drug-Free Housing; Halfway Houses; Treatment, History of*)

JEROME H. JAFFE

OXYCODONE

Oxycodone is one of the most widely used OPIOID ANALGESICS in the United States, and it is usually used in conjunction with the analgesics aspirin or acetaminophen. The combinations have proven effective and are, in some ways, superior to oxycodone alone, since they permit a lower

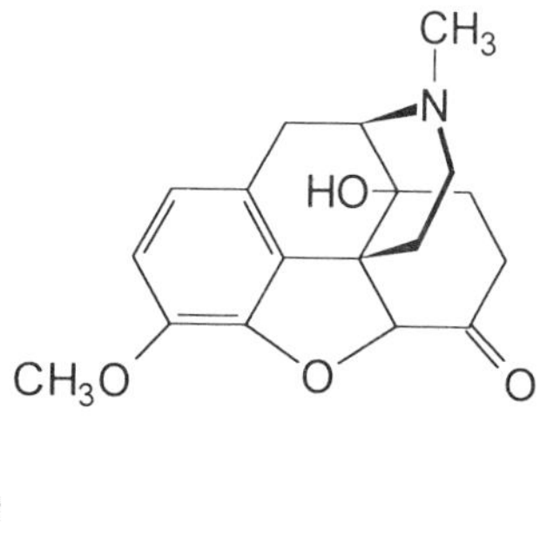

Figure 1
Oxycodone

dose of the opioid—and are therefore less likely to produce constipation, drowsiness, and nausea. Oxycodone is a derivative of OXYMORPHONE, the relationship being the same as that between CODEINE and MORPHINE. Like codeine, oxycodone is metabolized to oxymorphone, which is assumed to be responsible for its activity. Pharmacologically, the actions of oxycodone and oxymorphone are quite similar to those of morphine, so toxicity and ADDICTION can occur.

BIBLIOGRAPHY

JAFFE, J. H., & MARTIN, W. R. (1990). Opioid analgesics and antagonists. In A. G. Gilman et al. (Eds.), (1990). *Goodman and Gilman's the pharmacological basis of therapeutics*, 8th ed. New York: Pergamon.

GAVRIL W. PASTERNAK

OXYMORPHONE Oxymorphone is a potent semisynthetic OPIOID ANALGESIC derived from thebaine, one of the twenty ALKALOIDS occurring naturally in OPIUM. It is approximately fivefold more potent than MORPHINE and has very similar actions and side effects. It is used to treat moderate to severe PAIN. Oral formulations are not available in the United States, but it is available by injection or by rectal suppository. Like morphine, continued use of

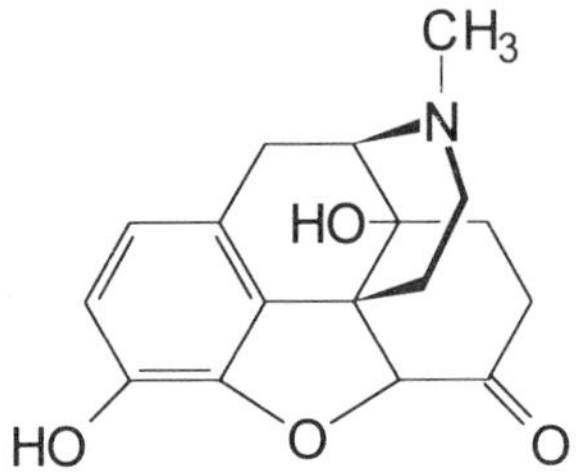

Figure 1
Oxymorphone

oxymorphone leads to TOLERANCE AND PHYSICAL DEPENDENCE. It is interesting that oxymorphone shares the same basic chemical structure as the ANTAGONISTS NALOXONE and NALTREXONE, the only difference being the substituent on the nitrogen. Neither naloxone nor naltrexone have analgesic activity; in contrast to oxymorphone, they are instead capable of blocking opiate actions.

BIBLIOGRAPHY

JAFFE, J. H., & MARTIN, W. R. (1990). Opioid analgesics and antagonists. In A. G. Gilman et al. (Eds.), *Goodman and Gilman's the pharmacological basis of therapeutics*, 8th ed. New York: Pergamon.

GAVRIL W. PASTERNAK

P-Q

PAIN: BEHAVIORAL METHODS FOR MEASURING ANALGESIC EFFECTS OF DRUGS

Pain is a sensation produced by potentially harmful stimuli, such as intense heat, stretching, cutting, or chemical irritation. The ways in which information about these stimuli is carried to the brain and the interpretation that results are very complex. Pain sometimes occurs in the absence of a harmful stimulus, such as in phantom-limb pain (where the limb has long been missing). In other instances, pain is not even felt, although harmful stimuli are present. Thus pain is both a sensation and a response to that sensation. The response to pain can vary depending on the individual and the circumstances. Given this complexity, it is not surprising that pain can be modified in many ways—by a variety of drugs, by hypnosis, and by stimulation such as acupuncture.

PAIN TRANSMISSION

The transmission of pain involves two systems—an ascending and a descending neural system. Ascending neural systems carry information about potentially harmful stimuli from peripheral nerves to the spinal cord and from there to the brain, where information about the emotional and psychological aspects of painful stimuli is incorporated. In addition, the perception of painful stimuli is altered by descending neural systems, which send information from the brain back to the spinal cord. Pain transmission can be altered at any point in this loop. Drugs such as aspirin (an analgesic) relieve pain by reducing pain sensitivity in the periphery. Local anesthetics such as lidocaine (Xylocaine) and procaine (Novocaine) relieve pain by blocking nerve conduction in specific areas. Morphine and other opioids (narcotics) alter pain transmission by interfering with the processing of painful stimuli in the spinal cord and the brain.

MORPHINE AND OTHER OPIOIDS IN HUMAN PATIENTS

Among all the drugs that relieve pain, opium and its derivative morphine, are certainly the best known. When morphine is given to patients who are experiencing severe pain, they often say the pain is less intense or that it no longer exists. Other patients say the pain is still present, but it just does not bother them. Thus, morphine affects both the sensation of pain and the patient's response to the painful stimulus. It is generally believed that morphine acts in both the spinal cord and the brain. In the spinal cord, morphine inhibits the flow of information about painful stimuli from the spinal cord to the brain. In the brain, morphine alters pain perception by altering activity in the descending pain-control system. In addition to relieving pain, morphine-like drugs produce a sense of pleasure (or euphoria) in some patients. Morphine and other opioids are the most effective drugs known for the relief of pain. Although their usefulness is sometimes limited by the fact that they can produce DEPENDENCE, this is generally not a problem in clinical settings.

NONOPIOID ANALGESICS

Although the opioids are considered the most effective drugs for the treatment of pain, THC (delta⁹-TETRAHYDROCANNABINOL), the active constituent of MARIJUANA, has some pain-relieving properties, but it is not as effective as morphine in this respect. Very large doses of drugs such as ALCOHOL and the BARBITURATES also appear to relieve pain; however, these effects do not represent true analgesia, since they only occur at doses of alcohol and the barbiturates that produce a loss of consciousness. Thus, the organism's lack of response to painful stimuli is simply an inability to respond.

STUDIES IN LABORATORY ANIMALS

To determine whether a newly developed compound has pain-relieving properties, scientists use behavioral procedures developed in laboratory animals. In general, these procedures measure the time it takes an organism to respond to a painful stimulus, first when no drug is present and then after a drug is given. Morphine and other opioids consistently alter this and other measures of pain perception. For example, morphine increases the time it takes an animal to remove its tail from a warm water bath, as illustrated in Figure 1. It takes about 2 seconds for the monkey to remove its tail from a warm water bath if morphine is not given. A small amount of morphine increases tail-removal time to about 8 seconds; larger amounts of morphine increase the time to as much as 20 seconds. Alteration in pain perception also depends on the intensity of the painful stimulus. If the water in the bath is very hot, only very large amounts of morphine will increase the time it takes animals to withdraw their tail, whereas a lesser amount of morphine will increase response time at lower temperatures. Similarly, some drugs such as BUPRENORPHINE are most effective in relieving pain when the pain is mild. Since buprenorphine also produces less dependence than morphine, it may be a very useful drug for treating mild forms of pain. By combining data about the pain-relieving effects of a drug with data about its likelihood to produce dependence, information is obtained about the usefulness of a new drug in a clinical setting.

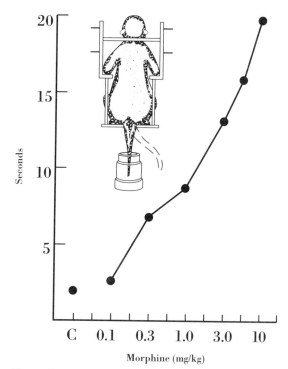

Figure 1
Pain Perception

(SEE ALSO: *Addiction: Concepts and Definitions; Pain: Drugs Used for; Opiates/Opioids*)

BIBLIOGRAPHY

BEECHER, H. K. (1959). *Measurement of subjective responses.* New York: Oxford University Press.

WASACZ, J. (1981). Natural and synthetic narcotic drugs. *American Scientist, 69,* 318–324.

LINDA DYKSTRA

PAIN: DRUGS USED FOR   Pain is a sensation unique to an individual. Its perception depends on the injury involved and the situation or context. A bruise obtained in a football game may not be appreciated at the time of the injury, yet in other circumstances the pain from a minor injury, such as stubbing a toe, may be overwhelming. The extent of the injury does not predict the amount of pain experienced. It is this wide variability that makes the treatment of pain difficult. Within the brain, there are two systems that can appreciate the sensation of pain. One deals with the objective component and tells the exact location of the injury and what type of injury it is. The other is more diffuse and comprises the "hurt." Many people have experienced both types of pain. Touching a hot object or stubbing a toe is quickly followed by the appreciation that an injury has occurred, followed an instant later by the pain. It is this second pain that contains the "suffering," the "hurt," and the elimination of this second pain is the goal of ANALGESIC therapy.

Pain encompasses a wide variety of sensations. The pain of a broken leg is quite different from that of abdominal cramping or a headache. Physicians have divided pains into three general categories. The first, and most common, is termed somatic pain. This results from tissue injury, such as a broken leg, metastases in the bone from cancer, muscle pulls, or ligament sprains. The second is termed visceral pain, and results from activation of pain fibers in internal organs, typically in the abdomen or chest. This category includes discomfort associated with gall bladder disease, peptic ulcers, or pancreatitis, to name a few. Unlike somatic pain, visceral pain is poorly localized. The most difficult pain to understand and to treat is deafferentation, or neuropathic pain, which is a consequence of injury to nerves. It is difficult for patients to describe these sensations, but they often use terms such as "burning," "shooting," or "electriclike." This type of pain is commonly seen in cancer patients where tumors invade nerve bundles. It also is seen with mild damage to nerves. The most common class of injury is the peripheral neuropathies. This collection of disorders results from a wide variety of causes; it affects nerves as they course through the body. The longest nerves are most sensitive to injury, which explains why this type of pain is most likely to develop in the feet. Diabetes is one of the most common causes. A special type of pain also falling in this classification is postherpetic neuralgia, a burning and/or shooting pain associated with *Herpes zoster*, known as shingles.

When considering pain, it is important to classify it as either acute (short-term) or chronic (long-term). The duration of many kinds of pain can be anticipated. The acute pain associated with surgery is usually limited in duration and, over the period of several days, decreases markedly. In contrast, the chronic pain associated with disseminated cancer can often be severe and persistent, actually increasing over time. Acute pain is associated with a number of very specific symptoms that are usually recognized by others—making it relatively easy to be believed. Patients may be pale and sweaty, the heart

may be beating rapidly, and they may be grimacing. Chronic pain is different; it is usually defined as pain that persists for six months or longer. Many of the signs we see acutely wear off during this time, despite the continued pain, leading some observers to conclude that the pain is minimal or even absent. This conclusion is incorrect and often leads to undertreatment and therefore unnecessary suffering. Despite the sophistication of modern medicine, the most accurate estimate of pain remains simply to ask the patient. Chronic pain may seem to have no cause, at times, may be difficult to evaluate or treat, and often requires specialists. Special pain clinics exist for such cases.

Pain medicines (analgesics) are often broken into three major groups. The first group comprises the most commonly used drugs—aspirin, acetaminophen, and related compounds; these drugs are effective for mild to moderate pain. The second group include the OPIOIDS (OPIATES). Some opioids are used for moderate pain while others are typically employed for more severe pain. Thirdly, there are a number of drugs used either for specific pain syndromes or in conjunction with the first two groups. The agents in this last group are termed adjuvant drugs.

The choice of analgesic is based on both the type of pain and its intensity. Most pain is treated in a standardized fashion. Initial therapy often utilizes aspirin, ibuprofen, or acetaminophen. These agents are available without prescription and can be very effective for mild to moderate types of pain. They have a number of properties that make them excellent analgesics. Their effectiveness against a wide variety of different types of pain and their oral dosage greatly enhance their utility. Unfortunately, these agents exhibit relatively low ceiling effects. This means that the maximal degree of analgesia that can be obtained by a drug can be limited, regardless of the dose. These drugs also reduce fevers and help with the muscle aches commonly associated with viral diseases, such as colds and influenza.

Typically, these agents act at the site of injury, leading to their classification as peripherally acting drugs as opposed to centrally acting drugs, such as the opiates, which work within the brain and spinal cord. These nonsteroidal anti-inflammatory drugs (NSAIDS) and aspirin work directly on the mechanisms of inflammation, which explains their effectiveness against arthritis. Ibuprofen became the first nonsteroidal drug approved for sale without a prescription, based on its long use and excellent safety record. Over the years, a number of additional drugs have been developed, many with analgesic potencies approaching those of morphine (Table 1). All these require prescriptions and carry risks greater than the drugs available over the counter. Side effects include a tendency to irritate the stomach and to interfere with the actions of platelets, a blood cell important in clotting; therefore, aspirin and the nonsteroidal anti-inflammatory drugs should be avoided in patients with ulcer disease, since the drugs can cause bleeding. Acetaminophen does not irritate the stom-

TABLE 1
Commonly Used Analgesics

	Average Oral Dose (in milligrams)	Frequency (in hours)	Comment
Nonnarcotics			
Aspirin	650	4–6	No prescription needed
Acetaminophen (Tylenol)	650	4–6	No prescription needed
Ibuprofen (Motrin)	200–400	4–6	No prescription needed
Fenoprofen (Nalfon)	200–400	4–6	Nonsteroidal anti-inflammatory
Diflunisal (Dolobid)	500–1000	8–12	Nonsteroidal anti-inflammatory
Naproxen (Naprosyn)	250–500	8–12	Nonsteroidal anti-inflammatory
Piroxicam (Feldene)	20–40	8–12	Nonsteroidal anti-inflammatory
Narcotics (opioids)			
Codeine	32–65	4–6	Often used in combination with aspirin or acetominophen (Tylenol#3)
Oxycodone	5–10	3–5	Often used in combination with aspirin (Percodan) or acetaminophen (Percocet)
Propoxyphene HCl	65–130	4–6	Used alone (Darvon) or in combination with aspirin and caffeine (Darvon Compound)
Propoxyphene napsylate	65–130	4–6	Used alone (Darvon-N) or with acetaminophen (Darvocet)
Morphine	30–60	4–6	Also available in slow-release formulations
Meperidine (Demerol)	50–100	3–5	Not very effective orally
Pentazocine (Talwin)	50–100	4–6	Partial agonist with tendency to produce unpleasant, subjective feelings; can precipitate withdrawal in dependent people
Hydromorphone (Dilaudid)	4–8	4–6	Very potent analgesic
Methadone (Dolophine)	5–20	8	Very effective analgesic; also used in maintenance programs
Levorphanol (Levo-Dromoran)	2–4	4–6	Potent analgesic used for cancer pain
Oxymorphone (Numorphan)	1–2	4–6	Only available by injection
Nalbuphine (Nubain)	10	4–6	Only available by injection
Butorphanol (Stadol)	2	4–6	Available by injection or by nasal inhalation; partial agonist; avoid using in dependent patients

ach and does not interfere with platelets—however, it has its own potential problems. Although it is one of the safest drugs available when used as directed, overdoses with acetaminophen can be very dangerous. Overdoses are associated with major damage to the liver which can be life-threatening. Care must be taken to use only the recommended doses of acetaminophen.

Opioids work within the brain and spinal cord to relieve the second pain—the hurt—described above. In this regard, they are amazing, since they take away pain without interfering with other sensations, unlike local anesthetics. It is this ability to selectively act on the hurt that makes them so valuable. A number of opioids are used for moderate pain (see Table 1). Of these, CODEINE is the most widely used, both alone and in combination with the nonopioids described above, followed by OXYCODONE. Both are usually used in combination with either aspirin or acetaminophen. The peripheral and central analgesics' work complement each other. If they work well together, they also bring with them the side effects of all the ingredients. Thus, both codeine and oxycodone produce constipation and sedation, along with occasional nausea, while the aspirin or acetaminophen have the problems noted above. Propoxyphene is another opioid used for mild to moderate pain. Like the others, it is most often used in combination with aspirin or acetaminophen. Standard doses are not much more effective than aspirin or acetaminophen alone, but at sufficiently high doses propoxyphene is an effective painkiller.

Pentazocine is a relatively unusual analgesic. It is an opioid indicated for moderate pain, but unlike morphine and codeine, which act primarily through mu receptors, pentazocine works in part through kappa receptors. Caution must be used when taking this agent along with other opioids, since it is a mixed AGONIST/ANTAGONIST and can precipitate WITHDRAWAL symptoms in dependent people. Many

TABLE 2
Comparison of Analgesics

	Action	Potential Side Effects	Special Comments
Nonnarcotic			
Aspirin	Relieves pain, reduces fever and inflammation, inhibits blood clotting	Gastrointestinal irritation, allergic reactions; reduced blood clotting may harm mother or fetus if taken during pregnancy; has been linked to Reye's syndrome in children	Especially effective for pain from inoperable cancer and dental surgery; inhibiting effect on bloodclotting can reduce incidence of heart disease
Acetaminophen	Relieves pain, reduces fever	Liver and kidney damage	Effective for mild-to-moderate pain but not for inflammation
Ibuprofen	Relieves pain, reduces inflammation and fever	Gastrointestinal irritation, allergic reactions	Especially effective for menstrual cramps as well as pain and inflammation in muscles and joints
Narcotic			
Codeine	Reduces pain, suppresses cough	Drowsiness, nausea, moderate physical and psychological dependence, constipation, respiratory depression	Effective for mild-to-moderate pain; especially effective against cough
Meperidine	Reduces pain and anxiety	Drowsiness, respiratory depression, high physical and psychological dependence	Effective for moderate-to-severe pain
Methadone	Reduces pain, alleviates heroin withdrawal symptoms	Drowsiness, respiratory depression, high physical and psychological dependence	Effective for severe pain

SOURCE: U.S. Department of Health and Human Services, Washington, D.C.

opioid addicts report an "allergy" to pentazocine when being treated by physicians, to avoid the possibility of withdrawal.

For more severe pain, a number of highly potent opioids are available (see Table 1). They include MORPHINE, hydromorphone, levorphanol, MEPERIDINE, and METHADONE. All are available orally. Morphine is now available in special slow-release formulations, which permit dosing as infrequently as every twelve hours. This is much more convenient for patients, particularly at night, when they no longer have to awaken to take their medicines. Special care must be taken when using these long-acting analgesics. Slow-release morphine, like methadone, may take days to reach stable levels in the blood. Thus, it can be difficult to adjust dosages without "overshooting"—which, if severe, can lead to OVERDOSES that may be life-threatening.

In hospitals, many patients receive opiates by injection or intravenously. Doses need to be adjusted to compensate for differing distributions and metabolism, but these changes are relatively straightforward for physicians working in the area of pain. Special devices are also available that permit patients to dose themselves, as needed, within specified guidelines. This approach is termed *patient controlled analgesia* (PCA). Even more sophisticated routes of administration are available. Some medications can be injected deep in the back, adjacent to the spinal canal (epidurally) where they can act primarily on the spinal cord. Localizing the medication to the spinal cord can minimize the side effects produced in the brain, such as nausea and respiratory depression.

The chronic use of opioids leads to a lessening of potency, which is termed *tolerance*. To overcome this, it may be necessary to increase the dose to maintain a constant effect. Furthermore, all patients taking sufficient quantities of drug for an extended time will become physically dependent—that is, they will experience some withdrawal if the drug is stopped. Very few patients taking opioids for medicinal purposes will ever become addicted—as the term is now used by psychiatrists. This distinction between the standard physiological responses of TOLERANCE/DEPENDENCE and ADDICTION is important, because fear of addiction should not interfere with the appropriate medical therapy of pain.

(SEE ALSO: *Abuse Liability of Drugs; Addiction: Concepts and Definitions; Controlled Substances Act; Opioids and Opioid Control; Pain: Behavioral Methods for Measuring Analgesic Effects of Drugs for; Tolerance and Physical Dependence*)

BIBLIOGRAPHY

INSEL, P. A. (1990). Analgesic-antipyretics and antiinflammatory agents: drugs employed in the treatment of rheumatoid arthritis and gout. In A. G. Gilman et al. (Eds.), *Goodman and Gilman's the pharmacological basis of therapeutics*, 8th ed. New York: Pergamon.

JAFFE, J. H. (1990). Drug addiction and drug abuse. In A. G. Gilman et al. (Eds.), *Goodman and Gilman's the pharmacological basis of therapeutics*, 8th ed. New York: Pergamon.

JAFFE, J. H., & MARTIN, W. R. (1990). Opioid analgesics and antagonists. In A. G. Gilman et al. (Eds.), *Goodman and Gilman's the pharmacological basis of therapeutics*, 8th ed. New York: Pergamon.

GAVRIL W. PASTERNAK

PAPAVER SOMNIFERUM   Poppy plants, of the genus *Papaver*, are long-stalked flowers of varying colors encompassing approximately 140 species. Of the many types of poppy plants, *Papaver somniferum* is known as the OPIUM poppy. It has white or blue-purple flowers and is widely cultivated in Asia, India, and Turkey, which supply much of the world's opium. Cultivation requires a tropical or subtropical climate without excessive rainfall. In the Northern Hemisphere, the plant flowers in late spring, after which the petals fall in a short time. This is followed by the rapid growth of the capsules (the plant's ovaries) for about two weeks. Incisions are carefully made in the capsule to obtain the milky juice, which is then dried as a gum that yields opium. The yield of opium can vary widely, but is typically about five pounds (2.25 kilograms) per acre.

The opium serves as a source of MORPHINE, CODEINE, and thebaine and is widely used in the production of important painkillers (ANALGESICS).

Typically, morphine comprises 10 percent of opium and most of the morphine used in medicine is obtained by purifying opium.

Illicit uses of opium are also widespread. In many parts of the world, opium is still smoked or eaten. Morphine extracted from opium may in turn be converted to HEROIN in clandestine laboratories. Heroin is the major opioid used illicitly in the United States. To prevent the collection and sale of opium for illicit conversion to heroin, new ways of processing the poppy plant have been developed. The most widely used consists of mowing the poppy fields before the pods are ripe enough to yield opium. The mowed stems, immature pods, and plant matter, referred to collectively as poppy "straw," are then shipped in bulk to large processing centers where the active AL-KALOIDS are extracted under careful supervision.

Other species of *Papaver* also contain alkaloids that can be converted into potent opioids. For example, *Papaver bractiatum* contains high concentrations of thebaine, which can be used to produce compounds several hundred times more potent than morphine.

(SEE ALSO: *Asia, Drug Use in; Crop-Control Policies; Golden Triangle as Drug Source; International Drug Supply Systems; Pain, Drugs Used for; Opioids and Opioid Control*)

BIBLIOGRAPHY

JAFFE, J. H., & MARTIN, W. R. (1990). Opioid analgesics and antagonists. In A. G. Gilman et al. (Eds.), *Goodman and Gilman's the pharmacological basis of therapeutics*, 8th ed. New York: Pergamon.

GAVRIL W. PASTERNAK

PARAPHERNALIA, LAWS AGAINST

Drug paraphernalia are articles that facilitate or enable the use of illicit drugs, such as hypodermic syringes for HEROIN or pipes for smoking MARIJUANA. Laws prohibiting possession and use of paraphernalia have been adopted in every state of the United States, despite substantial constitutional objections to them.

The first drug-paraphernalia laws, prohibitions against possessing opium pipes, were enacted by western states in the late nineteenth century as part of broad statutory efforts to suppress opium smoking by CHINESE immigrants. During the first third of the twentieth century, some states, in conjunction with an attempt to criminalize nonmedical use of OPIATES and COCAINE, also prohibited possession of hypodermic syringes without a medical prescription. By 1972, when the NATIONAL COMMISSION ON MARI-

HUANA AND DRUG ABUSE conducted a survey of state drug laws, about twenty states had adopted some type of drug-paraphernalia prohibition.

Commercialization of drug paraphernalia, especially through so-called head shops, in the early 1970s triggered a new generation of paraphernalia prohibitions, many of which criminalized sale as well as possession. These laws aimed at comprehensive bans on drug-related devices or articles intended for use with illicit drugs.

The drug-paraphernalia industry responded to the enactment of these laws by challenging their constitutionality on vagueness and overbreadth grounds. In most cases, courts struck the laws down as unconstitutionally vague—first, because they applied to objects that had lawful as well as unlawful uses, these laws failed to provide fair notice of the prohibited conduct; and second, the lack of explicit standards left police with discretion to enforce these laws in an arbitrary and discriminatory manner.

In 1979, the U.S. Drug Enforcement Administration (DEA) stepped into the fray. In an attempt to assist states and localities in drafting laws that would withstand constitutional scrutiny and at the same time effectively combat the drug-paraphernalia trade, the DEA drafted a Model Drug Paraphernalia Act (MDPA). Unlike prior state laws, the MDPA explicitly requires prosecutors to prove the defendant knew that the alleged paraphernalia would be used with illegal drugs. The addition of the so-called intent requirement was designed to alleviate the fair-warning concern associated with the earlier generation of statutes. In addition, the MDPA attempts to provide a more specific definition of drug paraphernalia by listing objects included within the category and by providing factors that judges should consider in determining whether an object falls within the definition. Finally, the act prohibits placement of an advertisement when one knows, or "reasonably should know," that it is intended to promote the sale of objects "designed or intended for use as drug paraphernalia.'"

A majority of states have adopted the MDPA or an equivalent statute, but its constitutionality has yet to be ruled on by the U.S. Supreme Court. In 1982, however, the Court upheld a local ordinance that required businesses to obtain a license in order to sell articles designed to be used with illegal drugs. Although this law did not involve a criminal statute prohibiting sale or possession of paraphernalia, most lower courts have subsequently upheld criminal laws modeled after the MDPA against vagueness and overbreadth challenges.

In the wake of the HIV/AIDS epidemic, another feature of traditional drug-paraphernalia laws has become controversial. In an effort to reduce the risk of transmission of the HUMAN IMMUNODEFICIENCY VIRUS (HIV) and other blood-borne diseases among needle-sharing illicit-drug users, state and local public-health authorities have sought to establish clean-needle exchange programs, usually through hospitals and clinics. To implement these programs, which have proven highly successful in reducing the risk of HIV transmission, lawmakers have had to repeal the paraphernalia laws and prosecutors have agreed not to enforce them *in this context*.

In general, drug-paraphernalia laws represent a type of drug legislation aimed mainly at declaring and symbolizing society's intolerance of illicit-drug use. Like other symbolic uses of criminal law, however, these laws are subject to highly discretionary enforcement and can have unintended costs.

(SEE ALSO: *Legal Regulation of Drugs and Alcohol; Needle and Syringe Exchanges and HIV/AIDS; Parents Movement; Prevention Movement; Substance Abuse and AIDS*)

BIBLIOGRAPHY

CORCORAN, A. M., & HELM, J. (1973). Compilation and analysis of criminal drug laws in the 50 states and five territories. In *Technical papers of the second report of the National Commission on Marihuana and Drug Abuse*. Washington, DC: U.S. Government Printing Office.

GOSTIN, L. (1991). The needle-borne HIV epidemic: Causes and public health responses. *Behavioral Science and the Law, 9*, 287.

GUINAN, M. D. (1983). The constitutionality of anti-drug paraphernalia laws—The smoke clears. *Notre Dame Law Review, 58*, 833.

RICHARD J. BONNIE

PARASITE DISEASES AND ALCOHOL

A long historical awareness exists regarding the association between heavy ALCOHOL use and an increased risk for or severity of symptoms caused by infectious diseases. In the United States, this awareness began in the medical literature of the late 1700s. It continues to be expanded in ongoing research. Historically, most of these diseases were viral and bacterial and caused death—such as tuberculosis. Some intestinal diseases were also noted, especially cholera, an acute epidemic infectious disease caused by *Vibrio cholerae* (a gram-negative bacillus) producing a soluble toxin in the intestinal tract, with profuse watery diarrhea, extreme loss of fluid and electrolytes, and a state of dehydration and collapse—death often following.

Modern research in immunosuppressed humans and animals has isolated a protozoan parasite *Cryptosporidium parvum*, which also affects the gastrointestinal tract. In immunocompetent hosts the disease is self-limiting, and recovery is accompanied by resistance to reinfection. *Cryptosporidium* is, however, common in patients with acquired immunodeficiency syndrome (AIDS); it has been noted in 16 to 50 percent of cases but is rarely manifested in HIV-positive people before loss of CD4 cells. Research with alcohol and COCAINE in AIDS-compromised animals indicated lessened resistance to *Cryptosporidium*. This was true as well with similar AIDS-compromised animals having colonies of trophozoites (a vegetative protozoan) of *Giardia muris* infecting the small intestine.

(SEE ALSO: *Complications; Substance Abuse and AIDS*)

BIBLIOGRAPHY

ALAK, J. I. B., ET AL. (1993). Humoral immune responses and resistance to *Cryptosporidium parvum* in a murine model of AIDS after chronic ethanol consumption. *Advances in the Biosciences, 86*, 331–334.

PETRO, T. M., DARBAN, H., & WATSON, R. R. (1993). Suppression of resistance to *Giardia muris* and cytokine production in a murine model of acquired immune deficiency syndrome. *Regional Immunology, 4*, 409–414.

WATSON, R. R. (1993). Resistance to intestinal parasites during murine AIDS: Role of alcohol and nutrition in immune dysfunction. *Parasitology, 107*,

RONALD ROSS WATSON

PAREGORIC

A camphorated OPIUM tincture; tinctures of opium are alcoholic extracts of opium, widely used in the treatment of diarrhea. Paregoric contains powered opium, anise oil, benzoic acid, camphor, glycerin, and diluted alcohol. With only 0.4 milligrams per milliliter of MORPHINE in 45 percent alcohol, it is more dilute than opium tincture—and the taste of the camphorated formula is generally disliked, helping to minimize excessive use or abuse.

Although paregoric is not indicated for bacterial or parasitic causes of diarrhea, it can be very helpful for other causes. Taken orally, it effectively slows down the gastrointestinal transit of wastes and enhances resorption of fluid from the intestine. Doses that effectively treat diarrhea typically do not cause euphoria or analgesia; however, excessive doses can be abused and can lead to DEPENDENCE.

(SEE ALSO: *Dover's Powder; Laudanum; Opiates/ Opioids*)

BIBLIOGRAPHY

JAFFE, J. H., & MARTIN, W. R. (1990). Opioid analgesics and antagonists. In A. G. Gilman et al. (Eds.), *Goodman and Gilman's the pharmacological basis of therapeutics*, 8th ed. New York: Pergamon.

GAVRIL W. PASTERNAK

PARENTS MOVEMENT The parents drug-prevention movement emerged in the latter half of the 1970s in response to the greatest escalation in drug use by children and adolescents in the history of the world. It originated with a number of people, who founded several different national organizations to lead the prevention movement.

In August 1976, an Atlanta mother, Marsha Keith Mannat Schuchard, Ph.D., and her husband, discovered at their daughter's thirteenth birthday party that she and most of her friends were using drugs. In response, they organized the nation's first parent peer group. Such groups consist of parents whose children are friends. The parents establish family and social guidelines to which they agree to adhere in order to protect their children and help them avoid unhealthy and destructive behaviors during adolescence. In a very short time, the young people whose parents formed this first parent peer group stopped using drugs and returned to the productive behaviors in which they had engaged before they became

involved with drugs. Dr. Schuchard later wrote about this experience in *Parents, Peers and Pot*, which the NATIONAL INSTITUTE ON DRUG ABUSE published and distributed free to more than 1 million people during the 1980s.

In the fall of 1977, a group of concerned Atlanta citizens formed NATIONAL FAMILIES IN ACTION. Founders included Keith Schuchard and Sue Rusche. Mrs. Rusche later became the organization's executive director. This organization called attention to the social and environmental factors that seemed to promote the use of illicit drugs. Its purpose is two-fold: (1) to replace commercial and societal messages that glamorize drug use with accurate information based on scientific research about drug effects; (2) to help people put this information to use by organizing community-based prevention groups. At the time of its founding, National Families in Action responded to the explosion of "head" shops that appeared to target children and teenagers as potential customers. (Drug users called themselves "heads"—"acid heads," "pot heads," "coke heads," and so on.) Head shops sold books and magazines that taught people how to use drugs. They also sold toys and gadgets to assist and enhance drug taking. The materials head shops sold were called drug PARAPHERNALIA. In January 1978, National Families in Action succeeded in getting the Georgia legislature to pass the nation's first laws banning the sale of drug paraphernalia.

At about the same time, Otto and Connie Moulton, of Danvers, Massachusetts, founded COMMITTEES OF CORRESPONDENCE. Their goal is to alert citizens to the activities of the drug culture and to drug-policy organizations that advocate the decriminalization and legalization of MARIJUANA, COCAINE, HEROIN, PCP, and other illicit drugs. They began sending out packets, called "Otto Bombs," detailing information about the local, state, and federal lobbying activities of pro-legalization organizations such as the National Organization for the Reform of Marijuana Laws (NORML), whose board and advisory board then consisted of many drug paraphernalia manufacturers and publishers of materials supporting or glamorizing drug use. Patterned after the original Committees of Correspondence, founded to uphold the rights of American colonists before and during the Revolutionary War, the modern-day version seeks to uphold the right of citizens to be drug free. A newsletter presents information from researchers and doctors refuting medial and scientific claims made by legalization proponents. The Committees of Correspondence also tracks the lobbying efforts of organizations that advocate legalizing drugs as they emerge. including the Drug Abuse Council in the 1970s and the DRUG POLICY FOUN-DATION in the 1990s.

In April 1978, Thomas "Buddy" Gleaton, Ed.D., invited Keith Schuchard and Sue Rusche to address the fourth annual Southeast Regional Drug Conference he held for drug-education professionals at Georgia State University, where he taught. He also invited various federal officials. Many accepted, particularly from the NATIONAL INSTITUTE ON DRUG ABUSE (NIDA). The NATIONAL PARENTS RESOURCE IN-STITUTE FOR DRUG EDUCATION (PRIDE) was founded in the summer of 1978, following this conference.

Publicity generated by the passage of Georgia's drug paraphernalia laws, by the fourth Southeast Drug Conference, and, later, by the publication of *Parents, Peers and Pot*, brought requests for help from parents throughout the United States. They wanted to form groups to ban paraphernalia sales in their cities, towns, and states, and to prevent substance abuse in their families and communities. For the next several years, leaders from National Families in Action, PRIDE, Committees of Correspondence, and other national organizations, along with leaders of groups from various states, traveled across the nation, helping parents form prevention groups. A contract that NIDA awarded to Pyramid made much of this work possible. Pyramid hired parent group leaders as consultants and paid their expenses to travel to communities that sought their help in organizing groups.

One of the first groups to form outside Georgia was Naples (Florida) Informed Parents, led by Pat and Bill Barton. The Florida leaders joined those from Georgia and Massachusetts to help parents in other states form similar groups. By 1979, hundreds, perhaps thousands, of parent groups had organized across the nation. In January 1979, Senator Charles Mathias (D-MD) held congressional hearings on the harmful effects of marijuana; he invited many parent-group leaders to Washington to testify. They, in turn, took advantage of this opportunity to meet and discuss the need to form a Washington-based parents' organization that could represent their interests with both Congress and the federal agencies that were making and implementing national drug policy. They agreed to meet at the fifth annual Southeast Regional Drug Conference, now known as the

PRIDE conference, at Atlanta in the spring of 1979. There, they founded the National Federation of Parents for Drug-Free Youth. Pat and Bill Barton were elected as the group's copresidents; a Maryland parent group leader, Joyce Nalepka, later became its executive director.

In the summer of 1980, the presidential election year, parent groups worked hard to get drug-abuse-prevention policy on the agendas of presidential candidates. After the election, the National Federation of Parents for Drug-Free Youth led a massive letter-writing campaign to President-elect Ronald Reagan, asking him to bring Carlton Turner, Ph.D., to the White House as his drug policy adviser. Dr. Turner, of the University of Mississippi, was responsible for growing all marijuana used in scientific research throughout the United States. He had devoted much time to educating parents at various conferences about the pharmacological effects of marijuana on the brain and body, and had earned their trust. President Reagan selected Dr. Turner as his drug adviser. Shortly after the inauguration, Dr. Turner helped the Federation arrange for parent-group leaders to brief First Lady Nancy Reagan on the prevention movement and to enlist her support for their cause. She not only responded positively but served informally as the national spokesperson for the drug-prevention movement. A few years later, President Reagan appointed parent-group leader Donald Ian Macdonald, M.D., a pediatrician from Florida, to serve as head of the Alcohol, Drug Abuse and Mental Health Administration (ADAMHA), the federal agency in the Department of Health and Human Services that was responsible for substance-abuse and mental-health research and services. One of Dr. Macdonald's legacies is the CENTER FOR SUBSTANCE ABUSE PREVEN-TION, then called the Office for Substance Abuse Prevention (OSAP), which he created during his tenure at ADAMHA. Congress formally authorized OSAP in the Anti-Drug Abuse Act of 1986.

Through this kind of concerted effort, the parent movement was able to place key policymakers in the federal government to emphasize and implement their goals—to prevent substance abuse before it starts; to help drug users quit; and to find treatment for those who are addicted and need help in quitting. The parent movement was the first leg of the national prevention effort. It is generally credited with developing and carrying out strategies that reversed those policies of the 1970s that seemed to increase drug use throughout that decade. These strategies included outlawing head shops and the sale of drug paraphernalia, stopping the decriminalization/legalization of marijuana, and insisting that producers of drug education materials replace "responsible" use messages with "no-use" messages. As a result of these strategies, the parent movement is thought to have been a key component in the 1980s for reversing the escalation in drug use by children, adolescents, and young adults that began in the 1960s—and for initiating the reduction in drug use that took place between 1979 and 1993.

(SEE ALSO: *Education and Prevention; Marihuana Commission; Partnership for a Drug-Free America; Prevention; Prevention Programs*)

SUE RUSCHE

PARTNERSHIP FOR A DRUG-FREE AMERICA

The Partnership for a Drug-Free America is a nonprofit coalition of the U.S. communications industries; its mission is to help reduce demand for illegal drugs by using the media to change the attitudes that affect drug trial and experimental (nonaddicted) use. The key officers of the organization are James E. Burke, chairman; Thomas A. Hedrick, Jr., president; Richard D. Bonnette, executive director; and Robert L. Caruso, chief financial officer.

The partnership was founded by Richard T. O'Reilly in early 1986 as a project of the American Association of Advertising Agencies. It was based on the idea by Philip Joanou, chairman of Dailey & Associates in Los Angeles, that the disciplines of marketing could be used effectively and efficiently over time to help "unsell" illegal drugs. The hypothesis was that prevention could be viewed as trying to affect individual decisions to buy or use illegal drugs in the same way that individual decisions to buy or use legal products and services are affected—except in reverse. Rather than using media messages on the *benefits* of a product, the partnership set out to reduce drug trial by building awareness of the *risks* and danger of using illegal drugs.

The Partnership's early strategy was based on a concept developed by Dr. Mitchell S. Rosenthal, president of the PHOENIX HOUSE treatment programs in New York. He theorized that the epidemic levels of drug use and addiction in the early 1980s was caused by a process of "normalization"—to both the use and users of illegal drugs—since the mid-1960s. According to Dr. Rosenthal, we could not achieve significant progress in "the war on drugs" until we reversed that process and "denormalized" individual and subcultural attitudes toward illegal drugs.

The three primary functions of the partnership are (1) to understand consumer attitudes that affect the trial and use of illegal drugs; (2) to develop messages targeted to specific demographic groups; and (3) to deliver those messages to the public through all forms of the media, but primarily public-service announcements. These functions, managed by a small full-time staff, have been accomplished through the volunteer efforts of research firms, advertising agencies, production groups, and the media. As of the end of 1992, more than 300 antidrug print and broadcast messages had been delivered, at no cost to the partnership and valued at more than 50 million dollars. Since the program's launch in March 1987, the media have donated more than 1.5 billion dollars in advertising time and space.

The partnership's prevention messages are targeted primarily to preteens and young teens, inner-city youth, and also parents, peers, and siblings, who are viewed as the key influencer groups. The focus of the messages is on building perceptions of risk and social disapproval, promoting resistance skills, and reinforcing a consistent tone of social denormalization in regard to illegal drugs. Overall media efforts are directed at achieving the goal of 1 million dollars a day in donated time and space. This results in the delivery of approximately one antidrug message per household per day. All major national media are visited personally by partnership staff to monitor the program. State and local media programs are also developed and supported through staff and volunteer efforts.

The organization's tracking research is funded by the NATIONAL INSTITUTE ON DRUG ABUSE (NIDA) and directed by the Gordon S. Black Corporation. The annual Partnership Attitude Tracking Survey (PATS) uses a centrally located sampling to evaluate attitudes toward illegal drugs among more than 8,500 preteens, teens, and adults. This research, along with other major NIDA studies and especially the HIGH SCHOOL SENIOR SURVEY done by the Institute for Social Research, suggests that since 1986 attitudes to illegal drugs have been changing. Furthermore, the surveys indicate that the partnership's messages have been a major source of information (among others) that helped effect these changing attitudes.

It is difficult to establish a scientifically conclusive cause-and-effect relationship between the partnership's efforts and U.S. trends in drug-use behavior. Many components are necessary—particularly community efforts—to reduce demand for illegal drugs, and it is unlikely that any one component is sufficient to the task. It is also imperative to note the importance of timing in this media effort, since the media are most effective in accelerating trends that are already in place. The media play a large role in American society and therefore in the lives of the children growing up in that society. The Partnership is mounting a very significant communications effort to influence the way Americans think about illegal drugs.

(SEE ALSO: *Advertising; Prevention: Shaping Mass Media Messages to Vulnerable Groups; Prevention Movement; Prevention Programs*)

THOMAS A. HEDRICK, JR.

PASSIVE SMOKING   *See* Nicotine; Tobacco: Medical Complications.

PATENT MEDICINES   *See* Over-the-Counter Medication.

PCP   *See* Phencyclidine (PCP).

PEER PRESSURE   *See* Adolescents and Drug Use; Causes of Substance Abuse; Lifestyle and Drug-Use Complications.

PEMOLINE   Although not structurally similar to the AMPHETAMINES, pemoline has similar PSYCHOMOTOR STIMULANT effects but only minimal effects on the cardiovascular system. Pemoline is often used therapeutically (despite being less effective than amphetamine or METHYLPHENIDATE) in the treatment of ATTENTION DEFICIT/hyperactivity disorder (AD-

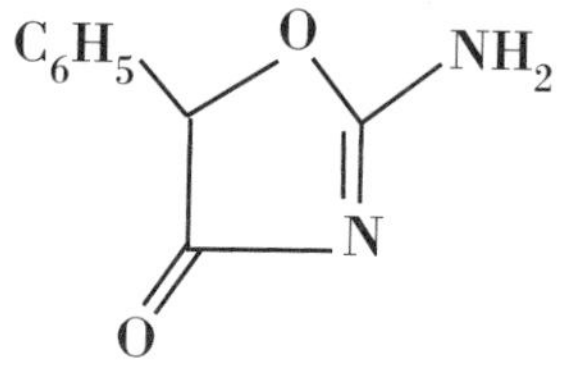

Figure 1
Pemoline

HD)—a syndrome that first becomes evident during childhood and is characterized by excessive activity and difficulty in maintaining attention. Pemoline has the advantage of a long half-life, which means that dosing can be once daily, but clinical improvement can be delayed by three to four weeks after initiation of pemoline therapy. In addiction, the likelihood for abuse of pemoline appears substantially less than that of the amphetamines.

MARIAN W. FISCHMAN

PERCODAN   *See* Oxycodone.

PERSONALITY AS A RISK FACTOR FOR DRUG ABUSE   The term *personality* refers to those relatively enduring aspects of attitudes, feelings, responses, and behaviors that permit us to recognize a particular person whom we have known over time. It is, in a way, a fingerprint of an individual's psychological makeup—the framework of how the individual thinks and acts. Psychiatrists believe that this framework arises out of childhood, powerfully shaped by the actions of parenting and the other social and environmental factors on a complex set of genetic and other biological givens. It is then further molded throughout one's development to achieve more or less lasting form in adolescence and early adulthood.

In the nineteenth century, we said that some people had willpower or a strong character; now we might refer to their good coping skills or to their ego strength—different ways of describing global measures of effective functioning. Current terms for more specific descriptors of personality might include the poles of introversion–extroversion, or approach–avoidance, as well as others.

There is a long tradition linking personality, or character, to alcohol and other substance use and abuse. In the popular imagination, the old usage of "alcoholic" or "drug fiend" conveyed images of weakness, untrustworthiness, and/or viciousness; more sophisticated imagery, "oral character," conveyed ideas of dependency and neediness—analogies to the greedy infant at the breast. Unfortunately, such simple postulates break down in the presence of the complexities of the real world: Not all substance abusers are frightening "drug fiends"; neither are they necessarily dependent, needy, demanding "oral characters."

The explanation for substance abuse is not found purely in the drug. Most adults are able to drink socially without becoming alcoholics; some of us are repeatedly exposed to opiates (e.g., after surgery) without becoming addicts. Clearly, the impact of personality on alcohol and other drug use depends on a variety of factors—the social context, the specific drug, and the stage of involvement with the drug. Is the individual brought up as a rich kid in the suburbs or poor in an inner-city ghetto? Is the person black or white? Do drugs and drug users surround the individual, and are they seen as normative, or are they considered dangerous, rare, and deviant? Is the drug a relatively weak reinforcer such as marijuana, or is it a powerful stimulant such as cocaine? Is the individual experimenting in the early stages of use, struggling with long-term dependency, or dominated by the pangs of withdrawal and craving? Although a number of predictor factors for substance abuse are known, such as age, sex, religiosity, and parental drug use, we do not know why only *some* of those at risk become drug dependent. Personality is another likely predictor of who will try a particular drug, who will continue to use it or abuse it, the success of the struggle with abstaining, and so forth.

As the preceding indicated, early thinking was that excessive drinking (alcohol) and smoking (tobacco) were linked to early childhood experiences of suckling and satiation, of hunger satisfied by taking something in through the mouth that resulted in blissful sleep. That this may, at least at times, be true was seen in one patient who had first been addicted to alcohol and then to a series of barbiturates and other sedatives; he said plaintively, "Doc, I could become addicted to orange juice if it gave me a dreamless sleep." Unfortunately, just as the thumb fails to provide milk, most drugs do not ultimately provide the desired end—the continuing sense of pleasure and/or relief. It was assumed that individuals who had had difficulty in the earliest stages of development might be particularly prone to some kinds of addiction—to depressive drugs, such as alcohol, sedatives, or opiates, which provide dreamy reverie states or sleep—and that difficulty in later stages of development might predispose to use of activating drugs, such as the stimulant amphetamines or cocaine.

Ongoing clinical experience and changing theories led investigators to focus additionally on aggression and on regulation of feelings. For example, many addicts appear to have difficulty distinguishing anxiety and anger, and they experience strong feelings as overwhelming, leading to loss of control. The drug may substitute, both pharmacologically and symbolically, for the parent—to "magically" help the individual maintain control. It has also been noted that many addicts appear not to have learned from their parents how to recognize, evaluate, and appropriately respond to danger. Many, or all, of these additional factors may operate at once: Individuals may be trying to satisfy primitive impulses and needs; there may be a defect in the recognition and control of feeling states; and they may be struggling to adapt to a stressful environment. A particular drug may, for a particular individual, transiently resolve these issues. Heroin may satiate, dampen, and control aggression—and provide relief from environmental pressures—for the moment. Amphetamines or cocaine may provide orgasmic pleasure, in the form of a "rush," as well as provide a sense of control and omnipotence. A patient who was dependent on amphetamines was panicked at the thought of dental anesthesia: "I can't stand the idea of not being in control, of being put to sleep. It's why I take the pills, to stay awake, to know what's happening."

Many individuals who misuse drugs will misuse many different kinds of drugs—the polydrug abusers. There are also people who, even after extensive experimentation with a variety of drugs, will choose to use and/or abuse a single drug or class of drugs—such as opiates, sedatives, or stimulants. It has been suggested that such individuals are driven to seek a particular drug experience, since the various drugs indeed have differing physiological and psychological effects.

Some studies lend support to this notion of particular personality contributions to drug preference.

For example, opiates tend to bolster withdrawal (from others) and repression (not acknowledging reality) by inducing a state of decreased motor activity, underresponsiveness to external situations, and reduction of perceptual intake. Such a state is conducive to reinforcing fantasies of omnipotence, magical wish-fulfillment, and self-sufficiency, but both sexual drive and aggression are diminished. In addition, there is evidence that opiate addicts are, in general, more severely impaired in terms of their ability to function in the ordinary world; they are less able to cope with the activities of daily living. In contrast, amphetamines elevate scores on autonomous functioning and sense of confidence; there is a feeling of heightened perceptual and motor abilities accompanied by a strengthened sense of potency and self-regard. These effects appear to serve the user's need to feel active and potent in the face of an environment perceived as hostile and threatening—and also to deny underlying fears of passivity.

It is important to remember that all of us have some quirks, that we do not always handle all kinds of stress equally well, that we all have some weaknesses in our personalities, some defects in our characters. These may predispose some of us to drug use and to particular drug choice. Others have significant defects in development, disordered adaptations to the real world in which we are expected to function; they may choose a particular drug or drugs to help them adapt to their difficulties—to make up, in a sense, for what is lacking within them. They are in effect choosing and self-administering their own medicine. This has been referred to as the *self-medication* theory of drug abuse. Certainly, drugs are capable of dramatically reversing painful emotional states; they can mute or free us from unmanageable feelings and provide some with the feeling that "It's the only time I've ever felt normal." Unfortunately, these effects are short-lived; side effects and the complications of physical dependance, tolerance, and withdrawal become prominent and even dominate the chronic user, who has become a substance abuser.

Be cautioned: These studies were done on people who had already been using illicit drugs for many years—who had been immersed in the "drug world" of copping (getting the drug), fearing detection and detention, and living with the altered state of consciousness induced by their drug of choice. These studies and others like them can tell us only of a correlation, not a causal relationship, where personality style or defect results in or leads to drug use/abuse. There are, however, some longitudinal studies that have followed schoolchildren for enough years to have seen some of them enter the drug world. In general, they show remarkable agreement in the descriptions of those children who become seriously involved with drugs. They are the opposite of the stereotype of the Eagle Scout (who is "thrifty, loyal, brave, clean, and reverent"); instead, they may be characterized as impulsive, with difficulty tolerating feelings and delaying gratification, and as possessing an antisocial personality style given to breaking rules, oppositional behavior, risk taking, and sensation seeking. These personality characteristics are present before immersion in the drug culture and are altered as the individual moves from initial use to continuing use, to the transition from use to abuse, to cessation or control of abuse—and, all too often, to relapse.

Be further cautioned: These findings may have been true at the time of the studies but may prove to be specific to that moment of history and no longer true. Zinberg (1984) has pointed out that the setting in which one takes a drug, and therefore the meaning of the drug-induced experience, is continually changing:

> Chronic users [of marijuana], those that began using prior to 1965, were observed to be more anxious, more antisocial, and more likely to be dysfunctional than were the naive subjects who were just beginning to use marijuana in 1968.... By the late 1960s, drug use was being experienced as a more normative choice ... in the early 1970s, controlled marijuana users could not possibly have been described as individuals driven to drug use by deep-seated, self-destructive, unconscious motives [p. 174].

An alternative view that has been suggested is that a series of otherwise accidental environmental reinforcers may so interact as to result in drug use in the absence, or the limited availability, of otherwise more necessary and pleasurable commodities. Experiments have shown just such development of "excessive behavior" in both animals and humans during conditions of deprivation—of not enough water or food—but they have not yet demonstrated such a role in the induction to drug use.

Despite these cautions, it appears that PERSONALITY is a contributor that predisposes some to substance use and abuse. Different personalities are likely to make differing contributions to drug use, depending on the particular drug, the historical moment, the social surround, and the other determinants of use. Although it is still difficult to demonstrate more than generalities about the personality of addictive behaviors, the construct of addictive personality(ies) may be "theoretically necessary, logically defensible, and empirically supportable" (Sadava, 1978). Without such a construct—which includes the characteristic response patterns of the individual, the symbolic meaning of the experience to the individual (while recognizing that this may be retrospective rationalization), as well as the specifics of the particular drug's pharmacology—it will be difficult to explain the variation in drug use among individuals with apparently comparable life experiences.

(SEE ALSO: *Adjunctive Drug Taking; Causes of Substance Abuse; Conduct Disorder and Drug Use; Coping and Drug Use; Families and Drug Use; Vulnerability As Cause of Substance Abuse*)

BIBLIOGRAPHY

KANDEL, D. B., (ED.). (1978). *Longitudinal research on drug abuse: Empirical findings and methodological issues*. Washington, DC: Hemisphere.

SADAVA, S. W. (1978). Etiology, personality and alcoholism. *Canadian Psychological Review, 19*, 1998–2014.

VAILLANT, G. E. (1983). *The natural history of alcoholism*. Cambridge: Harvard University Press.

ZINBERG, N. E. (1984). Drug, set and setting: The basis for controlled intoxicant use. New Haven: Yale University Press.

WILLIAM A. FROSCH

PERSONALITY DISORDER

The concept of *personality* refers to the set of relatively stable and characteristic behaviors that individuals display in perceiving and responding to the environment, along with a particular way of thinking about themselves. These patterns of behavior and self-perception are called personality traits. They are manifested in a variety of social interactions in day-to-day living, and their diversity is extensive. When these traits become exaggerated, inflexible, and maladaptive, they begin to impair social functioning and can cause subjective distress. Different constellations of maladaptive traits are clinically diagnosed as personality disorders. Frequently, individuals identified as having a personality disorder do not see themselves as others see them, do not recognize the annoyance their behavior engenders in those around them, and hence do not seek to change their behaviors unless there are significant social repercussions. The characteristic traits of a personality disorder typify the individual's long-term functioning and are generally recognizable by adolescence.

In psychiatry, clusters of certain personality traits are recognized in the DIAGNOSTIC AND STATISTICAL MANUAL *of Mental Disorders- 3rd ed.-revised* as constituting particular personality disorders. There is some overlap in the traits of some of the following identifiable personality disorders.

Paranoid	suspicious, mistrustful, hypervigilant, easily offended, unfeeling toward others
Schizotypal	odd and eccentric behavior, speech, and manner of thinking; withdrawn and isolated
Narcissistic	exaggerated sense of self-importance, feelings of entitlement to special favors, exploitation of others, lack of empathy, response of rage to criticism, disregard for social conventions
Histrionic	dramatic, emotional, erratic, with displays of seductive behavior; attention-seeking
Antisocial	antisocial behavior in many areas of life: lying, theft, violence, substance abuse, sexual promiscuity, spouse and child abuse, inconsistent work, legal conflicts; impulsivity and lack of remorse for antisocial acts
Borderline	unstable mood, behavior, relationships, and self-image; impulsive, self-destructive acts (e.g., suicide attempts, substance abuse); chronic feelings of emptiness, intolerance for being alone

Avoidant	timid, extreme sensitivity to real or imagined rejection, socially withdrawn, poor self-esteem
Dependent	avoidance of taking responsibility for their lives and a striving to get others to look after them; passive, submissive, with low self-esteem, and discomfort when alone
Obsessive-compulsive	perfectionist, orderly, inflexible, indecisive, constricted emotions, obstinate, overly conscientious
Passive-aggressive	resistance to demands for adequate social and occupational performance indirectly through procrastination, inefficiency, stubbornness, forgetfulness; frequent fault-finding with others

The origins of personality disorders are not well understood, but they clearly can be thought of as reflecting the contributions of genetic, constitutional (temperament), environmental (upbringing, relationships), sociocultural and maturational (psychological development) factors. The need for, and modalities of, treatment of personality disorders varies and can include psychotherapeutic and pharmacologic interventions.

(SEE ALSO: *Attention Deficit Disorder; Causes of Substance Abuse: Psychological (Psychoanalytic) Perspective; Comorbidity and Vulnerability; Conduct Disorder and Drug Use; Epidemiology of Drug Abuse; Personality As a Risk Factor for Drug Abuse; Vulnerability As Cause of Substance Abuse*)

BIBLIOGRAPHY

HIRSCHFELD, R. M. A. (1986). Personality disorders. In A. J. Frances & R. E. Hales (Eds.), *Psychiatry update— The American Psychiatric Association annual review* (Vol. 5). Washington, DC: American Psychiatric Press.

PERRY, J. C., VAILLANT, G. E., & GUNDERSON, J. G. (1989). Personality disorders. In H. I. Kaplan & B. J. Sadock (Eds.), *Comprehensive Textbook of Psychiatry* (5th ed., Vol. 2). Baltimore, MD: Williams & Wilkins.

MYROSLAVA ROMACH
KAREN PARKER

PET SCANNER   *See* Imaging Techniques.

PEYOTE   Peyote (or peyotl) is the common name for the cactus *Lophophra williamsii* or *Anhalonium lewinii*, which is found in the southwestern United States and northern Mexico. Although there are many compounds found in the cactus, some of which may be PSYCHOACTIVE, the principal HALLUCINOGENIC substance found in peyote is MESCALINE. As the other psychoactive substances may make some contribution to the PSYCHEDELIC experience, there may be some slight difference in the behavioral effects produced by taking peyote and pure mescaline, but the overall effects of peyote are very similar to those produced by mescaline.

Peyote, one of the oldest psychedelic agents known, was used by the Aztecs of pre-Columbian Mexico who considered it magical and divine. Its use spread to other Native American groups who used it to treat various illnesses, as a vehicle to communicate with the spirits, and in highly structured tribal religious rituals. For these rituals, the dried tops of the cactus—the buttons—are chewed or made into a tea. Since peyote may cause some initial nausea and vomiting, the participant may prepare for the ceremony by fasting prior to eating the buttons. Peyote is usually taken as part of a formalized group experience and over an extended period of time; the peyote ceremonies may take place at night and around a communal fire to increase the hallucinogenic effects and visions.

(SEE ALSO: *Ayahuasca; Dimethyltryptamine; Psilocybin*)

BIBLIOGRAPHY

EFRON, D. H., HOLMSTEDT, B., & KLINE, N. S. (EDS.). (1979). *Ethnopharmacologic search for psychoactive drugs.* New York: Raven Press.

DANIEL X. FREEDMAN
R. N. PECHNICK

PHARMACODYNAMICS

The study of the mechanism of drug actions is called pharmacodynamics. Most (but not all) drugs exert their action by binding to specific RECEPTORS. This binding may initiate changes that lead to the characteristic effects of the drug on body functions.

A central question in drug therapy (medication) is the proper dose of the drug that produces a desired action without many harmful side effects. To clarify this problem, pharmacologists analyze the relationship between dose and response. Most dose-response curves are sigmoidal (shaped like an S). The log-dose-response can be viewed as having four parameters: potency, slope, maximal efficacy, and variability. Potency describes the strength of drug effects. It is usually employed to calculate relative strengths among drugs of the same class. Slope is the central part of the curve that is approximately straight. It is used to analyze drug concentration (dose) from the observed corresponding responses. Maximal efficacy, or simply "efficacy," is the greatest effect produced by the drug. This is one of the major characteristics of a drug. Efficacy and potency of a drug are not necessarily correlated, and the two characteristics should not be confused.

Many drugs, including drugs of abuse, produce TOLERANCE—when it becomes necessary to take progressively larger doses to achieve the same drug effect. In some cases, the brain and other tissues on which a drug acts undergo adaptive changes (neuroadaptations) that tend to offset the drug effect. When a drug that produces neuroadaptation is withdrawn, the brain and other tissues have to readapt, because they are no longer balanced by drug effect. The adaptation produces a variety of signs and symptoms called withdrawal syndrome. The severity of this syndrome depends on the degree of adaptive changes in the nervous system—which, in turn, depends on the dose and the duration of exposure to the drug. The particular characteristics of the withdrawal syndrome depend on the pharmacological effects of the drug(s) and typically are opposite to the drug effects. For example, MORPHINE constricts the pupil; the morphine withdrawal symptom includes pupil dilation.

Most drugs of abuse produce pleasant effects in humans. For example, some people use AMPHETAMINES or other stimulants (e.g., COCAINE) to achieve a sense of well-being and euphoria. Some people use DEPRESSANTS—ALCOHOL, OPIOIDS, or TRANQUILIZERS—to relax. Still others use either stimulants or depressants to relieve boredom or reduce anxiety or pain. The common feature is that people use drugs because somehow the drug is rewarding to the user, either by producing a feeling of well-being (e.g., euphoria, elation) or by taking away a negative feeling (e.g., anxiety).

(SEE ALSO: *Addiction: Concepts and Definitions; Drug Interaction and the Brain; Drug Metabolism; ED50; LD50*)

USOA E. BUSTO

PHARMACOGENETICS (Genetic Factors and Drug Metabolism)

Researchers in pharmacogenetics study the genetic differences that cause individual variation in the body's response to or disposition of drugs. The source of most such variation is the genetically determined diversity of enzymes involved in plant metabolism—and more specifically in DRUG METABOLISM (since most drugs are derived from plants). Unlike the receptors (at nerve endings), whose variability is constrained by an essential and intricate function, variability among drug-metabolizing enzymes seems to have been encouraged during evolution. The most common of these enzymes are the cytochromes P450; located in the liver, they intercept drugs and other foreign chemicals passing from the gastrointestinal tract into the general circulation. Our capacity to metabolize the vast array of chemicals we are exposed to is due largely to the multiplicity of cytochrome P450 enzymes, each of which is able to metabolize compounds of diverse chemical structure.

These enzymes are thought to have evolved as a result of plant–animal "warfare." As animals ate evolving plants, the plants began to produce a variety of chemicals, some toxic, that deterred the predators long enough for the plant to reproduce. Animals adapted as well and one adaptation was the production of various P450 enzymes to detoxify plant chemicals; during the course of evolution, variants of individual enzymes were produced and retained in the human population. Today, atypical response to modern drugs is often a consequence of mutations that occurred and continue to occur during P450 evolution.

Debrisoquine hydroxylase (called CYP2D6) and mephenytoin hydroxylase (CYP2C18) are two widely studied P450 enzymes that are polymorphic—the variants occur more frequently than can be accounted for by recent mutations. In both cases, the variant enzymes are inactive. As a result, two distinct phenotypes exist: extensive metabolizers and poor metabolizers.

DEBRISOQUINE HYDROXYLASE (CYP2D6) POLYMORPHISM

This polymorphism was first discovered because of exaggerated response in a few individuals to debrisoquine, a drug used in the treatment of high blood pressure. The individuals were subsequently shown to have an impaired ability to metabolize the drug via hydroxylation, compared with other members of the study population. The molecular basis for the CYP2D6 poor metabolizer phenotype is the mutation or deletion of the gene that normally directs the synthesis of the CYP2D6 enzyme. Phenotyping is performed by administering a low dose of a drug that is metabolized by CYP2D6 (e.g., debrisoquine, dextromethorphan, or sparteine) and then measuring the relative amounts of unchanged drug and metabolite excreted in urine. Genotyping methods with greater than 95 percent accuracy are also available using polymerase chain-reaction-based amplification of DNA (deoxyribo-nucleic acid) in a peripheral blood sample.

The frequency of poor metabolizers varies with ethnic origin and ranges from approximately 7 percent of Caucasian to 1 to 2 percent of Asian (Chinese) and African (black) populations. It has often been noted that the substrates of CYP2D6 tend to be drugs that act on the central nervous system (e.g., AMPHETAMINES, tricyclic ANTIDEPRESSANTS, neuroleptics, and oral OPIATES such as CODEINE), but CYP2D6 is also important in metabolizing a number of cardiovascular drugs (e.g., beta-blocking agents and antiarrhythmics).

The consequences of CYP2D6 phenotype are illustrated in Figure 1. In this simplified representation, poor metabolizers exhibit higher drug levels and negligible metabolite levels in their plasma when compared with extensive metabolizers who were given the same dose. The potential clinical consequences for a particular drug depend on the relative pharmacological activity of the parent drug and metabolite. For some drugs (e.g., debrisoquine, perhexiline), the phenotypic difference in plasma levels of the active parent drug is so great that it exceeds any other sources of variation in response, and poor metabolizers are susceptible to exaggerated effects or side effects. For other drugs, the metabolite may be the active form. The pharmacology of the metabolite may be qualitatively similar to the parent (e.g., codeine conversion to MORPHINE), or the activity may be different (e.g., dextromethorphan conversion to dextrorphan). In either case, it is the extensive metabolizers who are expected to experience the greater drug effect. More often, the relative activity of parent drug and metabolite is not known. One of the implications of such phenotypic differences in active drug levels is that CYP2D6 activity may be a risk factor

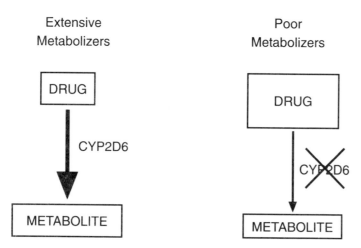

Figure 1
Consequences of CYP2D6 Phenotype

for people who abuse drugs that depend on this enzyme for their elimination from the body or for their conversion to pharmacologically active metabolites.

CYP2D6 has recently been shown to exist at very low levels in the brain. Variation in the central metabolism of drugs that appear to be bioactiviated by CYP2D6 (e.g., codeine conversion to morphine) may be of critical importance in their action, since it would determine the formation of a highly active metabolite in the vicinity of the neuronal receptor targets.

MEPHENYTOIN HYDROXYLASE (CYP2C18) POLYMORPHISM

This polymorphism is unrelated to the CYP2D6 polymorphism, and it affects the metabolism of a narrower range of drugs (the antimalarial drug proguanil and the anticonvulsants mephenytoin, diazepam, hexobarbital, omeprazole). The frequency of poor metabolizers is 2 to 5 percent in Caucasians and as high as 20 percent in Asian (Japanese and CHINESE) populations. The molecular defect resulting in inactive CYP2C18 is not yet known, and only phenotyping techniques can distinguish extensive metabolizers and poor metabolizers. To date, the clinical consequences of this polymorphism remain to be demonstrated.

(SEE ALSO: *Alcohol; Causes of Substance Abuse: Genetic Factors; Drug Interaction and the Brain; Vulnerability As Cause of Substance Abuse*)

BIBLIOGRAPHY

CHOLERTON, S., DALY, A. K., & IDLE, J. R. (1992). The role of individual human cytochromes P450 in drug metabolism and clinical response. *Trends in Pharmacology, 13*, 434–439.

EICHELBAUM, M., & GROSS, A. S. (1990). The genetic polymorphism of debrisoquine/sparteine metabolism—Clinical aspects. *Pharmacological Therapy, 46*, 377–394.

GONZALEZ, F. J., & NEBERT, D. W. (1990). Evolution of the P450 gene superfamily: Animal plant "warfare," molecular drive and human genetic differences in drug oxidation. *Trends in Genetics, 6*(6), 182–186.

VON MOLTKE, L. L., ET AL. (1994). Cytochromes in psychopharmacology. *Journal of Clinical Psychopharmacology, 14*(1), 1–4.

S. VICTORIA OTTON

PHARMACOKINETICS: GENERAL

Pharmacokinetics describes quantitatively the various steps of drug disposition in the body including absorbtion of drugs, distribution of the drugs to various organs, and their elimination by excretion and biotransformation. The rates of these processes are important in characterizing the fate of a medication in the body.

The actual percentage of a drug contained in a drug product that enters the circulation unchanged after its administration, combined with the rate of entry into the body, determines the *bioavailability* of a drug.

Once absorbed, most drugs are carried from their site of action and elimination by the circulating blood. Some drugs simply dissolve in serum water, but many others are carried bound to proteins, especially albumin. Plasma *protein binding* influences the fate of drugs in the body, since only the free (unbound) drug reaches the site of drug action. This interaction with binding sites is reversible.

The intensity of drug action is most frequently related to the concentration of the drug at the site of action. The duration of drug effect is related to the persistence of its presence at this site. The time to reach maximum drug concentrations (or peak effects) is usually referred to as t_{max}.

Whenever the fate of a drug in the body is described by pharmacokinetic parameters, a model of the body is assumed. The fundamental principles of pharmacokinetics are based on the most elementary model. The body is considered a single compartment. Distribution of the drug is considered uniform. The "volume" in which the drug is distributed is referred to as the *volume of distribution* (Vd). It is typically expressed in liters per kilogram (L/kg).

Elimination of the drug is assumed to be exponential. The *rate of elimination* of a drug is usually described by its *half-life* ($t_{1/2}$), which is the time required for 50 percent elimination of the drug. This is typically expressed in hours (h). Another way to express drug elimination is the *clearance*, which represents the volume of drug cleared from the body per unit of time. This is usually expressed in milliliters per minute per kilogram (ml/min/kg) but can also be expressed in liters per hour per kilogram (L/h/kg).

An effect of a single dose of a drug may be characterized by its latency, the time needed for drug concentrations to reach maximum levels (t_{max}). Mag-

TABLE 1
Pharmacokinetic Parameters of Opioids

Drug	Dosage/Route (mg)	Bioavail-ability (F) (%)	Protein Binding (%)	t_{max} (h)	Mean $t_{1/2}$ (h) (range)	Vd (L/kg)	Cl (ml/min/kg)
Butorphanol	2/IV	100 (IM)	80	0.75	3–4	5	385
Codeine	60 oral/IV	40–80 (oral)	7–53	1	3 (2.3–9.3)	2–6	15
Dextromethorphan		>50 (oral)	30–50	—	2–3 (estimated)	3–5	—
Heroin (3,6 diacelylmorphine) (see morphine)	4–16/IV	79 (oral)	—	—	3.0 min-utes	—	31
Buprenorphine	0.3/IV	40–90 (oral)	—	—	2–3	1–3	900–1,200 (ml/min)
Pentazocine	—	47 (oral)	65	—	4.5	7	17
Morphine	0.01/mg/kg	15–64 (oral) 100 (IM) 48 (rectal) 2 (epidural)	35	<1	2–4	3–4	12–21
Methadone	15–80	92 (oral)	40	<1	25 (13–47)	3.8	1.4
Meperidine	50–100/IM	50–60 (oral)	50–60	—	3–4	3–5	—
Propoxyphene	130	40–90 (oral)	—	1–2	2–15	—	—
Nalbuphine	—	16 (oral)	—	1–2	2–3	3–4	22
Naltrexone	—	5–40 (oral)	20	—	2–3	19	48

Vd = Volume of distribution   Cl = Clearance   IV = Intravenous   IM = Intramuscular

TABLE 2
Pharmacokinetic Parameters of Stimulant Drugs

Drug	Dosage/Route (mg)	Bioavail-ability (F) (%)	Protein Binding (%)	t_{max} (h)	Mean $t_{1/2}$ (h) (range)	Vd (L/kg)	Cl (ml/min/ kg)
Amphetamine	15–25/oral	—	23–26	1.25	14 (2–22)	6.1	0.2–0.6 (L/min)
Caffeine	1–5 mg/kg/ oral	100 (oral)	15–40	0.5–1	5 (1–10)	0.6	1
Cocaine	30–100/IV;IN	28–51 (IN)	7	0.5–1.5	0.8 (0.3–1.5)	2	11
Nicotine	0.25–2 (mg/kg/min)/ IV	30	5	—	2 (0.8–3.5)	1–2	18

Vd = Volume of distribution   Cl = Clearance   IV = Intravenous   IN = Intranasal

TABLE 3
Pharmacokinetic Parameters of CNS Depressants

Drug	Dosage/Route (mg)	Bioavail-ability (F) (%)	Protein Binding (%)	t_{max} (h)	Mean $t_{1/2}$ (h) (range)	Vd (L/kg)	Cl (ml/min/kg)
Alcohol (ethanol)	—	80 (oral)	—	<1	0.25	0.5	124 mg/kg/h
Benzodiazepines:							
Alprazolam	0.5–30/oral	90 (oral)	70	0.7–1.6	12 (6–18)	0.7–1.5	0.7–1.3
Bromazepam	0.25–3/oral	—	70	1	10–15	—	—
Chlordiazepoxide	20–50/oral IV, IM	100 (oral) PO or (IM)	95	0.5–3	10 (6–28)	0.3	0.5
Clobazam	10–20/oral	Good (oral)	90	1.3–1.7	25 (16–49)	0.9–1.8	0.36–0.63
Clonazepam	—	98	86	1–2	23 (20–80)	3.2	1.55
Clorazepate (see Desmethyldiazepam)	—	—	—	—	2.0	0.33	1.8
Desalkylflurazepam	—	—	—	1	75 (40–200)	22	4.5
Desmethyldiazepam	—	99	97	1–2	51 (51–120)	0.78	0.14
Diazepam	1–40/oral IM, IV	100 (oral) 50–60 (IM, rectal)	96	0.5–2	31 (14–61)	1 (0.9–3.0)	0.38–0.51
Flurazepam (see Desalkylflurazepam)	15–90/oral	—	97	—	—	—	—
Halazepam (see Desmethyldiazepam)	—	—	—	—	—	—	—
Lorazepam	2–4/oral	93 (oral) 90 (IM)	90	1.5	13 (8–25)	0.8–1.6	1 (0.8–1.3)
Midazolam	5–15/oral IV, IM	44 (oral)	95	0.3–0.7	2 1.4–5	0.8–17	6
Nitrazepam	15–30/oral	78 (oral)	87	2	26 (16–48)	1.2–2.7	0.86
Oxazepam	15–45/oral	97 (oral)	98	3 (0.5–8)	7 (5.1–13)	0.5–2.0	0.6–2.9
Prazepam (see Desmethyldiazepam)							
Temazepam	10–30/oral	>80 (oral)	98	0.8–4.7	12 (7–17)	1.3–1.5	1.0–3.4
Triazolam	0.25–1.0/oral	44 (oral) 53 s.l.	90	1.6	2.5 (2–5)	1.1	3.7–8.8

Vd = Volume of distribution   Cl = Clearance   IV = Intravenous   IM = Intramuscular   s.l. = Sublingual

nitude of peak effects and duration of action dosage and rates of absorption and elimination are influenced by these parameters. As dosage increases, latency is reduced and peak effect increased without change in the time of peak effect. Reduced elimination (long half-life, reduced clearance) results in an expected prolongation of drug effects and, in some cases, drug accumulation. Using more complex models than a single compartment model, physicians use pharmacokinetic data not only to characterize the fate of a drug in the body but also to calculate doses and frequency of drug administration for each

TABLE 4
Pharmacokinetic Parameters of Hallucinogens

Drug	Dosage/Route (mg)	Bioavail-ability (F) (%)	Protein Binding (%)	t_{max} (h)	Mean $t_{1/2}$ (h) (range)	Vd (L/kg)	Cl (L/min)
Marijuana (Δ⁹-tetrahydro-cannabinol)	0.5–30	8–24 (smoked) 4–12 (oral)	95–26	3–8 min	25 (19–57)	626(L)	0.2–1
Phencyclidine (PCP)	0.1–0.7/IV Inhaled	5–90	65	1.5	24 (7–51)	6.8	0.30 (0.14–0.77)

Vd = Volume of distribution   Cl = Clearance   IV = Intravenous

particular patient. This is important because there are wide variations among individuals in the absorption, distribution, and elimination of drugs.

Tables 1 through 4 are a summary of the available data on the kinetic properties of alcohol and other abused drugs. Some of the drugs of abuse included in this summary are illicit drugs (e.g., COCAINE) while others are effective pharmacological agents that have the potential to be abused (e.g., OPIOIDS).

Although some of the drugs included in the tables have been used for centuries (e.g., ALCOHOL, CAFFEINE), knowledge of their kinetics and metabolism is very recent and, in some cases, still incomplete. This is due partly to their complex metabolism and partly to the difficulties of studying drugs of abuse in humans.

The tables show the route of administration, the type of subjects used in the study, the doses used, and the most important kinetic parameters such as protein binding, half-life, volume of distribution, and clearance.

(SEE ALSO: *Drug Metabolism; Pharmacogenetics; Pharmacokinetics of Alcohol*)

USOA E. BUSTO

PHARMACOKINETICS: IMPLICATIONS FOR ABUSABLE SUBSTANCES

Pharmacokinetics is the study of the movements and rates of movement of drugs within the body, as the drugs are affected by uptake, distribution, binding, elimination, and biotransformation. An understanding of the biological basis of the clinical actions of abused drugs depends, in part, on knowledge of their neu-

rochemical and neuroreceptor actions that reinforce and sustain drug use (Hall, Talbert, & Ereshefsk, 1990). The pharmacokinetic properties of abusable substances represent a second important component of the database. The discipline of pharmacokinetics applies mathematical models to understand and predict the time course of drug amounts (doses) and their concentrations in various body fluids (Greenblatt, 1991, 1992; Greenblatt & Shader, 1985). Pharmacokinetic principles can be used to provide quantitative answers to questions involving the relationship of drug dosage and route of administration to the amount and time course of the drug present in systemic blood and at the receptor site of action.

Before an orally administered PSYCHOACTIVE DRUG can exert a pharmacological effect through its molecular recognition site in the brain, a number of events must take place (see Figure 1). The drug must reach the stomach and dissolve in gastric fluid. The stomach empties this solution into the proximal small bowel, which is the site of absorption of most medications. The drug must diffuse across the gastrointestinal mucosal barrier, reach the portal circulation, and be delivered to the hepatic (liver) circulation. (The liver detoxifies chemicals, including drugs.) Before reaching the systemic circulation, then, the absorbed drug must "survive" this initial exposure to the hepatic circulation—sometimes termed the "first-pass" through the liver (Greenblatt, 1993). After reaching the systemic blood, the drug is transported to the cerebral (brain) capillary circulation as well as to all other sites in the body that receive blood directly from the heart (cardiac output). The drug diffuses out of the cerebral capillary circulation, crosses the lipoidal (fatty) blood-brain bar-

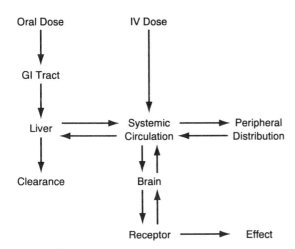

Figure 1

Schematic Representation of Physiological and Pharmacokinetic Events. These occur between administration of a centrally acting compound and the production of a pharmacological effect. If the medication is given orally, it must pass from the gastrointestinal (GI) tract to the portal circulation and to the liver before reaching the systemic circulation. Intravenous administration, however, yields direct access to the systemic circulation. Drugs of abuse may be taken by the intravenous route but are also taken by intranasal, intrabuccal, or inhalational routes, all of which will avoid the initial gastrointestinal-portal-hepatic exposure.

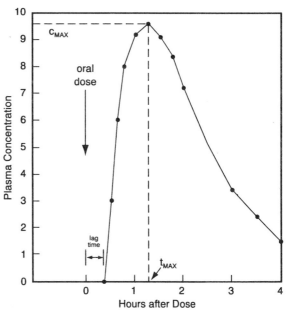

Figure 2

Schematic Plot of Plasma Concentration versus Time after Oral Dosage (given at time zero [arrow]). A lag time elapses between the time of administration and the beginning of appearance in the systemic circulation. Plasma levels then rise, reach a peak, and fall: c_{max} is the peak plasma concentration (9.6 units) and t_{max} is the time of peak concentration (1.25 hours after dosage).

rier, and reaches the extracellular water surrounding the neuroreceptor site of action. Only then is the drug available to interact with its specific molecular recognition site.

All of these processes take time, and some may serve as obstacles that delay or prevent the drug from reaching its site of action. Pharmacokinetic models incorporate the physiology of these processes, and can allow rational prediction of important clinical questions: How much drug reaches the brain? How fast does it get there? How long does it stay there?

DRUG ABSORPTION

The term *lag time* refers to the time elapsing between ingestion of an oral medication and its first appearance in the systemic circulation (see Figure 2). For most drugs, it generally falls between 5 and 45 minutes. For ethanol (drinking ALCOHOL, which

is also called ethyl alcohol), however, the lag time may be very short, because the drug is already a liquid at the time it is ingested, and a significant component of absorption probably occurs across the gastric mucosa as well as in the proximal small bowel (Frezza et al., 1990). The physicochemical features of the drug contribute importantly to the time necessary for dissolution and therefore to the lag time. All else being equal, drugs in solution have shorter lag times than those administered in suspension form; they are, in addition, more rapidly absorbed than capsule preparations and, finally, tablet preparations. For any given solid dosage form, lag time and absorption rate are likely to be shorter if the drug particles are more finally subdivided. Sustained-release (time-release) drug formulations are deliberately prepared to have long lag times and slow absorption rates, thereby avoiding drug effects associated with the peak concentration.

Absorption rate refers to the time necessary for the drug to reach the systemic circulation once the absorption process actually begins. Pharmacokinetic models can be applied to assign a half-life value to the process of absorption. Values of absorption half-life tend, however, to be of low statistical stability, and it is increasingly common to characterize the absorption process using the observed peak plasma concentration (c_{max}) and time of peak concentration (t_{max}). The t_{max} is actually a composite of the lag time plus the time necessary to reach peak concentration once absorption starts (Figure 2). In general, fast absorption implies a high value of c_{max} and a short value of t_{max}; slow absorption implies a long t_{max} and a low c_{max}. Again, sustained-release drug preparations are deliberately formulated to produce long lag times and slow absorption, thereby delaying and reducing the c_{max} after an oral dose. Drug absorption tends to be slower when medications are taken during or just after a meal, rather than in the fasting state (before a meal, on an empty stomach).

For these reasons, the ethanol in alcoholic beverages is relatively rapidly absorbed after oral ingestion. The popular lore that alcohol has a greater effect when taken on an empty stomach probably has a physiological basis, since peak concentrations will be higher and earlier when alcohol is taken in the fasting state. BENZODIAZEPINE derivatives (tranquilizers) clearly are not primary drugs of abuse and are seldom subject to misuse by the great majority of patients; however, benzodiazepines may be taken for nontherapeutic purposes by some substance abusers (Woods, Katz, & Winger, 1987, 1992; Shader & Greenblatt, 1993). The preference of specific benzodiazepines by drug abusers appears to be closely related to their rate of absorption. That is, rapidly absorbed benzodiazepines, leading to relatively high values of c_{max} shortly after dosage, appear to be preferred by drug abusers. The benzodiazepine diazepam (Valium), for example, is much more rapidly absorbed than is oxazepam (Serax or Serenid). In controlled laboratory settings, diazepam is more easily recognized as a potentially abusable substance by experienced drug users, and it is also preferred by this group to oxazepam (Griffiths et al., 1984a, 1984b). This preference also appears to be supported by epidemiological studies of PRESCRIPTION DRUG misuse (Bergman & Griffiths, 1986).

Some orally administered medications reach the systemic (blood) circulation in small or even negligible amounts relative to the dose ingested. In-

complete absorption from the gastrointestinal tract sometimes explains this. However, oral medications may be poorly available to the systemic circulation even if they are well absorbed. This is explained by the phenomenon termed *presystemic extraction*, which results from the unique anatomy and physiology of the gastrointestinal circulation (Greenblatt, 1993). Orally administered medications are absorbed into the portal rather than systemic circulation (Figure 3), and portal blood drains directly into the liver. Many drugs that are avidly metabolized in the liver may therefore undergo substantial biotransformation before reaching systemic blood. Some drugs may also be metabolized by the gastrointestinal (GI) tract mucosa. First-pass hepatic metabolism together with GI tract metabolism is collectively termed presystemic extraction. COCAINE, for example, is not favored as a drug of abuse by the oral route, because

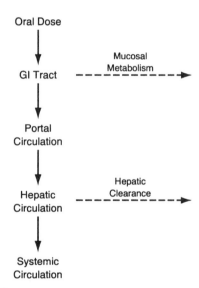

Figure 3

Possible Mechanisms of Presystemic Extraction. Orally administered medications may undergo metabolism as they pass through the gastrointestinal tract mucosa (dashed arrow), which contains significant amounts of Cytochrome P450-3A4. Mucosal metabolism of cyclosporine appears to occur in humans (Kolars et al., 1991). Metabolism may also occur as drug present in portal blood passes through the hepatic circulation (dashed arrow); this is termed "first-pass" metabolism. The net extent of presystemic extraction depends on the combination of mucosal metabolism and first-pass metabolism.

of nearly complete presystemic extraction, allowing only small amounts of the intact drug to reach the systemic circulation (Jatlow, 1988; Jeffcoat et al., 1989).

DRUG DISTRIBUTION

The process of distribution is an important determinant of pharmacokinetic properties, as well as the time course of action, of most centrally acting drugs, including those that are subject to abuse. Drugs reversibly distribute not only to their site of action in the brain but also to peripheral sites such as adipose (fat) tissue and muscle, where they are not pharmacologically active (figure 1). Only a small fraction of the total amount of a psychotropic drug in the body goes to the brain. An even smaller fraction actually binds to the specific molecular recognition site (receptor). The extent of distribution of a psychotropic drug is determined in part by lipid (fat) solubility (how well a substance dissolves in oils and fats; lipophilicity), which is related to molecular structure and charge. Most psychotropic drugs are highly lipid-soluble. Drug distribution is also determined by some characteristics of the organism: the relative amounts of adipose and lean tissue, blood flow to each individual tissue, and the extent of drug that binds to plasma protein. The overall extent of drug distribution throughout the body can be quantified by the pharmacokinetic volume of distribution, which is a ratio—the total amount of drug present in the body divided by the concentration in a reference compartment, usually serum or plasma. Lipid-soluble psychotropic drugs, as well as drugs of abuse, typically have very large pharmacokinetic volumes of distribution, which may exceed body size by tenfold or more. Although the drug cannot actually distribute to a space larger than the body, low plasma concentrations resulting from extensive uptake into peripheral tissues can yield a large apparent pharmacokinetic volume of distribution (Figure 4).

Drug distribution influences both the onset and the duration of drug action—as well as the observed value of elimination HALF-LIFE. After an intravenous

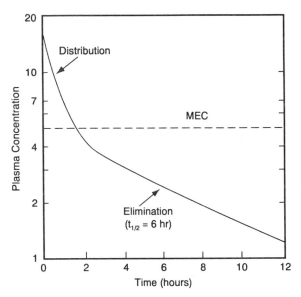

Figure 5
Plasma Concentrations of a Hypothetical Lipid-Soluble Drug after Intravenous Injection.
Disappearance from plasma is biphasic. The initial rapid phase is mainly due to drug distribution from central to peripheral compartments (see Figure 4). The slower phase of elimination is mainly due to clearance. For this drug, the elimination half-life in the postdistributive phase is 6 hours. If a plasma concentration of 5 units represents the minimum effective concentration (MEC) below which the drug exerts no detectable pharmacological effect, this drug in the dosage administered has a duration of action of approximately 2 hours.

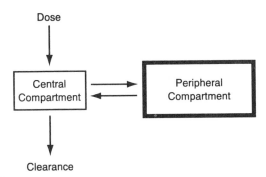

Figure 4
Schematic Diagram of the Two-Compartment Model. It is assumed that medications are administered into and cleared from the central compartment only, and that only the central compartment (which includes blood) is accessible to measurement. Reversible distribution occurs between central and peripheral compartments. For most psychotropic drugs, high lipid solubility favors distribution to the peripheral compartment, producing a large apparent pharmacokinetic volume of distribution.

(IV) injection, lipid solubility allows for the rapid crossing of the lipoidal blood-brain barrier, leading to a rapid onset of pharmacological action (drug effect). In behavioral terms, then, drug-taking produces immediate reinforcement. The duration of a drug's action, however, is determined mainly by the extent of its peripheral distribution. Plasma levels of lipid-soluble psychotropic drugs will decline rapidly and extensively after a single intravenous dose, because of peripheral distribution rather than elimination or clearance (Figure 5). A similar principle holds after oral administration of rapidly absorbed drugs (de Wit & Griffiths, 1991). Since duration of action after a single dose is determined more by distribution than by elimination or clearance, it is generally not accurate to equate elimination half-life and duration of action.

CLEARANCE AND ELIMINATION

The terms *clearance* and *elimination half-life* are commonly used to describe the bodily process of drug removal or disappearance. These two concepts are related but are not identical. Clearance is the most important, since it is a unique independent variable that best describes the capacity of a given organism to remove a given drug from its system. Clearance has units of volume divided by time—for example, milliliters/minute (ml/min) or liters/hour (L/h)—and is the total amount of blood, serum, or plasma from which a substance is completely removed per unit of time. Clearance is *not* identical either to the rate of drug removal or to the elimination half-life. For most psychotropic drugs, clearance is accomplished by the liver via processes of biotransformation that change the administered drug into one or more metabolic products (Figure 6); this is commonly called detoxification by the liver. The metabolites may appear in the urine, but the liver is still the organ that effects clearance. For drugs cleared exclusively by the liver, the numerical value of clearance cannot exceed hepatic blood flow.

Elimination half-life is described in units of time; it can be seen as the time necessary for the plasma concentration to fall by 50 percent after distribution equilibrium has been attained. The elimination phase of drug disappearance—at which time the concept of elimination half-life is applicable—may not be attained until completion of an initial phase of rapid drug disappearance resulting from periph-

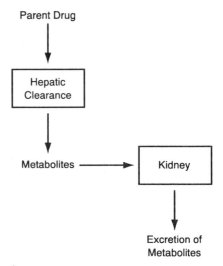

Figure 6
Psychotropic Drugs: Most, including drugs of abuse, are cleared via the liver by hepatic biotransformation to metabolic products. The metabolites may then be released into the circulation and excreted by the kidney.

eral distribution (see Figure 5). As discussed earlier, the duration of action of a single dose of a psychotropic drug is not necessarily related to its elimination half-life.

Pharmacokinetic theory yields the following relationship between a drug's elimination half-life, volume of distribution (Vd), and clearance:

$$\text{Elimination half-life} = \frac{0.693 \cdot Vd}{\text{clearance}}$$

The independent variables, appearing on the right side of the equation, are *Vd*, the physicochemically determined property reflecting the extent of distribution, and *clearance*, having units of volume divided by time, quantifying the capacity for drug removal. Elimination half-life is dependent on both of these. Note that a drug may have long elimination half-life, due either to a large Vd, a low clearance, or both.

PHARMACOKINETICS VERSUS PHARMACODYNAMICS

In contrast to pharmacokinetics, PHARMACODYNAMICS is the quantitative study of the time course of drug action. If drug distribution to the site of action occurs by passive diffusion from the systemic circulation, and if the intensity of drug action depends on the degree of RECEPTOR occupancy both in time and in quantity, then pharmacokinetics and pharmacodynamics are necessarily related. Kinetic-dynamic modeling, discussed in detail elsewhere (Greenblatt & Harmatz, 1993), addresses this relationship mathematically, by directly evaluating concentration versus effect. In the fields of psychopharmacology and substance abuse, kinetic-dynamic modeling is a major challenge, since (1) clinical drug effect (pharmacodynamic response) often is difficult to measure reliably and since (2) measured drug concentrations in systemic serum or plasma do not always parallel those at the central site of action. Nonetheless, recent advances in kinetic-dynamic modeling have significantly advanced our understanding of the relationship of the pharmacokinetics of psychotropic drugs to their pharmacodynamic effects.

IMPLICATIONS FOR TESTING OF URINE FOR SUBSTANCES OF ABUSE

Mandatory unannounced testing of urine samples for illegal drugs of abuse is conducted to detect and deter the use of these drugs, as well as to prevent potentially dangerous impairment of performance. The application of the fundamental principles of pharmacokinetics and pharmacodynamics, however, clearly indicates that urine testing is the wrong way to approach these objectives (Greenblatt, 1989; Greenblatt & Shader, 1990).

HEROIN, cocaine, and MARIJUANA, the principal illegal drugs of abuse, are subject to hepatic clearance, so urinary excretion is in the form of drug metabolites rather than the originally taken parent compounds (Agurell et al., 1986; Jatlow, 1988) (see Figure 6). As such, analytical methods for chemical testing of urine samples must be devised to detect these metabolites (Friedman & Greenblatt, 1986) (see Table 1). Screening IMMUNOASSAYS are notoriously insensitive, and many actual drug users will

escape detection by the screening test if the urine concentrations are below an arbitrary cutoff (Burnett et al., 1990). Negative tests can also be produced by dilution of urine via water loading (Lafolie et al., 1991) or by a variety of adulterants that interfere with analytical procedures (Schwarzhoff & Cody, 1993; Mikkelson & Ash, 1988). To complicate matters, immunoassays are nonspecific and have an unacceptably high false-positive rate. Most urine-testing programs deal with the false-positive problem by performing confirmatory tests on all positive results from the initial screening (Figure 7). However, even a positive test that is confirmed by gas chromatography/mass-spectroscopy does not conclusively identify that individual as a drug user. Positive urine tests may be produced by passive inhalation or dermal absorption, as (ironically) may occur in law-enforcement officials engaged in drug-enforcement activities (Baselt, Chang, & Yoshikawa, 1990; Elsohly, 1991). Recent evidence suggests that some nondrug-using individuals may excrete heroin metabolites resulting from foodstuffs (poppy-seed cake) or from endogenous metabolism (Hayes, Krasselt, & Mueggler, 1987; Mikus et al., 1994). Thus evaluation of the problems of analytical chemistry inherent in urine testing indicates that a negative test cannot rule out illegal drug exposure, nor can a positive test confirm it.

From a pharmacokinetic–pharmacodynamic standpoint, urine is an excretory product and not a body fluid. Urine concentrations of drug metabolites bear little relation to parent-drug concentrations in blood or at the site of action—the concentrations that actually determine pharmacodynamic effect (Osterloh, 1993). Even if chemically accurate, a "positive" urine test for a substance of abuse provides no useful information on the quantity of drug exposure, the duration or chronicity of exposure, or the pharmacodynamic effect of the drug at the time the urine sample was taken, or any time prior to or after that. A positive test does not confirm intoxication or im-

TABLE 1
Principal Urinary Metabolites of Potentially Abusable Drugs

Parent Drug	Urinary Metabolite
Marijuana (Tetrahydrocannabinol, THC)	11-nor-delta-9-THC-9-carboxylic acid
Cocaine	Benzoylecgonine
Heroin	Morphine glucuronide

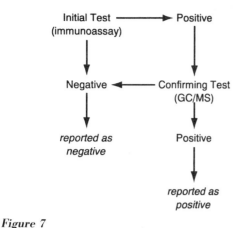

Figure 7

Urine-Testing Programs. Those for drugs of abuse typically use a two-tiered algorithm. An initial screening test is done with a relatively inexpensive, nonspecific, and insensitive immunoassay (such as enzyme-multiplied-, fluorescence-polarization-, or radioimmunoassay). If the initial test is negative, the result is reported as such, and no further testing is done. If the initial screen is positive, a second analysis is done on the same sample using a more accurate and specific method, such as gas-chromatography/mass-spectroscopy (GC/MS). If the confirmation test is negative, the result is reported as negative. If GC/MS confirms the initial screening test, the result is reported as positive.

pairment from that drug at any time, nor does a negative test rule them out. Thus, as a general rule, urine-testing programs are without adequate scientific foundation and cannot possibly attain the stated objectives (Greenblatt, 1989; Greenblatt and Shader, 1990; Sutherland, 1992). This does not mean that carefully controlled tests do not exist—for a discussion of this see DRUG TESTING AND ANALYSIS.

Detection and prevention of performance impairment in the workplace can, however, be achieved by the systematic testing of performance, using validated methods under properly controlled conditions. Such testing procedures would detect potentially dangerous impairment not only from illegal drugs of abuse but also from other causes, including use of legal substances (such as alcohol or antihistamines), sleep deprivation, other medical or psychiatric illness, or episodes of interpersonal stress. Chemical analysis of blood (not urine) could provide chemical confirmation for cases in which drug-induced performance impairment is suspected, provided a re-

search database is available to link blood concentrations to probable impairment, as exists in the case of alcohol (ethanol). Such an approach would provide a fair and direct method of coping with this problem.

COMMENT

A comprehensive approach to understanding the biological bases of substance abuse must combine the neurochemical and molecular mechanisms that underlie the behavioral effects of these drugs, as well as understanding their properties of absorption, distribution, and clearance. Advances were made in the 1980s and will continue to be made as research techniques in both disciplines become increasingly refined.

ACKNOWLEDGMENTS

Research supported by Grants DA-05258 and MH-34223 from the DEPARTMENT OF HEALTH AND HUMAN SERVICES.

The author is grateful for the collaboration of RICHARD I. SHADER, LAWRENCE G. MILLER, JEROLD S. HARMATZ, and DOMENIC A. CIRAULO.

(SEE ALSO: *Abuse Liability of Drugs: Testing in Humans; Benzodiazepines; Benzodiazepines: Complications; Pharmacokinetics of Alcohol; Psychomotor Effects of Alcohol and Drugs*)

BIBLIOGRAPHY

AGURELL, S., ET AL. (1986). Pharmacokinetics and metabolism of delta-one-tetrahydrocannabinol and other cannabinoids with emphasis on man. *Pharmacological Reviews, 38,* 21–43.

BASELT, R. C., CHANG, J. Y., & YOSHIKAWA, D. M. (1990). On the dermal absorption of cocaine. *Journal of Analytical Toxicology, 14,* 383–384.

BERGMAN, U., & GRIFFITHS, R. R. (1986). Relative abuse of diazepam and oxazepam: Prescription forgeries and theft/loss reports in Sweden. *Drug and Alcohol Dependence, 16,* 293–301.

BURNETT, D., ET AL. (1990). A survey of drugs of abuse testing by clinical laboratories in the United Kingdom. *Annals of Clinical Biochemistry, 27,* 213–222.

DE WIT, H., & GRIFFITHS, R. R. (1991). Testing the abuse liability of anxiolytic and hypnotic drugs in humans. *Drug and Alcohol Dependence, 28,* 83–111.

ELSOHLY, M. A. (1991). Urinalysis and casual handling of marijuana and cocaine. *Journal of Analytical Toxicology, 15,* 46.

FREZZA, M., ET AL. (1990). High blood alcohol levels in women: The role of decreased gastric alcohol dehydrogenase activity and first-pass metabolism. *New England Journal of Medicine, 322,* 95–99.

FRIEDMAN, H., & GREENBLATT, D. J. (1986). Rational therapeutic drug monitoring. *Journal of the American Medical Association, 256,* 2227–2233.

GREENBLATT, D. J. (1993). Presystemic extraction: Mechanisms and consequences. *Journal of Clinical Pharmacology, 33,* 650–656.

GREENBLATT, D. J. (1992). Pharmacokinetic principles in clinical medicine (Clinical Therapeutics Conference). *Journal of Clinical Pharmacology, 32,* 118–123.

GREENBLATT, D. J. (1991). Benzodiazepine hypnotics: Sorting the pharmacokinetic facts. *Journal of Clinical Psychiatry, 52* (No. 9, Supp.), 4–10.

GREENBLATT, D. J. (1989). Urine drug testing: What does it test? *New England Law Review, 23,* 651–666.

GREENBLATT, D. J., & HARMATZ, J. S. (1993). Kinetic-dynamic modeling in clinical psychopharmacology. *Journal of Clinical Psychopharmacology, 13,* 231–234.

GREENBLATT, D. J., & SHADER, R. I. (1990). Say "no" to drug testing. *Journal of Clinical Psychopharmacology, 10,* 157–159.

GREENBLATT, D. J., & SHADER, R. I. (1985). *Pharmacokinetics in Clinical Practice.* Philadelphia: Saunders.

GRIFFITHS, R. R., ET AL. (1984a). Relative abuse liability of diazepam and oxazepam: Behavioral and subjective dose effects. *Psychopharmacology, 84,* 147–154.

GRIFFITHS, R. R., ET AL. (1984b). Comparison of diazepam and oxazepam: Preference, liking and extent of abuse. *Journal of Pharmacology and Experimental Therapeutics, 229,* 501–508.

HALL, W. C., TALBERT, R. L., & ERESHEFSKY, L. (1990). Cocaine abuse and its treatment. *Pharmacotherapy, 10,* 47–65.

HAYES, L. W., KRASSELT, W. G., & MUEGGLER, P. A. (1987). Concentrations of morphine and codeine in serum and urine after ingestion of poppy seeds. *Clinical Chemistry, 33,* 806–808.

JATLOW, P. (1988). Cocaine: Analysis, pharmacokinetics, and metabolic disposition. *Yale Journal of Biology and Medicine, 61,* 105–113.

JEFFCOAT, A. R., ET AL. (1989). Cocaine disposition in humans after intravenous injection, nasal insufflation (snorting), or smoking. *Drug Metabolism and Disposition, 17,* 153–159.

KOLARS, J. C., ET AL. (1991). First-pass metabolism of cyclosporin by the gut. *Lancet, 338,* 1488–1490.

LAFOLIE, P., ET AL. (1991). Importance of creatinine analyses of urine when screening for abused drugs. *Clinical Chemistry, 37,* 1927–1931.

MIKKELSEN, S. L., & ASH, K. O. (1988). Adulterants causing false negatives in illicit drug testing. *Clinical Chemistry, 34,* 2333–2336.

MIKUS, G., ET AL. (1994). Endogenous codeine and morphine in poor and extensive metabolisers of the CYP2D6 (debrisoquine/sparteine) polymorphism. *Journal of Pharmacology and Experimental Therapeutics, 268,* 546–551.

OSTERLOH, J. (1993). Testing for drugs of abuse. *Clinical Pharmacokinetics, 24,* 355–361.

SCHWARZHOFF, R., & CODY, J. T. (1993). The effects of adulterating agents on FPIA analysis of urine for drugs of abuse. *Journal of Analytical Toxicology, 17,* 14–17.

SHADER, R. I., & GREENBLATT, D. J. (1993). Use of benzodiazepines in anxiety disorders. *New England Journal of Medicine, 328,* 1398–1405.

SUTHERLAND, R. (1992). Mandatory drug testing: Boon for public safety or launch of a witch-hunt? *Canadian Medical Association Journal, 146,* 1215–1220.

WOODS, J. H., KATZ, J. L., & WINGER, G. (1992). Benzodiazepines: Use, abuse, and consequences. *Pharmacological Reviews, 44,* 151–347.

WOODS, J. H., KATZ, J. L., & WINGER, G. (1987). Abuse liability of benzodiazepines. *Pharmacological Reviews, 39,* 251–419.

DAVID J. GREENBLATT

PHARMACOKINETICS OF ALCOHOL

The discipline known as pharmacokinetics deals with the way drugs are absorbed, distributed, and eliminated by the body and how these processes can be described in quantitative terms. The pharmacokinetics of alcohol (ethyl alcohol or ethanol) is an important issue in forensic toxicology and clinical medicine, when the amount of alcohol in the body is estimated from the concentration measured in a blood sample.

The Swedish scientist Erik M.P. Widmark (1889–1945) made pioneer contributions to knowledge about the pharmacokinetics of ethanol during the early decades of the twentieth century. Widmark observed that after the peak concentration in blood had been reached, the disappearance phase seemed to follow a near straight-line course, suggesting that the system for metabolizing alcohol was saturated (fully occupied), so that the amount of alcohol metabolized each hour did not depend on the amount in the blood. This situation is termed *a zero-order elimination process*. (Zero-order kinetics is contrasted with first-order kinetics, in which the metabolic system [e.g., the liver] is not saturated and in which the amount of drug metabolized per hour increases as the amount presented to the metabolic system increases.) Figure 1 (left frame) depicts zero-order elimination kinetics of ethanol after rapid intravenous infusion. Widmark used the Greek letter β to represent the negative slope of the disappearance phase and not the notation k_o used in Figure 1. The terminology and choice of symbols used in articles and books dealing with clinical pharmacokinetics are often confusing. Moreover, the concentrations of ethanol in blood and other body fluids are reported

using many different units, such as g% w/v, mg/dl, g/l, mmol/l; 21.7 mmol/l = 100 mg/dl = 1 g/l = 0.1 g% w/v.

Zero-order kinetics implies that the elimination rate of ethanol is independent of the BLOOD ALCOHOL CONCENTRATION (BAC) and therefore k_o should be the same regardless of the dose of ethanol administered; however, more recent studies have show that the slope of the BAC decay phase is steeper after larger doses of ethanol are ingested. Furthermore, when the BAC declines below about 10 mg/dl (0.01 g%, 2.17 mmol/l) the elimination curve of ethanol from blood flattens out and changes into a curvilinear decay profile.

Two different methods are described in the literature to portray the pharmacokinetics of ethanol. The method of choice seems to depend on the professional interests, the scientific background, and the training of those concerned. Specialists in foren-

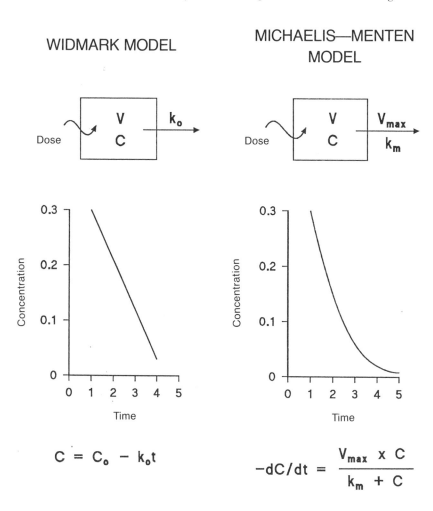

WIDMARK MODEL

MICHAELIS—MENTEN MODEL

$$C = C_o - k_o t$$

$$-dC/dt = \frac{V_{max} \times C}{k_m + C}$$

Figure 1

Elimination Kinetics of Ethanol Schematic diagram illustrating the elimination kinetics of ethanol. The left frame shows Widmark's zero-order model. The right frame shows Michaelis-Menten (MM) capacity-limited kinetics. An intravenous bolus dose of ethanol enters a volume V to produce a concentration C; k_o is the zero-order elimination rate constant; V_{max} is the maximum velocity of the reaction; and k_m is the Michaelis constant—the concentration of ethanol at half maximum velocity. Concentration-time profile are shown for zero-order and MM kinetics, and the mathematical expressions for the elimination rates are given.

sic medicine and toxicology, as well as other disciplines, favor the mathematical approach developed by Widmark. In contrast, scientists with their basic training in pharmacy and pharmacology prefer Michaelis-Menten (MM) kinetics, that is, saturable or capacity-limited enzyme kinetics. The MM model is depicted in Figure 1 (right frame) after intravenous input of ethanol. A pseudolinear phase is evident for most of the elimination profile, provided that the BAC remains sufficiently high ($>$ 10 mg/dl). At low substrate concentrations (C), a hockey-stick shape develops when data are plotted on cartesian graph paper. Accordingly, when C is much greater than k_m, the elimination rate approaches its maximum velocity; $-dC/dt = V_{max}$ (Figure 1, right frame). When C is less than k_m the elimination rate is proportional to the substrate concentration; $-dC/dt = (V_{max}/k_m) \cdot$ C and the MM equation collapses into first-order kinetics. This collapsing of the model is a consequence of capacity-limited kinetics and does not reflect any sudden change in the order of the biochemical reaction.

ETHANOL AS A DRUG

Ethanol differs from most other drugs in the way it is absorbed into the blood, metabolized in the liver, and how it enters the brain and produces its pharmacological effect. Ethanol (CH_3CH_2OH) has a molecular weight of 46.05, mixes with water in all proportions and carries only a weak charge; this means that the molecules of ethanol easily pass through biological membranes, including the blood-brain barrier. After absorption into the portal blood, ethanol passes through the liver, where enzymes begin the conversion into acetaldehyde and acetate. The end products of ethanol metabolism are carbon dioxide and water. The concentrations of ethanol in biological specimens depend on the dose ingested, the time after drinking, and the water content of the materials analyzed. The concentration-time profiles of ethanol and the pharmacokinetic parameters will differ depending on whether plasma, serum, urine, or saliva are the specimens analyzed. Several detailed reviews of ethanol pharmacokinetics are available and included in the bibliography.

Information about the absorption kinetics of ethanol is much less extensive than that about elimination kinetics. Unlike most other drugs, the dose of ethanol is not swallowed instantaneously because the drinking is usually spread over a period of time. For research purposes, however, ingestion of a bolus dose usually infers drinking times of five to fifteen minutes. The dosage form of ethanol, whether ingested as beer (3–6% w/v), wine (9–12% w/v), spirits (32–40% w/v), or as a cocktail (15–25% w/v) might influence the pharmacokinetic parameters. Absorption of ethanol starts in the stomach where about 20 percent of the dose can become absorbed. The remainder is absorbed from the upper part of the small intestine. The speed of absorption of alcohol depends to a large extent on the rate of gastric emptying, which varies widely among different subjects. Assuming that the rate of absorption from the gut is a first-order process, one can represent the entire concentration-time profile of ethanol with a single equation:

$$C = C_o(1 - e^{-kt}) - k_o t$$

Where C = BAC at some time t after administration
C_o = Initial back extrapolated BAC (see Figure 2)
k = First-order absorption rate constant
k_o = Zero-order elimination rate constant
t = Time after drinking

The peak BAC and the time of reaching the peak after drinking are important aspects of the absorption kinetics. Table 1 gives examples of these parameters after healthy men drank neat whiskey (40% v/v or 80 proof) on an empty stomach. The absorption of ethanol occurs more slowly from the stomach than from the intestine owing to the enormous difference in the absorption surface available. Factors that influence gastric emptying, such as food in the stomach before drinking, will alter the rate of absorption and the peak BAC reached. The absorption of ethanol occurs progressively during a drinking binge or spree, and studies have shown that the BAC fifteen minutes after the last drink has reached about 80 percent of the final peak BAC. Because of the saturation-type kinetics, the peak BAC and the area under the curve (AUC) increase more than expected from proportional increases in the dose. The slower the rate of delivery of ethanol to the liver the smaller the AUC for a given dose and vice versa. The systemic availability (bioavailability) of drugs like ethanol with dose-dependent kinetics should not be calculated from the ratio of AUC after oral and intravenous administration.

TABLE 1
Peak Blood Alcohol Concentration and Time to Reach the Peak after End of Drinking

Dose g/kg[1]	N	Peak BAC mg/dl		k_o mg/dl h		Time to peak (min)[3]			
		mean	(range)	mean	(range)[2]	10	40	70	100
0.34	6	56	(43–67)	12	(9–14)	5	1	—	—
0.51	16	74	(54–91)	13	(10–14)	11	3	1	1
0.68	83	92	(52–136)	13	(9–17)	33	26	21	3
0.85	44	120	(83–178)	15	(12–18)	13	24	7	—

Maximum concentration of ethanol in capillary (fingertip) blood and the time of reaching the peak after end of drinking. The zero-order rate of elimination of ethanol from blood (k_o) is also given. The subjects drank neat whiskey within 15–25 minutes after an overnight 10-hour fast.

[1]g ethanol/kg = 0.036 oz ethanol/kg.
[2]Zero-order elimination rate.
[3]Number of subjects reaching their peak BAC at 10, 40, 70 and 100 min., measured from end of drinking.

THE WIDMARK EQUATION

Figure 2 gives examples of the concentration-time profiles of ethanol obtained from oral and intravenous administration of a moderate dose. The ratio of the dose administered (D) to the initial extrapolated concentration of ethanol in blood (C_o) is the apparent volume of distribution (V_d) having dimensions L/kg. This defines the relationship between the concentration of ethanol spread over the body weight (in kilograms, kg) and the concentration in the blood.

$$C_o = D/(kg \times V_d)$$
$$D = C_o \times kg \times V_d \qquad [1]$$

Equation [1] is known as the *Widmark equation*; it is widely used to estimate alcohol in the body from measurements of alcohol in the blood. Widmark found that the average V_d for men was 0.68, with a range from 0.51–0.85, but in women the volume of distribution was less—with an average of 0.55 and a range of 0.44–0.66. These differences between the sexes stem from differences in body-tissue composition; proportionally, women carry more fat but less water than do men. Accordingly, women reach higher BACs than men if the same dose of ethanol is given according to body weight. A similar observation was made in studies of men with widely different ages, because body water decreases in the elderly. By dividing the dose of ethanol administered (g/kg) by the time needed to reach zero BAC (time$_o$) one obtains an estimate of the rate of clearance of ethanol from the body. This calculation neglects the nonlinear phase of ethanol elimination beginning at BAC below 10 mg/dl but does include the contribution from any first-pass metabolism occurring in the liver and gut.

If equation [1] is combined with the expression for zero-order elimination kinetics ($C = C_o - k_o t$) rearrangement gives equations 2 and 3:

$$D = kg \times V_d \times (C + k_o t) \qquad [2]$$

or

$$C = D/(kg \times V_d) - (k_o t) \qquad [3]$$

Equation [2] can be used to estimate the amount (dose D) of alcohol a person has consumed from knowledge of his or her BAC (C). Similarly, equation [3] allows estimating the BAC (C) that might exist after drinking a known amount of ethanol. For best results when using these equations, absorption and distribution of ethanol must be complete at the time of sampling blood. Owing to inter- and intra-individual variations in the pharmacokinetic parameters V_d and k_o the results obtained are subject to considerable uncertainty. This uncertainty should be allowed for when these calculations are made for legal purposes, for example, in trials concerned with DRIVING UNDER THE INFLUENCE of alcohol. A variability of ± 20 percent seems appropriate for most situations.

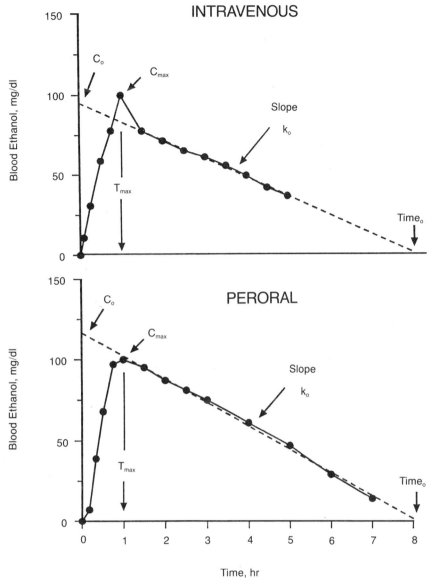

Figure 2
Examples of Concentration-Time Profiles of Alcohol Taken by Intravenous and Oral Routes of Administration
Examples of concentration-time profiles of ethanol obtained after intravenous infusion of 0.4 g ethanol/kg body weight in 15 minutes (upper part) and after ingestion of 0.8 g/kg (lower part). Several key pharmacokinetic parameters are shown.

RESEARCH ON ADH

The enzymes responsible for ethanol oxidation are mostly located in the liver, but recent research has focused on the existence of alcohol dehydrogenase (ADH)—the enzyme that transforms alcohol to acetaldehyde—in the gastrointestinal mucosa. Gastric ADH seems to be less effective in oxidizing ethanol in women (than in men) and in alcoholics (than in moderate drinkers). When a moderate dose of ethanol was ingested on an empty stomach, first-pass metabolism was negligible. This was explained by the ethanol bypassing gastric ADH, owing to rapid absorption occurring. However, the quantitative significance of gut metabolism in the overall disposal of ethanol remains controversial.

ELIMINATION RATES AND ENZYMES

Differences in the rate of disappearance of ethanol from blood might depend on genetic and environmental factors influencing an individual's catalytic activity of alcohol-metabolizing enzymes. In humans, the enzyme ADH occurs in multiple molecular forms, designated class I, II, and III. Class I enzymes are located mainly in the liver cytosol and have a low k_m for ethanol. Various isozymes (variations within a class) exist and β_1-ADH (class I) is predominant in Caucasians whereas β_2-ADH (class II) is the most abundant isozyme in Asians. The rate of ethanol elimination in the various racial groups is not much different from the variations seen within a single racial group in well-designed studies that allow for racial differences in body composition—the proportion of fat to lean body mass.

Alcoholics have a greater capacity to eliminate ethanol than do moderate drinkers. Disappearance rates from blood of 30 mg/dl/h are not uncommon—compared with a mean rate of only 15 mg/dl/h (range 8–20 mg/dl/h) in moderate drinkers. The liver microsomes contain enzymes capable of oxidizing ethanol as well as other drugs, organic solvents, and environmental chemicals. One particular form of the cytochrome P_{450} enzyme (denoted P450IIEI) metabolizes ethanol. This microsomal ethanol oxidizing system (MEOS) has a k_m of 40–60 mg/dl (8.7–13 mmol/l) compared with 2–5 mg/dl (0.4–1 mmol/l) for human ADH. More importantly, the P450IIEI isozyme becomes more active during prolonged exposure to ethanol—a process known as enzyme induction. Accordingly, because of continuous heavy drinking, alcoholics develop a high capacity for eliminating ethanol from the blood. Their enhanced capacity vanishes after a short period of abstinence, however, but liver disease (hepatitis, cirrhosis) in alcoholics does not seem to impair their ability to dispose of ethanol.

BEHAVIORAL EFFECTS OF ALCOHOL

Studies have shown that the behavioral effects of ethanol and its associated impairment of performance are more pronounced when the BAC is rising than when it is falling. This observation seems to depend, at least in part, on the distribution of ethanol between blood and tissue. The arterial blood concentration of ethanol is pumped to the brain and this exceeds the concentration measured in the venous blood, which is returning to the heart from skeletal muscles. This arterio-venous difference is most pronounced shortly after drinking; it decreases as ethanol diffuses equally into all body fluids. It seems

that this is not the whole story, because some evidence points to the development of acute cellular tolerance to ethanol's effects—an aspect of tolerance that quickly develops.

Despite extensive studies of ethanol pharmacokinetics spanning many years, there are still a number of unsettled issues and areas of debate. Two such issues are (1) the practical advantages of Michaelis-Menten kinetics as opposed to Widmark's zero-order model and (2) the role of gastric ADH in presystemic disposal of ethanol. The importance of blood source (artery, capillary, or vein) and the sampling site (arm or leg) on ethanol pharmacokinetics deserves further study, as does whether multicompartmental models should be invoked.

(SEE ALSO: *Accidents and Injuries from Alcohol; Addiction: Concepts and Definitions; Alcohol; Chinese Americans, Alcohol and Drug Use among; Drug Interactions and Alcohol; Drug Metabolism; Drunk Driving; Psychomotor Effects of Alcohol and Drugs; Vulnerability As Cause of Substance Abuse*)

BIBLIOGRAPHY

HOLFORD, N. H. G. (1987). Clinical pharmacokinetics of ethanol. *Clinical Pharmacokinetics 13*, 273–292.

VON WARTBURG, J. P. (1989). Pharmacokinetics of Alcohol. In K. E. Crow & R. D. Batt (Eds.), *Human Metabolism of Alcohol*. Boca Raton, FL: CRC Press.

WIDMARK, E. M. P. (1981). *Principles and applications of medicolegal alcohol determination*. Davis, CA: Biomedical Publications. (English translation of Widmark's 1932 monograph, in German)

WILKINSON, P. K. (1980). Pharmacokinetics of ethanol: A review. *Alcoholism: Clinical and Experimental Research 4*, 6–21.

A. W. JONES

PHARMACOLOGY In its broadest sense, pharmacology can be defined as the science dealing with interactions between living systems and molecules—in particular, chemicals (i.e., drugs)—usually introduced from outside the system. This definition also includes medical pharmacology, which is the science of drugs used to prevent, diagnose, and treat disease. Also included are the important roles played by chemicals in the environment that can cause disease, as well as the use of certain chemicals as molecular probes for the study of normal biochemistry and physiology. Toxicology is the branch of pharmacology that deals with the undesirable (i.e., toxic) effects of chemicals in biological systems.

(SEE ALSO: *Drug; Drug Metabolism; Drug Types; Pharmacodynamics; Pharmacokinetics; Poison*)

BIBLIOGRAPHY

BENET, L. Z., MITCHELL, J. R., & SHEINER, L. B. (1990). General principles. In A. G. Gilman et al. (Eds.), *Goodman and Gilman's the pharmacological basis of therapeutics*, 8th ed. New York: Pergamon.

NICK E. GOEDERS

PHARMACOTHERAPY *See* Treatment (for specific drugs); Treatment Types.

PHENCYCLIDINE (PCP) Although phencyclidine (PCP) and drugs of similar chemical structure (arylcyclohexylamines) are often called HALLUCINOGENS, they rarely produce HALLUCINATIONS, and the sensory distortions or apparent hallucinations that are produced are not the same type as LSD-induced hallucinations. Instead, phencyclidine belongs to a unique class of drugs called the dissociative anesthetics. Phencyclidine was developed in the 1950s as an anesthetic for veterinary medicine and later was tested in human surgical patients. There was great potential for PCP as an anesthetic, because it produced minimal effects on the heart and breathing was not suppressed. Unfortunately, the adverse side effects of PCP (e.g., dysphoria [unhappy, ill] and psychotic symptoms) led to a termination of the human clinical trials. The drug is no longer manufactured for veterinary use because supplies were diverted (stolen, highjacked) and the drug became widely abused in the 1970s. Ketamine, a drug chemically similar to PCP, is now used as a veterinary anesthetic and, in special cases, for anesthesia in humans. This drug is less powerful and shorter acting than PCP.

Phencyclidine abuse, mainly in pill form, peaked in the late 1970s and markedly declined throughout the 1980s and early 1990s. The most common route of administration in use in the 1990s is smoking. Phencyclidine is often added to MARIJUANA cigarettes, and it is commonly used while people are also drinking alcoholic beverages. Street names for PCP are "angel dust" or "crystal"; it is called "space base" when combined with COCAINE.

MECHANISM OF ACTION

Most investigators agree that the behavioral effects of PCP are mediated predominantly through RECEPTORS, which are proteins that are important for the normal functioning of cells within the body. Phencyclidine acts as an antagonist at the N-methyl-D-aspartate (NMDA) receptor-channel complex, which is one type of excitatory amino-acid receptor that is selectively activated by the agonists NMDA and GLUTAMATE. By definition, agonists produce stimulation while antagonists block the effects of agonists. When either glutamate or NMDA bind to the receptor, a channel within the cell membrane opens to allow sodium, calcium, and potassium ions to flow into and out of the cell. This movement of ions across the cell membrane causes a depolarization of the membrane which, if sufficiently large, causes the cell to fire. When the cell fires, an electrical charge passes along its membrane and NEUROTRANSMITTERS (chemicals that allow cells to communicate with each other) are released. Thus, glutamate and NMDA are important for normal cell-to-cell communication within the body.

PCP, as well as TCP, ketamine, dizocilpine (MK-801), and SKF 10,047 are representative of compounds that act as noncompetitive antagonists at the NMDA-receptor complex. The binding site for PCP resides within the channel and binding to this site physically prevents calcium and sodium ions from entering the cell while at the same time preventing potassium ions from leaving the cell. Blocking the movement of ions through the cell membrane in turn prevents the neuron from firing. In contrast to the noncompetitive antagonists, competitive antagonists such as CGS 19755, NPC 12626, CPP, and AP5 bind to the NMDA receptor itself without causing the ion channel to open. By simply occupying the receptor without activating it, competitive antagonists prevent NMDA from binding to and activating the receptor. Unlike noncompetitive antagonists, competitive NMDA-antagonist effects can be surmounted by higher doses of the agonist. However, the end result of both noncompetitive and competitive antagonists is a reduction of neuronal firing.

PHARMACOKINETICS AND METABOLISM

PCP use in humans occurs through several routes of administration including the intranasal (snorted), intravenous, oral, and inhalational (smoked) routes. When PCP is smoked in parsley cigarettes, approximately 70 percent of the total amount of PCP is inhaled. Of this amount, 38 percent is inhaled as PCP and 30 percent is inhaled as phenylcyclohexene, a by-product of PCP when it is heated. Peak blood concentrations of drug occur after only five to ten minutes, which is occasionally followed by a second peak one to three hours later. PCP is predominantly excreted in urine after intranasal, intravenous, and oral administration. The rate of PCP elimination through the kidneys depends on both urine pH and urine-flow rate. More specifically, PCP elimination occurs more rapidly when urine is acidic and when urine is passed rapidly.

DISCRIMINATIVE STIMULUS EFFECTS

One useful method of evaluating the pharmacological characteristics of PCP, as well as a variety of other drugs, is the drug-discrimination procedure. Typically, animals that are slightly food restricted are trained to respond for food on one lever after drug administration and on another lever after saline. On days when the drug is administered before the session, responding on the drug-associated lever results in food delivery while responding on the saline-associated lever does not. Conversely, on days when saline is administered before the session, responding on the saline-associated lever results in food delivery while responding on the drug-associated lever does not. After a number of training days, animals learn to reliably respond on the drug lever after the drug injection and on the saline lever after saline injection. Once this discrimination has been established, a number of test drugs can be administered to determine whether or not they produce effects similar to the training drug. Test drugs that substitute for the training drug (i.e., cause responses on the drug-as-sociated lever) are assumed to have discriminative stimulus effects that are similar to the training drug.

Using this procedure, several investigators showed that PCP and other noncompetitive antagonists produced similar discriminative stimulus effects in a number of different species (see Willetts, Balster, & Leander, 1990 for a review). These results suggested that the mechanisms of action of PCP and other noncompetitive antagonists, such as ketamine and dizocilpine, were similar. Furthermore, the discriminative stimulus effects of competitive antagonists such as CGS 19755, NPC 12626 and CPP were also similar to each other, which was again consistent with the notion that the mechanisms of action of competitive antagonists were similar. Given that competitive and noncompetitive antagonists both reduce neuronal firing, it was of interest to compare the discriminative stimulus effects of these two types of antagonists. In most species, the discriminative stimulus effects of competitive and noncompetitive antagonists were very different from each other.

Another difference between the competitive and noncompetitive antagonists was their abilities to antagonize the discriminative stimulus effects of NMDA. While both types of antagonist were effective in blocking the convulsant and lethal effects of NMDA, competitive antagonists in general were much more effective than noncompetitive antagonists in blocking the discriminative stimulus effects of NMDA. The noncompetitive antagonists partially antagonized NMDA but only at doses that produced substantial behavioral suppression. While most effects of NMDA were antagonized by both competitive and noncompetitive antagonists, the behavioral-suppressing effects of noncompetitive antagonists often interfered with their ability to antagonize the discriminative stimulus effects of NMDA.

Finally, another important finding with competitive and noncompetitive antagonists involved their interaction with other receptor systems. Studies showed that the discriminative stimulus effects of competitive antagonists such as CPP and NPA 12626 were similar to those produced by the BARBITURATE pentobarbital. Under certain conditions, the discriminative stimulus effects of PCP and pentobarbital were also similar. In addition to the interactions of NMDA antagonists with barbiturate receptors, some investigators have found similarities between PCP and ethanol (alcohol). These studies have proven to be important in describing both the similarities and differences between the noncompetitive and competitive NMDA-receptor antagonists.

TOLERANCE

Tolerance to a drug occurs when increasingly higher doses are needed to produce a specific effect or if drug effects diminish after repeated administration of the same dose of drug. It has not been possible to study tolerance to PCP in human subjects, but when interviewed, PCP users report that they increase the amount of PCP that they take over time (see Carroll, 1990). Another indicator of tolerance development is that burn patients treated with ketamine for pain often require higher doses over time. Although information from the clinic is useful, it is easier to study tolerance to ketamine, PCP, and similar drugs in animals. Laboratory studies with rats have shown that tolerance developed to the effects of PCP on food-reinforced responding, to the effects of PCP and dizocilpine on steroid hormone (adrenocorticotropin and corticosterone) release, and to the cataleptic effects of ketamine. Supersensitivity, the opposite of tolerance, occurs when repeated drug exposure produces a greater effect at a given dose. Some investigators have found that tolerance develops to some effects of PCP, such as head weaving, turning, and back pedaling, while supersensitivity occurs with other behaviors, such as sniffing, rearing, and ambulation. Although some scientists have hypothesized that PCP tolerance and supersensitivity are mediated through non-NMDA-receptor systems, others have suggested that PCP tolerance may be mediated through the NMDA receptor system. Repeated administration of dizocilpine, a PCP-like compound, produced a reduction in the number of NMDA receptors in rat brain, and that was correlated with tolerance to some of the behavioral effects produced by dizocilpine. Further studies will clarify the role of different receptor systems in the development of tolerance to the effects of PCP and related compounds.

Studies indicate that there are interactions between PCP and other drugs with respect to tolerance and supersensitivity of drug effects. For example, dizocilpine blocked the development of tolerance to morphine's analgesic (painkilling) effects, but it did not alter the analgesic effects when MORPHINE was administered acutely. Also, dizocilpine attenuated the development of tolerance to ethanol (ALCOHOL),

and it inhibited sensitization to amphetamine and cocaine (see *DHHS Fourth Triennial Report to Congress on Drug Abuse and Drug Abuse Research*, 1992 for further details).

DEPENDENCE

Physiological dependence on a drug is usually defined by a set of withdrawal signs that occur when steady use of the drug is discontinued. The withdrawal signs are typically the same for a given drug, and they follow a specific time course which is usually about six to forty-eight hours, depending on the drug. The withdrawal signs may be rapidly reversed after one administration of the drug.

Most of what we know about PCP dependence is from experimental studies with animals. There are only limited reports of PCP withdrawal effects in humans. In 1981, Tennant and colleagues studied 68 regular PCP users; they found that one third of them had sought treatment or medication to relieve the effects of PCP withdrawal. Withdrawal symptoms that they commonly reported were depression, drug craving, increased appetite, and increased need for sleep. Another way PCP dependence has been documented in humans is in studies of babies born to PCP-using mothers. Withdrawal signs that have been noted are diarrhea, poor feeding, irritability, jerky movements, high-pitched cry, and inability to follow a stimulus visually.

In laboratory studies with monkeys, similar signs of PCP withdrawal have been noted. Balster and Woolverton (1980) gave rhesus monkeys continuous access to PCP directly into the blood stream for 50 days, using an intravenous cannula system. The monkeys were trained to respond on a lever for an infusion of PCP. When PCP was replaced with a salt and water solution used to dissolve the drug (vehicle), withdrawal signs were noted, such as poor feeding, weight loss, irritability, bruxism (coughing), vocalizations, piloerection (hair standing up), tremors, less exploratory behavior in the cage, and poor motor coordination. The withdrawal syndrome began within four to eight hours, peaked between twelve and sixteen hours, and had disappeared by twenty-four to forty-eight hours. These results have been repeated in studies with rats. Some studies have reported PCP withdrawal effects after as little as two weeks of exposure. Thus, long-term use of the drug may not be necessary to produce physical dependence.

Recent studies with animals have shown that not only a short period of exposure to PCP but low doses of PCP result in withdrawal effects when drug administration is discontinued. Operant conditioning experiments are used as sensitive tests of drug-withdrawal effects in animals. In these experiments, animals are trained to respond on a lever or push a button or other device to obtain a food reward. At the same time they are allowed to self-administer drugs orally or intravenously. When drug access is removed, a decrease in operant responding for food is often seen, even when the drug dose is sufficiently low to produce no observable signs of withdrawal. These measures have also been used to demonstrate withdrawal effects from drugs such as cocaine, caffeine, and nicotine. When regular use of these drugs is discontinued there are no observable signs of withdrawal during abstinence. The most severe reductions in the operant behavioral baselines occur during the first forty-eight hours of drug withdrawal, a time during which physical signs occur when higher maintenance doses are used; however, the behavioral disruptions often last for long periods of time. During withdrawal, when animals will not respond on a lever for food, they readily consume hand-fed food. Thus, the decrease in feeding may not be due to illness but to a decrease in the motivation to work for food.

In the first study that demonstrated disruption in operant behavior during PCP withdrawal, Slifer and coworkers (1984) treated monkeys with continuous intravenous infusions for ten days. They were required to make 100 responses on a lever for each food pellet. When access to PCP was terminated, responding for food decreased substantially for up to seven days and did not return to normal levels until the monkeys were again allowed access to PCP. Similar results were found by other investigators using monkeys trained to self-administer orally delivered PCP. There was little difference in the results, depending on whether the PCP was self-administered or experimenter administered. In the monkey studies, there was only a weak relationship between dose and the severity of the withdrawal effect, but in rats, PCP dose, blood levels, and magnitude of the withdrawal effect were closely related. Recent studies have shown that there is cross-dependence between drugs that are chemically similar to PCP—such as

PCP and ketamine, dizocilpine, and the (+)isomer of SKF-10,047; however, cross-dependence was not demonstrated with either the racemate or (−)isomer of SKF-10,047 or with ethanol.

The PCP-withdrawal effect can be altered by changing schedules of reinforcement. In one study with monkeys, lever-press requirements or fixed ratios (FRs) for food were increased from 64 to 128 to 256 to 512 to 1024, and PCP-withdrawal effects were examined at each value. As the FR value increased, PCP withdrawal effects became more pronounced. At the two higher FRs, body weights declined and the severity of the withdrawal effect showed no further increases. To examine the effects of amount of food available, another experiment was conducted in which the FR was held constant at 1024 and the monkeys were either supplemented with 100 grams of hand-fed food or not. The amount of responding for earned food remained the same during supplemented and unsupplemented conditions, but when the effects of withdrawal were examined, a disruption in responding occurred only under the supplemented condition. When the monkeys had to earn their entire daily food ration, the withdrawal effect disappeared. These studies suggest that the severity of the PCP withdrawal effect is determined by the behavioral economics of food availability. The magnitude of PCP withdrawal increased as the price (FR of food) increased; but as the price became so high that body weight was lost, the PCP-withdrawal effect entirely disappeared. These data also suggest that PCP withdrawal is not necessarily an illness but a decreased level of motivation.

The use of drugs to treat the PCP-withdrawal syndrome has produced mixed results. When monkeys had access to orally delivered (+)SKF-10,047, the PCP-withdrawal-induced disruptions in food-maintained responding were reversed. This was not the case with (−)SKF-10,047 or the racemate (±)SKF-10,047. Injections of dizocilpine before PCP withdrawal, or two days into PCP withdrawal, greatly reduced or reversed, respectively, the disruptions in food-reinforced responding. Dizocilpine also dose-dependently reduced PCP self-administration. In contrast, while BUPRENORPHINE, a partial AGONIST at the mu-opiate receptor, also dose-dependently reduced PCP self-administration, it had no effect on PCP-withdrawal-induced disruptions in food-maintained responding. When PCP was self-administered concurrently with ethyl alcohol (ethanol) and then

PCP access was removed, PCP-withdrawal effects were as severe as when ethanol had not been available. Thus, ethanol did not alleviate the PCP withdrawal effect, although, as noted earlier, PCP and ethanol share discriminative stimulus effects (Grant et al., 1991). In other studies, PCP was self-administered concurrently with ethanol or caffeine. When PCP and the other drug were removed simultaneously, the withdrawal disruption was more severe than when PCP alone was withdrawn. (Further details of these withdrawal studies may be found in reviews by Carroll [1990] and by Carroll and Comer in the *DHHS Fourth Triennial Report to Congress on Drug Abuse and Drug Abuse Research*, 1992.)

REINFORCING EFFECTS

The reinforcing effects of a drug are determined by demonstrating that self-administration of the drug plus the solution it is dissolved in (vehicle) occurs in excess of self-administration of the vehicle alone. When drug-reinforced behavior is readily achieved in the animal laboratory, it is usually a good predictor that the drug has considerable abuse liability in the human population. The reinforcing effects of PCP have been studied using two animal models of self-administration, oral and intravenous. The intravenous route of self-administration requires the animal to make a specified number of responses on a lever or other manipulandum within a predefined time—then a fixed dose of the drug is delivered by an infusion pump via a catheter that is surgically implanted in a large vein that leads to the heart. Studies from various laboratories have demonstrated that intravenously delivered PCP functions as a reinforcer for rats, dogs, monkeys, and baboons.

Drugs that are chemically similar to PCP are also self-administered intravenously. These include drugs that have similar receptor-binding sites in the brain, such as ketamine, (+)SKF-10,047, dexoxadrol, and cyclazocine; and phencyclidine-like drugs that function as noncompetitive antagonists at the NMDA receptor, such as dizocilpine. Phencyclidine and dizocilpine self-administration is more reliably obtained when the animal has a history of self-administration of a drug with similar pharmacological or discriminative-stimulus effects. It has also been found that drugs that share discriminative-stimulus effect with PCP, such as (+)SKF-10,047, ketamine, PCE, TCP, and ethanol, are readily substituted for PCP in self-administration studies.

Oral PCP self-administration is established by presenting gradually increasing concentrations of PCP after the animal is given its daily food ration. After sufficient quantities of PCP are consumed, food is given after the drug self-administration session, and PCP consumption usually persists. This procedure provides a long-term stable baseline to examine variables that affect PCP-reinforced behavior. For example, alternative nondrug reinforcers, such as saccharin, reduce PCP-reinforced responding up to 90 percent of baseline if the FR for PCP is high or if the PCP concentration is very low. Free access to food decreases PCP self-administration, while even small reductions in the daily food allotment markedly increase PCP self-administration. Concurrent availability of ethanol also reduces PCP-reinforced responding.

A limited amount of information is available concerning drug pretreatment and PCP self-administration. Buprenorphine and dizocilpine pretreatment both resulted in dose-dependent decreases in PCP self-administration; however, potential treatment drugs such as fluoxetine and carbamazepine had no effect. Treatment with other drugs such as AMPHETAMINE or PENTOBARBITAL had a biphasic effect on PCP self-administration. Low doses of the pretreatment drugs increased PCP self-administration, and high doses decreased PCP self-administration.

TOXICITY

There is little evidence that long-term PCP use in adult humans (Luisada, 1981) and monkeys (see *DHHS Fourth Triennial Report to Congress on Drug Abuse and Drug Abuse Research*, 1992) results in any detectable organ or cellular damage. In monkeys that had been self-administrating PCP for eight years, tests of all organ systems, clinical chemistries, physical exams, and X rays revealed no differences between PCP-experienced and control animals that were the same age but had little drug experience. In humans, the form of toxicity most commonly associated with PCP use is a change in behavior. There have been a few accounts of bizarre and/or violent behavior associated with PCP use. Such reports have diminished since the preferred route of self-administration has shifted from oral (pill) to inhalation, which offers the users an ability to more carefully control the dose.

In monkeys, PCP produces a calming, tranquilizing effect. The immediate effects in humans are not seen in the hospital or clinic. Instead, the PCP user arrives in the emergency room several hours after PCP use, possibly while suffering acute withdrawal effects. Approximately twelve to fifteen hours after PCP was last taken, monkeys become agitated, violent, and aggressive. It is possible that many of the early reports of human violence and the PCP-related homicides were related to the withdrawal effects. It is necessary to determine the time course of unusual behavior and important to know the time of drug intake, although this is difficult to establish because the patient often loses memory of the drug-taking event.

Another unusual aspect of PCP toxicity is that users often complain of unpleasant effects long after chronic use has stopped. These reports could be caused by the fact that PCP is highly fat soluble and becomes stored for long periods of time in the body fat. During periods of weight loss, there is subsequent mobilization of fat-stored PCP into blood and brain tissues. Recent laboratory research with rats supports this hypothesis by demonstrating the ability of food deprivation to increase PCP levels in blood and brain (Coveney & Sparber, 1990).

Increasing data has become available on the effect of drugs of abuse on the offspring of dependent mothers, and it appears that the offspring of PCP users may be more vulnerable to the adverse effects of PCP than their adult counterparts. Golden and coworkers (1987) studied ninety-four PCP-exposed newborns and ninety-four nonexposed as controls; they found neurological abnormalities such as abnormal muscle tone and depressed reflexes in the exposed group. Another study followed twelve exposed infants for eighteen months and found a high percentage of medical problems (Howard et al., 1986). At six months the infants were irritable and hyperresponsive, and later they showed varying degrees of abnormalities in fine-motor, adaptive, language and social skills. A recent study of the offspring of 47 PCP abusers and 38 nonusers found that neurological dysfunction was common in the infants of PCP-abusing mothers (Howard, Beckwith, & Rodning, 1990). There was greater apathy, irritability, jitters, and abnormal muscle tone and reflexes. Follow-up interviews at six and fifteen months, using the Gesell Developmental Exam, revealed poor language development and a lower developmental quotient in general; however, the long-term outcome for PCP-exposed newborns is unknown.

TREATMENT

There are currently no PCP ANTAGONISTS that are useful for treatment of PCP OVERDOSE. Symptomatic treatment may be given for suppressed breathing rates, fever, high blood pressure, and increased salivation. Convulsions are treated with DIAZEPAM. Elimination of the drug may be enhanced by making the urine more acidic and/or pumping stomach contents. Attempts to minimize environmental stimuli have helped to control violent and self-destructive behavior. Psychiatric care may be needed for an extensive psychotic phase that may follow overdose (see Jaffe, 1989).

(SEE ALSO: *Abuse Liability of Drugs: Testing in Animals; Addiction: Concepts and Definitions; Adjunctive Drug Taking; Aggression and Drugs; Fetus: Effect of Drugs on the; Phencyclidine (PCP): Adverse Effects; Research, Animal Model: Drug Discrimination Studies; Tolerance and Physical Dependence*)

BIBLIOGRAPHY

CARROLL, M. E. (1990). PCP and Hallucinogens. *Advances in Alcohol and Substance Abuse, 9,* 167–190.

CONVENEY, J. R., & SPARBER, S. B. (1990). Delayed effects of amphetamine or phencyclidine: Interaction of food deprivation, stress and dose. *Pharmacological and Biochemical Behavior, 36,* 443–449.

GOLDEN, N. L., ET AL. (1987). Neonatal manifestations of maternal phencyclidine exposure. *Perinatal Medicine, 15,* 185–191.

GRANT, K. A., ET AL. (1991). Ethanol-like discriminative stimulus effects of noncompetitive n-methyl-d-aspartate antagonists. *Behavioural Pharmacology, 2,* 87–95.

HOWARD, J., BECKWITH, L., & RODNING, C. (1990). Adaptive behavior in recovering female phencyclidine/polysubstance abusers. *NIDA Research Monograph, 101,* 86–95.

JAFFE, J. H. (1989). Psychoactive substance abuse disorders. In H. I. Kaplan and B. J. Sadock (Eds.), *Comprehensive textbook of psychiatry,* 5th ed. Baltimore: Williams & Wilkins.

SLIFER, B. L., BALSTER, R. L., & WOOLVERTON, W. L. (1984). Behavioral dependence produced by continuous phencyclidine infusion in rhesus monkeys. *Journal of Pharmacology and Experimental Therapies, 230,* 339–406.

WILLETTS, J., BALSTER, R. L., & LEANDER, J. D. (1990). The behavioral pharmacology of NMDA receptor antagonists. *Tr. Pharmacol. Sci. 11,* 423–428.

MARILYN E. CARROLL
SANDRA D. COMER

PHENCYCLIDINE (PCP): ADVERSE EFFECTS

Widely known as PCP, PHENCYCLIDINE is an important drug of abuse in the United States. It produces a unique type of hallucinatory effect and is used both by smoking and ingestion. Persons under the influence of PCP experience mood changes, perceptual distortions, and feelings of dissociation from their surroundings. Since their judgment is impaired, they may take unnecessary risks. They may become unpredictable and violent. In certain individuals, PCP use, especially if repeated often, can result in the production of a mental disturbance referred to as PCP psychosis. It is not, however, known with certainty whether PCP itself, or a combination of factors involved in the lifestyle of PCP abusers, is the cause of brain damage or of long-term behavior impairment that also sometimes occurs in PCP abusers.

HISTORY

The use and abuse of PCP began in the United States in the middle 1960s, when it was primarily taken by ingestion; but the real epidemic of PCP abuse occurred in the 1970s, when smoking and insufflation ("snorting") became the more common forms of use (Burns & Lerner, 1976). Illicit synthesis of PCP is not very difficult for an experienced chemist, so PCP and its abuse spread rapidly, peaking about 1978. After 1980, its prevalence declined—however, PCP abuse continues to occur. National Institute on Drug Abuse surveys show that more Americans have experimented with PCP than with heroin, and the prevalence of recent use of PCP is about the same as with heroin, so it remains a significant public-health problem (National Institute on Drug Abuse, 1991).

PSYCHOLOGICAL EFFECTS OF PCP

The psychological effects of PCP abuse can be discussed under three headings: (1) the effects accompanying acute intoxication, (2) the personality disturbances that can sometimes develop in PCP abusers, especially when associated with chronic use, and (3) the possible neurobehavioral toxicity that might result from chronic use.

SIGNS AND SYMPTOMS OF PCP INTOXICATION

Low Dose. Dreamy carefree state, mood elevation, heightened or altered perception, impaired judgment, partial amnesia.

Intermediate Dose. Inebriation, motor incoordination, dissociation and depersonalization, confusion and disorientation, perceptual distortions and preoccupation with abnormal body sensations, diminished pain sensitivity, partial amnesia, and sometimes exaggerated mood swings and panic.

High Dose. Catatonia, "blank stare," drooling, nystagmus (eye-rolling), delirium and hallucinations, psychotic behavior, severe motor incoordination, total amnesia.

ACUTE PCP INTOXICATION

As with all drugs, the effects of PCP depend on the dose that is taken. The section above lists the typical effects of PCP at various doses. PCP abusers usually adjust their dosage to experience only the low-dose effects. High-dose effects are similar to a mild type of dissociative anesthesia.

Experienced drug abusers can readily distinguish the experience of PCP intoxication from that produced by other drugs such as MARIJUANA, MESCALINE, and LYSERGIC ACID DIETHYLAMIDE (LSD). Users typically report a feeling of dissociation from the environment and abnormal body sensations and body image. The perceptual distortions often cause things to appear far away or abnormal in size. Compared to LSD, the effects of PCP are not very PSYCHEDELIC.

The most dangerous effects of PCP intoxication arise from the impaired judgment and altered perceptions that occur. People can engage in risk-taking behavior and harm themselves or others. DRIVING, swimming, or other activities requiring coordination and good judgment become extremely dangerous. Someone on PCP may also engage in casual but high-risk sexual behaviors. PCP users experience profound mood swings—where what begins as a pleasant experience can turn into panic and terror—and their behavior is unpredictable. Sometimes these "bad trips" can lead to violent and uncharacteristic behaviors with disastrous results. In cases of high-dose intoxication, users can experience a toxic psychotic episode with DELIRIUM, profound HALLUCINATIONS, and paranoia. In cases of severe overdose, seizures, stroke, or kidney failure may lead to death (Burns & Lerner, 1976).

LONG-TERM USE

In persons who abuse PCP in large amounts over a long period, or in those who have psychological problems that make them especially vulnerable, a chronic psychosis may develop. This PCP psychosis is evident even when abusers are not high on PCP, and it may be quite difficult to treat. The symptoms of PCP psychosis differ considerably from person to person, but patients may show many features of SCHIZOPHRENIA, including the appearance of a thought disorder, paranoid ideation, hallucinations, mood changes, and aberrant behavior. These patients often require psychiatric hospitalization and treatment with ANTIPSYCHOTIC medications.

In research studies where PCP has been given repeatedly to animals, it has been possible to show the development of PHYSICAL DEPENDENCE (e.g., Balster & Woolverton, 1980). The doses required for dependence are quite high, so it may be that dependence in human PCP abusers is difficult to develop. There have been some clinical reports of withdrawal effects in heavy PCP abusers, but these do not appear to be present in most individuals needing treatment for PCP abuse. There are no specialized treatment methods for PCP abusers, and since many PCP abusers also abuse other drugs and/or alcohol, they are usually helped by the same counseling and psychotherapy programs that are used for other forms of drug abuse.

NEUROPSYCHOLOGICAL AFTEREFFECTS OF PCP ABUSE

It is not known for certain whether or not PCP causes brain damage or long-term neurological or behavioral impairment in chronic abusers. Although some PCP abusers develop neurobehavioral impairment, controlled experiments of the type that would need to be carried out to show that PCP alone was the cause of the problems have not been done. PCP abusers typically abuse many other drugs in addition to PCP, which may contribute to their problems, and they may have lifestyles and health habits that lead to neuropsychological dysfunction. For example, while under the influence of PCP, they may be involved in an accident resulting in brain injury, so the risk factors that accompany PCP abuse may be responsible for the clinical problems sometimes seen in abusers. It should be pointed out that PCP was used in humans for medical research for a number of years, and ketamine—a close analog of PCP—is, even in the early 1990s, given to thousands of patients. No legacy of neuropsychological impairment is seen in these individuals.

Does this mean that chronic PCP abuse does not cause neuropsychological impairment? Certainly, PCP—like all drugs—must be considered as a possible source of neural damage. In animal testing, it was found that even a single injection of a fairly high dose of PCP produced reversible pathomorphological changes in neurons of the cingulate and retrosplenial cortex in the brains of rats (Olney, Labruyere, & Price, 1989). Although it is not known if PCP produces these effects in humans, it is possible that it does and that this could lead to adverse health effects. Another possibly important basis for concern comes from studies which show that PCP, and related drugs, impair learning and memory in various animal models. PCP's ability to do this may be greater than for other classes of drugs of abuse, possibly due to PCP's ability to interfere with specific brain mechanisms for learning that involve N-methyl-D-aspartate (NMDA) RECEPTORS.

PCP AND VIOLENCE

Many people associate the abuse of PCP with violence and aggression, so this concern deserves special mention. Those under the influence of PCP often behave erratically and exercise poor judgment. These effects of PCP could certainly lead to violent behavior, and there are certainly numerous examples of extremely violent acts being performed by persons under the influence of PCP. This raises the question of whether PCP is uniquely associated with the production of violence and aggression: Is someone intoxicated with PCP more likely to be violent than someone who is intoxicated with COCAINE or alcohol?

Unfortunately, the answer to this question is not known. A great deal of criminal conduct in the United States is certainly carried out by people under the influence of alcohol or drugs. In addition, the public often associates drug use they do not understand with criminal and violent behavior. Every new drug epidemic is greeted with public concern that this drug causes violence. There is also the common practice of criminal attorneys using the defense of diminished capacity, because of drug use, to lessen the responsibility that their clients might bear for criminal conduct. All these factors undoubtedly contribute to the public attention focused on the relationship of PCP to violence.

Few good research studies have attempted to determine the specific role that PCP abuse may have in crime and violence. In one study (Wish, 1986) of nearly 5,000 arrestees in New York City in 1984 who agreed to leave a urine specimen for drug analysis, it was found that 56 percent tested positive for at least one drug of abuse. For those who had used PCP recently, most had committed robbery, not bizarre violent offenses. In fact, assault was more common among arrestees who had not used PCP than among those who had. These results support the conclusion that PCP may be no more likely to cause violence than some other drugs of abuse—but, clearly, more research on this question is needed.

The NATIONAL INSTITUTE ON DRUG ABUSE estimates that as many as 6 million Americans have tried PCP at least once. The very large majority of these occasions of PCP use were not associated with violent acts; however, if some users prone to violence take PCP and are faced with a threatening situation, they may act unpredictably and violently. Although there is no scientific evidence that PCP actually increases muscular strength, PCP users unmindful of their own potential safety or injuries can be a formidable risk, so law enforcement personnel are on guard against these dangerous situations. Alternatively, it should not be assumed that most people who abuse PCP will become violent—nor should every inexplicable act of violence be casually or speculatively attributed to PCP abuse.

(SEE ALSO: *Addiction: Concepts and Definitions; Amphetamine; Complications: Mental Disorders; Crime and Drugs; Tolerance and Physical Dependence*)

BIBLIOGRAPHY

BALSTER, R. L., & PROSS, R. S. (1978). Phencyclidine: A bibliography of biomedical and behavioral research. *Journal of Psychedelic Drugs, 10*(1), 1–15. A comprehensive list of publications on PCP before 1978.

BALSTER, R. L., & WOOLVERTON, W. L. (1980). Continuous access phencyclidine self-administration by rhesus monkeys leading to physical dependence. *Psychopharmacology, 70,* 5–10.

BURNS, R. S., & LERNER, S. E. (1976). Perspectives: Acute phencyclidine intoxication. *Clinical Toxicology, 9,* 477–501.

CARROLL, M. E. (1985). PCP: The dangerous angel. In S. H. Snyder (Ed.), *Encyclopedia of Psychoactive Drugs,* Vol. 8. New York: Chelsea House. This volume is intended for a general reader and covers a broad range of topics related to PCP abuse.

CLOUET, D. H. (ED). (1986). *Phencyclidine: An update.* National Institute on Drug Abuse Research Monograph 64. DHHS Publication No. (ADM)86-1443. The proceedings of a scientific/medical conference on PCP, containing information on a wide variety of topics.

FELDMAN, H. W., AGAR, M. H., & BESCHNER, G. M. (EDS.). (1979). *Angel dust: An ethnographic study of PCP users.* Lexington, MA: Lexington Books. This book presents an ethnographic study of the use patterns, health consequences, and lifestyles of a number of PCP-using social groups in four cities. This study is important, because it does not focus on abusers who are in treatment.

NATIONAL INSTITUTE ON DRUG ABUSE. (1991). *National household survey on drug abuse: Main findings 1990.* DHHS Publication No. (ADM)91-1788. Washington, DC: U.S. Government Printing Office.

OLNEY, J. W., LABRUYERE, J., & PRICE, M. T. (1989). Pathological changes induced in cerebrocortical neurons by phencyclidine and related drugs. *Science, 244,* 1360–1362.

POLLARD, J. C., UHR, L., & STERN, E. (1965). *Drugs and phantasy: The effects of LSD, psilocybin and sernyl on college students.* Boston: Little, Brown. A comparison of the effects of PCP and other hallucinogenic drugs given to normal students under carefully controlled conditions.

WISH, E. (1986). PCP and crime: Just another drug? In D. H. Clouet (Ed.), *Phencyclidine: An update.* National Institute on Drug Abuse Research Monograph 64. DHHS Publication No. (ADM)86-1443.

ROBERT L. BALSTER

PHENOBARBITAL This is the prototypic BARBITURATE central nervous system (CNS) DEPRESSANT. It is prescribed and sold as Luminal and was introduced into clinical medicine in 1912. It was used for a long period as a SEDATIVE-HYPNOTIC drug but has now largely been replaced by the much safer BENZODIAZEPINES.

Phenobarbital's long duration of action makes it useful for treating many forms of general and partial seizure disorders, such as epilepsy. Chronic use can result in TOLERANCE AND PHYSICAL DEPENDENCE, so it is classified as a Schedule III drug in the CONTROLLED SUBSTANCES ACT. Chronic treatment with phenobarbital can increase the activity of certain liver enzymes that metabolize other drugs. Thus a potential side effect is that other drugs (e.g., steroids, oral anticoagulants, digitoxin, beta-blockers, oral contraceptives, phenytoin, and others) are metabolized more quickly—and their effectiveness is reduced. Combinations of phenobarbital and other CNS depressants, such as ALCOHOL (ethanol), can lead to severe motor impairment and reduced breathing.

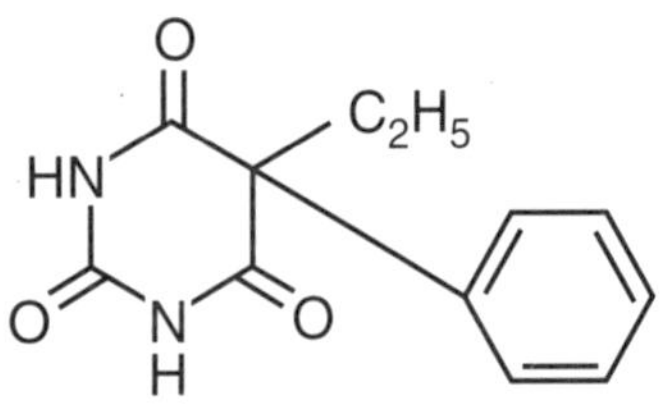

Figure 1
Phenobarbital

(SEE ALSO: *Drug Metabolism; Drug Interactions and Alcohol*)

BIBLIOGRAPHY

HARVEY, S. C. (1975). Hypnotics and sedatives. In L. S. Goodman & A. Gilman (Eds.), *The pharmacological basis of therapeutics*, 5th ed. New York: Macmillan.

SCOTT E. LUKAS

PHOENIX HOUSE Founded in 1967, Phoenix House was one of several second-generation THERAPEUTIC COMMUNITY (TC) programs that developed what are now recognized as "traditional" TC methods from the treatment approach originated at SYNANON by Charles Dederich. Phoenix House provides drug-free residential and outpatient treatment for adults and adolescents, plus intervention and prevention services. It operates fifteen units in three states, with a residential treatment capacity in excess of 1,600. It is one of the largest nongovernmental, nonprofit drug-abuse service agencies and has a 1-800-COCAINE substance-abuse information and referral service.

The six heroin addicts who founded Phoenix House in a Manhattan tenement apartment were guided by former addict counselors who had been trained in TC methods at Synanon and at the U.S. Naval Hospital in Oakland, California, working under the direction of Mitchell S. Rosenthal, M.D. A psychiatrist who had started the Oakland Naval Hospital's TC program, Rosenthal was then deputy commissioner for rehabilitation in New York City's Addiction Services Agency. From the program created at Phoenix House, he built a citywide drug-treatment system that, by 1971, included more than a dozen separate residences and a network of storefront centers. Privatized in 1972, Phoenix House was reorganized as a nonprofit agency operated by the Phoenix House Foundation, with Rosenthal as its president.

A period of contraction occurred during the 1970s, which was followed by accelerated growth and the expansion of services. Over the years, Phoenix House methodology has been adapted to the needs of adolescents, working men and women, criminal offenders, the homeless, single women with children, and other special populations. It now operates programs in correctional institutions and a shelter for homeless families, as well as the Phoenix Academies—residential high schools where adolescents in treatment recapture opportunities for higher education and careers.

Although many of the traditional TC practices observed at the adult, long-term residential-treatment facilities of Phoenix House have been modified at its other units, all programs adhere to the basic principles of TC treatment, relying on self-help methods and group process, and viewing substance abuse as a disorder affecting the whole person. All treatment is comprehensive, involves individualized treatment planning, and is designed to achieve the integration of clients into society as drug-free, productive, and socially responsible members.

The long-term residential treatment units of Phoenix House include an induction center and four other facilities in New York City; a treatment center in South Kortright, New York; units at two New York State correctional facilities; and treatment centers at Santa Ana and Turlock in California. Phoenix Academies are located in Yorktown, New York, Rockleigh, New Jersey, and Los Angeles, Santa Ana, and Descanso, California.

In New York City, Phoenix House also operates an evening outpatient program for adults, a day school and after-school intervention program for adolescents, an intervention program for homeless parents, a short-term residential program for working men and women, and a prevention unit that brings drug-education courses to some 35,000 schoolchildren each year.

Effectiveness of Phoenix House treatment was demonstrated by a long-term study (DeLeon, 1984) of treatment outcome that showed the strong connection between treatment success and time in treatment. Although success was greatest for program "graduates" (who completed 18 to 24 months of residence), substantial gains were also registered by nongraduates, including those who remained for only 3 months. The study showed that 75 percent of graduates achieved what researchers term a *best success* (with no posttreatment incident of narcotic use or use of a pretreatment drug of choice), while more than 90 percent sustained positive changes in behavior. Among "nongraduates" in treatment for 12 to 14 months, 38 percent achieved a best success and 55 percent sustained positive behavioral change.

(SEE ALSO: *Daytop Village; Prisons and Jails: Drug Treatment in; Project Return Foundation: Treatment Types: An Overview; Treatment Types: Self-Help and Anonymous Groups*)

BIBLIOGRAPHY

DeLeon, G. (1984). *The therapeutic community: Study of effectiveness*. National Institute on Drug Abuse. Treatment Research Monograph Series, DHHS Pub. No. (ADM) 84-1286. Washington, DC.

KEVIN McENEANEY

PHYSICAL DEPENDENCE A state, produced by repeated or prolonged drug exposure, in which the presence of drug in the body is required to maintain normal physiological function. This state is recognizable only by the occurrence of a withdrawal reaction when the drug is removed, that is reversed when the drug is again administered. Such dependence is believed to result from adaptive changes in the nervous system, opposite in direction to the drug effects, which offset these effects when drug is present, and produce a "drug-opposite" effect in its absence. Physical dependence is not synonymous with addiction, and can occur in nonaddicted persons.

(SEE ALSO: *Addiction: Concepts and Definitions; Disease Concept of Alcoholism and Drug Abuse; Tolerance and Physical Dependence*)

HAROLD KALANT

PILL POPPING *See* Slang and Jargon.

PLANTS, DRUGS FROM Humans have used their local plants for medicinal effects since prehistoric times. They gathered and ate plants and noticed the effects that some offered—whether therapeutic, mind-altering, or toxic. From trial and error they fashioned associations between cause and effect, keeping certain mushrooms, roots, barks, leaves, or berries for certain situations—the treatment of accidents, ill health, childbirth, coughs, fevers, rashes, and so on. Over the centuries, they established herbal medicine, as it is now called; they had also found certain plants that produced immediate and mind-altering effects, many of which were relegated to religious ritual. By the nineteenth century, Europeans had developed the science of chemistry to the point where the activator in many plants could be isolated and concentrated.

If experimentation with plant materials has led to cures, such as quinine for malaria or digitalis for heart disease, it has also led to the discovery of unpleasant effects or the discovery of poisons. From the literally thousands of substances that have been self-administered over the centuries, only a few continued to be used for nonmedicinal purposes. Even fewer have given rise to serious problems of chronic use and dependence. The legal and readily available drugs that are found naturally in plants (e.g., NICOTINE, CAFFEINE) or are derived from plants (e.g., ALCOHOL) will be described here first, because the use and abuse of these drugs is more widespread than all the other abused drugs combined. The health prob-

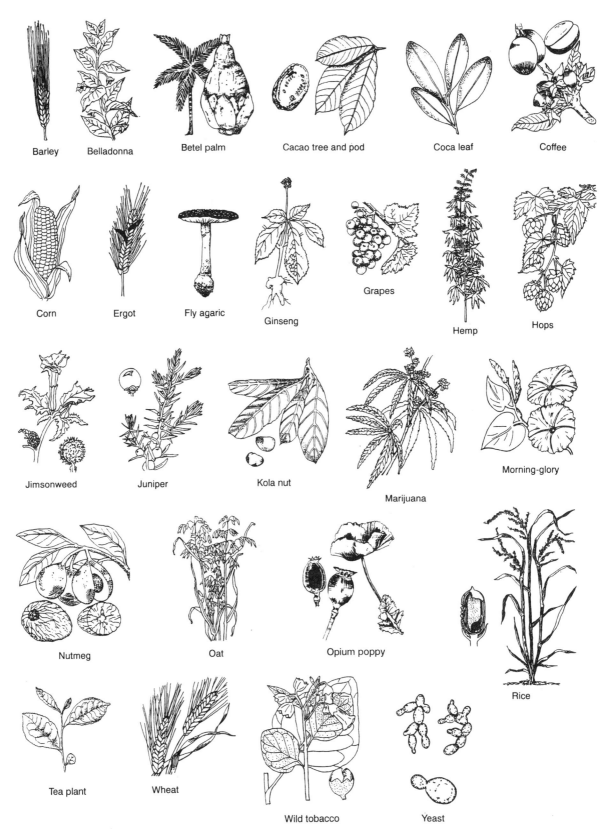

Barley   Belladonna   Betel palm   Cacao tree and pod   Coca leaf   Coffee

Corn   Ergot   Fly agaric   Ginseng   Grapes   Hemp   Hops

Jimsonweed   Juniper   Kola nut   Marijuana   Morning-glory

Nutmeg   Oat   Opium poppy   Rice

Tea plant   Wheat   Wild tobacco   Yeast

Figure 1
Some of the Plants Used in Making Drugs and
Alcoholic Beverages.

lems associated with the chronic use of alcohol and TOBACCO are, therefore, a very serious problem in our society, not only because of the large number of people who suffer and die each year from the direct toxic effects of these drugs but also because of the costs—the absenteeism from work and the unnecessary health-care cost. The illegal drugs will be discussed next; although the illicit use of MARIJUANA, COCAINE, OPIOIDS, and PSYCHEDELICS remains a major social, legal, financial, and health problem in the United States today, the proportion of the population physically dependent on these drugs is actually relatively low—only a small fraction of a percent. Finally, it is important to note that people often do not restrict their drug use to a single type. Alcohol users typically smoke cigarettes and may sometimes use other drugs as well. HEROIN users may also smoke and consume alcohol, marijuana, coffee or COLAS, and, in some instances, various STIMULANTS. Multiple drug use is, therefore, a relatively common occurrence.

ALCOHOL

Alcohol is perhaps the most widespread drug in use. It forms naturally by the fermentation process of plant materials and has been produced on purpose since at least neolithic times, when grains were first farmed, harvested, stored, and processed into gruels, porridges, puddings, and so forth. Often these spoiled, forming a fermented base. Alcohol is made as well from other starchy or sugary plant materials, such as fruits, canes, roots, and such. Fermentation (also called anaerobic respiration, or glycolysis) is the chemical process by which living cells, such as yeast, use sugar in the absence of air to produce part or even all of their energy requirements. In fermentation, sugar molecules are converted to alcohol and lactic acid. BEER, wine, and cheese production, as well as certain modern commercial processes require the fermentation by specific kinds of yeast, bacteria, and molds.

Ethyl alcohol, also called ethanol, is the type of alcohol that is usually produced for human consumption. In its pure form, alcohol is a clear liquid with little odor. People drink it primarily in three kinds of beverages: (1) beers are made from grains through brewing and fermentation and normally contain from 3 to 8 percent alcohol; (2) wines are fermented from fruits, such as grapes, and naturally contain from 8 to 12 percent alcohol (up to 21%

when fortified by adding more ethanol); (3) beverages or spirits DISTILLED from a fermented base, such as whiskey, gin, or vodka, contain about 40 to 50 percent alcohol, on average (often expressed in proof, so that 40% equals 80 proof; 50% is 100 proof).

NICOTINE AND TOBACCO

TOBACCO is a tall, herbaceous plant, the leaves of which are harvested, cured, and rolled into cigars, shredded for use in cigarettes and pipes, and processed for chewing or snuff. Tobacco has become a commercial crop in almost all tropical countries as well as in many temperate ones. The main source of commercial tobacco is *Nicotiana tabacum*, although *Nicotiana rustica* is also grown and is used in Asian tobaccos. Tobacco has been developed to yield a wide range of morphologically different types, from the small-leaved aromatic tobaccos to the large, broad-leaved cigar tobaccos. Tobacco is native to South America, where it was used in a drink for ritual purposes long before inhaling the smoke of the dried plant material was first documented by the Maya more than 2,000 years ago. Tobacco was then traded and grown in Central America; it moved into Mexico and the Caribbean and eventually into North America by about 800 A.D. The Arawaks of the Caribbean smoked tobacco, and during Columbus's voyage of 1492, he found the Arawaks smoking loosely rolled cigars. The Spanish took tobacco seeds to Europe, where Jean Nicot, France's ambassador to Portugal, sent tobacco to Paris in 1560 and gave the plant its genus (*Nicotiana*). In England, Sir Walter Raleigh began the popularization of pipe smoking in 1586, and the cultivation and consumption of tobacco spread with each voyage of discovery from Europe. Two kinds of tobacco were traded between Europe and America: "Spanish," from the West Indies and South America, and "Virginia," from the British plantations in their colony of Virginia. Despite its popularity in England, King James I forbade its production there since he vehemently disapproved of tobacco. Europeans at first smoked their tobacco in pipes, and later in cigars. It was often provided free to drinkers of coffee in coffee houses and cafés, as was the new product sugar. (Both remain strongly associated with coffee drinking.) Cigarettes spread in popularity only after the Crimean War (1854–1856), and their spread was especially aided by the first cigarette-making machine, developed in the United States in 1881.

NICOTINE is the most powerful ingredient of the tobacco plant, found primarily in the leaves. Nicotine is an extremely poisonous, colorless, oily alkaloid that turns brown upon exposure to the air. Nicotine can affect the central nervous system, resulting in respiratory failure and general paralysis. Nicotine can also be absorbed through the skin. Only two to three drops—less than 50 milligrams—of the pure alkaloid placed on the tongue can be rapidly fatal to an adult. A typical cigarette contains 15 to 20 milligrams of nicotine; however, the actual amount that reaches the bloodstream (and, therefore, the brain) through normal smoking is only about 1 milligram. Nicotine is responsible for most of the short-term as well as the long-term effects of smoking and plays a major role in the reinforcing properties.

CAFFEINE

CAFFEINE is an odorless, slightly bitter, alkaloid chemical found in coffee beans, tea leaves, and kola nuts, and several other plants used by humans such as cacao (CHOCOLATE) and maté (a South American holly used as a popular drink). In small amounts, caffeine acts as a mild stimulant and is harmless to most people. In large amounts, however, caffeine can result in insomnia, restlessness, and cardiac irregularities.

Tea. Tea is the beverage made when the processed leaves of the TEA plant are infused with boiling water. Native to Southeast Asia, the tea plant, *Camellia sinensis*, is a small, shrub-like evergreen tree that belongs to the family Theaceae. The seeds of the tea plant contain a volatile oil, and its leaves contain the chemicals caffeine and tannin. Although second to coffee in commercial value, tea ranks first as the most often consumed beverage. More than 50 percent of the world's population drink some form of tea every day. Many also use tea medicinally, as a stimulant. The tea plant originated in the region encompassing Tibet, western China, and northern India. According to ancient Chinese legend, the emperor Shen-Nung learned how to brew the beverage in 2737 B.C., when a few leaves from the plant accidentally fell into water he was boiling. Tea leaves began to be processed in China (dried, smoked, fermented, pressed, etc.) and were sold in cakes of steamed leaves, as powder, or in leaf form. Tea was

introduced by Chinese Buddhist monks into Japan (9th to 13th centuries), where the preparation and consumption of tea developed into the ritual tea ceremony called *cha no yu*. Tea culture then spread into Java, the Dutch East Indies, and other tropical and subtropical areas. British merchants formed the East India Company (1600–1858) and introduced teas from China and India into England, the American colonies, and throughout the British Empire.

Coffee. The COFFEE bean is the world's most valuable legal agricultural commodity. In 1982, for example, the coffee-importing bill for the United States alone was 2.537 billion dollars. Of the many varieties of the genus *Coffea* (family Rubiaceae) known to exist, only two species have significant commercial importance—*C. arabica* and *C. robusta* together constitute 99 percent of production. Coffee is native to the Ethiopian highlands and has been cultivated and brewed in Arab countries for centuries. The drink was introduced into Europe in the mid-seventeenth century and European colonial plantations were established in Indonesia, the West Indies, and Brazil, soon making coffee cultivation an important element in imperialist economies. Today, Latin America and Africa produce most of the world's coffee. The United States is the largest importer, having broken with the British tea tradition during the Revolutionary War to maintain the new American drink of coffee instead (purchased from non-British sources).

MARIJUANA

MARIJUANA is the common name given to any drug preparation derived from the hemp plant, CANNABIS SATIVA. Two varieties of this plant are *Cannabis sativa* variety *indica* and variety *americana*. The several forms of this drug are known by various names throughout the world, such as *kif* in Morocco, *dagga* in South America, and GANJA in India. HASHISH refers to a dried resinous substance that exudes from the flowering tops of the plant (also known as *charas* in Asia). In Western culture, cannabis preparations have acquired a variety of slang names, including grass, pot, tea, reefer, weed, and Mary Jane or MJ. Cannabis has been smoked, eaten in baked goods, and drunk in beverages. In Western cultures, marijuana is prepared most often from the dried leaves and flowering shoots of the plant as a tobaccolike mixture that is smoked in a pipe or rolled into a cigarette. As one of the oldest known drugs, cannabis was acknowledged as early as 2700 B.C. in a Chinese manuscript. Throughout the centuries, it has been used both medicinally and as an intoxicant. The major psychoactive component of this drug, however, was not known until the mid-1960s. This ingredient is TETRAHYDROCANNABINOL, commonly known as THC. PSYCHOACTIVE compounds (cannabinoids) are found in all parts of the male and female plant, with the greatest concentrations found in the flowering tops. The content of these compounds varies greatly from plant to plant, depending on genetic and environmental factors.

COCAINE

COCAINE is an ALKALOID drug found in the leaves of the coca plant, the common name of a shrub, *Erythroxylon coca*, of the coca family, Erythroxylaceae. Coca is densely leaved and grows to heights of 8 feet (2.5 m). It is cultivated in its native South America but also in Africa, Southeast Asia, and Australia for the narcotic alkaloids of its leaves, particularly cocaine. Whole or powdered dried leaves, usually mixed with lime (calcium carbonate), have been chewed by the people of what is now Colombia, BOLIVIA, and Peru for centuries, to dull the sense of hunger and to lessen fatigue. The coca shrub should not be confused with the cacao tree, the source of cocoa and chocolate.

Cocaine was first used in Western medicine as a local anesthetic. In 1884 it was used by Carl Koller, an ophthalmologic surgeon. Historically, the chief medical use for cocaine has been as a local anesthetic, especially for the nose, throat, and cornea, because of its effectiveness in depressing nerve endings. This has been largely replaced by less toxic, synthetic local anesthetics. Used systemically, cocaine stimulates the central nervous system, producing feelings of excitation, elation, well-being, enhanced physical strength and mental capacity, and a lessened sense of fatigue. It also results, however, in increases in heart rate, blood pressure, and temperature, and its use can result in death. Cocaine use became popular because of its stimulating properties. In Western countries, it is frequently ingested by sniffing its fine white powder, often called snow. It is sometimes injected intravenously, although repeated injections can result in skin abscesses, hepatitis, and the spread of AIDS. Cocaine can also be inhaled (smoked) once it has been converted to its free-base form; some preparations of freebase co-

caine are known as rock, or crack. CRACK-cocaine gained popularity in the late 1980s and early 1990s, because it is relatively inexpensive as a single dose, (e.g., $10 to $20 per "hit"); usually smoked in a special pipe, it produces an extreme euphoria as it is rapidly absorbed from the lungs and carried by the blood directly to the brain.

OPIUM

OPIUM is a drug obtained from the juice of the immature seed pods of the oriental poppy, *Papaver somniferum*. There are over 20 natural alkaloids of opium, including CODEINE and MORPHINE. Morphine is the largest component and it contributes most significantly to opium's physiological effects. HEROIN (diacetylmorphine) was derived from morphine and is the most important drug synthesized from opium's natural alkaloids. As a folk medicine, opium has been used to relieve pain, reduce such drives as hunger and thirst, induce sleep, and ease anxiety and depression. Opium and some of its derivatives are highly addictive, and their use has led to abuse and serious drug problems. Drugs from opium or derived from opium are still used widely in medicine, despite the development of synthetic opioid drugs such as MEPERIDINE (Demerol). The therapeutic effects of the opioids include PAIN relief, suppression of the cough reflex, slowing of respiration, and slowing of the action of the gastrointestinal tract. Opium's constipating effect led to its initial use, in the form of paregoric, in treating diarrheas and dysenteries. The main producers and exporters of opium are located in India and Turkey. About 750 tons (680 metric tons) of opium are annually needed to meet medical uses worldwide.

Opioids have been used since ancient times, both for medicinal purposes and for pleasure. Opium was taken orally, as a pill or added to beverages, for centuries in the Middle East, India, and Asia. Addiction did not become a wide problem until the practice of opium smoking was introduced by the British from India into China in the late seventeenth century (in an effort to gain a trade opening to the "closed" empire of China). China attempted to deal with the problem by restricting the cultivation and importation of opium in the nineteenth century. This led to the Opium Wars (1839–1842), since the opium trade became highly profitable to the British East India Company. Britain won over China, and opium was sold to the Chinese through treaty ports until the twentieth century.

In Europe and North America in the eighteenth century, opioids became widely used as most effective and reliable analgesics (painkillers). Heroin was developed in Germany in the 1890s and used from 1898 as a cough suppresser and analgesic with the hope that it would not lead to addiction, as did morphine (from which it was derived). From the first year or two after introduction, some clinicians agreed that it did not show addictive properties. A few even suggested that it might be useful in treating people addicted to morphine. Within a few years it became clear that, like morphine, its use could lead to addiction comparable in gravity to that of morphine.

On the street, opium is seen as a dark brown chunk of gum (from the pod of the opium poppy) or in dried powdered form. It is smoked, eaten, and drunk or injected as a solution for medicinal and recreational purposes. Indian and Chinese immigrants brought the practices with them, but the number of users is not great. During the early phases of addiction, opium produces a feeling of euphoria or well-being. With time, one may become dependent through physical and emotional factors. Tolerance develops and larger and larger doses of the drug are required to produce the same effect. If denied access to the drug, an addict will experience severe withdrawal symptoms; sudden withdrawal in a heavily dependent person has occasionally been fatal.

MESCALINE

PEYOTE, or mescal, is the common name of the small spineless cactus Lophophora williamsii, found in the southwestern United States and north-central Mexico. Peyote is used in Native American religious rituals, primarily for its HALLUCINOGENIC effects. At the end of the nineteenth century, Arthur Heffter demonstrated that MESCALINE (3,4,5-trimethoxy-phenethylamine) was responsible for peyote's pharmacological effects. Mescaline is related to the AMPHETAMINES. When ingested, it can produce HALLUCINATIONS, frequently of a visual nature, characterized by vivid colors, designs, and a distorted space perception. It stimulates the autonomic nervous system and can cause nausea, vomiting, sweating, tachycardia (rapid heartbeat), pupillary dilation, and

anxiety. The use of peyote in Native American ritual, referred to as Peyotism, was documented by Europeans in the sixteenth century. The modern practice of the peyote-based religion began in the late nineteenth century, was widely practiced by Native Americans in the southwestern United States, and was incorporated as the Native American Church in 1918. This church claimed more than 200,000 members in the 1960s. From the church member's point of view, peyote symbolizes spiritual power; the peyote "button"—the dried top of the cactus—is eaten as a sacrament to induce a hallucinogenic trance (of a few hours duration) for communion with God.

PSILOCYBIN

PSILOCYBIN is the active substance contained in the fruiting bodies of the Psilocybe mexicana mushroom (called the MAGIC MUSHROOM); it is a potent hallucinogen that can cause psychological disturbances. Taken orally or injected, the drug produces effects similar to those of the chemically unrelated LSD (LYSERGIC ACID DIETHYLAMIDE), and cross-tolerance has been experienced between psilocybin, LSD, and mescaline. The use of psilocybin is illegal in the United States, except for the direct consumption of mushrooms by a few religious groups as part of their ritual.

OTHER SUBSTANCES

Throughout the world, many other natural plant substances are used for mind- and mood-altering effects. These include the use of the KAVA root (Piper methysticum) for an intoxicating drink in the South Pacific; indole-containing snuff (distilled from indigo, genus Indigofera) among the Amazonian Indians of Brazil; KHAT leaves of a bush indigenous to East Africa containing an amphetamine-like drug (cathinone); BETEL NUT derived from the betel palm (Areca catechu) and widely used throughout the Pacific rim; and FLY AGARIC (a toxic mushroom, Amanita muscaria) among the Uralic-speaking tribes of Siberia.

(SEE ALSO: Alcohol: History of Drinking; Asia, Drug Use in; Ginseng; Ibogaine; Jimsonweed; Morning Glory Seeds; Nutmeg; Opioids and Opioid Control: History; Tobacco: Industry)

BIBLIOGRAPHY

BLOOM, F. E. (1988). Neurobiology of alcohol action and alcoholism. *Annual Review of Psychiatry, 8*, 347–360.

JAFFE, J. H. (1990). Drug addiction and drug abuse. In A. G. Gilman et al. (Eds.), *Goodman and Gilman's the pharmacological basis of therapeutics*, 8th ed. New York: Pergamon.

JAFFE, J. H., & MARTIN, W. R. (1990). Opioid analgesics and antagonists. In A. G. Gilman et al. (Eds.), *Goodman and Gilman's the pharmacological basis of therapeutics*, 8th ed. New York: Pergamon.

SCHULTES, R. E. (1981). Coca in the northwest Amazon. *Journal of Ethnopharmacology, 3*, 173–194.

SCHULTES, R. E. (1969). Hallucinogens of plant origin. *Science, 163*, 245–247.

NICK E. GOEDERS

POISON   A substance that, when introduced into the body in relatively small quantities, causes destruction or malfunction of some tissues and organs. Depending on the quantity in the body (the dose), a poison can kill. The word poison usually implies that a substance has no healthful use and is to be considered dangerous even in small quantities. Most common household substances are poisonous, including bleach, ammonia, drain cleaners, paint supplies, and so on.

SUBSTANCES CAUSING DEATHS FROM ACCIDENTAL POISONING

DRUGS
    Analgesics and antipyretics
    Sedatives and hypnotics
    Tranquilizers
    Antidepressants
    Other psychotropic agents
    Other drugs acting on nervous system
    Antibiotics and other antimicrobial agents
    Cardiovascular drugs
    Hormones
    Hematological agents
    Other drugs
OTHER SUBSTANCES
    Alcohols
    Cleaning and polishing agents and paint
    Petroleum products
    Pesticides
    Corrosives and caustics
GASES
    Utility gas
    Carbon monoxide
    Nitrogen oxides
    Freon
    Other gases

In the practice of medicine, many useful DRUGS, such as antibiotics for treating infections or antihypertensive drugs for treating high blood pressure, can be poisonous or toxic in higher doses. Almost all drugs that are abused can be poisonous or toxic; some, even at relatively low doses.

A few drugs that are commonly used in medicine in small amounts to produce important therapeutic effects are also used in other contexts as poisons. For example, the drug warfarin is used medically as an anticoagulant (to increase the time it takes blood to clot), an important effect for people who have had strokes or heart-valve replacement—but warfarin is also used as rat poison, because when rats eat it in large amounts they die soon after from massive hemorrhages. The same "mustard gas" (nitrogen mustards) that, as poison gas, caused much death and suffering in World War I, actually has medical use in the treatment of certain leukemias. Similarly, a series of extremely potent chemicals were developed during World War II as "nerve gases" for warfare, which act by flooding the body with excess acetylcholine (a body substance necessary for synaptic transmission), causing muscle paralysis and death. Consequently, close chemical relatives of some of the most potent nerve gases ever developed are being used to treat medical disorders, such as myasthenia gravis, in which there is not enough acetylcholine in nerve endings.

Treatment of someone who has been poisoned may require removal of the poison from the body (e.g., with the use of a stomach pump for ingested poisons), administration of an antidote if one exists, or simply support in repairing the damage done to the body. Many cities have a telephone "hot line" or poison-control center number where information about poisons, antidotes, and actions to take in case of poisoning can be obtained; often, they will alert emergency medical service (EMS) units to arrive in mere minutes. In case of a poisoning, including a drug overdose, it is essential to call for expert medical help as quickly as possible to minimize damage to the victim.

(SEE ALSO: *Complications: Medical and Behavioral Toxicity Overview; Drug Types; Inhalants; Methanol*)

BIBLIOGRAPHY

KLAASSEN, C. D. (1990). Principles of toxicology. In A. G. Gilman et al. (Eds.), *Goodman and Gilman's the pharmacological basis of therapeutics*, 8th ed. New York: Pergamon.

MICHAEL J. KUHAR

POISON CONTROL CENTERS *See* Appendix I, Volume 4.

POLICY ALTERNATIVES

This section includes two articles that introduce the reader to some of the issues surrounding public debate on the decriminalization of, or the legalization of, drugs. *Prohibition of Drugs: Pro and Con* is a short summary of the diverse opinions, expressed continually over the last 80 years, about the optimal way to deal with the reality that psychoactive drugs exist; that many people like the effects of those drugs; that some who use them do so to excess; that some are necessary for medical purposes; and that the substances themselves can be toxic not just for the user but for others who are affected by the user's behavior. The second article, *Safer Use of Drugs*, takes the view that society can reduce the toxic personal and social effects of drugs by informing potential users about how the risks of drug use can be minimized.

The argument that harm from drug use can be reduced by teaching people how to use drugs safely is viewed by many experts as counterproductive and likely to foster drug use. The *Partnership for a Drug-Free America* has developed its media campaign on the premise that the decision to try a drug is powerfully driven by two specific attitudes: perception of risk and social disapproval. This premise is supported by data emerging from the national *High School Senior Survey* (*Monitoring the Future* study) that the likelihood of drug use, especially initial experimentation, goes down as appreciation of the risks associated with drug use goes up. The more a young person feels that drugs are socially acceptable and/ or not dangerous, the more she or he is likely to try them. It is difficult to imagine an educational process that can teach "ways of using drugs safely" without simultaneously communicating a message of tolerable risk and a degree of social acceptability.

Prohibition of Drugs: Pro and Con

The history of U.S. social and legal policy in regard to psychoactive and intoxicating drug use has been characterized by periodic shifts, strong ideological presuppositions, and deep disappointment. Any analysis of current policy and the debate about drug legalization must recognize the historical roots of current policy that affect the various positions in the debate.

A brief historical note may help place the current discussion of drug policy in the United States in perspective. To borrow a phrase from Ecclesiastes, there is nothing new under the sun. Those engaged in the current, often heated, discussions about national drug policy often act as if their concerns, insights, and positions about intoxication, drug use, and society are unique to our age. A cursory review of history indicates that the debate on the meaning and effects of alcohol and other drug use on morals, public safety, productivity, and health is at least as old as written language. Some of the earliest recorded civilizations struggled with the issue and often adopted laws and policies that attempted to strictly regulate or prohibit the use of alcohol and other drugs.

Often these laws were based on a culture's perspective on the will of the divine or combined with basic civil codes. For example, the Torah appears to be very concerned with excessive alcohol use. It was seen as leading to gross immorality. The Christian New Testament holds similar views particularly on the excess use of alcohol. The theme seems to be one of avoiding all things that harm the body or one's relationship with God and moderation even in all things that are good. The Koran takes a very strong prohibition stand against alcohol and all intoxicating substances. Since much of modern Western civilization derives from these religious traditions, they continue to influence public thinking and policy. From a less theocentric perspective, many ancient civil codes also struggled with the regulation or prohibition of intoxicating chemicals. For example, the Romans seemed especially concerned that slaves and women not use alcohol and forbade its use by them. The concern appeared to be that alcohol would make slaves less productive and more difficult to control and that it would also lead to female sexual impurity. Chinese emperors prohibited the use of opium among their subjects. In addition, during the sixteenth and seventeenth centuries when tobacco use began to spread around the world, many societies, including the Ottoman Empire, Great Britain, Russia, and Japan, initially tried prohibiting the substance.

These ancient and more recent laws and codes show that the regulation or prohibition of socially perceived harmful substances is not new to our age, nor is the range of views on the negative consequences of regulation or prohibition and what would constitute a more effective, less harmful policy.

Among the many legacies that underpin the present discussion of drug policy in contemporary society are four at times overlapping and sometimes contradicting philosophical and cultural traditions. The first is the basic American heritage of individual liberty and limited government interference with any variety of human choice, even if that choice is harmful to the individual making the decision and morally repugnant to the majority of society. This position was eloquently argued by British philosopher and economist John Stuart Mill (1806–1873) in his essay *On Liberty* (1859). It perhaps finds contemporary expression in such a social phenomenon as the pro-choice movement and in the proponents of the legalization or decriminalization of drugs.

A second major social tradition is rooted in the moral utilitarian view of government that is also a part of the nation's heritage. The utilitarian perspective, also argued by Mill in his book *Utilitarianism* (1863), emphasized that government had a legitimate right to prohibit the behaviors that actually caused real harm to others. From this viewpoint, government had the right and responsibility to protect the common welfare by legally prohibiting individuals from engaging in behavior that was demonstrably harmful, not to themselves (which would have been an interference with liberty), but to other citizens.

The moral utilitarian perspective was an important underlying element in many of the late nineteenth- and early twentieth-century social-reform movements that culminated in the many state laws prohibiting narcotics and other drug use and the national HARRISON NARCOTICS ACT OF 1914 and the Volstead Alcohol PROHIBITION Act of 1920. The utilitarian perspective was that narcotics and alcohol use caused real harm to others and society in general in the form of family poverty, crime, violence, and health-care costs.

A third social tradition that has influenced U.S. drug policy is commercialism. There is ample evidence that through the nineteenth century, U.S. society had a strong commercial attitude toward alcohol use and the use of a variety of powerful drugs. As has been documented by historians, merchandise catalogs, as well as a variety of traveling entrepreneurs, legally distributed OPIUM, BARBITURATES, and COCAINE as wonderful cure-alls for the ills of the human condition. These merchants were an organized, respected part of the commercial establishment. Perhaps based on British narcotics commercialism, there has always been a commercial attitude toward alcohol and drug distribution in the United States. From the commercial perspective, alcohol and drugs are a wonderful commodity. They are often rapidly metabolized, highly addictive, and easily distributed. However, by the end of the nineteenth century, this rather freewheeling distribution of drugs caused a widespread public reaction that became incorporated into a variety of health- and social-reform movements.

A fourth significant element in the development of national alcohol and drug policy is a public-health perspective. As was noted, at the turn of the twentieth century, the United States was in the midst of major social and health reforms. After the passage of the 1906 Pure Food and Drug Act, a host of public-health-based government bureaus and regulations emerged, focusing on improving the quality of meats and other foods and requiring the accurate labeling of drugs. In addition, the American Medical Association initiated major reforms in the medical profession, eliminating over-the-counter narcotic drug advertisements in their journal and supporting the licensing of physicians as the only legitimate prescriber of many drugs. The public-health reform movements attempted to decommercialize drug distribution and make drug use a medical, not commercial, decision. The passage of the Harrison and Volstead acts probably represented a significant triumph of the moral utilitarian and public-health perspectives.

Following the Harrison Act and further legislation, the U.S. government instituted various bureaus and departments to carry out law enforcement and antidrug educational programs. Any review of the education programs of the Bureau of Narcotics would tend to conclude that they primarily constituted a heavy dose of propaganda with little basis in scientific fact. The federal proclivity for restricting the availability of drugs and arresting users and dealers continued strongly through the 1960s. During the decades following the Harrison Act and until the 1960s, the media and government were fairly united in their opposition to drug use, and there were few questions about the efficacy of drug laws or the social policy upon which those laws were based.

In the 1960s, U.S. society experienced the coming of age of the first of the baby boomers—those born between 1946 and 1958. By their sheer numbers, a proportion of this generation challenged the traditional socialization mechanisms of society and significantly questioned traditional assumptions, rationales, explanations, and authority. In a drive for generational self-discovery, drug use, particularly as a means to alter consciousness, became a part of the youth movement of the late 1960s and the 1970s. Most of the baby boomers who used drugs explored the use of MARIJUANA and HALLUCINOGENS, but over the same years HEROIN use was increasing in inner cities across the country; crime, too, was increasing. Despite the declaration of a "war on drugs" by the Nixon administration in 1970–1971, national surveys conducted during the 1970s and early 1980s showed annual increases in almost all types of drug use among high school seniors, household residents, and criminal-justice populations. The one exception was heroin, the major target of the Nixon drug war. Heroin use levels declined and then remained stable, but COCAINE use rose in the late 1970s and the 1980s, as did marijuana use among young people. By 1985, more than 20 percent of U.S. adults had taken drugs illegally, and for persons aged 18 to 34 more than 50 percent had done so.

Perhaps because of the fundamental changes in national drug-using behavior that occurred during this period, the modern movement to legalize drugs began. The basis of the argument was that (1) many of the drugs that were then illegal were not as harmful as government and media propaganda portrayed them to be; (2) drugs such as marijuana were relatively less harmful than alcohol and tobacco; and (3) the use of marijuana was a generational choice. In fact, the 1978 NATIONAL HIGH SCHOOL SENIOR SURVEY showed that in the prior thirty days, a higher proportion of seniors had smoked marijuana than

had smoked tobacco. By 1979, the media and American households were holding serious discussion about the legalization of marijuana, moving toward the BRITISH SYSTEM of heroin maintenance, and considering the legalization of cocaine as a nonaddictive stimulant. Social political movements such as NORML were organized to achieve passage of laws decriminalizing marijuana use. With the tacit support of the Carter administration, there were a few states, such as Alaska, that decriminalized the possession of small amounts of marijuana for personal use. Even the director of the National Institute on Drug Abuse in the late 1970s, Robert Dupont, appeared to accept the likelihood that marijuana would be decriminalized. However, in 1977, in reaction to growing marijuana use by young people and a perception that government itself was being tolerant of drug use, groups of parents organized a grassroots campaign to buttress the resistance to drug law liberalization. By 1978, the PARENTS MOVEMENT had become a force to be considered, and their views had ready access to the White House policy office. The apparently about-to-be-successful national movement to legalize many drugs in the 1970s came to an abrupt end with the 1980 election of President Ronald W. Reagan.

Corresponding with the election of President Reagan, there was a general conservative shift in national consciousness. First Lady Nancy Reagan, who made drug use among young people one of her prime topics of concern, was a welcome speaker at annual national meetings of the parents groups. The public debate on legalization during the early 1980s was also affected by increasing evidence of the physical and psychological consequences of drug use, declining illegal drug use among high school students, decreasing use among household members, and, maybe, the initiation of maturation among the baby boomers. During the 1980s, U.S. policy was characterized by the increasing intolerance of drug addiction or even recreational drug use. On an official level, this came to be called ZERO TOLERANCE.

According to the official federal policy of the 1980s, the assumption was that to a large extent drug use was an individual choice that could be affected by raising the cost of drug use to the users. It was believed that if enforcement reduced the availability of drugs, thus raising their prices, and increased the consequences of use by increasing the severity and certainty of punishment, individuals would choose to say no to illegal drug use. During the 1980s, funding shifted from a balance between demand reduction (treatment) and supply reduction (enforcement), to one primarily focusing on enforcement. The federal government became disengaged from a primary responsibility for treatment, while at the same time it increased its involvement in enforcement. The change in support was not dramatic at first. The total federal budget for all demand-side and supply-control activities was about 1.5 billion dollars in 1981, with about two-thirds allocated to law enforcement and supply control. This amount escalated sharply, starting in 1986, when President Reagan redeclared a "war on drugs." By 1989, the total had reached 6.7 billion dollars, with two-thirds allocated to controlling drug supply. The resources escalated still further during the Bush administration, reaching 12.2 billion dollars in fiscal year 1993.

By the end of the 1980s, the national drug-abuse policy of zero tolerance with a heavy focus on enforcement without any comparable increase in support for treatment began receiving critical reviews from policymakers, public administrators, clinicians, and academic researchers. These critical reviews were generally based on civil libertarian and public health harm-reduction perspectives. The key points made by national policy critics were:

1. About two-thirds of all felony arrestees in major metropolitan areas were currently using cocaine.
2. A large proportion of all criminal charges were drug charges. This had resulted in a significant expansion of prisons and the proportion of the population incarcerated. All of this had occurred at a very high economic cost.
3. The high profits from the drug trade were funding international terrorism and resulting in a rapidly increasing rate of violence in American urban areas.
4. Because of the vast amount of cash generated in the drug trade, there was an extraordinary amount of corruption at every level of each branch of government.
5. In an attempt to reduce illegal drug use, draconian laws focusing on search and property seizures had been passed that undermined hard-won civil rights.
6. Treatment availability for the poor had been reduced, with many cities reporting monthlong waiting lists for publicly funded treatment slots.

All of these real consequences have resulted in a major reinvigoration of the interest in legalizing or decriminalizing drug use. Those who argue for legalization come from a wide variety of professions and ideological positions, but they all essentially believe that U.S. society has reached the point where it can no longer afford to enforce existing law. There simply are not enough police, courts, prosecutors, or jail cells, nor is there the sense of justice that will allow U.S. society to enforce laws that have been broken by more than 20 percent of U.S. citizens.

In summary, the zero-tolerance just-say-no policy of the 1980s had come to be viewed by critics as resulitng in a virtual saturation of the criminal-justice and prison system with drug law offenders, the undermining of crucial civil rights, and the decreasing availability of drug treatment for the poor accompanied by increasing violence in high drug-trafficking areas and large-scale public corruption. Many critics came to view drug laws as contrary to the very basis of a libertarian civil government. These critics saw the war on drugs declared in the 1980s and continued to the present as inimical to civil liberty. In addition to the civil libertarian perspective, there are many critics of current drug-prohibition policy that focus on a public-health harm-reduction perspective. From this perspective, current policy is not reducing the public-health harm caused by drug use. Strict law enforcement and reduction in treatment availability has resulted in denying treatment to those being personally harmed by drug abuse. The public-health-reduction model emphasizes that drug abuse and addiction are the product of a complex set of psychological, sociological and economic variables that are very little affected by the threat of prison. This perspective argues that the best way to reduce the personal and public-health harm of drug use would be to increase drug education and prevention, increase drug-treatment availability, and reduce the harm caused by drug abuse by providing clean needles and, perhaps, decriminalizing use—thus significantly reducing the cost of drugs and the associated CRIME.

Although there are very few detailed legalization proposals, those who advocate decriminalization generally argue that national policy should move toward an approach in which the distribution of drugs such as marijuana, cocaine, and heroin would not be governed by criminal law but by governmental regulations that controlled the manufacture, distribution, and use of these substances so that they would go only to those already addicted or be dispensed under very regulated conditions. Advocates of this policy believe that the movement of drug policy from criminal law to regulatory restrictions would result in the relatively easy availability of drugs and inexpensive access to them for those who are addicted, thus resulting in a significant reduction in corruption and violence as well as an increasing willingness on the part of addicts to enter treatment. This, it is asserted, would relieve the severe overcrowding of the criminal-justice system. At the same time, it is argued, because of strict regulation, this policy change would more effectively protect young people as well as public health and safety than the current policy (see Nadelmann, 1988; Wisotsky, 1991).

Critics of the legalization perspective do not question many of the basic judgments of the consequences of the 1980s national policy, but they do severely question the assumptions upon which legalization is based. Those who are opposed to drug legalization often draw upon the moral utilitarian and public-health perspectives. They make the following arguments:

1. During the 1980s and continuing into the early 1990s, drug use, by all measures, significantly decreased among high school and college students as well as in the general population.

2. It is naive to assume that increasing availability, lowering cost, and reducing legal consequences will have no effect on the incidence and prevalence of marijuana, cocaine, and heroin use. From this perspective, it is argued that once these drugs are legalized, even though regulated, they will enter the arena of advocacy through free speech and thus the realm of market creation and expansion through advertising. Alcohol use, which is severely regulated and illegal for those under twenty-one years of age, is initiated in junior high school. In addition, about a third of high school seniors report being drunk each month. In most states, tobacco cannot be sold to minors, but smoking among junior high school students is common. These facts imply that regulation to make a drug available to one age group actually makes it available to all age groups.

3. The resulting increase in use in society and broadening of the societal base of use will result in detrimental health, behavioral, and economic consequences that will far outweigh any proposed benefit of legalization.

4. There is no broad societal base for legalizing drugs. Surveys among high school seniors clearly show that a large majority oppose the legalization of drugs—even the legalization of marijuana. Traditionally liberal countries such as Switzerland and SWEDEN have tried relaxing drug laws and were forced to modify their positions by their citizens, who daily had to experience the consequences of wide drug availability. Additionally, in a referendum in November 1991, Alaskans voted to rescind a marijuana legalization law passed in the 1970s and recriminalized marijuana possession. In a democracy, governmental policy cannot ignore the voice of the public. Finally, in the first presidential debate of the 1992 election, one of the few things that all three candidates agreed on was that drugs absolutely should not be legalized and that the criminal justice system plays a useful role in forcing users into treatment. Dr. Joycelyn Elders, the first Surgeon General in the Clinton Administration, was criticized for merely suggesting that the issue of legalization should be debated.

5. In these times of concern with HIV infection and AIDS, it may be hard to conceive of popular or governmental support for any policy that may increase injecting-drug use.

6. While the costs of drug law enforcement and incarceration of offenders may seem high, it is a misconception to assume that those incarcerated are all petty first-time violators of the drug laws. DiIulio (1993) asserts that ". . . in 1991 more than 93 percent of all state prisoners were violent offenders, repeat offenders (one or more prior felony convictions) or violent repeat offenders." He suggests that the vast majority of "drug" criminals were not convicted of simple possession, but of sale or manufacture. In short, most people would probably want to have these offenders behind bars even if the antidrug laws did not exist.

Many of those opposed to legalizing drugs, such as former Secretary of Health, Education and Welfare Joseph A. Califano, Jr., and Mathea Falco, a former Carter administration official, argue that the existing policy should be drastically modified to increase the availability of treatment and educational and economic opportunities in societal groups with high drug-use rates. Specifically, what is called for is an increase in treatment availability in the criminal-justice system, either through diversion or probation to treatment or through the provision of therapeutic services in jails and prisons, as well as a major increase in the availability of publicly funded treatment slots in the United States. It is argued that every dollar invested in treatment results in several dollars saved in terms of other social costs, including crime.

Some who oppose drug legalization believe that the current discussion has subtly eroded the public's will to fight illegal drug use. From this perspective, the only way to retain the *reduction* in general societal drug use that occurred during the 1980s is to retain a vigorous enforcement of drug laws. The advocates of strict law enforcement have taken note of the most recent high school surveys that indicate an increase in drug use among students, and they believe that this increase reflects the weakening of the war on drugs in the current administration and a kind of backdoor legitimization, a demoralizing discussion of the failure of drug policy. Previous drug policy leaders such as William J. Bennett argue that national drug policy during the 1980s was effective in reducing drug use in the general youth and adult population by making use morally, socially, and legally unacceptable and that the current discussion is making drug use more acceptable, resulting in recent increases in use (Bennett & Walters, 1995a, 1995b; Rosenthal, 1995). Bennett and Walters do not believe that support of treatment programs is a useful investment, and they would leave it to state governments to decide to exactly what degree treatment should be supported.

While it may be very difficult to reconcile the extremes of the drug legalization debate, there is some common ground that could emerge into a broadly acceptable public policy. Many involved in the current drug policy debate share a common belief that there is a need for increasing drug education, prevention and treatment availability, as well as expanding economic opportunities.

Some of those on both sides have strongly endorsed the need to restore the balance between interdiction and treatment in favor of treatment. Ignoring federal responsibility for treatment has been disastrous. Both sides would probably agree that a crucial priority for the federal administration would be to provide treatment availability for all those who seek it and to incorporate drug-abuse treatment into national health-care policy. In addition, many on both sides of the debate would probably also agree as to the convincing need of addressing basic questions of educational and economic opportunity, as well as that of institutionalized racism, which may function as societal underpinnings of drug-use epidemics.

(SEE ALSO: *Anslinger, Harry J. and U.S. Drug Policy; Crime and Drugs; Opioids and Opioid Control: History; Prevention Movement; Prohibition; Temperance Movement; U.S. Government Agencies*)

BIBLIOGRAPHY

BENNETT, W. J., & WALTERS, J. P. (1995a). Renewing the war on drugs. *Washington Times*, February 10.

BENNETT, W. J., & WALTERS, J. P. (1995b). Why aren't we attacking the supply of drugs? *Washington Times*, February 9.

BRECHER, E. M. (1972). *Licit and illicit drugs*. Boston: Little, Brown.

CALIFANO, J. A., JR. (1995). It's drugs, stupid. *New York Times Magazine*, January 29.

DIIULIO, J. J., JR. (1993). Cracking down. *New Republic*, May 10.

DUPONT, R. S. (1979). *Marihuana:* A review of the issues regarding decriminalization and legalization. In G. M. Beschner & A. S. Friedman (Eds.), *Youth drug abuse*. Lexington, MA: Lexington Books.

INCIARDI, J. A. (1992). *The War on Drugs II*. Mountain View, CA: Mayfield.

INCIARDI, J. A., & MCBRIDE, D. C. (1991). The case against legalization. In J. A. Inciardi (Ed.), *The drug legalization debate*. Newbury Park, CA: Sage Publications.

JOHNSTON, L. D., O'MALLEY, P. M., & BACHMAN, G. (1988). *Illicit drug use, smoking, and drinking by America's high school students, college students, and young adults 1975–1987* (National Institute on Drug Abuse, DHHS 89–1602). Washington, D.C.: U.S. Government Printing Office.

MCBRIDE, D. C. (1991). The case against legalization. In J. A. Inciardi (Ed.), *The drug legalization debate*. Newbury Park, CA: Sage Publications.

MILL, J. S. (1863). *Utilitarianism.* (Frazer's Magazine). London: Parker, Son and Bourn.

MILL, J. S. (1859). *On liberty*. Boston: Atlantic Monthly Press. (Reprint 1921).

NADELMANN, E. A. (1988). The case for legalization. *Public Interest, 92,* 3–31.

National Drug Control Strategy. (1995). The White House. Washington, DC.

REUTER, P. (1992). *Hawks ascendant: The punitive trend of American drug policy.* Santa Monica, CA: RAND.

ROSENTHAL, A. M. (1995). The cruelest hoax. *New York Times,* January 3.

TREBACH, A. S., & INCIARDI, J. A. (1993). *Legalize it? Debating American drug policy.* Washington, DC: American University Press.

WISOTSKY, S. (1991). Beyond the war on drugs. In J. A. Inciardi (Ed.), *The drug legalization debate.* Newbury Park, CA: Sage Publications.

YOUNG, J. H. (1961). *The toadstool millionaires: A social history of patent medicines in America before federal regulation.* Princeton, NJ: Princeton University Press.

DUANE C. MCBRIDE

Safer Use of Drugs   People commonly use drugs in safe ways, that is, nonabusively. Safe use means that drug use does not significantly impair health or interfere with social or economic functioning. For example, most users of alcohol consume that drug in moderation, not to the point of extreme intoxication, during specified hours, and for specified purposes, such as relaxation after daily work or promotion of social interchange.

Any drug can be used or abused, although some drugs and some ways of introducing them into the body may favor safe use. In general, less potent forms of drugs taken by mouth are more likely to be associated with safe use, whereas more potent forms taken parenterally (that is, introduced other than by way of the intestinal track) are less likely to be associated with safe use.

It is difficult to discuss the safe use of illegal drugs, because foes of those substances regard them as "drugs of abuse" that cannot possibly be consumed in nonabusive ways. This attitude is unhelpful. Whether a drug is used or abused has little to do with whether a drug is legal or illegal; it depends, rather, on the relationship an individual forms with it. One can as easily find examples of abusive use of legal drugs (TOBACCO, ALCOHOL, and OVER-THE-COUNTER medications) as of safe use of illegal ones. Take for example, the majority of coffee drinkers in our society who are addicted to the CAFFEINE in coffee (meaning they will have a withdrawal reaction on sudden cessation of intake). Many of these people also experience adverse effects on health as a result of their coffee addiction (cardiac arrhythmias, stomach and intestinal problems, irritation of the urogenital tract, tremors, insomnia, mood swings, and more). Many users of MARIJUANA, however, consume that drug moderately and occasionally, without suffering ill effects on health or behavior.

By observing safe use of drugs throughout the world—from Native Americans who use HALLUCINOGENIC PLANTS ritually to the many people who have figured out how to enjoy alcohol, tobacco, and caffeine nonaddictively and nonabusively—one can draw up a list of suggestions for users to increase the likelihood of safe use.

1. Know that the substance you are using is a drug or contains a drug.
2. Know how it affects your mind and body and what the risks are of moderate to excessive use.
3. Use lower potency (dilute) forms of drugs rather than higher potency (concentrated, refined) forms.
4. It is always safer to take drugs by mouth rather than by other routes of administration.
5. If the substances are illegal, it is important to know your sources in order to avoid adulterated, toxic, or misrepresented products.
6. Limit frequency of use by defining appropriate occasions and purposes for use. Regular, especially daily, use of any psychoactive drug commonly leads to loss of desired effects (tolerance) and to dependence.
7. Do not use any drug without good reason or just to go along with the crowd.
8. Seek advice about drugs from books and from people who know from experience what their real benefits and risks may be.
9. Reactions to drugs are strongly shaped by dose, mind set (expectations) and setting. Pay attention to these variables to reduce the risk of bad reactions.

Clearly, it is in society's interest to discourage the unsafe use of drugs. It is also in society's interest to foster the safe use of drugs by those who are inclined to use them. Of course, abstinence is a sure way to avoid problems, but there is no reason to think that most people will choose it in regard to drugs any more than they choose it in regard to sex. Therefore, providing good education about ways of using drugs safely should be a priority along with encouraging abstinence.

In addition, government drug policy should not work against safe use. Strongly prohibitionist policies may drive out of circulation dilute, natural forms of drugs, while encouraging the growth of black markets in concentrated, refined, and adulterated forms. This has certainly been the case with coca leaf and COCAINE. Coca leaf, with a low abuse potential and significant medical usefulness, has disappeared from our world, as powder and CRACK-cocaine have become more available—a change that has favored unsafe use rather than safe use. It would therefore be in society's interest to make dilute, low-potency forms of natural drugs more available.

(SEE ALSO: *Drugs from Plants; Education and Prevention; Partnership for a Drug-Free America; Prevention Movement*)

ANDREW T. WEIL

POLYDRUG ABUSE This term refers to the common observation that individuals who are considered drug abusers often abuse more than one type of drug. Almost all drug abusers smoke NICOTINE cigarettes and a large proportion consume alcoholic beverages, but many of them do not consider the co-occurrence of these two forms of drug use as an instance of polydrug abuse.

There are several types of polydrug abusers. They include those who abuse two or more substances but with a definite preference for one; only when they are not able to get supplies of their preferred drug do they abuse other types of drugs. These other types of drugs may either be from the same pharmacological class (e.g., HEROIN abusers may abuse other NARCOTICS as CODEINE or Demerol) or from different pharmacological classes (e.g., STIMULANT abusers—such as COCAINE abusers—may also use heroin, a narcotic). Some polydrug abusers do not necessarily have a favorite drug but instead may select different drugs for consumption at different times (e.g., stimulants in the morning, SEDATIVES at night) or under different conditions.

Polydrug abuse can also refer to the consumption of a drug to counteract an unpleasant effect produced by another drug or by withdrawal from another drug. For example, individuals who take enough stimulants to become highly agitated and aroused may take a tranquilizer to counteract the unpleasant side effects. Finally, polydrug abuse can refer to the consumption of different drugs simultaneously (e.g., speedballs). The assumption is that the different drugs in combination constitute more than the sum of their individual parts, producing a unique, highly reinforcing effect.

(SEE ALSO: *Barbiturates: Complications; Drug Abuse Warning Network; Drug Interactions and Alcohol; Prescription Drug Abuse; Sedatives: Adverse Consequences of Chronic Use*)

CHRIS-ELLYN JOHANSON

POPPY/OPIUM POPPY See Opium; *Papaver somniferum.*

POT See Marihuana; Slang and Jargon.

POVERTY AND DRUG USE Popular stereotype has it that drug use is more prevalent among the poor. In fact, a lack of money—in itself—does not seem to be associated with drug use. Empirical research has found, however, that in the United States, a number of attitudes, behaviors, and conditions linked to drug use also are linked to poverty, thus creating a situation that encompasses more than a lack of money. The study of poverty and drugs in the United States is complicated by the complexity of poverty as a conceptual category and by methodological problems in the measurement of drug use.

Webster's *Tenth New Collegiate Dictionary* defines poverty as "the state of one who lacks a usual or socially acceptable amount of money or material possessions." The sociological definition focuses on the relational aspect of poverty: Poor people are those who are at the bottom of a hierarchy of social stratification. Such a system is marked by unequal distribution of resources and income and also by differences in prestige, lifestyle, and values. In a review of the literature about poverty, Rossi and Blum (1969) listed the "critical features" of poverty—attitudes, behaviors, or conditions that are believed to distinguish poor from nonpoor people. Poor people often are categorized by unemployment or intermittent employment, low-status and low-skill jobs, unstable family and interpersonal relationships, low involvement in the community, alienation from the larger society, low aspirations, and individual feelings of helplessness. They noted that poverty was correlated with divorce and unhappy marriage, illegitimacy, low rates of voting, dropping out of school, high arrest rates and incidence of mental disorders, poor physical health, an increased incidence of mental disorders, and high mortality rates. Rossi and Blum concluded that poor people differed quantitatively, but not qualitatively, from nonpoor people; that is, the differences in their attitudes, their behaviors, and their conditions were differences of degree, not kind.

When studying the relationship of poverty to drug use, the literature is quite devoid of attempts to use the multidimensional conception of poverty. Instead, researchers have tended to choose one critical feature and look at its relationship to the use of specific drugs. Such studies have examined the association of U.S. drug use with income (e.g., National Institute on Drug Abuse, 1991); educational attainment (e.g., Zucker & Harford 1983); educational success (e.g., Schulenberg et al. 1994); employment (e.g., National Institute on Drug Abuse, 1991); HOMELESSNESS (e.g., McCarty et al. 1991); and neighborhood (e.g., Peterson and Harrell 1992). The results of these studies are largely inconclusive, thereby pointing not to a simple correlation between poverty (or poverty-linked attitudes, behaviors, and conditions) and drug use, but to more subtle pathways of direct and indirect effects.

One reason it is difficult to reach any valid generalizations about poverty and drug use is that poor people are a highly heterogeneous group. They live in all regions of the United States, in both rural and urban areas. Although poor people are more likely to be young and to belong to ethnic or racial minority groups, they are represented in all age and ethnic groups.

Much of the 1980s and 1990s sociological literature on poverty (as well as popular concern, as evidenced in the media) has focused on the concept of an American underclass (Mincy, Sawhill, & Wolf, 1990)—a population caught in an intergenerational cycle of poverty, isolated from mainstream society, living in an urban ghetto, and at risk for a number of social ills, including drug use. It should be noted that only a small proportion of poor people lead lives fitting this description. Many poor people are poor for only a short time (Duncan, 1984). That segment of the poor population described as an underclass, however, does appear to be at higher risk for use of some drugs, particularly CRACK-cocaine (Peterson & Harrell, 1992). The finding of an association between underclass status and drug use does not explain the mechanism by which it comes to exist. Researchers have speculated that increased use of crack-cocaine since its introduction in the mid-1980s may not be a direct effect of poverty upon the individual, but rather a result of increased availability; that is, a person living in certain urban, inner-city areas would simply have easier access to crack-cocaine than would a person living elsewhere (Lillie-Blanton, Anthony, & Schuster, 1993).

Collecting information about poverty and drug use has proved to be methodologically problematic. For example, surveys of drug use often are based on household samples. Those who are poor are less likely to live in stable households and more likely to live in extended or amorphous households—both situations that would result in their being excluded from such a survey. Some reporting of drug use comes from testing of arrestees, and this may introduce a bias in the estimation of the amount of drug use by people who are poor. In addition, many studies focus on certain drugs (e.g., crack-cocaine, HEROIN) and not others (e.g., MARIJUANA, COCAINE), and this may tend to misrepresent the extent of drug use among poor people as compared to its extent among people of the middle and upper classes.

Perhaps the greatest impact of poverty in relation to drugs is that on prevention and treatment. With private inpatient and outpatient treatment costing thousands of dollars and the long waiting lists for admission to publicly funded programs, impoverished drug users are less likely to obtain access to treatment. The heterogeneity of the poor and the lack of an empirical association between income level and drug use imply that making the poor the object of a targeted prevention and treatment effort might not be successful. Instead, the extant research on poverty and drug use suggests that policy efforts be directed at ensuring that lack of money does not become a barrier to participation in prevention and treatment programs.

ACKNOWLEDGMENT

We wish to acknowledge the contribution of KAREN CLARKE, who helped review the literature.

(SEE ALSO: *Alcohol: History of Drinking; Ethnicity and Drugs; Families and Drug Use; Homelessness and Drugs; Vulnerability As Cause of Substance Abuse*)

BIBLIOGRAPHY

DUNCAN, G. J. (1984). *Years of poverty, years of plenty.* Ann Arbor, MI: Institute for Social Research, The University of Michigan.

LILLIE-BLANTON, M., ANTHONY, J. C., & SCHUSTER, C. R. (1993). Probing the meaning of racial/ethnic group comparisons in crack cocaine smoking. *Journal of the American Medical Association, 269* (8), 993–997.

McCARTY, D., ET AL. (1991). Alcoholism, drug abuse, and the homeless. *American Psychologist, 11,* 1139–1148.

MINCY, R. B., SAWHILL, I. V., & WOLF, D.A. (1990). The underclass: Definition and measurement. *Science, 248,* 450–453.

NATIONAL INSTITUTE ON DRUG ABUSE. (1991). *National Household Survey on Drug Abuse: Main findings 1990.* Washington, DC: U.S. Government Printing Office. (DHHS Publication No. [ADM] 91-1788).

PETERSON, G. E., & HARRELL, A. V. (1992). Introduction: Inner-city isolation and opportunity. In A. V. Harrell and G. E. Peterson, (Eds.). *Drugs, crime, and social isolation.* Washington, DC: The Urban Institute Press.

ROSSI, P. H., & BLUM, Z. D. (1969). Class, status, and poverty. In D. P. Moynihan (Ed.), *On understanding poverty: Perspectives from the social sciences.* New York: Basic Books.

SCHULENBERG, J., ET AL. (1994). High school educational success and subsequent substance use: A panel analysis following adolescents into young adulthood. *Journal of Health and Social Behavior, 35,* 45–62.

ZUCHER, R. A. & HARFORD, T. C. (1983). National study of the demography of adolescent drinking practices in 1980. *Journal of Studies on Alcohol, 44,* 974–985.

NORA JACOBSON
MARGARET E. ENSMINGER

PREGNANCY AND DRUG DEPENDENCE: OPIOIDS AND COCAINE

During the 1980s, increasing numbers of pregnant drug-dependent women went to medical facilities—some to receive ongoing prenatal care, but others only to deliver their babies without the benefit of any prenatal care. Such women fear the threat of confrontation with legal authorities. The general lack of women-oriented drug-treatment programs contributes to this major health problem—addiction in pregnancy. It has also contributed to increased medical and social maladies and mortality in such mothers and their infants.

The 1990 NATIONAL HOUSEHOLD SURVEY ON DRUG ABUSE estimated that almost 50 percent, approximately 29 million of the 60 million women of childbearing age, used an illicit drug at least once in their lifetimes. In 1988, one study reported for the United States an annual occurrence rate (prevalence) of 11 percent, resulting in an estimated 375,000 drug-exposed births; these data cannot be applied to the entire country, since they were collected from a limited number of mainly urban hospitals—and the frequency, amount, type, and duration of drugs used were unavailable. The basis is also unclear for the reported estimates of 50,000 to 100,000 cocaine-exposed babies born each year. The occurrence of drug abuse among pregnant women varies widely in local studies—from 7.5 percent in Rhode Island, to 14.8 percent in Pinellas County, Florida, to 17 to 31 percent in a Boston hospital. These local rates cannot be used to estimate the prevalence of drug abuse among pregnant women in the United States; they can only provide data for averages.

As a result of the uncertainty among data sources, in 1992, the NATIONAL INSTITUTE ON DRUG ABUSE (NIDA) began a national hospital-based study known as the *National Pregnancy and Health Survey.* This survey collected data on the prevalence of licit and illicit drug use by pregnant women, limited data on infant birth weight, and the duration of hospital stay. The results were released in late 1994 and the summary tables are included here. Additional surveys in progress include the *National Maternal and Health Survey* conducted by the National Center for Health Statistics, which will collect data on drug-abusing women who had a live birth, stillbirth, or an infant who died before one year of age, and the *National Longitudinal Survey,* which collects data on the frequency of marijuana and cocaine use during pregnancy by women who have given birth to a child since 1986.

OPIOIDS

Due to preexisting conditions and ongoing active drug use, the opioid-dependent woman frequently suffers from chronic ANXIETY and DEPRESSION. So-

TABLE 1
Estimated Use of Selected Substances during Pregnancy: Total U.S.

Substance	Percentage		Population (in Thousands)	
	Estimate	95% C.I.*	Estimate	95% C.I.
Illicit drug use and nonmedical use of psychotherapeutics				
Any illicit drug use[1]	5.5	4.2–7.2	220.9	168.1–289.2
Marijuana	2.9	1.9–4.5	118.7	77.1–181.6
Cocaine	1.1	0.8–1.7	45.1	30.5–66.6
Crack	0.9	0.6–1.4	34.8	22.3–54.3
Other cocaine	0.3	0.1–0.7	12.7	6.0–26.6
Methamphetamine	0.1	0.0–0.4	4.5	1.2–17.3
Heroin	0.1	0.0–0.4	3.6	0.8–17.1
Methadone	0.1	0.0–0.4	3.4	0.8–14.2
Inhalants	0.3	0.1–0.7	12.1	5.1–28.6
Hallucinogens	0.2	0.1–0.6	8.7	3.0–25.1
Nonmedical use of any psychotherapeutics[2]	1.5	1.0–2.3	61.2	40.1–93.2
Amphetamines	0.0	0.0–0.3	1.2	0.1–13.4
Sedatives	0.3	0.1–0.8	10.3	3.4–30.8
Tranquilizers	0.0	0.0–0.3	1.9	0.3–12.9
Analgesics	1.2	0.8–1.9	48.7	30.2–78.4
Alcohol	18.8	16.2–21.7	756.9	653.4–872.7
Cigarettes	20.4	18.5–22.4	819.7	744.0–901.0
Medical use of any psychotherapeutics[3]	10.2	7.7–13.6	412.3	308.2–546.3
Amphetamines	0.3	0.2–0.7	13.4	6.4–28.0
Sedatives	3.6	2.3–5.6	144.1	91.3–225.7
Tranquilizers	1.4	0.8–2.4	55.4	31.2–97.9
Analgesics	7.6	5.6–10.2	305.2	226.2–408.9

*Confidence Index
[1]Use of marijuana, cocaine (all forms), methamphetamine, heroin, methadone, inhalants, hallucinogens, or nonmedical use of psychotherapeutics during pregnancy.
[2]Nonmedical use of any prescription amphetamines, sedatives, tranquilizers, or analgesics during pregnancy.
[3]Medical use of any prescription amphetamines, sedatives, tranquilizers, or analgesics during pregnancy.
SOURCE: National Institute on Drug Abuse, 1994.

cial problems, such as POVERTY, HOMELESSNESS, involvement in an abusive or battering relationship, and ALCOHOLISM, may overwhelm her ability to cope with life activities. She usually lacks confidence and hope for the future and has extreme difficulty with interpersonal relationships, especially with men. One study found that 83 percent of addicted women were raised in households marked by parental drug or alcohol abuse, 67 percent of those women had been sexually assaulted, 60 percent had been physically assaulted, and almost 100 percent of the women wished that they were someone else as they were growing up. In addition to these problems, the treatment and resolution of their addiction is a complex biopsychosocial matter which requires understanding and patience. Addiction is a chronic, progressive, relapsing disease, and one cannot expect a smooth and rapid recovery. It should not be surprising, therefore, that the lifestyle of the pregnant addict has a profound influence on her psychological, social, and physiological well-being.

She may have several other children who are currently not living with her, but instead with a relative or in placement. Drug-dependent women are fre-

TABLE 2

Estimated Percentage and Population of Infants Exposed *in utero* to Selected Substances: Total U.S.

Substance	Percentage		Population (in Thousands)	
	Estimate	95% C.I.*	Estimate	95% C.I.
Illicit drug use and nonmedical use of psychotherapeutics				
Any illicit drug use[1]	5.4	4.1–7.1	222.0	169.4–289.6
Marijuana	2.9	1.9–4.4	118.7	77.6–180.6
Cocaine	1.1	0.7–1.6	45.1	30.4–66.8
Crack	0.9	0.5–1.3	34.8	22.2–54.5
Other cocaine	0.3	0.1–0.7	12.7	6.0–26.7
Methamphetamine	0.1	0.0–0.4	4.5	1.2–17.3
Heroin	0.1	0.0–0.4	3.6	0.8–17.1
Methadone	0.1	0.0–0.4	3.4	0.8–14.4
Inhalants	0.3	0.1–0.7	12.1	5.1–28.7
Hallucinogens	0.2	0.1–0.6	8.7	3.0–25.3
Nonmedical use of any psychotherapeutics[2]	1.5	1.0–2.3	62.2	40.4–95.7
Amphetamines	0.0	0.0–0.3	1.2	0.1–13.6
Sedatives	0.3	0.1–0.8	10.3	3.4–30.8
Tranquilizers	0.0	0.0–0.3	1.9	0.3–13.1
Analgesics	1.2	0.7–2.0	49.8	30.5–81.1
Alcohol	18.6	16.0–21.6	762.0	654.1–883.1
Cigarettes	20.2	18.4–22.2	826.1	749.9–907.9
Medical use of any psychotherapeutics[3]	10.1	7.5–13.5	414.1	307.6–552.2
Amphetamines	0.3	0.2–0.7	13.4	6.4–28.0
Sedatives	3.6	2.2–5.6	146.0	91.6–230.6
Tranquilizers	1.4	0.8–2.4	55.4	31.1–98.3
Analgesics	7.5	5.5–10.1	305.2	224.9–411.1

*Confidence Index
[1]Use of marijuana, cocaine (all forms), methamphetamine, heroin, methadone, inhalants, hallucinogens, or nonmedical use of psychotherapeutics during pregnancy.
[2]Nonmedical use of any prescription amphetamines, sedatives, tranquilizers, or analgesics during pregnancy.
[3]Medical use of any prescription amphetamines, sedatives, tranquilizers, or analgesics during pregnancy.
SOURCE: National Institute on Drug Abuse, 1994.

quently intelligent, although the average level of high school achievement is usually at the tenth-grade level. Housing situations are frequently chaotic, and plans for the impending birth of the child may not have been considered.

It is well known that medical complications impact many drug-involved pregnancies; the most frequently encountered complications include anemia, various infections such as pneumonia, hepatitis, urinary tract infections, and sexually transmitted diseases. The women are at risk for human immu-nodeficiency virus (HIV) disease culminating in acquired immunodeficiency syndrome (AIDS).

The HIV disease has been increasingly linked to drug usage. The practice of sharing contaminated needles to inject HEROIN or COCAINE, the practice of prostitution to buy drugs, or the direct sex-for-drugs transaction associated with "crack" smoking have all contributed to this serious international health crisis. Currently, the spread of HIV is less linked to homosexual spread and more to heterosexual transmission and intravenous drug abuse. Although the exact risk

TABLE 3
Estimated Use of Selected Substances during Pregnancy, by Race/Ethnicity: Percentage, Total U.S.

| Substance | Race/Ethnicity | | | | | |
| | White, non-Hispanic | | Black, non-Hispanic | | Hispanic | |
	Estimate	95% C.I.*	Estimate	95% C.I.	Estimate	95% C.I.
Illicit drug use or nonmedical use of psychotherapeutics						
Any illicit drug use[1]	4.4	2.8–6.7	11.3	8.2–15.4	4.5	2.9–7.0
Marijuana	3.0	1.6–5.4	4.6	3.0–7.1	1.5	0.5–3.9
Cocaine	0.4	0.1–1.1	4.5	2.9–7.0	0.7	0.2–2.7
Crack	0.3	0.1–1.1	4.1	2.6–6.5	0.1	0.0–2.1
Other cocaine	0.1	0.0–0.6	0.7	0.2–2.1	0.6	0.1–2.8
Methamphetamine	0.1	0.0–0.7	0.1	0.0–1.9	(no)	(no)–(no)
Heroin	0.0	0.0–0.7	0.2	0.0–1.7	(no)	(no)–(no)
Methadone	0.1	0.0–0.5	0.2	0.0–1.7	(no)	(no)–(no)
Inhalants	0.2	0.1–0.8	0.4	0.1–1.8	0.4	0.1–2.7
Hallucinogens	0.1	0.0–0.6	0.3	0.1–1.7	0.7	0.2–2.2
Nonmedical use of any psychotherapeutics[2]	1.1	0.6–2.0	3.1	1.7–5.4	1.8	0.8–3.8
Amphetamines	0.0	0.0–0.6	(no)	(no)–(no)	(no)	(no)–(no)
Sedatives	***	***–***	1.0	0.3–3.5	0.4	0.1–1.8
Tranquilizers	***	***–***	(no)	(no)–(no)	0.2	0.0–1.7
Analgesics	1.0	0.5–1.9	2.2	1.2–4.1	1.3	0.5–3.3
Alcohol	22.7	19.3–26.4	15.8	11.5–21.3	8.7	5.4–13.7
Cigarettes	24.4	21.7–27.2	19.8	14.9–25.8	5.8	3.8–8.6
Medical use of any psychotherapeutics[3]	11.2	8.0–15.5	10.4	7.2–15.0	6.2	3.9–9.7
Amphetamines	0.3	0.1–0.9	0.4	0.1–1.8	0.3	0.1–1.7
Sedatives	4.0	2.2–7.0	4.4	2.0–9.5	1.4	0.5–4.2
Tranquilizers	1.1	0.7–2.0	2.9	0.8–9.6	0.9	0.3–3.3
Analgesics	8.3	5.8–11.7	7.7	5.6–10.7	4.4	2.7–6.9

*Confidence Index

[1]Use of marijuana, cocaine (all forms), methamphetamine, heroin, methadone, inhalants, hallucinogens, or nonmedical use of psychotherapeutics during pregnancy.

[2]Nonmedical use of any prescription amphetamines, sedatives, tranquilizers, or analgesics during pregnancy.

[3]Medical use of any prescription amphetamines, sedatives, tranquilizers, or analgesics during pregnancy.

, ***– Low precision, no estimate reported.

(no) No observations, C.I. not computed.

0.0 Estimate <0.05, rounded to 0.0 with valid C.I.

SOURCE: National Institute on Drug Abuse, 1994.

of an infected mother passing the disease to her offspring is not precisely known, it is estimated that approximately 25 to 30 percent of infants exposed in this fashion will actually contract AIDS. Counseling in an effort to prevent HIV infection, therefore, forms an essential part of services that must be offered to pregnant substance-abusing women or women involved in relationships with addicted men.

Nutritional deficiencies associated with drug addiction are due largely to the lack of proper food intake, which may result in iron and folic-acid deficiency anemias. Toxic responses to narcotics may

TABLE 4
Estimated Use of Selected Substances during Pregnancy, by Race/Ethnicity: Population (Thousands), Total U.S.

| | Race/Ethnicity | | | | | |
| | White, non-Hispanic | | Black, non-Hispanic | | Hispanic | |
Substance	Estimate	95% C.I.*	Estimate	95% C.I.	Estimate	95% C.I.
Illicit drug use or nonmedical use of psychotherapeutics						
Any illicit drug use[1]	113.1	72.8–174.0	75.0	54.2–102.4	28.1	18.0–43.3
Marijuana	77.5	42.5–139.5	30.8	19.9–47.2	9.1	3.4–24.2
Cocaine	9.2	3.0–27.9	30.0	19.3–46.2	4.4	1.2–16.4
Crack	7.0	1.7–28.4	27.2	17.1–42.9	0.6	0.0–12.6
Other cocaine	3.0	0.6–14.8	4.4	1.4–14.0	3.8	0.8–17.1
Methamphetamine	3.7	0.8–17.6	0.8	0.1–11.0	(no)	(no)–(no)
Heroin	0.8	0.0–15.8	1.3	0.1–11.6	(no)	(no)–(no)
Methadone	2.2	0.3–14.2	1.2	0.1–11.1	(no)	(no)–(no)
Inhalants	6.4	2.1–19.6	2.9	0.7–12.1	2.7	0.4–16.7
Hallucinogens	2.2	0.3–15.9	2.2	0.4–11.4	4.3	1.3–13.7
Nonmedical use of any psycho-therapeutics[2]	27.8	14.7–52.1	20.4	11.4–36.1	11.0	5.1–23.4
Amphetamines	1.2	0.1–15.0	(no)	(no)–(no)	(no)	(no)–(no)
Sedatives	***	***–***	6.6	1.8–23.3	2.2	0.4–10.8
Tranquilizers	***	***–***	(no)	(no)–(no)	1.0	0.1–10.6
Analgesics	24.9	12.2–50.6	14.8	7.8–27.6	7.8	3.0–20.1
Alcohol	588.6	501.0–686.5	105.0	76.6–141.4	53.6	33.3–84.7
Cigarettes	632.9	564.4–706.8	131.6	98.9–171.7	35.6	23.5–53.3
Medical use of any psychotherapeutics[3]	291.9	208.8–402.5	69.4	47.6–99.6	38.5	24.4–60.1
Amphetamines	8.9	3.5–22.4	2.5	0.5–11.7	2.0	0.4–10.6
Sedatives	102.9	57.6–181.3	29.4	13.3–63.2	8.7	2.8–26.2
Tranquilizers	29.6	17.0–51.5	19.0	5.4–63.6	5.8	1.7–20.1
Analgesics	215.1	150.4–304.1	51.5	36.9–71.1	27.0	16.9–42.5

*Confidence Index
[1]Use of marijuana, cocaine (all forms), methamphetamine, heroin, methadone, inhalants, hallucinogens, or nonmedical use of psychotherapeutics during pregnancy.
[2]Nonmedical use of any prescription amphetamines, sedatives, tranquilizers, or analgesics during pregnancy.
[3]Medical use of any prescription amphetamines, sedatives, tranquilizers, or analgesics during pregnancy.
, ***– Low precision, no estimate reported.
(no) No observations, C.I. not computed.
0.0 Estimate <0.05, rounded to 0.0 with valid C.I.
SOURCE: National Institute on Drug Abuse, 1994.

contribute to malnutrition by interfering with the body's ability to absorb or utilize nutrients. Abnormalities result because of the high incidence of altered function of the intestine, liver, and pancreas; malnutrition is often related to the presence of liver disease (since nausea causes addicts to eat infre-

quently or to vomit). Low sugar levels in the bloodstream or certain vitamin (B_6, thiamine) and mineral (magnesium) deficiencies may cause seizures in both alcoholics and drug addicts. Hepatitis, a viral infection of the liver, often accompanies the abuse of injectable drugs; it causes addicts to eat infrequently—

due to fatigue, swollen liver, nausea, and vomiting—which in turn diminishes the intake of nutrients, vitamins, minerals, and trace elements. Consequently, intensive diet therapy is needed in correcting drug and alcohol addiction—to balance fluids, electrolytes, trace elements, minerals, and vitamins—especially in acutely ill patients.

In addition to many potential medical problems, the lifestyle of some pregnant addicts becomes burdensome. To meet the high cost of maintaining a drug habit, she may often indulge in robbery, forgery, the sale of drugs, and/or prostitution. Because most of her day may be consumed by the activities of either obtaining drugs or using drugs, she spends most of her time unable to function in society's usual activities. She may have intermittent periods of normal alertness and well-being, but for most of the day, she will be either "high" or "sick." The high (euphoric) state will keep her sedated or tranquilized, absorbed in herself, and incapable of fulfilling familial responsibility. The sick (withdrawal) state is generally characterized by craving for more drugs, malaise, nausea, tearing, perspiration, tremors, vomiting, diarrhea, and cramps. Since hormonal changes in pregnancy manifest some of these symptoms in nondrug users, the sick state may be more frequent or intensified for addicts.

IMPACT OF MATERNAL OPIOID USE ON FETAL WELFARE

Opioid dependence in the pregnant woman is not only overwhelming to her own physical condition but also dangerous to that of the fetus (and eventually to the newborn infant). Because of her lifestyle, and because she may fear calling attention to her drug habit, the pregnant addict often does not seek prenatal care. Obstetrical complications associated with heroin addiction include miscarriages, premature separation of the placenta, infection of the membranes surrounding the FETUS, stillbirth, retardation of the growth of the fetus, and premature labor.

Because no quality control exists for street drugs, doses and substances used to stretch the dose may cause repeated episodes of underdose, withdrawal, and/or overdose. Maternal narcotic withdrawal has been associated with the occurrence of stillbirth. Severe withdrawal is associated with increased muscular activity, thereby increasing the rates of me-

tabolism and oxygen consumption; during maternal withdrawal, fetal activity also increases, as does the oxygen need of the fetus. The oxygen reserve in the placenta may not be able to supply the extra oxygen needed by the fetus. During labor, contractions further inhibit the blood flow through the uterus. If labor coincides with withdrawal symptoms in the mother, the fetus will also withdraw. Since uterine blood flow will vary at this time, and less oxygen will be delivered to the fetus, fetal death may occur.

COCAINE

Cocaine is known to cause many medical complications in adult users, including heart attacks, irregular heart beats, rupture of major blood vessels, strokes, fevers, seizures, infections, as well as a range of psychiatric disorders. The medical impact of cocaine on human pregnancy must consider all associated variables such as poverty, homelessness, inadequate prenatal and postpartum care, deficient nutrition, varying types of cocaine usage, multiple drug use, sexually transmitted diseases, and the possible presence of toxic chemicals that are mixed with or used to process cocaine.

Suppression of maternal appetite with inadequate nutritional intake is well recognized in cocaine "binging." Many cocaine users admitted for treatment may have at least one vitamin deficiency (B_1, B_6, C). Correction of these vitamin deficiencies is important during pregnancy so that essential chemicals (neurotransmitters) that transmit messages in the brain can be replenished.

Cocaine's chemical properties (low molecular weight and high solubility) allow it to cross the placenta easily and enter the fetus. The passage from maternal circulation to the fetus is enhanced by the injection or smoking of cocaine. In addition, because of acid/base balance issues and low levels of certain enzymes, which usually metabolize the drug, accumulation of cocaine in the fetus occurs. Furthermore, the "binge" pattern commonly associated with cocaine use may lead to even higher levels of cocaine in the fetus. Transfer of cocaine appears to be greatest in the first and third trimesters of pregnancy. Cocaine has a very potent ability to constrict blood vessels. A deleterious effect of this blood vessel constriction is fetal deprivation of essential nutrients and decreases in the amount of fetal oxygen. In addition to an acute oxygen deprivation, long time use of cocaine may produce a chronic decrease in nu-

TABLE 5
Estimated Use of Selected Substances during Pregnancy, by Age: Percentage, Total U.S.

| | Age in Years | | | | | |
| | Under 25 | | 25–29 | | 30 and older | |
Substance	Estimate	95% C.I.*	Estimate	95% C.I.	Estimate	95% C.I.
Illicit drug use or nonmedical use of psychotherapeutics						
Any illicit drug use[1]	5.7	3.5–9.3	5.1	3.4–7.6	5.5	3.9–7.6
Marijuana	3.5	1.7–7.1	2.4	1.3–4.2	2.8	1.8–4.3
Cocaine	0.4	0.1–1.3	1.6	0.9–3.0	1.6	0.9–2.8
Crack	0.2	0.0–0.8	1.3	0.7–2.6	1.3	0.7–2.5
Other cocaine	0.2	0.0–1.3	0.5	0.1–1.5	0.3	0.1–1.1
Methamphetamine	0.2	0.0–1.2	0.1	0.0–1.2	(no)	(no)–(no)
Heroin	(no)	(no)–(no)	0.1	0.0–1.4	0.2	0.0–1.0
Methadone	0.1	0.0–0.8	0.1	0.0–1.4	0.1	0.0–1.0
Inhalants	0.3	0.1–1.2	0.3	0.1–1.3	0.3	0.0–1.7
Hallucinogens	0.0	0.0–1.1	0.2	0.0–1.2	0.4	0.1–1.3
Nonmedical use of any psychotherapeutics[2]	1.7	1.0–2.8	1.2	0.5–2.6	1.5	0.7–3.4
Amphetamines	0.1	0.0–0.8	(no)	(no)–(no)	(no)	(no)–(no)
Sedatives	0.5	0.2–1.7	0.2	0.0–1.2	(no)	(no)–(no)
Tranquilizers	0.1	0.0–0.9	0.1	0.0–1.3	(no)	(no)–(no)
Analgesics	1.1	0.6–2.1	0.9	0.3–2.4	1.5	0.7–3.4
Alcohol	12.4	9.6–15.8	21.8	17.7–26.5	24.0	20.2–28.2
Cigarettes	21.9	18.5–25.8	19.4	16.5–22.8	19.3	15.3–24.0
Medical use of any psychotherapeutics[3]	10.1	6.9–14.4	10.1	7.6–13.3	10.6	6.9–15.9
Amphetamines	0.1	0.0–0.8	0.8	0.3–2.1	0.2	0.0–1.0
Sedatives	5.0	2.9–8.5	2.9	1.6–5.0	2.5	1.3–4.5
Tranquilizers	1.0	0.3–3.4	2.6	1.4–4.9	0.9	0.3–3.0
Analgesics	6.5	4.4–9.6	7.9	5.6–11.1	8.6	5.6–12.9

*Confidence Index
[1]Use of marijuana, cocaine (all forms), methamphetamine, heroin, methadone, inhalants, hallucinogens, or nonmedical use of psychotherapeutics during pregnancy.
[2]Nonmedical use of any prescription amphetamines, sedatives, tranquilizers, or analgesics during pregnancy.
[3]Medical use of any prescription amphetamines, sedatives, tranquilizers, or analgesics during pregnancy.
, ***– Low precision, no estimate reported.
(no) No observations, C.I. not computed.
0.0 Estimate <0.05, rounded to 0.0 with valid C.I.
SOURCE: National Institute on Drug Abuse, 1994.

trients and oxygen, leading to diminished growth of the fetus.

The use of cocaine by the mother may also affect the course of labor. CRACK (smokable cocaine in its base form) also appears to increase directly contractions of the uterus and may thus precipitate the onset of premature labor. Higher rates of early pregnancy loss and third-trimester separations of the placenta appear to be major complications of maternal cocaine use. Increased blood pressure and increased body temperature caused by cocaine may be responsible for early fetal loss and later separation of the placenta. The latter is hazardous to the fetus and the mother because of bleeding, shock, and the chance

TABLE 6
Estimated Use of Selected Substances during Pregnancy, by Age: Population (Thousands), Total U.S.

	Age in Years					
	Under 25		25–29		30 and older	
Substance	Estimate	95% C.I.*	Estimate	95% C.I.	Estimate	95% C.I.
Illicit drug use or nonmedical use of psychotherapeutics						
Any illicit drug use[1]	91.5	55.5–148.6	55.5	37.2–82.1	73.9	52.8–102.9
Marijuana	55.3	26.7–112.4	25.5	14.2–45.6	37.8	24.5–58.0
Cocaine	6.4	1.9–21.4	17.6	9.3–32.9	21.1	11.8–37.6
Crack	2.6	0.5–13.1	14.1	7.0–28.2	18.1	9.6–33.7
Other cocaine	3.8	0.7–20.6	5.0	1.5–16.1	3.8	1.0–15.0
Methamphetamine	3.3	0.6–18.7	1.1	0.1–13.3	(no)	(no)–(no)
Heroin	(no)	(no)–(no)	0.8	0.0–15.0	2.8	0.6–13.8
Methadone	1.4	0.1–12.8	0.8	0.0–15.0	1.2	0.1–13.7
Inhalants	5.2	1.4–19.4	3.2	0.7–13.8	3.6	0.6–22.7
Hallucinogens	0.6	0.0–18.1	2.2	0.4–12.9	5.9	1.9–18.2
Nonmedical use of any psycho-therapeutics[2]	27.4	16.7–44.9	12.9	5.8–28.3	20.9	9.5–45.6
Amphetamines	1.2	0.1–13.0	(no)	(no)–(no)	(no)	(no)–(no)
Sedatives	8.4	2.6–26.6	1.9	0.3–12.7	(no)	(no)–(no)
Tranquilizers	0.9	0.0–14.4	1.0	0.1–13.5	(no)	(no)–(no)
Analgesics	18.0	9.5–33.8	9.9	3.7–26.2	20.9	9.5–45.6
Alcohol	197.4	153.1–252.2	235.6	191.5–286.7	324.0	273.3–380.7
Cigarettes	349.3	295.2–410.3	209.9	177.8–246.2	260.5	207.2–323.6
Medical use of any psychotherapeutics[3]	160.2	110.6–228.8	109.4	82.3–144.1	142.6	92.6–215.0
Amphetamines	2.4	0.4–12.9	8.4	3.0–22.8	2.7	0.5–13.7
Sedatives	80.1	46.5–135.9	30.8	17.5–53.7	33.2	17.2–61.3
Tranquilizers	15.2	4.2–54.5	28.2	14.7–53.3	12.0	3.5–40.5
Analgesics	103.9	69.8–152.8	85.5	60.1–120.4	115.8	75.6–174.5

*Confidence Index

[1]Use of marijuana, cocaine (all forms), methamphetamine, heroin, methadone, inhalants, hallucinogens, or nonmedical use of psychotherapeutics during pregnancy.

[2]Nonmedical use of any prescription amphetamines, sedatives, tranquilizers, or analgesics during pregnancy.

[3]Medical use of any prescription amphetamines, sedatives, tranquilizers, or analgesics during pregnancy.

, ***– Low precision, no estimate reported.

(no) No observations, C.I. not computed.

0.0 Estimate <0.05, rounded to 0.0 with valid C.I.

SOURCE: National Institute on Drug Abuse, 1994.

of death for both, if an emergency cesarean section is not performed.

The major fetal effect of cocaine is retardation of growth, resulting in smaller than normal babies at the time of birth. Although animal studies suggest that cocaine may cause malformations of the fetus, data from studies in humans are contradictory. Some reports have shown an increased chance of abnormalities of the heart, limbs, and urinary tract, but others show no differences; studies in humans have not included large populations, and good scientific methods have not been utilized to control for many other factors that may contribute to abnormalities. Studies like these are very difficult to design for human populations.

It is currently thought that the incidence of malformations in infants as a result of cocaine taken by pregnant women is very low and that those that do occur are the result of disruption in the fetal blood vessels due to the constriction that occurs. This vessel constriction diminishes blood supply, which causes organs to malform at varying stages of fetal development. Abnormalities have been observed in the intestines, the kidneys, and the extremities.

RECOMMENDATIONS TO AMELIORATE THE EFFECTS OF DRUGS ON WOMEN AND THEIR CHILDREN

Despite the increased use of other drugs of abuse, such as cocaine, opioid abuse continues to be a major problem in the United States. Numerous investigators have reported the extremely high incidence of obstetrical and medical complications among street addicts, as well as the increase in medical conditions and death among their newborn infants.

Insufficient data exist for measuring the long-term effects of maternal drug usage. Controversy exists on how best to prevent and treat the adverse effects of addiction. It now seems clear, however, that providing comprehensive multidisciplinary drug-treatment services and prenatal care for addicts will significantly reduce the medical and psychological conditions and the death rate in both mothers and infants. Recommendations for treatment for drug-dependent women are multifaceted. The pregnant woman who abuses drugs must be designated as high risk; she warrants specialized care in a perinatal center where she can be provided with comprehensive addictive and obstetrical care and psychosocial counseling. Care must be provided in a supportive, proactive, and nonjudgmental fashion. The women must know that sharing of confidential information with health-care providers will *not* render them liable to criminal prosecution under state law statutes that define drug addiction in pregnancy as a form of fetal abuse.

Treatment of addiction in pregnancy may involve voluntary drug-free THERAPEUTIC COMMUNITIES, outpatient or day treatment, and, in narcotic-dependent women, METHADONE MAINTENANCE. The pregnant drug-dependent woman should be evaluated in a hospital setting where a complete history and physical examination may be performed and targeted laboratory tests carried out to evaluate her overall health status. Opioid dependent women should receive appropriate methadone maintenance, with support from an extensive medical and psychosocial network. Psychosocial counseling should be provided by experienced social workers who are aware of the medical needs, as well as the social and psychological needs, of these women. The pregnant woman addicted to BARBITURATES or major tranquilizers along with opioids should be medically withdrawn during her second trimester in a setting that furnishes appropriate monitoring of fetal well-being.

Maternal–infant attachment should have special emphasis. Parenting skills of these women need to be strengthened in an effort to nullify the anticipated (assumed) increase in child neglect and abuse that occurs in this population. Social and medical support should not end with the hospitalization. An outreach program, incorporating public health nurses and community workers, should be established. The ability of the mother to care for the infant after discharge from the hospital should be assessed by frequent observations in the home and clinic settings. Mechanisms by which to follow and supervise the infant's course after discharge from the hospital must be developed.

The major impact of comprehensive care, coupled with methadone maintenance for opioid-dependent women, has been the reduction of perinatal illness and mortality and the reduction of rates of low birthweight in offspring. Increases in birthweights, in themselves, have dramatically reduced illness and mortality for drug-exposed infants and children (mortality rates for low-birthweight newborns are forty times that of the full-term infants of normal weight).

Moreover, it is known that low-birthweight infants contribute greatly to the population of infants who test as mentally retarded (IQ of 70 or below), as well as those who will have great difficulty in school because they are "poor learners." These handicapped individuals will be unable to compete fully in our increasingly complex society. In addition, the incidence of cerebral palsy and lethal malformations are increased in low-birthweight infants. Emotional disturbances, social maladjustments, and visual and hearing deficits are also increased. With the increasing number of addicted women, custodial facilities for their mentally and neurologically deficient infants may be necessary if programs do not deal with prevention and treatment during pregnancy.

(SEE ALSO: *Addicted Babies; Complications; Fetal Alcohol Syndrome; Fetus: Effects of Drugs on the; Injecting Drug Users and HIV; Opioid Complications and Withdrawal; Substance Abuse and AIDS*)

BIBLIOGRAPHY

FINNEGAN, L. P., & KANDALL, S. R. (1992). Maternal and neonatal effects of drug dependence in pregnancy. In J. Lowinson et al., *Comprehensive textbook of substance abuse*, 2nd ed. Baltimore: Williams & Wilkins.

HADEED, A. J., & SIEGEL, S. R. (1989). Maternal cocaine use during pregnancy: Effect on the newborn infant. *Pediatrics, 84,* 205.

KALTENBACH, K., & FINNEGAN, L. P. (1988). The influence of the neonatal abstinence syndrome on mother–infant interaction. In E. J. Anthony & C. Chiland (Eds.), *The child in his family: Perilous development: Child raising and identify formation under stress.* New York: Wiley-Interscience.

ZUCKERMAN, B., ET AL. (1989). Effects of maternal marijuana and cocaine use on fetal growth. *New England Journal of Medicine, 320,* 762.

LORETTA P. FINNEGAN
MICHAEL P. FINNEGAN
GEORGE A. KANUCK

PRESCRIPTION DRUG ABUSE

Unfavorable responses to medical treatments—addiction to prescribed drugs or to those used in treatments—is termed *iatrogenic*. A wide array of medicines can be associated with addiction or abuse in some people. Such drugs include the OPIOIDS, antihistamines, anticholinergics, and steroids, among others—but the most common are those prescribed for psychological problems.

Some drugs acting on the mind have a low potential for abuse and dependence, for example, the ANTIPSYCHOTICS, antidepressants, and lithium salts. Others, such as the BARBITURATES and AMPHETAMINES, have a high potential.

BARBITURATES

Although barbiturates are more or less obsolete as tranquilizers and sleeping tablets, addiction to them is still encountered. TOLERANCE AND PHYSICAL DEPENDENCE can rapidly occur during therapy—and abrupt withdrawal can result in a severe and life-threatening withdrawal state. Studies in abusers show them to greatly prefer barbiturates to the BENZODIAZEPINES, which have replaced them pharmacologically and are discussed below. MEPROBAMATE, a carbonate used as a tranquilizer, is similar in many ways to the barbiturates, including its abuse potential.

Clinically, patterns of nonmedical use of nonopioids vary greatly; large quantities can be injected into a vein or muscle, often producing abscess formation. Other users take large amounts by mouth, on a binge or spree basis, the most popular being pentobarbital, amylbarbital, quinalbarbital, and Tuinal—the amylbarbital/quinalbarbital combination. Some users become permanently intoxicated and totally engrossed in maintaining their supply, licit or illicit. POLYDRUG use in combination with amphetamines or opioids is common.

Withdrawal can be hazardous, with the risk of SEIZURES or psychotic features, when discontinuing chronic usage of 500 milligrams a day or more. Withdrawal DELIRIUM (similar to DELIRIUM TREMENS, DTs) is common and often difficult to treat; a chronic state with HALLUCINATIONS may ensue.

BENZODIAZEPINES

The benzodiazepines supplanted the barbiturates because they seemed to be at least as effective, with few side effects and less likelihood of producing addiction. Benzodiazepines are preferred to placebo by drug abusers but vary in this regard; for example, diazepam (Valium) and lorazepam (Ativan) seem more likely to be taken than is oxazepam (Serax or Serenid). Benzodiazepines have been abused in various countries at various times. They have been injected as the main drug of abuse or as part of a polydrug-abuse pattern. Abusers of alcohol may also abuse benzodiazepines, finding that with drug interaction a potentiation occurs, that is, the combination is particularly powerful. Most benzodiazepine abuse is with drugs obtained legally from a number of complaisant prescribers, but the very heavy user may have to resort to illicit sources of supply. About 50 percent of abusers of benzodiazepines were introduced to the drug within the medical context.

Within polydrug abuse, the benzodiazepine is used to eke out the supply of opioid or to ease the crash from the high euphoria of COCAINE use. Patterns of usage and beliefs about the possible effects of benzodiazepine use vary widely among hard-drug abusers, but, generally speaking, benzodiazepines are viewed as potential drugs of dependence in their own right and not as relatively innocuous adjuncts.

It is fairly uncommon for patients started on benzodiazepines for therapeutic purposes to increase their dosage steadily. Nevertheless, since benzodiazepine use is widespread, high-dose users are seen fairly often. It is unclear why some patients escalate their dosage, whereas most remain at therapeutic levels indefinitely.

AMPHETAMINES

Amphetamines are stimulants, which raise mood, increase the sense of well-being, energy, and alertness, and decrease appetite. Some few users, paradoxically, become the opposite—drowsy, anxious, and irritable.

Normal-dose usage was typically prescribed; an obese, middle-aged, mildly depressed housewife took two or three doses every day as a pick-me-up, a mild stimulant and appetite suppressant. (Some weak physical dependence ensued from such use, mainly seen as sleep changes on withdrawal.) With the discouragement of such indications, usage by physicians and patients has fallen off. Another obsolete use was as a vigilance-enhancer in those who felt the need to keep awake for excessively long periods, such as medical interns or long-distance truck drivers. Few people progressed from iatrogenic oral misuse to intravenous abuse.

Intravenous amphetamine produces euphoria, similar to but more sustained than that following the use of cocaine. After a few hours, the effects wear off, leaving the abuser feeling exhausted, drowsy, and depressed. Clandestine laboratories manufacturing amphetamine are still at work. Their preferred substance is METHAMPHETAMINE, which can be synthesized easily. Since intravenous use of methamphetamine is usual, and tolerance quickly occurs, larger and more frequent doses become required to achieve the desired effect. Toxic effects supervene, with repetitive face and hand movements and stereotyped behavior—for example, the user assembling and dismantling mechanical objects. A full-blown paranoid type of psychosis may develop, with loss of reality and delusions of persecution. Individual susceptibility to these toxic effects varies greatly. Polydrug abuse of amphetamines is common; co-administration of amphetamine with heroin ("speedball") or a barbiturate is believed to optimize the pleasurable effects while minimizing the toxic ones.

APPETITE SUPPRESSANTS

Appetite suppressants cover a range of compounds, from the decongestant phenylpropanolamine (often available without prescription), to powerful amphetamine analogues (chemical variants). Most are stimulant, although one, fenfluramine, is quite sedative. As with the amphetamines, patterns of use and abuse vary a great deal, from chronic daily ingestion of a therapeutic dose to binge or spree use of large quantities. As a general rule, the more amphetaminelike the appetite suppressant, the more likely is abuse.

Trying to stop the use of appetite suppressants may be difficult for abusers, because of withdrawal symptoms of tiredness, dysphoria (discomfort), or frank depression. These problems and growing doubts about sustained effectiveness (for their original dietary purposes) have led many doctors to cease prescribing them.

(SEE ALSO: *Iatrogenic Addiction*)

BIBLIOGRAPHY

ATOR, N. A., & GRIFFITHS, R. R. (1987). Self-administration of barbiturates and benzodiazepines: A review. *Pharmacology, Biochemistry, and Behavior, 57,* 391–398.

CLAYTON, R. R., VOSS, H. L. & LOSCUITO, L. A. (1987). *Main findings: 1985 National Household Survey on Drug Abuse.* Rockville, MD: National Institute on Drug Abuse.

WESSON, D. R. & SMITH, D. E. (1977). *Barbiturates: Use, misuse, and abuse.* New York: Human Sciences Press.

M. A. LADER

PREVENTION   This section contains articles describing the evolution and ongoing activities of "grassroots" efforts to empower families and communities to prevent substance abuse through education, mutual support, and lobbying efforts. These articles are representative, but by no means all-inclusive, of prevention and education activities that have been organized at the community level—they serve merely as an introduction. Included in this section are: *Community Drug Resistance; National*

Families in Action; National Federation of Parents for Drug-Free Youth/National Family Partnership (NFP); Prevention of Alcoholism: The Ledermann Model of Consumption; Shaping Mass-Media Messages to Vulnerable Groups. For an overview of this important and influential movement see the entries, *Parents Movement* and *Prevention Movement.* Also see *Education and Prevention* and the section on *Prevention Programs.*

Community Drug Resistance   By the 1980s, many urban neighborhoods in the United States became seriously debilitated by the departure of middle-class residents to the suburbs, the influx of illegal immigrants, growing unemployment rates, weak family structures, and a host of other underclass problems. In the mid-1980s, the proliferation of cheap CRACK cocaine, used mainly by inner-city adolescents and young adults, transformed a bad situation into a desperate one. For some residents, this new upsurge in drug use was the last straw, they got angry and began looking for ways to reclaim their neighborhoods and their children.

The fear and anger fueled by COCAINE use had already breathed new life into the community anticrime movement of the 1970s. Social scientists and policymakers had concluded that community anticrime programs were unlikely to arise spontaneously in poor, crime-ridden neighborhoods where they were most needed—and, indeed, could not even be implemented successfully by professional organizers (Rosenbaum, 1986). But research on citizen antidrug programs has found that, actually, they are most likely to arise in poor neighborhoods, where drug activity is most common (Davis et al., 1991).

TYPES OF COMMUNITY ANTIDRUG PROGRAMS

Weingart (1992) proposed a typology for understanding citizen antidrug organizations. He defines four types of programs: (1) Law-enforcement enhancement, (2) civil justice, (3) treatment and prevention, and (4) community building. Weingart argues that community antidrug efforts are overwhelmingly dominated by the first category of program—those that aim to complement the activities of law-enforcement agencies.

Law-Enforcement Enhancement.   Block-watch programs train participants to observe drug activity from their homes and to report it—usually to a designated member of the block-watch organization—in as much detail as possible (descriptions of suspects, locations of drug caches, license numbers of buyers). That person relays the information periodically to a designated police-liaison officer and, in return, the police-liaison officer reports back to the organization on the form of action taken as a result of the complaints.

Citizen patrols are commonly used programs that enhance law-enforcement efforts. Patrols vary in the degree of confrontation they use with drug dealers. Some simply observe and call their base or the police when drug sales are spotted; others have gone as far as obviously photographing or otherwise harassing drug dealers. Experience suggests that just a few individuals patrolling can be effective in removing drug activity from a neighborhood, although it has proven difficult to maintain residents' commitments to participation over extended periods of time.

Civil-Justice Efforts.   These involve bringing suits against drug dealers in civil court for actionable nuisances (the noise and violence that accompany drug activity) and bringing suits against property owners, demanding abatement of a nuisance (drug selling) at a particular location. These are by far the most common type of civil action, and many major cities have made enforcement of nuisance-abatement laws a priority of the city attorney's office. Civil actions against property owners seem to be highly effective in abating drug sales at the targeted location, usually through eviction of drug sellers (and other occupants of their apartments). Questions remain about whether these actions simply displace the problem to another locale and whether they violate the rights of property owners and (at times) innocent tenants (Smith, Davis, & Hillenbrand, 1992).

Treatment and Prevention Programs.   Such programs often rely on the voluntary efforts of drug abusers, their families, and their neighbors to help one another. Community-treatment programs range in size from those that are part of national organizations—like COCAINE ANONYMOUS and NARCOTICS ANONYMOUS—to small grass-roots programs, which are often church affiliated. Local drug-education efforts are usually citywide rather than neighborhood based, and they often work through the schools. Other prevention programs offer neighborhood youths supervised recreational activities for self enhancement and as an alternative to the drug culture.

Community Building. Finally, some community-building efforts try to unite local residents against drugs through vigils, rallies, and marches. These kinds of activities are common in major U.S. cities. Other community groups have fought back against drugs by eliminating signs of disorder—such as uncollected trash and graffiti; by enhancing the appearance and safety of the neighborhood—installing better street lighting or clearing refuse and planting flowers; or by demanding that local officials raze the abandoned buildings and clear the refuse-filled empty lots used by drug abusers and drug dealers.

LINKS TO POLICE ORGANIZATIONS

Some community antidrug efforts (most notably Black Muslim patrols in Washington, D.C., in Brooklyn, N.Y., and elsewhere) have been mounted without the involvement of the police. Such efforts are relatively rare, however. In the vast majority of programs, potential activists have found a willing ally in the police. At the time that citizens were attempting to organize against drugs, many local police departments were in the process of undergoing a conversion to a community-oriented approach to law enforcement, which invited citizen participation (Skogan, 1990).

Davis et al. (1991) report that the police played a critical role in the maintenance, and sometimes the formation, of community antidrug programs. Although police administrators are normally the first to state that they cannot mobilize a neighborhood against drugs, they often facilitate incipient organizations by donating space and speakers for meetings, by acting as advocates with other city agencies (sanitation, building inspection, etc.), by providing training and backup for patrols, and by bringing together leaders from different neighborhoods to cross-pollinate ideas. (See Garofalo & McCleod, 1986, for a similar conclusion about the police role in earlier citizen anticrime programs.)

RISKS OF VIGILANTISM

The same war on drugs that promotes the vigorous enforcement and prosecution of drug cases may also have adverse consequences, including the erosion of personal freedoms and the promotion of vigilantism. Since about 1980, the Supreme Court has responded to public outcries for tougher action against drug dealers by upholding cases that permit broader latitude in surveillance and search activities. Furthermore, very aggressive citizen efforts (such as the Black Muslim patrols and the Guardian Angels) have been criticized for harassment and violent assaults against drug dealers.

PROGRAM EVALUATION

Because community antidrug programs are new, the media have been the main source of information about their activities; however, a handful of studies have been undertaken to explore the implementation and impact of community antidrug programs. Davis et al. (1991) examined the kinds of communities that have spawned antidrug efforts and the effects the programs have had on residents' perceptions of crime and disorder. Contrary to extant theories of community organizing—which suggest that resident programs against crime can only be mounted successfully in middle-class areas—the investigators found that antidrug initiatives were more common in low-income neighborhoods, even after taking into account the fact that such neighborhoods had more drug activity. In addition, the study looked closely at four of these initiatives. Residents in neighborhoods served by the programs reported lower fear of crime and greater neighborhood satisfaction than residents of comparable nearby areas without programs.

Rosenbaum and his colleagues studied the initiation of a national demonstration program called Community Responses to Drug Abuse (CRDA). Using federal funds, ten communities in nine U.S. cities implemented a variety of antidrug projects, including closing drug houses and creating drug-free zones. The researchers interviewed participants, observed program activities, and analyzed records at the ten sites. Rosenbaum et al. (1992) report that the local community organizations accomplished a great deal with limited funds. A crucial lesson learned by the organizations was that enforcement activities provide only a limited solution to the drug problem. The most effective strategies involved broader partnerships with other agencies and institutions, such as churches and schools.

Finally, the federal Community Partnership Demonstration Program, funded by the Office for Substance Abuse Prevention, provides assistance to more than 250 programs for the prevention of substance abuse and now the CENTER FOR SUBSTANCE ABUSE PREVENTION (CSAP) allows local organizations considerable discretion to shape their own initiatives in combating drug and alcohol abuse.

(SEE ALSO: *Crime and Drugs; Education and Prevention; Gangs and Drugs; Prevement Movement*)

BIBLIOGRAPHY

DAVIS, R., SMITH, B., LURIGIO, A., & SKOGAN, W. (1991). *Community response to crack: Grassroots anti-drug programs.* Final report to the National Institute of Justice. New York: Victim Services Agency.

GAROFALO, J., & McLEOD, M. (1986). *Improving the effectiveness of neighborhood watch programs.* Unpublished report to the National Institute of Justice from the Hindelang Criminal Justice Research Center, State University of New York at Albany.

ROSENBAUM, D. (1986). The theory and research behind neighborhood watch: Is it a sound fear and crime reduction strategy? *Crime & Delinquency, 33,* 103–134.

ROSENBAUM, D. ET AL. (1992). *The community response to drug abuse national demonstration program: Final process evaluation report.* Report submitted to the National Institute of Justice. Chicago: Center for Law and Justice, University of Illinois.

SKOGAN, W. G. (1990). *Disorder and decline: Crime and the spiral of decay in American cities.* New York: The Free Press.

SMITH, B, DAVIS, R., & HILLENBRAND, S. (1992). *Ridding private sector housing of drug havens.* Report to the National Institute of Justice. Washington, DC: American Bar Association.

WEINGART, S. (1992). A typology of community responses to drugs. In R. Davis et al. (Eds.), *Drugs and community.* Springfield, IL: Charles C. Thomas.

ROBERT C. DAVIS
ARTHUR J. LURIGIO

National Families in Action In November 1977 a group of concerned citizens in Atlanta, Georgia, troubled by the emergence of commercial and environmental pressures that seemed to encourage people to use addictive drugs, formed National Families in Action. These commercial and environmental pressures coincided with an escalation in drug use among children and young adults to the highest levels in the history of the world. The organization's founders—parents, doctors, law-enforcement officials, political leaders, educators, business leaders, and others—sought to replace the glamorization of drug use with accurate, reliable information based on scientific research about drug effects.

Initially, National Families in Action targeted the drug-PARAPHERNALIA industry. If drugs such as MARIJUANA and COCAINE were illegal, the group reasoned, it made no sense to allow the sale of implements to enhance their use. Three months after its founding, National Families in Action got the Georgia legislature to pass the nation's first laws prohibiting the sale of drug paraphernalia. Publicity surrounding this event brought calls from people across the United States who wanted to organize similar groups to ban drug-paraphernalia sales in their communities. They also wanted to educate their families and communities about the harmful effects of drugs, to prevent drug use before it started, to help users stop, and to find treatment for those who couldn't stop using drugs. The organization published a manual, *How to Form a Families in Action Group in Your Community,* which helped many thousands of groups organize. In addition, members traveled throughout the United States to help families organize community-based, substance-abuse-prevention groups.

National Families in Action established a drug-information center, collecting articles from medical and scientific journals about all aspects of substance abuse, including research about drug effects, prevention of use, intervention, and treatment. It also collects articles from newspapers and magazines about drug policy and the emergence and growth of the grass-roots prevention movement in the United States (and, increasingly, abroad). In addition, the collection houses publications of the drug-paraphernalia industry and organizations that advocate drug legalization. National Families in Action's drug information center contains more than 500,000 documents on substance abuse. The center answers questions from people throughout the world who call or write for information.

In 1982, National Families in Action began publishing *Drug Abuse Update,* a quarterly digest containing abstracts of articles collected at the center. In 1990, the organization introduced *Drug Abuse Update for Kids,* written for children in elementary and middle schools. It publishes other drug-education materials as well, including a curriculum about drugs and the brain titled *You Have the Right to Know*. From 1984 to 1990, National Families in Action's executive director, Sue Rusche, wrote a twice-weekly column on substance abuse that was syndicated by King Features to more than 100 newspapers throughout the nation.

In 1990, National Families in Action received a demonstration grant from the CENTER FOR SUBSTANCE ABUSE PREVENTION to help families who live in two Atlanta public housing developments prevent substance abuse in their communities. Called Inner City Families in Action, this project was named one of eleven exemplary programs in the United States in 1993. The program trains parents to teach *You Have the Right to Know* to neighbors, friends, and children. It helps parents obtain needed skills to complete their education and enter the work force. It also helps parents form Families in Action groups to seek treatment for loved ones who are addicted to drugs, to engage children in productive activities, and to prevent substance abuse in their communities.

Throughout its history, National Families in Action has developed numerous networks and national coalitions to advance the field of substance-abuse prevention. These include the Prevention, Intervention and Treatment Coalition for Health (PITCH), an association of community-based prevention organizations that serve many different ethnic and cultural groups throughout the nation. Through its advocacy efforts, PITCH helped bring about the creation of a new federal agency, the Substance Abuse and Mental Health Services Administration, to further develop the prevention field. National Families in Action is increasingly called upon to help citizens from other nations develop prevention groups.

Along with other national prevention organizations, National Families in Action has played a pivotal role in driving drug use down since 1979.

(SEE ALSO: *Education and Prevention; Marihuana Commission*)

SUE RUSCHE

National Federation of Parents for Drug-Free Youth/National Family Partnership (NFP)
A large number of parent-group leaders, who had previously organized drug-prevention groups in their states and local communities, formed the National Federation of Parents for Drug-Free Youth in the spring of 1979. With the assistance of national organizations in Atlanta, Massachusetts, and elsewhere, these leaders had organized prevention groups of parents in response to the greatest escalation in drug use by American adolescents in the history of the world. They organized to protect children by striving to prevent drug use before it began, by helping young drug users to stop, and by obtaining treatment for those who couldn't stop by themselves.

During the 1970s, legislatures in eleven states decriminalized MARIJUANA. During this same period, an explosion of head shops proliferated throughout the United States. These places, which sold PARAPHERNALIA to enhance drug use, targeted their products to children and teenagers. The national decriminalization discussion produced rhetoric that ignored or played down the harmful effects of drugs, and this rhetoric spilled over into drug-education materials, which counseled the "responsible use" of drugs that were both dangerous and illegal. Song lyrics and films in the adolescent culture tended to reinforce the popularity and acceptance of drug use. These factors appeared to contribute to, if not actually drive, the astonishing escalation in adolescent drug use throughout the 1970s.

Parent groups organized to prevent children from entering the drug culture and to rescue those who already had, taking aim at the drug-paraphernalia industry and fighting decriminalization. By 1979, however, it had become clear that action at local and state levels was not enough. Representation at the national level was critical, particularly in light of the fact that a federal bill to decriminalize marijuana was gaining support from members of Congress. Parent-group leaders formed the National Federation of Parents for Drug-Free Youth to represent their interests in Washington.

The first order of business was to defeat the pending federal decriminalization bill, which would have removed criminal penalties for the possession of up to an ounce of marijuana. Federation volunteers bought 1-ounce jars of parsley to demonstrate that an ounce was not an insignificant amount, and to reinforce the fact that an ounce of marijuana can yield from forty to sixty "joints." They delivered these jars to each member of Congress, educating senators and representatives about the high levels of marijuana and other drug use that decriminalization in some states appeared to have produced among young people, and asking them to vote against the federal decriminalization bill. The Federation succeeded in this effort. Congress voted the bill down and rejected decriminalization for good.

Shortly afterward, the Federation led a letter-writing campaign to newly elected President Reagan, asking him to place leaders sympathetic to parent-groups' concerns in important drug-policy roles in his administration. In addition, the Federation brought parent-group leaders from communities across the United States to Washington to brief First Lady Nancy Reagan about their efforts. As a result, Mrs. Reagan became an informal spokesperson for the Federation and its work.

When Drug Enforcement Administration agent Enrique Camarena was brutally murdered while on duty in MEXICO, the Federation's Virginia chapter conceived a Red Ribbon campaign to honor the slain agent. The chapter wanted to express support for law-enforcement officers nationwide who put their lives on the line every day to enforce the nation's drug trafficking laws. The initial campaign developed into Red Ribbon Week, held annually each October. During this week, schools and communities across the nation celebrate the Red Ribbon campaign for drug-free communities.

In 1993, the Federation refocused its mission and scope, reincorporating under the name of National Family Partnership (NFP).

The NFP conducts training for parents, youth, and community leaders to help them organize prevention groups. Along with other national preven-

tion organizations, the NFP has contributed to the reduction in drug use that has occurred since 1979.

(SEE ALSO: *Education and Prevention*; *Marihuana Commission*; *Parents Movement*)

SUE RUSCHE

Prevention of Alcoholism: The Ledermann Model of Consumption

The Ledermann model of alcohol consumption is an important concept for anyone who wishes to understand efforts undertaken in the second half of the twentieth century to prevent heavy drinking and alcoholism. The point of departure for this concept is a set of observations about how ALCOHOL consumption is distributed in human societies.

Many have thought of this distribution as occurring in two parts. First, there is the great mass of "normal" drinkers; their drinking might be plotted as a bell-shaped curve, with a few people drinking no more than a sip in a year, an increasing number drinking greater amounts than a sip but less than the average amount, and then a declining number drinking more than the average amount, until the graph reaches the normal drinkers who drink much more than the average, and these are relatively few in number. Second, there is a much smaller number of "abnormal" drinkers; their drinking distribution also might be plotted as a bell-shaped curve, but this curve is shifted to the right of the distribution for

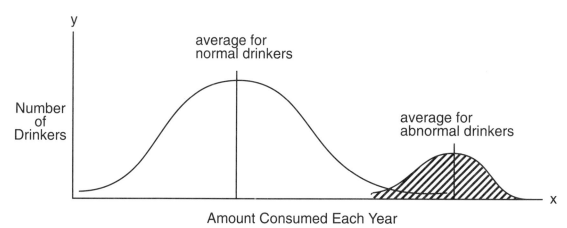

Figure 1
Two "Bell-shaped" Distributions

normal drinkers. Figure 1 shows this two-distribution concept of normal and abnormal drinking, with the number of drinkers on the y axis and the amount consumed on the x axis.

Sully Ledermann, a French demographer, thought of this problem in relation to a single distribution that was not bell shaped or normal in its distribution. He imagined that drinking ought to be plotted in relation to a single curve, with a shape that is known as "lognormal" and without a categorical distinction between normal and abnormal drinkers. The shape is known as lognormal because the natural logarithms of individual consumption rather than actual consumption values are normally distributed. If Ledermann is correct, then the majority of individuals within a society will drink relatively modest amounts of alcohol and a small proportion will drink large quantities, but this will appear in an asymmetric or "skewed" distribution curve with a longer tail to the right of the average alcohol-consumption level (see Figure 2). To the right of the curve, there should be no bump, which would be caused by the presence of an abnormal-drinkers category, distinct from the category of normal drinkers.

One of the important discussion points raised by Ledermann in relation to alcohol consumption has to do with the prevention and the reduction of heavy drinking. Categorical distinctions between normal and abnormal drinkers make it possible to focus prevention and intervention efforts on the abnormal drinkers. One implication of the Ledermann model, however, is that efforts can be focused on the great mass of people who drink modestly as well as on the heavier drinkers; in so doing, dramatic reductions can be brought about in the average amount of alcohol consumed and in the proportion of people who are very heavy drinkers. This difference in approach is part of an important late-twentieth-century debate about how societies can best organize to reduce the hazards of alcohol use.

BACKGROUND

Ledermann (1915–1967) first proposed his single-distribution hypothesis in a French publication entitled *Alcool, Alcoolisme, Alcoolisation* (1956). In a second report published in 1964, Ledermann attempted to test and confirm the validity of his theory by using empirical data on drinking behavior from multiple studies. Born in Algeria, Ledermann spent most of his career in Paris, at the National Institute of Demographic Studies (INED) and the University of Paris. A prolific researcher, he was interested in the distribution of alcohol consumption within society; this interest developed out of a broader effort to identify the reasons for the lower average longevity of the people in France, relative to that of the people in other European countries. Increasingly, he came to believe that a close connection existed between the average or per capita level of alcohol consumption within a society and the prevalence of excessive drinkers at risk for alcohol-related injury or death, and that this relationship could be described mathematically.

Ledermann argued that the lognormal distribution of alcohol consumption resulted from the tendency of individuals to develop and change their

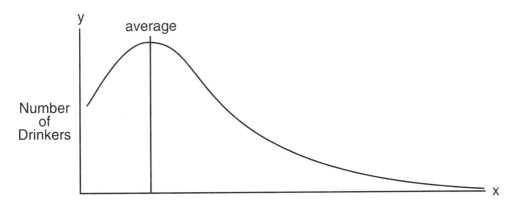

Figure 2
One Lognormal Distribution

drinking habits according to a "boule de neige" (snowball) mechanism driven by social pressures. What does this idea imply? The Norwegian scientist Ole-Jørgen Skog noted that, in general, lognormal distributions tended to result from the exponential (multiplicative) combination of behaviors (1985). On an individual level, this means that persons will tend to increase or decrease their frequency of a behavior by an amount proportional to the initial frequency with which they perform it. For example, we might expect that a person currently consuming 30 liters of alcohol per year would perceive an increase of 6 liters as being comparable to an increase of 1 liter by an individual who currently consumes 5 liters. Such phenomena grow exponentially, in snowball fashion, and tend to distribute according to a lognormal function within populations. Ledermann believed that the snowball effect was caused by the operation of social pressures within drinking environments. This notion implies that the drinking behaviors of individuals within a particular social environment or "drinking culture" are tightly interrelated, such that changes in the alcohol consumption level of some individuals are very likely to induce changes in the consumption level of others. Skog and other scholars have elaborated upon this rudimentary social-interaction hypothesis in an effort to understand how shifts in the drinking habits of one sector may rapidly diffuse throughout the entire population.

The Ledermann model provides a simple formula for estimating the distribution of alcohol use in any homogeneous population of drinkers (that is, any population in which the average consumption level does not vary significantly across subgroups). In addition to assuming lognormality with his model, Ledermann also hypothesized that the proportion of drinkers consuming more than 365 liters of absolute alcohol (ethanol) annually was small and invariant across populations, because such high consumption levels (1 liter per day) would quickly have lethal effects. With this constant determined, he could establish mathematically the full distribution of alcohol consumption within a population, knowing only the per capita or average consumption level. Knowledge of the distribution of alcohol consumption yields three important additional insights. First, one can estimate the proportion of heavy or excessive alcohol users in the population. This value is frequently defined as the percentage of drinkers consuming 10 centiliters or more of absolute alcohol per day, but the threshold of excessive use can theoretically be established at any level of consumption. Second, the total amount of alcohol consumed by heavy users can be estimated. Third, and most important, the effect of changes in average consumption on the proportion of excessive drinkers in the population can be predicted. This final corollary of the model is perhaps the most controversial, because it suggests that the prevalence of excessive alcohol use within a society might be manipulated by restrictions on alcohol availability or other preventive efforts designed to reduce the general level of consumption in the population. The implications of the Ledermann model for alcohol-control policy and other public health efforts were carefully elucidated in a monograph by Finnish scholar Kettil Bruun and an international body of colleagues (1975).

Ledermann's hypotheses have been the object of intense scrutiny and debate in the four decades since they were first proposed. Many researchers have examined the "fit" between the lognormal distribution and data obtained from actual populations of drinkers, with mixed results. Significant deviations from expectations of the model have been demonstrated in some cases; in other populations, the distribution closely approximated lognormality. Ledermann's assumption of constancy across populations of the proportion of heavy drinkers who consume 365 liters or more of alcohol annually has been severely challenged. In general, these critiques have weakened the deterministic character of Ledermann's original formulation, without challenging the basic assertion that there is a close connection between average alcohol consumption in the population and the prevalence of excessive or "at risk" drinkers. Other critics, however, have argued that this relationship is tautological—that is, a high per capita alcohol-consumption level logically requires there be a high proportion of heavy drinkers, and a low per capita consumption level logically requires there be a low proportion of heavy drinkers. As Skog (1985) pointed out, in mathematical terms this logical limitation is expressed as a Markov's inequality: The proportion of the population with a consumption level that is higher than an established limit cannot exceed the ratio between per capita consumption of the population and the limit. Results obtained from actual drinking populations, however, indicate that the relationship between these phenomena is much stronger than the Markov's inequality implies, thus suggesting that social interaction or additional factors may indeed have an important influence on the prevalence of heavy drinking behavior. The debate over these issues is unresolved, but it is clear that Ledermann's ideas have served as a major stimulus in the effort to understand the relationship between the "drinking culture" of a society and the prevalence of excessive alcohol use.

(SEE ALSO: *Addiction: Concepts and Definitions; Advertising and Alcohol Use; Alcohol: History of Drinking; Disease Concept of Alcoholism and Drug Abuse; Legal Regulation of Drugs and Alcohol; Prevention; Social Costs of Alcohol and Drug Abuse*)

BIBLIOGRAPHY

BRUUN, K., ET AL. (1975). *Alcohol control policies in public health perspective* (Vol. 25). Helsinki: Finnish Foundation for Alcohol Studies.

DUFFY, J. C. (1986). The distribution of alcohol consumption—30 years on. *British Journal of Addiction, 81,* 735–741.

LEDERMANN, S. (1964). *Alcool, alcoolisme, alcoolisation: Mortalité, morbidité, accidents due travail* (Institut National d'Etudes Démographiques, Trav. et Doc., Cah. No. 41). Paris: Presses Universitaires de France.

LEDERMANN, S. (1956). *Alcool, alcoolisme, alcoolisation: Données scientifiques de caractère physiologique, economique et social* (Institut National d'Etudes Démographiques, Trav. et Doc., Cah. No. 29). Paris: Presses Universitaires de France.

SKOG, O-J. (1985). The collectivity of drinking cultures: A theory of the distribution of alcohol consumption. *British Journal of Addiction, 80,* 83–99.

BRYAN M. JOHNSTONE

Shaping Mass-Media Messages to Vulnerable Groups

The early history of drug-abuse PREVENTION campaigns is one of limited success and sometimes spectacular failures. In the mid-1970s, the NATIONAL INSTITUTE ON DRUG ABUSE (NIDA) suspended support for research in this area altogether—in the belief that such campaigns might actually be teaching young people how to use drugs. However, more recent theoretical and methodological developments in studies of communication and of prevention campaigns have led to substantially greater evidences of success.

Early research assumed immediate and direct effects of media on reactive audiences. According to K. A. Neuendorf (1991), two underlying assumptions were (1) that audiences needed only to be informed in order to achieve prevention goals and (2) that audiences could be addressed as a single mass, composed of self-aware consumers of public information for whom prevention-campaign designers could create messages "palatable to one and all."

Evaluations of the media campaigns, however, soon indicated that these assumptions were not warranted. General, information-only messages disseminated to a mass audience are likely to have little effect. Rather, campaigns that are based on theories of persuasive communication, are designed for specific audiences, and have pragmatic goals experience greater success. Moreover, current research in the development of more sophisticated public campaigns contains an added dimension known as formative research. In formative research, messages are pretested on a sample of the target audience to determine if the message is clear, credible, and likely to get the desired response. Reviews of mass-media-based drug-abuse prevention efforts have led to the recommendation by Flay and Sobel (1983) that more emphasis be placed on formative research in the laboratory before proceeding to expensive program dissemination and field evaluations: "Only after it has been established that an efficacious communication product has been developed is it worth disseminating it" (p. 26).

The role of communication in the design of health campaigns received more attention with increased recognition that intervention strategies are at the heart of prevention campaigns and that communication is at the heart of intervention. Planners of contemporary public campaigns seek to design messages that will appeal specifically to members of carefully defined target audiences.

Recently, more effective, large-scale campaigns have been found to bring about substantial changes in public opinion. According to Flay (1983), who studied forty antismoking campaigns, these campaigns involved extensive use of the media and are targeted at various segments of the general public over extended periods of time. A well-known example is the work by the PARTNERSHIP FOR A DRUG-FREE AMERICA, a private-sector coalition of groups in advertising, media, and public communication. The project involves thousands of professionals who voluntarily create, produce, and disseminate antidrug advertisements across the nation. A drawback for the prevention professional is that unless the partnership funds the effort or money and time constraints are of little concern, the approach may be too expensive. A second approach, which may be appropriate for more moderate campaigns, although still not inexpensive, involves mass-media campaigns targeted at individuals. Direct changes in individual attitudes and behavior have been brought about through mass-media messages that employ formative research and are precisely targeted to at-risk individuals. A number of these campaigns have been supplemented by interpersonal approaches as described below.

THEORETICAL BACKGROUND AND CURRENT RESEARCH

Assumptions about the nature of human information processing have evolved from early assumptions based on rational models of human behavior to more complex models involving both cognitive and affective forces. A few years ago a panel of prominent psychologists and communication researchers reviewed changing beliefs about the function of cognition and affect in communication. The panel's conclusions suggest a more complex view of humans as sometimes operating on subroutines, such as "drive home from work," of which they are only dimly aware and sometimes selecting communication stimuli, such as music or television programs, to help them manage their moods.

This perspective is consistent with what is known of humans as biological creatures who employ attention processes shaped by millions of years of evolution. In primeval times, a slight change in the coloration of a leaf or the presence of a subtle odor could lead to life-threatening or life-sustaining experiences. These changes triggered arousal responses. To the present day, humans continue to respond to novelty with alertness and to familiarity with relaxation.

The level of arousal at which humans operate most comfortably, the optimal level of arousal, varies among individuals and significantly affects the kinds of stimuli to which they most likely will attend and how they will respond to those stimuli. Individuals who operate at high levels of arousal prefer varied, novel, and complex sensations and experiences, and are willing to take both physical and social risks in order to engage in such experience.

A number of studies have indicated a biobehavioral basis for this sensation-seeking behavior (e.g., Bardo, Nieswander, & Pierce, 1988). Among these are biochemical measures, including monoamine oxidase (MAO)—an enzyme that regulates the levels of MONOAMINE NEUROTRANSMITTERS in brain NEURONS. Lower levels of platelet MAO tend to be associated with higher levels of sensation seeking. Although there is evidence of high heritability of the sensation-seeking trait, the environment also accounts for a significant portion of its variance.

The sensation-seeking scale (SSS) was developed by Zuckerman (1978) to measure this trait. Four ten-item subscales make up the total instrument: boredom susceptibility (BS), disinhibition (DIS), experience seeking (ES), and thrill and adventure seeking (TAS). Under established procedures for determining levels of need for sensation, approximately half of any given age group falls into the sensation-seeking category.

Substantial research indicates that a strong connection exists between substance use by ADOLESCENTS and young adults and SENSATION seeking. Zuckerman and others found that 74 percent of college undergraduates scoring high on the SSS had used one or more drugs; only 23 percent of those scoring low on the scale had done so. Evaluating a variety of personality measures, Segal and Singer (1976) found that the SSS provided the most discrimination between users and nonusers. Use of specific drugs, such as AMPHETAMINES, MARIJUANA, HASHISH, and LSD correlated strongly with sensation seeking. Donohew, Palmgreen, & Lorch (1991) have found highly significant differences in alcohol and other drug use among junior and senior high school students (see Table 1).

The explanation behind the positive relationship between sensation seeking and substance use is twofold. First, high sensation seekers obtain the direct neurological stimulation from the substance itself. The effect of the drug (e.g., stimulant versus depressant) does not seem to be important but simply its ability to alter states of consciousness. Second, the illegality or surreptitiousness associated with substance use provides a further source of stimulation.

The research supporting the relationship between substance use and sensation seeking offers strong justification for using the SSS to target prime at-risk groups and for designing anti-drug-use messages and programs to reach them. An approach to the de-

TABLE 1
Sensation Seeking and Drug Use among Junior and Senior High School Students[a]

	Junior High LSS n = 658	Junior High HSS n = 565	Senior High LSS n = 450	Senior High HSS n = 420
Marijuana	6.2	24.7	13.0	38.3
Cocaine	.6	3.4	1.8	6.6
Liquor	20.7	58.3	28.5	67.1
Beer	24.9	58.3	38.0	70.5
Uppers	1.4	14.9	2.0	14.8
Downers	.5	11.2	1.6	6.6

[a]These figures represent percentages of students at the 7th- through 12th-grade levels who indicated using particular drugs at least once during the last 30 days.
[b]LSS = low sensation seekers; HSS = high sensation seekers.

sign of health campaigns employed by Donohew and associates (in press) based on this relationship is leading to increased effectiveness. The approach features three components:

1. Designing messages to appeal to high sensation seekers and, where possible, offering exciting alternatives to drug use.
2. Placing the messages in TV and other programming that high sensation seekers are likely to be audiences for.
3. Advertising a hot line for viewers to call to learn about exciting alternatives to drug use (in order to meet sensation needs in more prosocial ways).

Data analysis from a recent field experiment (Donohew, Palmgreen, & Lorch, 1994) indicates that approximately two-thirds of callers to an advertised information hot line were high sensation seekers and that high sensation seekers were more likely to recall the antidrug messages.

Supplementing media messages with an interpersonal communication component is a current trend in campaigns. Often viewers are offered a hot line number to call for more information that will result either in "mediated-interpersonal" contact with a health professional over the telephone or in direct interpersonal contact with existing community agencies and organizations. Sometimes seminars, discussion groups, or support groups can be created as a result of the campaign. Also, media messages can be coupled with interpersonal programs, such as

school-based programs, that provide greater individual contact and message reinforcement.

(SEE ALSO: *Advertising; Comorbidity and Vulnerability; Conduct Disorder and Drug Use; Education and Prevention; Prevention Movement; Vulnerability As Cause of Substance Abuse*)

BIBLIOGRAPHY

BARDO, M. T., NIESWANDER, J., & PIERCE, R. (1988). Effects of opiate and dopaminergic drugs on novelty preference behavior. *Society for Neuroscience Abstracts, 14,* 683–689.

DONOHEW, L., PALMGREEN, P., & LORCH, E. (1994). Attention, need for sensation, and health communication campaigns. *American Behavioral Scientist, 38*(2), 310–322.

DONOHEW, L., ET AL. (EDS.). (1991) *Persuasive communication and drug abuse prevention.* Hillsdale, NJ: Erlbaum.

DONOHEW, L., ET AL. (EDS.). (1988). *Communication, social cognition and affect.* Hillsdale, NJ: Erlbaum.

FLAY, B. R. (1983). State-of-the-art in mass media and smoking behavior. Paper presented at NCI workshop on the role of mass media in smoking prevention and cessation, Washington, DC.

FLAY, B. R., & SOBEL, J. R. (1983). The role of mass media in preventing adolescent substance abuse. In T. J. Glynn et al., (Eds.), *Preventing adolescent drug abuse:*

Intervention strategies. Washington, DC: NIDA Research Monograph Series (47).

NEUENDORF, K. A. (1991). Health images in the mass media. In E. B. Ray & L. Donohew (Eds.), *Communication and health.*

RICE, R. E., & ATKIN, C. K. (1989). (Eds.) *Public communication campaigns.* Newbury Park, CA: Sage.

ROGERS, E. M., & STOREY, J. D. (1987). Communication campaigns. In C. R. Berger & S. H. Chaffee (Eds.), *Handbook of communication science* Beverly Hills, CA: Sage.

SEGAL, B., & SINGER, J. L. (1976). Daydreaming, drug and alcohol use in college students: A factor analytic study. *Addictive Behaviors, 1,* 227–235.

ZUCKERMAN, M. (1978). *Sensation-seeking: Beyond the optimal level of arousal.* Hillsdale, NJ: Erlbaum.

LEWIS DONOHEW
NANCY GRANT HARRINGTON

PREVENTION MOVEMENT

The grass-roots drug-prevention movement began in the mid-1970s with parents who were concerned about the health and safety of their children. During that decade, drug use among U.S. adolescents escalated from relatively low levels to the highest levels in history. Some young drug users were addicted and needed treatment. Others were in trouble with drugs but had not yet become addicted. Some were dying of drug overdoses, and many were being killed in alcohol- and drug-related automobile crashes.

Many social and environmental factors appeared to contribute to the escalation in the use of alcohol and other drugs among the young. Parents organized to address these factors in order to prevent drug use among young people. Youth groups also formed to help prevent substance abuse among their peers.

Media groups soon organized in response to parents' concerns about the glamorization of drug use on television, in films, and in song lyrics that influenced young people. A few years later, the advertising community initiated a campaign to design and air commercials with strong antidrug messages targeted to children and adolescents.

A drug-related tragedy on the aircraft carrier U.S.S. *Nimitz,* in which many young servicemen were killed during routine practice maneuvers, brought the MILITARY into the prevention movement. It reinstituted universal drug testing to ensure that such an event would never happen again. Such testing was started in the early 1970s but was discontinued in 1975. The business community adopted drug-testing policies similar to those initiated by the military to prevent drug use in the workplace, particularly in jobs that involved public safety.

Educators and researchers concerned about drug use in primary and secondary schools and on college campuses developed school-based approaches to drug prevention. The law enforcement community added its voice to the prevention effort through community policing programs. It also joined forces with the education community through efforts such as DARE (Drug Abuse Resistance Education), in which police officers teach DARE's drug-education curriculum to students in elementary, middle, and high schools.

Local, state, and national political leaders created policies and allocated resources to stem the flow of drugs into the country and to help people prevent substance abuse in their families and communities.

Specific ethnic and cultural groups created prevention groups as well. They focused on strengthening their communities through a renewed appreciation of their heritages and building on the resiliency that had enabled them to survive the long-term, debilitating effects of racism and poverty.

Seeing the opportunity to contribute its considerable strength and human resources, the faith (religious) community also initiated drug-prevention programs. When researchers established the links between substance abuse and the transmission of HIV/AIDs, the AIDS-prevention community joined hands with the substance-abuse-prevention community.

Partnerships and coalitions formed to bring all parts of the community together to develop and implement substance-abuse-prevention strategies collaboratively.

The grass-roots substance-abuse prevention movement continues to expand. Each component that joins it seeks to create communities in which individuals and families can live healthy lives free of substance abuse and the problems it generates.

THE PARENT MOVEMENT

Parents initiated the prevention movement in response to the escalation of drug use among teenagers throughout the 1970s. Surveys conducted by the government indicate that in 1962, just seventeen years before drug use peaked, less than 2 percent of the entire U.S. population had tried any illicit drug. By 1979, when the use of most drugs peaked, 24 million Americans used drugs regularly. Seventy percent of young adults (ages 18–25), 65 percent of high school seniors, and 34 percent of youth (ages 12–17) had tried an illicit drug. Rates of ALCOHOL and TOBACCO use were even higher. Ninety-five percent of young adults, 93 percent of seniors, and 70 percent of youths had tried alcohol; 83 percent of young adults, 74 percent of seniors, and 54 percent of youths had tried cigarettes. One in nine seniors smoked marijuana daily.

Several social and environmental factors appeared to be contributing to this escalation in drug use among young people.

Between 1972 and 1978, eleven states decriminalized MARIJUANA. The political rhetoric that accompanied the decriminalization effort tended to deny or minimize the harmful effects of marijuana and other drugs. Governmental action that equated penalties for marijuana possession with those for traffic violations tended to reinforce this belief. Most people thought that state legislatures would not make marijuana more available through decriminalization if it was truly harmful.

Prevention research in the early 1970s persuaded some that drug education did not reduce drug use, and government funding for drug education materials ceased. This created a vacuum that was filled by decriminalization advocates and those who stood to profit from increased drug sales. A great deal of the educational materials available throughout the 1970s taught people how to "use drugs responsibly" rather than what scientists were learning about the harmful effects of drugs. These materials tended to promote drug use rather than prevent it.

As states decriminalized marijuana, an industry emerged to assist people in their drug taking. It manufactured drug PARAPHERNALIA—toys and gadgets designed to enhance drug use—and sold it in "head shops," places where so-called "pot heads," "acid heads," "coke heads," and other drug users could buy implements to help them take drugs. Head shops also sold promotional materials and "starter kits" targeted to young aspiring drug users. By 1977, some 30,000 head shops were conducting business across the nation.

Each of these factors helped increase drug use among children and teenagers. Parents organized to help young people who were using drugs stop doing so through education, prevention, counseling, or treatment. They also sought to prevent nonusers from starting, and to reinforce those who decided not to use drugs by emphasizing the desirability of living drug-free lives. Groups that led this effort include the NATIONAL PARENTS RESOURCE INSTITUTE ON DRUG EDUCATION (PRIDE), NATIONAL FAMILIES IN ACTION, the NATIONAL FEDERATION OF PARENTS FOR DRUG-FREE YOUTH, the American Council on Drug Education, and COMMITTEES OF CORRESPONDENCE, as well as state groups such as Texans War on Drugs, Florida Informed Parents, Tennessee Families in Action, and Alaskans for Drug-Free Youth. There also are thousands of local groups in cities, towns, and counties across the country.

Parent groups targeted the social and environmental factors they felt were contributing to the escalation in drug use among young people and developed strategies to address those factors. They did this by establishing clear definitions. They defined drug abuse to include all illegal drugs and all legal drugs and substances used illegally. The latter include alcohol and tobacco for those under the legal purchasing age, medicines, glue, gasoline, and other substances. Then, parent groups set clear goals: to prevent use before it starts, to persuade users to stop, and to help those who can't stop, find treatment. For alcohol, the goals were slightly different: to prevent use before the legal drinking age, to persuade those who choose to drink when they reach the legal drinking age to follow low-risk drinking guidelines, and to help those who are addicted to alcohol find treatment.

To achieve these goals, parent groups developed several strategies. They mounted an intensive effort to secure the passage of laws to ban the sale of drug paraphernalia. Over a four-year period they succeeded in getting such laws passed in several communities and states. Today, nearly every state has such laws. Challenges to the paraphernalia laws were brought by the National Organization for the Reform of Marijuana Laws (NORML), which argued that they were unconstitutional. Many of NORML's board members were members of the drug paraphernalia industry. However, after several conflicting rulings issued by federal district and appeals courts across the nation, in the early 1980s the U.S. Supreme Court upheld these laws as constitutional. This ended the joint effort of NORML and the paraphernalia industry to defeat the paraphernalia laws.

NORML also led the marijuana decriminalization movement. Between 1972 and 1978, the organization persuaded eleven states to decriminalize marijuana. The parent prevention movement stopped this effort: parents prevented additional states from decriminalizing after 1978, and defeated a federal effort to decriminalize marijuana nationwide. In some states that had decriminalized, such as Alaska, parents worked to recriminalize the drug after surveys showed that marijuana use among young people was considerably higher in these states than in states that had not decriminalized.

Parent groups placed primary focus on ensuring that drug-education materials convey a no-use message rather than recommending the "responsible use" of drugs that are both illegal and harmful. They did this by going to the medical and scientific literature and insisting that drug-education materials reflect what was reported in that literature about drug effects.

These strategies seemed to have contributed to the peak and then steady decline in drug use among adolescents, young adults, and the entire population since parents initiated the prevention movement. The MONITORING THE FUTURE study (HIGH SCHOOL SENIOR SURVEY), conducted by the government annually since 1975, shows a direct correlation between rates of use and young people's belief that drugs are harmful. The more students who believe a specific drug will hurt them, the fewer students who use that drug. Sadly, for reasons not yet fully understood, the steady rise of high school students who perceived drugs to be harmful leveled off and began to decline in the early 1990s. As a result, the steady decline in drug use over fourteen years reversed in 1993, and student drug use once again began rising.

A reemergence of calls for drug decriminalization and legalization worries prevention advocates. This effort is being led once again by NORML and by the DRUG POLICY FOUNDATION, an organization that emerged from NORML, many of whose leaders were once active in NORML. The glamorization of drug use in song lyrics and films is also reappearing, as are claims by legalization proponents that people can use highly addictive drugs "responsibly." Whether these shifts are contributing to the turnaround in drug use is not yet clear. Nor is it clear whether the rise in drug use among high school students is a temporary aberration or a permanent trend. Nonetheless, the prevention movement is redoubling its efforts to ensure that drug abuse resumes and sustains its downward trend.

THE MOVEMENT AGAINST DRUNK DRIVING

At the same time the parent drug-prevention movement was targeting illicit drug use and the problems it generated among young people, another group of parents and families took aim at the problem of *drunk driving* and the devastation it was creating on the highways, particularly among young people. At the time, deaths from alcohol-related crashes were so prevalent that drunk-driving crashes had become the leading cause of death among adolescents. Families of many young people whom drunk-driving crashes had killed organized groups such as MOTHERS AGAINST DRUNK DRIVING (MADD), REMOVE INTOXICATED DRIVERS (RID), and STUDENTS AGAINST DRIVING DRUNK (SADD) to stop the carnage on the highways. As with the parent-led drug-free movement, parents who led the anti-drunk-driving movement first raised the nation's awareness about the problem and then developed strategies to address it.

Among the many contributions this movement has made, perhaps the most significant deals with the age at which young people may legally purchase and consume alcohol. For many years the legal drinking age in every state was twenty-one. During the Vietnam War, however, when young men aged eighteen and over were drafted into the military, most states lowered their legal drinking ages to eighteen in recognition of the fact that if young men were old enough to fight for their country, they ought to be old enough to drink. The unanticipated consequence of this action, however, was a further lowering of the age at which young teenagers and even preadolescents were able to purchase alcohol, albeit illegally. This led to the appalling rise in the number of young people who were killed in drunk-driving crashes.

As anti-drunk-driving groups tried to persuade state legislatures to return the legal drinking age to twenty-one, their efforts were consistently defeated, year after year, by the alcohol industry, which had considerably more dollars to spend lobbying against such an action. In many states, drug-free parents groups joined forces with MADD, RID, and other parent-led anti-drunk-driving groups, but to no avail. What broke the log jam was MADD's strategy of advocating a federal bill that would deny federal highway funds to states that refused to raise the drinking age to twenty-one. Although the alcohol industry succeeded in defeating the federal effort for several years in a row, the anti-drunk-driving forces finally overwhelmed the industry, and Congress passed the federal bill.

Faced with the loss of federal highway funds, nearly all states have raised the drinking age to twenty-one. The U.S. Department of Transportation estimates changes in the drinking-age laws had saved some 13,000 teenage lives by the mid-1990s. Furthermore, when MADD, RID, and similar groups first organized, some 52,000 Americans were killed on the highway each year. About half of those deaths were due to alcohol-related crashes. By the mid-1990s, highways deaths were down to 41,000 per year, and drunk-driving deaths had been reduced to 17,700 per year.

These groups continue to work to reduce drunk driving and the problems it generates by advocating better enforcement of existing laws, passage of new laws, and ways to mandate repeat DUI (DRIVING UNDER THE INFLUENCE) offenders into treatment.

THE MEDIA

In response to parental concerns about the glamorization of drug and alcohol use in films and on television, several groups were organized to address these concerns. The family of actor Paul Newman founded the Scott Newman Center in memory of Newman's only son, Scott, who died of an overdose of alcohol and drugs. The center bestows awards on producers and writers who create television programs and films that contain strong no-use messages and that enhance the public's understanding of substance-abuse issues and ways of dealing with them successfully. The Entertainment Industries Council developed programs to work with filmmakers to educate them about substance abuse and to deglamorize drug use in movies. The National Academy of Television Arts and Sciences implemented strategies to enhance the industry's awareness of the impact it could have in reducing substance abuse through the power and reach of the mass media.

In the early 1980s, advertising and public-relations agencies formed the PARTNERSHIP FOR A DRUG-FREE AMERICA. These agencies volunteer their talent and time to create and produce antidrug commercials targeted to young people. The Partnership solicits free air time and space in the electronic and print media to place these commercials and ads. Thus far, it has secured several billion dollars' worth of media placement for the antidrug messages it creates.

ETHNIC AND CULTURAL GROUPS

The introduction of CRACK in the mid-1980s made this form of COCAINE cheap and plentiful. It brought illicit drug use and addiction into poverty-stricken communities that had previously avoided massive illicit drug use. Consequently, members of the African-American, Hispanic and Latino, Native American, and Asian-American communities organized to prevent drug use, drug addiction, and drug-related crime. The passage of the Anti-Drug Abuse Act of 1986 assisted this effort. With it, Congress made demonstration grants available to local groups to prevent substance abuse among youth at high risk of becoming involved with illicit drugs.

This movement has mounted intensive efforts to confront the consequences of poverty and racism. One consequence is to have made poor communities more vulnerable to drug use and the health and social problems it creates. Ethnic and cultural groups have organized to confront these problems, helping addicts find treatment and reclaiming their communities from drug dealers. They also have addressed other environmental factors, taking aim at the tobacco and alcohol industries' efforts to target ethnic communities for increased consumption of their products. They have defeated the introduction of new brands of cigarettes and alcoholic beverages targeted to African Americans and Hispanics and Latinos. Campaigns to eliminate the disproportionately high numbers of alcohol and tobacco billboards located near schools and churches in inner-city neighborhoods have resulted in outright bans of such ADVERTISING in Baltimore, Maryland, and Cincinnati, Ohio.

COMMUNITY PARTNERSHIPS

In the mid-1980s, the Robert Wood Johnson Foundation invited communities to submit proposals to establish community partnerships bringing together parents, young people, schools, ethnic and cultural groups, religious institutions, businesses, local governing bodies, and social and civic organizations to reduce substance abuse. So many communities responded to the foundation's invitation that the government provided 100 million dollars in assets seized from drug smugglers, making even more funds available to communities to establish partnerships to prevent substance abuse and related problems.

RESULTS

These four distinct groups—parents, the media, ethnic and cultural groups, and community partnerships—comprise the cornerstones of the grass-roots prevention movement that has been working to prevent substance abuse since the 1970s. Many other segments of the population have joined the prevention movement and have made important contributions to the overall reductions in substance abuse. The grass-roots prevention effort has brought about policy shifts at all levels of government that have reduced both demand and supply. Collectively, these efforts have resulted in significant, sometimes astonishing, declines in drug use:

- The number of Americans who currently use illicit drugs was reduced from 24 million in 1979 to 11 million in 1992. Current use of cocaine, which peaked in 1985, dropped from 5.8 million to 1.3 million in 1992. Alcohol-related traffic deaths were reduced from 26,000 to 17,700 per year. The surgeon general estimated that by preventing people from starting to smoke and by persuading smokers to stop, we will have saved a total of 3 million lives by the year 2000.

- Between 1979 and 1992, lifetime drug use declined dramatically: among youth, from 34 to 17 percent; among seniors, from 65 to 44 percent; and among young adults, from 70 to 52 percent.

- Lifetime alcohol use declined as well: among youth, from 70 to 39 percent; among high-school seniors, from 93 to 88 percent; and young adults, from 95 to 86 percent.

- Lifetime tobacco use also has been reduced: among youth, from 54 to 34 percent; among seniors, from 74 to 62 percent; and among young adults, from 83 to 69 percent.

Current (past month) drug use declined significantly as well, in some cases even more dramatically than lifetime use, as Table 1 shows.

This is not to suggest that the problem of substance abuse is over in the United States. Clearly it is not. Although cocaine use has not reversed and alcohol use continues to decline among high school students, the rise in other drug use detected by the

TABLE 1
Drug Use by Age Group, 1979 and 1992

Drug	Age Group	1979 (percent)	1992 (percent)
Any illicit drug	young adults	37.1	13.0
	high-school seniors	38.9	14.4
	youth	17.6	6.1
Marijuana	young adults	35.4	11.0
	high-school seniors	36.5	11.9
	youth	16.7	4.0
Cocaine (1985–1992)	young adults	9.3	1.8
	high-school seniors	5.7	1.3
	youth	1.4	0.3
Alcohol	young adults	75.9	59.2
	high-school seniors	71.8	51.3
	youth	37.2	15.7
Cigarettes	young adults	42.6	31.9
	high-school seniors	34.4	27.8
	youth	12.1	9.6

SOURCE: High School Senior Survey, 1993.

TABLE 2
Drug Use by Age Group, 1992 and 1993

Drug	Age Group	1992	1993
Any illicit drug	high-school seniors	14.4%	18.3%
Marijuana	high-school seniors	11.9	15.5
Cigarettes	high-school seniors	27.8	29.9

SOURCE: High School Senior Survey, 1994.

1993 Monitoring the Future Survey sounds the alarm, as Table 2 shows.

Furthermore, deaths directly and indirectly related to substance abuse—deaths that could have been prevented by changing behavior—remain unacceptably high, as Table 3 shows.

Nonetheless, the significant reductions in drug use and in drug-related deaths since the mid-1970s that have resulted from the grass-roots prevention movement, and from the additional public and private efforts it stimulated, suggest that we know what to do. If the nation is wise enough to continue the effort by encouraging, financing, and expanding the prevention effort, there is no reason why we cannot further reduce substance abuse, substance-abuse-related deaths, and all the other problems to which substance abuse contributes among all people of all ages in all communities.

(SEE ALSO: *Accidents and Injuries from Alcohol; Adolescents and Drug Use; Advertising and Alcohol Use; Advertising and the Alcohol Industry; Advertising and Tobacco Use; Education and Prevention; Ethnicity and Drugs; Industry and Workplace, Drug Use in; Marihuana Commission; National Household Survey on Drug Abuse*)

SUE RUSCHE

TABLE 3
Causes of Death in the United States, 1993

Cause of Death	Number
Tobacco	400,000
Alcohol	100,000
Illicit drugs	20,000
AIDS	221,000
Firearms	35,000

SOURCE: *Journal of the American Medical Association*, November 10, 1993.

PREVENTION PROGRAM DIRECTORY
See Appendix, Volume 4.

PREVENTION PROGRAMS   These articles describe specific programs being carried out in schools and communities throughout the United States in the continuing effort to prevent experimentation with and use of alcohol and drugs. Included here are articles on *"Here's Looking at You"*; *Life Skills Training; Napa Project, Revisited; Ombudsman Program; PRIDE; Project SMART; Talking with Your Students about Alcohol; the Waterloo Smoking Prevention Project.*

The overview entries on the *Parents Movement* and the *Prevention Movement* and the individual articles in the section entitled *Prevention* provide a framework for the articles in this section. For an extensive listing of other organizations engaged in similar efforts and of other programs now being used, see also *Education and Prevention* and the directory in the Appendix, Volume 4.

"Here's Looking at You" Series   The "Here's Looking at You" (HLAY) program grew out of the work done by Clay Roberts and Douglas Goodlett for their master's degrees in the 1970s. By 1978, Mr. Roberts and other health-education specialists at Seattle's Educational Service District No. 121 (ESD–121) had created a full alcohol-education curriculum for kindergarten through twelfth grades, designed mainly for delivery in fifteen to twenty class presentations each year, complete with multimedia support materials. The federal grant that funded the project required an evaluation by outside researchers, for which a contract was concluded in 1978 with Armand L. Mauss and colleagues at Washington State University, Pullman, Washington.

Encouraging results from this preliminary evaluation led to more extensive funding for three years by the NATIONAL INSTITUTE ON ALCOHOLISM AND ALCOHOL ABUSE (NIAAA) for HLAY demonstration projects and their evaluations at several sites around the nation. The only longitudinal evaluation was done by Mauss and colleagues in the Seattle and Portland (Oregon) areas (Hopkins et al., 1988; Mauss et al. 1988). In 1981, Roberts and some of his colleagues left ESD–121 to form Roberts, Fitzmahan, & Associates, a private corporation in Seattle, which completed four major revisions of the original program. In 1993, R., F., & A. merged with the Comprehensive Health Education Foundation (C.H.E.F), which since 1974 has produced and distributed the curricula and materials developed by the Roberts teams.

From the beginning, HLAY has been based on an educational theory involving both cognitive and affective elements: knowledge (information), attitudes, self-esteem, decision-making skills, and other social skills (Mooney et al., 1979). In subsequent versions of the program (HLAY-2 and HLAY-2000, with updates of the latter), strenuous efforts have been made to improve the educational strategy in light of ongoing psychosocial research and program evaluations. The program components as of 1992 fall into three basic categories: information, social skills, and "bonding." A two-pronged "inoculation" strategy—stressing both "risk factors" and "protective factors"—runs through these three categories. In both design and delivery, HLAY is one of the most thorough and sophisticated school-based programs in the United States, and also one of the most widely used.

The underlying theoretical basis that has evolved for this program is recognizable by social scientists as combining elements of both "rational choice" and "control" theories. In layman's terms, the program rests on the assumption that schoolchildren will be far less likely to use alcohol or other drugs if they are (1) given full and reliable information about the properties of chemical substances and the consequences of using them; (2) trained in self-control, decision making, and other social skills (including Refusal Skills®); and (3) assisted in feeling positive about themselves and in bonding with friends, families, schools, and communities. Many of these outcomes would obviously be desirable in other arenas of youth and health as well.

Evaluations of the early versions (HLAY-1 and HLAY-2) have generally found little positive program impact on reported student *use* of alcohol or other drugs, although consistent evidence exists for impact on knowledge and occasional evidence of impact on attitudes, self-esteem, and decision-making skills (Swisher et al., 1985; Hopkins et al., 1988; Kim, 1988; Green & Kelley, 1989). Evaluations of other school-based drug/alcohol programs have also generally produced findings of doubtful impact on actual behavior, except perhaps in the case of lessened tobacco smoking (e.g., Moscowitz, 1989; Tobler, 1986).

One evaluation of HLAY-2000 has measured positive program impact on reported actual *use* of alcohol or other drugs (DuBois et al., 1989). This evaluation, covering grades 1 to 6, found evidence of positive impact on knowledge, self-esteem, and refusal skills but no evidence of impact on actual substance *use*, except in the case of chewing tobacco in grades 1 to 3. Other unpublished evaluations of HLAY-2000 (Bubl, 1988; Barrett, 1989) have not measured program impact on actual use of drugs or alcohol but have shown some evidence of impact at various grade levels on knowledge, self-esteem, coping, decision-making and refusal skills, and making friends.

It is inherently difficult, of course, to prevent or change undesirable behavior through any classroom curriculum—given the wider and emotionally powerful influences of home, peers, and community—which create a countervailing mode, especially in the case of alcohol (Goodstadt, 1986; Lohrmann & Fors, 1986; Mauss et al., 1988).

Even where classroom programs might have a beneficial impact, it is difficult to measure with much sensitivity, given the present stage of evaluation technology and especially when such measurement depends on the self-reports of children. HLAY-2000 has attempted to take the wider influences into account, focusing greater emphasis on family and on peers. Yet its effectiveness in reducing actual *use* of substances remains to be demonstrated through adequate evaluation.

(SEE ALSO: *Education and Prevention; Parents Movement; Prevention*)

BIBLIOGRAPHY

BARRETT, C. J. (1989). Substance abuse prevention: A program evaluation. Unpublished MSW Thesis, Southern Connecticut State University.

BUBL, J. E. (1988). Evaluation of the "Here's Looking at You, 2000" curriculum in the rural Marion County, Oregon, schools. Unpublished MS Thesis, University of Oregon.

DuBois, R. ET AL. (1989). Here's Looking at You 2000: Program report and evaluation, 1988–1989 school year. Philadelphia: Corporate Alliance for Drug Education (CADE).

GOODSTADT, M. S. (1986). School-based drug education in North America. What is wrong? What can be done? *Journal of School Health, 56,* 278–281.

GREEN, J. J., & KELLEY, J. M. (1989). Evaluating the effectiveness of a school drug and alcohol prevention curriculum: A new look at Here's Looking at You-Two. *Journal of Drug Education, 19*(2), 117–132.

HOPKINS, R. H., ET AL. (1988). Comprehensive evaluation of a model alcohol education curriculm. *Journal of Studies on Alcohol, 49*(1), 38–50.

KIM, S. (1988). A short- and long-term evaluation of "Here's Looking at You" alcohol education program. *Journal of Drug Education, 18*(3), 235–240.

LOHRMANN, D. K. & FORS, S. W. (1986). Can school-based educational programs really be expected to solve the adolescent drug abuse problems? *Journal of Drug Education, 16,* 327–339.

MAUSS, A. L., ET AL. (1988). The problematic prospects for prevention in the classroom: Should alcohol education programs be expected to reduce drinking by youth? *Journal of Studies on Alcohol, 49*(1), 51–61.

MOONEY, C. ET AL. (1979). Here's Looking at You: A school-based alcohol education project. *Health Education, 10*(6), 38–41.

MOSCOWITZ, J. M. (1989). The primary prevention of alcohol problems: A critical review of the research literature. *Journal of Studies on Alcohol, 50*(1), 54–88.

SWISHER, J. D., NESSELROADE, C., & TATANISH, C. (1985). Here's Looking at You-Two is looking good: An experimental analysis. *Journal of Humanistic Educational Development, 23,* 111–118.

TOBLER, N. S. (1986). Meta-analysis of 143 adolescent drug prevention programs: Quantitative outcome results of program participants compared to a control or comparison group. *Journal of Drug Issues, 16*(4), 537–567.

ARMAND L. MAUSS

Life Skills Training

Toward the end of the 1970s, an approach to drug-abuse prevention called Life Skills Training (LST) was initiated. This approach differed from the type of information programs conducted by many schools until that time. Instead of students being taught a collection of facts about drugs and the dangers of using them, they were taught general skills for living happier and healthier lives. Studies testing the LST approach have been conducted since 1980, and they provide evidence that teaching life skills can help adolescents avoid becoming involved with drugs.

BEYOND HEALTH KNOWLEDGE

The LST program was developed because conventional approaches to TOBACCO, ALCOHOL, and drug abuse had not worked (Botvin, 1986). Most schools around the country had courses in health education, tobacco education, alcohol education, or drug education. In these courses, students typically have been taught that using tobacco, alcohol, MARIJUANA, or other drugs is bad for their health, and they may find out how or why they are dangerous. Sometimes students are given detailed information about how these substances affect the body, how long the effects last, and even how people use them. Many tobacco-, alcohol-, and drug-education programs have tried deliberately to scare students by pointing out how many people die each year from drug abuse.

It is widely believed that if students really knew how harmful smoking, drinking, or using drugs is, they would not do it. However, numerous studies have found that teaching facts or using scare tactics does not work. Students in educational programs that make use of these approaches still end up using drugs. It is now clear that teaching facts is not enough, and the LST program was therefore designed to go beyond merely providing students with facts about the harmful effects of using drugs.

TARGETING THE CAUSES OF DRUG ABUSE

Many prevention programs do not work because they do not deal with the real causes of drug abuse (U.S. Public Health Service, 1986). Although we still need to learn more about what leads to drug abuse and how it develops, much is already known. This knowledge about the causes of drug abuse and the theories that researchers have developed to explain it provide the foundation upon which successful prevention programs, such as LST, are based. At this point, most drug-prevention experts agree that drug abuse does not have a single cause. Many different factors cause individuals to first try one or more drugs, and then they gradually become both physically and psychologically dependent on them (Schinke, Botvin, & Orlandi, 1991).

Perhaps the most important influence that leads to drug use among young people comes from the world around them. Individuals who use drugs generally have friends, older brothers and sisters, or parents who also use drugs. Under these circumstances, they not only have easy access to one or more drugs, but they also constantly get the message that drug use is normal and acceptable, fun and exciting, a way of dealing with stress and anxiety, something that is grown-up and cool. ADOLESCENTS may be influenced to use drugs both because they see people they look up to (such as athletes and entertainers) using them and because the popular media all too often "glorifies" tobacco, alcohol, and drug use. In addition, adolescents are often pressured to use drugs by their friends and lured by advertisers trying to sell cigarettes or alcoholic beverages.

According to research studies, some individuals are more likely than others to succumb to such pressures. Generally, they are the people who have low self-esteem, are unable to handle anxiety, feel that their lives are controlled by outside forces, are not sufficiently able to stand up for their rights to say no to unreasonable requests, have positive attitudes to drug use, and believe that most people use drugs. Finally, the values and leisure activities of individuals who use drugs appear to be different from those of nonusers. For example, adolescents who use drugs tend to get lower grades in school, are more likely than nonusers to engage in antisocial behaviors, and are less likely to participate in organized extracurricular activities such as sports or clubs.

HOW AND WHEN DRUG INVOLVEMENT DEVELOPS

Most people start using drugs during their early teenage years or slightly before (Schinke, Botvin, & Orlandi, 1991). This is the time when they are experimenting with a wide range of behaviors and life-style patterns as part of the natural process of growing up, becoming more independent, and discovering their own identity. They typically try alcohol or cigarettes first; later, they may try marijuana or another illegal drug. Contrary to what adults might like to think, more than half of all adolescents try one or more of these substances. Most individuals who try drugs do not use them more than a few times, but those who do run the very real risk of developing a compulsive pattern of use characterized by increases in both the frequency and amount of drug use and possibly development of drug dependence.

LIFE SKILLS TRAINING PROGRAM

The LST program (Botvin, 1989) was developed after careful consideration of many factors found to be associated with the initiation of adolescent drug abuse. Designed to help young people develop an increased sense of personal control, it teaches them basic personal and social skills that are relevant to the everyday problems confronting them. The problem of drug abuse is addressed within this larger context of acquiring basic life skills and enhancing personal and social competence. Because the LST approach was intended for middle or junior high school students, it focuses on preventing use of the so-called gateway substances of tobacco, alcohol, and marijuana.

The main objectives of the LST program are: (1) to provide students with the information and skills they need to resist social pressures to use drugs; (2) to decrease potential motivations for using drugs by helping students develop greater autonomy, self-esteem, self-mastery, and self-confidence; (3) to enable students to cope effectively with anxiety, particularly anxiety induced by social situations; (4) to increase students' knowledge of the immediate negative consequences of drug use and provide them with accurate information concerning the prevalence rates of tobacco, alcohol, and marijuana use; and (5) to promote the development of attitudes and beliefs supportive of a life-style that excludes drug use.

The LST program was developed to be used with students who attend either middle school (grades 6–8) or junior high school (grades 7–9), but because the pressures and opportunities to use drugs typically increase throughout the middle and junior high school years, the present LST curriculum is a three-year program. It consists of fifteen class periods during the first year, ten booster sessions in the second year, and five booster sessions in the third year. The booster sessions, which are intended to reinforce the material taught in the first year of the program, focus on the demonstration and practice of the life skills that form the foundation of this prevention approach. The LST program contains the following five components, each of which consists of two to six sessions.

Knowledge and Information (Four Sessions). Session 1 provides general information about cigarette smoking, with emphasis on the prevalence of cigarette smoking among teenagers and adults, and on its declining social acceptability. Session 2 focuses on the immediate physiological consequences of smoking, with the help of biofeedback apparatus to demonstrate the effect of cigarette smoking on heart rate, carbon monoxide in expired air, and hand steadiness. The third and fourth sessions are like the first session except that they concentrate on alcohol and marijuana respectively. Students are given the necessary information to enable them to counter common myths and misconceptions about alcohol and marijuana.

Decision Making and Independent Thinking (Four Sessions). The first two sessions deal with making decisions effectively and responsibly. Students examine how they currently make important decisions and are taught an effective strategy for doing so. The sessions also identify tactics that others may use to influence one's decisions. Sessions 3 and 4 introduce the techniques used by advertisers to influence the decisions students make as consumers. Students discuss the purpose of advertising and review the use of deceptive advertising techniques. Cigarette and liquor advertisements are highlighted as examples of how these deceptive techniques may be used. Students are taught to identify the techniques and to avoid falling prey to the appeals used to persuade them to smoke cigarettes and drink alcoholic beverages.

Self-directed Behavior Change (Two Sessions). The third component of the LST program focuses on self-image and self-improvement. During this component, an eight-week self-improvement project is begun in order to help students improve one aspect of their lives. Session 1 explains what self-image is, how it is formed, and how it can be improved through a self-improvement plan. During Session 2, students select a skill or behavior that they would like to improve or change and identify a long-term (eight-week) goal and a series of short-term (weekly) goals. Students complete reports so that they can follow their weekly progress as they learn to shape their own behavior.

Coping with Anxiety (Two Sessions). The fourth component involves learning to cope with anxiety and features explicit instructions on how to use specific anxiety-reduction techniques. Students are invited to discuss their own experiences with anxiety, the symptoms of anxiety, and the common situations that produce anxiety. Three anxiety-reduction techniques are taught: a basic relaxation exercise, a deep-breathing exercise, and the use of mental rehearsal. A twelve-minute relaxation audiotape is used to introduce the students to the relaxation exercise in the classroom. Students are encouraged to practice these techniques at home and to integrate them into their everyday lives as active coping strategies. They are also made aware of their own anxiety-producing thought sequences, shown how these sequences influence their reaction to anxiety-producing situations, and taught the importance of controlling them.

Social Skills (Six Sessions). Sessions 1 and 2 of this largest of the five components cover communication skills, define verbal and nonverbal communication, and present specific guidelines for avoiding misunderstandings. Session 3 is designed to help students overcome shyness. Students are taught the skills needed to initiate social contacts, give and receive compliments, and begin, maintain, and end conversations. Session 4, which focuses on the skills needed to maintain opposite-sex relationships, begins with a discussion of attraction. It teaches students how to initiate and sustain conversations and how to plan dates and social activities. In the last two sessions, which involve assertiveness training, students discuss the reasons for not being assertive and learn the benefits of appropriate assertiveness. General verbal and nonverbal assertive behaviors are taught, and students are asked to apply them to situations involving peer pressure to smoke, drink, or use marijuana.

During the 1980s, LST was tested by Botvin and his colleagues in eight separate studies that involved more than 25,000 students from over 150 schools in New York and New Jersey. Most of the studies focused on cigarette smoking, but several also examined the impact of LST on alcohol and marijuana use. The LST approach typically produced reductions of 50 percent to 80 percent in new smoking, drinking, and marijuana use after the first year of the program (Botvin & Tortu, 1988), but booster sessions appear to be necessary to maintain these initial prevention effects. Studies have demonstrated that the LST program can be effectively implemented by adult providers and peer leaders. Not surprisingly, it was found that the effectiveness of the LST program was related to how thoroughly it was implemented. Students whose teachers conducted the program carefully and completely demonstrated lower rates of drug use than did students whose teachers either deviated from the program or taught only part of it.

The early evaluation studies of the LST program involved predominantly white, middle-class students. More recent research has begun to examine its effectiveness with inner-city minority students. The promising results so far suggest that it may work with a relatively broad spectrum of students. For example, three separate studies that tested the impact of the program on cigarette smoking indicated that it can significantly reduce smoking among both Hispanic and black youth. The school-based LST model has also been modified so it can be used with children living in homeless shelters in New York City.

Research is currently under way to determine the extent to which the LST approach is effective in reducing risk for HIV infection. Studies are also being conducted to investigate the long-term effectiveness of this type of prevention strategy with tobacco, alcohol, and marijuana as well as to determine the extent to which it is effective with other illicit drugs. Notwithstanding the need for additional research, it is clear from the past twelve years of testing that LST may be one of the especially promising substance-abuse prevention programs that is currently available.

(SEE ALSO: *Advertising and Alcohol Use; Advertising and Tobacco Use; Coping and Drug Use; Prevention*)

BIBLIOGRAPHY

BOTVIN, G. J. (1993). *Life skills training: Teacher's manual.* Princeton, NJ: Princeton Health Press.

BOTVIN, G. J. (1990). Substance abuse prevention: Theory, practice and effectiveness. In J. Wilson & M. Tonry (Eds.), *Crime and justice.* "Drugs and Crime" Special Issue, vol. 13. Chicago: University of Chicago Press.

BOTVIN, G. J. (1986). Substance abuse prevention research: Recent developments and future directions. *Journal of School Health, 56,* 369–374.

BOTVIN, G. J., & TORTU, S. (1988). Preventing adolescent substance abuse through life skills training. In *Fourteen ounces of prevention: A casebook for practitioners.* Washington, DC: American Psychological Association.

SCHINKE, S., BOTVIN, G. J., & ORLANDI, M. A. (1991). Substance use in children and adolescents: Evaluation and intervention. *Developmental and Clinical Psychology and Psychiatry,* vol. 22. Newberry Park, CA: Sage.

U.S. PUBLIC HEALTH SERVICE. (1986). *Drug abuse and drug abuse research: The second triennial report to Congress from the secretary* (DHHS Publication No. ADM 87-1486). Washington, DC: U.S. Government Printing Office.

GILBERT J. BOTVIN

Napa Project, Revisited   During the early 1970s, the field of drug-abuse PREVENTION turned strongly toward school-based affective and alternatives programs. Many in the field at that time became convinced that these programs would strengthen children's interpersonal competencies and positive social connections, make schools more hospitable and rewarding places for them to spend time, and increase their sense of efficacy and contribution. We were sure that such programs, in contrast to the informational drug-education programs that preceded them, would deal with root causes of substance abuse, especially inadequate socialization and inadequate opportunities for meaningful experience, involvement, and accomplishment.

Despite the excitement about affective programs, funding for them was hard to find. Well-designed and -implemented programs remained scarce. Prevention was still the stepchild in the drug-abuse field, the rationale for affective programs was not widely understood, and their proponents felt handicapped by a lack of research evidence attesting to their effectiveness. It was a chicken-or-egg bind; little funding resulted in inadequately implemented, poorly evaluated programs, which resulted in little evidence of effectiveness, which perpetuated the funding problem.

The Napa Project was designed to demonstrate the promise of school-based affective and alternatives programs. It was oriented toward exemplary strategies for elementary and junior high school students because interventions in senior high seemed to be too late. The hope was to see each strategy implemented in high-quality fashion and in fertile circumstances, for periods measured in semesters and years, not days or weeks. The goal was to assess the strategies' effects on a range of student outcomes. Further, in addition to implementing and evaluating the strategies individually, there was reason to assess them in several combinations and sequences, in recognition that significant effects might not be attainable with any one strategy alone.

The Napa Project was conducted between 1978 and 1983, in close collaboration with the Napa Unified School District in northern California. All the studies were done in the Napa schools, which served a largely white, middle- and working-class community on the periphery of the San Francisco Bay area. The intervention and evaluation costs of the project were supported by a large, multiyear grant from the NATIONAL INSTITUTE ON DRUG ABUSE (NIDA).

THE SEVEN PREVENTION STRATEGIES

Underlying selection of seven prevention strategies for the Napa Project was a theoretical model linking them to improvements in classroom and school environments, and then to positive changes

in students' competencies, values, and attitudes (see Figure 1). Derived from the work of the Jessors (Jessor & Jessor, 1975) and Fishbein (Ajzen & Fishbein, 1977; Schlegel et al., 1977), the causal model held that as students' satisfaction with self, peers, and school increased, and as they perceived their peers to have more positive attitudes toward school, their own attitudes toward drug use would become less accepting and they would perceive the norms of their peers to be similarly antidrug. This was intended to decrease both intentions to use drugs and actual drug use.

Several criteria were used in selecting strategies. Each had to be consistent with a general approach emphasizing school and peer-group influences, and individual competencies and attitudes. Each had to be implementable at moderate cost and without major changes in the basic goals or methods of public schooling. Each needed to be representative of then-current affective and alternatives programs or, in one case, of what was then the newly emerging "social influence" approach to drug education.

The project development began with a search for established, packaged strategies that met the above criteria. Where none were found, project staff developed a strategy using material from existing programs and curricula.

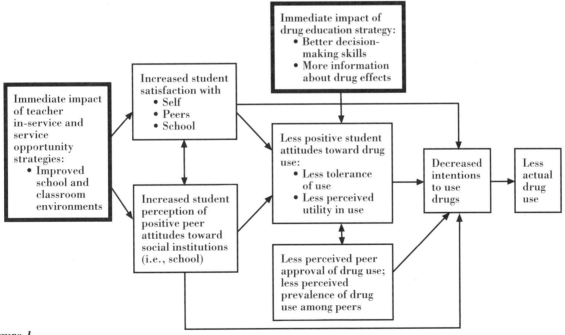

Figure 1
Hypothesized Change Model

Four of the seven strategies that ultimately were selected involved in-service teacher-training programs. These programs focused on helping teachers become comfortable and proficient with instructional practices that were believed to influence students' attitudes toward school, their self-concepts, and their development of social competencies. In these programs, the topic of drug use was hardly mentioned. Rather, the programs were designed to improve teachers' classroom management and instructional skills, and to enhance the overall classroom learning climate. The four strategies and the grade levels at which they were implemented are listed below.

- Magic Circle—teachers were trained to lead structured class meetings designed to build a sense of connection and community, and to foster social and academic development (grades 3–4).
- Effective Classroom Management (ECM)-Elementary—teachers were taught communication skills, discipline techniques, and self-concept enhancement techniques for use throughout the school day (grades 4–6).
- Effective Classroom Management (ECM)-Junior High—communication, discipline, and self-concept enhancement skills were adapted for teaching in the junior high environment (grades 7–9).
- Jigsaw—teachers were taught to organize classrooms into cooperative learning groups of five or six students, in which each student was given the responsibility of teaching an essential piece of the regular curriculum to the other group members (grades 4–6).

Two alternatives strategies were offered as elective academic courses to junior high students. In these courses, students were taught skills and provided with opportunities for helping peers or younger children. The courses did not address the topic of drug use; instead, they sought to teach social competencies and to enhance self-esteem. The alternatives strategies were the following:

- Cross-age tutoring—students regularly tutored younger children in reading or other academic subjects (grades 8–9).

- Operating a school store—students ran a school store on their campus, selling school supplies and snacks, while learning relevant business skills in a related academic course (grades 8–9).

The final strategy was a drug education course that taught social competencies and drug information to seventh graders. In the course, students were taught Maslow's (1980) framework for understanding motivation; learned a systematic decision-making process; analyzed techniques used in commercial advertising; learned assertiveness skills for dealing with peer pressure; and practiced setting personal goals. Toward the end of the course, students were provided with information about tobacco, alcohol, and marijuana, in response to their written questions. They also applied the social skills in considering drug-use issues.

IMPLEMENTATION OF THE STRATEGIES

The Napa Project employed a three-member program staff to oversee and support implementation of the strategies. All three individuals had prior experience as classroom teachers and as in-service trainers. The primary role of the program staff was not to provide services directly to students, except in the delivery of the drug-education course and in one trial of the cross-age tutoring course. Rather, with the other strategies, their role was to provide training, coaching, and consultation to volunteer teachers in the participating schools.

Evaluation data collected from students and teachers during initial testing of the strategies were used to revise curricula and procedures for subsequent implementation. The ECM and drug-education strategies were substantially revised, and the other strategies were modified in minor ways.

In trials of the in-service strategies, teachers who volunteered to participate were trained by program staff during nine to twelve weekly two-hour workshops. Several times during and after training, one of the trainers observed each teacher's use of the in-service skills in the classroom, and provided additional encouragement and guidance. All of the in-service courses combined lectures, discussions, readings, simulations, and practice exercises. At each training session, previously taught skills were reviewed, implementation problems were discussed, and new skills were introduced and practiced. All teachers who completed the training received a stipend, and graduate credit was offered.

The first version of the cross-age tutoring course was offered each semester to eighth and ninth graders, and was taught by a junior high school teacher assisted by a program staff member. The second version was offered to eighth graders only and was taught by program staff. The class met daily during the entire semester. Tutors traveled to nearby elementary schools to work one on one or in small groups with younger students. They also met as a group to refine skills, discuss problems, and plan schedules. Program staff closely monitored the tutors' activities at the elementary schools. Tutors received grades and academic credit for their participation in the course.

The school store class met daily and was taught by a junior high school teacher with assistance from a program staff member. Teaching methods included lectures, demonstrations, self-guided learning modules, experiential activities, simulations, and role-playing. Students volunteered some of their own time to work in the store. Each student participated in most aspects of store operations, including sales, marketing, and accounting.

The twelve sessions of the drug-education course were taught by a program staff member once a week in social studies classes. Instruction included lectures, whole-class and small-group discussions, demonstrations, experiential activities, role-playing, and audiovisual presentations.

EVALUATION METHODS AND RESULTS

The seven strategies were evaluated individually and in certain combinations in twelve separate studies. All studies assessed the implementation of the strategies as well as their effects on students. All studies used research designs with assessments before and after the intervention, and follow-up testing to detect any delayed effects. Most studies used random assignment of students, classrooms, or schools; in the few where assignment was nonrandom, pretesting generally showed good initial equivalence between conditions. A wide array of student outcomes were measured, using reliable instruments and carefully controlled assessment procedures. The designs

of most (but not all) studies held up well over time; loss of subjects from the studies occurred at moderate levels and was comparable across conditions in most studies.

Process and outcome evaluations were conducted by the project's full-time four-person research staff. General information about the studies can be found in Schaps et al. (1984), which lists twenty-eight publications describing the various studies.

None of the strategies was shown to be effective. The four in-service strategies and the two alternatives strategies had no systematic effects on students' perceptions of classroom climate; attitudes toward self, peers, or school; attendance; academic achievement; perceptions of peer group norms; or drug-related attitudes, intentions, or behaviors. Moreover, this lack of effects could not be readily explained by poor implementation of the strategies. Implementation of the alternatives strategies was generally satisfactory. Although implementation of the in-service strategies did vary greatly from teacher to teacher, and was found to be inadequate in many classrooms, no effects were found even for the subgroups of students had who the greatest exposure to the strategies, or who were in classrooms where the strategies were best implemented, or who received combinations of strategies over two or three years.

In the first two of three studies of the drug-education course, no effects were found for the eighth graders or for seventh grade boys, but positive effects were found on seventh grade girls' drug knowledge, perceived peer attitudes toward use, and their use of alcohol, cigarettes, and marijuana. But, inexplicably, no pattern of positive effects for girls was found in a third study.

WHY THE LACK OF EFFECTS?

What might account for the strategies' ineffectiveness? Project staff at the time decided that the strategies were based on an inadequate theory of substance-abuse prevention. Outcome data from the studies consistently showed that the seven strategies did *not* foster constructive social attitudes, norms, and competencies. And some longitudinal data suggested that even if the strategies had influenced social norms and attitudes, the ultimate effects on substance abuse might have been negligible. The data showed that positive attitudes toward school and academic self-esteem were *not* strongly associated with later avoidance of substance abuse.

In the mid-1900s, it still appears that the Napa Project's strategies and underlying theory were inadequate. Other evaluations of school-based affective programs have shown similarly disappointing results, and some reviewers have since written off all school-based educational programs as ineffective approaches to prevention.

Nevertheless, the failings of Napa's strategies and theory may well have been inadequacies of scope and depth, not of direction. That is, Napa may well have been on a potentially fruitful track in seeking to promote socially constructive norms, attitudes, and competencies—and reduce substance abuse—by fundamentally altering students' experience of schooling. But those who designed Napa may have grossly underestimated the scope and substance of the needed changes, and also the resources and processes needed to enact those changes. Even the classrooms in which the best implementation of Napa's strategies was observed may not have differed much from "ordinary" ones.

Why might even Napa's "high implementing" classrooms have fallen short of providing what is needed to promote students' development? Because, as designed, Magic Circle, ECM, and Jigsaw all failed to address the content of the day-to-day curriculum or the broad, long-term goals of schooling. Moreover, each strategy emphasized only classroom pedagogy, or discipline, or climate. Consequently, none of the strategies put forward an encompassing educational vision or framework, and none addressed more than a modest subset of the variables now thought to be involved in achieving such a vision.

With what is known today about the complexity of making fundamental change in education, it can also be argued that Napa's support to teachers was narrow and shallow. True, in providing twelve workshops over a period of several months, supplemented by one or two individualized consultations with a trainer, Napa offered teachers more support than most prevention programs of its time. Today, however, it is recognized that no change in meaningful ways, most techers need several years of focused staff development; regular opportunities for planning, reflection, and problem solving with peers; congruent instructional and curricular materials; encouragement from school and central office administrators; supportive assessment practices; and protection from conflicting demands for change.

Napa's two alternatives strategies are subject to a similar criticism: limited scope. Working as a tutor or a staff member of a school store for 30–45 minutes several days a week may be enjoyable and satisfying, but it may have no effect on the broader experience of schooling unless a school changes in other ways. This is what the data suggest. Students seemed to like participating in the alternatives activities, and to feel that they got something valuable from the experience, but their more general attitudes toward school, learning, and peers, and their attitudes toward self, were little affected.

SUMMARY

In summary, it may be that the types of school programs which will effectively promote positive development and inhibit problem behaviors will be very broad in conception. Such programs may simultaneously address issues of school structure, climate, policy, curriculum, pedagogy, assessment, and more. They may focus on the dual goals of (1) creating a challenging, accessible curriculum and (2) a sense of community for all children, characterized by a strong sense of shared purpose and values, with by stable, supportive connections among students and adults. The most significant effects of such changes may be less on *intra*personal attitudes and tendencies, and more on *inter*personal norms. The preventive benefits of such communities may lie in their capacities to hold and guide children more effectively as they develop.

Perhaps, then, the Napa Project was so limited in the changes it sought and in the assistance it provided, that the "success" it achieved was too puny to make a meaningful difference in the development of children. Perhaps the school-reform efforts of the 1990s, guided by more comprehensive theories of learning and development, and better understanding of what is involved in making meaningful change, will accomplish what Napa hoped to, but did not know how to, achieve.

(SEE ALSO: *Education and Prevention; Prevention Movement*)

BIBLIOGRAPHY

AJZEN, I., & FISHBEIN, M. (1977). Attitude-behavior relations: A theoretical analysis and review of empirical research. *Psychological Bulletin, 84,* 888–918.

FLAY, B., & PETRAITIS, J. (1993). The theory of triadic influence: A new theory of health behavior with implications for preventive interventions. In G. Albrecht (Ed.), *Advances in medical sociology*. Vol. 4: *A reconsideration of models of health behavior change*. Greenwich, CT: JAI Press.

FULLAN, M. (1993). *Change forces: Probing the depths of educational reform*. London: Falmer Press.

HANSEN, W. (1990). Theory and implementation of the social influence model of primary prevention. In K. Rey, C. Faegre, and P. Lowery (Eds.), *OSAP prevention monograph—3: Prevention research findings: 1988*. Rockville, MD: Office of Substance Abuse Prevention, U.S. Department of Health and Human Services.

HAWKINS, D., & CATALANO, R. (1987). The Seattle Social Development Project: Progress report on a longitudinal prevention study. Paper presented at the National Institute on Drug Abuse Science Press Seminar, Washington, DC.

JESSOR, R., & JESSOR, S. (1975). Adolescent development and the onset of drinking. *Journal of Studies on Alcohol, 36*, 27–51.

MASLOW, A. (1980). *Motivation and personality*. New York: Harper & Row.

MOSKOWITZ, J. (1989). The primary prevention of alcohol problems: A critical review of the research literature. *Journal of Studies in Alcohol, 50*(1), 54–88.

SCHAEFFER, G., MOSKOWITZ, J., MALVIN, J., & SCHAPS, E. (1983). School-related attitudes and drug involvement: Testing a causal model using latent variables and longitudinal data. Technical report. Napa, CA: Pacific Institute for Research and Evaluation.

SCHAPS, E., MOSKOWITZ, J., MALVIN, J., & SCHAEFFER, G. (1984). *The Napa drug abuse prevention project: Research findings*. DHHS Publication No. (ADM) 84-1339. Washington, DC: U.S. Department of Health and Human Services.

SCHLEGEL, R., CRAWFORD, C., & SANBORN, M. (1977). Correspondence and mediational properties of the Fishbein model: An application to adolescent alcohol use. *Journal of Experimental Social Psychology, 13*, 412–430.

ERIC SCHAPS

Ombudsman Program Ombudsman is a word of Swedish origin that can be loosely translated as "a helping person." The Ombudsman program is a drug-abuse prevention program geared to students in grades five through nine. The program was developed by the Drug Education Center, located in Charlotte, North Carolina, and is based on the assumption that the most effective way to prevent adolescent alcohol and other drug (AOD) abuse is through the promotion of individual personal growth. This is attempted via enhancements of self-esteem, social skills, and the empowerment of students in a group project that seeks to help others. Students meet once or twice per week (depending on the course module chosen) during regular classroom hours. In addition, the program activities are designed to be integrated into academic subject areas. Either a trained facilitator or a classroom teacher who has been trained by a certified Ombudsman trainer directs the program.

The Ombudsman program was one of the first drug-abuse prevention programs in the United States to be funded by the NATIONAL INSTITUTE ON DRUG ABUSE (NIDA) in 1977. The purpose of the NIDA grant was to fully develop and evaluate the outcome of this program. Prior to 1977, many drug-abuse prevention programs were based on drug information and often tainted by "scare tactics." Officials at NIDA hoped to promote a shift away from these tactics, and this was part of the reason for supporting an adolescent drug-abuse prevention program based on a personal growth-oriented approach. In 1979 and again in 1984, the Ombudsman program was recognized by the U.S. Office of Education.

The program has three phases. The first, Self-Awareness, involves a series of exercises that encompass activities for building self-esteem. The purpose of these activities is to foster the development of self-worth and respect for others. The second phase, Group Skills, gives students an opportunity to foster communication, positive group interaction, and refusal/resistance decision-making skills. Information on the effects of drugs is taught in this phase. During the third and last phase, students apply the knowledge and skills they have gained in the program by planning and carrying out a project that helps others within their own community or at school. Ombudsman program activities are experiential, utilize cooperative learning techniques, and appeal to a variety of learning styles.

Ombudsman program outcomes have been evaluated by using the Student Attitudinal Inventory (Kim, 1981c). Short-term evaluation results indicate that the program can affect seven high-risk student attitudinal factors closely related to adolescent drug-using behavior: negative social attitude, rebelliousness, low valuing of school, poor student-teacher relationship, perception of incohesive family relationship, low self-esteem, and attitudes favoring drug use. It has also been learned that the program is more effective among younger than among older students. Finally, data from one long-term evaluation suggest that there is a greater proportion of students who no longer use drugs (i.e., who gave up experimenting with drugs) among the students trained in the Ombudsman program than among those who have not participated in the program.

(SEE ALSO: *Education and Prevention; Prevention*)

BIBLIOGRAPHY

KIM, S. (1983). The short-term effect of a national prevention model on student drug abuse. *Journal of Primary Prevention*, 118–128.

KIM, S. (1981a). An evaluation of the Ombudsman primary prevention program on student drug abuse. *Journal of Drug Education*, *11*(1), 27–36.

KIM, S. (1981b). How do we know whether a primary prevention program on drug abuse works or does not work? *The International Journal of the Addictions*, *16*(2), 359–365.

KIM, S. (1981c). Student Attitudinal Inventory for outcome evaluation of adolescent drug abuse prevention programs. *Journal of Primary Prevention*, *2*(2), 91–100.

SEHWAN KIM
CARL SHANTZIS
JONNIE MCLEOD

PRIDE (National Parents Resource Institute for Drug Education) Thomas "Buddy" Gleaton, Ed.D., and Marsha Keith Manatt Schuchard, Ph.D., founded PRIDE in Atlanta, Georgia, in 1978. PRIDE's purpose is to help parents form groups to protect their children from becoming involved with MARIJUANA and other drugs. The organization is based on the following fundamental principles: (1) drug abuse is a health issue; (2) the family is the greatest bulwark against adolescent drug use; (3) families need help from the rest of the community to steer young people safely through the many temptations and dangers that confront them every day.

Initially, PRIDE based its group model on parent peer groups initiated by Dr. Schuchard and her fam-

ily. A parent peer group encourages parents to get to know and link up with the parents of their children's friends. They establish social guidelines for their children to which they all agree to adhere, and they try to create positive alternatives for young people to prevent them from engaging in unhealthy and destructive behaviors during adolescence. Dr. Schuchard's handbook, *Parents, Peers and Pot*, published and distributed by the National Institute on Drug Abuse, outlines how to form parent peer groups.

PRIDE later expanded its parent peer-group model to include larger groups of parents who wanted to work for change throughout their communities to prevent drug abuse among young people. The organization offers training to parents across the nation. PRIDE also added a youth component, training junior high and high school students and encouraging them to take a stand against drug use. In both cases, the essence of the PRIDE philosophy is to help parents and young people reverse adolescent peer pressure that encourages negative behaviors, and use it as a force to persuade young people to adopt positive behaviors.

The PRIDE Drug Use Survey has helped thousands of local school systems determine the extent of ALCOHOL, marijuana, and other drug use among students in elementary, middle, and high school. A large data base allows PRIDE to spot early trends in the rise or fall of various drugs used by students. A systemwide survey of Atlanta public school students in 1994 demonstrated a shocking correlation between drug use and possession of guns and other weapons. The more involved a student is with drugs, the more likely he or she is to possess a weapon. If this early indicator holds true for students in other school systems, it will provide even more reason to intensify efforts to prevent drug use among students, in order to free them from violence as well as drug abuse and addiction.

PRIDE sponsors an annual conference on drug abuse, bringing experts from many related disciplines together to educate young people, parents, and educators about drug effects and to highlight the prevention work of parent and youth groups, including the PRIDE teams of students across the nation. The conference has grown from a small event to one that attracts thousands of participants.

First Lady Nancy Reagan and the first ladies of several nations hosted a PRIDE conference during the Reagan administration. The visiting first ladies returned home to help parents in their own countries initiate parent groups.

As the United States devoted more resources to developing the discipline of substance-abuse prevention, and as grass-roots organizations such as PRIDE and others increasingly demonstrated that PREVENTION reduces drug use and abuse, European and other nations became intensely interested in learning about the American prevention experience. PRIDE has done much to foster this interest, and the PRIDE conference increasingly draws participants from other nations to learn about American grass-roots prevention techniques and processes.

Along with other national prevention organizations, PRIDE has had a major influence in reducing substance abuse in the United States since 1979, when the use of most drugs peaked.

(SEE ALSO: *Parents Movement; Prevention Programs; U.S. Government Agencies*)

SUE RUSCHE

Project SMART Project SMART was begun in 1981 in Los Angeles by Drs. C. Anderson Johnson, Brian R. Flay, William B. Hansen, and John W. Graham as a pioneering effort to scientifically test programs for preventing experimental and habitual use of multiple substances. Originally the name stood for Self-Management and Resistance Training, but since then, it has come to stand more generally for the programs created by this University of Southern California research team, and for the programs used in many of their projects.

As children enter middle school and junior high school, they begin experimenting with BEER and wine. Many of these young people also try cigarettes and hard liquor. Some follow this by experimenting with MARIJUANA and they develop patterns of habitual and heavy use of ALCOHOL and TOBACCO. It was the goal of Project SMART researchers to interrupt the usual pattern of experimentation and habituation by presenting innovative programs that provided students with skills for overcoming situations that might promote use. Project SMART provides instructions for classroom teachers on how to prevent experimentation with and regular use of alcohol, tobacco, marijuana, and other drugs. Originally, there were two sets of Project SMART materials. One set

focused on teaching students self-management skills. The other set provided students with social pressure resistance skills for dealing with peer pressure to use substances as well as skills for avoiding pressure from television, movies, music, adults, and ADVERTISING that might make substance use attractive. Both sets of materials included information about the consequences of alcohol, tobacco, and marijuana use.

These two curricula represented two different ways of thinking about what causes young people to experiment with substances. Self-management training came from the idea that young people use drugs to help them handle the challenges of growing up. This approach was based on the hypothesis that young people who lack the ability to manage their lives may experiment with alcohol or drugs as an alternative to handling difficult situations. The goal was to increase the skills that are important to being successful so that substance use would not be seen as a practical alternative. The training was applied to making decisions, handling STRESS, improving self-esteem, and increasing a person's ability to set and achieve goals. Students were taught how to identify and manage their stress through relaxation training; how to increase their self-esteem through finding positive qualities about themselves and others; and how to make well-thought-out decisions by mastering a process for identifying problems, thinking of alternatives, and weighing consequences. They learned how to set and achieve goals through practicing personal goal setting.

Training in resistance skills came from the idea that experimentation with alcohol, tobacco, and other drugs was related to social pressure. It was hypothesized that young people had to deal with offers, threats, and dares to use drugs and that they experimented with alcohol and other drugs in order to fit in with their peer group. Thus young people who had the skills necessary for refusing offers in a way that allowed them to still be accepted were thought to be able to avoid experimentation. In this program, students learned to identify different types of peer pressure to use drugs, including "friendly," "hot," and "silent" pressures. They were then taught some simple but effective strategies for refusing offers and resisting pressure, such as saying "No, thanks," giving a reason or excuse, repeating the same response to continued pressure (the "broken record" technique), walking away, avoiding the situation, giving the cold shoulder, changing the subject, and/or using strength in numbers. Students practiced these techniques in role plays. The program also informed students about the real rates of use among their peers, which were nearly always lower than what students originally expected.

Both programs included a session in which students were videotaped making a personal commitment to use the skills they had been taught in order to remain drug-free. Students who had been trained in self-management skills described what they would do instead of using drugs, while students who had received resistance training described how they would respond if someone offered them a drug. Finally, in both programs, admired and respected classmates were used as peer facilitators. The peer facilitators helped conduct both programs' small-group activities and were trained to demonstrate the skills that students needed to master.

In 1982 and 1983, the two programs were presented to two groups of students in the Los Angeles Unified School District (Hansen et al., 1988). A third group of students received no special Project SMART program. Students who participated in the project filled out surveys that asked them to report on their use of alcohol, tobacco, and marijuana. After being pretested, students were given the program in the seventh grade and then completed surveys in the eighth and ninth grades. The students who had not been given any special program also completed surveys at the same time.

At the end of this project, it was found that students who had received training in how to resist pressure were less likely to use all three substances. Of the group not involved in any program, 17 percent who were nonsmokers at the beginning of the project had started using cigarettes by the ninth grade. By comparison, 29 percent of originally nonsmoking students who had received training in self-management skills and 12 percent of students who had received training in the social-pressure-resistance skills had started using cigarettes by the ninth grade. Among students who were originally nondrinkers, in schools that had received no program, 13 percent reported having had two or more drinks in the month before they were tested in ninth grade. Among originally nondrinking students who had been trained in self-management, 17 percent reported drinking this much, whereas only 5 percent of originally nondrinking students who had been

trained in resisting social pressures reported drinking this much by ninth grade. By the ninth grade, 12 percent of originally nonusing students from schools that had received no program reported using marijuana. In the schools in which self-management skills had been taught, 22 percent of originally nonusing students reported using marijuana compared to 13 percent of originally nonusing students in schools where resistance to social pressure had been taught.

The program materials for Project SMART have continued to evolve. Numerous research projects have added variations to what students are taught, and the actual methods for teaching the resistance skills have been revised and refined as the people who created and delivered the programs learned more about how young people think and found better ways to get the message of the program across.

In 1985, Drs. William B. Hansen and John W. Graham started a new project that emphasized a new strategy for prevention termed normative education. This program component was designed to establish a social norm that was intolerant of alcohol, tobacco, and other drug use. Normative education was based on that part of the original Project SMART that gave young people feedback about the rates of substance use among their peer group. In addition, the program encouraged young people to discuss openly with their parents and their friends the appropriateness of drinking in a number of situations. For example, they made judgments about getting drunk on graduation night, about young people drinking toasts at weddings, and about young persons drinking alcohol when saddened by the loss of a pet. The responses demonstrated that the majority of young people answering felt that even low levels of alcohol use were almost always inappropriate in these situations. Among additional normative-education activities that were added, one activity called for groups of students to make up lyrics to a rap or folk song that addressed drinking and other drug use. Another activity involved playing a game in which teams tried to guess the most common answers to such questions as: What is one word that describes people who get drunk? The answers had been obtained by previously surveying the students and tallying their most common responses. These activities all reinforced the naturally intolerant norm for alcohol and drug use that can be found among nearly all young adolescents. Both the new and old programs made use of peer facilitators, as had the original Project SMART, and both programs included information about the consequences of using alcohol, tobacco, and marijuana. In addition, also included in both sets of materials were homework assignments given to students to be completed with their parents.

A test was conducted that compared students who had participated in the normative-education program with students who had participated in the program that taught resistance to peer pressure. The two groups had received their program in the seventh grade and results tallied in the eighth grade were analyzed and published (Hansen & Graham, 1991). The results showed that each group of students had benefited from the program it received. Students who were given normative education expected greater intolerance of alcohol and drug use among their peers than did the other group. Students who had been taught how to resist pressure were much more capable than the students of the other group of refusing when tested in a situation in which a fellow student pretended to offer them a can of beer. However, the students who had established conservative group norms were less likely to drink alcohol, get drunk (see Figure 1), develop problem behaviors in relation to alcohol, use marijuana (see Figure 2), and/or smoke tobacco.

Ultimately, Project SMART became a curriculum guide that included the best components of all the research projects that contributed to it. The program now consists of two parts: the basic program that is delivered to students in the first year of middle or junior high school and a booster program that is given the following year. Some Project SMART program materials have been given different names, such as Project STAR and Project I-STAR. Except for being based on the most up-to-date versions of the curriculum, these programs are identical to Project SMART. The success of the curriculum was reproduced in a large study conducted in the greater Kansas City area. In this region, students from schools that received the program exhibited reduced rates of alcohol, tobacco, and marijuana use compared to students from schools that received no special program (Pentz et al., 1989).

(SEE ALSO: *Adolescents and Drug Use; Advertising; Education and Prevention; High School Senior Survey; Partnership for a Drug-Free America; Prevention*)

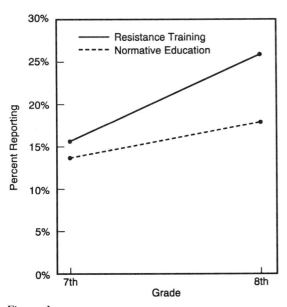

Figure 1
Percent of Students Who Reported Having Been Drunk

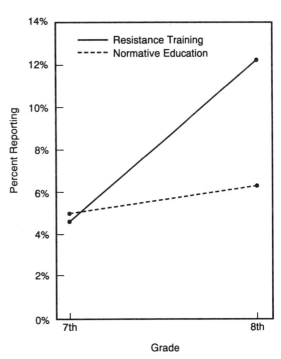

Figure 2
Percent of Students Who Reported Using Marijuana

BIBLIOGRAPHY

HANSEN, W. B., & GRAHAM, J. W. (1991). Preventing alcohol, marijuana, and cigarette use among adolescents: Peer pressure resistance training versus establishing conservative norms. *Preventive Medicine, 20,* 414–430.

HANSEN, W. B., ET AL. (1988). Affective and social influences approaches to the prevention of multiple substance abuse among seventh grade students: Results from Project SMART. *Preventive Medicine, 17,* 135–154.

PENTZ, M. A., ET AL. (1989). A multi-community trial for primary prevention of adolescent drug abuse: Effects on drug use prevalence. *Journal of the American Medical Association, 261,* 3259–3266.

WILLIAM B. HANSEN

Talking With Your Students About Alcohol (TWYSAA) This is an alcohol education curriculum for use in schools. Three levels of the curriculum are available: Level One for grades five and six; Level Two for grades seven and eight; and Level Three for grades nine and ten. TWYSAA is designed to be presented once at each level, so students receive instruction every other year. The major goals of the program are to increase the number of students who choose to abstain from drinking ALCOHOL, to delay the age at which students begin to drink alcohol (if they do begin drinking), and to reduce high-risk drinking.

TWYSAA is part of a series of alcohol-education programs designed by the Prevention Research Institute in Lexington, Kentucky. The original program, Talking With Your Kids About Alcohol, was a ten-hour course for parents—aimed at educating them about how to make low-risk drinking choices for themselves and giving them guidance on how to teach their children about alcohol. TWYSAA was the second program developed for the series; it is meant to be used as a complement to the course for parents. Schools that wish to implement the TWYSAA curriculum are first required to make Talking With Your Kids About Alcohol available to parents.

AN OVERVIEW OF THE PROGRAM

Several hours of instruction are provided at each level of TWYSAA. Activities include teacher presentations, class discussions, role-playing, group games, and training in assertiveness. Homework is assigned that encourages discussion of family attitudes about drinking and parental expectations about children's drinking behavior.

TWYSAA is concerned with influencing the students' drinking behavior not just in the present (during childhood and adolescence) but throughout their entire lives. The authors realized that, once children are no longer in school, their opportunities to receive in-depth education about alcohol would be very limited. TWYSAA teaches children how to estimate their own personal biological risk of developing ALCOHOLISM—based on their family history and individual physiological factors. Students also learn how factors such as age, gender, fatigue, illness, pregnancy, menstrual period, and medication must be considered in making drinking choices. Safety issues, such as DRUNK DRIVING are discussed, and information is given about alcohol's negative effects on the cognitive skills needed for success in school. Students are encouraged to honor the parental and religious values of their households as well as the legal prohibition (in the U.S.) against purchasing alcohol or drinking before age 21. They also learn and practice ways to "turn down a drink without turning off a friend."

TEACHER TRAINING AND MATERIALS

The TWYSAA course is usually taught as a part of the school's health-education curriculum. Teachers prepare for course instruction by attending a three- or four-day training program. The Prevention Research Institute provides these training sessions at various locations; it will also bring the training program to a location if a sufficient number of teachers want to be trained. At the training session, teachers receive all the course materials including slides to use in classroom presentations, a detailed teacher's guide with lesson plans, and printed materials that may be reproduced for students.

EVALUATION STUDY

In 1987, TWYSAA was named an "exemplary program" by the U.S. Office for Substance Abuse Prevention (OSAP). An evaluation of the curriculum was conducted by the Prevention Research Institute over a three-year period in nine schools in Kentucky and Ohio. Drinking behaviors and attitudes were measured before students took the TWYSAA course, immediately after the course, then one and two years later. A group of students (a control group) who did not take the course were also studied for comparison. The findings from this study indicated that TWYSAA had successfully achieved each of its major goals. Compared to students who had not taken the course, more of the TWYSAA students chose to abstain from alcohol; those who began drinking started later; and fewer TWYSAA students drank heavily.

(SEE ALSO: *Education and Prevention; Prevention*)

BIBLIOGRAPHY

DAUGHERTY, R., & O'BRYAN, T. (1990). *Talking With Your Students About Alcohol*, 5th ed. Lexington, KY: Prevention Research Institute.

NATIONAL ASSOCIATION OF STATE ALCOHOL AND DRUG ABUSE DIRECTORS (NASADAD) & NATIONAL PREVENTION NETWORK (NPN). (1987). *20 prevention programs: Helping communities help themselves*. Rockville, MD: National Clearinghouse for Alcohol and Drug Information.

O'BRYAN, T. (1989). Talking With Your Kids About Alcohol. *Adolescent Counselor*, Apr/May, pp. 54–55.

PREVENTION RESEARCH INSTITUTE. (1988). *Second report: Talking With Your Students About Alcohol*. Lexington, KY: Author.

DIANNE SHUNTICH

Waterloo Smoking Prevention Project

In 1979, the Waterloo Smoking Prevention Project represented one of the first rigorous efforts to evaluate a "social influences" approach to smoking prevention. Based in Waterloo, Canada, this project made use of a school-based curriculum to help students become aware of the social pressures to smoke and to practice ways of resisting those pressures.

THE GRADE 6 CURRICULUM AND LATER BOOSTER SESSIONS

The first curriculum component of the Waterloo Project consisted of two sessions in Grade 6 that were intended to provide information on the consequences of SMOKING. This was done with a method

pioneered by the ancient Greek philosopher Socrates, who posed questions and then used the answers and discussion to shed light on difficult problems. In the Waterloo sessions, the Socratic method was used to stimulate the development in students of beliefs, attitudes, and intentions regarding smoking. Information obtained during the discussion was repeated in later work by the instructors and also via videotapes, poster making, and class discussions, so as to increase students' understanding and recall of the material.

The second and probably most important component of the project was the focus on the social influences that cause one to smoke (e.g., family, media, peers) and the development of skills to resist such pressures. Ideas were again elicited from students and repeated in a variety of ways. Students then practiced using the skills by role-playing what they could do when someone wanted them to smoke.

The third component of the program concerned decision making and public commitment. Students were asked to pull together the information learned earlier and to consider the social consequences of smoking in their own social environment. Each student then made a decision about smoking and announced it to the rest of the class, along with the reasons for the decision. "Booster" sessions were used to strengthen the students' skills. After the sixth-grade curriculum, students in Waterloo schools were given two booster sessions in Grade 7 and one booster session in Grade 8. All curriculum sessions were delivered by advanced graduate-school students who were on the research staff.

EVALUATION

The Waterloo Project research team completed a very rigorous experiment to evaluate the short-term and long-term impact of this smoking-prevention curriculum and its booster sessions in grades 6 to 8. Out of twenty-two participating schools, eleven were designated at random to receive the Waterloo Project curriculum; the other eleven schools did not use any social-influences curriculum.

After tracking virtually all of the students in the participating schools, the research team used questionnaires to ask them whether and when they had started to smoke tobacco. There seemed to be a beneficial impact of the program before students reached Grade 9: Students who had received the curriculum were less likely to have started smoking. These early effects were not maintained during the high school years, however: The smoking levels of students who had received the curriculum were just as high as those of students who had not received it.

The value of the social-influences approach in preventing the onset of regular smoking by the end of high school needs further study. Results from the Waterloo Project and from other studies suggest that program effects obtained in junior high school might gradually decay during the following years and totally disappear by Grade 12. This kind of outcome may mean that high school booster sessions are necessary. Results of reported high school interventions suggest that social-influence curricula could be effective with high school students, although so far effects have been small (see Flay, 1985). Booster sessions were also recommended by an expert advisory panel convened by the National Cancer Institute (Glynn, 1989).

The apparent lack of effects of social-influence programs in preventing students from smoking by the time they reach Grade 12 should not be overinterpreted. First, it is possible that boosters in early high school years might help to maintain substantial early effects. Second, there is a much better understanding today than there was in the late 1970s and early 1980s of the essential components of effective prevention programs (Glynn, 1989). These improvements might well mean that current versions of social-influence programs might produce more durable effects. Third, society at large has changed since 1979, and social values are now more supportive of nonsmoking.

One surprising result of the Waterloo Project was the large difference observed between the students who remained in school and those who dropped out—DROP-OUTS were more likely to smoke. This difference cannot be explained by differences in age or by pretest differences in smoking risk or experience; it is, moreover, similar to results obtained from a Minnesota study (Pirie, Murray, & Luepker, 1988). Indeed, this difference may have been underestimated in both the Minnesota study and the Waterloo Project because school leavers were followed up less successfully than the students still in school. The high rates of smoking by early school leavers or dropouts warrant special attention by future researchers. Prevention programs for youth who leave

or drop out of school will need to go beyond the school setting. Early cessation programs also need to be developed for this group; these will entail helping young-adult smokers to consider quitting, something that most current smoking-cessation programs do not include.

Additional studies of the long-term effects of social-influence prevention interventions are necessary. Strong conclusions cannot be drawn from the only two long-term follow-up studies conducted to date. Although the short-term effects of the social-influences approach appear very promising, interpretation of the long-term value of these programs must await further long-term follow-up studies.

(SEE ALSO: *Education and Prevention; Prevention*)

BIBLIOGRAPHY

BEST, J. A., ET AL. (1984). Smoking prevention and the concept of risk. *Journal of Applied Psychology, 14,* 257–273.

FLAY, B. R. (1985). Psychosocial approaches to smoking prevention: A review of findings. *Health Psychol, 4,* 449–488.

FLAY, B. R., ET AL. (1989). Six-year follow-up of the first Waterloo school smoking prevention trial. *American Journal of Public Health, 79*(10), 1371–1376.

FLAY, B. R., ET AL. (1985). Are social psychological smoking prevention programs effective? The Waterloo study. *Journal of Behavioral Medicine, 8,* 37–59.

FLAY, B. R., ET AL. (1983). Cigarette smoking: Why young people do it and ways of preventing it. In P. McGrath & P. Firestone (Eds.), *Pediatric and adolescent behavioral medicine.* New York: Springer-Verlag.

GLYNN, T. J. (1989). Essential elements of school-based smoking prevention programs: Research results. *Journal of School Health, 59,* 181–188.

PIRIE, P. L., MURRAY, D. M., & LUEPKER, R. V. (1988). Smoking prevalence in a cohort of adolescents, including absentees, dropouts, and transfers. *American Journal of Public Health, 78*(2), 176–178.

BRIAN R. FLAY

PRISONS AND JAILS Prisons serve as the principal form of punishment in the United States. The arrest rate is highest in the world of the general population, 400 per 100,000, and for black people, it is over 3,000 per 100,000. Today the nation's correctional facilities are so overcrowded that many prisoners are released prematurely in order to accommodate newly incarcerated persons. Federal prison facilities built for 31,000 inmates are holding more than 56,600, while the states are holding 533,000 in cells built for 436,000 (Lipton, Falkin, & Wexler, 1992). This represents an increase of 55% in total inmates since 1981. In such conditions, the median incarceration period is two to three years, and few inmates are serving their complete sentences. Apart from a general increase in crime rates, much of this crowding is due to the public outcry against drug-related crimes and the resultant tougher sentencing practices that have been enacted against the committers of these crimes and against repeat offenders (Wexler et al., 1992). Most states have chosen to respond to prison crowding by accelerating the construction of new prisons, but in 1992 dollars this costs 70,000 to 100,000 dollars per bed space. The diversion of offenders into community treatment programs and an increased emphasis on preventative measures need to be pursued.

Especially since the advent of CRACK use in the mid-1980s, drug-dependent offenders have been responsible for a disproportionate amount of crime as compared to nonusers. About 75 percent of the nation's prisoners have used drugs (Bureau of Justice Statistics, 1991). Many persons are actively engaged in the use of drugs around the time of their arrests. Current urinalysis surveys of persons arrested in twenty-two major U.S. cities indicated that as many as 70 percent tested positive for cocaine and 20 percent tested positive for heroin (National Institute of Justice, 1989). Other studies have shown that active heroin use accelerates the user's crime rate by a factor of 4 to 6 and that the crime is at least as, and usually more, violent than that of the nonuser. Crack-related crime is at least as frequent as HEROIN-related crime and it is certainly more violent. Drug-using felons are also often repeat offenders (Lipton et al., 1992).

It has become imperative to find ways to keep offenders from reverting to crime and to avoid spending more money on new jails. Recently, intensive substance-abuse treatment programs have become an important part of the corrections approach in prisons because of accumulating evidence that treatment is capable of reducing recidivism rates (Wexler,

1994). To date, 15 percent of U.S. inmates are involved in drug treatment programs, and efforts are being made to increase this proportion substantially.

(SEE ALSO: *Crime and Drugs; Drug Use Forecasting Program; Prisons and Jails: Drug Treatment in; Shock Incarceration and Boot Camp Prisons; Treatment Alternatives to Street Crime; Treatment in the Federal Prison System*)

BIBLIOGRAPHY

BUREAU OF JUSTICE STATISTICS (1991). *Drugs and crime facts, 1991.* Rockville, MD: U.S. Department of Justice, Office of Justice Programs.

LIPTON, D. S., FALKIN, G. P., & WEXLER, H. K. (1992). Correctional drug abuse treatment in the United States: An overview. *Drug abuse treatment in prisons and jails* (NIDA Research Monograph 118). Rockville, MD: U.S. Department of Health and Human Services.

NATIONAL INSTITUTE OF JUSTICE (1989). Drug use forecasting update. *NIJ Reports*, July/August.

WEXLER, H. K. (1994). Progress in prison substance abuse treatment: A 5-year report. *Journal of Drug Issues, 24*(2), 361–372.

WEXLER, H. K., BLACKMORE, J., & LIPTON, D. S. (1991). Project REFORM: Developing a drug abuse treatment strategy for corrections. *Journal of Drug Issues, 21*(2), 473–495.

WEXLER H. K., MAGURA, S., BEARDSLEY, M. M., & JOSEPHER, H. (1994). ARRIVE: An AIDS prevention model for high-risk parolees. *International Journal of the Addictions, 29* (3), 363–388.

HARRY K. WEXLER

PRISONS AND JAILS: DRUG TREATMENT IN

Since the late 1980s in the United States, there has been a shift in the direction of the corrections field—a movement away from the model emphasizing security and control and toward a model emphasizing rehabilitation and treatment. Looking for ways to reduce recidivism and to control overcrowding (and recognizing the close connection between substance abuse and crime), correctional authorities have begun expanding prison-based drug-treatment programs. This movement is based on a growing body of research; it supports intensive prison-based drug-treatment programs as being well suited for incarcerated drug abusers and an effective means of controlling recidivism (Falkin, Wexler, & Lipton, 1992).

HISTORICAL CONTEXT

Until the late 1980s, the idea of rehabilitating prisoners was considered largely futile. This strong belief that corrections could not rehabilitate the offender was fueled by research studies that concluded that "nothing works" to reduce recidivism (Lipton, Martinson, & Wilks, 1975). Treatment in prisons was seen as an ill-fated attempt by liberals to help irredeemable offenders. In the 1970s, rehabilitation was replaced by a retributive corrections philosophy of "just deserts," in which deterrence and infliction of punishment formed the primary goals. This change in sentencing philosophy resulted in a dramatic climb in prison populations, without a commensurate decline in crime.

One reason for the increasing numbers of individuals requiring incarceration was the general population increase and the ensuing increase in serious drug involvement. Public outcry against the sharply rising crime rates of the early 1970s led politicians to call for more certain and severe sentences through the enactment of determinant sentencing and persistent felony offender laws. Public concern in 1986 regarding the spread of CRACK-cocaine also created a demand for action. Legislators responded by mandating tougher sentences against drug dealers and users. As a result, the nation's prisons were filled with serious drug-abusing offenders, many of them recidivists.

It is evident that, for criminals, incarceration is not adequate either as a deterrent or as a means of controlling recidivism. The majority of inmates, especially the most serious offenders, have severe lifestyle problems that are manifested most significantly by chronic substance abuse. Without receiving appropriate treatment while they are in prison, a high percentage of them will relapse into drug use after their release and will be sent back to prison. Research findings that have demonstrated reductions in recidivism for inmates who participated in drug treatment while in prison generated considerable interest at the federal level and contributed to support for drug treatment for prisoners (Wexler, 1994).

RECENT DEVELOPMENTS

The aforementioned factors set the stage in the mid-1980s for the enactment of several federal laws that appropriated millions of dollars for drug enforcement, prevention, education, and treatment. In particular, interest in correctional rehabilitation for drug-abusing offenders was reflected in the Anti-Drug Abuse Act of 1986, whose substantial funding for substance-abuse treatment included a large amount for correctional drug treatment.

A team of researchers at the National Development and Research Institutes (NDRI) reexamined correctional rehabilitation and drug-abuse treatment in prisons across the country. From this study, a set of guiding principles emerged for effective rehabilitation of drug-abusing offenders, and of offenders generally (Wexler, Lipton, & Johnson, 1988). Based on the available scientific evidence, it appears that drug treatment in correctional settings can curb recidivism provided the programs have the following central features: (1) treatment services are based on a clear and consistent treatment philosophy; (2) an atmosphere of empathy and safety prevails; (3) a committed, qualified treatment staff has been recruited; (4) there are clear and unambiguous rules of conduct; (5) ex-offenders and ex-addicts are used as role models, staff, and volunteers; (6) peer role models and peer pressure are used; (7) relapse-prevention programs are provided; (8) there is continuity of care while prisoners are in custody and a program of aftercare when they return to the community; (9) evaluations of treatment have been integrated into the design of the program; and (10) integrity of the treatment program is maintained, as well as autonomy, flexibility, and openness (Wexler et al., in press). With the proper program elements in place, treatment programs can achieve a significantly greater reduction in recidivism than can be obtained by continuing a policy of imprisonment without adequate treatment.

These findings and principles were shared with the staff of the Bureau of Justice Administration (BJA), who were charged with guiding the administration of the expenditures for correctional drug treatment. BJA's strategy included funding an array of technical-assistance projects to guide the implementation of this part of the law. One of these projects was Project REFORM (Comprehensive State Department of Corrections Treatment Strategy for

Drug Abuse project). During the five years of its operations (1987–1991), eleven participating state departments of correction were enabled to develop state plans and implement many substance-abuse initiatives. When the BJA funding of REFORM was completed, the CENTER FOR SUBSTANCE ABUSE TREATMENT (CSAT) established Project RECOVERY (Technical Assistance and Training Services to Demonstration Prison Drug Treatment Programs) to continue these technical-assistance activities for eighteen months (1991–1992) in a total of fourteen states. Participants in Projects REFORM and RECOVERY believed that a primary goal of corrections was the reduction of recidivism—that is, intervention into the lives of offenders to prevent a return to prior patterns of criminal behavior (Wexler & Lipton, 1993). A large number of drug-treatment programs were implemented by the states that participated in these two projects. Many of these treatment programs are in the process of being evaluated and early outcome data show significant decreases in recidivism.

From the late 1980s to the early 1990s, parallel changes in substance-abuse initiatives were occurring at the federal level in the Bureau of Prisons (BOP). In the late 1970s, BOP had called for the establishment of drug-treatment programs in all federal prisons. From lack of resources, these were low-intensity programs, with an emphasis on drug education, self-help, and group psychotherapy. In the early 1990s, however, while continuing but improving these established treatment programs, BOP attempted to develop an intensive and comprehensive drug-abuse strategy that had a multilayered approach to programming. The BOP approach included one level of drug education, three levels of treatment (outpatient counseling, comprehensive residential programs, and residential pilot programs), and one level of transitional services (community reentry and continuing care in the community) (Human Resource Division, 1991). An evaluation in 1991 of BOP's progress found that intensive residential programs specifically designed for inmates were substantially underenrolled, typically running at less than half capacity. The programs relied on voluntary participation and there was little incentive to join (Murray, 1992). (State programs often offer an improved chance of parole in exchange for participation, but parole was abolished in the federal system by the Sentencing Reform Act of 1984). Because

BOP did not utilize recovering counselors who could serve as credible rehabilitated role models, it was difficult to convince prospective participants that treatment could work. This reduced the effectiveness of BOP's recruitment efforts and might weaken long-term outcomes.

The future of drug treatment in prisons nevertheless appears promising. There is growing governmental support for establishing correctional substance-abuse treatment programs. Two important guidebooks about drug treatment in prisons that reflect this shift in federal policy have been produced by federal agencies: *Intervening with Substance-Abusing Offenders* (National Institute of Corrections, 1991) and *Establishing Substance Abuse Treatment Programs in Prisons* (Wexler et al., in press). This important shift in correctional policy is also gaining public support at the state level. For example, in 19xx Texas voters endorsed a billion-dollar bond package for the construction of new correctional treatment facilities that will serve 14,000 inmates with significant substance-abuse problems. During periods of rapid expansion in this field, experienced practitioners advise that extra attention be paid to maintaining the quality of treatment efforts through the inclusion of support structures and the adequate training of staff.

(SEE ALSO: *Amity, Inc.; Coerced Treatment for Substance Offenders; Prisons and Jails; Therapeutic Communities; Treatment in the Federal Prison System*)

BIBLIOGRAPHY

FALKIN, G. P., WEXLER, H. K., & LIPTON, D. S. (1992). Drug treatment in state prisons. In D. R. Gerstein & H. J. Harwook (Eds.), *Treating drug problems* (Vol. 2). Washington, DC: National Academy Press.

HUMAN RESOURCE DIVISION. (1991). Despite new strategy, few federal inmates receive treatment. *Report to the Committee on Government Operations, House of Representatives.* Washington, DC: United States General Accounting Office.

LIPTON, D., MARTINSON, R., & WILKS, J. (1975). *The effectiveness of correctional treatment.* New York: Praeger.

MURRAY, D. W. (1992). Drug abuse treatment programs in the Federal Bureau of Prisons: Initiatives for the 1990s. *Drug abuse treatment in prisons and jails* (NIDA Research Monograph 118). Rockville, MD: U.S. Department of Health and Human Services.

NATIONAL INSTITUTE OF CORRECTIONS. (1991). Intervening with substance-abusing offenders: A framework for action. *The report of the National Task Force on Correctional Substance Abuse Strategies.* Washington, DC: U.S. Department of Justice.

WEXLER, H. K. (1994). Progress in prison substance abuse treatment: A 5-year report. *Journal of Drug Issues,* 24(2), 361–372.

WEXLER, H. K. (1992). Overview of correctional drug treatment evaluation research. *The Psychotherapy Bulletin,* 27(1), 25–27.

WEXLER, H. K., & LIPTON, D. S. (1993). From REFORM to RECOVERY: Advances in prison drug treatment. In J. Inciardi (Ed.), *Drug treatment and criminal justice.* Sage Criminal Justice System Annuals, Vol. 29. Beverly Hills: Sage Publications.

WEXLER, H. K., LIPTON, D. S., & JOHNSON, B. D. (1988). *A criminal justice system strategy for treating drug offenders in custody.* Washington, DC: National Institute of Justice, Issues and Practices.

WEXLER, H. K., ET AL. (in press). *Establishing substance abuse treatment programs in prisons: A practitioner's handbook.* Washington, DC: Center for Substance Abuse Treatment.

HARRY K. WEXLER

PRISONS AND JAILS: DRUG USE AND AIDS IN

From the time it began, in the early 1980s, the epidemic of HIV (HUMAN IMMUNODEFICIENCY VIRUS) infection that causes AIDS (acquired immune deficiency syndrome) has steadily spread throughout the prisons and jails in the United States. When there is such an ongoing epidemic of an infectious disease, numbers and statistics are continually changing. This article describes the status of the epidemic through the early years of the 1990s.

By 1991, more than 5,000 cases of AIDS had been reported among prisoners. In some correctional systems, AIDS was the leading cause of death (Vlahov, 1992). The prevalence of AIDS in the U.S. federal correctional setting (202 cases per 100,000) tends to exceed the prevalence in the general community (17 per 100,000), largely because such prisons contain a greater proportion of individuals who possess the socio-demographic and behavioral characteristics that placed them at higher risk for HIV exposure (Wexler et al., 1992). For example, many inmates have histories of intravenous (IV) drug use.

Because AIDS risk behavior is endemic among criminal-justice clients in the United States, it is generally held that potential exists for the spread of HIV through this population and back into the community. Hence, much attention has been directed recently toward AIDS in prisons.

GEOGRAPHIC VARIATION AMONG PRISONS

HIV testing (seroprevalence surveys) has demonstrated considerable geographic variation across correctional systems. Seropositive rates have been lowest (less than 1.0%) among entrants to prisons in the Midwest and have been highest in the mid-Atlantic states (Vlahov, 1992). These rates seem to reflect rates similar to those for IV drug users in the correctional facilities' respective communities. For example, 28 percent of a sample of New York State prison admissions from December 1987 to January 1988 had histories of IV drug use, with an overall HIV seropositive rate of 17.4 percent (Wexler et al., 1992). This is similar to the high seroprevalence rates seen in IV drug users in New York City, where most of the inmates lived before sentencing. Despite geographic variations, several surveys have indicated that HIV seroprevalence is higher among incarcerated females than males, among racial/ethnic minorities than whites, and among inmates over 25 years of age (Vlahov, 1992).

RISK FACTORS FOR INMATES

The major risk factor for HIV infection and AIDS within the correctional population is IV drug use prior to incarceration. An investigation conducted by the New York State Prison system in 1989 estimated that at least 95 percent of the inmates diagnosed with AIDS had a history of IV drug use, as compared with the 3 percent who acquired the disease strictly through homosexual contact (Vlahov, 1992). Because IV drug use and homosexual activity do continue to occur within institutions, affecting a proportion of inmates, it is commonly feared that correctional facilities will serve as an "amplifying reservoir" of HIV infection that may then spread to the community. However, transmission of the virus while in prison is found only infrequently. By the early 1990s, studies found a stabilization rather than an increase of prevalence among male inmates entering prison (Vlahov, 1992).

This temporal stabilization of seroprevalence in conjunction with the rarity of intraprison transmission still cannot belie the fact that HIV infection remains a major prison health problem and a potential community crisis, if parolees resume IV drug use and high-risk sexual practices. Responses to the issue of AIDS in correctional settings have included (1) inmate risk education, (2) screening for the HIV antibody, (3) and the segregation of seropositive inmates in order to prevent transmission and to initiate treatment for related health complications. Even though IV drug use remains the major HIV risk behavior, treatment for drug abuse as a preventative measure is used infrequently for inmates or parolees. Efforts are currently being made to develop comprehensive AIDS prevention models for former IV drug users released to parole (Wexler et al., 1994).

(SEE ALSO: *Injecting Drug Users and HIV; Substance Abuse and HIV/AIDS*)

BIBLIOGRAPHY

MAGURA, S., ROSENBLUM, A. R., & JOSEPH H. (1991). AIDS risk among intravenous drug-using offenders. *Crime and Delinquency, 37,* 86–100.

MCBRIDE, D. C., & INCIARDI, J. A. (1990). AIDS and the IV drug user in the criminal justice system. *Journal of Drug Issues, 20,* 267–280.

VLAHOV, D. (1992). HIV-1 infection in the correctional setting. *Drug Abuse Treatment in Prisons and Jails* (NIDA Research Monograph 118). Rockville, MD: U.S. Department of Health and Human Services.

WEXLER, H. K., MAGURA, S., BEARDSLEY, M. M., & JOSEPHER, H. (1994). ARRIVE: An AIDS education/relapse prevention model for high risk parolees. *International Journal of the Addictions, 29,* 361–386.

HARRY K. WEXLER

PROBATION *See* Coerced Treatment for Substance Offenders.

PROBLEM DRINKING *See* Addiction: Concepts and Definitions.

PRODUCTIVITY: EFFECTS OF ALCOHOL ON

ALCOHOL is the most commonly used and abused drug in the United States. In 1991, approximately 50 percent of all 125 million employed workers in the United States had taken at least one drink, and about 6 percent reported they had been drinking heavily (five or more drinks on five or more occasions) during the past month. Heavy drinking is more than four times as prevalent among male workers than it is among female workers, and it is most prevalent in male-dominated, semiskilled, transient occupations such as construction and transportation.

Alcohol can affect productivity in various ways. The relevant physiological effects of alcohol include intoxication, hangovers, WITHDRAWAL (abstinence syndrome) after long-term heavy use, and residual physical, mental, or social disabilities due to abuse or chronic dependence. The most important effects of intoxication—clumsiness, sleepiness, difficulty in processing new information or communicating ideas—impair physical safety and cognitive capability. Both effects can lead to poor performance, absenteeism, and job loss. Hangovers or periods of withdrawal can have similar results. Liver and heart damage, stroke, and irreparable injuries are the most common physical and mental disabilities. The most common social disability is withdrawal of trust by associates.

The consequences for economic productivity are measured not by taking them individually but by statistically estimating the overall loss of wage-earning capacity attributable to alcohol abuse and dependence. These losses are computed in two forms. *Morbidity cost* is the annual loss of earnings by individuals who are impaired by alcohol compared to the earnings of unimpaired people with similar demographic characteristics. According to the most recent estimate of this loss in the United States, one fourth of working-age men and one twentieth of working-age women were so impaired, thus averaging a 4 percent loss of earnings potential, or a total of a 35-billion dollar loss in income reduction in 1993. *Mortality cost* is the present value of the lost lifetime earnings of the nearly 100,000 individuals (two thirds of them male) who are estimated to die annually because of alcohol use—one fourth in traffic crashes, one fifth from liver disease, one eighth from homicide or suicide, one tenth from other accidents, and the remainder from esophageal cancer and a wide variety of other toxic effects. The average expected value of future earnings lost was about 33 billion dollars in 1993.

Morbidity and mortality costs account for well over half the estimated economic burden of alcohol-related illness. However, morbidity and mortality cost estimates involve complex econometric modeling procedures and use survey data from many sources. Model results have differed by as much as 200 percent for morbidity costs and 25 percent for mortality costs.

(SEE ALSO: *Accidents and Injuries from Alcohol; Complications: Medical and Behavioral Toxicity Overview; Economic Costs of Alcohol Abuse and Alcohol Dependence; Industry and Workplace; Drug Use in; Social Costs of Alcohol and Drug Abuse*)

DEAN R. GERSTEIN

PRODUCTIVITY: EFFECTS OF DRUGS ON

Concern about drug use in the U.S. workforce has focused on the most common illicit drugs—COCAINE and MARIJUANA—although also common is the nonmedical use of TRANQUILIZERS, SEDATIVES, and STIMULANTS. In 1990, about 7 percent of employed workers had used an illicit drug in the past month, according to national surveys. Illicit drugs are used at higher rates by men than by women and also at higher rates by low-paid workers in transient occupations than by other workers.

Laboratory studies show that typical single doses of marijuana effect small temporary impairments in performing complex tasks, whereas typical single doses of cocaine may effect small temporary enhancements—especially when the performance of subjects is impaired by fatigue. To the extent that generalization is possible, sedatives and tranquilizers are similar in their effects to marijuana. Using illicit drugs during off hours is much more common than doing so while on the job. The effects of hangovers, post intoxication fatigue, or withdrawal from a chronic run of use may be significant for productivity, as may also be the potential accumulation of longer term disabilities, including social mistrust.

Productivity loss due to drugs is estimated by comparing the earnings of problem users with those of other people with similar demographic characteristics. The total income losses are now estimated at about 10 billion dollars annually, with a large frac-

tion of this estimate being attributable to the nonmedical use of sedatives and tranquilizers. Productivity losses account for about one sixth of the total estimated economic burden of drug problems.

(SEE ALSO: *Industry and Workplace, Drug Use in; Productivity: Effects of Alcohol on; Social Costs of Alcohol and Drug Abuse*)

DEAN R. GERSTEIN

PROFESSIONAL CREDENTIALING

A host of health-care professionals provide treatment for substance-abuse disorders. They include, but are not limited to, physicians, psychologists, social workers, nurses, pastors, and addiction or drug-abuse counselors. Institutions and programs that train these professionals are accredited, and the individuals, after undergoing the training, may obtain credentials from a professional or state body. In this context, one must define the terms *accreditation* and *credential* and examine the role of each in protecting the interests of the consumer of substance-abuse treatment.

In the United States, there are two forms of educational accreditation—institutional accreditation, which began in the late 1790s in New York State, and professional accreditation, which began in the first years of the twentieth century. Accreditation is a voluntary, self-regulating process designed to evaluate the strengths and weaknesses of an educational institution. Institutional accreditation and professional accreditation have a pattern in common. It involves: (1) preparation of a detailed and objective self-study by the institution or professional program that outlines and evaluates objectives, activities, and achievements; (2) an on-site visit by a team of peers that provides expert evaluation and offers suggestions for improvement; and (3) a subsequent review and decision by a central governing commission or board to award or deny accreditation. The location of the institution determines which one of the six regional accreditation organizations will accredit it. An exception to this regional pattern is made for institutions with programs of a specialized nature, such as trade and technical education, rabbinical and Talmudic education, and the like. National accreditation bodies accredit these programs.

The U.S. Secretary of Education and the Commission on Recognition of Post-Secondary Accreditation (CORPA) recognize both regional and national accreditation organizations—that is, they accredit the accreditors. Professional accreditation is carried out, in the main, by organizations formed by members of the profession. For example, the American Psychological Association accredits doctoral programs in psychology. These specialized accreditation bodies operate nationally. Within each field, one accrediting agency is recognized by the Committee on Post-Secondary Education. These recognized agencies come together to form the Assembly of Specialized Accrediting Bodies, which works on issues of common interest to those involved in professional education. The counseling function is recognized within several professions that undergo accreditation, but the subspecialty of substance-abuse or addictions counseling is not at present independently recognized within this framework (as of 1995).

Although accreditation applies to programs or institutions and does not cover substance-abuse counseling, *credentials* apply to individuals and do cover this subspecialty. Institutions that offer training in substance-abuse counseling design their programs to meet the requirements outlined by the state or by potential employers so that graduates can obtain *certification*. Graduates must then pass tests certifying that they have a specific level of proficiency in the theoretical and practical aspects of substance-abuse treatment. For example, in Michigan the Department of Public Health and other interested organizations initiated a program for the professional development of counselors that is based on education, experience, supervised practical training, professional recommendation, testing and review, ethics, and residence. Michigan requires that persons undergo a three-tier testing process covering the theoretical and practical aspects of substance-abuse treatment to become certified addictions counselors (CACs). The first test covers fundamental knowledge of substance-abuse counseling; the second, applications to specific populations; and the third, the oral presentation of a case. Certification is for a specific term and renewal requires additional education. Once certified, a person may provide addiction treatment in states other than the one that awarded certification, through a reciprocity agreement that covers states with membership in the International Certification Reciprocity Consortium.

In addition to certification by the state, certification may also be obtained through professional or-

ganizations. For example, the American Society of Addiction Medicine, under the auspices of the American Medical Association, certifies physicians who wish to treat substance abuse. The association offers courses that review topics in addiction theory and practice, examines candidates who wish to obtain credentials, and certifies their advanced knowledge and skills in this area. Other professional associations such as the American Psychological Association are currently developing procedures and mechanisms for providing substance-abuse-treatment credentials to their members who supply mental health services in this area.

Both accreditation and certification work to improve the quality of the education and specialty training that individuals receive and to assure the quality of the services provided. As a safeguard, consumers of substance-abuse services may determine whether the professional delivering the services was trained in a program accredited by the appropriate professional organization in a university or college accredited by the appropriate regional accrediting board. Consumers may also determine whether the professional holds credentials as a substance-abuse counselor, since these credentials certify that a person has met certain educational requirements and displayed the level of knowledge and skill deemed necessary in the profession.

(SEE ALSO: *American Academy of Psychiatrists in Alcoholism and Addictions; American Society of Addiction Medicine*)

M. MARLYNE KILBEY
AMY L. STIRLING

PROHIBITION OF ALCOHOL

The Eighteenth Amendment to the Constitution of the United States prohibited the "manufacture, sale and transportation of intoxicating liquors." The amendment, passed by Congress in 1917, was written to become effective one year after its ratification by the states. The amendment outlawed only the manufacture, transport, and sale of liquor; it did not criminalize the possession of ALCOHOL for personal use, nor did it make purchase of liquor from bootleggers a criminal offense, nor did it define what was meant by "intoxicating" liquors. To implement the amendment, Congress passed the National Prohibition Act, better known as the Volstead Act. The Volstead Act was crafted to allow supplies of alcohol to be produced and transported for scientific and other commercial purposes. It also defined an intoxicating liquor as any beverage containing more than 0.5 percent alcohol. It could have set the permissible level higher and allowed, for example, the production, transportation, and sale of BEER, but it did not. Prohibition became effective in 1920. A Prohibition Bureau was established within the Treasury Department to carry out the provisions of the law. Under the Volstead Act, Treasury agents could obtain a search warrant only if they could prove that alcohol was being sold, thus precluding searches of individual homes no matter how much liquor might be there. Some wealthy people, given the ample notice that Prohibition was coming, laid in enough alcoholic beverages to last them through most of the following decade. The law also had the effect of allowing manufacture for personal use. Such home production sometimes became part of a cottage industry contributing to the supplies distributed by bootleggers. Even committed Prohibitionists appeared to believe that the public would not tolerate any effort to criminalize the act of drinking itself. The Volstead Act, unlike some state laws, permitted the manufacture of beer as long as the beer contained no more than 0.5 percent alcohol (near beer).

Given the common belief that Prohibition failed utterly to alter the consumption of alcohol or its adverse effects on health, it is appropriate to ask, To what extent did the law reduce alcohol use in the United States? First, there is no question that it succeeded in eliminating 170,000 saloons, even if it did not change the attitudes of most Americans about the morality of drinking. And, while some writers have asserted that drunkenness actually increased during Prohibition, most available records point to the opposite conclusion (Aaron & Musto, 1981; Lender & Martin, 1987). The most consistent findings on the impact of Prohibition come from statistics on medical problems known to be linked to alcohol consumption, especially excessive alcohol consumption. Among these problems were hospital admissions for alcoholism and admissions to state mental institutions for alcoholic dementia and alcoholic psychosis. Striking decreases were observed in New York and Massachusetts, two states that did not have restrictions on alcohol consumption prior to 1920. Massachusetts state mental hospital admissions for alcoholic psychosis fell from 14.6 per

100,000 in 1910, to 6.4 in 1922, and were 7.7 in 1929; in New York, such admissions fell from 11.5 in 1910, to 3.0 in 1920, rising again to 6.5 in 1931 (Aaron & Musto). Deaths from alcohol-related diseases also fell. National statistics showed that the number of deaths from cirrhosis, about 14.8 per 100,000 in 1907, were only 7.9 in 1919, 7.1 in 1920, and did not rise above 7.5 during the 1920s. There were decreases in arrests for drunkenness and in the costs of jailing public inebriates. Commander Evangeline Booth of the Salvation Army asserted that not only had drinking fallen off sharply, especially among the poor, but there were fewer broken homes because of wages lost to drinking or violence related to drinking.

Aaron and Musto state, "Observers . . . have been unanimous in concluding that the greatest decreases in consumption occurred in the working class. . . . In large measure, intoxicants priced themselves out of the market" (Aaron & Musto 1981, p. 165). A quart of beer or a quart of gin were five to six times more expensive in 1930 than they were prior to Prohibition. Prohibition defenders asserted that instead of purchasing liquor in saloons, workers were putting their earnings into cars and refrigerators. Admittedly, the impact on alcohol consumption was greatest in the early years of Prohibition. As bootlegging increased in the late 1920s, medical problems linked to alcohol use began to rise again, but they did not reach the high levels experienced before 1920. Other data on per capita alcohol consumption immediately after repeal in 1934 indicated that there must have been a drastic decline in average alcohol consumption during the Prohibition years. Undoubtedly, crime associated with bootlegging increased. Many bootleggers became quite wealthy. Some who were involved in illegal activities prior to Prohibition used the wealth flowing from bootlegging to extend and further develop organized criminal enterprises, some of which later became involved with trafficking in illicit drugs. One of the most notorious of the figures associated with organized crime was Al Capone, who came to national attention as a result of his Chicago-based criminal activities. Aaron and Musto point out, however, that organized rackets existed in large cities before Prohibition and that the homicide rate increased most sharply between 1900 and 1910.

Unquestioned, also, is the unreliable quality of bootlegged liquor, much of which was produced by diverting or hijacking industrial alcohol. Some industrial alcohol could simply be flavored and sold as scotch, gin, or bourbon. Much of it, however, had been mixed with METHANOL (methyl alcohol) or other chemicals to render it undrinkable—denatured. Bootleggers hired chemists to remove the denaturants by redistillation ("washing"). Inadequate processing, which was not uncommon, produced a liquor that could be toxic or even lethal. The liquor produced in England and Canada and smuggled in by ship or truck was of a higher quality. One smuggler who brought in such quality liquor, Bill McCoy, has given us a term still used to describe an authentic product—the "real McCoy."

The continued criticism of Prohibition and the frustration of enforcing the Volstead Act led many of their advocates to become increasingly defensive and hostile to those not seen as supporters. Concern for the drunkard sharply diminished. According to Lender and Martin, "Many crusaders began labelling rehabilitation as nothing more than a waste of time and energy; prohibition, they promised would make such work unnecessary" (Lender & Martin 1987, p. 159). Groups interested in treatment declined. The Association for the Study of Inebriety dissolved in the mid-1920s. Volstead Act advocates became more hostile toward alcoholics as criticism of Prohibition increased. Some suggested amending the Act to make drinking itself a criminal offense. One such suggestion came from an official in the Prohibition Unit of the Treasury Department, Harry J. ANSLINGER, then the Assistant Commissioner of Prohibition. Thus the nineteenth-century concerns of the TEMPERANCE MOVEMENT for the physical and spiritual health of alcoholics turned, in the 1920s, to calls for stiffer jail terms, or even exile, for chronic alcoholics. In the context of these attitudes, the harsh penalties that were then being meted out under the leadership of the Treasury Department for mere possession of illicit drugs become somewhat more comprehensible.

The enforcement of the Volstead Act had been vested in the Treasury Department's Prohibition Unit within the Internal Revenue Bureau. The first National Prohibition Administrator, the head of the Prohibition Unit, was John F. Kramer. The Narcotics Division, headed by Levi G. Nutt, a pharmacist by training, was part of the Prohibition Unit. The Narcotics Division became an independent unit in the Treasury Department in 1930 when the Prohibition Unit was transferred to the Department of Justice.

Harry J. Anslinger was appointed first Commissioner of Narcotics.

Despite growing criticism, Prohibition, according to Aaron and Musto, was still alive and well when Herbert C. Hoover was elected president by a large margin in 1928. An overwhelming majority of both houses of Congress and nearly all the state governors supported the Eighteenth Amendment. Even opponents of Prohibition did not realistically expect to see it repealed. But the onset of the Great Depression in 1929 dramatically changed the situation. Opponents of Prohibition no longer argued for its repeal because of its demoralizing effects on civil liberty but argued instead that the revival of the liquor industry would provide jobs and tax revenue. In the 1932 campaign for the presidency, Franklin D. Roosevelt promised to repeal Prohibition. Almost immediately after his inauguration, he had changes introduced in the Volstead Act to legalize the sale of beer.

In 1933, the Twenty-First Amendment to the Constitution was ratified. It was brief and to the point: "Section 1. The Eighteenth Article of Amendment to the Constitution of the United States is hereby repealed." The federal government, however, retained responsibility to regulate and tax beverage alcohol and to prevent its illegal production. Section 2 of the Amendment allowed the states to continue Prohibition under state laws if they so desired. Some states did so; many states adopted alcohol beverage control laws (ABC laws). These were intended to curb the abuses that had characterized the production and sale of alcohol prior to prohibition. Among other provisions, ABC laws restricted the hours when alcohol could be sold (to make taverns and bars less attractive) and banned liquor sales on Sundays and election days. Some ABC laws created state-operated monopolies for the sale of packaged beverages. The various federal laws dealing with control of alcohol remained the responsibility of various federal agencies. It was not until 1972 that they were brought together and responsibility for overseeing them was assigned to a single agency—the Bureau of Alcohol, Tobacco, and Firearms (BATF) in the Department of the Treasury.

(SEE ALSO: *Alcohol: History of Drinking; Harrison Narcotics Act of 1914; Legal Regulation of Drugs and Alcohol; Tax Laws and Alcohol; Temperance Movement; Treatment, History of*)

BIBLIOGRAPHY

AARON, P., & MUSTO, D. F. (1981). Temperance and prohibition in America: A historical overview. In M. Moore & D. Gerstein (Eds.), *Alcohol and public policy: Beyond the shadow of Prohibition.* Washington, DC: National Academy Press.

LENDER, E. M. & MARTIN, J. K. (1987). *Drinking in America.* New York: Free Press.

MUSTO, D. F. (1987). *The American disease: Origins of narcotic control.* Expanded ed. New York: Oxford University Press.

TICE, P. M. (1992). *Altered states: Alcohol and other drugs in America.* Rochester: The Strong Museum.

JEROME H. JAFFE

PROHIBITION OF DRUGS: PRO AND CON *See* Policy Alternatives.

PROJECT RETURN FOUNDATION, INC.

Project Return Foundation, Inc., a nonprofit, nonsectarian, multipurpose human-services agency, operates five New York City residential drug-free (RDF) THERAPEUTIC-COMMUNITY (TC) programs: Olympus, Discovery, Exodus, Genesis, and Sponsorship. The agency was founded in 1970 as a self-help and community center for substance abusers by two recovering addicts, Carlos Pagan and Julio Martinez. Later, Martinez became director of the New York State Division for Substance Abuse Services.

Under the leadership of its current president, Jane Velez, the agency has diversified significantly and now provides services to battered women; addicted women with serious and persistent mental illness and their infants; parolees with substance-abuse problems; and homeless men and women with substance-abuse problems living in New York City shelters. Project Return also operates an outreach, anti-AIDS education/prevention program, a medically supervised, drug-free outpatient program, and a modified TC-oriented health-related facility for substance abusers who are HIV + and symptomatic. The latter service is administered jointly by Project Return Foundation, Inc., Samaritan Village, and H.E.L.P., Inc. In total, nearly 1,000 men and women receive daily treatment and rehabilitative services through programs administered by Project Return Foundation, Inc.

The foundation's five RDF TCs are located in the west Bronx. The Olympus, Discovery, Genesis, Exodus, and Sponsorship programs are colocated at 1600 Macombs Road, Bronx, NY 10452. In 1992, the all-male Olympus program, with its 249 clients, was divided into two programs (Olympus and Discovery) of roughly equal size. Both programs will service men and women. The forty-eight-bed Genesis program, part of a federal waiting-list reduction initiative, will eventually be relocated and enlarged to eighty beds.

UNIQUE FEATURES

All of Project Return's RDF TC programs are run according to the same clinical principles—they provide comprehensive, holistic, individualized treatment and rehabilitative services to the residents through interdisciplinary treatment teams. Interdisciplinary teamwork spans the entire length of stay in the TC programs, from admissions to discharge. Interdisciplinary teams are involved in the formulation and modification of the Preliminary and Master Treatment Plans, the provision of day-to-day services, Aftercare Planning, the management of any exceptional clients with special abilities and/or problems (HIV +, mentally ill chemical abusers [MICA], physical handicaps), requests for Work-Out and Live-Out, Discharge Planning, and Aftercare Services.

The teams are comprised of roughly equal numbers of substance-abuse counselors, almost all of whom are recovering former members of TCs, and professional specialists, including a physician, a (part-time) psychiatrist, nurses, Ph.D.-licensed clinical psychologists, social workers, family therapists, vocational-rehabilitation counselors, a recreational and an art therapist, legal counselors, HIV health educators/coordinators, and admissions counselors. A number of the professional staff are also recovering, and increasing numbers of the substance-abuse counselors have opted to return to school to pursue higher education.

Another unique feature of the RDF TCs is the wide array of clients the agency purposefully admits to its programs, including MICA or dual-diagnosis clients, homeless substance abusers, pregnant women (up to the sixth month of pregnancy), clients with serious learning disabilities, clients with physical disabilities, and gay and lesbian substance abusers. All these subpopulations are integrated into the mainstream

TC program. Their unique needs are addressed through a host of different specialty groups (e.g., an incest survivors group, an HIV+ support group) and individual sessions with staff specialists.

Project Return's TC programs have developed a number of innovative programs. These include the ten-week Relapse Prevention Program, the Incest Survivors Group, the Domestic Violence Group, the Socialization Group, the development and utilization of an Intensive Care Management Group to monitor the progress of exceptional clients, and the use of Behavioral Criteria to monitor clients' progress in moving throughout the various levels of the program.

Project Return Foundation, Inc. is unusual in the extent to which it has opened itself to interactions with the outside world. For example, the agency has accepted interns and externs from a number of universities, including Long Island University, New York University, Columbia, Adelphi, Fordham, and Yeshiva. In addition, a number of major research studies have been undertaken at Project Return Foundation, Inc. Several are in progress and more are being planned for the future. Visits from government officials and treatment staff from programs throughout the United States and from other countries, such as Canada, the United Kingdom, Australia, Sweden, Germany, Argentina, and Hungary are common occurrences. All of the RDF TCs work with community advisory boards and each is deeply involved in Project Return's Family Association.

SUCCESSES

Project Return Foundation, Inc. boasts one of the lowest early departure rates in all of New York State and has increased the number of graduates steadily during the late 1980s and early 1990s. It has demonstrated that it is possible to professionalize TCs and still retain the heart and soul of the TC movement. It also has consistently demonstrated a capacity to treat successfully the two most common characteristics of modern addiction—multiple substance abuse (including crack and the abuse of alcohol) and various comorbidities (including MICA, HIV+, tuberculosis, and diabetes).

(SEE ALSO: *Substance Abuse and AIDS; Treatment Types; Vulnerability As Cause of Substance Abuse: Sexual and Physical Abuse*)

BIBLIOGRAPHY

CARROLL, J. F. X. (1992). The evolving American therapeutic community. *Alcoholism Treatment Quarterly*, 9(3/4), 175–181.

JANE VELEZ
JEROME F. X. CARROLL

PROPOXYPHENE *d*-Propoxyphene (Darvon®) is an OPIOID drug that is structurally related to METHADONE. It is used clinically to produce analgesia when the level of PAIN is not severe. Its popularity rests largely on the belief that propoxyphene is less likely to cause addiction than CODEINE, a drug that is also used for relief of moderate levels of pain. Propoxyphene is typically used in combination with aspirin or acetaminophen. Its ANALGESIC effects are synergistic with those of aspirin and other nonsteroidal anti-inflammatory agents.

When it was introduced into clinical medicine in the early 1960s, propoxyphene was not subject to special narcotic regulatory control. This fact may explain its early popularity, which was probably due to clinicians' unrealistic fears about the addictive potential of codeine and to the inconvenience of prescribing it under the narcotic regulations that were in effect before the CONTROLLED SUBSTANCES ACT of 1970 was passed.

Although propoxyphene has only one-half to two-thirds the potency of codeine, it has been used to control symptoms of the opioid WITHDRAWAL syndrome. It is not commonly abused because it produces unpleasant toxic effects at high doses.

(SEE ALSO: *Opiates/Opioids*)

BIBLIOGRAPHY

JAFFE, J. H. & MARTIN, W. H. (1990). Opioid analgesics and antagonists. In A. G. Gilman et al. (Eds.), *Goodman and Gilman's the pharmacological basis of therapeutics*, 8th ed. New York: Pergamon.

JEROME H. JAFFE

PROJECT SMART/STAR *See* Prevention; Prevention Programs.

PSILOCYBIN This is an indole-type HALLU-CINOGEN, found naturally with another hallucinogen in a variety of mushrooms—the most publicized being the Mexican or MAGIC MUSHROOM, *Psilocybe mexicana*, as well as other *Psilocybe* and *Conocybe* species. These mushrooms have long been consumed by Native Americans, especially in Mexico and the southwestern United States, as part of religious rites.

Psilocybin produces effects similar to LYSERGIC ACID DIETHYLAMIDE (LSD), but it is less potent and is metabolized in the body to form psilocin, another hallucinogenic compound. Both of these compounds have been synthesized in clandestine laboratories and made available on the streets.

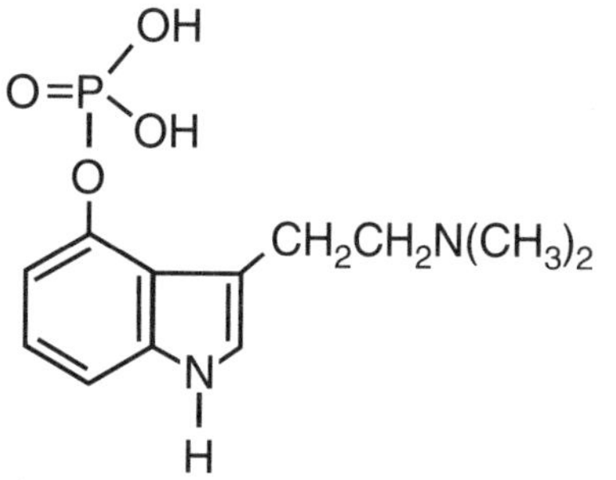

Figure 1
Psilocybin

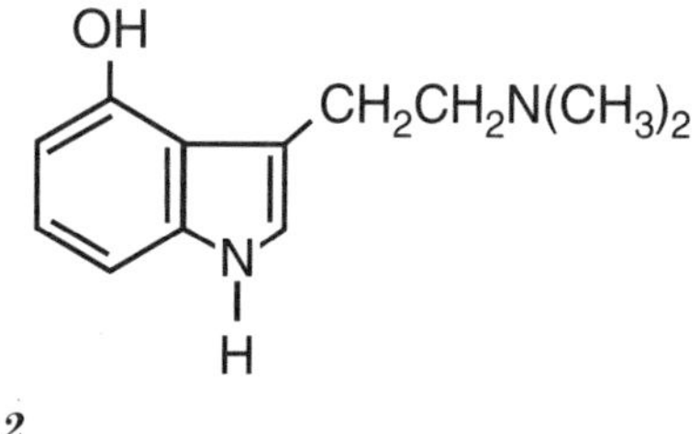

Figure 2
Psilocin

(SEE ALSO: *Hallucinogenic Plants; Peyote; Plants, Drugs from*)

BIBLIOGRAPHY

WEIL, A. (1972). *The natural mind.* Boston: Houghton Mifflin.

DANIEL X. FREEDMAN
R. N. PECHNICK

PSYCHEDELICS *See* Hallucinogens; Lysergic Acid Diethylamide (LSD) and Psychedelics.

PSYCHIC DEPENDENCE *See* Addiction: Concepts and Definitions.

PSYCHOACTIVE *Psychoactive* is a general term that came into use about 1961. It describes a substance that affects the central nervous system, producing changes in mental activity and/or behavior. A psychoactive substance or process may affect the way an individual thinks or the manner in which the environment is perceived or experienced; it may change the behavior of an individual in a given situation.

Psychoactive drugs can be used nonmedically, to induce altered states of consciousness (e.g., HALLUCINOGENS, such as LSD), or clinically, to treat mental illness (e.g., ANTIPSYCHOTICS to treat SCHIZOPHRENIA).

(SEE ALSO: *Psychopharmacology*)

NICK E. GOEDERS

PSYCHOACTIVE DRUG Any of a group of drugs (called psychotropic from 1948 until 1961) that act upon the central nervous system, producing changes in mental activity and/or behavior. Psychoactive drugs are among the most widely used group of pharmacologically active agents, with extremely important clinical applications, including anesthesia for surgery and ANALGESIA for relief of pain.

Psychoactive drugs are used to suppress disorders of movement and to treat ANXIETY, mania, DEPRESSION, or SCHIZOPHRENIA. Often, drugs used primarily to treat disorders in peripheral organs can also affect the central nervous system (e.g., beta-blocking agents, used to treat high blood pressure or disorders of heart rhythm, or steroid hormones used to control inflammation). Psychoactive drugs are also taken by people in a nonmedical context to combat fatigue (e.g., CAFFEINE or AMPHETAMINE), to improve mood (e.g., ALCOHOL, COCAINE, or HEROIN), or as elements of cultural rituals (wine or PEYOTE at religious ceremonies).

NICK E. GOEDERS

PSYCHOANALYSIS Psychoanalysis is an analytic technique originated by Sigmund Freud (1856–1939), an Austrian neurologist. It has been altered by his students and their students, in turn, throughout the twentieth century. Psychoanalysis is a theory of the way the mind works: (1) Sequences of thoughts are determined—they do not occur by chance; (2) Much of our thinking takes place out of awareness—it is unconscious and not easily recovered; (3) The experiences of early childhood, particularly those with important caretakers, continue to have an impact (often unconsciously) on our daily lives; (4) Feelings, both sexual and aggressive, are present at birth and affect behavior. The theory helps us understand something of the addicts' complex motivations and of their inner experience and behaviors.

Psychoanalysis is also a method: It attempts to understand mental processes by free association (following thoughts wherever they lead without selection or censoring) and by the analysis of dreams, fantasies, and behaviors. Psychoanalysts apply this method as a therapy or treatment for certain forms of mental disability.

(SEE ALSO: *Causes of Substance Abuse: Psychological [Psychoanalytic] Perspective*)

BIBLIOGRAPHY

BRENNER, C. (1973). *An elementary textbook of psychoanalysis.* New York: International Universities Press.

WILLIAM A. FROSCH

PSYCHOLOGICAL DEPENDENCE *See* Addiction: Concepts and Definitions.

PSYCHOLOGICAL TREATMENT FOR SUBSTANCE ABUSE *See* Treatment; Treatment Types.

PSYCHOMOTOR EFFECTS OF ALCOHOL AND DRUGS Alcohol and other drugs of abuse can alter normal behavior in a deleterious way. Epidemiological studies have shown that 50 percent or more of all single-vehicle traffic fatalities in the United States are associated with the use of ALCO-

HOL. The risk of a driver causing an accident increases progressively the more that BLOOD ALCOHOL CONTENT (BAC) increases past 0.4 grams per liter (g/l). At BACs of 1.0 g/l, the risk is tenfold, and with BACs of 1.5 g/l, the risk is almost thirtyfold compared to nonalcohol conditions. The same phenomenon applies to accidents in which pedestrians are killed by drunk drivers.

PSYCHOMOTOR PERFORMANCE

Most behavioral tasks are complex processes in which information sampling and its processing, motor responses, and sensorimotor coordination are involved. A decrement in any part of this system leads to impaired performance. Numerous studies describe techniques used to assess the psychomotor functions of people under the influence of chemicals with the potential for impairing performance. The vastness of the range of behavioral activities, however, makes it unlikely that any one, or even a small number, of tests could completely describe the impairing properties of alcohol and other drugs under all conceivable circumstances.

A way to approach this problem is to isolate the main variables of performance into smaller entities and measure the effects separately with a set of relevant tests. Since psychomotor behavior consists of external stimuli and a rational response to them, a simplified chain of events can be divided into a sensory part (detection of stimulus), a central part (complex processing of the sensory information), and a motor part (overt behavior or motor reaction to the stimulus).

It is sometimes difficult to select the most sensitive, accurate psychomotor test for various agents that impair performance. Sets of tests have been used—for example, in studies on the likelihood of bus drivers to have traffic accidents. The capabilities that best characterized the drivers with low accident records were constant and keen attention, adequate information processing, and the absence of hasty reactions. Eye-to-hand coordination was less important, and simple reaction times represented the poorest correlation to safe driving. Although it is logical to choose a set of tests that cover the most important variables, in most tests there is an overlap among several skills. Alcohol, drugs, and their combinations, moreover, may impair these integrated variables to a varying extent in different individuals. Because of this, one cannot predict or give exact numerical data for the amount of impairment associated with a single variable of the system affected. Nor does impairment in one sensitive test mean that the overall performance is severely impaired. In practice, it may not be important to know whether the accident of a drunken driver resulted from impaired attention rather than from poor motor coordination or slowed reactions, when all these skills were more or less affected.

CONFOUNDING FACTORS IN PSYCHOMOTOR TESTING

Substance abuse is commonly, but not necessarily, associated with an acquired TOLERANCE; this means that after repeated administration, a given dose of a drug produces a decreased effect and larger doses become needed to obtain the effects observed with the original dose. Deleterious psychomotor effects are usually easy to detect when large single doses are taken by people who have not yet acquired tolerance to the effects of drugs. The question becomes more complex when the user who takes small doses acquires significant tolerance to them because of regular use.

For any skilled performance, a large variation is observable among individuals. Thus some people may, by nature, have slower reactions or poorer information-processing capacities, and their best performance in the respective tests may be clearly worse compared with that of more capable subjects—even when the more capable are under the influence of performance-impairing drugs. The decremental drug effect can be similar in both cases, but the more capable subjects can afford it because of their better reserves. It is consequently difficult to define safe and unsafe doses of any agent.

Other factors that may influence psychomotor behavior include motivation, learning, adaptation to the task, and drowsiness. Paying the subjects according to how well they perform might improve motivation and performance, and this might skew the test results. (Such a motivational enhancer is not always mentioned in the research reports.) Impairment of performance may not be detected in tasks of short duration in a stimulating environment, whereas deleterious effects can be documented in monotonous tasks of long duration. Transposed to normal life situations, this observation may explain why an inebriated driver can get through a difficult driving test without any significant errors but cannot handle a surprising event after several hours of monotonous driving on a highway.

ALCOHOL

An alcohol dose affects the central nervous system (CNS)—the predominant effect being a depression of central functions. This means that the higher the dose of alcohol, the more the CNS is depressed. The most highly integrated brain functions are involved first; when the brain cortex is released from its functions of integrating and control, processes related to judgment and behavior occur in a disorganized fashion and the proper operation of behavioral tasks becomes disrupted.

The effects of alcohol are biphasic, and the phases depend on the dose and the rate of administration. With higher alcohol concentrations, central depressant effects dominate. Low concentrations seem to stimulate various functions by inhibiting the control mechanisms. This is seen in animal studies as decreased motor activity with large doses of alcohol and increased activity with small doses. In humans, very small doses of alcohol do not necessarily impair performance, and the tension-relieving effects of alcohol can sometimes be seen in some tests of short duration. However, there is no reason to overestimate this occasionally observed pseudostimulant effect of alcohol; in actuality, alcohol impairs various skills that are needed to cope with everyday routines.

Several investigators have demonstrated that alcohol does induce a larger decrease in test performances requiring hand–eye coordination, whereas simple tests of cognitive ability show less of a decrease. When more complex cognitive functions are studied, however, low to moderate BACs (0.3–1 g/l) impair sensory tasks and sensorimotor skills less than they do complex cognitive behavior, such as performing two tasks simultaneously ("divided attention"). It thus seems that alcohol impairs the rate of information processing by slowing the ability to switch attention from one to another sensory input to motor control, without significantly impairing sensory motor functions as such. In fact, moderate BACs (less than 1 g/l) are not associated with dramatic changes in such basic neurophysiological mechanisms as neuromuscular transmission or the conduction velocity of motor nerves. Alcohol effects are thus better seen in situations where the information load is increased and highly integrated functions are needed for the task.

It is well known that the muscles of the eye and eye movements often easily reflect the CNS depression caused by alcohol. One of the most sensitive signs is the appearance of lateral nystagmus; small twitches or vibration in the position of the eye are seen when the person looks to the side. The angle of the gaze at which the nystagmus appears correlates with the alcohol dose: On average, a BAC of 0.5 g/l induces nystagmus at a 45-degree angle of deviation, whereas a BAC of 1 g/l produces nystagmus even at the 35-degree angle. Also, saccadic eye movements (from one fixation point to another) become slower with BACs of 0.8 g/l to 1 g/l. All this indicates that people who are drunk have a narrower sector of intact vision than people who are sober. Visual information becomes disrupted if eyes must be turned to the side to detect stimuli, or if eyes must be moved quickly from one point to another.

Several types of tests measure skilled performance in tasks related to driving behavior. *Tracking tasks* involve hand-to-eye coordination, and the task is to keep an object on a prescribed path by controlling its position through turning a steering wheel. Impairment of performance is seen at BACs of as little as 0.7 milligrams per milliliter (mg/ml). *Choice reaction task* refers to a situation where aural or visual stimuli (or both) need response according to rules that necessitate mental processing before giving the answer. In traffic, driving requires a division of attention between a tracking task and surveillance of the environment. When a driver must process information from more than one source concomitantly—by adding sudden reaction tasks to the tracking task—very low BACs are sufficient to produce significant impairment of performance.

Clinical tests for drunkenness include many simple tasks that are easy to measure even in field conditions. These can be divided into three subtests. (1) Motor subtests consist of measuring a person's ability to walk along a straight line with eyes open and closed; maintain a steady turning gait; fit the tips of index fingers together with eyes closed, and collect small objects (e.g., matches) from the floor. (2) *Vestibular* subtests assess the person's body sway, with eyes open and closed, and nystagmus. (3) *Mental* subtests assess the driver's ability to subtract backward, orientation as to time, and overall behavior.

The performance in each subtest is graded from 0 to 3, but these clinical tests are not very sensitive to small BACs (nystagmus exlcuded), and there is great individual variation. The use of these clinical tests for drunkenness in field conditions has greatly diminished since portable BREATHALYZERS became available. The tests are most useful in situations where one has to decide whether to take a blood test for detection of other drugs when no alcohol is found in the driver's breath. Unfortunately, tests developed to detect alcohol effects are less sensitive to the effects of BENZODIAZEPINES and other CNS depressant drugs.

DRUG–ALCOHOL INTERACTIONS

It is well known that large doses of CNS-active drugs impair various of the functions and interact at least additively with alcohol, thereby resulting in heavy sedation or unconsciousness. This effect suggests that even small doses of alcohol may impair performance when taken together with correctly prescribed CNS-active drugs such as anxiolytics, ANTIPSYCHOTICS, ANTIDEPRESSANTS, and OPIOIDS. The deleterious interaction is most obvious when single doses are taken. The issue becomes more complex in chronic alcohol abuse when acquired tolerance of varying extent has developed. Such an adaptation often decreases the expected pharmacological actions of other psychoactive drugs (an effect termed *cross tolerance*).

ALCOHOL AND BENZODIAZEPINES

Taken orally, benzodiazepines have a low acute toxicity. Low doses taken with alcohol (ethanol) may impair skilled performance. A specific benzodiazepine antagonist (flumanzenil) effectively cancels the share of benzodiazepines in mixed intoxications.

Although the risk of a driver having an accident while under the influence of alcohol increases progressively as the BACs increase, a study of the epidemiology and psychomotor effects of benzodiazepines and alcohol are not clear in this respect. One might expect their combined action to be potent, but this has not been documented. Under experimental conditions, a person's tolerance to the drug has been found to minimize or cancel the expected enhanced action of the benzodiazepine in combination with alcohol.

ALCOHOL AND CANNABIS

With chronic (long-term) use of *Cannabis* (MARI-JUANA), a person may acquire a tolerance to its effects. However, tests show the combined effects of ethanol and cannabis to be detrimental to skilled performance. This interaction is potentiative and multidimensional, resulting partly from the fact that *Cannabis* shows a peculiar increase of effect with time that is unrelated to plasma-*Cannabis* levels.

ANTI-ALCOHOL DRUGS

This category generally covers both drugs used to diminish motivation for drinking plus those that cancel (as antagonists) alcohol intoxication. Although the list of possible antagonists is long—and includes AMPHETAMINES and CAFFEINE—no convincing antagonism has been documented. Therefore, no pharmacological agent exists to cancel out the psychomotor effects of alcohol to allow sober performance.

(SEE ALSO: *Accidents and Injuries from Alcohol; Addiction: Concepts and Definitions; Blood Alcohol Concentration; Driving, Alcohol, and Drugs; Driving Under the Influence; Drunk Driving*)

BIBLIOGRAPHY

DORIAN, P., ET AL. (1983). Amitriptyline and ethanol: Pharmacokinetic and pharmacodynamic interaction. *European Journal of Clinical Pharmacology, 25,* 325–331.

MATTILA, M. J. (1990). Alcohol and drug interactions. *Annals of Medicine, 22,* 363–369.

MOSCOWITZ, H. (1984). Attention tasks as skilled performance measures of drug effect. *British Journal of Clinical Pharmacology, 18,* 51S–61S.

O'HANLON, J. F. & DEGIER, J. J. (EDS.). (1986). *Drugs and driving,* London: Taylor & Francis.

STARMER, G. A., & BIRD, K. D. (1984). Investigating drug-ethanol interactions. *British Journal of Clinical Pharmacology, 18,* 27S–35S.

MAURI J. MATTILA
ESKO NUOTTO

PSYCHOMOTOR STIMULANT   This term is used to describe drugs that act as central nervous system (CNS) stimulants. Such drugs generally are appetite suppressants, decrease sleep and fatigue, increase energy and activity, and at higher doses can cause convulsions and death.

Ingestion typically results in increased wakefulness and a decreased sense of fatigue, increased speech and motor activity, alertness, and, frequently, elevation of mood. Many of the drugs in this class have a potential for abuse, with reports of euphoria at higher doses. Although users often report improved performance on physical and mental tasks, this is rarely the case, but they do restore performance that has been impaired by fatigue.

Prolonged use of most of these drugs can result in tolerance to many of their effects. Repeated high doses can result in distorted perception and overt psychotic behavior.

(SEE ALSO: *Amphetamine; Cocaine; Tolerance and Physical Dependence*)

MARIAN W. FISCHMAN

PSYCHOPHARMACOLOGY   Psychopharmacology is that branch of science that involves the study of the effects of interactions between drugs that affect the central nervous system (i.e., PSYCHOACTIVE DRUGS) and living systems. Behavioral and neurobiological effects as well as the mechanisms of actions and side effects of drugs are often examined. Clinical psychopharmacological investigations include the effects of drugs used in treating psychiatric disorders (such as ANXIETY, DEPRESSION, SCHIZOPHRENIA, and mania) as well as other dysfunctions within the central nervous system (such as movement disorders, Alzheimer's disease). Also included are the effects of psychoactive drugs used nonmedically to induce altered states of consciousness, to improve mood, or to otherwise affect the mental status and/or behavior of the individual. Preclinical studies of psychoactive drugs using ANIMAL models are also an important aspect of psychopharmacology.

Psychopharmacology is an interdisciplinary field of science. Psychopharmacologists may be psychologists with extra training in PHARMACOLOGY, pharmacologists with special training in psychology and behavior, or physicians who obtain extra training in relevant disciplines.

BIBLIOGRAPHY

MELTZER, H. Y. (ED.). (1987). *Psychopharmacology: The third generation of progress.* New York: Raven Press.

NICK E. GOEDERS

PSYCHOSTIMULANT   *See* Drug Types; Psychomotor Stimulant.

PSYCHOTHERAPY   *See* Treatment; Treatment Types.

PSYCHOTROPIC SUBSTANCES CONVENTION OF 1971   The 1971 Convention on Psychotropic Substances extended the international drug control system to cover mood-altering substances such as stimulants (e.g., AMPHETAMINES), SEDATIVE-HYPNOTICS (e.g., BARBITURATES), and HALLUCINOGENS (e.g., LSD and MESCALINE). It limited the use of these substances to medical and scientific purposes, and it did not cover ALCOHOL or TOBACCO. As of November 1994, 132 governments were party to the convention.

GENERAL PROVISIONS

The manufacture, trade, and distribution of psychotropic substances are subject to licensing, record keeping, and reporting. The convention generally permits governments great flexibility in applying the provisions to meet their particular needs, because it recognizes that psychotropic substances are widely used in medical practice to treat mental and physical disorders. In addition, the convention includes provisions for the prevention of abuse and for the treatment and rehabilitation of drug addicts. Because of the convention, a substance abuser may receive treatment, education, aftercare, and rehabilitation as an alternative or in addition to punishment.

A patient may not obtain any of the substances regulated under the convention without a medical prescription, although exceptions are allowed under certain circumstances, when licensed pharmacists may supply small quantities of the substances that are less likely to be abused. In addition, the convention set forth precautions to be taken to ensure that the distribution of psychotropic substances conformed to sound medical practice. An example of such practice is the proper labeling of retail packages to include adequate directions for use and warnings, if necessary.

A party to the convention may prohibit exportation of psychotropic substances from its country. It may also notify other parties, through the United Nations secretary-general, that it prohibits the import of schedule II, III, or IV substances into its country.

All signatories must provide detailed annual statistical reports on the production, trade, and consumption of psychotropic substances to the International Narcotics Control Board (INCB), the central authority that was established to coordinate control of the illegal manufacture and use of narcotics. The reports for substances in schedules I and II must be more detailed than those for substances in schedules III and IV, which are not as rigidly regulated.

SCHEDULES OF PSYCHOTROPIC SUBSTANCES

A psychotropic substance is assigned to one of four schedules by balancing the drug's potential for abuse and the threat it poses to the public health against its therapeutic benefits. The placement of a drug in one of the schedules affects its trade, manufacture, distribution, and use. Hallucinogens and other drugs that are of no—or severely limited—medical use are placed in schedule I. Schedule I substances, the most stringently regulated of the four schedules, may only be used for scientific and limited medical purposes in government-operated licensed establishments. The manufacture, trade, distribution, and possession of these substances require special licensing or authorization from the government. The amounts of these substances that may be supplied, imported, and exported are limited, even for authorized uses, and records of their use must be kept.

Schedule II drugs, such as METHAQUALONE and amphetamines, possess a high potential for abuse and limited medical usefulness, and therefore they are subject to tighter controls over their production and trade than substances in schedule III and schedule IV. Governments must issue special import and export authorizations before these drugs can be traded internationally. Experience has shown that placement of a substance in schedule II severely reduces its use.

Schedule III and schedule IV have been assigned to such drugs as depressants, sedative hypnotics, anxiolytics, barbiturates, and minor tranquilizers. Individuals and businesses involved in the manufacture, trade, and distribution of schedule III and schedule IV psychotropic substances must have licenses from the government. They must maintain records of the manufacture and wholesale trade, import, and export of these substances. The World Health Organization (WHO) has designated several drugs in schedule IV, including BENZODIAZEPINES such as diazepam (Valium®) and alprazolam (Xanax®), "essential drugs" that governments must assure are available for medical purposes.

ROLE OF THE WORLD HEALTH ORGANIZATION

The Convention on Psychotropic Substances allows the United Nations Commission on Narcotic Drugs to add substances to its schedules, and also to transfer or remove them. WHO recommends what it considers to be the appropriate placement of drugs within schedules. A party to the convention may ask the United Nations secretary-general to recommend that WHO place other drugs under control. WHO reviews substances to determine whether they have the "capacity to produce a state of dependence and central nervous system stimulation or depression resulting in hallucination or disturbances in motor function or thinking or behavior or mood" and whether they pose a risk to public health. WHO must make known in great detail the criteria it applied in evaluating a psychotropic substance for control.

The evaluations WHO makes are based on scientific and medical criteria, but in deciding whether to accept or reject WHO's recommendations, the United Nations Commission on Narcotic Drugs may consider social, economic, and political issues. A two-thirds majority vote is required, however, before the Commission may alter or amend a schedule. If, because of exceptional circumstances, a party cannot apply the provisions of the convention to a newly added substance, it may notify the United Nations secretary-general and obtain permission to satisfy only minimal control requirements.

SIGNIFICANCE OF THE CONVENTION

The 1971 Convention on Psychotropic Substances recognized that the abuse of mood-altering substances, like the abuse of narcotic drugs, could have harmful effects, at the same time that it acknowledged that psychotropic drugs provide important medical and scientific benefits. Through the treaty drawn up at the Convention, the international community took another step in the cooperative effort to curtail drug abuse while preserving the availability of psychotropic substances for legitimate medical use.

(SEE ALSO: *International Drug Supply Systems; Single Convention on Narcotic Drugs; WHO Expert Committee on Drug Dependency*)

BIBLIOGRAPHY

BEAN, P. (1974). *The social control of drugs.* New York: Wiley.

BRUUN, K., PAN, L., & REXED, I. (1975). *The gentlemen's club: International control of drugs and alcohol.* Chicago: University of Chicago Press.

REXED, B., ET AL. (1984). *Guidelines for the control of narcotic and psychotropic substances.* Geneva: World Health Organization.

WORLD PEACE THROUGH LAW CENTER. (1975). *International drug control* (prepared for the Sixth World Conference of the Legal Profession, sponsored in part by the U.S. Department of Justice). Washington, DC: Author.

ROBERT T. ANGAROLA

PUBLIC INTOXICATION   Before the seventeenth century, public intoxication was not, by itself, a crime in England. Drunkenness was punishable as a criminal offense only if it resulted in some form of breach of the peace or disorderly conduct. In 1606, however, in England, simple public intoxication was first made a criminal offense. This English precedent was reflected in some laws in the American colonies as well as in the United States in the city, county, and state laws enacted after the American Revolution. By the early 1960s, about two million arrests occurred annually for simple public intoxication, representing about 33 percent of all arrests in the United States.

Since then, remarkable changes have occurred in the handling of public intoxication. Through efforts initially in the courts and later through federal and state legislation, important steps have been taken to transfer the handling of public intoxication from the criminal-justice system to more humane and effective public-health care. The major stumbling blocks to further progress have been the lack of adequate funding and uncertainty about the most effective way of treating alcohol abuse and alcoholism.

INITIAL COURT CHALLENGES

Beginning in 1964, lawyers argued that derelict alcoholics could not lawfully be punished for their public intoxication on two independent grounds. First, they argued that these derelict alcoholics did not have the *mens rea* (Latin, guilty mind or intent) required for conviction of a crime, because their public intoxication was a symptom of the disease of alcoholism. Second, they argued that punishing an alcoholic for exhibiting the symptoms of that disease in public was cruel and unusual punishment, prohibited by the U.S. Constitution.

In lower court cases, these arguments prevailed. In 1968, however, in the case of *Powell v. Texas*, the U.S. Supreme Court handed down a split decision on this issue. Four justices found that it would be cruel and unusual punishment to convict Powell, an admitted alcoholic, for simple public intoxication. Four other justices determined that the matter should be left to the states and should not be decided on a constitutional level. The ninth and controlling justice determined that, because Powell had a home, he could properly be held responsible for being intoxicated in public and thus was appropriately convicted. This left open the question of whether a derelict alcoholic, without a home, could also be convicted.

ENACTMENT OF FEDERAL STATUTES

Faced with a stalemate in the Supreme Court, advocates for reforming the public-intoxication laws turned to Congress. In spite of a large number of federal public health statutes, none referred explicitly to the problems of intoxication and ALCOHOLISM. Congress responded by enacting the Alcoholic Rehabilitation Act of 1968, which recognized alcoholism as a major health and social problem, and recommended handling public intoxication as a

health problem rather than as a law-enforcement matter. This was followed by the Comprehensive Alcohol Abuse and Alcoholism Prevention, Treatment, and Rehabilitation Act of 1970, which created the NATIONAL INSTITUTE ON ALCOHOL ABUSE AND ALCOHOLISM to administer all alcoholism programs and authority assigned to the U.S. Department of Health, Education, and Welfare (now the U.S. Department of Health and Human Services). These new federal laws for the first time provided a national focus for handling public intoxication on a public-health basis.

CHANGES IN THE STATE STATUTES

Before the court cases and the federal statutes, simple public intoxication constituted a criminal offense throughout the United States. Following the dramatic legal developments in the courts and in Congress, state and local laws rapidly began to change. Initially in the District of Columbia and in Maryland, and subsequently throughout other parts of the country, the criminal statutes prohibiting simple public intoxication were repealed and replaced with new laws establishing detoxification programs for intoxicated persons and rehabilitation programs for chronic alcoholics. By the early 1990s, more than 67 percent of the states had revised their laws to reflect this change in approach.

THE CURRENT STATUS

Existing federal and state laws now provide a firm foundation for handling public intoxication as a public-health problem rather than as a matter for the criminal-justice system. Relatively little additional change can be accomplished solely by further litigation or legislation.

With these legal and legislative hurdles overcome, two additional obstacles have arisen to impede further progress. First, the competition for federal and state health funds has become intense. Other important health needs, including basic health care for the needy and treatment for people with acquired immunodeficiency sydrome (AIDS), have made it very difficult for public officials to devote adequate resources for the expansion of public-health programs to include public intoxication and alcoholism. The problem has been compounded by a lack of any clearly effective method for the prevention or treatment of intoxication and alcoholism. A low rate of rehabilitation has led many public health officials to conclude that scarce public resources are more effectively devoted to other illnesses, especially communicable diseases. Unless there is additional investment, the police will remain deeply involved in identifying and responding to intoxicated individuals, and their response will not necessarily be limited to transporting the individual to a sobering-up station.

Progress in the prevention and treatment of intoxication and alcoholism has therefore been slow, in spite of the major changes made in the courts, the Congress, and state and local legislative bodies. Unless and until the American public places a higher priority on the handling of public intoxication as a public-health matter or medical science finds more effective methods to prevent and treat this problem, this situation is unlikely to change.

(SEE ALSO: *Alcohol: History of Drinking; Detoxification; Homelessness, Alcohol, and Other Drugs; Temperance Movement; Treatment: Alcohol; Treatment, History of*)

BIBLIOGRAPHY

PRESIDENT'S COMMISSION ON LAW ENFORCEMENT AND ADMINISTRATION OF JUSTICE. (1967). *Task force report: Drunkenness.* Washington, DC: Author.

SECRETARY OF HEALTH, EDUCATION, AND WELFARE. (1971). *First special report to the U.S. Congress on alcohol & health,* Chapter VII: The legal status of intoxication and alcoholism. DHEW Publication No. (HSM) 72–9099. Washington, DC: Author.

PETER BARTON HUTT.

QUAALUDE *See* Methaqualone.

QUITTING SMOKING *See* Nicotine Delivery Systems for Smoking Cessation; Tobacco.